130

Advanced
PHYSICS
for You

Keith Johnson
Simmone Hewett
Sue Holt
John Miller

...ive support of: Patrick Organ, Richard Bradford,
...mer, Martin Gregory, Tim Akrill, Andy Raw,
...Jackson, George Snape, Nicola Thomas,
...Devereux, Michael Johnson, Julie Thornton,
...ie, Tim Egginton, Darren Forbes, Jane Cope,
...r, Averil Macdonald, Tricia New, Ann Johnson,
...l Deehan, Paul Hadfield, Marilyn Miller

OXFORD
UNIVERSITY PRESS

UNIVERSITY PRESS

Great Clarendon Street, Oxford, OX2 6DP, United Kingdom

Oxford University Press is a department of the University of Oxford.
It furthers the University's objective of excellence in research, scholarship, and
education by publishing worldwide. Oxford is a registered trade mark of Oxford
University Press in the UK and in certain other countries

First published in 2000   This edition published in 2015

British Library Cataloguing in Publication Data
Data available

978 140 852737 5

10 9 8 7 6

Paper used in the production of this book is a natural, recyclable product made
from wood grown in sustainable forests. The manufacturing process conforms to
the environmental regulations of the country of origin.

Printed in India by Multivista Global Pvt. Ltd

Everything should be made as simple as possible, but not simpler.
Albert Einstein

There is no higher or lower knowledge, but one only, flowing out of
experimentation.
Leonardo da Vinci

I do not know what I may appear to the world, but to myself
I seem to have been only a boy playing on the seashore, and
diverting myself in now and then finding a smoother pebble or
a prettier shell than ordinary, while the great ocean of truth
lay all undiscovered before me.
Sir Isaac Newton

Introduction

Advanced Physics for You is designed to help and support you during your advanced Physics course. No matter which exam course you are following, this book will help you to make the transition to A-level.

The book is carefully laid out so that each new idea is introduced and developed on a single page or on two facing pages. Words have been kept to a minimum and as straightforward as possible, with clear diagrams, and a cartoon character called 'Phiz'.
Pages with a red triangle in the top corner are the more difficult pages and can be left at first.

Each important fact or new formula is clearly printed in **heavy type** or is in a coloured box.
There is a summary of important facts at the end of each chapter, to help you with revision.

Worked examples are a very useful way of seeing how to tackle problems in Physics. In this book there are over 200 worked examples to help you to learn how to tackle each kind of problem.

At the back of the book there are extra sections giving you valuable advice on study skills, practical work, revision and examination techniques, as well as more help with mathematics.

Throughout the book there are 'Physics at Work' pages. These show you how the ideas that you learn in Physics are used in a wide range of interesting applications.

There is also a useful analysis of how the book covers the different examination syllabuses, with full details of which pages you need to study, on the website at: www.oxfordsecondary.co.uk/advancedforyou

At the end of each chapter there are a number of questions for you to practise your Physics and so gain in confidence. They range from simple fill-in-a-missing-word sentences (useful for doing quick revision) to more difficult questions that will need more thought.

At the end of each main topic you will find a section of further questions, mostly taken from actual advanced level examination papers.
For all these questions, a 'Hints and Answers' section at the back of the book gives you helpful hints if you need them, as well as the answers.

We hope that reading this book will make Physics more interesting for you and easier to understand.
Above all, we hope that it will help you to make good progress in your studies, and that you will enjoy using **Advanced Physics for You**.

Keith Johnson
Simmone Hewett
Sue Holt
John Miller

Contents

Chapters marked * are usually not in AS, only in A-level.

For more details of which sections you need to study for your particular examination, see page 506, and download the relevant 'Specification Map' from www.oxfordsecondary.co.uk/advancedforyou

Electricity and Magnetism

Quantum Physics

Nuclear Physics

Modern Physics

Extra Sections

1 Basic Ideas

How fast does light travel? How much do you weigh?
What is the radius of the Earth?
What temperature does ice melt at?

We can find the answers to all of these questions by measurement.
Speed, mass, length and temperature are all examples of **physical quantities**.
Measurement of physical quantities is an essential part of Physics.

In this chapter you will learn:
- the difference between 'base' and 'derived' units,
- how you can use units to check equations,
- how to use 'significant figures',
- how to deal with vectors.

Measuring temperature …

… and time …

Stating measurements

All measurement requires a system of units.
For example: How far is a distance of 12?

Without a unit this is a meaningless question. You must always give a measurement as **a number multiplied by a unit**.

For example:

12 m means 12 multiplied by the length of one metre.
9 kg means 9 multiplied by the mass of one kilogram.

But what do we mean by one metre and one kilogram?
Metres and kilograms are two of the seven internationally agreed **base units**.

… and weight.

Base quantities and units

The Système International (SI) is a system of measurement that has been agreed internationally. It defines 7 base quantities and units, but you only need six of them at A-level.

The 7 base quantities and their units are listed in the table:

Their definitions are based on specific physical measurements that can be reproduced, very accurately, in laboratories around the world.

The only exception is the kilogram. This is the mass of a particular metal cylinder, known as the prototype kilogram, which is kept in Paris.

Base quantity		Base unit	
Name	Symbol	Name	Symbol
time	t	second	s
length	l	metre	m
mass	m	kilogram	kg
temperature	T, θ	kelvin	K
electric current	I	ampere	A
amount of substance	n	mole	mol
luminous intensity *(Not used at A-level)*		candela	cd

Derived units

Of course, we use far more physical quantities in Physics than just the 7 base ones. All other physical quantities are known as *derived quantities*. Both the quantity and its unit are derived from a combination of base units, using a defining equation:

Example 1
Velocity is defined by the equation:

$$\text{velocity} = \frac{\text{distance travelled in a given direction (m)}}{\text{time taken (s)}}$$

Both distance (ie. length) and time are base quantities. The unit of distance is the metre and the unit of time is the second. So, from the defining equation, the derived unit of velocity is **metres per second**, written **m/s** or **m s^{-1}**.

Example 2
Acceleration is defined by the equation:

$$\text{acceleration} = \frac{\text{change in velocity (m s}^{-1})}{\text{time taken (s)}}$$

Again, combining the units in the defining equation gives us the derived unit of acceleration.
This is **metres per second per second** or **metres per second squared**, written **m/s^2** or **m s^{-2}**.

What other units have you come across in addition to these base units and base unit combinations?
Newtons, watts, joules, volts and ohms are just a few that you may remember.
These are special names that are given to particular combinations of base units.

Example 3
Force is defined by the equation:

force = mass (kg) $\times$ acceleration (m s^{-2})   (see page 55)

The derived unit of force is therefore:
kilogram metres per second squared or **kg m s^{-2}**.
This is given a special name: the **newton** (symbol **N**).

The table below lists some common derived quantities and units for you to refer to.

Some of the combinations are quite complicated. You can see why we give them special names!

Physical quantity	Defined as	Unit	Special name
density	mass (kg) $\div$ volume (m^3)	kg m^{-3}	
momentum	mass (kg) $\times$ velocity (m s^{-1})	kg m s^{-1}	
force	mass (kg) $\times$ acceleration (m s^{-2})	kg m s^{-2}	newton (N)
pressure	force (kg m s^{-2} or N) $\div$ area (m^2)	kg m^{-1} s^{-2} (N m^{-2})	pascal (Pa)
work (energy)	force (kg m s^{-2} or N) $\times$ distance (m)	kg m^2 s^{-2} (N m)	joule (J)
power	work (kg m^2 s^{-2} or J) $\div$ time (s)	kg m^2 s^{-3} (J s^{-1})	watt (W)
electrical charge	current (A) $\times$ time (s)	A s	coulomb (C)
potential difference	energy (kg m^2 s^{-2} or J) $\div$ charge (A s or C)	kg m^2 A^{-1} s^{-3} (J C^{-1})	volt (V)
resistance	potential difference (kg m^2 A^{-1} s^{-3} or V) $\div$ current (A)	kg m^2 A^{-2} s^{-3} (V A^{-1})	ohm (Ω)

▷ Homogeneity of equations

We have seen that all units are derived from base units using equations. This means that in any correct equation the base units of each part must be the **same**. When this is true, the equation is said to be homogeneous. Homogeneous means 'composed of identical parts'.

Example 4
Show that the following equation is homogeneous: $\text{kinetic energy} = \frac{1}{2} \times \text{mass} \times \text{velocity}^2$

From the table on page 7:
Unit of kinetic energy = joule = $\text{kg m}^2 \text{ s}^{-2}$
Unit of $\frac{1}{2} \times \text{mass} \times \text{velocity}^2 = \text{kg} \times (\text{m s}^{-1})^2 = \text{kg m}^2 \text{ s}^{-2}$ (Note: $\frac{1}{2}$ is a pure number and so has no unit.)

The units on each side are the same and so the equation is homogeneous.

This is a useful way of checking an equation. It can be particularly useful after you have rearranged an equation:

Example 5
Phiz is trying to calculate the power P of a resistor when he is given its resistance R and the current I flowing through it.
He cannot remember if the formula is: $P = I^2 \times R$ or $P = I^2 \div R$.
By checking for homogeneity, we can work out which equation is correct:

Using the table on page 7:   Units of P = watts (W) = $\text{kg m}^2 \text{ s}^{-3}$
Units of $I^2 = \text{A}^2$
Units of R = ohms (Ω) = $\text{kg m}^2 \text{ A}^{-2} \text{ s}^{-3}$

Multiplying together the units of I^2 and R would give us the units of power. So the first equation is correct.

One word of warning. This method shows that an equation could be correct – but it doesn't prove that it is correct!

Can you see why not? Example 4 above is a good illustration. The equation for kinetic energy would still be homogeneous even if we had accidentally omitted the $\frac{1}{2}$.

▷ Prefixes

For very large or very small numbers, we can use standard prefixes with the base units.
The main prefixes that you need to know are shown in the table:

Example 6
a)   Energy stored in a chocolate bar   = 1 000 000 J

= 1×10^6 J
= 1 megajoule
= 1 MJ

b)   Wavelength of an X-ray   = 0.000 000 001 m

= 1×10^{-9} m
= 1 nanometre
= 1 nm

Prefix	Symbol	Multiplier
tera	T	10^{12}
giga	G	10^9
mega	M	10^6
kilo	k	10^3
deci	d	10^{-1}
centi	c	10^{-2}
milli	m	10^{-3}
micro	μ	10^{-6}
nano	n	10^{-9}
pico	p	10^{-12}
femto	f	10^{-15}

▷ The importance of significant figures

What is the difference between lengths of 5m, 5.0m and 5.00m?

Writing 5.00m implies that we have measured the length more precisely than if we write 5m.

Writing 5.00m tells us that the length is accurate to the nearest centimetre.
A figure of 5m may have been rounded to the nearest metre.
The actual length could be anywhere between $4\frac{1}{2}$m and $5\frac{1}{2}$m.

The number 5.00 is given to three **significant figures** (or 3 **s.f.**).

To find the number of significant figures you must *count up the total number of digits, starting at the first non-zero digit, reading from left to right*.
The table gives you some examples:

It shows you how a number in the first column (where it is given to 3 s.f.) would be rounded to 2 significant figures or 1 significant figure.

In the last example in the table, why did we change the number to 'standard form'?
Writing 1.7×10^2 instead of 170 makes it clear that we are giving the number to two significant figures, not three.
(If you need more help on standard form look at the *Check Your Maths* section on page 473.)

3 s.f.	2 s.f.	1 s.f.
4.62	4.6	5
0.005 01	0.005 0	0.005
3.40×10^8	3.4×10^8	3×10^8
169	1.7×10^2	2×10^2

Significant figures and calculations

How many significant figures should you give in your answers to calculations?
This depends on the precision of the numbers you use in the calculation. Your answer cannot be any more precise than the data you use. This means that you should round your answer to the same number of significant figures as those used in the calculation.

If some of the figures are given less precisely than others, then round up to the *lowest* number of significant figures.
Example 7 explains this.

Make sure you get into the habit of rounding all your answers to the correct number of significant figures. You may lose marks in an examination if you don't!

Example 7
The swimmer in the photograph covers a distance of 100.0 m in 68 s.
Calculate her average speed.

$$\text{speed} = \frac{\text{distance travelled}}{\text{time taken}} = \frac{100.0 \text{ m}}{68 \text{ s}} = 1.470\,588\,2 \text{ m s}^{-1}$$

This is the answer according to your calculator.
How many significant figures should we round to?

The distance was given to 4 significant figures. But the time was given to only 2 significant figures. Our answer cannot be any more precise than this, so we need to round to 2 significant figures.

Our answer should be stated as: $\underline{1.5 \text{ m s}^{-1} \text{ (2 s.f.)}}$

▷ Vectors and scalars

Throwing the javelin requires a force. If you want to throw it a long distance, what two things are important about the force you use?

The javelin's path will depend on both the **size** and the **direction** of the force you apply.

Force is an example of a *vector* quantity.
Vectors have both size (magnitude) and direction.

Other examples of vectors include: velocity, acceleration and momentum. They each have a size and a direction.

Quantities that have size (magnitude) but *no* direction are called **scalars**. Examples of scalars include: temperature, mass, time, work and energy.

The table shows some of the more common vectors and scalars that you will use in your A-level Physics course:

Look back at the table of base quantities on page 6.
Are these vectors or scalars? Most of the base quantities are scalars.
Can you spot the odd one out?

Scalars	Vectors
distance	displacement
speed	velocity
mass	weight
pressure	force
energy	momentum
temperature	acceleration
volume	electric current
density	torque

▷ Representing vectors

Vectors can be represented in diagrams by arrows.

- **The length of the arrow represents the magnitude of the vector.**

- **The direction of the arrow represents the direction of the vector.**

Here are some examples:

Force vectors
A horizontal force of 20 N:

Using the same scale, how would you draw the vector for a force of 15 N at 20° to the horizontal?

A vertical force of 10 N:

Velocity vectors
A velocity of 30 m s^{-1} in a NE direction:

Using the same scale, how would you draw the vector for a velocity of 15 m s^{-1} in a NW direction?

A velocity of 20 m s^{-1} due west:

▷ Vector addition

What is 4 kg plus 4 kg? Adding two masses of 4 kg *always* gives the answer 8 kg. Mass is a scalar. You combine scalars using simple arithmetic.

What about 4 N plus 4 N? Adding two forces of 4 N can give any answer between 8 N and 0 N.
Why do you think this is?
It's because force is a vector. When we combine vectors we also need to take account of their *direction*.

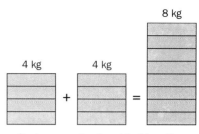

Scalars are simply added together.

Often in Physics we will come across situations where two or more vectors are acting together. The overall effect of these vectors is called the **resultant**. This is the single vector that would have the same effect.
To find the resultant we must use the directions of the 2 vectors:

Vectors acting along the same straight line

Two vectors acting in the *same* direction can simply be added together:

$$\text{Resultant} = F_1 + F_2$$

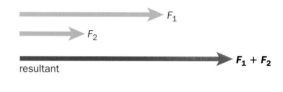

If the vectors act in *opposite* directions, we need to take one direction as positive, and the other as negative, before adding them:

$$\text{Resultant} = F_1 + (-F_2) = F_1 - F_2$$

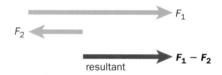

Example 8
Phiz is standing on a moving walkway in an airport.
The walkway is moving at a steady velocity of 1.50 m s^{-1}.
a)   Phiz starts to walk forwards along the walkway at 2.00 m s^{-1}.
     What is his resultant velocity?

   Both velocity vectors are acting in the same direction.
   Resultant velocity $= 1.50$ m s$^{-1} + 2.00$ m s^{-1}
   $\qquad\qquad\qquad = 3.50$ m s^{-1} in the direction of the walkway.

b)   Phiz then decides he is going the wrong way. He turns round and starts to run at 3.40 m s^{-1} in the opposite direction to the motion of the walkway. What is his new resultant velocity?

   The velocity vectors now act in opposite directions.
   Taking motion in the direction of the walkway to be positive:
   Resultant velocity $= +1.50$ m s$^{-1} - 3.40$ m s^{-1}
   $\qquad\qquad\qquad = -1.90$ m s^{-1} (3 s.f.)

   As this is negative, the resultant velocity acts in the opposite direction to the motion of the walkway. He moves to the left.

▷ Combining perpendicular vectors

To find the resultant of two vectors (X, Y) acting at 90° to each other, we draw the vectors as adjacent sides of a rectangle:

The resultant is the **diagonal** of the rectangle, as shown here:

The **magnitude** (size) of the resultant vector R can be found using Pythagoras' theorem:

$$R^2 = X^2 + Y^2$$

The **direction** of the resultant is given by the angle θ:

$$\tan \theta = \frac{\text{opposite}}{\text{adjacent}} = \frac{Y}{X} \qquad \therefore \; \theta = \tan^{-1}\left(\frac{Y}{X}\right)$$

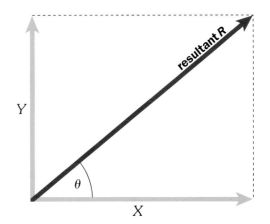

You can also find the resultant using an accurate scale drawing. The magnitude of the resultant can be found by measuring its length with a ruler. The direction can be measured with a protractor. For more on scale drawing, see page 14.

Example 9
Two tugs are pulling a ship into harbour. One tug pulls in a SE direction. The other pulls in a SW direction. Each tug pulls with a force of 8.0×10^4 N. What is the resultant force on the ship?

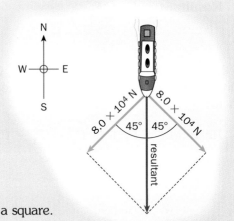

The two forces act at 90° to each other. Using Pythagoras' theorem:

Magnitude of resultant $= \sqrt{(8.0 \times 10^4 \text{ N})^2 + (8.0 \times 10^4 \text{ N})^2}$

$= \sqrt{1.28 \times 10^{10}} \text{ N}$

$= \underline{1.1 \times 10^5 \text{ N}}$ (2 s.f.)

Since both tugs pull with the same force, the vectors form adjacent sides of a square. The resultant is the diagonal of the square. So it acts at 45° to each vector. The resultant force must therefore act <u>due south</u>.

Example 10
A man tries to row directly across a river.
He rows at a velocity of 3.0 m s^{-1}.
The river has a current of velocity 4.0 m s^{-1} parallel to the banks.
Calculate the resultant velocity of the boat.

The diagram shows the two velocity vectors.

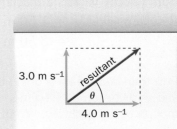

We can find their resultant using Pythagoras' theorem:

Size of resultant $= \sqrt{(3.0 \text{ m s}^{-1})^2 + (4.0 \text{ m s}^{-1})^2} = \sqrt{25} \text{ m s}^{-1} = 5.0 \text{ m s}^{-1}$

Direction of resultant: $\tan \theta = \dfrac{\text{opposite}}{\text{adjacent}} = \dfrac{3.0}{4.0} \qquad \therefore \; \theta = \tan^{-1}\left(\dfrac{3.0}{4.0}\right) = 37°$

So the resultant velocity is <u>5.0 m s^{-1}</u> at <u>37°</u> to the bank.

▷ Vector subtraction

The diagram shows the speed and direction of
a trampolinist at two points during a bounce:

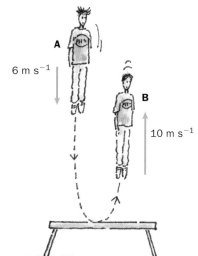

What is the trampolinist's change in **speed**
from A to B?

Change in speed = new speed − old speed
                 = 10 m s^{-1} − 6 m s^{-1}
                 = 4 m s^{-1}

What about his change in **velocity**?
To find the change in a vector quantity
we use vector subtraction:

Change in velocity = new velocity − old velocity
                   = 10 m s^{-1} **up** − 6 m s^{-1} **down**

Remember, with vectors we must take account of the direction.
In this example let us take the upward direction to be positive,
and the downward direction to be negative.

We can then rewrite our equation as:

Change in velocity = +10 m s^{-1} − (−6 m s^{-1})
                   = +10 m s^{-1} + 6 m s^{-1}
                   = +16 m s^{-1}

So the change in velocity is 16 m s^{-1} in an **up**ward direction.

Can you see that subtracting 6 m s^{-1} downwards is the same as
adding 6 m s^{-1} acting upwards?
***Vector subtraction is the same as the addition of a vector of
the same size acting in the opposite direction.***

Two negatives make a positive.

Example 11
A boy kicks a ball against a wall with a horizontal velocity of 4.5 m s^{-1}.
The ball rebounds horizontally at the same speed.
What is the ball's change in velocity?

Although the speed is the same, the velocity has changed. Why?
Velocity is a vector, so a change in direction means a change in
velocity.

Let us take motion towards the wall as positive,
and motion away from the wall as negative.

Change in velocity = new velocity − old velocity
                    = (−4.5 m s^{-1}) − (+4.5 m s^{-1})
                    = −4.5 m s^{-1} − 4.5 m s^{-1}
                    = −9.0 m s^{-1} (2 s.f.)

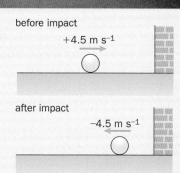

before impact
+4.5 m s^{-1}

after impact
−4.5 m s^{-1}

So the change in velocity is 9.0 m s^{-1} in a direction away from the wall.

▷ Combining vectors using scale drawing

If more than one vector acts on an object we can find the
resultant vector by drawing each vector consecutively
head to tail. The resultant **R** will be a line joining the start
to the final end point. This works for any number of
vectors acting at a point.

If each vector is drawn carefully to scale, using a protractor
to ensure the correct angles, then the resultant can be found
by taking measurements from the finished diagram.

Notice that it doesn't matter in which order you choose to
redraw the vectors. You always end up with the same
resultant. Two of the six possible arrangements for vectors
a, *b*, *c* are shown here:
Try out the other combinations for yourself.

What if, after drawing each vector head to tail, you end up back
where you started?
If this happens then the **resultant is zero**. This will be the
case for a system of forces acting on an object **in equilibrium**
(see pages 32 and 36).

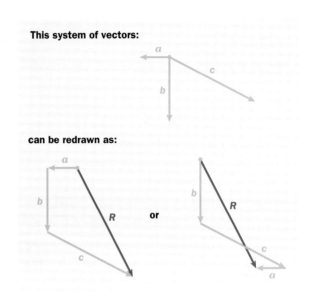

This system of vectors:

can be redrawn as:

or

Vector triangles

We can redraw the diagram on page 12 for two
perpendicular vectors using the head-to-tail technique:
Can you see that this gives the same resultant?

Any two vectors can be combined in this way.
The resultant becomes the third side of a **vector triangle**.
This is also a useful technique where the vectors are not
perpendicular to each other.

If the vectors' lengths and directions are drawn accurately to scale,
the resultant can be found by measuring the length of the resultant
on the diagram with a ruler.
The direction is measured with a protractor.

The method also works for **vector subtraction**.
Again the vectors are drawn head to tail, but the vector
to be subtracted is drawn in the reverse direction.
The diagrams show the resultant of the addition and
subtraction of two vectors **P** and **Q**:

If you prefer a more mathematical solution, lengths and angles
can be calculated using trigonometry. For a right-angled triangle,
Pythagoras' theorem and sine, cosine or tangent will give you
the magnitude and direction of the resultant (see also page 474).
Other triangles can be solved using the sine or cosine rule.

Although a mathematical solution does not require a
scale diagram, you should always draw a sketch using the
head-to-tail method to ensure you have your triangle the
right way round.

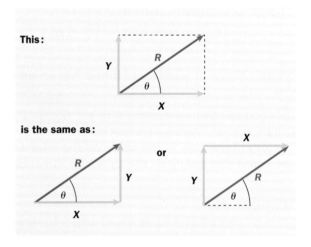

This:

is the same as:

or

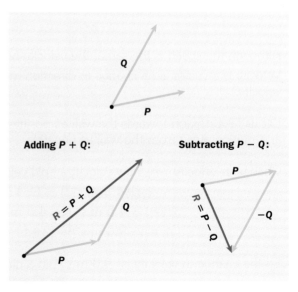

Adding P + Q:

Subtracting P – Q:

14

▷ Resolving vectors

We have seen how to combine two vectors that are acting at 90°, to give a single resultant. Now let's look at the reverse process.

We can **resolve** a vector into two **components** acting at right angles to each other. The component of a vector tells you the effect of the vector in that direction.

So how do we calculate the 2 components?
Look at the vector V in this diagram:

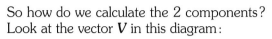

We can resolve this vector into two components, V_1 and V_2, at right angles to each other:

   V_1 acts at an angle θ_1 to the original vector.
   V_2 acts at an angle θ_2 to the original vector. $(\theta_1 + \theta_2 = 90°)$

To find the size of V_1 and V_2 we need to use trigonometry:

$$\cos \theta_1 = \frac{adjacent}{hypotenuse} = \frac{V_1}{V} \qquad \text{Rearranging this gives:} \quad V_1 = V \cos \theta_1$$

$$\cos \theta_2 = \frac{adjacent}{hypotenuse} = \frac{V_2}{V} \qquad \text{Rearranging this gives:} \quad V_2 = V \cos \theta_2$$

So to find the component of a vector in any direction you need to *multiply by the cosine of the angle between the vector and the component direction*.

Example 12
A tennis player hits a ball at 10 m s^{-1} at an angle of 30° to the ground.
What are the initial horizontal and vertical components of velocity of the ball?

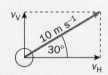

Horizontal component: $v_H = 10 \cos 30° = \underline{8.7 \text{ m s}^{-1}}$ (2 s.f.)

The angles in a right angle add up to 90°.
So the angle between the ball's path and the *vertical* = 90° − 30° = 60°

Vertical component: $v_V = 10 \cos 60° = \underline{5.0 \text{ m s}^{-1}}$ (2 s.f.)

Example 13
A water-skier is pulled up a ramp by the tension in the tow-rope. This is a force of 300 N acting horizontally. The ramp is angled at 20.0° to the horizontal.
What are the components of the force from the rope acting (a) parallel to and (b) perpendicular to the slope?

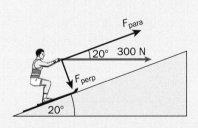

The angle between the rope and the parallel slope direction = 20°
So, the angle between the rope and the perpendicular direction = 90° − 20° = 70°

a) Component of force **parallel** to ramp $F_{para} = 300 \text{ N} \times \cos 20° = \underline{282 \text{ N}}$ (3 s.f.)

b) Component of force **perpendicular** to ramp $F_{perp} = 300 \text{ N} \times \cos 70° = \underline{103 \text{ N}}$ (3 s.f.)

15

▷ Combining non-perpendicular vectors by calculation

On the previous page you learnt how to resolve a vector into two perpendicular components. You can also use the same technique to combine multiple vectors without the need for scale drawings.

The trick is to first resolve each vector into its components. Often it's convenient to resolve horizontally and vertically (but you can use any convenient mutually perpendicular directions).

All the horizontal components can then be added together (taking account of their direction).
Similarly, all the vertical components can be added.
The resultant horizontal and vertical vectors can then be combined using Pythagoras' theorem.

See how this works in the following example:

Example 14
Three forces act on a single point, as shown in the diagram:
Find the magnitude and direction of their resultant force.

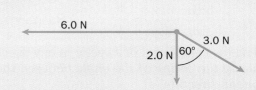

Step 1: Resolve the 3.0 N force into its horizontal and vertical components:

Horizontal component $= 3.0 \times \cos 30° = 2.6$ N
Vertical component $= 3.0 \times \cos 60° = 1.5$ N

Step 2: Replace the 3.0 N force with its components in the diagram (shown in black):

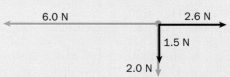

Step 3: Find the total force in the horizontal and vertical directions:

Total horizontal force $= 6.0 - 2.6 = 3.4$ N (to the left)
Total vertical force $= 2.0 + 1.5 = 3.5$ N (downwards)

Step 4: Finally, combine the total horizontal and vertical forces into a single resultant force:

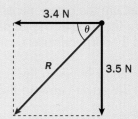

Using Pythagoras' theorem:
Magnitude of resultant $R = \sqrt{3.4^2 + 3.5^2} = 4.9$ N (2 s.f.)

$\tan\theta = \dfrac{\text{opposite}}{\text{adjacent}} = \dfrac{3.5}{3.4}$ $\therefore\ \theta = \tan^{-1}\left(\dfrac{3.5}{3.4}\right) = 46°$

So the resultant force is <u>4.9 N</u> acting at <u>46° below the original 6.0 N force</u>.

Summary

There are 6 base quantities that you need for A-level (time, length, mass, temperature, electric current, and amount of substance).
All other quantities are derived from these.

For an equation to be correct it must be homogeneous. This means that all the terms have the same units. But remember, a homogeneous equation may not be entirely right!

You should always give your numerical answers to the correct number of significant figures.

Scalars have size (magnitude) only.

Vectors have size (magnitude) and direction. Vectors can be represented by arrows.

When vectors are added together to find the resultant, we must take account of their direction.

For vectors acting along the same straight line we take one direction as positive and the other as negative.

To add two perpendicular vectors X and Y you can use Pythagoras' theorem: $R^2 = X^2 + Y^2$

A single vector can be resolved to find its effect in two perpendicular directions.
The component of a vector in any direction is found by multiplying the vector by the *cosine* of the angle between the vector and the required direction.

Subtracting a vector is the same as adding a vector of the same size, acting in the opposite direction.

▷ Questions

1. Can you complete the following sentences?
   a) Measurements must be given as a number multiplied by a ....
   b) Seconds, metres and kilograms are all .... units. Units made up of combinations of base units are known as .... units.
   c) Vector quantities have both .... and .... Scalars have only ....
   d) The single vector that has the same effect as two vectors acting together is called the ....
   e) The effect of a vector in a given direction is called the .... in that direction.
   f) The .... of a vector in any direction is found by mutiplying the .... by the .... of the angle between the vector and the required direction.

2. Rewrite each of the following quantities using a suitable prefix:
   a) 2 000 000 000 J
   b) 5900 g
   c) 0.005 s
   d) 345 000 N
   e) 0.000 02 m

3. The drag force F on a moving vehicle depends on its cross-sectional area A, its velocity v and the density of the air, ρ.
   a) What are the base units for each of these four variables?
   b) By checking for homogeneity, work out which of these equations correctly links the variables:
      i) $F = kA^2\rho v$
      ii) $F = kA\rho^2 v$
      iii) $F = kA\rho v^2$
      (The constant k has no units.)

4. In a tug-of-war one team pulls to the left with a force of 600 N. The other team pulls to the right with a force of 475 N.
   a) Draw a vector diagram to show these forces.
   b) What is the magnitude and direction of the resultant force?

5. Two ropes are tied to a large boulder. One rope is pulled with a force of 400 N due east. The other rope is pulled with a force of 300 N due south.
   a) Draw a vector diagram to show these forces.
   b) What is the magnitude and direction of the resultant force on the boulder?

6. A javelin is thrown at 20 m s^{-1} at an angle of 45° to the horizontal.
   a) What is the vertical component of this velocity?
   b) What is the horizontal component of this velocity?

7. A ball is kicked with a force of 120 N at 25° to the horizontal.
   a) Calculate the horizontal component of the force.
   b) Calculate the vertical component of the force.

8. Find the resultant of the following system of forces (not drawn to scale):
   a) by scale drawing,
   b) by calculation.

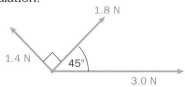

Further questions on page 114.

2 Looking at Forces

We use forces all the time, often without even noticing.
What forces are you using right now?
As you sit reading this book, forces keep you in the chair,
hold your chair together and allow you to turn the pages.

Forces are involved whenever objects interact. This is true
for objects as big as planets, or as small as atoms.

In this chapter you will learn:
- about the different types of forces that exist,
- what causes these forces,
- how to draw the forces acting on an object.

There are three basic types of force that exist in our universe:

Windsurfers need to balance forces carefully.

▷ Gravitational force

All objects with mass attract each other with a gravitational force.
So why don't you feel yourself being pulled towards other objects
all the time? We only really notice gravitational forces caused by
extremely massive objects such as the Earth.
It is the gravitational pull of the Earth that keeps your feet on the
ground and gives you weight. We will look at gravitational forces
in more detail in Chapter 8.

In conversation people often use the word 'weight' when they
really mean mass. So what is the difference? What do these
words actually mean?

Your mass is the amount of matter you contain.
Mass is measured in **kilograms (kg)**.

Your weight is the force of gravity pulling you down.
Weight, like all forces, is measured in **newtons (N)**.

*Isaac Newton is said to have 'discovered'
gravity when an apple fell on his head!*

Weight and mass are linked by this equation:

Weight	**=**	**mass**	**×**	**g**
(in N)		(kg)		(N kg^{-1})

In symbols:

$$W = m\,g$$

g is known as the ***gravitational field strength*** or
the ***acceleration due to gravity***.
On Earth $g = 9.81$ N kg^{-1} (or 9.81 m s^{-2}).
So every 1 kg has a weight of 9.81 N.
(In calculations it is sometimes rounded up to 10 N kg^{-1}.)

Example 1
A man has a mass of 70.0 kg. What is his weight (a) on Earth, where $g = 9.81$ N kg^{-1},
(b) on the Moon, where $g = 1.60$ N kg^{-1} and (c) on Jupiter, where $g = 26.0$ N kg^{-1}?

a) On Earth: weight $W = mg = 70.0$ kg $\times$ 9.81 Nkg^{-1} = <u>687 N</u> (3 s.f.)

b) On the Moon: weight $W = mg = 70.0$ kg $\times$ 1.60 Nkg^{-1} = <u>112 N</u> (3 s.f.)

c) On Jupiter: weight $W = mg = 70.0$ kg $\times$ 26.0 Nkg^{-1} = <u>1820 N</u> = 1.82×10^3 N (3 s.f.)

▷ Electromagnetic forces

Electromagnetic forces cause attraction and repulsion between positive and negative electric charges.
They also cause magnetic poles to attract and repel.

Opposite charges attract:

Opposite poles attract:

Like charges repel:

Like poles repel:

Hair-raising repulsion between electric charges !

Electromagnetic forces play a vital part in holding our world together. Our world is made of atoms. But what holds atoms together?

Electromagnetic attraction holds negatively charged electrons to the positive nucleus. Electromagnetic forces hold molecules together, and control all chemical reactions.
You may have already come across the links between electricity and magnetism if you studied the electric motor and generator at GCSE. We will look at electromagnetic effects in more detail in Chapter 18.

▷ Nuclear forces

You might not have met this last group of forces. They are only experienced by sub-atomic particles. There are two types:

- **Weak nuclear force**
  These are the forces involved in the radioactive decay of atoms. We will look at this in more detail in Chapters 25 and 27.

- **Strong nuclear force**
  We have said that electromagnetic forces hold electrons to the nucleus, but what holds the nucleus together? Why does the repulsion between positive protons in the nucleus not force the nucleus apart? The answer is that they are held together by the strong nuclear force.
  As its name suggests, the strong nuclear force is the strongest force of all. Unlike electromagnetic force, it only acts over a very short range to bind neighbouring nucleons together.

The CERN research facility in Geneva. Scientists at CERN are carrying out experiments to find out more about weak and strong nuclear forces.

But what about simple pushing and pulling forces?
Where do they fit in? At first glance it is hard to see how the forces you use when pushing a trolley or pulling a rope can be one of these three types. Can you work out which type they are?

All the forces involved when objects are in direct contact are **electromagnetic** forces.
For example, your pushing force is due to repulsion between the electrons of the atoms in your hand and those in the object!

We will look in more detail at these everyday contact forces on the next few pages.

▷ 'Normal' contact forces

The force of gravity pulls you towards the centre of the Earth.
So what stops you from sinking into the ground?
The answer is the 'normal' contact force, sometimes called the 'normal' reaction. This is a force that exists wherever two solid surfaces are in contact.
The word 'normal' means **at 90°** to the surfaces.

The diagram shows you the direction of the normal contact force for two surfaces in contact. Each surface feels an equal and opposite force.

What causes this force? Contact forces are due to the **electromagnetic** forces acting between the atoms and molecules of the two surfaces in contact. Pushing two surfaces together creates a repulsive electromagnetic force that we feel as the contact force. The strength of this force depends on the materials involved.
You can push your hand through tissue paper as the contact forces are weak. Pushing your hand through a brick wall is a bit trickier!

▷ Tension

When we stretch a wire or string it is in a state of tension.
When a material is stretched its molecules move further apart.
An attractive electromagnetic force between the molecules acts to pull them back together again, creating tension.

Compressive forces are produced by the opposite effect.
When a solid is squashed its molecules move closer together.
A net repulsive electromagnetic force tries to push them further apart. We will look at tension and compression in more detail in Chapter 13.

The diagram shows the tension forces in a wire supporting a picture frame. Tension forces, like all forces, come **in pairs**.
Two forces act along each section of the wire, trying to pull the atoms back together as the wire stretches.

But don't these forces just cancel out? Overall, all forces will cancel, but each tension force acts on a different object.
The forces at the top of the wire act on the nail. These forces need to be balanced by the other forces acting on the nail such as its weight, friction and contact forces between the wall and the nail.
The forces of tension at the bottom of the wire act on the picture frame to balance its weight.

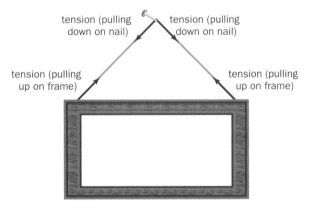

▷ Friction between solid surfaces

Friction tries to prevent motion between two surfaces in contact.
As it is trying to stop things moving, friction always acts parallel to the surfaces in contact and in the *opposite* direction to the motion (or the motion it's preventing).

What does the size of the frictional force depend on?
There are two key factors:
● the type of surfaces in contact, and
● how hard the surfaces are pressed together.

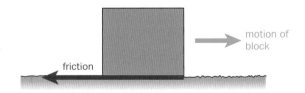

*Only the friction force acting on the box is shown.
All forces come in pairs so an equal but opposite
friction force acts on the ground.*

So what causes friction? You may think that some surfaces are perfectly smooth, but take a look at this magnified photograph:

Magnified photo of the surface of paper.

The page of this book feels smooth but, on a microscopic scale, the surface is very uneven. An object resting on the page is only actually in contact at a few points. The object's weight effectively acts on this very small contact area. This creates huge pressures that weld the surfaces together. We have to break these welds to slide something over the page. This is what creates friction.

As well as opposing motion, friction can cause surfaces to heat up and eventually wear away. But is friction always a bad thing? What would our world be like without it?

Think about how hard it is to walk over a surface such as ice where the friction is reduced. With no friction between you and the ground you could not walk forwards at all. One foot needs to push back on the ground without slipping as you step forwards. And how many objects around you would collapse without friction to hold screws and nails in place? Even your clothes would eventually unravel without friction to hold the fibres together! Without friction you could not even pick anything up to hide your embarrassment! Perhaps friction is not such a bad thing after all.

Loss of friction would be a problem!

Just look at how many ways you either use or try to reduce friction when you ride a bicycle:

You can reduce your friction with the air by crouching low, wearing a streamlined helmet and tight clothing, and even by shaving your legs!

Friction keeps your hands on the handlebars

The brakes operate by applying friction between the wheel rim and the brake blocks

Oil is used to reduce the friction in the gears and chain

Friction between the tyres and the road allows you to move forwards

Friction keeps your feet on the pedals

Motive force

We call a force that drives something forwards a 'motive force'. Motive forces are created by friction.

But doesn't the engine drive a car forwards? Not directly. The engine power turns the car's wheels. Without friction the wheels would just spin on the spot.

The diagram shows you the front wheel of a car on a road:

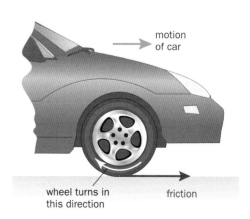

motion of car

wheel turns in this direction

friction

As the wheel turns, friction is created at the point of contact between the wheel and the road. At this point the wheel is moving backwards. To prevent this motion the friction force must act in the opposite direction. This is the forward force that drives the car.

▷ Fluid resistance (drag)

Liquids and gases are *fluids* – they both flow. When an object moves through a fluid, it feels some resistance to its motion. This is a type of friction. Friction in fluids is often called 'drag'.

So what causes drag when an object moves through a fluid? The fluid needs to be pushed out of the way as the object moves through it. Friction is created by contact between the molecules of fluid and those of the solid surface.
The faster an object moves through a fluid, the greater the drag as more molecules move past each other each second.

Drag also depends on the **viscosity** of the fluid. The viscosity of a fluid is a measure of how easily it will flow. Even air has some viscosity. This creates air resistance. Thick, sticky liquids such as treacle are very viscous. An object moving through treacle would suffer a lot of drag. Viscosity is very temperature dependent.
You may have noticed that honey and syrups flow more easily when hot. Engine oils thicken on cold days making it harder to start a car. In liquids, viscosity decreases with increasing temperature. Heat increases the speed of the liquid molecules, decreasing the intermolecular forces.
The opposite is true of gases. Gases become more viscous as they get hotter due to the greater frequency of intermolecular collisions.

The motion of real fluids is complex. But in simple terms we can treat the fluid as moving in one of two ways:

1. Streamline (laminar) flow

The diagram shows the flow of a fluid around a smooth object. Lines of flow are called **streamlines**. In streamline flow the movement of fluid is steady and the pattern of streamlines does not change.
This is an idealised model of flow. The fluid particles moving past a given point all follow the same path at the same speed. In streamline flow the drag is *proportional* to the velocity (see the next page).

Streamline flow is sometimes called **laminar flow**. Laminar flow means layered flow. When a fluid moves through a pipe, layers of fluid move parallel to each other at different speeds. Fluid particles next to the pipe stick to it and so are stationary. The next fluid layer slides over the first one so it flows a little faster. The fluid at the centre moves fastest. The layers all move steadily with no mixing.

2. Turbulent flow

Above a certain **critical speed**, all streamline flow becomes turbulent. The fluid motion is no longer steady. The speed and direction of the fluid particles moving past a given point in turbulent flow are constantly changing.
Turbulent flow occurs in most practical situations involving moving fluid. You can see turbulent flow in air when you blow out a candle. The rising smoke swirls and mixes in the turbulent air.

Turbulence creates additional drag. For turbulent flow, the drag is proportional to the *square* of the velocity. So doubling the speed of flow will give you four times the drag force!
The amount of turbulence also depends on shape. Hard-edged shapes will create turbulence at much lower speeds than smooth streamlined shapes such as aerofoils.

The 'streamlined' shape of a dolphin allows water to flow smoothly around it, reducing drag.

Fluid	T /°C	η /milli Pa s
hydrogen	0	8.4×10^{-3}
	27	9.0×10^{-3}
air	0	17.4×10^{-3}
	27	18.6×10^{-3}
water	20	1.00
	40	0.65
	100	0.28
mercury	20	1.55
blood	37	3–4
honey	20	2–10
tar	20	30 000

The SI unit of viscosity η is the **pascal second** (Pa s).

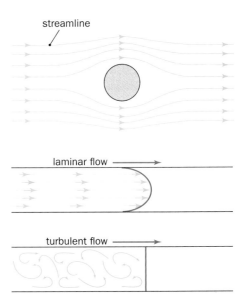

Car in a wind tunnel. The smoke-streams show streamline airflow over the car.

▷ Stokes' Law

The mathematics of viscous drag on irregular shapes is difficult. However, we can calculate the drag acting on a **smooth sphere** moving through a fluid using an equation worked out in 1851 by Sir George Stokes. This is known as **Stokes' Law**:

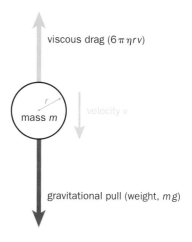

$$\boxed{\textbf{Viscous drag } = \ 6\pi \ \eta \ r \ v}$$

where

η = viscosity of the fluid (kg s^{-1} m^{-1})
r = radius of the sphere (m)
v = velocity of the sphere relative to the fluid (m s^{-1})

The viscous drag is measured in newtons (N).
The force acts in the opposite direction to the sphere's velocity.

Stokes' Law assumes the fluid is uniform throughout and that the flow is laminar. The equation is therefore only valid at relatively low speeds.

For a sphere falling through fluid, the upward drag force (together with any upthrust, see page 28) can be enough to balance the weight. The sphere then falls at a constant speed called its 'terminal velocity' (see page 45).

Example 2
At what speed does a marble of radius 8.00 mm and mass 7.00 g fall through oil of viscosity 0.985 kg s^{-1} m^{-1} so that the viscous drag force acting on it balances its weight? (Ignore any upthrust; $g = 9.81$ N kg^{-1})

When the forces are balanced, viscous drag $= W = mg$

Rearranging Stokes' equation: $v = \dfrac{\text{viscous drag}}{6\pi\eta r} = \dfrac{mg}{6\pi\eta r} = \dfrac{7.00\times10^{-3}\times9.81}{6\pi\times0.985\times8.00\times10^{-3}} = \underline{0.462 \text{ m s}^{-1}}$ (3 s.f.)

▷ Lift

Try this quick experiment. Hold a strip of paper just below your lips and blow across its surface. What happens?
You might expect the paper to be forced downwards but in fact it lifts up! Lift forces are essential for flight. To maintain steady flight, an aircraft must produce enough lift to balance its weight.

So what causes lift? The surprising answer is that lift is caused by the viscosity of the air. As you blow across the paper strip, a laminar airflow follows the downward curve of the paper as the bottom layer of air sticks to the paper surface. This forces the air downwards. The downward force on the air is matched by an equal upward force on the paper, as described by Newton's Third Law (see page 56). This upward force is the lift.

lift

The wings of birds and many aircraft have a special shape called an aerofoil. As the wing moves through the air, the air is dragged around the curve of the wing and forced downward, resulting in an upward lift on the wing.

A key factor in wing lift is the **angle of attack**. The steeper the angle, the greater the downward component of force on the air around the wing. This creates greater lift. If the angle of attack becomes too great, though, the airflow becomes turbulent, leading to a sudden loss of lift. This can cause an aircraft to 'stall' in mid-air.

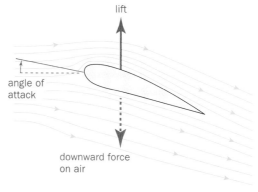

angle of attack

downward force on air

▷ Free-body force diagrams

We have seen that forces are caused by the interaction of two objects.
So forces always come in pairs.
Can you remember how we represent forces in diagrams?
Forces are vector quantities, so they are drawn as vector arrows (see Chapter 1).
This can be straightforward, but in a single physics problem you may have to deal with several interacting objects and many different forces.

Even a simple example such as a man standing on the Earth involves four forces. Can you work out what these are?

1. The Earth exerts a gravitational force of attraction on the man (his weight).
2. The man exerts a gravitational force of attraction on the Earth.
3. The Earth exerts a normal contact force on the man.
4. The man exerts a normal contact force on the Earth.
   (Remember: 'normal' means at 90°)

So how can we simplify this?
For most situations we only need to draw a **free-body force diagram**.
This is a diagram that shows all the forces acting on *just one object*.
Here are the free-body force diagrams for the man, and for the Earth:

normal
contact force
of Earth on man

gravitational force of
Earth on man

Free-body force diagram for the man.

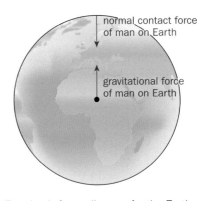

normal contact force
of man on Earth

gravitational force
of man on Earth

Free-body force diagram for the Earth.

Here are some examples of the free-body force diagrams that you will need to draw to solve mechanics problems:

1. Balanced beams

This is the free-body force diagram for a beam resting on two supports. The diagram shows only the forces *acting on the beam*:

The weight of a uniform beam is taken to act at its centre.
(See Chapter 3 for more about *centre of mass*.)

If an object is placed on the beam, it will exert a normal contact force on the beam. This force is numerically equal to the object's weight.

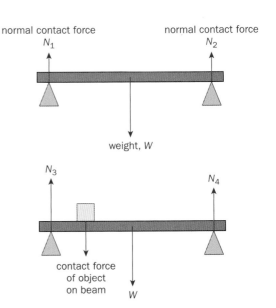

normal contact force
N_1

normal contact force
N_2

weight, W

N_3

N_4

contact force
of object
on beam

W

2. Suspended objects

Here is the free-body force diagram for a
bungee jumper at the bottom of a jump:

It shows the two forces **acting on the person**
at that instant.

Which of these two forces is larger at this point?
Why does the person bounce back up?

3. Objects resting against rough and smooth surfaces

The words 'rough' and 'smooth' have a particular meaning in Physics.
If a surface is described as 'rough' we need to take account of friction.
If a surface is described as 'smooth' we can treat it as friction-free.
This is often done to make a problem easier.

So how does this affect our free-body force diagrams?
As an example take a look at this free-body force diagram for a ladder
resting against a wall. It shows only the forces **on the ladder**:

In this example the ground is a 'rough' surface but the wall is 'smooth'.
At the point of contact with the smooth wall there is only a normal
contact force.
At the rough ground, though, we need to show two forces acting.
These are the normal contact force and the friction force.

The normal contact forces act at 90° to each surface.
How do we decide the direction of the friction force? Think about
which way the ladder would move if it started to slip. The bottom of
the ladder would slip away from the base of the wall. We draw the
friction force in the opposite direction to oppose this motion.

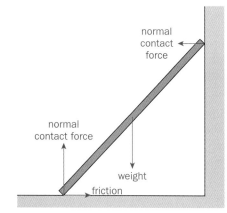

A ladder leaning against a wall.

4. Moving objects

Here you also need to remember to include any drag forces.

a) Free-body force diagram for a car on a level road:

Where on the car do the contact forces and friction
with the road actually act? Notice that even though
the car is in contact with the road at each of the
four wheels, we can simplify this to show a single
overall contact force and a single motive force.

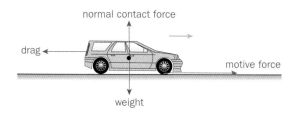

b) Free-body force diagram for a car on a slope:

Check that you are happy with the direction of these
forces. Weight acts vertically downwards.
The normal contact force acts at 90° to the road.
The motive force acts along the slope in the direction
that the car is moving. The drag force acts parallel
to the road and in the opposite direction to the motion.

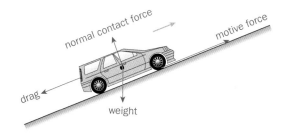

▷ Pressure

People sometimes confuse pressure and force. Pressure tells us how concentrated a force is. It is the *force per unit area*:

$$\text{pressure} = \frac{\text{force (in N)}}{\text{area (in m}^2)} \quad \text{or} \quad p = \frac{F}{A}$$

The SI unit of pressure is the **pascal (Pa)**. **1 Pa = 1 N m^{-2}**.
In the formula **F** is the 'normal' force, at 90° to the surface.

Why do camels have large flat feet? By spreading their weight over a large area they reduce the pressure that they exert on the ground. This means they are less likely to sink into the sand.

Sharpened knives cut more easily for the opposite reason.
A sharp edge has a smaller area. This creates a greater pressure when you press down, making it easier to cut through things.

Is pressure a scalar or a vector? Although its calculation involves force, which you know is a vector, pressure is a *scalar* quantity. By definition, its direction must always be at 90° to the surface it acts on.

Why is it easier to walk over soft snow in skis rather than boots? Skis increase your area of contact, reducing the pressure on the snow.

Example 3
A man can safely lie on a bed of nails by spreading his weight over 750 nails. The man weighs 600 N and the area of each nail is 8.0×10^{-6} m^2. Calculate the average pressure on each nail.

Average weight acting on each nail $= \dfrac{600 \text{ N}}{750 \text{ nails}} = 0.80$ N per nail

Average pressure on each nail, $p = \dfrac{F}{A} = \dfrac{0.80 \text{ N}}{8.0 \times 10^{-6} \text{ m}^2}$

$\qquad\qquad\qquad\qquad = 1.0 \times 10^5 \text{ N m}^{-2}$
$\qquad\qquad\qquad\qquad = 1.0 \times 10^5 \text{ Pa}$

Fluid pressure

You cannot feel it, but the air in our atmosphere is exerting a huge pressure on you. At sea level the weight of air above you exerts a pressure on your body of over 100 000 Pa! This is about the same as the pressure produced by 8 people standing on the area of this book. So why aren't you crushed? Luckily the atmospheric pressure is balanced by the pressure inside your body.

Did you know that we need atmospheric pressure to breathe? To breathe in, your diaphragm moves down, creating more space in your chest. This lowers your internal pressure so that the higher atmospheric pressure outside your body pushes air into your lungs.

Why does it become more difficult to breathe at higher altitudes? At high altitude there is less weight of air pushing down above you. This reduces the atmospheric pressure.

Underwater, the pressures on you become even greater. This is due to the added weight of water pushing down above you. Submarines need thick hulls to withstand the enormous pressures deep underwater.

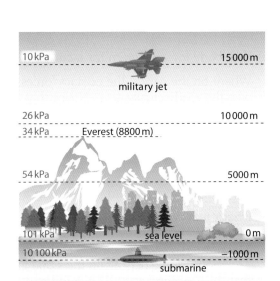

10 kPa	15 000 m
	military jet
26 kPa	10 000 m
34 kPa	Everest (8800 m)
54 kPa	5000 m
101 kPa	sea level — 0 m
10 100 kPa	−1000 m
	submarine

Calculating fluid pressure

We can calculate the amount of pressure a fluid exerts by applying our basic definition of pressure.
Think about the cylinder of fluid in the diagram:

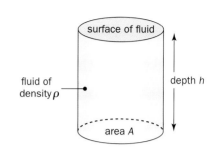

Volume of cylinder = area of base × height = $A h$

Mass of fluid in cylinder = volume × density = $A h \rho$

Force acting on area A = weight of fluid = mass × g = $A h \rho g$

Pressure exerted on the base, $p = \dfrac{\textbf{force (weight)}}{\textbf{area}} = \dfrac{A h \rho g}{A} = h \rho g$

So to find the pressure p due to fluid of density ρ at depth or height h you can use the equation:

pressure	=	height	×	density	×	gravitational field strength
(Pa)		(m)		(kg m^{-3})		(N kg^{-1})

or $\boxed{p = h \rho g}$

This equation only applies to **incompressible fluids**, as we have assumed that the density ρ does not change with depth.
This is a good approximation for most liquids.
With gases the density tends to increase with depth, as gases compress under their own weight.

Surprisingly, the pressure in a fluid does not depend on the shape or area of the container or on the total mass of the fluid.
Fluid pressure depends only on depth and density.
In the diagram, the pressure at depth h is **the same** in all three containers.

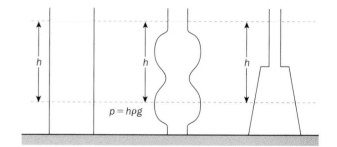

$p = h \rho g$

The key points to remember about fluid pressure are:

- Pressure increases with depth and all points at the same depth are at the same pressure.

- Pressure acts equally in all directions. This means that the forces produced will always be at right angles to any surface in contact with the fluid.

Example 4
What is the total pressure exerted on a diver at a depth of 80.0 m under the sea?
Density of seawater ρ_{sea} = 1030 kg m^{-3}.
Atmospheric pressure = 1.01×10^5 Pa. Take g = 9.81 N kg^{-1}.

Pressure due to seawater = $h \rho g$ = 80.0 × 1030 × 9.81 = 8.08×10^5 Pa

Total pressure on diver = pressure due to water + atmospheric pressure
$$= 8.08 \times 10^5 + 1.01 \times 10^5 = \underline{9.09 \times 10^5 \text{ Pa}}$$

▷ Upthrust

Have you ever tried to push an inflated balloon or beach ball underwater? It's not easy! You can feel the water pushing it back up. The force you feel is called **upthrust**.
The upthrust increases as you push more of the ball under water. Upthrust occurs in liquids and gases. It is the force that keeps ships and hot-air balloons afloat.

So what causes upthrust?

Upthrust occurs because, as we saw on page 27, the pressure in a fluid increases with depth. When you push an object underwater, the forces that the water pressure exerts at the base of the object are greater than those exerted at the top.
The upthrust is the resultant upward force.

What affects the amount of upthrust?

Think about the submerged block in the diagram. The block has top and bottom faces of area A and the density of the fluid is ρ:

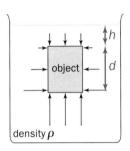

Fluid pressure on the top face = $h\rho g$
So downward force on the top face = area × pressure = $Ah\rho g$

Fluid pressure on the bottom face = $(h+d)\rho g$
So upward force on the bottom face = area × pressure = $A(h+d)\rho g$

The upthrust is the resultant upward force:
Upthrust = upward force on bottom face − downward force on top face
$$= A(h+d)\rho g - Ah\rho g$$
$$= Ad\rho g$$

Now let's calculate the weight of the fluid 'displaced' or pushed out of the way by the block:

Mass of fluid displaced = volume × density = $(A \times d) \times \rho$
Weight of fluid displaced = mass × gravitational field strength = $Ad\rho g$
This is exactly the same value as the upthrust.

> **When an object is completely or partly immersed in a fluid, it experiences an upthrust equal to the weight of the fluid displaced.**

This is known as **Archimedes' principle** after the Greek mathematician who discovered it when he noticed his bath water overflowed as he got in. Legend has it that he was so excited by his discovery he ran naked down the street shouting 'Eureka' ('I have found it')!

Archimedes' principle is true for *all fluids* (that is, all liquids *and* all gases), and it works for any shape. You can prove this for yourself with the simple apparatus shown in the diagram:
Notice that the upthrust can be found either by measuring the weight of the water displaced by the object or from the change in the reading on the spring balance.

So upthrust depends on the **volume of the object submerged** and the **density of the fluid**.
These determine the weight of fluid displaced by the object.

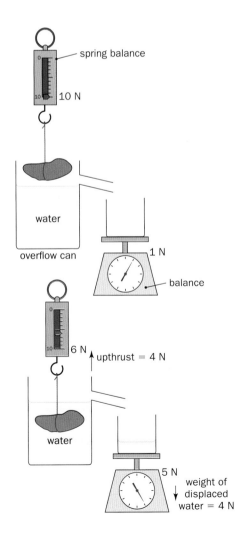

▷ Floating and sinking

Ocean liners weighing many thousands of tonnes can float because of the huge amount of water that their hulls displace. For an object to float it must displace its own weight of the surrounding fluid.

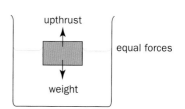

If an object floats: **upthrust = weight**
If an object sinks: **upthrust < weight**

A simple way to decide if something will float or sink is to compare its overall density (see page 184) with that of the surrounding fluid:

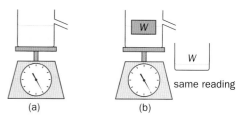

- If an object is **denser** than the liquid, it will weigh more than the same volume of liquid. So the upthrust will be less than the weight and it will sink.

- If an object is **less dense** than the liquid, the fluid it displaces when fully submerged will weigh more than the object. The upthrust will be greater than the weight, which will make the object float up to the surface. Once it's partially above the surface, the amount of fluid it displaces reduces, reducing the upthrust, until the weight and upthrust are balanced.

- If an object has exactly the **same density** as the liquid, it will stay still, neither sinking nor floating upwards, as no resultant force acts on it.

Look at the table here and decide which substances will sink in water. What will float on mercury? Will ice float on petrol?

Substance			Density	
Solid	Liquid	Gas	kg m^{-3}	g cm^{-3}
gold			19 000	19
	mercury		14 000	14
lead			11 000	11
iron			8000	8
	water		1000	1
ice			920	0.92
	petrol		800	0.80
		air	1.3	0.0013

Example 5

A cargo ship has a rectangular hull 22 m wide and 180 m long. When loaded it floats with the bottom of the hull 5.0 m below the water surface. What is the total weight of the ship? (Density of seawater = 1030 kg m^{-3}; g = 9.81 N kg^{-1})

From Archimedes' principle, upthrust = weight of water displaced
$$= \text{mass of water} \times g$$
$$= (\text{volume} \times \text{density}) \times g$$
$$= (22 \times 180 \times 5) \times 1030 \times 9.81 = 2.0 \times 10^8 \text{ N (2 s.f.)}$$

As the ship is floating, upthrust = weight of ship = $\underline{2.0 \times 10^8 \text{ N}}$.

Example 6

A glass of water contains a 3.0 cm cube of ice. Calculate:
a) the volume of water that the floating ice cube displaces, and
b) the volume of water in the melted ice cube.
Use the density values for ice and water given in the table above.

a) Weight of displaced water = upthrust = weight of ice cube
Since **weight = m g**, mass of displaced water = mass of ice cube = density × volume
$$= 920 \times 0.03^3 = 0.025 \text{ kg}$$
∴ volume of displaced water = mass ÷ density = 0.025 ÷ 1000 = $\underline{2.5 \times 10^{-5} \text{ m}^3}$

b) The mass will not change as the ice melts. So mass of melted ice = 0.025 kg.
Volume of melted ice = mass ÷ density = 0.025 ÷ 1000 = $\underline{2.5 \times 10^{-5} \text{ m}^3}$

So the entire melted ice cube fits into the same space as the part of the floating cube that was below the surface! This is why the water level stays exactly the same as the ice melts. Try this out for yourself.

▷ Physics at work: Automotive forces

Hydraulic brakes

Liquids are almost incompressible. If you put a section of liquid under pressure, the pressure is transmitted to all parts of the liquid. This is known as **Pascal's principle** after the man who discovered it. This idea is used in all hydraulic systems, such as those used in car brakes:

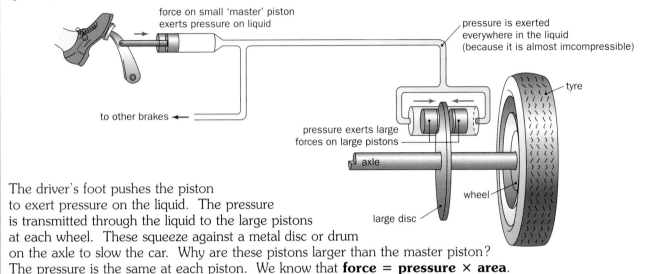

The driver's foot pushes the piston
to exert pressure on the liquid. The pressure
is transmitted through the liquid to the large pistons
at each wheel. These squeeze against a metal disc or drum
on the axle to slow the car. Why are these pistons larger than the master piston?
The pressure is the same at each piston. We know that **force = pressure × area**.
So doubling the area with the same pressure gives twice the force at the brakes that you apply to the pedal.
Hydraulic brakes also prevent you swerving due to uneven application of the braking force at the wheels, as the pressure is transmitted **equally** to each wheel.

Racing car physics

Did you know that at high speeds the fins, wings and spoilers on a modern Formula 1 car produce more down-force than the weight of the car? In theory it could drive upside down along a ceiling!

The huge down-forces help to push the car's tyres on to the track, improving grip and traction. These forces are created in the same way that an aeroplane wing generates lift.
The car's shape forces the air streaming past it to move upwards.
The upward force on the air equals the downward force on the car.
The cars are also designed to reduce air turbulence, as turbulence increases drag, which acts to slow the car down.

A key factor affecting an F1 car's acceleration is the grip of the tyres. Racing cars use smooth tyres called 'slicks'. These maximise the area of contact between the tyre and the road. This produces high friction to grip the road, increasing the motive force or traction.

So why do conventional tyres have grooves cut into them?
Racing slicks are only safe to use in dry road conditions.
If water comes between the tyre and the road, the friction is reduced.
The tyres lose their grip and the car can 'hydroplane' – the driver loses control of the car as it glides over a film of water.
Grooves are cut into a tyre to help prevent this. The tread patterns of a modern tyre are mathematically designed to take out as much water from under the tyre as possible, improving its grip on the road.
An F1 'wet tyre' can disperse 65 litres of water per second!

Formula 1 racing tyres must withstand far larger forces than standard car tyres. Their special rubber is designed for grip rather than durability.

Summary

The 3 basic types of forces are gravitational, electromagnetic and nuclear (strong and weak).
Weight and mass are linked by the equation: **weight** (N) = **mass** (kg) × **g** (N kg^{-1})

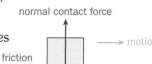

Friction is the force that tries to stop things moving. Friction acts parallel to the surfaces in contact, while a normal contact force acts at 90° to two surfaces in contact.

Friction in fluids is called *drag*. Fluid flow is smooth or *streamlined* at low velocity. At higher velocities, flow becomes *turbulent*. Turbulence increases drag.

Stokes' Law: For a smooth sphere of radius r moving at velocity v in a fluid of viscosity η, **viscous drag = 6π η r v**

Forces always come in pairs, but free-body force diagrams show all of the forces acting on just *one* object.

$$\textbf{pressure (Pa)} = \frac{\textbf{force (N)}}{\textbf{area (m}^2\textbf{)}}$$ In a fluid: **pressure** (Pa) = **depth** (m) × **density** (kg m^{-3}) × **g** (N kg^{-1})

An object immersed in a fluid feels an upthrust equal to the weight of the fluid it displaces.
This is known as **Archimedes' principle**. If an object floats, the upthrust must equal the object's weight.

▷ Questions

(Take g = 9.81 N kg^{-1} on Earth)

1. Can you complete these sentences?
   a) The .... of an object is the amount of matter it contains. It is measured in ....
   b) Weight is the force of .... pulling down on an object. Weight is measured in ....
   c) .... tries to prevent motion between surfaces in contact. Friction in fluids is known as ....
   d) The pressure in a fluid increases with the .... and .... of the fluid. The viscosity of a fluid depends on its ....
   e) The upward force on an object immersed in a fluid is called .... The upthrust is equal to the .... of the fluid displaced.

2. Which weighs more, a 100 kg mass on Earth or a 120 kg mass on Venus (g_{Venus} = 8.87 N kg^{-1})?

3. A brick has a mass of 3.00 kg. Its dimensions are 30 cm × 10 cm × 7.0 cm. Calculate the force and the pressure exerted by the brick on the table when it rests on:
   a) the smallest face, b) the largest face.

4. Explain the following:
   a) Drawing pins have pointed ends.
   b) A ladder would be useful if you had to rescue someone who had fallen through thin ice.
   c) Tractors have large tyres.

5. Draw a free-body force diagram for:
   a) a ladder resting at an angle between a rough wall and rough ground;
   b) the same ladder as in (a) but this time with a man standing $\frac{1}{4}$ of the way up from the bottom.

6.

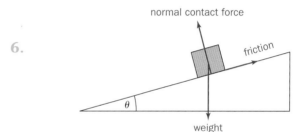

The block in the diagram weighs 120 N. It will start to slip down the slope when the component of its weight acting down the slope equals the maximum frictional force. If the block slips when the angle θ = 35.0°, what is the maximum frictional force?

7. A swimming pool is 1.0 m deep at one end and 3.5 m deep at the other. What change in pressure would you experience if you swam from one end to the other along the bottom of the pool? (Density of water = 1000 kg m^{-3})

8. A ship displaces 4500 m^3 of seawater. The ship weighs 5.0 × 10^7 N. Will it float or sink? (Density of seawater = 1030 kg m^{-3})

9. A buoy is tied to the seabed by a strong chain. At high tide it is completely submerged.
   a) Draw a free-body force diagram showing the **three** forces acting on the submerged buoy.
   b) The buoy has a volume of 1.8 m^3. Calculate the upthrust on it if the density of seawater is 1030 kg m^{-3}.
   c) If the buoy has a mass of 200 kg, calculate the tension in the chain.

Further questions on page 114.

3 Turning Effects of Forces

Why are door handles usually on the opposite edge of the door to the hinges? Try pushing open a door near the hinge, and then at the other edge, to feel the difference for yourself.

By pushing further from the hinge you produce a much larger turning effect.

When you need to undo a tight nut, a spanner with a long handle is more effective than a short one, for the same reason.

By applying the force further from the pivot you create a greater turning effect.

In this chapter you will learn:
- how to calculate the turning effect of a force,
- the conditions needed for an object to be balanced or 'in equilibrium',
- the meaning and significance of the centre of gravity.

Conditions for equilibrium

If an object is stationary, all the forces acting on the object must cancel out. We say that the object is **in equilibrium**.

But is the reverse true? If there is no overall force acting on an object, does this mean it is in equilibrium?

To answer this question for yourself, look at these two diagrams:

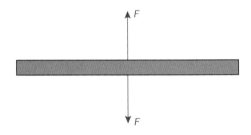

In both cases, the force up equals the force down, and so they cancel out.

In the first diagram, the rod would remain stationary.
It is in equilibrium.

But what happens to the second rod when we apply these forces?

Although the forces cancel out, they don't act along the same line. The two forces will create a turning effect or **torque**. The rod will start to rotate clockwise, so it is not in equilibrium.

From this example you should be able to see that if an object is in equilibrium, **two** things must be true:

1. there is no net force acting in any direction, *and*

2. there is no net turning effect about any point.

The size of the turning effect produced by a force is called the **moment of the force**.

▷ Calculating moments

Try this simple experiment. Hold a ruler horizontally and hang a weight on the ruler near your hand. Try to keep the ruler horizontal as you move the weight further from your hand. How does the turning effect change?

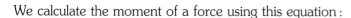

Now try it with a heavier weight.
You can feel that the turning effect or moment depends on the size of the force (the weight) **and** the distance from your hand.

We calculate the moment of a force using this equation:

> **moment of a force** = **force** × **perpendicular distance** from the force
> about a point   to that point
>   (N m)   (N)   (m)

Moments are measured in **newton-metres**, written **N m**.

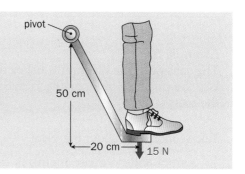

Example 1
Calculate the moment of the pushing force on the pedal in the diagram:

Moment = force × *perpendicular* distance to pivot

         = 15 N × 0.20 m

         = 3.0 N m

Couples

When two forces that are equal in size and opposite in direction do not act along the same straight line, we say they form a **couple**. A couple has no resultant force. It only produces a turning effect. Your hands on a steering wheel provide a couple to turn the wheel:

The turning effect of a couple is sometimes called a **torque**.

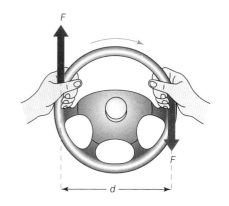

We calculate the moment (torque) of a couple using this equation:

> **couple** = **magnitude of one force** × **perpendicular distance** between
>   the two forces forming the couple
>   (N m)   (N)   (m)

or

> **couple = F × d**

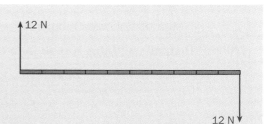

Example 2
Calculate the couple produced by the forces acting on the metre rule in the diagram:

Couple = F × d = 12 N × 1.00 m = 12 N m

Compare this with the answer you get by calculating the **total moment** about **any** point on the metre rule.
(To do this, find the moment of each force separately and add them together.)
What do you notice? You should find that you always get the **same** answer of 12 N m.

▷ Principle of moments

Moments can be clockwise or anti-clockwise.
Look at the photograph of the children on the seesaw:

The weight of the child on the left produces a moment that tries to turn the seesaw anti-clockwise.
The weight of the child on the right produces a clockwise moment.

If the seesaw is balanced, what does this tell us about these two moments? For an object to be in equilibrium (balanced) there must be no overall turning effect about any point.
The clockwise and anti-clockwise moments must cancel out.

This is summed up by **the principle of moments**:

> When an object is in equilibrium:
>
> **sum of the clockwise moments**   = **sum of the anti-clockwise moments**
> about any point   about the same point

Example 3
The seesaw in the diagram is balanced.
Use the principle of moments to calculate the weight W.

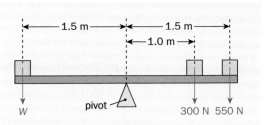

Taking moments about the pivot:

$$\text{sum of anti-clockwise moments} = \text{sum of clockwise moments}$$
$$W \times 1.5\ \text{m} = (300\ \text{N} \times 1.0\ \text{m}) + (550\ \text{N} \times 1.5\ \text{m})$$
$$W \times 1.5\ \text{m} = 300\ \text{N m} + 825\ \text{N m}$$
$$W \times 1.5\ \text{m} = 1125\ \text{N m}$$
$$\therefore\ W = \frac{1125\ \text{N m}}{1.5\ \text{m}} = \underline{750\ \text{N}}\quad (2\ \text{s.f.})$$

Example 4
The diagram shows the forces acting on your forearm when you hold a weight with your arm horizontal. Your elbow joint acts as a pivot:

The clockwise moments produced by the weight of your arm and the weight in your hand must be balanced by an anti-clockwise moment from your biceps muscle.
Use the principle of moments to calculate the force exerted by your biceps, F_B.

Taking moments about the elbow:

$$\text{sum of clockwise moments} = \text{sum of anti-clockwise moments}$$
$$(20\ \text{N} \times 0.16\ \text{m}) + (60\ \text{N} \times 0.32\ \text{m}) = (F_B \times 0.040\ \text{m})$$
$$22.4\ \text{N m} = F_B \times 0.040\ \text{m}$$
$$\therefore\ F_B = \frac{22.4\ \text{N m}}{0.040\ \text{m}} = \underline{560\ \text{N}}\quad (2\ \text{s.f.})$$

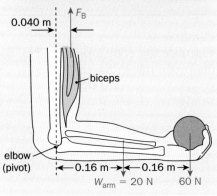

Note that any reaction force at the pivot would have **no** moment about this point, as the perpendicular distance is zero.

▷ Centre of gravity (centre of mass)

Imagine you are carrying a heavy ladder on your shoulder.
How would you position it so that it balanced?
You would rest the **centre** of the ladder on your shoulder.
This point is called its **centre of gravity**.
Although gravity is pulling down on *every* part of the ladder, the whole weight of the ladder can be treated **as if it acts at this point**.
You can certainly feel the whole weight on your shoulder!

What is the difference between centre of gravity and centre of mass?
The **centre of gravity** of an object is the point at which we can take its entire **weight** to act.
The **centre of mass** of an object is the point at which we can take its entire **mass** to be concentrated.

(In places where the gravitational field strength is uniform, the centre of gravity and the centre of mass are at the same point.
This is true for all objects near the surface of the Earth.)

Locating centres of gravity (centres of mass)

For uniform, symmetrical objects the centre of gravity is at the geometric centre.
The diagrams show you the position of the centre of gravity G for some common shapes:

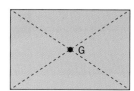

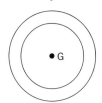
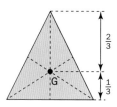

An object will always balance if you support it at its centre of gravity.
Check this for yourself with something simple like a ruler or a book.

Why does a plumb-line always hang vertically?
If you hang an object up so that it can swing, it will always come to rest with its centre of gravity directly below the point of support.
Why is this?
If the line of the weight does not go through the pivot, then there is a moment to make the object turn. It turns until there is no moment.

For irregular shapes you can use this idea to find their centre of gravity:

Example 5 Finding the centre of gravity of an irregular lamina
A lamina is a thin, flat shape. In this example our irregular lamina is an elephant shape cut from flat card.

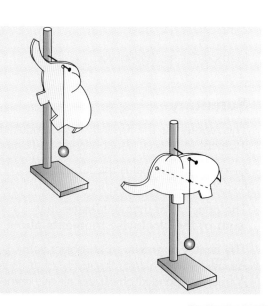

Hang the card from a pin, so that it can swing freely:
When it comes to rest, use a plumb-line to draw a vertical line on the card from this support.
The centre of gravity must be somewhere on this line.

Now hang the card from a different point. Use the plumb-line again to draw a second vertical line. The centre of gravity must be on this line as well. So the centre of gravity is where the lines cross.

How can you check this? You can either draw a third line, which should cross at the same place, or try to balance the card at this point.

▷ Solving equilibrium problems

You can find one or more unknown forces acting on an object in equilibrium by following these 3 steps:

1. Draw a free-body force diagram for the object. Remember to show the weight of the object acting at its centre of gravity.

2. Calculate moments about one or more points. Read the tips:

3. Resolve the forces to find the total force acting in a convenient direction (eg. vertically or horizontally). This must be zero.

By rearranging the equations that these steps produce, you will be able to find the unknown forces.

Tip 1: *If there are **two unknown forces**, take moments about a point through which one of the unknown forces acts. As this unknown force has no moment about this point, it will not appear in your equation (see Example 6).*

Tip 2: *Remember, if **only three non-parallel forces** act on an object, these must form a closed vector triangle when drawn head to tail (see page 14).*

Example 6
A diver weighing 650 N stands at the end of a uniform 2.0 m long diving board of weight 200 N. What are the reaction forces at the supports A and B, if the board is balanced as shown in the diagram?

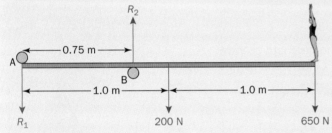

Take moments about **A**:   clockwise moments = anti-clockwise moments

$$(200 \text{ N} \times 1 \text{ m}) + (650 \text{ N} \times 2 \text{ m}) = R_2 \times 0.75 \text{ m}$$
$$200 \text{ N m} + 1300 \text{ N m} = R_2 \times 0.75 \text{ m}$$
$$1500 \text{ N m} = R_2 \times 0.75 \text{ m}$$
$$\therefore R_2 = \frac{1500 \text{ N m}}{0.75 \text{ m}} = \underline{2000 \text{ N}}$$

To find R_1 you have a choice:

Method 1. Resolve vertically:
$$R_2 = R_1 + 650 \text{ N} + 200 \text{ N}$$
$$2000 \text{ N} = R_1 + 850 \text{ N}$$
$$\therefore R_1 = 2000 \text{ N} - 850 \text{ N} = 1150 \text{ N}$$
$$= \underline{1200 \text{ N}} \text{ (2 s.f.)}$$
Method 1 is usually easier.

Method 2. Take moments about **B**:
anti-clockwise moments = clockwise moments
$$R_1 \times 0.75 \text{ m} = (200 \text{ N} \times 0.25 \text{ m}) + (650 \text{ N} \times 1.25 \text{ m})$$
$$R_1 \times 0.75 \text{ m} = 862.5 \text{ N m}$$
$$\therefore R_1 = \frac{862.5 \text{ N m}}{0.75 \text{ m}} = \underline{1200 \text{ N}} \text{ (2 s.f.)}$$

Summary

The moment is the turning effect produced by a force:
Moment of a force about a point (in N m) = force (N) × perpendicular distance from the force to the point (m)

A couple (or torque) is the turning effect due to two equal forces acting in the opposite direction:
Couple (in N m) = magnitude of one force (N) × perpendicular distance between the two forces in the couple (m)

The principle of moments states that, for an object in equilibrium:
sum of the clockwise moments about any point = sum of the anti-clockwise moments about that point

The centre of gravity of an object is the point at which its entire weight appears to act.
An object will balance if supported at its centre of gravity.
A freely suspended object will always come to rest with its centre of gravity directly below the point of suspension.

▷ Questions

1. Can you complete these sentences?
   a) The moment of a force about a point is equal to the force multiplied by the .... distance from the .... to that point.
   b) Moments are measured in ....
   c) The principle of moments states that, if an object is in ...., the total .... moments are .... to the total .... moments.
   d) The centre of gravity is the point through which the whole .... of the object seems to act.
   e) A couple consists of two forces of equal .... acting in .... directions. A couple has no resultant force. It produces only a .... effect.

2. a) A mechanic applies a force of 200 N at the end of a spanner of length 20 cm. What moment is applied to the nut?
   b) Would it be easier to undo the nut with a longer or shorter spanner? Explain your answer.

3. The diagrams show rulers balanced at their centres of gravity.
   What are the missing values X, Y and Z?

   a)

   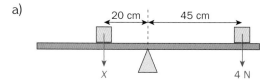

   b)

   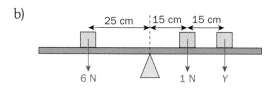

   c)

   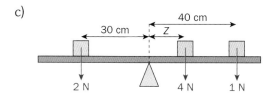

4. Explain the following:
   a) A Bunsen burner has a wide, heavy base.
   b) Racing cars are low, with wheels wide apart.
   c) A boxer stands with his legs well apart.
   d) A wine glass containing wine is less stable.
   e) Lorries are less stable if carrying a load.

Further questions on page 115.

5. Calculate the size of these couples:
   a)

   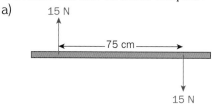

   b)

   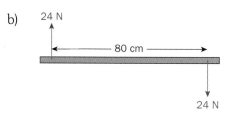

6. The uniform rulers in the diagrams are balanced. The rulers are 1.0 m long and weigh 1.0 N. Find the missing values X and Y.
   a)

   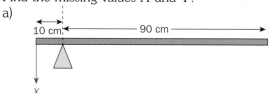

   b)

   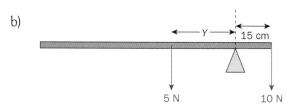

7. Two men are carrying a uniform ladder of length 12 m and weight 250 N. One man holds the ladder 2.0 m from the front end and the other man is 1.0 m from the back of the ladder.
   a) Draw a free-body force diagram for the ladder.
   b) Calculate the upward contact forces that each man exerts on the ladder.

8. The diagram shows a woman using a lever to lift a heavy object. The lever consists of a uniform plank of wood pivoted as shown. The plank is 3.0 m long and weighs 200 N. The object weighs 1200 N.

   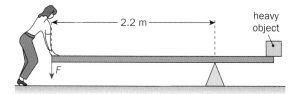

   a) Draw a free-body force diagram showing all the forces acting **on the plank**.
   b) Calculate the downward force F she needs to apply to keep the plank balanced.
   c) What is the reaction force at the pivot?

4 Describing Motion

We live in a world full of movement. Humans, animals and the many forms of transport we use are obvious examples of objects designed for movement. This chapter is about the Physics of motion.

In this chapter you will learn:
- how to describe motion in terms of distance, displacement, speed, velocity, acceleration and time,
- how to use equations that link these quantities,
- how to draw and interpret graphs representing motion.

▷ Distance and displacement

Distance and displacement are both ways of measuring how far an object has moved. So what is the difference?

Distance is a scalar. Displacement is a *vector* quantity (see page 10). Displacement is the distance moved in a particular direction.

The snail in the picture moves from A to B along an irregular path:
The *distance* travelled is the total length of the dashed line.

But what is the snail's *displacement*?
The magnitude of the displacement is the length of the straight line AB. The direction of the displacement is along this line.

▷ Speed and velocity

The speed of an object tells you the distance moved per second, or the *rate of change of* distance:

$$\textbf{average speed} = \frac{\textbf{distance travelled (m)}}{\textbf{time taken (s)}}$$

Speed is a scalar quantity, but velocity is a vector.
Velocity measures the rate of change of *displacement*:

$$\textbf{average velocity} = \frac{\textbf{displacement (m)}}{\textbf{time taken (s)}}$$

Both speed and velocity are measured in **metres per second**, written **m/s** or **m s⁻¹**. With velocity you also need to state the direction.

Using these equations you can find the *average* speed and the *average* velocity for a car journey.
A speedometer shows the actual or *instantaneous* speed of the car. This varies throughout the journey as you accelerate and decelerate.

So how can we find the instantaneous speed or velocity at any point? The answer is to find the distance moved, or the displacement, over a very small time interval. The smaller the time interval, the closer we get to an instantaneous value (see also page 40).

Military jet
450 m s⁻¹

Racing car
60 m s⁻¹

Cheetah
27 m s⁻¹

Sprinter
10 m s⁻¹

Tortoise
0.060 m s⁻¹

Example 1
The boat in the diagram sails 150 m due south and then 150 m due east.
This takes a total time of 45 s.
Calculate (a) the boat's average speed, and (b) the boat's average velocity.

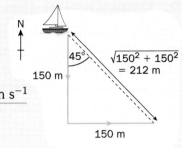

a) average speed $= \dfrac{\text{distance travelled}}{\text{time taken}} = \dfrac{150\text{ m} + 150\text{ m}}{45\text{ s}} = \dfrac{300\text{ m}}{45\text{ s}} = \underline{6.7\text{ m s}^{-1}}$

b) average velocity $= \dfrac{\text{displacement}}{\text{time taken}} = \dfrac{212\text{ m}}{45\text{ s}} = \underline{4.7\text{ m s}^{-1}}$ in a SE direction

▷ Acceleration

Acceleration is the rate of change of velocity:

$$\textbf{acceleration} = \frac{\textbf{change in velocity (m s}^{-1})}{\textbf{time taken (s)}}$$

Acceleration is measured in metres per second per second, or
metres per second squared, written **m/s²** or **m s⁻²**.
It is a vector quantity, acting in a particular direction.

The change in velocity may be a change in *speed,* or *direction*, or both.
If an object is slowing down, its change in velocity is negative.
This gives a negative acceleration or 'deceleration'.

▷ Indicating direction

It is important to state the direction of vectors such as displacement,
velocity and acceleration. In most motion problems you will be dealing
with motion in a straight line (*linear motion*).
In this case you can use + and − signs to indicate direction.
For example, with horizontal motion, if you take motion to the **right** as
positive, then:

-3 m	means a displacement of 3 m to the left
$+8$ m s^{-1}	means a velocity of 8 m s^{-1} to the right
-4 m s^{-2}	means an acceleration of 4 m s^{-2} to the left
	(*or* a deceleration of an object moving towards the right)

The runner is going round the curve at a constant speed.
So how can he also be accelerating?
His velocity is changing because his direction is changing.

The sign convention you choose is entirely up to you.
In one question it may be easier to take *up* as positive whereas in
another you might use *down* as positive. It doesn't matter as long as
you keep to the **same** convention for the entire calculation.

Example 2
A ball hits a wall horizontally at 6.0 m s^{-1}. It rebounds horizontally at 4.4 m s^{-1}.
The ball is in contact with the wall for 0.040 s. What is the acceleration of the ball?

Taking motion *towards* the wall as positive:

$$\begin{aligned}\text{change in velocity} &= \text{new velocity} - \text{old velocity} \\ &= (-4.4\text{ m s}^{-1}) - (+6.0\text{ m s}^{-1}) \\ &= -10.4\text{ m s}^{-1}\end{aligned}$$

$$\text{acceleration} = \frac{\text{change in velocity}}{\text{time taken}} = \frac{-10.4\text{ m s}^{-1}}{0.040\text{ s}} = \underline{-260\text{ m s}^{-2}}$$

Negative, therefore in a direction *away from* the wall.

▷ Displacement–time graphs

The diagram shows a graph of displacement against time, for a car:

What type of motion does this straight line represent?
The displacement increases by equal amounts in equal times. So the object is moving at **constant velocity**.

You can calculate the velocity from the graph:

$$\text{Velocity from 0 to A} = \frac{\text{displacement}}{\text{time taken}}$$

$$= \text{gradient of line 0A}$$

The steeper the gradient, the greater the velocity.

Velocity is a vector so the graph also needs to indicate its direction. Positive gradients (sloping upwards) indicate velocity in one direction. Negative gradients (sloping downwards) indicate velocity in the opposite direction.

How would you draw a displacement–time graph for a stationary object? If the velocity is zero, the gradient of the graph must also be zero. So your graph would be a horizontal line.

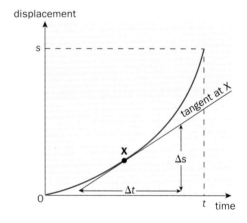

The gradient of a displacement–time graph gives us the velocity.

This second graph is a curve. How is the velocity changing here? The gradient of the graph is gradually increasing. This shows that the velocity is increasing. So the object is **accelerating**.

We could find the *average* velocity for this motion by dividing the total displacement s by the time taken t ($= s/t$).

But how do we find the actual (instantaneous) velocity at any point? The instantaneous velocity is given by the gradient at that point. This is found by drawing a tangent to the curve and calculating its gradient ($= \Delta s/\Delta t$; see the labels on the graph; see page 476).

What would the gradient of a *distance*–time graph represent? In this case the gradient would give the scalar quantity, *speed*.

▷ Velocity–time graphs

This graph shows the motion of a car travelling in a straight line:

It starts at rest, speeds up from 0 to A, travels at constant velocity (from A to B), and then slows down to a stop (B to C).
What does the gradient tell us this time?

$$\text{Gradient of line 0A} = \frac{\text{change in velocity}}{\text{time taken}} = \text{acceleration}$$

The steeper the line, the greater the acceleration.

A positive gradient indicates acceleration. A negative gradient (eg. BC) indicates a negative acceleration (deceleration).

Straight lines indicate that the acceleration is constant or uniform.

If the graph is curved, the acceleration at any point is given by the gradient of the tangent at that point.

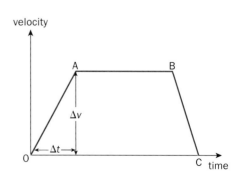

The gradient of a velocity–time graph gives us the acceleration.

The **area** under a velocity–time graph also gives us information. First let's calculate the displacement of the car during its motion.

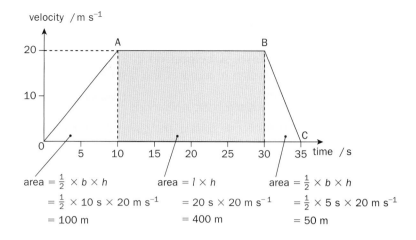

From 0 to A:
displacement $\begin{aligned} &= \text{average velocity} \times \text{time taken} \\ &= (\tfrac{1}{2} \times 20 \text{ m s}^{-1}) \times 10 \text{ s} \\ &= 100 \text{ m} \end{aligned}$

From A to B:
displacement $\begin{aligned} &= \text{average velocity} \times \text{time taken} \\ &= 20 \text{ m s}^{-1} \times 20 \text{ s} \\ &= 400 \text{ m} \end{aligned}$

From B to C:
displacement $\begin{aligned} &= \text{average velocity} \times \text{time taken} \\ &= (\tfrac{1}{2} \times 20 \text{ m s}^{-1}) \times 5 \text{ s} \\ &= 50 \text{ m} \end{aligned}$

Compare these values with the areas under the velocity–time graph. (They are marked on the diagram.) What do you notice?
In each case *the area under the graph gives the displacement*.

This also works for non-linear velocity–time graphs:

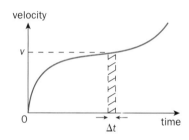

Area of shaded strip $= v \times \Delta t$ = average velocity $\times$ time interval
$\qquad\qquad\qquad\qquad\qquad\quad$ = displacement in this interval

Adding up the total area under the curve would give us the total displacement during the motion.

What does the area under a *speed*–time graph represent?
This gives us the *distance* moved.

Example 3
The velocity–time graph represents the motion of a stockcar starting a race, crashing into another car and then reversing.

a) Describe the motion of the car during each labelled section.
b) What is the maximum velocity of the car?
c) At which point does the car crash?
d) Does the car reverse all the way back to the start point?

a) 0 to A: The car accelerates.
   A to B: The car is moving at constant velocity.
   B to C: The car rapidly decelerates to a standstill.
   C to D: The car is not moving.
   D to E: The velocity is increasing again but the values are negative. Why? The car is starting to reverse.
   E to F: The car is now reversing at constant velocity.
   F to G: The car decelerates and stops.

b) From the graph, maximum velocity = 15 m s^{-1}.

c) The car crashes at point B. This causes the rapid deceleration.

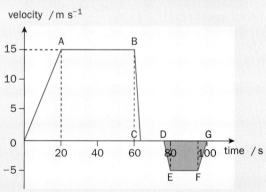

d) The area under the positive part of the velocity–time graph (shaded blue) gives the forward displacement of the car.
   The negative area (shaded red) gives the distance that the car reversed.
   As the red area is smaller than the blue area, we can see that the car did not reverse all the way back to the start point.

▷ Equations of motion

These are 4 equations that you can use whenever an object moves with **constant, uniform acceleration** in a straight line.
The equations are written in terms of the 5 symbols in the box:

s = **displacement** (m)
u = **initial velocity** (m s^{-1})
v = **final velocity** (m s^{-1})
a = **constant acceleration** (m s^{-2})
t = **time interval** (s)

They are derived from our basic definitions of acceleration and velocity.
Check your syllabus to see if you need to learn these derivations, or whether you only need to know how to use the equations to solve problems.

Derivations

From page 39,   acceleration $= \dfrac{\text{change in velocity}}{\text{time taken}} = \dfrac{\text{final velocity} - \text{initial velocity}}{\text{time taken}}$

Writing this in symbols: $a = \dfrac{v - u}{t}$

So $at = v - u$ which we can rearrange to give

$$\boxed{v = u + at}$$ . . . . . (1)

From page 38,   average velocity $= \dfrac{\text{displacement}}{\text{time taken}}$

If the acceleration is constant, the average velocity during the motion will be halfway between u and v. This is equal to $\frac{1}{2}(u + v)$.
Writing our equation for velocity in symbols:

$$\tfrac{1}{2}(u + v) = \dfrac{s}{t}$$ which we can rearrange to give

$$\boxed{s = \tfrac{1}{2}(u + v)\,t}$$ . . . . . (2)

Using equation (1) to replace v in equation (2):

$$s = \tfrac{1}{2}(u + u + at)\,t$$
$$\therefore \; s = \tfrac{1}{2}(2u + at)\,t$$ which we can multiply out to give

$$\boxed{s = ut + \tfrac{1}{2}at^2}$$ . . . . . (3)

From equation (1), $t = \dfrac{v - u}{a}$

Using this to replace t in equation (2):

$$s = \tfrac{1}{2}(u + v)\,\dfrac{(v - u)}{a}$$
$$\therefore \; 2as = (u + v)(v - u)$$
$$\therefore \; 2as = v^2 - u^2$$ which we can rearrange to give

$$\boxed{v^2 = u^2 + 2as}$$ . . . . . (4)

Note:

- You can use these equations only if the acceleration is **constant**.

- Notice that each equation contains only 4 of our five '$suvat$' variables.
  So if we know any 3 of the variables we can use these equations to find the other two.

42

Example 4
A cheetah starts from rest and accelerates at 2.0 m s^{-2} due east for 10 s.
Calculate   a) the cheetah's final velocity,
            b) the distance the cheetah covers in this 10 s.

First, list what you know: s = ?
                            u = 0 (= 'from rest')
                            v = ?
                            a = 2.0 m s^{-2}
                            t = 10 s

a) Using equation (1):   **$v = u + at$**
                          v = 0 + (2.0 m s^{-2} × 10 s) = <u>20 m s^{-1}</u> due east

b) Using equation (2):   **$s = \frac{1}{2}(u + v)\,t$**
                          $s = \frac{1}{2}$ (0 + 20 m s^{-1}) × 10 s = <u>100 m</u> due east

Or you could find the displacement by plotting a velocity–time graph
for this motion. The magnitude of the displacement is equal to
the area under the graph. Check this for yourself.

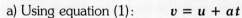

Example 5
An athlete accelerates out of her blocks at 5.0 m s^{-2}.
a) How long does it take her to run the first 10 m?
b) What is her velocity at this point?

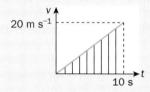

First, list what you know: s = 10 m
                            u = 0
                            v = ?
                            a = 5.0 m s^{-2}
                            t = ?

a) Using equation (3):   **$s = ut + \frac{1}{2}at^2$**
                          $\therefore$ 10 m = 0 + ($\frac{1}{2}$ × 5.0 m s^{-2} × t^2)   So $t^2 = \dfrac{10\ \text{m}}{2.5\ \text{m s}^{-2}}$ = 4.0 s^2 $\therefore t =$ <u>2.0 s</u>

b) Using equation (1):   **$v = u + at$**
                          v = 0 + (5.0 m s^{-2} × 2.0 s) = <u>10 m s^{-1}</u> (2 s.f.)

Example 6
A bicycle's brakes can produce a deceleration of 2.5 m s^{-2}.
How far will the bicycle travel before stopping, if it is moving
at 10 m s^{-1} when the brakes are applied?

First, list what you know: s = ?
                            u = 10 m s^{-1}
                            v = 0
                            a = −2.5 m s^{-2} (negative for *deceleration*)

Using equation (4):   **$v^2 = u^2 + 2as$**
                       0 = (10 m s^{-1})^2 + (2 × −2.5 m s^{-2} × s)
                       0 = 100 m^2 s^{-2} − (5.0 m s^{-2} × s)   So $s = \dfrac{100\ \text{m}^2\,\text{s}^{-2}}{5.0\ \text{m s}^{-2}}$ = <u>20 m</u> (2 s.f.)

▷ Vertical motion under gravity

Free fall

An object is in 'free fall' if the only force acting on it is gravity.
So a falling object can only be in free fall if there is no air resistance.
All objects in free fall accelerate downwards at the same rate.

The acceleration does not depend upon the mass of the object.
This was supposedly first demonstrated by Galileo when he dropped
a light stone and a heavy stone from the Leaning Tower of Pisa.
Since the stones felt little air resistance they both hit the ground at
the same time.
Try dropping two small objects with different masses for yourself.
Which hits the ground first? If air resistance is negligible they should
accelerate at the same rate and land together.

Generally we treat the **acceleration due to gravity** g on Earth
as a constant. Accurate measurements show that **g = 9.81 m s⁻²**.
This actually varies slightly from place to place. The value of g
decreases as you go further away from the surface of the Earth (see
Chapter 8).

g is also known as the **gravitational field strength**, measured in
newtons per kilogram, N kg⁻¹. So **g = 9.81 N kg⁻¹** (see page 18).

To simplify calculations we sometimes use $g = 10$ m s⁻² $= 10$ N kg⁻¹.

Experiment to measure the acceleration of free fall, g

A small steel ball feels very little air resistance. By timing its fall
through a known distance, you can calculate a value for the
acceleration due to gravity, g.
To eliminate your reaction time and get accurate measurements,
you need to use an electronic timer, like the one in the diagram:
When the switch is moved quickly to B, it simultaneously releases
the ball and starts the timer. When the ball hits the trapdoor the
circuit is broken at C and the timer stops.

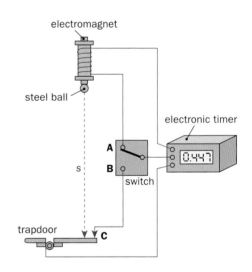

On page 42, we derived the equation $s = ut + \frac{1}{2}at^2$
You can measure the distance the ball falls, s, and the time taken, t.
The initial velocity u is zero. The acceleration a is the acceleration
due to gravity, g.
Putting this into the equation gives: $\quad s = 0 + \frac{1}{2}gt^2$

Rearranging this gives: $\quad g = \dfrac{2s}{t^2}$

You can use this equation to find g from a single drop of the ball.

How could you get a better value? You could repeat the experiment
for a range of different heights and plot a graph:

Compare $\quad s = \frac{1}{2}g \; t^2 + 0 \quad$ with the equation for a straight line:
$\qquad\quad\; y = m \; x \; + \; c \quad$ (see *Check Your Maths*, page 478)

Plotting s on the y-axis and t^2 on the x-axis should give a straight line
through the origin (since the intercept $c = 0$).
How can you then find g from the graph? The gradient m of the
straight line must be $\frac{1}{2}g$.
Why is this a good method? How does it show if one result is wrong?

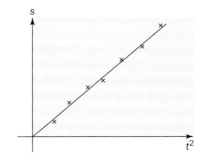

▷ Terminal velocity

On Earth, falling objects always experience some **air resistance**. Often we can ignore this, but at high speeds air resistance can become significant. Skydivers rely on air resistance for a safe landing.

Think about the forces acting on a skydiver during a jump. When she first jumps from the plane, her weight makes her accelerate downwards. See 0–A on the graph below.
Her weight remains fixed, but what happens to the air resistance on her? The air resistance is not constant; it increases with speed (page 22).

Eventually the air resistance and weight will balance out. As there is no net force acting, she stops accelerating. At this point the skydiver is travelling at her maximum possible speed or **terminal velocity** (A–B).

air resistance, F

0 to A
W > F
(acceleration)

weight, W

When the parachute is first opened (point B), the air resistance it creates at this high speed will be greater than the skydiver's weight. In which direction does the net force act at this point? This upward force makes her slow down rapidly, reducing the air resistance.

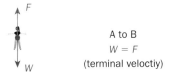

F

W

A to B
W = F
(terminal velocity)

Eventually the air resistance will balance the weight again, at a new lower terminal velocity (C–D). This allows her to land safely.

Here is the velocity–time graph for this motion:

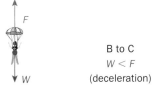

F

W

B to C
W < F
(deceleration)

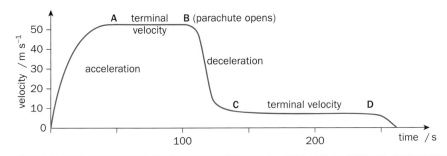

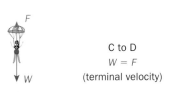

F

W

C to D
W = F
(terminal velocity)

Example 7
Phiz flips a coin into the air. Its initial velocity is 8.0 m s^{-1}.
Taking $g = 10$ m s^{-2} and ignoring air resistance, calculate:
a) the maximum height h that the coin reaches,
b) the velocity of the coin on returning to his hand,
c) the time that the coin is in the air.

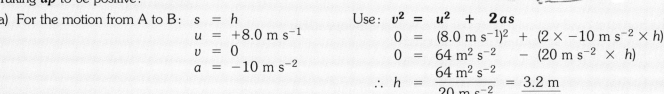

Taking **up** to be positive:

a) For the motion from A to B:
$\quad s = h$
$\quad u = +8.0$ m s^{-1}
$\quad v = 0$
$\quad a = -10$ m s^{-2}

Use: $v^2 = u^2 + 2as$
$\quad 0 = (8.0 \text{ m s}^{-1})^2 + (2 \times -10 \text{ m s}^{-2} \times h)$
$\quad 0 = 64 \text{ m}^2 \text{ s}^{-2} - (20 \text{ m s}^{-2} \times h)$
$\quad \therefore h = \dfrac{64 \text{ m}^2 \text{ s}^{-2}}{20 \text{ m s}^{-2}} = \underline{3.2 \text{ m}}$

b) The acceleration is the same going up and coming down. If the coin decelerates from 8.0 m s^{-1} to 0 m s^{-1} on the way up, it will accelerate from 0 m s^{-1} to 8 m s^{-1} on the way down. The motion is symmetrical. So the velocity on returning to his hand is $\underline{8.0 \text{ m s}^{-1} \text{ downwards}}$.

c) For the motion from A to B:
$\quad u = +8.0$ m s^{-1}
$\quad v = 0$
$\quad a = -10$ m s^{-2}
$\quad t = ?$

Use: $v = u + at$
$\quad 0 = 8.0 \text{ m s}^{-1} + (-10 \text{ m s}^{-2} \times t)$
$\quad \therefore t = \dfrac{8.0 \text{ m s}^{-1}}{10 \text{ m s}^{-2}} = \underline{0.8 \text{ s}}$

The coin will take the same time to move from B to A. So total time in the air $= \underline{1.6 \text{ s}}$.

▷ Projectiles

So far we have looked at motion in a straight line. We can also use the motion equations (from page 42) with objects projected or thrown through the air at an angle.

What examples of projectiles can you think of?
Missiles and cannon balls are sometimes used in exam questions.
An understanding of projectiles is also important in many sports.
How many sports can you think of that involve either the movement of objects through the air, or athletes hurling themselves through the air?
Football, golf, darts, netball, high jump and long jump are just a few examples.

For example, think about what happens when a ball is thrown at a velocity v at an angle θ to the ground:

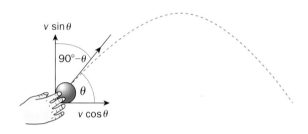

We can resolve this initial velocity into two components.
What are components? (See page 15 if you need a reminder.)

Initial **horizontal** velocity $= v \cos \theta$

Initial **vertical** velocity $= v \cos (90° - \theta) = v \sin \theta$

Ignoring air resistance, the only force acting on the ball during its flight is gravity.
This gives a downward acceleration that **only** affects the vertical component of velocity.

In the horizontal direction there is no acceleration. What does this tell you about the horizontal velocity during the flight?
The horizontal velocity remains constant (assuming there is no air resistance).

The overall effect is that the projectile will follow a curved or **parabolic** path through the air.
The distance the ball travels will depend on the horizontal velocity and the time of flight.
The time of flight depends on the vertical velocity.

For information on spreadsheet modelling of projectile motion, see page 483.

The horizontal and vertical motions of an object are **independent** and can be treated **separately** in calculations.
Try this simple experiment to convince yourself:

Experiment to show the independence of horizontal and vertical motion

Set up a ruler and two coins as shown in the diagram:
Press on the ruler and tap the end so that coin A falls vertically while coin B is projected sideways.

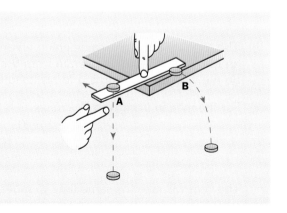

Listen to the coins. Do they hit the ground at the same time?

The vertical acceleration of each coin is exactly the same, even though one is moving horizontally as well as vertically.
This means that they hit the floor at exactly the same time.

Example 8
A stone is thrown over the edge of a cliff with a horizontal velocity
of 15 m s^{-1}. The cliff is 1.5×10^2 m high.
Ignoring any air resistance and taking $g = 9.81$ m s^{-2}, calculate:
a) the time it takes for the stone to reach the ground,
b) the distance it lands from the foot of the cliff,
c) the magnitude of the stone's velocity when it hits the ground.

a) First, look at the **vertical** motion.
   Taking **down** to be positive: $s = +150$ m
$$u = 0$$
$$a = +9.81 \text{ m s}^{-2}$$
$$t = ?$$

Use: $$s = ut + \tfrac{1}{2}at^2$$
$$150 \text{ m} = 0 + (\tfrac{1}{2} \times 9.81 \text{ m s}^{-2} \times t^2)$$
$$\therefore 150 \text{ m} = 4.905 \text{ m s}^{-2} \times t^2$$
$$\therefore t^2 = \frac{150 \text{ m}}{4.905 \text{ m s}^{-2}} = 30.6 \text{ s}^2$$
$$\therefore t = \sqrt{30.6 \text{ s}^2} = \underline{5.5 \text{ s}} \quad (2 \text{ s.f.})$$

b) Then look at the **horizontal** motion.
   As there is no horizontal acceleration, the horizontal velocity of 15 m s^{-1} does not change during the motion.

$\therefore$ Horizontal distance $=$ horizontal velocity $\times$ time of flight $= 15$ m s^{-1} $\times$ 5.5 s $= \underline{83 \text{ m}}$ (2 s.f.)

c) We need to find the resultant of the horizontal and vertical components of velocity at impact:
   Vertical velocity: $v_V = u + at = 0 + (9.81 \text{ m s}^{-2} \times 5.5 \text{ s}) = 54 \text{ m s}^{-1}$
   Horizontal velocity: $v_H = 15$ m s^{-1} (unchanged)

Using Pythagoras' theorem, resultant velocity $= \sqrt{(54 \text{ m s}^{-1})^2 + (15 \text{ m s}^{-1})^2}$
$$= \sqrt{3141} \text{ m s}^{-1}$$
$$= \underline{56 \text{ m s}^{-1}} \quad (2 \text{ s.f.}) \text{ in the direction shown:}$$

Example 9
A golfer hits a ball so that it moves off with a velocity of 26 m s^{-1}
at 30° to the horizontal.
Ignoring any air resistance and taking $g = 9.81$ m s^{-2},
calculate:
a) the time that the ball is in the air, and
b) the horizontal distance the ball travels.

a) First, look at the **vertical** motion from A to B.
   The angle to the vertical $= 90° - 30° = 60°$

   Taking **up** to be positive: $v = 0$ (since it is moving horizontally at B)
$$u = 26 \text{ m s}^{-1} \times \cos 60° = 13 \text{ m s}^{-1}$$
$$a = -9.81 \text{ m s}^{-2}$$
$$t = ?$$

Use: $v = u + at$
$$0 = 13 \text{ m s}^{-1} + (-9.81 \text{ m s}^{-2} \times t)$$
$$\therefore t = \frac{13 \text{ m s}^{-1}}{9.81 \text{ m s}^{-2}} = 1.3 \text{ s} \qquad \text{So the time from A to C is twice this, ie. } \underline{2.6 \text{ s}} \quad (2 \text{ s.f.})$$

b) Then look at the **horizontal** motion from A to C.
   Horizontal component of velocity $= 26$ m s^{-1} $\times \cos 30° = 22.5$ m s^{-1}. This is constant during the motion.
   Horizontal distance $=$ horizontal velocity $\times$ time of flight $= 22.5$ m s^{-1} $\times$ 2.6 s $= \underline{59 \text{ m}}$ (2 s.f.)

47

▷ Effect of friction on projectiles

On the previous pages we assumed there was no friction acting on the projectiles.
We treated the path of a projectile as a smooth, symmetrical curve or parabola.
But what happens to real projectiles? Air resistance produces a significant force that we cannot always ignore.
Take a look at the time-lapse photograph below of a motorcycle stunt rider flying through the air:

The green line shows the parabolic path the motorcycle would take in the absence of friction. The red line follows its actual path.
Air resistance reduces both the horizontal distance or **range** and the **maximum height** reached. The path is no longer a parabola.
The object falls more steeply than it rises.

Effect on horizontal motion

We saw on page 46 that, with no friction or air resistance, the horizontal motion of a projectile remains constant. In reality, drag forces will oppose the motion, leading to a horizontal deceleration.

There is a complex relationship between drag and speed.
At low speeds the drag is proportional to velocity but at higher speeds drag becomes related to velocity squared (see page 22).
In general the drag, and hence the horizontal deceleration, will decrease as the horizontal velocity decreases. This leads to a non-linear fall in the horizontal velocity of the projectile. This will reduce the range of the projectile.
With sufficient time in the air, the horizontal velocity could fall to zero. The projectile would then fall vertically.

Effect on vertical motion

When we ignore friction, the vertical motion of a projectile is only affected by the force due to gravity. This produces a constant downward acceleration g throughout the motion. Friction affects the upward and downward motion differently. Can you think why? Which way do the vertical forces act during the projectile's flight?

Friction always acts in the opposite direction to the motion.
As a projectile moves upwards, the forces due to gravity and friction both act downwards. They work together to decelerate the projectile, bringing it to rest more quickly and reducing the maximum height reached. Friction causes the decrease in vertical velocity to become non-linear, as the drag force is not constant. Drag decreases as the projectile slows down.

On the way back down, friction and weight act in opposite directions. The projectile accelerates downwards but at a decreasing rate, as the upward drag force increases with increasing speed. Drag and weight may eventually balance so that the projectile falls at its terminal velocity (see page 45).

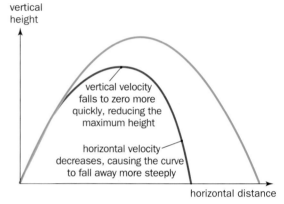

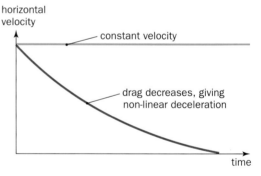

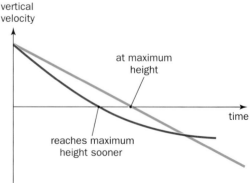

The three graphs show the motion of a projectile initially fired upwards at an angle to the ground.
The green lines show friction-free motion.
Red lines show how this changes when friction acts.

48

▷ Physics at work: Speed and stopping distance

Fast cars

In 2014, the Hennessey Venom became the world's fastest production road car, reaching 270.49 mph on the Space Shuttle landing strip in Florida. This beat the previous record holder, the Bugatti Veyron, by just 0.63 mph!

The Hennessey Venom.

So what limits the top speed of a car? On page 45 you saw how a falling object will accelerate until it reaches its maximum speed, or terminal velocity. This is due to the increase in the drag force as the object speeds up. Exactly the same effect limits the speed of a car. Even with the accelerator pressed to the floor, there is a maximum driving force available from the engine. This will accelerate the car rapidly at first but as the speed increases the drag force on it will also increase. Eventually the drag force becomes large enough to balance the motive force and so it is no longer possible to accelerate. The graph shows the shape of the velocity–time graph for a car reaching its maximum speed:

Thrust SSC.

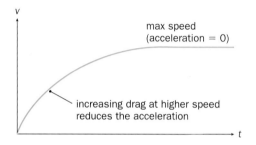

max speed
(acceleration = 0)

increasing drag at higher speed
reduces the acceleration

Bloodhound SSC.

So the key to high speed is minimising drag by using streamlined shapes, as well as maximising engine power. This has been taken to extremes in vehicles such as Thrust SSC. SSC stands for 'supersonic car'! Thrust SSC holds the official land speed record for a wheeled vehicle, achieving an incredible 763.035 mph using a mix of car and aircraft technology. Thrust's driver, Andy Green, is now involved in designing Bloodhound SSC, a rocket-powered car being designed to break the 1000 mph barrier!

Stopping distances

When you learn to drive you need to learn the Highway Code to pass your Driving Theory Test. The Highway Code includes a table of typical stopping distances for cars travelling at different speeds. A section of this is shown below. Notice that the total stopping distance is made up of two parts. The **thinking distance** is the distance travelled before the driver reacts and starts to brake. You move a surprising distance after spotting a hazard before you even press the brake pedal. The **braking distance** is the distance travelled while decelerating to a stop once you've pressed the brake. What deceleration and reaction time do these figures assume? We can find out using our equations of motion. Using the 70 mph (31 m s^{-1}) figures:

$$v^2 = u^2 + 2as$$
$$0 = (31 \text{ m s}^{-1})^2 + (2 \times a \times 75 \text{ m}) \quad \text{So} \quad a = -\frac{(31 \text{ m s}^{-1})^2}{(2 \times 75 \text{ m})} = \underline{-6.4 \text{ m s}^{-2}} \text{ (negative as it is a deceleration)}$$

$$\text{Reaction time} = \frac{\text{thinking distance}}{\text{average speed}} = \frac{21 \text{ m}}{31 \text{ m s}^{-1}} = \underline{0.68 \text{ s}}$$

30 mph   9 m   14 m

50 mph   15 m   38 m

Note that these are only typical values. In practice the actual distances would depend on factors such as your attention, road surface, weather and the vehicle's condition.

70 mph   21 m   75 m

Thinking distance   Braking distance

Summary

Distance and speed are scalar quantities. Displacement, velocity and acceleration are vectors. They have direction.

$$\text{average speed (m s}^{-1}) = \frac{\text{distance travelled (m)}}{\text{time taken (s)}}$$

$$\text{average velocity (m s}^{-1}) = \frac{\text{displacement (m)}}{\text{time taken (s)}}$$

$$\text{acceleration (m s}^{-2}) = \frac{\text{change in velocity (m s}^{-1})}{\text{time taken (s)}}$$

Motion graphs:

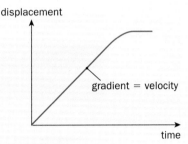

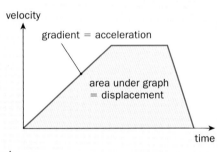

The 4 equations of motion:
Use them only if the acceleration is constant.

$$v = u + at$$
$$s = \tfrac{1}{2}(u + v)\,t$$
$$s = ut + \tfrac{1}{2}at^2$$
$$v^2 = u^2 + 2as$$

Remember to choose a sign convention to indicate the direction of vector quantities. When using these equations with projectiles, we treat the horizontal and vertical motions completely independently.

If we can ignore air resistance, all falling objects accelerate downwards at the same rate, known as the acceleration due to gravity, g. On Earth, $g = 9.81$ m s^{-2}.

Terminal velocity is the maximum speed a moving object reaches.
This is caused by air resistance (drag) preventing further acceleration.

▷ Questions

(Take $g = 9.81$ m s^{-2})

1. Can you complete these sentences?
   a) Velocity is the rate of change of .... It is measured in ....
   Acceleration is the rate of change of .... It is measured in ....
   b) A change in velocity can be a change in .... or ...., or both.
   c) Velocity is given by the .... of a displacement–time graph. The steeper the gradient, the .... the velocity.
   d) Acceleration is given by the gradient of a graph of .... against ....
   The area under a velocity–time graph gives the ....
   e) Negative values of displacement or velocity on a graph indicate motion in the opposite ....
   f) An object is in free fall if the only force acting on it is .... Most falling objects will experience friction or air .... This .... with increasing velocity until the air resistance upwards balances the weight downwards. At this point the object is said to be falling at its .... velocity

   g) An object projected sideways and one allowed to fall vertically from the same height will both hit the ground at the .... time.
   The .... and .... motions of a projectile are completely independent.

2. A toy car moves round a circular track of radius 1.0 m at a steady speed of 0.60 m s^{-1}.

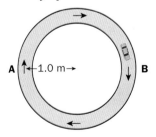

   For the motion from A to B, calculate:
   a) the distance moved along the track,
   b) the time taken,
   c) the car's displacement,
   d) the car's average velocity.

3. A car is moving along a road at 8.0 m s^{-1}. It then speeds up to 20 m s^{-1} in 4.0 s.
   a) What is the car's acceleration?
   b) If the car then decelerates at 4.0 m s^{-2}, how long does it take to stop?

4. Sketch displacement–time graphs for each of the following objects:
   a) a stationary object,
   b) a lift moving up, stopping then coming down,
   c) a ball falling to the ground.

5. Can you match each of these velocity–time graphs to the description of the motion?

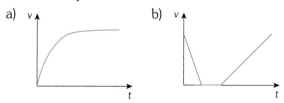
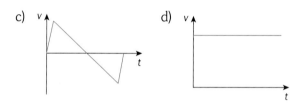

   i)   a ball thrown up into the air and then caught again
   ii) a skydiver reaching terminal velocity
   iii) a sprinter running in a straight line at steady speed
   iv) a car slowing down, stopping at traffic lights, then moving off.

6. The graph shows the motion of a car travelling along a straight road:

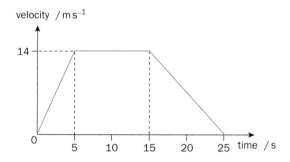

   a) What is the initial acceleration of the car?
   b) What is the deceleration produced by the brakes?
   c) How far does the car travel in the first 5 s?
   d) How far does the car travel in total?

Further questions on page 116.

7. A runner accelerates from rest at 3.4 m s^{-2} for 3.0 s. She then maintains a steady velocity for the rest of the 100 m race.
   a) How far does she travel during the acceleration?
   b) What velocity does she reach?
   c) How long in total does she take to run 100 m?

8. The graph shows the motion of a lift starting at rest and initially moving upwards:

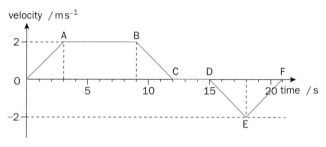

   a) What is the acceleration of the lift between 0 and A?
   b) What happens to the lift between B and C?
   c) How high is the lift above its starting point at C?
   d) What happens between D and F?
   e) What is the overall displacement from the starting point by the end of the motion?

9. A sandbag is dropped from a height of 150 m, from a hot-air balloon that is moving upwards with a velocity of 5.0 m s^{-1}. Ignore air resistance.
   a) What is the initial velocity of the sandbag?
   b) How long will the bag take to reach the ground?

10. A pencil is knocked horizontally off the edge of a desk at 2.0 m s^{-1}. The desk is 65 cm high.
    a) How long does it take the pencil to reach the floor? Ignore air resistance.
    b) What horizontal distance does the pencil travel during this time?

11. An aid parcel is released from a plane flying horizontally at 60.0 m s^{-1}, at a height of 1000 m.
    a) What are the horizontal and vertical components of the parcel's initial velocity?
    b) How long does the parcel take to hit the ground? Ignore air resistance.
    c) At what horizontal distance should the plane be from the target when the parcel is released?
    d) How would air resistance affect your answers to (b) and (c)?

12. A golf ball is hit at 26 m s^{-1} at 45° to the ground. Ignore air resistance.
    a) How long is the ball in the air?
    b) How far does the ball travel horizontally?
    c) What is the maximum height the ball reaches?
    d) How would air resistance affect your answers to (a), (b) and (c)?

5 Newton's Laws and Momentum

In the previous chapter we looked in detail at the motion of objects. But what causes motion? A simple answer is **force**. We will now look more closely at the links between force and motion.

In this chapter you will learn:
- about Newton's three laws of motion,
- the meaning of the terms inertia, momentum and impulse,
- how to use Newton's laws and the related principle of conservation of momentum to solve problems.

Linking force and motion

So what is the link between force and motion?

Force is needed to start things moving. If an object is not moving, it will stay still as long as there is no **resultant** force acting on it.

Remember, no resultant force does not necessarily mean that there are no forces acting at all. It means that any forces acting on the object are balanced and cancel each other out.

For example, a book will stay in the same place on a table. Two equal but opposite forces act on the stationary book: its weight and the normal contact force. To get the book to move we need a resultant force to act in some direction.

So a force will get things moving, but does motion stop if the force is removed?
Our ideas have changed over the last 2000 years:

Aristotle thought a continuous force is needed to keep things moving. Our own experience often seems to support this. Cars and bicycles will slow down and stop when we turn off the engine or stop pedalling. But is this because there is no force acting?

In fact the opposite is true. Cars and bicycles only stop because there **is** a force acting. What is this force? Friction. Without friction the car or bicycle would just keep on going!

Later experiments carried out by **Galileo** showed that force causes *changes* in an object's motion. Forces are needed:
- to start and to stop motion,
- to change an object's speed,
- to change an object's direction.
But constant force is **not** needed to maintain motion.

Isaac Newton published his 3 laws of motion in 1687. They explain the links between force and motion, and are described on the next few pages.

(These laws break down at speeds close to the speed of light, but for everyday objects they form the basis of all mechanics.)

Aristotle (384 – 322 BC)

Galileo Galilei (1564 – 1642)

Isaac Newton (1642 – 1727)

▷ Newton's First Law

> **If there is no resultant force acting on an object,**
> - **if it is at rest, it will stay at rest**
> - **if it is moving, it keeps on moving at a constant velocity (a constant speed in a straight line).**

On Earth it is difficult to create a friction-free environment to test this law. You may have seen a simple demonstration using an air-track. Given a small push to get it going, a glider on the track will carry on moving with almost constant speed for some time.

The same idea is used in space flight. Once in deep space, rockets do not need to use their engines continuously. A short blast gets the spacecraft moving in the right direction. It then continues through friction-free space until the engines are used again to alter its course.

Astronauts are tethered to their spacecraft. With no friction present, a small push could send them moving away into space.

▷ Inertia

Newton's first law tells us that objects have in-built resistance to any change in their motion.

A stationary object only starts to move when you apply a resultant force.

A moving object keeps moving at a steady speed in a straight line. To change the speed or direction you need to apply another resultant force.

This reluctance to change velocity is called **inertia**.
The inertia of an object depends on its mass.
A bigger mass needs a bigger force to overcome its inertia and change its motion.

You have experienced your own inertia when travelling in a car.

- What overcomes your inertia as the car accelerates?
You can feel the car seat pushing against your back to move you forwards.

Inertia depends on mass.

- What happens as you turn a corner?
Your body tries to keep moving in a straight line, while the car turns the corner. You feel as if you are being pushed into the side of your seat.

- What happens when the car decelerates?
Your inertia keeps you moving and you lurch forwards.
If the car brakes suddenly and you are not properly restrained, your inertia can be deadly! You will keep moving forwards at high speed until the steering wheel or windscreen forces a change in your motion!

Car safety is discussed in 'Physics at work' on page 65.

Inertia can damage your health!

▷ Momentum

You wouldn't want to get in the way of a charging elephant or a speeding bullet. Both are hard to stop, but for different reasons. What are these?
The elephant is moving relatively slowly but it has great **mass**.
A single bullet is light. It is hard to stop because of its high **velocity**.

2000 kg moving at 5 m s⁻¹.

The mass and velocity of an object determine its **momentum**:

| **momentum, p = mass, m × velocity, v** | or | $p = m v$ |
| (kg m s⁻¹)   (kg)   (m s⁻¹) | | |

The greater an object's momentum, the more force needed to stop it.

Momentum is a vector quantity so we need to specify its direction.
This is always the same direction as the velocity of the object.

0.02 kg moving at 400 m s⁻¹.

> *Example 1*
> Calculate the momentum of an elephant of mass 2000 kg moving at 400 m s⁻¹, the speed of a bullet!
>
> momentum, $p = m v$ = 2000 kg × 400 m s⁻¹ = <u>800 000 kg m s⁻¹</u>
>
> Definitely one not to get in the way of!

▷ Newton's Second Law

Newton realised that when a resultant force acts on an object, its velocity changes (ie. its speed and/or direction changes).

This must mean that *force causes a change in momentum*.
This is summarised by Newton's second law of motion:

> **The rate of change of momentum of an object is directly proportional to the resultant force acting on it.**
> **The change in momentum takes place in the direction of that force.**

We can write this as:

$$\text{resultant force} \propto \frac{\text{change in momentum}}{\text{time taken}}$$

('∝' means 'is proportional to')

or, in symbols: $\quad F = k \dfrac{(mv - mu)}{\Delta t}$

Δ means 'a change in'

where:

● u and v are the initial and final velocities,

● Δt ('delta-tee') is the time taken for the change.

● k is a constant of proportionality.
   The unit of force, the newton N, is carefully defined so that $k = 1$. This means that:

> **One newton (1 N) is the resultant force needed to cause a rate of change of momentum of 1 kg m s⁻¹ each second.**

So Newton's second law can be stated as:

$$F = \frac{m\,v - m\,u}{\Delta t}$$ or $$F = \frac{\Delta(m\,v)}{\Delta t}$$ . . . . (1)

If the mass of the object does not change, we can simplify this to:

$$F = \frac{m\,(v - u)}{\Delta t} = \text{mass} \times \frac{\text{change in velocity}}{\text{time taken}} = \text{mass, } m \times \text{acceleration, } a$$

So, **if the mass is constant**, Newton's second law becomes:

Force	**=**	**mass**	**×**	**acceleration**
(N)		(kg)		(m s^{-2})

or $$F = m\,a$$ . . . . (2)

Do you recognise this equation? This is the form of Newton's second law that you would have used at GCSE.
It tells us that force is directly proportional to acceleration. The greater the resultant force acting, the greater the acceleration of an object. The acceleration is in the same direction as the resultant force.

If no resultant force acts on an object, equation (2) tells us that there is no acceleration. Can you see how this links back to Newton's first law? With no resultant force the object must continue at the same velocity or remain at rest.

GCSE seems a long time ago!

Equation (2) also gives us an alternative definition of the newton:

> **One newton (1 N) is the resultant force needed to give a mass of 1 kg an acceleration of 1 m s^{-2}.**

Weight

In Chapter 2 we used the equation $W = m \times g$ to calculate the weight W of an object of mass m. Can you see that this is really a version of Newton's second law (equation 2 above)?

Weight W is the force of gravity acting on the mass and g is often called the acceleration due to gravity.

Example 2

thrust T

A rocket of mass 1000 kg carries 800 kg of fuel at take-off.
If the thrust T from the engines is 22 000 N, calculate:
a) the initial acceleration, and
b) the acceleration just as all the fuel is used up, assuming the thrust has remained constant.
Take $g = 10$ N kg^{-1} throughout for simplicity.

a) Total mass initially $= 1000$ kg $+ 800$ kg $= 1800$ kg

Resultant upward force $= T - W = T - mg = 22\,000$ N $- (1800$ kg $\times 10$ N kg$^{-1}) = 4000$ N

From equation (2), acceleration, $a = \dfrac{F}{m} = \dfrac{4000\text{ N}}{1800\text{ kg}} = 2.2$ m s^{-2}

weight W

b) Resultant upward force $= T - W = T - mg = 22\,000$ N $- (1000$ kg $\times 10$ N kg$^{-1}) = 12\,000$ N

Just as the fuel is used up, acceleration, $a = \dfrac{F}{m} = \dfrac{12\,000\text{ N}}{1000\text{ kg}} = 12$ m s^{-2}

▷ Newton's Third Law

In Chapter 2 we looked at forces acting between two objects. Newton realised that forces cannot exist in isolation. *Forces always act in pairs.* This is Newton's third law of motion:

> **If an object A exerts a force on an object B, then B exerts an equal but opposite force on A.**

This is sometimes written as:

> **To every action force there is an equal but opposite reaction.**

The term *opposite* in these definitions means *in the opposite direction.*

You exert a force on the trolley but the trolley exerts an equal force back on you.

So why don't these forces just cancel out, with no effect?

This is the most common misconception about Newton's third law forces. The most important thing to remember is that the two forces act on **different** objects.

The motion of an object is determined solely by the forces acting on it. This is why the free-body force diagrams that you learned about on page 24 are so useful when solving problems. Remember those diagrams show all the forces acting on just **one** object.

You can test Newton's third law for yourself in some simple experiments:

Example 3

Two people on ice skates stand facing each other:

If Adam pushes Ben away from him, Ben will move off backwards. But what happens to Adam? Does he remain stationary?

No. He also moves backwards because, during the push, Ben will exert an equal but opposite force on him.

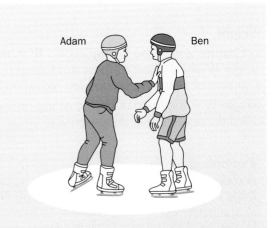

Adam   Ben

The force of Adam   =   The force of Ben
    on Ben   on Adam
 (to the right)   (to the left)

Can you see that **changing round the words** on the left-hand side tells you what the words should be on the right-hand side?

Example 4

Set up two force-meters as shown in the diagram:

One measures the force that you exert on the table when you pull. The other force-meter measures the force that the table exerts on you.

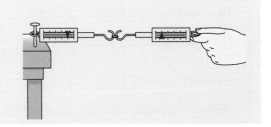

As you pull, what do you notice about the two readings? The readings on both meters are identical.

Identifying Newton's third law pairs

Spotting a Newton's third law pair of forces is not always easy. The diagram shows the 2 forces acting on a person standing on Earth:

normal contact force
N

weight W

The forces are equal in size and opposite in direction but they are **not** a Newton pair. Why not?

Remember that the two forces in a Newton pair always act on **different** objects.
But here the upward force is the contact force of the Earth **on the person**, and the weight is the force of gravity **on the person**.
So both of these forces act on the **same** object, the person.

The diagram is a free-body force diagram **for the person** (as on page 24). The Newton's third law pairs for those two forces are not shown here. What would the two matching forces be?
The matching force for N is the contact force of the person **on the Earth**; for W it is the gravitational pull of the person **on the Earth** (see Example 5). These forces are irrelevant if we are only concerned with the motion of the person.

To help you identify Newton pairs you can use the following checklist. Each force in a Newton's third law pair:

- has the same magnitude (size),
- acts along the same line but in opposite directions,
- acts for the same time,
- acts on a **different** object,
- is of the same type (eg. two contact forces, or two gravitational forces),
- can be identified by changing round the words as in Example 3.

Example 5
The diagrams show the forces acting when a gymnast balances on a beam. They have been drawn as free-body force diagrams, showing all the forces acting on (a) the gymnast, (b) the beam, (c) the Earth.

Can you identify each of the 4 Newton third law pairs? Remember that each force in a pair will be acting on a **different** object.

The matching pairs are numbered and colour coded to help you:

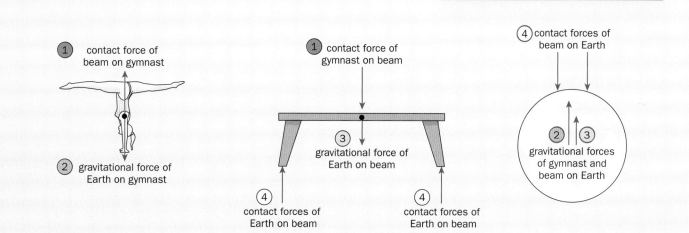

1 contact force of beam on gymnast

2 gravitational force of Earth on gymnast

1 contact force of gymnast on beam

3 gravitational force of Earth on beam

4 contact forces of Earth on beam

4 contact forces of Earth on beam

4 contact forces of beam on Earth

2 3 gravitational forces of gymnast and beam on Earth

▷ Newton's Laws: More worked examples

Example 6
A glider weighing 8.0 kN is being towed horizontally through the air
at a steady speed. The diagram shows the forces acting on the glider.
If the drag force D is 600 N, calculate:
a) the tension T in the tow-line,
b) the lift force L on the glider.

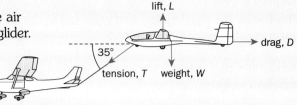

a) Steady horizontal speed means no horizontal acceleration.
So resultant horizontal force = 0

∴ total force acting to the left = total force acting to the right
$$T \cos 35° = D = 600 \text{ N}$$
$$\therefore T = \frac{600 \text{ N}}{\cos 35°} = 732.5 \text{ N} = 7.3 \times 10^2 \text{ N} \quad (2 \text{ s.f.})$$

Components of T:

b) Travelling horizontally so there is no vertical acceleration
So resultant vertical force = 0

∴ total force acting upwards = total force acting downwards
$$L = W + T \cos 55°$$
$$= 8000 \text{ N} + (730 \text{ N} \times \cos 55°) = 8420 \text{ N} = 8.4 \times 10^3 \text{ N} \quad (2 \text{ s.f.})$$

Example 7
The diagram shows the forces acting on a person in a stationary lift.
The person has a mass of 70 kg. Taking $g = 10$ N kg^{-1},
calculate the contact force R when:
a) the lift is at rest,
b) the lift is accelerating upwards at 1.0 m s^{-2},
c) the lift is accelerating downwards at 2.0 m s^{-2},
d) the lift is ascending at a steady speed.

a) As the person is not moving, the resultant force on him is zero.
∴ contact force R acting upwards = weight W acting downwards
$$= \text{mass} \times g$$
$$= 70 \text{ kg} \times 10 \text{ N kg}^{-1} = \underline{700 \text{ N}}$$

b) Accelerating **up**wards. Applying Newton's second law:
resultant **up**wards force $(R - W)$ = mass × acceleration
$$R - 700 \text{ N} = (70 \text{ kg} \times 1.0 \text{ m s}^{-2})$$
$$\therefore R = 70 \text{ N} + 700 \text{ N} = \underline{770 \text{ N}}$$

c) Acceleration **down**wards. Applying Newton's second law:
resultant **down**wards force $(W - R)$ = mass × acceleration
$$700 \text{ N} - R = (70 \text{ kg} \times 2.0 \text{ m s}^{-2})$$
$$\therefore R = 700 \text{ N} - 140 \text{ N} = \underline{560 \text{ N}}$$

d) Constant speed implies no acceleration, so resultant force = 0
∴ contact force acting upwards = weight acting downwards = mass × g = 70 kg × 10 N kg^{-1} = $\underline{700 \text{ N}}$

Can you see why you feel heavier as a lift accelerates upwards, and lighter as it accelerates downwards?
Normally you feel a contact force from the floor equal to your weight. As the lift accelerates upwards, you
experience a greater contact force, and so you feel heavier. The opposite is true for downward acceleration.

Example 8

A car of mass 1200 kg tows a caravan of mass 1800 kg along a level road. They are linked by a rigid tow-bar. The car accelerates at 1.80 m s⁻², pushed by a motive force of 5600 N. The resistive force on the car due to friction and air resistance is 65.0 N. Calculate:
a) the resistive force F on the caravan,
b) the tension T in the tow-bar.

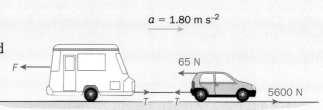

a) The tension in the tow-bar pulls equally on the car and on the caravan. If we start by looking at the car and caravan together we can ignore tension, as the tension forces balance out. The car and the caravan must have the same acceleration as they are linked by a rigid bar.

Applying Newton's second law ($F = m\,a$) to the car and the caravan together:

$$\textbf{Force} \ = \ \textbf{mass} \ \times \ \textbf{acceleration}$$
$$(5600\text{ N} - 65\text{ N} - F) = (1200\text{ kg} + 1800\text{ kg}) \times 1.80\text{ m s}^{-2}$$
$$5535\text{ N} - F = 3000\text{ kg} \times 1.80\text{ m s}^{-2}$$
$$5535\text{ N} - F = 5400\text{ kg m s}^{-2} = 5400\text{ N}$$
$$\therefore \ F = 5535\text{ N} - 5400\text{ N} = \underline{135\text{ N}} \quad (3\text{ s.f.})$$

b) We can find the tension in the tow-bar by applying Newton's second law to either the car or the caravan. We will do both here. Always start by drawing free-body force diagrams for each one:

Method 1: Forces on **car** (apply $F = ma$)
$$5600\text{ N} - 65\text{ N} - T = 1200\text{ kg} \times 1.80\text{ m s}^{-2}$$
$$5535\text{ N} - T = 2160\text{ kg m s}^{-2} = 2160\text{ N}$$
$$\therefore \ T = 5535\text{ N} - 2160\text{ N} = 3375\text{ N} = \underline{3.38 \times 10^3\text{ N}} \quad (3\text{ s.f.})$$

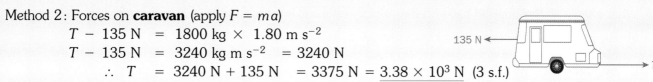

Method 2: Forces on **caravan** (apply $F = ma$)
$$T - 135\text{ N} = 1800\text{ kg} \times 1.80\text{ m s}^{-2}$$
$$T - 135\text{ N} = 3240\text{ kg m s}^{-2} = 3240\text{ N}$$
$$\therefore \ T = 3240\text{ N} + 135\text{ N} = 3375\text{ N} = \underline{3.38 \times 10^3\text{ N}} \quad (3\text{ s.f.})$$

Example 9

A battery-powered wheelchair can provide a motive force of 170 N. The person and wheelchair have a combined weight of 1000 N and the total resistive force acting is 20 N.
What is the maximum angle θ of the ramp that will allow the wheelchair to travel up it at a steady speed?

Steady speed means no acceleration, and so no resultant force along the slope.

So: total force = total force
  up slope   down slope
$$F = R + W\sin\theta$$
$$170\text{ N} = 20\text{ N} + (1000\text{ N} \times \sin\theta)$$
$$\therefore \ \sin\theta = \frac{170\text{ N} - 20\text{ N}}{1000\text{ N}} = 0.15$$
$$\therefore \ \theta = \sin^{-1}0.15 = \underline{8.6°} \quad (2\text{ s.f.})$$

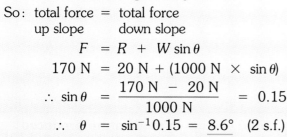

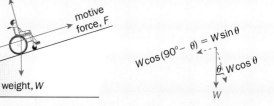

▷ Impulse

Why do you 'follow through' in sports such as tennis that involve hitting a ball? By following through you keep the force acting on the ball for a **longer time**. To see the advantage of this we need to refer back to our first version of Newton's second law, as given on page 55:

$$\text{resultant force} = \frac{\text{change in momentum } \Delta\,(m\,v)}{\text{time taken, } \Delta t}$$

Rearranging this gives:

Force × time = change in momentum
  (N)   (s)   (kg m s⁻¹)

This tells us that the greater the force on an object and the longer it acts for, the greater the change in the object's momentum.

By following through when you hit a tennis ball you increase the time that the force acts for. This produces a greater change in the ball's momentum.
Similarly, by drawing your hands backwards when you catch a ball you can reduce the sting. Can you see why? The momentum of the ball is reduced over a longer time, reducing the force on your hands.

The quantity **'force × time'** is known as **impulse**. Impulse measures the effect of a force. Impulse is measured in newton seconds (N s). These are equivalent to the units of momentum (kg m s⁻¹):

$1\,\text{N s} = 1\,\text{kg m s}^{-1}$

Example 10
a) What is the impulse produced by a force of 4 N acting for 3 s?
   Impulse = F × t = 4 N × 3 s = <u>12 N s</u>

b) What effect will this impulse have on the velocity of (i) a 2 kg mass and (ii) a 3 kg mass?

   (i) An impulse of 12 N s will produce a change in momentum p of 12 kg m s⁻¹.
        Momentum $p = m\,v$
        So, if the mass m is 2 kg, the change in velocity v must be <u>6 m s⁻¹</u> in the direction of the force.

   (ii) If the mass m is 3 kg, the change in velocity v must be <u>4 m s⁻¹</u> in the direction of the force.

Force–time graphs

The graph shows how the force applied to a golf ball by a club varies with time:

What does the area under the curve tell us?

Area of the shaded strip = F × Δt

= impulse produced during the
  small time interval Δt

We could divide the whole area under the curve into tiny strips and find the impulse produced in each interval. Adding together the area of each strip would then give us the total impulse. So:

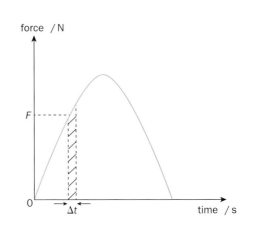

force / N

F

0   Δt time / s

Area under a force–time curve = total impulse acting = total change in momentum produced

Example 11
The graph shows the force acting on a tennis ball of mass 60 g
during a return shot.
a) What is the impulse on the ball?
b) If the ball reaches the player with velocity of 22 m s^{-1} moving
   to the left, what is the velocity of the return shot to the right?

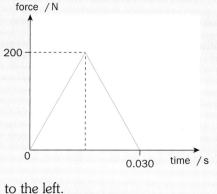

a) Impulse = area under curve = area of triangle
   = $\frac{1}{2}$ × base × height = $\frac{1}{2}$ × 0.030 s × 200 N = <u>3.0 N s</u>

b) An impulse of 3.0 Ns gives a change in momentum of 3.0 kg m s^{-1}
   in the direction of the force, ie. to the right.
   Initial momentum of ball = mv = 0.060 kg × 22 m s^{-1} = 1.3 kg m s^{-1} to the left.

   Remembering that momentum is a vector quantity, and taking motion to the right as positive:

   Change in momentum = new momentum − old momentum

   +3.0 kg m s^{-1} = new momentum − (−1.3 kg m s^{-1})

   ∴ new momentum, p = 3.0 kg m s^{-1} − 1.3 kg m s^{-1} = 1.7 kg m s^{-1}

   ∴ new velocity = $\frac{p}{m}$ = $\frac{1.7 \text{ kg m s}^{-1}}{0.060 \text{ kg}}$ = <u>28 m s^{-1} to the right</u>

Example 12
A helicopter hovers by forcing a column of air downwards.
The force down on the air produces an equal force up on the
helicopter (Newton's third law) which must balance its weight.

What is the maximum weight of the helicopter if the rotors force
1000 kg of air downward at 15 m s^{-1} each second?

Change in momentum of air in 1 s = mv = 1000 kg × 15 m s^{-1} = 15 000 kg m s^{-1}

Force acting on air = $\frac{\text{change in momentum}}{\text{time taken}}$ = $\frac{15\,000 \text{ kg m s}^{-1}}{1 \text{ s}}$ = 15 000 kg m s^{-2} = 15 000 N

∴ Upward force on helicopter = maximum weight = 15 000 N = <u>1.5 × 10^4 N</u> (2 s.f.)

Example 13
Water leaves a hose pipe of cross-sectional area 3.2 × 10^{-4} m^2 with
a horizontal velocity of 3.0 m s^{-1}. It strikes a vertical wall and
runs down it without rebounding. What is the force exerted on the wall?
(Density of water ρ_w = 1000 kg m^{-3})

The water travels at 3 m s^{-1}, so in 1 s a 3 m jet of water leaves the pipe.

Volume of water hitting the wall in 1 s = length × area of jet
                                        = 3 m × 3.2 × 10^{-4} m^2 = 9.6 × 10^{-4} m^3

∴ Mass of water hitting the wall in 1 s = volume × density
                                        = 9.6 × 10^{-4} m^3 × 1000 kg m^{-3} = 0.96 kg

Horizontal momentum before hitting wall = mv = 0.96 kg × 3.0 m s^{-1} = 2.88 kg m s^{-1} each second
Horizontal momentum after hitting wall = 0

Force of wall on water = $\frac{\text{change in momentum}}{\text{time taken}}$ = $\frac{0 - 2.88 \text{ kg m s}^{-1}}{1 \text{ s}}$ = −2.88 kg m s^{-2} = −2.88 N

∴ By Newton's third law, force of water on wall = 2.88 N = <u>2.9 N</u> (2 s.f.)

▷ Conservation of momentum

Suppose you can just reach the bank from a boat:

Why are you likely to get wet if you try to step out? What happens to the boat as you step forward? We can explain why the boat moves away from the bank using Newton's laws.

As you step from the boat, the boat exerts a contact force on you to propel you forwards. This must be equal but opposite to the force exerted by you on the boat (Newton's third law). Both of these forces act for the same time.

Now apply Newton's second law ($F = \Delta(mv) / \Delta t$) to both you and the boat. Since the force and the time are the same, both you and the boat have the same change in momentum, but in **opposite** directions. So as you gain momentum towards the bank, the boat gains equal momentum away from the bank.

But what is the overall change in momentum here?
Remember that momentum is a vector quantity.
Equal changes in opposite directions cancel out.
Consequently, the overall change in momentum is always zero.
This is known as the **principle of conservation of momentum**:

> Whenever objects interact, their **total momentum** in any direction remains **constant**, provided that no external force acts on the objects in that direction.

Investigating conservation of momentum

You can investigate the principle of conservation of momentum using a linear air track:

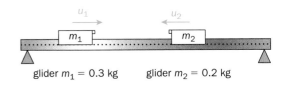

glider m_1 = 0.3 kg   glider m_2 = 0.2 kg

The air track helps to eliminate the external force of friction. Using light gates to measure the speed of the gliders, you can set up simple experiments to find their total momentum before and after collisions. You can use magnets or springs to repel the gliders away from each other.

Some typical results are given in this table:
Here motion to the right is taken as positive, and motion to the left as negative.

	Before the interaction				After the interaction			
	Velocity u_1 of glider 1 / ms^{-1}	Velocity u_2 of glider 2 / ms^{-1}	$m_1 u_1$ / kg ms^{-1}	$m_2 u_2$ / kg ms^{-1}	Velocity v_1 of glider 1 / ms^{-1}	Velocity v_2 of glider 2 / ms^{-1}	$m_1 v_1$ / kg ms^{-1}	$m_2 v_2$ / kg ms^{-1}
1	0.50	0	0.15	0	0.30	0.30	0.09	0.06
2	0.50	−0.50	0.15	−0.10	−0.30	0.70	−0.09	0.14
3	0	0	0	0	−0.40	0.60	−0.12	0.12
4	0.40	0.20	0.12	0.04	0.30	0.35	0.09	0.07

How do these results support the principle of conservation of momentum?
If you look carefully you can see that, in each case:

$$m_1 u_1 + m_2 u_2 = m_1 v_1 + m_2 v_2$$

So:

> **Total momentum before the interaction = total momentum after the interaction**

We use this equation in problems involving collisions or explosions.

We often use the principle of conservation of momentum to solve problems where objects collide or explode apart.

Some examples include: snooker balls, a rocket during lift-off, a collision between two cars and a bullet being fired from a gun.

Snooker players use the conservation of momentum instinctively with every shot.

The downward momentum of the hot gases results in upward momentum of the rocket.

Example 14 **Collisions**

a) A car of mass 1000 kg moving at 20 m s^{-1} collides with a car of mass 1200 kg moving at 5.0 m s^{-1} in the same direction. If the second car is shunted forwards at 15 m s^{-1} by the impact, what is the velocity v of the first car immediately after the crash?

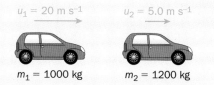

Taking motion to the right as positive:

$$\textbf{Momentum before collision} = \textbf{momentum after collision}$$
$$(1000 \text{ kg} \times 20 \text{ m s}^{-1}) + (1200 \text{ kg} \times 5 \text{ m s}^{-1}) = (1000 \text{ kg} \times v) + (1200 \text{ kg} \times 15 \text{ m s}^{-1})$$
$$20\,000 \text{ kg m s}^{-1} + 6000 \text{ kg m s}^{-1} = (1000 \text{ kg} \times v) + 18\,000 \text{ kg m s}^{-1}$$
$$\therefore 8000 \text{ kg m s}^{-1} = 1000 \text{ kg} \times v$$
$$\therefore v = \underline{8.0 \text{ m s}^{-1}} \text{ (to the right, since the answer is positive)}$$

b) If the cars collide head on at the same speeds as in (a), what would their combined velocity v be after the collision if they stick together on impact?

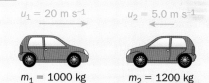

Taking motion to the right as positive:

$$\textbf{Momentum before collision} = \textbf{momentum after collision}$$
$$(1000 \text{ kg} \times 20 \text{ m s}^{-1}) - (1200 \text{ kg} \times 5 \text{ m s}^{-1}) = (1000 \text{ kg} + 1200 \text{ kg}) \times v$$
$$20\,000 \text{ kg m s}^{-1} - 6000 \text{ kg m s}^{-1} = 2200 \text{ kg} \times v$$
$$\therefore 14\,000 \text{ kg m s}^{-1} = 2200 \text{ kg} \times v$$
$$\therefore v = \underline{6.4 \text{ m s}^{-1}} \text{ (to the right, since the answer is positive)}$$

Example 15 **Explosions**

a) A bullet of mass 10 g is fired at 400 m s^{-1} from a rifle of mass 4.0 kg. What is the recoil velocity v of the rifle?

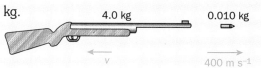

Before the gun is fired there is no momentum. The gun and the bullet are not moving. Taking motion to the right as positive:

$$\textbf{Momentum before explosion} = \textbf{momentum after explosion}$$
$$0 = (0.010 \text{ kg} \times 400 \text{ m s}^{-1}) + (4.0 \text{ kg} \times v)$$
$$0 = 4.0 \text{ kg m s}^{-1} + (4.0 \text{ kg} \times v)$$
$$\therefore v = \underline{-1.0 \text{ m s}^{-1}} \text{ (to the left, since the answer is negative)}$$

b) The bullet is fired into a block of wood of mass 390 g resting on a smooth surface. If the bullet remains embedded in the wood, calculate the velocity v that the block moves off at.

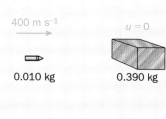

$$\textbf{Momentum before collision} = \textbf{momentum after collision}$$
$$(0.010 \text{ kg} \times 400 \text{ m s}^{-1}) + 0 = (0.010 \text{ kg} + 0.390 \text{ kg}) \times v$$
$$4.0 \text{ kg m s}^{-1} = 0.400 \text{ kg} \times v$$
$$\therefore v = \underline{10 \text{ m s}^{-1}} \text{ (to the right, since the answer is positive)}$$

▷ Conservation of momentum in two dimensions

The photograph shows some spectacular fireworks.
Notice how symmetrically they explode.

As a rocket approaches its maximum height it slows to a stop and
its momentum falls to zero. If the firework then explodes, the total
momentum must still be zero. Momentum must be conserved in
every direction. So for every shower of sparks thrown out in a
given direction there must be a shower with equal momentum
in the opposite direction, creating a beautiful starburst effect.

Remember, momentum is a vector quantity. In situations involving
multiple directions, we can resolve the momentum of the interacting
objects in two perpendicular directions. We can then apply the
law of conservation of momentum in each direction.

See how this works in the following examples:

Example 16

Two cars collide at a junction. The white car has a mass of 1500 kg. Before
the collision it was moving east at 18.0 m s^{-1}. The red car has a mass of
1200 kg. It was travelling at $52.0°$ to the white car at 12.0 m s^{-1}.

If the cars stick together on impact, at what velocity does the wreckage travel
immediately after the crash?

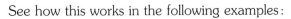

Before the collision we can resolve the momentum E–W and N–S:

Total momentum in E–W direction $= (1500 \times 18.0) + (1200 \times 12.0 \cos 52.0°)$
$ = 3.59 \times 10^4$ kg m s^{-1}
Total momentum in N–S direction $= (0) + (1200 \times 12.0 \cos (90 - 52.0°))$
$ = 1.13 \times 10^4$ kg m s^{-1}

After the collision, the total momentum in each direction is unchanged.
If R is the resultant momentum, then, using Pythagoras' theorem:

$$R^2 = (3.59 \times 10^4)^2 + (1.13 \times 10^4)^2 \quad \therefore R = 3.76 \times 10^4 \text{ kg m s}^{-1}$$

$$\tan \theta = \frac{1.13 \times 10^4}{3.59 \times 10^4} \quad \therefore \theta = 17.5°$$

Finally, dividing the momentum by the combined mass of the cars gives the velocity $= \dfrac{3.76 \times 10^4}{(1200 + 1500)} = 13.9$ m s^{-1}

So the wreckage moves off at 13.9 m s^{-1} at $17.5°$ N of E.

Example 17

In a game of snooker, the cue ball strikes a stationary red ball off-centre.
The balls move apart at the angles shown in the diagram, with the cue
ball moving at 0.94 m s^{-1}. Both balls have the same mass m.

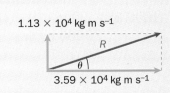

a) At what speed v_R does the red ball move off?
b) What was the initial speed v_I of the cue ball?

a) **Before** the collision there is no momentum perpendicular to the motion of the cue ball.
   After the collision, the components of momentum of the two balls in this direction must cancel out:

$$m \times v_R \sin 32° = m \times 0.94 \sin 58° \quad \therefore v_R = \frac{0.94 \sin 58°}{\sin 32°} = 1.5 \text{ m s}^{-1}$$

b) Horizontal momentum **before** collision $=$ horizontal momentum **after** collision

$$m v_I = (m \times v_R \cos 32°) + (m \times 0.94 \cos 58°)$$
$$v_I = (1.5 \cos 32°) + (0.94 \cos 58°) = 1.8 \text{ m s}^{-1}$$

▷ Physics at work: Car safety

Seat belts and crumple zones

Seat belts save lives! In a collision with a stationary object, the front of your car stops almost instantly. But what about the passengers?

Unfortunately they will obey Newton's first law and continue moving forwards at constant velocity until a force changes their motion. Without a seat belt, this force will be provided by an impact with the steering wheel or windscreen. This can cause you serious injury, even at low speeds.
A seat belt allows you to decelerate in a more controlled way, reducing the forces on your body. In an accident it is designed to stretch about 25 cm. This allows the restraining force to act over a longer time.

Newton's second law ($F = \Delta(mv)/\Delta t$, see page 55) tells us that the longer the time Δt taken to reduce a passenger's momentum $\Delta(mv)$, the smaller the force F needed to stop them.

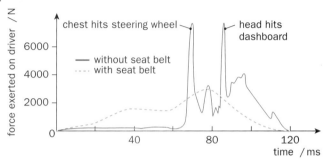

Look at the graph. It compares the forces acting on a driver involved in a collision with, and without, a seat belt:

Newton's second law also explains why modern cars are designed with **crumple zones**.
The passenger cell is designed as a rigid cage to maximise passenger protection by transmitting forces away around the roof and floor of the car.
But crumple zones are deliberately built into the car's bodywork. By crumpling, the car takes a longer time to come to rest. Again this means a lower rate of change of momentum and a smaller force acting on the passengers.
So, although cars suffer significant damage in even a minor collision, you are far more likely to walk away from an accident.

Sections of a car are designed to crumple in an accident.

Airbags

Airbags are designed to provide a cushion between your upper body and the steering wheel or dashboard. This can reduce the pressure exerted on you by more than 80% in a collision.

Crash detection, inflation and deflation of the airbag (to allow the driver to see again) must all take place in less time than it takes you to blink!

Without an airbag, your head would hit the steering wheel about 80 milliseconds after an impact. To be of any use, the crash must be detected and the airbag inflated in less than 50 ms!
So the bag is inflated explosively. This could be dangerous if it inflated accidentally under normal driving conditions.

So what triggers the airbag? Why does it inflate during a collision and not whenever you brake hard?

In a collision you are brought to rest very rapidly, say in 0.10 s. Even at low speeds this produces high deceleration. For example, at 20 mph (9 m s^{-1}):

$$\text{acceleration} = \frac{\Delta v}{t} = \frac{0 - 9 \text{ m s}^{-1}}{0.10 \text{ s}} = -90 \text{ m s}^{-2} = -9.2\, g \quad \text{(where } g = \text{acceleration due to gravity)}$$

This means that your seat belt must be able to exert a force over 9 times your body weight without snapping!

Compare this with the deceleration during an emergency stop, even at high speed.
On page 49 we used data from the Highway Code to calculate the typical deceleration of a car from 70 mph.
The answer of 6.4 m s^{-2} (0.65 g) is 14 times smaller than the deceleration in even a low-speed collision.
An acceleration sensor can easily detect the difference between the two, and only activate the airbag in a collision.

Summary

Newton's First Law:

If there is no resultant force acting on an object,
- if it is at rest, it will stay at rest;
- it it is moving, it keeps on moving at a constant velocity (at constant speed in a straight line).

Newton's Second Law:

The rate of change of momentum of an object is directly proportional to
the resultant force acting on it:
The change of momentum takes place in the direction of that force.
If the mass is constant, then **Force = mass × acceleration** or $F = m\,a$

$$F = \frac{\Delta(m\,v)}{\Delta t}$$

Newton's Third Law:

If an object A exerts a force on an object B, then B exerts an *equal* but *opposite* force on A.

The inertia of an object is its reluctance to change velocity. Inertia increases with increasing mass.

The **momentum** p of an object depends on its mass m and velocity v: $p = m\,v$
It is a vector and is always in the same direction as velocity.
The principle of conservation of momentum states that, whenever objects interact,
their total momentum remains constant, provided that no external force acts on the objects, so:

Total momentum before the interaction = total momentum after the interaction

Impulse measures the effect of a force:
Impulse (N s) = force × time = change in momentum produced (kg m s^{-1})
The area under a force–time curve gives the total impulse acting,
which equals the total change in momentum produced.

The unit of force is the **newton**. The newton can be defined in two ways:
- 1 newton is the resultant force needed to cause a rate of change of momentum of 1 kg m s^{-1} each second, or
- 1 newton is the resultant force needed to give a mass of 1 kg an acceleration of 1 m s^{-2}.

▷ Questions

(Take $g = 9.81$ m s^{-2})

1. Can you complete these sentences?
   a) A .... is needed to change the motion of an object.
   Forces can .... or decelerate an object, to change the object's .... (ie. speed or direction).
   b) The momentum of an object depends on its .... and .... Momentum is measured in ....
   c) The equation $F = ma$ is only valid in situations where .... is constant.
   d) One newton is the resultant force needed to give a mass of .... an acceleration of ....
   e) Forces in a Newton's third law pair have equal .... but act in opposite ....
   f) Impulse depends on .... and .... It is measured in ....
   The impulse of a force equals the change in .... produced. Total impulse can be found from the area under a .... – .... graph.
   g) In collisions and explosions, total .... remains constant, provided that no external .... acts.

2. Calculate the momentum of:
   a) an elephant of mass 2000 kg moving at 3 m s^{-1},
   b) a bullet of mass 20 g moving at 300 m s^{-1}.

3. What is the force needed to give a train of mass 25 000 kg an acceleration of 4 m s^{-2}?

4. A car of mass 1500 kg accelerates away from traffic lights. The car produces a constant motive force of 9000 N. The initial drag force opposing the motion is 1500 N.
   a) What is the initial acceleration?
   b) Why does acceleration gradually decrease even though the force from the engine is unchanged?

5. A grasshopper of mass 2.4×10^{-3} kg jumps upwards from rest. By pushing with its legs for 1.2 ms it achieves a take-off speed of 0.65 m s^{-1}. Calculate:
   a) its upward momentum,
   b) the force exerted by its legs.

6. Explain, with examples, how to identify a Newton's third law pair of forces.

7. A man is standing on a chair resting on the ground.
   a) Draw 3 free-body force diagrams, for
      i) the man,
      ii) the chair, and
      iii) the Earth,
      showing all the forces acting on each object.
   b) Show clearly which of these forces are Newton pairs of equal but opposite forces.

8.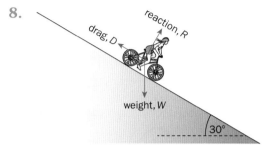

The diagram shows the forces acting on a cyclist freewheeling down a hill. Her total mass, including the bicycle, is 79 kg. She accelerates from rest to 6.0 m s^{-1} in 3.0 s.
   a) What is her acceleration?
   b) Calculate the weight W.
   c) By applying Newton's second law along the slope, find the drag force D.
   d) After a few seconds she reaches a steady speed down the slope. What is the drag force now? Why has it changed?

9. An empty lift has a mass of 1200 kg. It is supported by a steel cable attached to the roof. Draw a diagram showing the two forces acting on the lift.

   Find the tension in the lift cable when the lift is:
   a) ascending at a steady speed,
   b) accelerating upwards at 1.50 m s^{-2},
   c) accelerating downwards at 2.00 m s^{-2}.

10. During a serve, a tennis racket is in contact with the ball for 25 ms. The 60 g ball leaves the racket with a horizontal velocity of 31 m s^{-1}. If its initial horizontal velocity was zero, calculate:
    a) the change in momentum of the ball,
    b) the impulse from the racket,
    c) the average force exerted on the ball.

11. A car moving at 25 m s^{-1} collides with a wall. The driver of mass 65 kg is brought to rest by his seat belt in 0.20 s. Calculate:
    a) the driver's change in momentum,
    b) the average force exerted by the seat belt on the driver.
    c) How many times greater is this force than the driver's own weight?

12.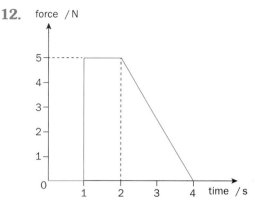

The graph shows the force acting on a 3.0 kg mass over a period of 4.0 s.
   a) Use the graph to calculate the total impulse.
   b) What change in momentum does this impulse produce?
   c) If the mass was initially at rest, what is its velocity after the 4 s?

13. During a hailstorm 0.54 kg of hailstones fall on to a flat roof in 6.0 s without rebounding. The hail hits the roof at 12 m s^{-1}. Calculate:
    a) the change of momentum each second,
    b) the force exerted on the roof.

14. A glider of mass 200 g is moving at 0.60 m s^{-1} along an air track. It collides with a second stationary glider of mass 250 g.
    If the gliders stick together on impact, calculate their new combined speed.

15. A car of mass 1500 kg travelling at 12 m s^{-1} collides head-on with a lorry moving at 20 m s^{-1}. The lorry has a mass of 9000 kg. If the collision reduces the speed of the lorry to 15 m s^{-1}, what is the car's velocity after the impact? In which direction is this?

16. A rocket of total mass 3500 kg is moving at 250 m s^{-1} through space. When the booster rockets are fired, 1200 kg of burnt fuel is ejected from the back of the rocket at 20 m s^{-1}. What is the new speed of the rocket?

17. A cannon fires a cannonball of mass 55 kg at 35 m s^{-1}. The cannon recoils at 2.5 m s^{-1}.
    a) What is the mass of the cannon?
    b) If the cannonball becomes embedded in a target of mass 600 kg, at what speed does the target move immediately after impact?

18. A car of mass 1200 kg, moving east at 16.2 m s^{-1}, collides with a lorry of mass 3500 kg moving north at 10.4 m s^{-1}. As a result of the collision, the car is shunted at 12.0 m s^{-1} in a direction 75° N of E. What is the velocity of the lorry immediately after the crash?

Further questions on page 117.

6 Work, Energy and Power

Holding a heavy set of weights above your head may feel like hard work, but in a Physics sense you are doing no work at all!
You are doing work only when you move the weights up.
In Physics, work is done only **when a force moves**.

In this chapter you will learn:
- how to calculate work and power,
- the equations for kinetic and potential energy,
- how to use the principle of conservation of energy to solve problems.

Work and energy

We can calculate the work done when a force moves using this equation:

Work done, W = force, F × displacement, s
in the direction of the force
(joules)   (newtons)   (metres)

Note that although force and displacement are vectors, work is a **scalar** quantity.

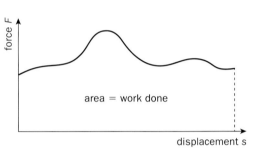

If the force F varies with displacement s, the work done can be found from the area under an F–s graph.

What if the force and displacement are not in the same direction, as in the diagram? In this case you need to resolve the force to find the component acting in the direction of the displacement:

Work done, $W = F \cos \theta_1 \times$ displacement, s

What about the vertical component of the force, $F \cos \theta_2$?
This does no work, because there is no movement in the vertical direction.

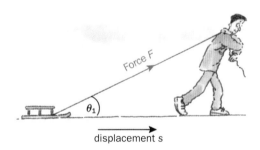

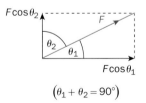

$(\theta_1 + \theta_2 = 90°)$

Doing work requires energy. Energy is often defined as the ability to do work. Energy is also a scalar quantity.

Doing work involves a transfer of energy from one form to another. For example, when you use your muscles to move an object, some of the stored energy in your body is transferred to kinetic (movement) energy.
The amount of work done tells you how much energy has been transferred from one form to another:

Work done = Energy transferred

A joule is a small unit of energy.

Both work and energy have the same unit, the joule (J):

1 joule is the work done (or energy transferred) when a force of 1 N moves through a distance of 1 m (in the direction of the force).

▷ Power

If two cars of the same weight drive up the same hill, they do the same amount of work. But what if one car is more powerful than the other?

Although the total energy transferred in reaching the top is the same for both cars, the more powerful car will get there **faster**.
Power is a measure of how fast work is done, or energy is transferred:

$$\textbf{Power, } \textbf{\textit{P}} = \frac{\textbf{work done, } \Delta \textbf{\textit{W}}}{\textbf{time taken, } \Delta \textbf{\textit{t}}} = \frac{\textbf{energy transferred, } \Delta \textbf{\textit{E}}}{\textbf{time taken, } \Delta \textbf{\textit{t}}}$$

If the work rate is not steady, these equations give the **average** power.

The unit of power is the watt, W.

1 watt = 1 joule per second

In her muscles, stored energy is transferred to kinetic energy. The more powerful the athlete, the faster she reaches her top speed.

Combining the equations for power and work done gives us another equation for calculating power:

$$\text{Power} = \frac{\text{work done}}{\text{time taken}} = \frac{\text{force} \times \text{displacement}}{\text{time taken}}$$

$$= \text{force} \times \text{average velocity (in the direction of the force)}$$

Or, in symbols:

$$\begin{array}{ccc} \textbf{\textit{P}} = & \textbf{\textit{F}} \times & \textbf{\textit{v}} \\ \text{(W)} & \text{(N)} & \text{(m s}^{-1}) \end{array}$$

Example 1
A man weighing 650 N runs up the hill in the diagram:
It takes him 15 s to reach the top.
Calculate:
a) the work he has done,
b) the power output of his legs.

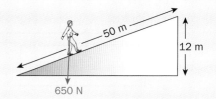

a) The force is the man's weight. This acts vertically so the displacement needed is the **vertical** height climbed.

Work done = force × displacement in direction of force
$$= 650 \text{ N} \times 12 \text{ m}$$
$$= 7800 \text{ J}$$

b) **Power** $= \dfrac{\textbf{work done, } \Delta \textbf{\textit{W}}}{\textbf{time taken, } \Delta \textbf{\textit{t}}} = \dfrac{7800 \text{ J}}{15 \text{ s}} = 520 \text{ J s}^{-1} = \underline{520 \text{ W}}$

Example 2
A speed boat is travelling at a steady speed of 15 m s^{-1}.
The resistance to its motion as it moves through the water is 1800 N.
Calculate:
a) the force F exerted by the boat's propeller on the water,
b) the power output of the engine.

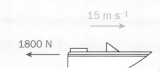

a) By Newton's third law, the force F exerted **by the boat on the water** = force exerted **by the water on the boat**. The boat is not accelerating so this force must be balanced by the resistance to motion. So $F = \underline{1800 \text{ N}}$.

b) Power, $\textbf{\textit{P}} = \textbf{\textit{F}} \times \textbf{\textit{v}} = 1800 \text{ N} \times 15 \text{ m s}^{-1} = \underline{27\,000 \text{ W}}$

▶ Types of energy

How many different types of energy can you think of? Heat energy, light energy, chemical energy, kinetic energy, sound energy and electrical energy are just some of the labels you may have come across. In fact there are only two basic types of energy: kinetic energy and potential energy.

Phiz using elastic PE on his Physics homework. What other energy types can you spot?

Kinetic energy (KE or E_k)

Kinetic energy is the energy an object has because of its *motion*.
We can calculate the kinetic energy an object has, using this equation:

$$\text{Kinetic energy} = \tfrac{1}{2} \times \text{mass} \times \text{speed}^2$$
$$\text{(joules)} \qquad \text{(kg)} \qquad \text{(m s}^{-1}\text{)}^2$$

or

$$E_k = \tfrac{1}{2}\,m\,v^2$$

Potential energy (PE or E_p)

'Potential' means 'stored'. Potential energy is the energy *stored* in an object due to its position, state or shape. Potential energy exists in various forms. Some examples are given in the table:

In mechanics we are usually concerned with **gravitational potential energy** (GPE). This is the energy gained by an object when it is lifted up against the force of gravity.
We can calculate the amount of gravitational potential energy that an object gains, using this equation:

Type of potential energy	Examples of where stored
Gravitational PE	Water held behind a dam
Chemical PE	Food and fuels
Elastic PE	Stretched spring
Electrical PE	Electric field
Nuclear PE	Mass of an atomic nucleus

$$\begin{array}{ccccc} \text{Change in gravitational} & = & \text{mass} & \times & \text{gravitational} & \times & \text{change in} \\ \text{potential energy} & & & & \text{field strength} & & \text{height} \\ \text{(joules)} & & \text{(kg)} & & \text{(N kg}^{-1}\,or\,\text{m s}^{-2}) & & \text{(m)} \end{array}$$

Or, in symbols:

$$\Delta E_p = m\,g\,\Delta h$$

Notice that this equation gives you the *change* in potential energy, not an absolute value. You can choose to take potential energy to be zero at any convenient point, usually ground level.

Where do we get these equations from?

KE equation

The diagram shows a force F acting on an object so that it accelerates from rest to a velocity v over a distance s.

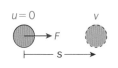

Using one of the motion equations (page 42):

$$v^2 = u^2 + 2as = 0 + 2as \qquad \therefore\ a = \frac{v^2}{2s}$$

$$\begin{aligned} \text{KE gained} = \text{work done} &= F \times s \\ &= ma \times s \quad \text{(since } F = ma) \\ &= m\frac{v^2}{2s} \times s \\ &= \tfrac{1}{2}mv^2 \end{aligned}$$

PE equation

The diagram shows an object of mass m lifted vertically through a height Δh.

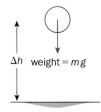

The force acting vertically is the object's weight (mg).

$$\begin{aligned} \text{PE gained} &= \text{work done against gravity} \\ &= \text{force} \times \text{distance} \\ &= mg \times \Delta h \end{aligned}$$

▷ Interchange between KE and PE

Why do roller-coasters always start by lifting the cars to the highest point on the track? Is it just to build up your anticipation?

At the top of the track the cars have their maximum gravitational potential energy. A motor is needed to lift the cars up, but from then on they can freewheel along the track. As the cars accelerate downwards the stored potential energy is transferred to kinetic energy.

What happens to the speed of the cars as they climb up again? The cars slow down as some of the KE is transferred back to PE. There is a continual interchange between PE and KE during the ride.

You can test this for yourself using a heavy ball on a smooth curved track. Release the ball and its PE is transferred to KE as it accelerates. The ball continues to roll until all the KE is transferred back to PE again.

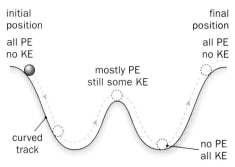

If there is no friction, how far would the ball travel along the track before it first stops moving?
When it stops it has no KE. All the energy must have been transferred back to PE. As this must be the same PE it started with, the ball would roll until it reaches a point on the track which has the same height as the release height (if there is no friction).

With no friction, the ball will roll until it reaches its initial height, regardless of the shape of the track.

The interchange between PE and KE forms the basis of many examination questions. Here are two typical examples:

Example 3
A tennis ball is dropped from a height of 2.0 m.
Ignoring air resistance, at what speed does it hit the ground?
(Take $g = 10$ m s^{-2})

As the ball falls, its PE is transferred to KE:

$$\textbf{PE lost} = \textbf{KE gained}$$
$$m\,g\,\Delta h = \tfrac{1}{2}\,m\,v^2$$
$$\cancel{m} \times 10 \text{ m s}^{-2} \times 2.0\text{ m} = \tfrac{1}{2} \times \cancel{m} \times v^2$$
$$\therefore \quad v^2 = 40 \text{ m}^2\text{ s}^{-2} \qquad \text{So} \quad v = \sqrt{40 \text{ m}^2\text{ s}^{-2}} = \underline{6.3 \text{ m s}^{-1}} \text{ (2 s.f.)}$$

Notice that you do **not** need to know the mass of the ball as the mass m cancels out on each side of the equation.

Example 4
In the roller-coaster in the diagram, the cars are released from rest at a height of 50 m.
Ignoring friction, what is the speed of the cars at
(a) the bottom and (b) the top of the loop?
(Take $g = 10$ m s^{-2})

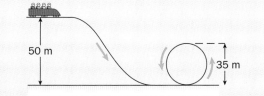

a) At the bottom of the loop the cars are 50 m below their initial height.

$$\textbf{PE lost} = \textbf{KE gained}$$
$$m\,g\,\Delta h = \tfrac{1}{2}\,m\,v^2$$
$$\cancel{m} \times 10 \text{ m s}^{-2} \times 50 \text{ m} = \tfrac{1}{2} \times \cancel{m} \times v^2$$
$$\therefore \quad v^2 = 1000 \text{ m}^2\text{ s}^{-2}$$
$$\text{So} \quad v = \sqrt{1000 \text{ m}^2\text{ s}^{-2}} = \underline{32 \text{ m s}^{-1}} \text{ (2 s.f.)}$$

b) At the top of the loop the cars are just 15 m below their initial height.

$$\textbf{PE lost} = \textbf{KE gained}$$
$$m\,g\,\Delta h = \tfrac{1}{2}\,m\,v^2$$
$$\cancel{m} \times 10 \text{ m s}^{-2} \times 15 \text{ m} = \tfrac{1}{2} \times \cancel{m} \times v^2$$
$$\therefore \quad v^2 = 300 \text{ m}^2\text{ s}^{-2}$$
$$\text{So} \quad v = \sqrt{300 \text{ m}^2\text{ s}^{-2}} = \underline{17 \text{ m s}^{-1}} \text{ (2 s.f.)}$$

▷ Conservation of energy

On the previous page we looked at the interchange between kinetic energy and potential energy. But often kinetic energy is lost without any obvious increase in potential energy.

For example, what happens to the kinetic energy when a parachutist lands or a car crashes? Where does your kinetic energy go when a bicycle slows down when you stop pedalling?

In all of these cases the energy is transferred to the surroundings, which become warmer. We say that the energy is *dissipated*.

In the case of the bicycle this is a gradual process through friction. In a car crash there is a sudden transfer that causes heating of the crumpled metal and the surrounding air:

The surroundings are often said to gain 'heat energy'. The correct term for this is **internal energy**. Internal energy is the name given to the total KE and PE of all the atoms or molecules of a substance. We will look at this in more detail in Chapter 14.

The main point here is that the energy does not just disappear. Energy cannot be used up. It is always transferred to other, often less obvious, forms. The **principle of conservation of energy** is one of the most fundamental rules in Physics:

> **Energy can be transferred from one form to another, but it cannot be created or destroyed.**
> **The total amount of energy always stays the same.**

You can solve many problems in mechanics by applying the principle of conservation of energy.
Always remember to include any work done or energy transferred to internal energy in overcoming frictional forces.
Friction is the most likely reason for apparent energy 'loss' from a system.

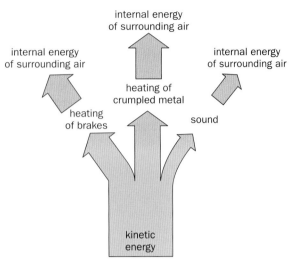

An energy transfer diagram, or 'Sankey' diagram, of the energy transfers in a car crash.
The width of the arrow represents the amount of each type of energy.

Example 5
A skier of mass 75 kg accelerates from rest down the slope in the diagram. At the bottom of the slope he is travelling at 40 m s⁻¹. What is the average frictional force opposing his motion down the slope? (Take $g = 9.8$ m s⁻²)

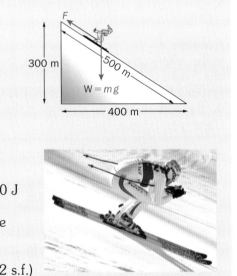

In travelling down the slope, the energy is conserved, so:
PE lost = KE gained + work done against friction

PE lost = $mg\,\Delta h$ = 75 kg × 9.8 m s⁻² × 300 m = 220 500 J
KE gained = $\frac{1}{2}mv^2$ = $\frac{1}{2}$ × 75 kg × (40 m s⁻¹)² = 60 000 J
∴ Work done against friction = 220 500 J − 60 000 J = 160 500 J

Work done against friction = frictional force × distance in direction of force
160 500 J = F × 500 m

∴ $F = \dfrac{160\,500 \text{ J}}{500 \text{ m}}$ = 321 N = 3.2 × 10² N (2 s.f.)

▷ Efficiency

A machine is any device that transfers energy from one form to another. Cars, electric motors and our bodies are all examples of machines.

Unfortunately, when energy is transferred, not all of it is transferred in a useful way. So what happens to the rest of the transferred energy? Usually some energy is 'wasted' as internal (heat) energy.

The proportion of energy that is *usefully* transferred is called the **efficiency** of the machine:

$$\text{efficiency} = \frac{\textbf{useful energy output}}{\textbf{total energy input}} = \frac{\textbf{useful power output}}{\textbf{total power input}}$$

Efficiency is a ratio. It has no units. Efficiency can also be expressed as a percentage, by multiplying by 100%.

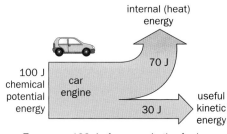

For every 100 J of energy in the fuel, only 30 joules are transferred usefully. The engine's efficiency is 0.30 or 30%.

Example 6
A water wheel is powered by 3600 kg of water falling through a height of 5.0 m each minute.
If the power output is 1200 W, what is the efficiency of the wheel?
(Take $g = 10 \text{ m s}^{-2}$)

PE lost by falling water = energy gained by the wheel

PE lost in 1 minute $= mg\,\Delta h = 3600 \text{ kg} \times 10 \text{ m s}^{-2} \times 5.0 \text{ m} = 180\,000 \text{ J}$

$\therefore$ Power provided $= \dfrac{\text{energy transferred, } \Delta E}{\text{time taken, } \Delta t} = \dfrac{180\,000 \text{ J}}{60 \text{ s}} = 3000 \text{ W}$

Efficiency $= \dfrac{\text{power output}}{\text{power input}} = \dfrac{1200 \text{ W}}{3000 \text{ W}} = \underline{0.40}$ (or 40%)

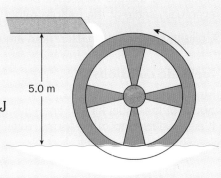

▷ Linking kinetic energy and momentum

A moving object has kinetic energy and momentum. Both of these depend on just the mass m and velocity v of the object. So how are kinetic energy and momentum related?

We can derive an expression that links these quantities by combining our equations for kinetic energy and momentum:

Kinetic energy, $E_k = \frac{1}{2}m v^2$

Momentum, $p = m v$

Rearranging the momentum equation gives: $v = \dfrac{p}{m}$

Substituting for v in the kinetic energy equation:

$$E_k = \tfrac{1}{2}mv^2 = \tfrac{1}{2}\,m \times \frac{p^2}{m^2}$$

Finally, cancelling the m's:

$$\boxed{E_k = \frac{p^2}{2m}}$$

This links the kinetic energy E_k of a mass m to its momentum p.

*So to **double** the momentum we need **4 times** the kinetic energy!*

▷ Elastic and inelastic collisions

In Chapter 5 we saw that when objects collide, the total *momentum* is always conserved (if there is no external resultant force), so:

Total momentum before = total momentum after

When we consider energy, the result may be different.
The *total* energy is also always conserved, but the kinetic energy may not be.
It depends on whether the collision is **elastic** or **inelastic**.

Elastic collisions

These are collisions in which kinetic energy is conserved:

Total KE before collision = total KE after collision

In a gas, the collisions between the molecules are elastic.
What would happen if they weren't elastic? Repeated collisions would slow the gas molecules down, so they would then eventually settle at the bottom rather than filling their container!

Collisions between snooker balls are very nearly elastic.

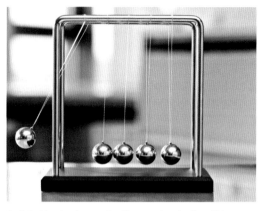
In this Newton's cradle, the moving ball collides elastically with a stationary one of the same mass. The total KE and momentum are transferred from ball to ball.

Inelastic collisions

Most collisions are inelastic.
Some of the initial kinetic energy is apparently 'lost'.
It is transferred to other forms, usually internal (heat) energy.

For inelastic collisions:

Total KE before collision > total KE after collision

Crash barriers and crumple zones of cars are specifically designed to collide *in*elastically, to absorb the kinetic energy in a crash.

In an inelastic collision, KE is 'lost'.

Example 7
A gas molecule of mass 2.0×10^{-26} kg moving at 600 m s^{-1} collides with a stationary molecule of mass 10×10^{-26} kg.
The first molecule rebounds at 400 m s^{-1}.
Is this collision elastic or inelastic?

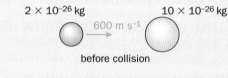

Taking motion to the right as positive:

Total momentum before collision = total momentum after collision

$$(2.0 \times 10^{-26} \text{ kg} \times 600 \text{ m s}^{-1}) = (2.0 \times 10^{-26} \text{ kg} \times -400 \text{ m s}^{-1}) + (10 \times 10^{-26} \text{ kg} \times v)$$

$$\therefore \quad 1200 \text{ kg m s}^{-1} = -800 \text{ kg m s}^{-1} + (10 \text{ kg} \times v)$$

$$\therefore \quad 1200 \text{ kg m s}^{-1} + 800 \text{ kg m s}^{-1} = 10 \text{ kg} \times v$$

$$\therefore \quad v = \frac{2000 \text{ kg m s}^{-1}}{10 \text{ kg}} = 200 \text{ m s}^{-1}$$

Total KE *before* collision $= \frac{1}{2}mv^2 = \frac{1}{2} \times 2.0 \times 10^{-26} \text{ kg} \times (600 \text{ m s}^{-1})^2 = 3.6 \times 10^{-21}$ J

Total KE *after* collision $= [\frac{1}{2} \times 2.0 \times 10^{-26} \text{ kg} \times (400 \text{ m s}^{-1})^2] + [\frac{1}{2} \times 10 \times 10^{-26} \text{ kg} \times (200 \text{ m s}^{-1})^2] = 3.6 \times 10^{-21}$ J

Since the total KE is the same before and after the collision, the collision is elastic.

Summary

Work done, W = force, F × displacement in the direction of the force, s
If the force acts at an angle θ to the displacement: $W = F\cos\theta \times s$

Energy is the ability to do work. Work and energy are measured in joules (J). Work done = energy transferred.

Power is the rate of doing work.   Power, $P = \dfrac{\text{work done, } \Delta W}{\text{time taken, } \Delta t} = \dfrac{\text{energy transferred, } \Delta E}{\text{time taken, } \Delta t}$ or $P = Fv$
1 watt = 1 J s^{-1}

The two types of mechanical energy are kinetic energy ($E_k = \frac{1}{2}mv^2$) and potential energy ($\Delta E_p = mg\,\Delta h$).
Energy is always conserved, though it is not always transferred in useful forms.

Efficiency $= \dfrac{\text{useful energy output}}{\text{total energy input}} = \dfrac{\text{useful power output}}{\text{total power input}}$ (× 100%)

If kinetic energy is conserved in a collision, the collision is elastic.
In inelastic collisions, some of the initial kinetic energy is transferred to other forms.
Momentum p is always conserved if no resultant external force is acting.

$E_k = \dfrac{p^2}{2m}$

▷ Questions

(Take g = 9.8 m s^{-2})

1. Can you complete these sentences?
   a) .... is the ability to do work. Both work and energy are measured in ....
   b) 1 J is the work done when a force of .... moves through a distance of .... in the direction of the force.
   c) Power is a measure of the .... of energy transfer. It is measured in .... 1 watt = 1 .... per ....
   d) The two main types of energy are .... energy and .... energy.
   e) The total KE and PE of the molecules of a substance is called its .... energy.
   f) In an elastic collision .... energy is conserved. If some energy is transferred to other forms, a collision is described as ....

2. A cyclist is travelling along a level road at 8.0 m s^{-1}. Her legs provide a forward force of 150 N. How much work does she do in 10 s?

3. A car with a 54 kW engine has a top speed of 30 m s^{-1} on a horizontal road.
   Calculate:
   a) the force driving the car forward at the top speed,
   b) the resistance to the car's motion at this speed.
   (Hint: When moving at its top speed, the car has zero acceleration.)

4. A lift and passengers weigh 15 000 N. Ignoring friction, what is the power of the motor needed to move the lift up at a steady speed of 3.0 m s^{-1}?

5. A cyclist of mass 100 kg is freewheeling down a hill inclined at 10° to the horizontal.
   He keeps his brakes on so that he travels at a steady speed of 6.0 m s^{-1}. The brakes provide the only force opposing the motion.
   Draw a diagram showing the forces acting on the cyclist. Calculate:
   a) the component of his weight acting down the hill,
   b) the braking force,
   c) the work done by the brakes in 3 s.

6. A pump lifts 200 kg of water per minute through a vertical height of 15 m. Calculate the output power. Calculate the input power rating of the pump if it is:
   a) 100% efficient,   b) 65% efficient.

7. A tennis ball of mass 60 g is dropped from a height of 1.5 m. It rebounds but loses 25% of its kinetic energy in the bounce.
   Ignoring any air resistance, calculate:
   a) the velocity at which the ball hits the ground,
   b) the velocity at which it leaves the ground,
   c) the height reached on the rebound.

8. A 10 kg child slides down a 3.0 m long slide inclined at 30° to the horizontal. A friction force of 35 N acts along the slide. Calculate:
   a) the potential energy lost in reaching the bottom,
   b) the work done against friction on the slide,
   c) the child's kinetic energy at the bottom of the slide,
   d) the child's velocity at the bottom of the slide.

Further questions on page 118.

7 Circular Motion

What do CD players, satellites, spin-dryers, fairground rides and the hammer thrower in the photograph have in common? They all use circular motion.

In this chapter you will learn:
- why an object moving in a circle must be accelerating,
- the equations we use to describe circular motion,
- how to use these equations with horizontal and with vertical circles.

What makes an object move in a circle?

Do you remember Newton's first law (see page 53)?

This says that an object continues to move in a *straight* line unless a resultant force acts on it.
So to make something move in a circle we need a force.

For example, the hammer thrower in the photograph makes the hammer move in a circle using the pull of the wire:

In which direction does the force act?

Your own experiences of circular motion may lead you to the wrong answer here.

Imagine yourself on the 'chairoplane' ride in the photograph:

What force do you feel as you swing round in a circle?
It *feels* as if you are being pushed outwards.
People often talk, *wrongly*, about an outwards or 'centrifugal' force.

In fact, the resultant force on you is *in*wards!
Your body is trying to obey Newton's first law, and so your body tries to keep moving in a straight line.
What stops you going in a straight line is a force pulling you inwards, *towards the centre of the circle*.
This force is provided by the tension in the chains supporting your seat.

To make an object move in a circle you always need a resultant force towards the centre of the circle.

This is called a **centripetal force**.

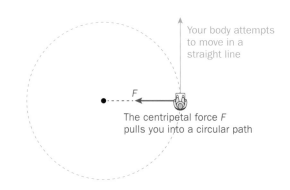

Your body attempts to move in a straight line

The centripetal force *F* pulls you into a circular path

Although a centripetal force is needed to keep an object moving in a circle, it does **no** work on the object. Can you see why?

At any point, the object is moving in a direction along a tangent to the circle. The force is always at 90° to this, towards the centre of the circle.
So there is no movement in the direction of the force, and so no work is done.

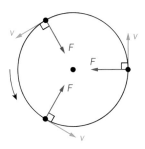

Force and velocity are always at right angles.

What if the centripetal force stops?

Think about the hammer thrower again.
The centripetal force comes from the pull of the wire.
When the wire is released, the centripetal force suddenly stops.
What happens to the hammer?

Newton's first law gives us the answer. Without a force to keep changing its direction, an object will carry on moving in a straight line. This will be along a tangent to the circle.

So the hammer flies off in a straight line in the direction it was heading when it was released. You can see the importance of timing your release carefully!

Where does the centripetal force come from?

The centripetal force is not some new force acting on the object. Centripetal force is just the name we give to the resultant force acting on the object in a direction towards the centre of the circle. Here are some examples:

The Earth in orbit
The Earth orbits the Sun along an almost circular path. What provides the centripetal force here?

The Sun's gravitational pull on the Earth provides the centripetal force that keeps us in orbit.
In a similar way, the Moon is kept in orbit by the gravitational pull of the Earth.

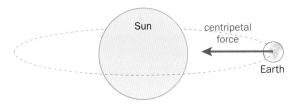

An electron orbiting the nucleus
An electron needs a centripetal force to keep it in orbit around the nucleus. What provides the centripetal force here?

The force towards the centre of the circle comes from the attraction between the positively-charged nucleus and the negatively-charged electron.

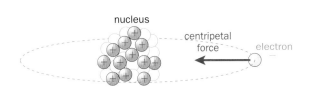

A car on a banked track
The diagram shows the forces acting on a car moving in a horizontal circle round a track:
Which two of these forces contribute towards the centripetal force?

The normal reaction and the force of friction both have a component towards the centre of the circle.
These combine together to give the resultant centripetal force that is needed to make the car go round the curve.

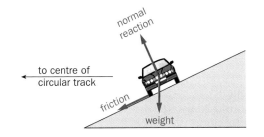

▷ Measuring rotation

Measuring angles

You will be used to measuring angles in degrees, but angles can also be measured in another unit — the **radian**. This is the unit we use for circular motion.

The diagram shows a Ferris wheel with 1 radian marked on it:
To turn through one radian you need to move a distance round the circle equal to its radius r.
Using the diagram, roughly how many radians are there in a circle?

We can calculate this number exactly using the definition of the radian:

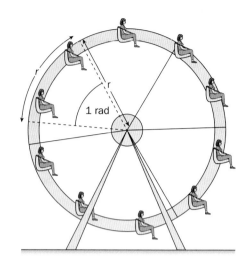

$$\text{angle } \theta \text{ in radians} = \frac{\text{length of arc, } s}{\text{radius, } r} \quad \text{or} \quad \theta = \frac{s}{r}$$

For a full circle (360°), the arc length is the circumference ($2\pi r$).

∴ For a full circle in radians $\theta, = \dfrac{s}{r} = \dfrac{2\pi r}{r} = 2\pi \quad (\approx 6.28 \text{ rad})$

So:

$$360° = 2\pi \text{ radians}$$

Angles in radians (rad) are often written as multiples of π.

Look back at the Ferris wheel diagram.
We can now calculate how many degrees there are in 1 radian:

$$1 \text{ radian} = \frac{360°}{2\pi} \approx 57°$$

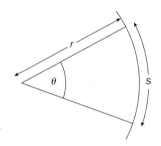

Example 1

A toy train moves round a circular track of radius 0.50 m.
a) How many radians has the train turned through in moving 1.4 m along the track?
b) What is this angle in degrees?

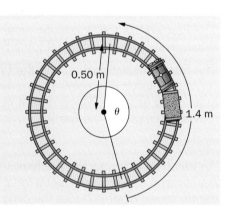

a) Angle in radians $= \dfrac{s}{r} = \dfrac{1.4 \text{ m}}{0.50 \text{ m}} = \underline{2.8 \text{ rad}} \quad (2 \text{ s.f.})$

b) Angle in degrees $= 2.8 \text{ rad} \times \dfrac{360°}{2\pi \text{ rad}} = \underline{160°} \quad (2 \text{ s.f.})$

Time period and frequency

For circular motion, the **time period T** is the time for one complete rotation. Time period is measured in seconds.

The **frequency f** is the number of rotations per second:

$$\text{frequency, } f = \frac{\text{number of rotations}}{\text{time taken}}$$

$$T = \frac{1}{f} \quad \text{and} \quad f = \frac{1}{T}$$

Frequency is measured in units of per second (s^{-1}).
This unit is also given the special name **hertz (Hz)**.
One rotation per second = 1 Hz.

78

Measuring angular speed

The photo shows a girl on a roundabout:

How can we describe how fast she is moving?
There are two ways:

1. We can calculate her **linear speed**, v, at any instant.
   This is the speed she would move off at, along a tangent
   to the circle, if there was no centripetal force.

2. We can also calculate the angle θ she turns through
   in a given time.
   This is called her **angular speed**, ω:

$$\text{angular speed, } \omega = \frac{\text{angle turned through, } \theta \text{ (rad)}}{\text{time taken, } t \text{ (s)}} \quad \text{or} \quad \omega = \frac{\theta}{t}$$

Angular speed is measured in radians per second (rad s^{-1}).
ω is sometimes called the 'angular velocity'.
It is also called the 'angular frequency'.

We can also calculate the angular speed if we know
the time period T or the frequency f of the motion.

One revolution is 2π radians. This takes T seconds.

So $\omega = \dfrac{\theta}{t} = \dfrac{2\pi}{T} = 2\pi f$ (since $\dfrac{1}{T} = f$), so: $\boxed{\omega = \dfrac{2\pi}{T}}$ and $\boxed{\omega = 2\pi f}$

Linking v and ω

The skaters in the picture all turn through an angle of 360° (2π rad)
together. They all have the same *angular* speed, ω.
But which skaters have the fastest *linear* speed v?

The skaters on the outside cover the greatest distance.
They must move at a higher speed than those near the centre, to
keep up. For one revolution:

$$\text{speed, } v = \frac{\text{distance}}{\text{time}} = \frac{\text{circumference}}{\text{time period}} = \frac{2\pi r}{T}$$

But $\dfrac{2\pi}{T} = \omega$, so: $\boxed{v = r\,\omega}$

So linear speed v increases with the radius r of the circle.

Example 2
The minute hand on a watch is 6.40 mm long. Calculate: (a) its frequency,
(b) its angular speed, and (c) the speed of its free end.

a) Time period T for 1 revolution = 1 hour = 60×60 s = 3600 s.

   So $f = \dfrac{1}{T} = \dfrac{1}{3600 \text{ s}} = 2.78 \times 10^{-4}$ Hz (3 s.f.)

b) Angular speed $\omega = 2\pi f = 2\pi \times 2.78 \times 10^{-4}$ Hz = 1.75×10^{-3} rad s^{-1}

c) Linear speed $v = r\,\omega = 6.40 \times 10^{-3}$ m $\times 1.75 \times 10^{-3}$ rad s^{-1} = 1.12×10^{-5} m s^{-1}

▷ Centripetal acceleration

How can you accelerate yet keep moving at constant speed?
The answer is to move in a circle!
Because your direction keeps changing, your *velocity* must also
be changing. This means that you are accelerating.
This is known as **centripetal acceleration**.

The fighter pilot in the photographs experiences very high
accelerations during a turn. This can reduce the blood flow
to the brain. The acceleration is sometimes measured in *g*'s,
where 1 *g* is 9.81 m s^{-2}. Accelerations above 4*g* can lead to
tunnel vision and unconsciousness.

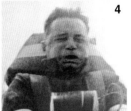

Facial distortion due to increasing g-force.

The faster the plane and the tighter the turn, the greater the
acceleration. In fact:

$$\textbf{centripetal acceleration, } \boldsymbol{a} = \frac{\textbf{speed}^2}{\textbf{radius}}$$

or

$$\boldsymbol{a} = \frac{\boldsymbol{v}^2}{\boldsymbol{r}}$$

Since $v = r\omega$,
we can also write $a = \dfrac{v^2}{r} = \dfrac{r\omega \times \cancel{r}\omega}{\cancel{r}} = r\omega^2$ so:

$$\boldsymbol{a} = \boldsymbol{r}\,\boldsymbol{\omega}^2$$

See page 84 for a derivation of the
centripetal acceleration formula.

Remember that acceleration is a vector quantity.
Which direction is it in?
The acceleration is in the same direction as the centripetal force
that causes it. This is *towards the centre of the circle*.

Example 3
A car moves at 10 m s^{-1} around a bend of radius 50 m.
What is its centripetal acceleration?

$$a = \frac{v^2}{r} = \frac{(10 \text{ m s}^{-1})^2}{50 \text{ m}} = \frac{100 \text{ m}^2 \text{ s}^{-2}}{50 \text{ m}} = \underline{2.0 \text{ m s}^{-2}}$$

▷ Centripetal force

Newton's second law tells us that: force = mass × acceleration.

We have two equations for centripetal acceleration.
Multiplying them by the mass m of the object moving in a circle
gives us two equations for the resultant centripetal force:

$$F = \frac{m\,v^2}{r}$$ and $$F = m\,r\,\omega^2$$

Example 4
Helen Sharman was the first Briton in space. She orbited Earth
once every 91.8 minutes in the Mir space station.
Mir had a mass of 21 100 kg and an orbit radius of 6750 km.
Calculate   a) Helen's angular speed in orbit,
            b) the centripetal force on Mir (provided by the pull of gravity).

Helen Sharman with Russian astronauts.

a) Angular speed, $\omega = \dfrac{\theta}{t} = \dfrac{2\pi}{T} = \dfrac{2\pi \text{ rad}}{91.8 \times 60 \text{ s}} = \underline{1.14 \times 10^{-3} \text{ rad s}^{-1}}$

b) Centripetal force, $F = mr\omega^2 = 21\,100 \text{ kg} \times 6750 \times 10^3 \text{ m} \times (1.14 \times 10^{-3} \text{ rad s}^{-1})^2 = \underline{1.85 \times 10^5 \text{ N}}$

▷ Moving in horizontal circles

The car in the photograph is skidding out of control as it tries to turn a corner. Why?
There is not enough centripetal force to keep it moving in a circle.

The diagram shows the forces acting on a car cornering on a level road. What provides the centripetal force?
The only force with a component towards the centre of the circle is the frictional force between the tyres and the road.
So friction provides the centripetal force.

The higher the speed and the tighter the bend, the greater the centripetal force needed.
We can use the equations on the opposite page to calculate the frictional force needed to corner safely:

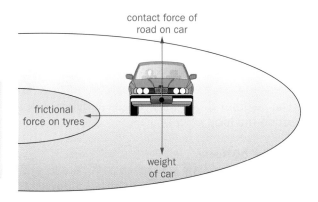

Example 5
What is the centripetal force needed for a car of mass 1000 kg to follow a curve of radius 50.0 m at 18.0 m s^{-1} ($\approx$ 40 mph)?

$$F = \frac{m\,v^2}{r} = \frac{1000 \text{ kg} \times (18.0 \text{ m s}^{-1})^2}{50.0 \text{ m}} = \underline{6480 \text{ N}}$$

Worn tyres and wet or icy road conditions will reduce the friction available. How could you take the corner safely with less friction?
Slow down!

Circling at an angle

The diagram shows the forces acting on a person on a 'chairoplane' ride.
What provides the resultant centripetal force here?
It comes from the tension T in the chain. Although the tension is not horizontal, it has a component towards the centre of the circle.
The component towards the centre is $T\cos\theta$.
So:

$$\text{Centripetal force} = T\cos\theta = \frac{m\,v^2}{r}$$

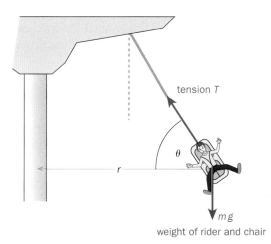

What about the vertical component of the tension?
This component balances the weight of the chair and rider.

What happens as the ride speeds up?
The chairs swing out further. Can you see why?

The faster the ride, the greater the centripetal force needed.
Swinging further out increases the component of tension acting towards the centre of the circle.
The tension force must also increase to balance the weight.

Similar forces are needed for aircraft banking.
Why must the pilot tilt the plane to turn it in a circle?

The lift force on the plane always acts at 90° to the wings.
Tilting the plane gives a component of lift towards the centre of the circle.
This is the resultant centripetal force needed to turn the plane.

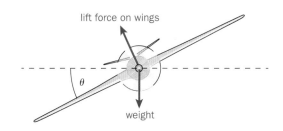

▷ Moving in vertical circles

Have you ever looped the loop on a roller-coaster ride?
What stops you falling out? As long as the speed is high enough
you will stay in contact with your seat.
To understand why, we need to look at the resultant centripetal
force on the riders.

To keep you moving in a circle there must be a resultant
centripetal force ($= mv^2/r$) acting towards the centre of the circle.
What provides this force?

Only two forces act on a rider moving round the loop:

1. The rider's **weight mg**. This does not change.

2. The **contact force R** that the seat exerts on the rider.
   This varies in size as the car goes round the circle.

The diagram shows these forces at different positions
round the loop:

At the bottom
The weight and the contact force R_B are in opposite directions.
Which is the larger force? We need a resultant force towards the
centre, to provide the centripetal force. So the contact force must
be bigger than the weight.
At the bottom:

$$R_B - mg = \frac{mv^2}{r}$$

At the top
Both the weight and the contact force R_T act downwards towards
the centre of the circle. Together they provide the centripetal force:

$$R_T + mg = \frac{mv^2}{r}$$

At the side
The contact force R_S alone provides the centripetal force.
Can you see why? The weight acts vertically downwards
so it has no component towards the centre of the circle.
At the side:

$$R_S = \frac{mv^2}{r}$$

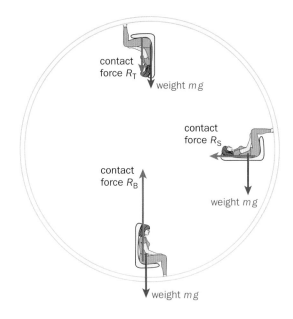

Can you see that the contact force is much larger at the bottom
than at the top of the loop? At the bottom of the loop the seat
pushes up on you with a large force. This is why you feel yourself
being pushed into your seat more at the bottom.

Failing to loop the loop!

As long as the ride is fast enough, it is impossible for you to fall out.
But what happens if the ride slows down?
If you are moving more slowly, the centripetal force needed is smaller.
This means that the contact force from the seat will be smaller.

If the contact force at the top drops to zero, you will *just* make the loop.
Any slower than this and your weight will be larger than the
required centripetal force. If this happens you will fall out of your seat!

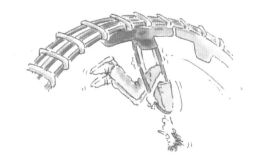

Example 6
A roller-coaster has a vertical loop of radius 12 m.
The cars hurtle round the loop at 14 m s^{-1}. Calculate:
a) the centripetal force needed on a passenger of mass 60 kg,
b) the contact forces on the passenger at the top and bottom of the loop.
c) At which point in the loop does the passenger feel heaviest?

14 m s^{-1}
←—12 m—→

a) Centripetal force, $\mathbf{F} = \dfrac{\mathbf{m}\,\mathbf{v}^2}{\mathbf{r}} = \dfrac{60 \text{ kg} \times (14 \text{ m s}^{-1})^2}{12 \text{ m}} = \underline{980 \text{ N}}$ (2 s.f.)

b) Taking $g = 10$ N kg^{-1}, the passenger's weight $= \mathbf{mg} = 60$ kg $\times$ 10 N kg$^{-1} = 600$ N
   Using the notation on the opposite page,

   At the top: $R_T + mg = \dfrac{mv^2}{r}$ At the bottom: $R_B - mg = \dfrac{mv^2}{r}$

   $R_T + 600 \text{ N} = 980 \text{ N}$ $R_B - 600 \text{ N} = 980 \text{ N}$

   $\therefore R_T = 980 \text{ N} - 600 \text{ N} = \underline{380 \text{ N}}$ $\therefore R_B = 980 \text{ N} + 600 \text{ N} = \underline{1580 \text{ N}}$

c) We judge our weight from the size of the contact force on us.
   The passenger feels heaviest when the contact force is largest. This is at the bottom of the loop.

Example 7
What is the minimum speed for a roller-coaster with a vertical
loop of radius 12 m?

At the top of the loop: $R_T + mg = \dfrac{mv^2}{r}$

When $R_T = 0$, $\cancel{m}g = \dfrac{\cancel{m}v^2}{r}$

Rearranging this gives: $v^2 = rg = 12 \text{ m} \times 10 \text{ m s}^{-2}$

$\therefore v = \sqrt{120 \text{ m}^2 \text{ s}^{-2}} = \underline{11 \text{ m s}^{-1}}$ (2 s.f.)

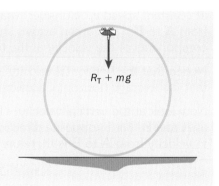
$R_T + mg$

▷ Physics at work: Weightlessness and artificial gravity

To be truly 'weightless' you must be at a place where there is no
gravitational field. So why do astronauts appear to be weightless?

We sense our weight from the size of the contact force pushing against us.
Imagine standing in a stationary lift. The upward contact force that you
feel equals your weight downwards. But what if the cable snaps?
Both you and the lift fall freely with the same acceleration due to gravity.
You feel weightless as you no longer feel the floor pushing up on you.

An astronaut in orbit is like a person in a free-falling lift.
The spacecraft and the astronaut fall freely around the Earth with the same
centripetal acceleration (see page 80). The astronaut feels weightless
because there is no contact force between her and the spacecraft.

To create 'artificial gravity' a spacecraft could be made to rotate.
The astronauts would then feel the centripetal force exerted on them by
the wall of the spacecraft. This would create the sensation of weight.
The smaller the radius, the faster the rotation needed to create enough
force. So to avoid making its occupants nauseous, a rotating spacecraft
would need to be very large. At present such a design is just too expensive.

▷ Physics at work: Bobsleigh bends

Why does a bobsleigh move up the wall of the run when it goes round a corner?

There must be a resultant centripetal force on the bobsleigh for it to follow a circular path. Where does this force come from?

There is very little friction on the ice, so the main forces on the bobsleigh are its weight and the normal contact force. By going up the wall, the normal contact force has a component that acts towards the centre of the circle. This provides the centripetal force.

The faster the bobsleigh takes the bend, the higher it goes. This gives a greater horizontal component, which increases the centripetal force.

normal contact force

weight

▷ Derivation of the centripetal acceleration formula

The diagram shows a particle moving along a circular path of radius r at constant speed v. Consider what happens as the particle moves from **A** to **B**. The green arrows show the horizontal and vertical components of the velocity at the two points.

Notice that the horizontal velocity is unchanged. This tells us that the average horizontal acceleration from A to B is zero.

Now look at the vertical velocity. This is the same size at A and B, but in opposite directions. The change in velocity from A to B is therefore $2v\sin\theta$ (upwards).

So vertical acceleration $= \dfrac{\text{change in velocity}}{\text{time taken}} = \dfrac{2v\sin\theta}{t}$

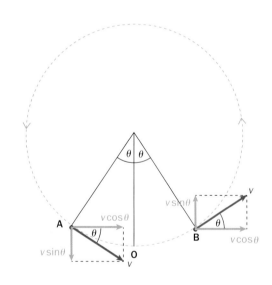

The distance travelled from A to B is the length of the arc. From our definition of the radian on page 78:

$$\text{length of arc} = \text{angle in radians} \times \text{radius} = 2\theta r$$

So the time t taken to move from A to B $= \dfrac{\text{distance}}{\text{speed}} = \dfrac{2\theta r}{v}$

Substituting this expression for t into our equation above for the vertical acceleration gives:

$$\text{vertical acceleration} = \dfrac{2v\sin\theta}{(2\theta r/v)} = \dfrac{v^2\sin\theta}{r\theta} \quad \text{(upwards)}$$

Maths box

As $\theta \to$ zero, $\sin\theta \to \theta$

Try this out for yourself on your calculator. (Remember to use radian mode.)

This is the acceleration for the particle moving from A to B. To find the instantaneous acceleration at a point O on the circle, we need to let the angle θ tend to zero so that A and B get closer and closer together. For small angles, $\sin\theta$ and θ will cancel. (See the *Maths box* and page 475.)
So our equation for acceleration a becomes:

$$a = \frac{v^2}{r}$$

The acceleration at O is upwards, towards the centre of the circle. This is our formula for centripetal acceleration on page 80.

Summary

To move an object in a circle there must be a resultant force on it, towards the centre of the circle.

This is called the centripetal force: $\quad F = \dfrac{m\,v^2}{r} \quad$ or $\quad F = m\,r\,\omega^2$

An object moving at steady speed in a circle is accelerating because its direction keeps changing.

This centripetal acceleration, a, is towards the centre of the circle: $\quad a = \dfrac{v^2}{r} \quad$ or $\quad a = r\,\omega^2$

Angles in circular motion are measured in radians (2π radians $= 360°$; 1 radian $\approx 57°$).

Angular speed ω is the angle turned through per second: $\quad \omega = \dfrac{\theta}{t}$

Angular speed ω and linear speed v are linked by the equation: $\quad v = r\,\omega$

Frequency f is the number of rotations per second (in Hz). Time period T is the time taken for one rotation.

$$\omega = 2\pi f \qquad \omega = \dfrac{2\pi}{T} \qquad\qquad T = \dfrac{1}{f} \qquad f = \dfrac{1}{T}$$

▷ Questions

1. Can you complete these sentences?
   a) An object cannot move in a circle unless there is a resultant force acting .... the centre of the circle. This is called a .... force.
   If this force is removed, the object will continue moving in a .... line, because of Newton's ...... law.
   b) The centripetal force causes a centripetal .... The bigger the linear ...... (in m s^{-1}) and the smaller the .... of the circle, the bigger the acceleration.
   c) The number of rotations in one second is known as the .... This is measured in .... The time for one rotation is called the ....
   d) One revolution is 360° or .... radians.

2. Convert these angles into radians:
   360°, 180°, 90°, 45°, 30°, 1°.

3. Convert these angles into degrees:
   2π rad, $\pi/2$ rad, $\pi/3$ rad, 1 rad.

4. A car wheel turns through 6 radians in 0.2 seconds. What is its angular speed?

5. A spin-dryer whirls clothes at an angular speed of 85 rad s^{-1}. The drum has a radius of 0.20 m. Calculate the speed of the clothes.

6. Complete this table:

Period /s	Frequency /Hz	Angular speed /rad s^{-1}
	25	
4.2		
		12

7. When a CD rotates at an angular speed of 10.5 rad s^{-1}, how long does one revolution take?

8. A car of mass 800 kg moves in a circle of radius 30 m at a speed of 8.0 m s^{-1}. Calculate:
   a) its centripetal acceleration,
   b) the centripetal force acting.

9.
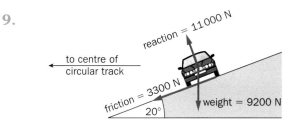

   The diagram shows the forces on a car on a banked track. What is the resultant centripetal force?

10. The Earth takes one year to orbit the Sun along a path of radius 1.5×10^{11} m. Calculate:
   a) the frequency of the Earth's orbit,
   b) the Earth's angular speed.
   c) At what linear speed is the Earth moving?

11. A pilot has a weight of 800 N. He flies his plane at 60 m s^{-1} in a vertical circle of radius 100 m.
   a) Calculate the centripetal force on the pilot.
   b) Draw a diagram to show the two forces on the pilot at (i) the top and (ii) the bottom of the loop.
   c) What is the contact force of the seat on the pilot at (i) the top and (ii) the bottom of the loop?
   d) Express your answers to (c) as multiples of the pilot's weight.

Further questions on page 119.

8 Gravitational Forces and Fields

Gravity is something we take for granted. It seems obvious to us that a dropped object will fall. But why? What is gravity? How do we explain it?
Isaac Newton is credited with 'discovering' gravity when an apple fell on his head. He suggested that the apple fell due to a force of attraction between the apple and the Earth.

One of Newton's greatest achievements was to extend this idea to *all* objects that have mass. He realised that ***all masses attract each other with a gravitational force***.

You know that gravity is pulling you towards the Earth as you read this. But did you realise that there is also a small force of attraction between you and this book (and *every* other object with mass in the Universe for that matter!)?

In this chapter you will learn:
- how to calculate the gravitational force,
- what we mean by gravitational field,
- about the motion of satellites in orbit.

▷ Gravitational force

All objects with mass attract each other with a gravitational force. The size of the force between any two objects depends on their masses and how far apart they are.
We can calculate the force F between two objects of mass m_1 and m_2 using Newton's law of gravitation:

$$F = \frac{G\, m_1\, m_2}{r^2}$$

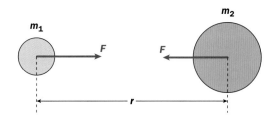

Newton wrote this equation for point masses. However, we can use the law with real objects if we assume that the mass of each object is concentrated at its centre of mass (see page 35). The distance r is measured between the centres of mass.

G is a constant of proportionality known as the **gravitational constant**. G has a value of $\mathbf{6.67 \times 10^{-11}\ N\ m^2\ kg^{-2}}$.

This is a very small number. What does this tell us about the size of gravitational forces?
Gravitational forces are very weak unless we are looking at objects with enormous mass such as stars and planets.

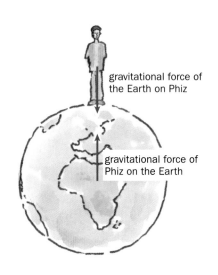

gravitational force of the Earth on Phiz

gravitational force of Phiz on the Earth

Remember that gravitational force is a mutual attraction. You attract the Earth with the same force that the Earth attracts you! However, this force has more effect on you because you are much lighter than the Earth.
Which of Newton's laws of motion is about this idea of equal but opposite forces? (See page 56.)

An inverse-square relationship

Newton's law of gravitation is an example of an **inverse-square law**.
Inverse means that as the separation of the masses **increases**, the gravitational force **decreases**.
In this case the force decreases in proportion to the **square** of the distance between the masses.

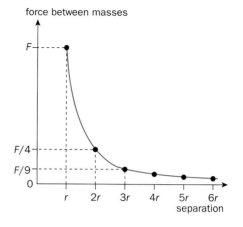

force between masses

So: 2 × distance   gives $\frac{1}{4}$ × gravitational force

3 × distance   gives $\frac{1}{9}$ × gravitational force

10 × distance   gives $\frac{1}{100}$ × gravitational force

Example 1

The gravitational force of attraction between two marbles placed with their centres 1.0 m apart is 8.00×10^{-14} N.
What would be the force of attraction if they were placed 4.0 m apart?

The separation is 4 times as great so the force must fall to $\frac{1}{(4)^2}$ or $\frac{1}{16}$ of the original force.

$\therefore$ New force of attraction $= \frac{1}{16} \times 8.00 \times 10^{-14}$ N $= \underline{5.0 \times 10^{-15}}$ N (2 s.f.)

Example 2

The Moon orbits the Earth along a path of radius 3.84×10^8 m.
The mass of the Moon, M_M, is 7.35×10^{22} kg.
The Earth's mass, M_E, is 6.00×10^{24} kg.
$G = 6.67 \times 10^{-11}$ N m^2 kg^{-2}

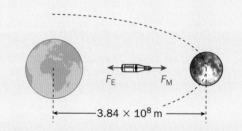

a) The gravitational force of attraction between the Earth and the Moon keeps the Moon in orbit. Calculate the size of this force.

b) A rocket of mass 42 000 kg is fired from Earth to the Moon. What is the net gravitational force on the rocket when it is 3.00×10^8 m from the centre of the Earth?

a) $F = \dfrac{G\, M_E\, M_M}{r^2} = \dfrac{6.67 \times 10^{-11} \times 6.00 \times 10^{24}\ \text{kg} \times 7.35 \times 10^{22}\ \text{kg}}{(3.84 \times 10^8\ \text{m})^2} = \underline{1.99 \times 10^{20}\ \text{N}}$ (3 s.f.)

b) Gravitational force between rocket and Earth:

$F_E = \dfrac{G\, M_E\, m}{r^2} = \dfrac{6.67 \times 10^{-11} \times 6.00 \times 10^{24}\ \text{kg} \times 42\,000\ \text{kg}}{(3.00 \times 10^8\ \text{m})^2} = 187\ \text{N}$

Gravitational force between rocket and Moon:

$F_M = \dfrac{G\, M_M\, m}{r^2} = \dfrac{6.67 \times 10^{-11} \times 7.35 \times 10^{22}\ \text{kg} \times 42\,000\ \text{kg}}{(3.84 \times 10^8\ \text{m} - 3.00 \times 10^8\ \text{m})^2} = 29.2\ \text{N}$

The net force on the rocket is the vector sum of these two forces:

$\therefore$ Net force $= 187\ \text{N} - 29.2\ \text{N} = \underline{158\ \text{N}}$ towards the Earth (3 s.f.)

At some point on its journey the net force on the rocket will be zero. At this point $F_E = F_M$.
Is this point closer or further from the Moon?

In fact, the forces balance at 3.45×10^8 m from Earth. Try working this out for yourself.

▷ Gravitational field

If a mass has a gravitational force on it, then we say it is in a **gravitational field**. We use **field lines** to show the gravitational field around an object with mass.
This is similar to the way we draw magnetic field lines to show the field round a magnet.

The diagrams show the Earth's gravitational field:

For spherical objects such as Earth the field is *radial*. All the field lines point towards the centre of mass, ie. the centre of the sphere.

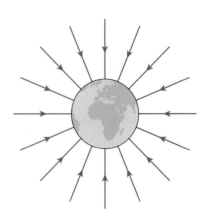

The field lines tell us two things:

- The arrows on the field lines indicate the **direction** of the field. This is the direction in which a mass in the field would be pulled.

- The spacing of the lines tells us about the **strength** of the field. The closer the field lines are together, the stronger the force.

  What happens to the strength of the field as we move away from Earth? The field lines get further apart, as the field gets weaker.

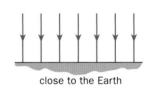

close to the Earth

Close to the Earth's surface there may be small variations in the gravitational field, but we usually ignore them. We treat the field as uniform. In this case we draw the field lines parallel to each other.

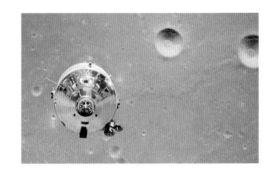

Houston, we have a problem . . .
In 1970 the Apollo 13 spacecraft suffered major system failure en route to the Moon. The fuel cells failed and oxygen was running out fast. The rockets had insufficient fuel to turn back so the crew used the gravitational field of the Moon to swing themselves safely back to Earth.

Gravitational field strength, *g*

You can measure the strength of the Earth's gravitational field at any point using a mass on a spring balance.
The stronger the field, the greater the gravitational force pulling the mass towards the Earth.
Of course heavier masses are pulled more than lighter ones.

This is why we state the strength of a field in terms of the force acting *on each kilogram* of mass in the field:

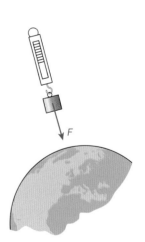

$$\text{gravitational field strength, } g = \frac{\text{force, } F \text{ (N)}}{\text{mass, } m \text{ (kg)}}$$ or $$g = \frac{F}{m}$$

where F is the force acting on a mass m in the field.
Gravitational field strength g is a **vector** quantity. It is measured in $N\,kg^{-1}$.

At the Earth's surface $g = \textbf{9.81 N kg}^{-1}$. We treat this value as a constant, although in fact it varies very slightly from place to place.

The gravitational field strength g (in $N\,kg^{-1}$) has the same value as the acceleration due to gravity g (in $m\,s^{-2}$) at that point (see page 44).

Calculating g in a radial field

How can we find the value of g at any point in a gravitational field? Taking a spring balance out into space is rather impractical!

We can derive an equation for g. Consider a small mass m in the gravitational field of a larger mass M, such as the Earth:

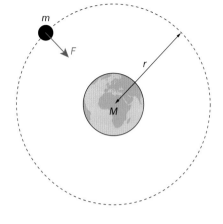

From our definition of gravitational field strength g on the opposite page, the force acting on the mass in the field is $F = m\,g$. Do you recognise this form of the equation? You have used this before to calculate the weight of objects. (See page 18)

This force is the result of gravitational attraction. So we can also calculate the size of this force using Newton's law of gravitation.

This tells us: $\quad F = \dfrac{GMm}{r^2}$

Since these are two ways of expressing the **same** force F, we can equate them, so:
$$\cancel{m}g = \frac{GM\cancel{m}}{r^2}$$

Cancelling the m's gives:

$$\boxed{g = \frac{GM}{r^2}}$$

This gives us the field strength g at any point around a mass M. Note that the distance r is measured from the centre of mass.

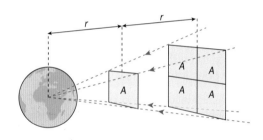

This equation is another **inverse-square** relationship. You can also see this by looking at the field lines. Remember that the spacing of the lines indicates the strength of the field. For example, at twice the distance away from Earth, the field lines are spaced over four times the area. This shows that the field is only $\frac{1}{4}$ as strong at this distance.

Example 3
Find the gravitational field strength at:
a) the surface, and
b) 1.0×10^6 m above the surface of a star of radius 1.2×10^6 m.
The mass of the star is 1.5×10^{30} kg.
$G = 6.67 \times 10^{-11}$ N m^2 kg^{-2}

a) At the surface, $\quad \mathbf{g} = \dfrac{GM}{r^2} = \dfrac{6.67 \times 10^{-11} \times 1.5 \times 10^{30} \text{ kg}}{(1.2 \times 10^6 \text{ m})^2} = \underline{6.9 \times 10^7 \text{ N kg}^{-1}}$ (2 s.f.)

b) Remembering that r is measured from the **centre** of the star:

At 1.0×10^6 m above the surface, $g = \dfrac{GM}{r^2} = \dfrac{6.67 \times 10^{-11} \times 1.5 \times 10^{30} \text{ kg}}{(1.2 \times 10^6 \text{m} + 1.0 \times 10^6 \text{ m})^2} = \underline{2.1 \times 10^7 \text{ N kg}^{-1}}$ (2 s.f.)

Example 4
Find the mass of the Earth, given that its radius is 6400 km and g at the surface is 9.81 N kg^{-1}. ($G = 6.67 \times 10^{-11}$ N m^2 kg^{-2})

$r = 6400$ km $= 6400 \times 10^3$ m

$\mathbf{g} = \dfrac{GM}{r^2} \quad$ and so: $\quad M = \dfrac{gr^2}{G} = \dfrac{9.81 \text{ N kg}^{-1} \times (6400 \times 10^3 \text{ m})^2}{6.67 \times 10^{-11} \text{ N m}^2 \text{ kg}^{-2}} = \underline{6.02 \times 10^{24} \text{ kg}}$ (3 s.f.)

▷ Satellites

Have you ever had the feeling that you were being watched?
Well, you'd be right! Several thousand **artificial satellites** circle
the Earth. These 'eyes in the sky' are continually watching over us.

Can you name our oldest satellite?
The original Earth satellite is the Moon. This is a **natural satellite**.
A satellite is any object held in orbit around a larger one by
gravitational attraction. The Earth itself is a satellite of the Sun.

Artificial satellites have many uses.
Observation satellites monitor the weather, map changes in land
use and vegetation, track pollution and watch over military targets.

Satellites are also used for space observation.
Orbiting telescopes can probe deep into space without interference
from the Earth's atmosphere.

We also increasingly rely on satellites for navigation and global
communications. Satellites are big business.

View from a weather satellite (in 'false colour').

Into orbit: falling through space

Why do satellites stay in orbit rather than falling back to Earth?
In fact, satellites are *continually falling*.
A rocket engine carries a satellite to the correct height and into a
path parallel to the Earth's surface. The engines are then stopped.
The only force then acting is the Earth's gravitational pull, so the
satellite goes into free fall.

As it was initially moving parallel to the Earth's surface, the satellite
falls along a curved path, like a projectile (see page 46).
As the Earth curves away beneath it, the satellite never hits the
ground, but continues round in its orbit.
The satellite must move at exactly the right velocity for its orbit.

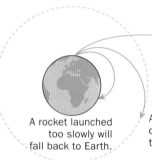

A rocket launched at
11 200 m s⁻¹ will have
enough energy to
escape the Earth's
gravity completely
(see page 96).

A rocket launched
too slowly will
fall back to Earth.

A satellite in orbit is in
continual free fall around
the Earth.

Types of orbit

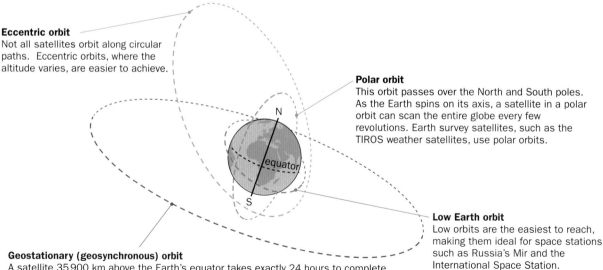

Eccentric orbit
Not all satellites orbit along circular
paths. Eccentric orbits, where the
altitude varies, are easier to achieve.

Polar orbit
This orbit passes over the North and South poles.
As the Earth spins on its axis, a satellite in a polar
orbit can scan the entire globe every few
revolutions. Earth survey satellites, such as the
TIROS weather satellites, use polar orbits.

Low Earth orbit
Low orbits are the easiest to reach,
making them ideal for space stations
such as Russia's Mir and the
International Space Station.

Geostationary (geosynchronous) orbit
A satellite 35 900 km above the Earth's equator takes exactly 24 hours to complete
one orbit. As the Earth spins, the satellite appears to stay fixed in the sky.
Geostationary orbits are used for communications satellites such as the INTELSAT
network. The METEOSAT weather satellites are also placed in geostationary orbit
to continuously monitor large areas of the globe.

Satellite calculations

Can you remember the equations for circular motion?
If not, look back at the previous chapter now (page 80).
For calculations on the motion of satellites, you will need to combine
the equations for gravitation with those for circular motion.

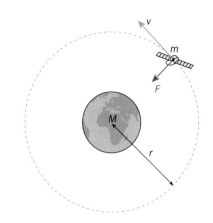

A centripetal force is needed to keep a satellite moving along a
circular path. Where does this force come from?
The only force acting on the satellite is Earth's gravity. The centripetal
force is provided by this gravitational attraction. For a satellite of
mass m in an orbit of radius r and velocity v about a planet of mass M:

Gravitational force of attraction = mass × centripetal acceleration

$$\frac{GMm}{r^2} = m \times \frac{v^2}{r}$$

Substituting $g = \dfrac{GM}{r^2}$ (from page 89) gives:

$$\not{m}g = \not{m}\frac{v^2}{r} \qquad \therefore \quad g = \frac{v^2}{r}$$

You will find these equations very useful for solving satellite problems where g varies with altitude.

Example 5
A spaceship orbits the Moon, travelling at 1460 m s^{-1}. How high is it
above the Moon's surface? The Moon has a mass of 7.35×10^{22} kg
and a radius of 1.74×10^6 m. ($G = 6.67 \times 10^{-11}$ N m^2 kg^{-2})

Gravitational force of attraction = mass × centripetal acceleration

$$\frac{GM\not{m}}{r^2} = \not{m} \times \frac{v^2}{\not{r}}$$

Cancelling and rearranging gives: $r = \dfrac{GM}{v^2} = \dfrac{6.67 \times 10^{-11} \times 7.35 \times 10^{22} \text{ kg}}{(1460 \text{ m s}^{-1})^2} = 2.30 \times 10^6$ m

Remember r is measured from the centre of mass:
$\therefore$ height above the surface = 2.30×10^6 m $- 1.74 \times 10^6$ m = $\underline{5.60 \times 10^5 \text{ m}}$ (3 s.f.)

Example 6
A geostationary satellite must orbit the Earth along a path of
radius 4.23×10^7 m. Use this information to calculate:
a) the speed of the satellite,
b) the gravitational field strength at this height, and
c) the mass of the Earth.
($G = 6.67 \times 10^{-11}$ N m^2 kg^{-2})

a) For a geostationary orbit, the period of rotation T = 24 hours = 24 × 60 × 60 s = 86 400 s
   Circumference of orbit = $2\pi r$

$$\textbf{Speed} = \frac{\textbf{distance}}{\textbf{time}} = \frac{2\pi r}{T} = \frac{2\pi \times 4.23 \times 10^7 \text{ m}}{86\,400 \text{ s}} = \underline{3.08 \times 10^3 \text{ m s}^{-1}} \text{ (3 s.f.)} \ (\approx 6000 \text{ mph})$$

b) Gravitational field strength $\boldsymbol{g} = \dfrac{\boldsymbol{v^2}}{\boldsymbol{r}} = \dfrac{(3080 \text{ m s}^{-1})^2}{4.23 \times 10^7 \text{ m}} = \underline{0.224 \text{ N kg}^{-1}}$ (3 s.f.)

c) $\boldsymbol{g} = \dfrac{\boldsymbol{GM}}{\boldsymbol{r^2}}$ and so: $M = \dfrac{gr^2}{G} = \dfrac{0.224 \text{ N kg}^{-1} \times (4.23 \times 10^7 \text{ m})^2}{6.67 \times 10^{-11}} = \underline{6.01 \times 10^{24} \text{ kg}}$ (3 s.f.)

▷ Gravitational potential, V

You have come across the word 'potential' before when we calculated changes in gravitational potential energy (ΔE_p) in Chapter 6.
Do you remember the equation for this? (See page 70.)
Close to the Earth's surface, $\Delta E_p = mg\Delta h$.
For this we assumed that g is a constant (= 9.81 N kg^{-1}).

However, when you calculate energy changes over large distances in space you cannot use this equation. Can you see why?
In space, the value of g changes as the distance from Earth changes.

So instead we use a quantity called **gravitational potential** (given the symbol V). The gravitational potential at any point tells us the potential energy *of each kilogram of mass* at that point.

The potential energy depends on the mass M which causes the gravitational field, and the distance r from the centre of mass:

$$V = -\frac{GM}{r}$$

G is the gravitational constant.
V is measured in **joules per kilogram** (J kg^{-1}). Like energy, gravitational potential is a scalar quantity. It has magnitude only.

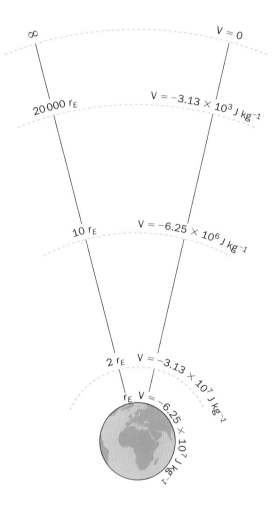

Why is V negative?

In Chapter 6 we said that you can choose the point at which potential energy is zero. This is usually ground level.

But what if we are comparing the potentials of a person standing on Earth and a person on Mars? What common point could we take as the zero here?
Gravitational potential is defined so that $V = 0$ *at an infinite distance away*.
But potential energy increases the higher we lift an object, i.e. the further we are from Earth. So how can V keep increasing and eventually reach zero? The answer is that V must have negative values.

Gravitational potential V at a point is defined as
the work done against gravity when a 1 kg mass is brought from infinity to that point.
This must be negative as gravity pulls the masses together. We only do work *against* gravity if the 1 kg mass is moved further away.

You can see from the diagram that V has its most negative value at the planet's surface. It becomes less negative as we move higher.
You can check these values using the equation for V.

Example 7
Calculate the gravitational potential V at 4.50×10^{20} m from the centre of a star of mass 8.45×10^{40} kg.
($G = 6.67 \times 10^{-11}$ N m^2 kg^{-2})

• $V = ?$

$\longleftarrow\quad 4.50 \times 10^{20}$ m $\quad\longrightarrow$

$$V = -\frac{GM}{r} = -\frac{6.67 \times 10^{-11} \text{ N m}^2 \text{ kg}^{-2} \times 8.45 \times 10^{40} \text{ kg}}{4.50 \times 10^{20} \text{ m}} = \underline{-1.25 \times 10^{10} \text{ J kg}^{-1}} \quad (3 \text{ s.f.})$$

Equipotentials

All points at the same height above the Earth must have the same gravitational potential. A line joining all the points with the same potential is called an **equipotential**.
This is like a contour line on a map.

The diagram shows the equipotential surfaces around the Earth:

For a radial field like this the equipotentials are concentric spheres. The gravitational field lines cross the equipotential surfaces at right angles. This means that the gravitational force acts at 90° to the equipotentials.
If an object such as a satellite moves along an equipotential its potential energy does not change. As there is no movement in the direction of the force, no work is done.

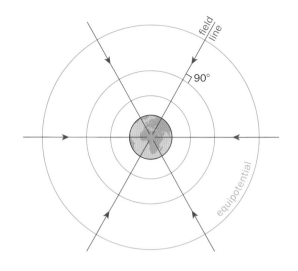

Energy changes

So how do we calculate the work done, or energy change, when an object moves from one point in a field to another?
First we need to find the change in gravitational potential.
This gives us the change in energy *per kilogram*. So to find the total energy change we need to multiply by the mass of the object:

> **Energy change** (J) = **change in gravitational potential** (J kg^{-1}) × **mass** (kg)

Example 8
Calculate the energy change when a rocket of mass 50 000 kg moves at steady speed from the surface to a height of 3.50×10^6 m above the Earth.
The Earth has a mass of 6.00×10^{24} kg and a radius of 6.40×10^6 m.
($G = 6.67 \times 10^{-11}$ N m^2 kg^{-2})

Remembering that the distance r is measured from the centre of mass:

$$V \text{ at the surface} = -\frac{GM}{r} = -\frac{6.67 \times 10^{-11} \text{ N m}^2 \text{ kg}^{-2} \times 6.00 \times 10^{24} \text{ kg}}{6.40 \times 10^6 \text{ m}} = -6.25 \times 10^7 \text{ J kg}^{-1}$$

$$V \text{ at } 3.50 \times 10^6 \text{ m} = -\frac{GM}{r} = -\frac{6.67 \times 10^{-11} \text{ N m}^2 \text{ kg}^{-2} \times 6.00 \times 10^{24} \text{ kg}}{(6.40 \times 10^6 + 3.50 \times 10^6) \text{ m}} = -4.04 \times 10^7 \text{ J kg}^{-1}$$

∴ Change in potential $\Delta V = (-4.04 \times 10^7 \text{ J kg}^{-1}) - (-6.25 \times 10^7 \text{ J kg}^{-1}) = 2.21 \times 10^7 \text{ J kg}^{-1}$

∴ Total energy change = ΔV × mass of rocket = 2.21×10^7 J kg^{-1} × 50 000 kg = $\underline{1.11 \times 10^{12} \text{ J}}$ (3 s.f.)
The energy change is positive. Work must be done against gravity to move the rocket away from Earth.

Example 9
A meteorite is attracted by the gravitational field of the Moon. The Moon has a mass of 7.7×10^{22} kg and radius 1.7×10^6 m. Neglecting the initial speed of the meteorite, calculate the speed at which it strikes the Moon's surface. ($G = 6.67 \times 10^{-11}$ N m^2 kg^{-2})

Kinetic energy gained = gravitational potential energy lost in moving from infinity to the surface:

$$\tfrac{1}{2} m v^2 = m \times \Delta V$$
$$\tfrac{1}{2} \cancel{m} v^2 = \cancel{m} \times \frac{GM}{r} = \frac{6.67 \times 10^{-11} \times 7.7 \times 10^{22}}{1.7 \times 10^6} = 3.02 \times 10^6$$
$$\therefore \quad v = \sqrt{2 \times 3.02 \times 10^6} = \underline{2.5 \times 10^3 \text{ m s}^{-1}} \quad (2 \text{ s.f.})$$

▷ Kepler's Laws of Planetary Motion

In 1543 Copernicus's 'heliocentric' (Sun-centred) model of the universe was published. In this new model the planets followed circular orbits around the Sun. The established church objected to this idea but many scientists were also skeptical, as the model did not quite fit the observed motion of the planets.

Johannes Kepler solved this problem by careful analysis of data collected by the great Danish astronomer Tycho Brahe. Kepler realised that Copernicus's model would work if the planets followed *elliptical*, rather than circular, orbits around the Sun.

An ellipse is a curve in which the sum of the distances from any point on the curve to two fixed points is constant. The fixed points are known as *foci*.
You can use this idea to draw an ellipse for yourself using a fixed length of string held by drawing pins at two points (the foci).
Pulling the string taut with a pencil and moving it round the pins will produce an ellipse. The lengths a and b must always add up to the fixed length of the string.
The closer the foci, the more circular the ellipse becomes.
A circle is just a special ellipse with both foci at the same point.

In the early 1600s, Kepler published three laws that described the observed motion of the planets:

Johannes Kepler (1571–1630), German mathematician and astronomer.

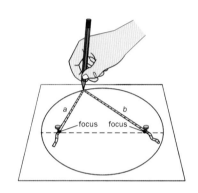

Kepler's First Law: Orbit shape

All planets move in elliptical orbits with the Sun at one focus.

Most planets have orbits that are nearly circular.
Kepler's realisation that the orbits were elliptical took careful study of the geometry. The diagrams on these pages exaggerate the elliptical shape for clarity.

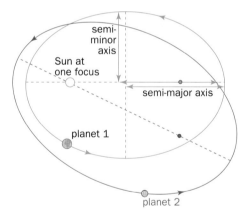

Kepler's Second Law: Orbit areas

An imaginary line between the Sun and a planet sweeps out equal areas in equal times.

For a circular orbit, this law tells us that a planet would have a constant angular velocity.
For an elliptical orbit, the area swept out in a given time remains constant as the orbit speed varies.
The shaded sectors in the diagram show the areas swept out by a line between the Earth and the Sun in one month:
Closer to the Sun, the gravitational pull is stronger and so the Earth moves more quickly. It moves along a longer path in a given time. The sector swept out is short but wide.
Further from the Sun, the Earth moves more slowly, covering a shorter path in a given time. The sector swept out is long but thin.
The shaded sectors have different shapes but the *same* area.

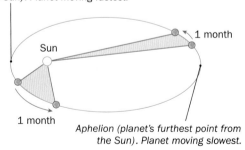

Perihelion (planet's closest point to the Sun). Planet moving fastest.

Aphelion (planet's furthest point from the Sun). Planet moving slowest.

Kepler's Third Law: Orbit period

The square of a planet's orbit period T is proportional to the cube of the average distance r between the planet and the Sun.

$$T^2 \propto r^3$$

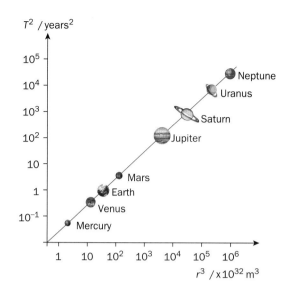

Plotting T^2 against r^3 for the planets gives a straight line:

The distance r is taken as the length of the semi-major axis of the planet's elliptical orbit.

Kepler's work was based on direct observations of the planets. However, his laws apply to all orbits, including that of the Moon and artificial satellites.

Kepler's results provided the foundation for Newton's work on gravitation. Newton wanted to **explain** the observed motion described by Kepler's laws. In 1666, aged just 24, Newton formulated his law of gravitation. Newton was then able to derive Kepler's third law by equating the centripetal force needed to keep a planet in orbit with the force due to gravity. Here's how this works:

Derivation of $T^2 \propto r^3$

Consider a planet of mass M_P moving along a circular orbit of radius r around the Sun, which has mass M_S:
The planet is orbiting with velocity v and takes a time T to complete one orbit. F is the resultant force acting on the planet.

Applying Newton's law of gravitation (see page 86), $F = \dfrac{GM_S M_P}{r^2}$

Applying the equation for centripetal force (see page 80), $F = \dfrac{M_P v^2}{r}$

Equating these two equations gives: $\dfrac{GM_S M_P}{r^2} = \dfrac{M_P v^2}{r}$ $\therefore v^2 = \dfrac{GM_S}{r}$

For a circular orbit, $v = \dfrac{\text{distance travelled}}{\text{time taken}} = \dfrac{2\pi r}{T}$ $\therefore \dfrac{4\pi^2 r^2}{T^2} = \dfrac{GM_S}{r}$

Rearranging, we get: $T^2 = \left(\dfrac{4\pi^2}{GM_S}\right) r^3$

The value of $\left(\dfrac{4\pi^2}{GM_S}\right)$ is a constant for all planets, and anything orbiting the Sun, so $T^2 \propto r^3$.

Example 10
Io and Europa are two of the moons of Jupiter. Io has an average orbit radius r_I of 4.22×10^8 m. Europa has an average orbit radius r_E of 6.71×10^8 m. If Io has an orbit period T_I of 1.77 days, how long does Europa's orbit take?

$T^2 \propto r^3$ $\therefore \dfrac{T^2}{r^3} = \text{constant}$ $\therefore \dfrac{T_I^2}{r_I^3} = \dfrac{T_E^2}{r_E^3}$ So $\dfrac{1.77^2}{(4.22 \times 10^8)^3} = \dfrac{T_E^2}{(6.71 \times 10^8)^3}$

$\therefore T_E^2 = \dfrac{1.77^2}{(4.22 \times 10^8)^3} \times (6.71 \times 10^8)^3 = 12.59$ $\therefore T_E = \sqrt{12.59} = \underline{3.55 \text{ days}}$

Linking V and g

The equations for the **gravitational field strength g** and the **gravitational potential V** at a distance r from the centre of a planet of mass M are closely related:

From page 89, $g = \dfrac{GM}{r^2}$

From page 92, $V = -\dfrac{GM}{r}$

So, at a given point, $\boldsymbol{g} = -\dfrac{\boldsymbol{V}}{\boldsymbol{r}}$

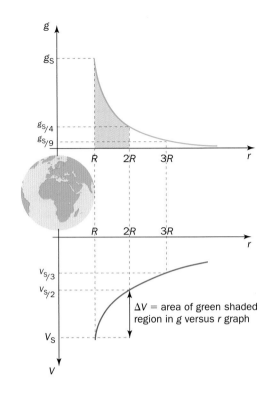

We can also link these variables through their two graphs:

The rate of increase of the gravitational potential with distance r is known as the **potential gradient, $\Delta V/\Delta r$**. Notice that the potential gradient (on the graph) gradually decreases.
The value of this gradient at any point gives the value of g, the gravitational field strength, at that distance.

$$\boxed{\boldsymbol{g} = -\dfrac{\boldsymbol{\Delta V}}{\boldsymbol{\Delta r}} = \textbf{potential gradient}}$$

Also, the area under the graph of g versus r gives the change in potential ΔV. For example, the area of the green shaded strip from R to $2R$ in the top graph is equal to the change in potential ΔV from R to $2R$ marked on the bottom graph.

ΔV = area of green shaded region in g versus r graph

Escape velocity

If you throw an object straight up into the air, it gradually slows to a stop and then falls back down. The faster you throw the object, the higher it goes. But how fast must you throw something for it to escape the pull of Earth's gravity completely and move off into space? The speed needed is called the **escape velocity**.

To calculate the escape velocity we need to work out the initial kinetic energy needed to do enough work against gravity to move from the planet's surface to an infinite distance:

At the planet's surface, $V = -\dfrac{GM}{R}$ At infinite distance, $V = 0$

Work done in moving to infinity = change in potential $\times$ mass m of the object
$$= \dfrac{GMm}{R}$$

This is the kinetic energy needed so, if v_E is the escape velocity:

$$\tfrac{1}{2} m v_E^2 = \dfrac{GMm}{R} \qquad \therefore \boxed{\boldsymbol{V_E} = \sqrt{\dfrac{\boldsymbol{2GM}}{\boldsymbol{R}}} = \sqrt{\boldsymbol{2gR}}} \quad \text{(substituting for } g = GM/R^2\text{)}$$

Notice that this depends on the mass and radius of the planet but not on the mass of the object.

> *Mathematicians may have spotted that the equation for g can be found by differentiating the equation for V with respect to r.*

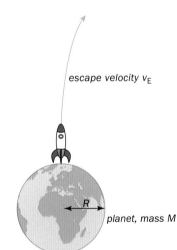

escape velocity v_E

planet, mass M

Example 11
How fast would you need to throw a ball for it to escape the pull of Earth's gravity, given that the mass of the Earth is 6.00×10^{24} kg and its radius is 6.38×10^6 m? ($G = 6.67 \times 10^{-11}$ N m^2 kg^{-2})

$$v_E = \sqrt{\dfrac{2GM}{R}} = \sqrt{\dfrac{2 \times 6.67 \times 10^{-11} \times 6.00 \times 10^{24}}{6.38 \times 10^6}} = 11\,200 \text{ m s}^{-1} \text{ (about 25\,000 miles per hour!)}$$

▷ Physics at work: Satellites and space probes

Space telescopes

Space telescopes have huge advantages over Earth-based observatories. From their position in space they are unaffected by the absorption and distortion caused by the Earth's atmosphere.

The Hubble Space Telescope (HST) was launched into low Earth orbit in 1990, orbiting Earth once every 97 min at an altitude of 600 km. The HST provided us with truly awe-inspiring images of our universe. The eXtreme Deep Field (XDF) image shown here was compiled from 10 years of images with a total exposure time of two million seconds! Each spot of light is a galaxy. The faintest galaxies in the image are one ten-billionth of the brightness detectable with the naked eye.

Hubble's successor, the Webb Space Telescope (see page 155), is on schedule for launch in 2018. Its light-gathering main mirror (constructed from 18 hexagonal segments) will have more than 7 times the area of the HST mirror, giving even greater resolution.

The Hubble eXtreme Deep Field (XDF) image was assembled from 342 separate exposures.

Global Positioning System (GPS)

We rely increasingly on GPS to navigate, whether using maps on our mobile phones or in-car 'sat navs'. So how do these work?

A network of 24 satellites orbiting 20 200 km above Earth is continuously transmitting radio time signals. Your GPS receiver picks these up and feeds the signal to a computer chip that can calculate the exact time taken for a satellite signal to reach it. By taking readings from four different satellites, the device can calculate its own three-dimensional position and velocity. This is known as **triangulation**.

The GPS system provides us with information accurate to within 10 m, anywhere in the world.

One of 24 NAVSTAR satellites in the GPS network.

Reaching Saturn

The Cassini space probe has been exploring Saturn and its moons since 2004. Cassini took nearly seven years to reach Saturn from Earth. The diagram shows its flight path. Why did it take such a circuitous route?

A probe uses the gravitational pull of other planets to swing itself towards its target. At each planetary fly-by a 'slingshot' effect increases the probe's speed. Fly-bys of Venus, Earth and Jupiter were needed to gain enough speed to overcome the enormous gravitational pull of the Sun and reach Saturn.

Saturn arrival
1 July 2004

orbit of Jupiter

orbit of Saturn

Venus swing-by
20 June 1998

orbit of Earth

Venus swing-by
21 April 1998

orbit of Venus

Earth swing-by
16 August 1998

launch
8 October 1997

Jupiter swing-by
30 December 2000

Summary

The universal constant of gravitation $G = 6.67 \times 10^{-11}$ N m^2 kg^{-2}

All masses attract each other with a gravitational force. Force $\boldsymbol{F} = \dfrac{\boldsymbol{G\, m_1\, m_2}}{\boldsymbol{r^2}}$

The **gravitational field strength g** at a point
is the force acting on each kilogram of mass in a gravitational field: $\boldsymbol{g} = \dfrac{\boldsymbol{F}}{\boldsymbol{m}}$ or $\boldsymbol{F} = \boldsymbol{m\,g}$
g is a vector quantity. It is measured in N kg^{-1}.

In a radial field: $\boldsymbol{g} = \dfrac{\boldsymbol{GM}}{\boldsymbol{r^2}}$ The 'potential gradient' $= -\dfrac{\Delta V}{\Delta r} = g$

The **gravitational potential V** at a point
is the potential energy of each kilogram of mass in a gravitational field.
V is a scalar quantity. It is measured in J kg^{-1}.

In a radial field: $\boldsymbol{V} = -\dfrac{\boldsymbol{GM}}{\boldsymbol{r}}$

V is taken to be zero at infinity.   Change in energy = change in gravitational potential $\times$ mass

All orbital motion is described by **Kepler's Laws**.
Orbital period T and radius r are linked by Kepler's 3rd Law: $\boldsymbol{T^2 \propto r^3}$

▷ Questions

You will need the following data:

$G = 6.67 \times 10^{-11}$ N m^2 kg^{-2}

$g = 9.81$ N kg^{-1} at the Earth's surface

Mass of the Earth $= 6.00 \times 10^{24}$ kg

Radius of the Earth $= 6.4 \times 10^6$ m

Mass of the Sun $= 2.00 \times 10^{30}$ kg

1. Can you complete these sentences?
   a) The gravitational force between two objects depends on their .... and their ....
   b) Newton's law of gravitation is an example of an inverse .... relationship.
      If the separation doubles, the force ....
   c) A gravitational .... is a region in which a mass feels a gravitational force.
      Field lines can indicate the .... and the .... of a field.
      Round the Earth, the shape of the gravitational field is ....
   d) The gravitational field strength g (in N kg^{-1}) has the same value as the .... due to gravity (in m s^{-2}) at that point.
      The value of g on the surface of the Earth is .... N kg^{-1}
   e) A satellite that takes exactly 24 hours to complete one orbit of the Earth is described as ....
   f) A line joining all points with the same potential is called an ....
   g) To find the energy change when an object moves from one point in a field to another, we multiply the change in .... by the .... of the object.

2. a) Using $W = mg$, calculate the weight of a 2.0 kg mass on Earth.
   b) Using Newton's law of gravitation, calculate the gravitational force on a 2.0 kg mass on the Earth's surface.
   c) What do you notice about your answers to parts (a) and (b)?

3. Calculate the gravitational force between:
   a) two protons of mass 1.67×10^{-27} kg separated by 1.40×10^{-14} m,
   b) a dog of mass 35 kg in the UK and a 60 kg kangaroo on the other side of the world in Australia,

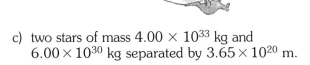

   c) two stars of mass 4.00×10^{33} kg and 6.00×10^{30} kg separated by 3.65×10^{20} m.

4. The Earth orbits the Sun along a path of radius 1.50×10^{11} m.
   a) Find the gravitational force exerted by the Earth on the Sun.
   b) Saturn has a mass 95 times the Earth's and an orbit radius 9.5 times the Earth's.
      Compared to Earth, does it exert a larger or smaller force on the Sun?

5. Two stars of mass 6.2×10^{32} kg and M kg are 1.4×10^{16} m apart. The net gravitational force is zero at P, 1.5×10^{15} m from M along the line between the two stars. What is the value of M?

6. a) Find g at 200 000 m above the Earth's surface.
   b) An astronaut has a mass of 70 kg. What is her weight (i) on Earth and (ii) at 200 000 m?

7. a) Calculate the value of g at points R, $2R$, $3R$, $4R$ and $5R$ from the centre of the Earth, where R is the Earth's radius.
   b) Use your values from part (a) to plot a graph of g against distance from the Earth, starting at the Earth's surface:

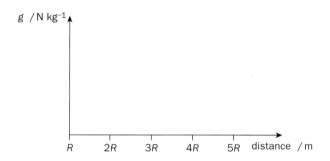

8. How far above the Earth's surface is g:
   a) one-quarter of its value at the surface,
   b) one-ninth of its value at the surface?

9. Find the radius of Mars, given that the gravitational field strength g at its surface is 3.73 N kg^{-1} and its mass is 6.42×10^{23} kg.

10. A mass of 5.00 kg feels a force of 16.6 N at a point 200 km above the surface of Mars.
    a) Calculate the value of g at this point.
    b) The radius of Mars is 3390 km. Calculate its mass.

11. A geostationary satellite orbits the Earth, travelling at 3080 m s^{-1}.
    a) What is its period of rotation?
    b) Calculate its height above the Earth's surface.

12. Calculate the velocity of Venus in its orbit around the Sun. The radius of its orbit is 1.08×10^{11} m.

13. How fast must a rocket be travelling to take off from the surface of a planet of mass 5.6×10^{26} kg and radius 6.0×10^{7} m and move off into space?

14. A galaxy orbits a black hole with an orbit radius of 2.2×10^{17} km and a period of 3.4×10^{9} years. Estimate the orbital period for a galaxy orbiting the same black hole at a distance of 3.6×10^{17} km.

15. a) Calculate the gravitational potential at the Earth's surface.
    b) At what height above the Earth does the potential fall to half this value?

16.

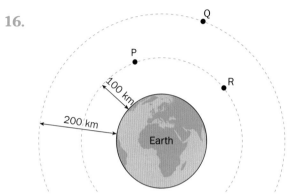

a) Calculate the gravitational potential at the points P and Q shown in the diagram.
b) How much energy is needed to move a rocket of mass 33 000 kg from P to Q?
c) Explain why no work is done as the rocket moves from P to R.

17. The gravitational potential 4.5×10^{7} m from the centre of a planet is -1.3×10^{8} J kg^{-1}.
    a) Calculate the mass of the planet.
    b) Why is the potential negative?

18. The Moon has a mass of 7.35×10^{22} kg and a radius of 1740 km.
    a) Calculate the gravitational potential at its surface.
    b) A probe of mass 100 kg is dropped from a height of 1 km on to the Moon's surface. Calculate its change in potential energy.
    c) If all the potential energy lost is converted to kinetic energy, calculate the speed at which the probe hits the surface.

19. A satellite of mass 19 500 kg orbits the Earth along a path of radius 6.9×10^{6} m. Calculate:
    a) the speed of the satellite,
    b) the kinetic energy of the satellite,
    c) the gravitational potential at this altitude,
    d) the change in potential energy in moving the satellite from the Earth's surface to its orbit,
    e) the minimum energy needed to place the satellite in this orbit. (Ignore friction with the atmosphere.)

Further questions on page 119.

9 Simple Harmonic Motion

We have already looked at one type of periodic motion, circular motion in Chapter 7.
We now look at a second type, where an object repeatedly moves backwards and forwards, or up and down, in a regular way.

The photograph shows a bungee jumper. At the end of his fall he bounces up and down about a central point before eventually stopping. This is an example of a periodic motion called **simple harmonic motion (SHM)**.

In this chapter you will learn:
- how to describe SHM using graphs and equations,
- about the energy transfers involved in SHM,
- the effects of damping and resonance.

Adrenaline-charged SHM!

Keeping time

Why do some old clocks use a pendulum?
A pendulum oscillates backwards and forwards with a regular beat. Even as the oscillation dies away, its **time period** (the time for one complete oscillation) stays the same.

Why? Because the amplitude of the swing also decreases. The pendulum moves a smaller distance at a slower speed, and the time stays the same.

This effect was first noticed by Galileo in 1581. He watched the lamps in Pisa's cathedral swinging backwards and forwards. The time for each swing stayed almost the same even as the swinging died down. Galileo realised that this regular motion could be used for time keeping.

All clocks rely on some type of periodic motion.
Oscillating pendulums are still used in many mechanical clocks. Modern quartz watches use the regular vibrations of quartz crystals and atomic clocks rely on the natural vibrations of atoms such as caesium.

When an object oscillates with constant time period, even if the amplitude varies, we say it is moving with simple harmonic motion (SHM).

The regular oscillations of a pendulum through a small angle are approximately simple harmonic.

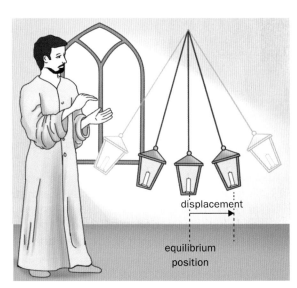

*Galileo noticed the lamps had constant time period. This is known as **isochronous** oscillation.*

SHM is a common type of motion.
The pistons in a car engine, a cork bobbing up and down on water and a vibrating guitar string all move with SHM.

Let's look in more detail at what happens during SHM.
Imagine yourself on the swing in the photograph:
We'll assume you swing through only a small angle (moving almost horizontally) so then your motion will be simple harmonic.

First think about your velocity

How does your **velocity** change as you move backwards and forwards?
At each end of the oscillation you are stationary for a moment. The swing speeds up as you move back towards the centre. Once you go past this point you start to slow down again.

Remember that velocity is a vector so we also need to consider its direction. We can take forward motion as positive and backward motion as negative.

Now let's think about your acceleration

In which direction are you **accelerating** during each swing? Starting at A in the diagram you speed up towards the centre. You are accelerating towards B, the centre.

Once past the middle you slow down until you stop again at C.

But deceleration is the same as negative acceleration, so decelerating away from B is the same as accelerating towards B. So you are still accelerating *towards the centre!*

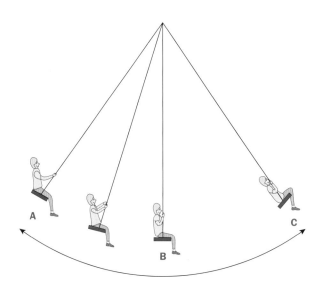

From point C you accelerate back towards the centre again. From B to A you decelerate to a stop. But again you are really accelerating towards B.

Can you see that throughout the motion your acceleration is *always* directed towards the central point, B?
B is called the **equilibrium position**.
It is where you come to rest when you stop moving.

How does the size of the acceleration change?

This is perhaps easier to see with another common example of SHM – a mass with a spring. The mass in the diagram is tethered between 2 identical springs. When you pull the mass to one side and let go, it oscillates backwards and forwards.

What pulls the mass back to the centre each time?
When the mass is displaced to the left, F_1 decreases and F_2 increases so the resultant pull of the springs is to the right. Similarly, when the mass is displaced to the right, the resultant pull is to the left, to the centre. The more you displace the mass, the greater the resultant force that pulls it back.

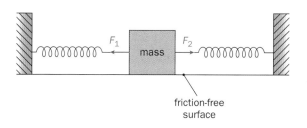

friction-free surface

How does this affect the acceleration?
Newton's second law tells us that the greater the force, the greater the acceleration. So the greater the displacement away from the centre, the greater the acceleration.

In summary, for an object moving with simple harmonic motion:

- **the acceleration is always directed towards the equilibrium position at the centre of the motion,**

- **the acceleration is directly proportional to the distance from the equilibrium position.**

▷ Graphs of SHM

One way to represent the simple harmonic motion of an object is to draw graphs of its displacement, its velocity and its acceleration against time.

What would a **displacement–time graph** look like for a simple pendulum? We can produce this graph directly using a pendulum pen like the one in the diagram:

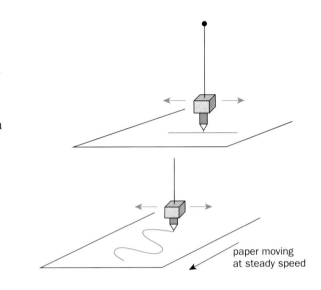

As the pen swings back and forth it draws, over and over, the same line on the paper underneath.
But what if we pull the paper along steadily beneath the pen? The pen would draw a regular wave. This is the shape of the displacement–time graph for SHM. The graph is *sinusoidal*. It is the shape of a sine or cosine curve.

paper moving at steady speed

Remember that displacement is a *vector* quantity. Displacements in one direction are taken as positive and those in the opposite direction are negative.

One complete oscillation means a movement from one extreme to the other and back again. The time this takes is called the **time period, T**. The number of oscillations in one second is called the **frequency, f**. Frequency is measured in hertz (Hz).

You have met these quantities before in the chapter on circular motion (page 78). Remember that:

$$f = \frac{1}{T} \qquad \text{and} \qquad T = \frac{1}{f}$$

The **amplitude A** of the motion is the maximum displacement. Note that this is measured from the *centre* of the oscillation.

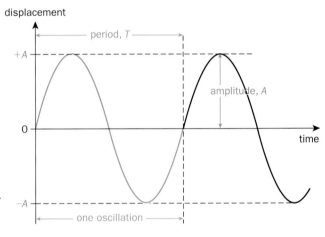

Comparing oscillations

Imagine two masses on springs bouncing up and down side by side. How can these oscillations differ?
They could have different periods and amplitudes.

Even if these are the same, how else can they differ?
The oscillations could be out of step. To describe how far out of step two oscillations are, we use the idea of **phase**.
Phase measures how far through a cycle the movement is.

By analogy with circular motion, one complete oscillation has a phase of **2π radians** (see page 78). The phase difference between two oscillations is usually given in radians:

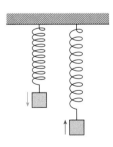

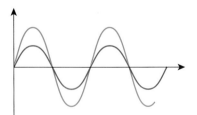

These oscillations are in step.
They are **in phase**.
The phase difference is **zero**.

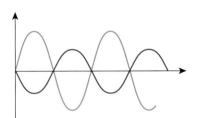

These oscillations are a $\frac{1}{2}$-cycle out of step. They are in **anti-phase**. The phase difference is **π rad** (180°).

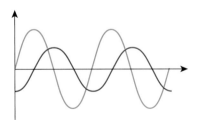

These oscillations are a $\frac{1}{4}$-cycle out of step.
The phase difference is **π/2 rad** (90°).

102

Linking displacement, velocity and acceleration

The picture shows a boy on a swing :

For small angles the swing moves with simple harmonic motion.

What will graphs of the boy's displacement, velocity and acceleration against time look like ?

We already know that the **displacement–time graph** is sinusoidal. This is the first graph:

At which point during the motion have we started timing ?
At $t = 0$ the displacement is zero. So timing must have started as the swing passed through the equilibrium position, B.

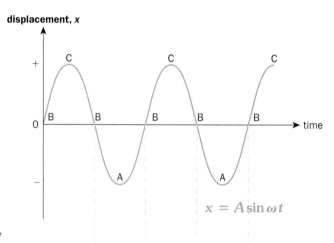

displacement, x

$x = A \sin \omega t$

Remember that **velocity** is the rate of change of displacement. So the velocity at any point is equal to the *gradient* of the displacement–time graph (see page 40).

Look at this displacement–time graph again.
Where is the gradient at its maximum ? The gradient is steepest when the displacement is zero. The gradient falls to zero as the swing reaches its maximum amplitude.

This tells us that the velocity is greatest at the centre of the motion, falling to zero at each extreme (as you already know).

The second graph shows the **velocity–time graph** for the motion :
The forward motion of the swing is taken as positive.
The backward motion is taken as negative.

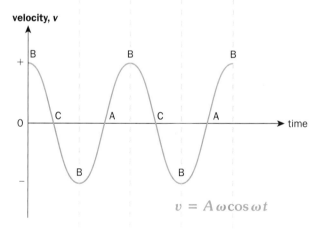

velocity, v

$v = A \omega \cos \omega t$

How can we use this velocity–time graph to work out the swing's **acceleration** ? Acceleration is the rate of change of velocity. So the acceleration is the *gradient* of the velocity–time graph.

Look at this velocity–time graph again.
The gradient is steepest when the velocity is zero.
This is at the maximum amplitude. The acceleration increases as the displacement increases. This was discussed on page 101.

This last graph shows the **acceleration–time graph** for the swing :
Acceleration to the right is positive and to the left is negative.

Notice that when the displacement is positive, the acceleration is negative. Can you see why ?
As the swing moves forwards from B to C and back again, the displacement is positive. The acceleration towards the centre is to the left. This is the negative direction.
As the swing moves from B to A and back again the displacement is now negative. The acceleration towards the centre is now to the right and so the acceleration is positive.

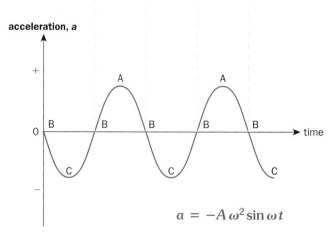

acceleration, a

$a = -A \omega^2 \sin \omega t$

Look at these three graphs carefully.
Think about the motion of the swing and make sure you understand how the three graphs are related.
And you thought swings were child's play!

▷ Equations for SHM

Linking acceleration a and displacement x

The most important equation for SHM links the acceleration a of an object with its displacement x from the centre of the oscillation.

The acceleration is **directly proportional** to the displacement, or:

$$a \propto x$$

We can write this as an equation by inserting a constant of proportionality:

$$\boxed{a = -(2\pi f)^2\, x}$$

f is the frequency in hertz (Hz). This is the number of oscillations per second. The expression $2\pi f$ is known as the 'angular frequency', ω.

Why is there a minus sign in the acceleration equation?
This is because the acceleration is always back towards the centre. So the acceleration is always in the opposite direction to the displacement, as discussed on page 103.

The maximum displacement is called the amplitude, A. Do not confuse this with the acceleration, a.

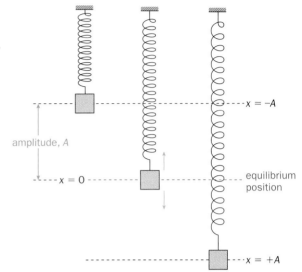

amplitude, A
$x = 0$
$x = -A$
equilibrium position
$x = +A$

Example 1
The pendulum of an old clock oscillates once every 2.0 s.
Calculate: a) its frequency,
            b) its acceleration when it is 50 mm from the midpoint.

a) frequency $f = \dfrac{1}{T} = \dfrac{1}{2.0\text{ s}} = \underline{0.50\text{ Hz}}$ (2 s.f.)

b) acceleration $a = -(2\pi f)^2\, x$

$$= -(2\pi \times 0.50\text{ Hz})^2 \times 0.05\text{ m}$$

$$= \underline{-0.49\text{ m s}^{-2}} \quad \text{(towards the centre)}$$

Linking displacement x and time t

The equation for acceleration can be solved mathematically to give two equations for the displacement x.
The solutions depend on the point at which timing starts.

You have already seen that the displacement–time graph for SHM is sinusoidal. So the equations for this shape of graph must be sines or cosines. In fact:

If timing starts at the **centre** of the oscillation: $x = A\sin 2\pi f t$
If timing starts at the **maximum displacement**: $x = A\cos 2\pi f t$

These equations give the displacement x after a time t.
A is the amplitude of the oscillation and f is the frequency.

The angles are in **radians** so you must have your calculator in radian mode when calculating the value of the sine or cosine.

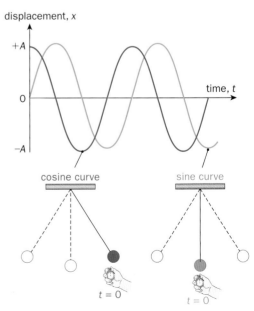

displacement, x
$+A$
0
$-A$
time, t
cosine curve
sine curve
$t = 0$
$t = 0$

Example 2
A plucked guitar string vibrates at 260 Hz with an amplitude of 2.0 mm.
The vibration is timed from when the string moves through the centre
of its oscillation.
Assuming the motion is SHM, find the displacement of the string after 0.52 s.

Since timing starts at the centre (so $x = 0$ when $t = 0$),

$$x = A \sin 2\pi f t$$
$$\therefore \quad x = (2.0 \times 10^{-3}\,\text{m}) \times \sin(2\pi \times 260\,\text{Hz} \times 0.52\,\text{s}) \quad \text{[Calculator in radian mode !]}$$
$$= (2.0 \times 10^{-3}\,\text{m}) \times 0.951$$
$$= \underline{1.9 \times 10^{-3}\,\text{m}} \quad \text{(2 s.f.)}$$

Linking velocity v and displacement x

If an object is moving with SHM, its velocity at any point can be
found using this equation:

$$\boxed{v = \pm 2\pi f \sqrt{A^2 - x^2}}$$

f is the frequency, A is the amplitude and x is the displacement
from the equilibrium position. Notice that the velocity can be
positive or negative depending on its direction.

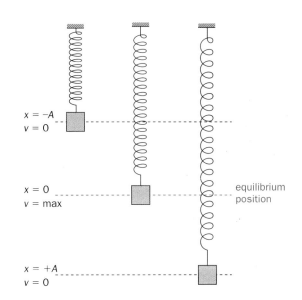

How does this equation tie in with what you already know about
the velocity during SHM?

Look at the equation. What happens to v as x gets smaller?
The value of v increases. So an object moving with SHM speeds
up as it moves towards the centre.

What value for v does the equation give us at the maximum
displacement? At the maximum displacement, x equals the
amplitude A. So the velocity here must be zero.

We can use this equation to find the **maximum** velocity during
simple harmonic motion.
What is the value of x when the object is moving fastest?
The maximum velocity is at the equilibrium position, where the
displacement is zero. So the equation becomes:

$$v_{max} = \pm 2\pi f \sqrt{A^2 - 0^2}$$

$$\boxed{v_{max} = \pm 2\pi f A}$$

Example 3
A baby in a bouncer bounces up and down with a time period of 1.2 s and
an amplitude of 90 mm. The motion can be assumed to be simple harmonic.
Calculate (a) the frequency of the bounces and (b) the baby's maximum velocity.

a) frequency $f = \dfrac{1}{T} = \dfrac{1}{1.2\,\text{s}} = \underline{0.83\,\text{Hz}}$ (2 s.f.)

b) maximum velocity $v_{max} = \pm 2\pi f A$
$$= \pm 2\pi \times 0.83\,\text{Hz} \times 0.090\,\text{m} = \underline{\pm 0.47\,\text{m s}^{-1}}$$

▷ The simple pendulum

A pendulum oscillates with SHM, provided the amplitude is small.
Try timing small oscillations for a simple pendulum.

What does the time period depend on?
There are two things you can vary: the mass of the bob and
the length of the string. Try changing the mass.
What effect does this have? Surprisingly, the mass has **no** effect
on the time period of a pendulum.
What effect does changing the **length** have? As the string gets
longer you should find that the time period increases.

In fact, the time period T of a pendulum can be found from:

$$T = 2\pi \sqrt{\frac{l}{g}}$$

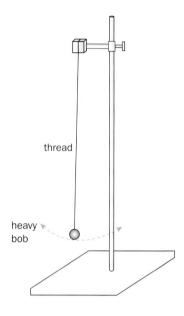

thread

heavy
bob

l is the length of the pendulum (in metres). This is measured
to the **centre** of mass of the bob.
g is the acceleration due to gravity (in m s^{-2}, see page 44).
Notice that you do not need to know the mass of the bob.

See page 483 for a spreadsheet model of this.

Example 4
Calculate the time period of a pendulum of length 0.75 m.
Take $g = 9.8$ m s^{-2}.

$$T = 2\pi \sqrt{\frac{l}{g}} = 2\pi \sqrt{\frac{0.75 \text{ m}}{9.8 \text{ m s}^{-2}}} = 2\pi \sqrt{0.076 \text{ s}^2} = \underline{1.7 \text{ s}}$$

$l = 0.75$ m

Measuring the acceleration due to gravity, *g*

We generally treat the acceleration due to gravity on Earth as a
constant. In fact g varies with your latitude (north–south) and
your height above sea level. Metal ore deposits in the Earth's crust
can also affect the value.
The actual values for g range from about 9.78 m s^{-2} at the equator
to 9.83 m s^{-2} at the North Pole.

How can you use a simple pendulum to measure g accurately?
Look back at the equation for the period of a pendulum.
You can measure the time period T with a stopwatch.
If you know the length of the pendulum, then you can rearrange
the equation to calculate g.

To get a reliable figure you should measure the time period for
several lengths of the pendulum. You can then find g
by plotting a graph. What should you plot?

Compare the equation for the period of a pendulum with the
equation for a straight line: $y = mx + c$ (see page 478):

Can you see that plotting T on the y-axis against $\sqrt{l}$ on the
x-axis should give a straight line, through the origin?

The gradient of this graph would equal $2\pi/\sqrt{g}$.

By finding the gradient (page 476) you can work out a value for g.

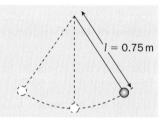

$$T = 2\pi \sqrt{\frac{l}{g}}$$

$$\therefore \quad T = \frac{2\pi}{\sqrt{g}} \times \sqrt{l}$$

$$y = m \quad x + c$$

gradient   intercept = 0

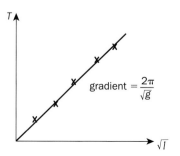

gradient $= \dfrac{2\pi}{\sqrt{g}}$

▷ A mass on a spring

Hang a mass on a spring, pull it down a small way and let go.
What happens?
The mass bounces up and down about its equilibrium position.
It oscillates with simple harmonic motion.

What affects the time period of this oscillation?
This time the **mass** does have an effect.
The greater the mass the slower it accelerates. This will increase
the time for each oscillation.

What else affects the period?
The other variable is the **stiffness** of the spring. A stiffer spring will
pull the mass back to its equilibrium position with more force.
This produces greater acceleration. If the mass moves faster
the time period will decrease.

In fact, the time period T for a mass m oscillating on a spring is
given by this equation:

$$T = 2\pi \sqrt{\frac{m}{k}}$$

The stiffness of the spring is represented by the **spring constant, k**
(see page 186). This is a constant for a particular spring.
The spring constant is measured in newtons per metre (N m^{-1}).

The open coil spring in this diagram can make a mass oscillate
horizontally over a friction-free surface. When compressed the
spring pushes to the right. When extended it pulls to the left.

Can we use the same equation to find the time period here?
The mass m and spring constant k are the same whether the
motion is horizontal or vertical. Since these are the only things
that the period depends on, we can use the same equation.

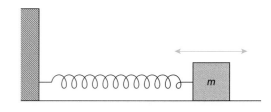

Restoring force ∝ displacement
$$F = -k\,x$$
∴ Acceleration ∝ displacement
So the mass oscillates with SHM.

See page 483 for spreadsheet modelling of this.

Example 5
At the bottom of a bungee jump the woman in the photograph
bounces up and down on the end of the bungee rope through
an amplitude of 2.0 m.
The woman's mass is 66 kg.
The rope has a spring constant of 240 N m^{-1}.

Assuming the bouncing is SHM, calculate:
a) the period of the oscillations,
b) the frequency of the bounces,
c) the maximum acceleration during a bounce.

a) period $T = 2\pi \sqrt{\dfrac{m}{k}} = 2\pi \sqrt{\dfrac{66 \text{ kg}}{240 \text{ N m}^{-1}}} = \underline{3.3 \text{ s}}$

b) frequency $f = \dfrac{1}{T} = \dfrac{1}{3.3 \text{ s}} = \underline{0.30 \text{ Hz}}$

c) The acceleration is greatest at the maximum displacement
from the centre, i.e. at $x = 2.0$ m

acceleration $a = -(2\pi f)^2\, x = -(2\pi \times 0.30 \text{ Hz})^2 \times 2.0 \text{ m} = \underline{-7.1 \text{ m s}^{-2}}$ (towards the equilibrium position)

▷ Energy in SHM

Imagine yourself back on the swing. At which points do you have your maximum and minimum kinetic energy?

At the maximum amplitude of your swing, you are stationary for a moment. At these two points your kinetic energy is zero. Your kinetic energy reaches its maximum at the centre as you speed up towards this point.

Remember that energy is always **conserved**. So where does the increase in kinetic energy come from?
Think about how your potential energy changes.

Your gravitational potential energy has its maximum value at the extremes and its lowest value at the centre. So, in moving from A to B in the diagram, your potential energy is transferred to kinetic energy. As you slow down from B to C, your kinetic energy is transferred back to potential energy again.

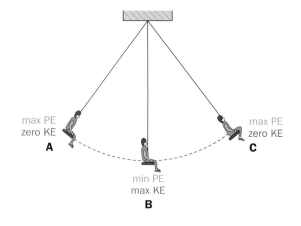

All simple harmonic motion involves this continual transfer between kinetic energy and potential energy.
The **total** amount of energy in the system remains constant.

To keep things simple we usually choose to take the potential energy to be zero at the equilibrium position.

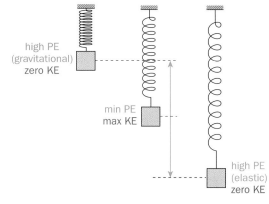

The graph shows how the potential energy and kinetic energy vary with displacement for an object moving with SHM:

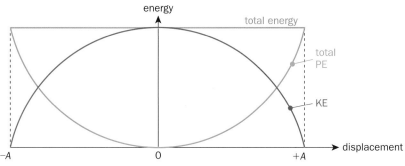

Energy and amplitude

The more energy you put into a system, the bigger the oscillations will be. Think about a mass on a spring. The more you pull it down, the more it oscillates. But is the relationship linear?
Does putting in twice the energy give you twice the amplitude?

To work this out let's look at the energy of an object moving with SHM as it moves through its equilibrium position.
At this point the object is moving at its maximum velocity and the total energy is all kinetic energy:

Total energy $= \text{KE} = \frac{1}{2}mv_{max}^2 = \frac{1}{2}m(2\pi f A)^2 = \frac{1}{2}m(4\pi^2 f^2)A^2$

Total energy $\propto$ **A^2**

So to make the system oscillate with **twice** the amplitude you need to give it **four** times the energy.

Energetic oscillations

▷ Damping

So far we have assumed that no energy is lost from an oscillating system and that it continues to oscillate indefinitely. What actually happens in practice?
If you leave a pendulum or a mass on a spring oscillating, it eventually slows down and stops. The time period stays the same as the amplitude gets smaller and smaller.

Why does this happen? Air resistance slows the object down. Energy is lost from the system in overcoming this friction. This effect is called **damping**.

The graphs show how damping affects both the displacement of an oscillating object and the total energy of the system:

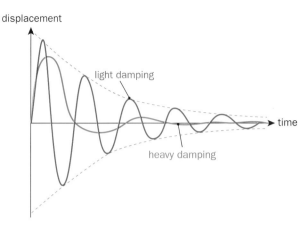

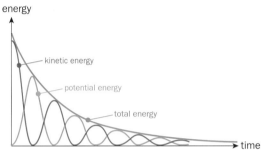

The time period is unchanged even though the displacement reduces, because the object slows down. So it takes the same time to cover the smaller distance in each oscillation.
How quickly the oscillations die away depends on the degree of damping. Air resistance provides **light damping**.
The oscillations take a long time to die away.
But imagine a pendulum moving through water. Water would cause **heavier damping**. Once released, the pendulum would take longer to return to its equilibrium position and it would hardly oscillate at all.

Damping is often applied to oscillating systems deliberately. In car suspension systems, for example, 'shock absorbers' or dampers are fitted to the suspension springs.
These quickly damp the vibrations when a car goes over a bump. Driving would be very uncomfortable without them!
The level of damping used in a car's suspension needs to stop the oscillations as quickly as possible. Damping that allows an object to move back to its equilibrium position in the *quickest* possible time, without oscillating, is called **critical damping**.

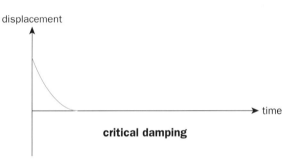

What happens if you increase the damping above this critical level?
If a system is **overdamped** it does not oscillate.
It takes a long time to return to its equilibrium position.
A pendulum moving through thick treacle would be overdamped.

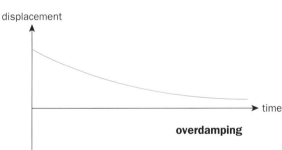

Overdamping is useful where rapid fluctuations need to be ignored. An example of this is a car fuel gauge. Overdamping stops the pointer oscillating as the fuel sloshes about in the tank.

Drums are lightly damped by the surrounding air.

Car suspensions use critical damping.

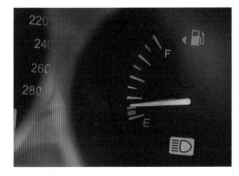

Fuel gauges are overdamped.

109

▷ Resonance

Imagine pushing a child's swing. How do you time your pushes to make it go higher? You push in time with the swing's natural movement. The amplitude of the vibration builds up when you match the frequency of your pushing force with the **natural frequency** of the swing:

You can *force* an object to vibrate at any frequency by applying an external driving force. For example, you can hold the seat of a swing and move it backwards and forwards at any frequency you choose. This is called a **forced vibration**.
But if you displace an object and then let it oscillate freely it will always oscillate at its own natural frequency. This is called a **free vibration**. All oscillating systems have their own natural frequency. When an external driving frequency is the *same* as the natural frequency, the amplitude builds up and up.
Energy is transferred from the driver to the vibrating object. This effect is called **resonance**.

Resonance can be destructive. Have you ever heard the violent vibrations of a washing machine at some spin speeds?
Or sat on a bus and felt it judder and rattle at a certain speed? If so, you have experienced resonance first hand.
The vibrations can build up to dangerous levels. Cartoons sometimes show glasses smashing when someone hits a high note. This can really happen! The frequency of the sound must match the natural frequency of the glass. The glass then resonates, vibrating more and more until it breaks.

Resonance can also be very useful. Every time you use a microwave oven you are using resonance (see opposite page). Tuning circuits in TV and radio sets work by resonating at the frequency of the station you select. Digital watches rely on the resonant vibration of a quartz crystal to keep time. And there are valuable medical applications, such as magnetic resonance imaging (see page 438) and the use of tuned ultrasound to break up kidney stones through resonant vibration.

Resonance and damping

Damping absorbs energy. This reduces the effect of resonance. The graph shows the effect of different levels of damping on the resonance peak.

You can see that, as the damping increases:

• the amplitude of the resonance peak decreases,

• the resonance peak gets broader,

• the 'resonant frequency' gets slightly lower, so the peak moves to the left on the graph.

Damping is used where resonance could be a problem.

A good example is the damping of buildings in earthquake zones. The foundations are designed to absorb energy.
This stops the amplitude of the building's oscillations reaching dangerous levels when an earthquake wave (see page 136) arrives.

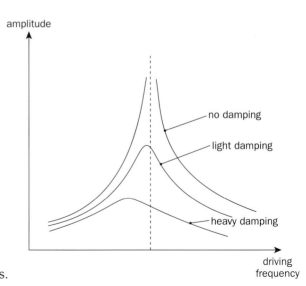

▷ Physics at work: Oscillations and resonance

The Tacoma Narrows bridge

In 1940 in Washington state, USA, the Tacoma Narrows suspension bridge collapsed just three months after opening. A steady crosswind caused swirling air, which set the bridge vibrating. By chance, these vibrations matched a natural frequency of the bridge.
The bridge resonated and started to oscillate with increasing violence. Eventually the amplitude of the oscillations was too great for the structure to stand and it collapsed :

Resonance has caused other bridge disasters. In 1850, over 200 French soldiers died when the bridge they were marching over resonated with their steps and collapsed. This is why soldiers always break step when they march over a bridge.

Foucault's pendulum

In 1851, French physicist Léon Foucault devised a simple way to show that the Earth is rotating. He hung a giant pendulum from the dome of the Pantheon in Paris. The pendulum wire of the replica in the photograph is 67 m long and supports a 28 kg lead bob.

As the pendulum oscillates, the plane of its swing appears to gradually rotate, eventually completing a full circuit. The hour markers around its edge track its progress.

A replica of Foucault's pendulum, installed in the Pantheon in 2010 after the wire on the original snapped.

So why does the pendulum rotate? If fact it doesn't.
As the pendulum swings backwards and forwards, the Earth rotates beneath it. To a viewer rotating with the Earth, this gives the impression that the pendulum's plane of oscillation rotates. Foucault's pendulum in Paris actually takes 32.7 hours to complete a full circuit. How do we explain this when a full rotation of the Earth is only 24 hours?
A pendulum at the North Pole would take exactly 24 hours to complete a full rotation as the spin of the Earth occurs directly beneath it. A pendulum at the equator would not appear to rotate at all as there is no 'twist' of the floor relative to its motion. At other points on the Earth, such as Paris, the angular speed at which the pendulum appears to rotate varies with latitude.

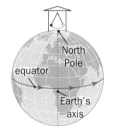
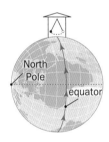

There are Foucault pendulums in many museums and public buildings around the world. The pendulums should slow down and stop due to damping caused by air resistance. But many, such as the one in the lobby of the United Nations building in New York, never slow down. They use electromagnets to provide a tiny force at just the right moment in each swing. This is in perfect resonance, so that energy is restored to keep the pendulum swinging.

How do microwaves cook food?

Microwave ovens use resonance. The frequency of the microwaves almost equals the natural frequency of vibration of a water molecule. This makes the water molecules in food resonate. This means they take in energy from the microwaves. As the water's internal energy increases, the food gets hotter.

The slight mismatch of frequencies is deliberate. It prevents all the energy being absorbed at the surface and allows the microwaves to penetrate deeper into the food.

Summary

An object moving with simple harmonic motion (SHM) must obey the defining equation: $a = -\omega^2 x$, where $\omega = 2\pi f$.
During SHM:
- the time period is constant,
- the acceleration is always directed towards the same point – this is the equilibrium position,
- the acceleration is directly proportional to the displacement from the equilibrium position.

During SHM, velocity $v = \pm 2\pi f \sqrt{A^2 - x^2}$ so that: $v_{max} = \pm 2\pi f A$

If timing starts at the centre of the oscillation: $x = A \sin \omega t$ $v = +A \omega \cos \omega t$ $a = -A \omega^2 \sin \omega t$
If timing starts at the maximum amplitude: $x = A \cos \omega t$ $v = -A \omega \sin \omega t$ $a = -A \omega^2 \cos \omega t$

A simple pendulum and a mass on a spring move with SHM: $T = 2\pi \sqrt{\dfrac{l}{g}}$ $T = 2\pi \sqrt{\dfrac{m}{k}}$

During SHM, energy is repeatedly transferred from kinetic to potential and back again.
The total energy remains constant. The total energy is proportional to the **square** of the amplitude.
Energy can be lost from an oscillating system through damping.

If the driving frequency = the natural frequency, the object gains energy and the amplitude increases.
This is called resonance. Damping reduces the effect of resonance.

▷ Questions

1. Can you complete these sentences?
   a) The time for one oscillation is called the time .... This stays the same, even if the size, or ...., of the swing decreases.
   b) The midpoint of the oscillation is called the .... position. The distance away from this point is called the ....
   c) During SHM the .... of the object is always directed towards the equilibrium position.
   d) The acceleration is directly proportional to the ....
   e) The velocity is greatest at the ....
   At the maximum displacement the velocity is ....

2. A pendulum has a time period of 3.0 s and an amplitude of 0.10 m.
   Calculate:
   a) its frequency,
   b) its maximum acceleration.

3. The displacement of a girl on a swing is given by the equation $x = A \sin 2\pi f t$.
   a) Sketch a graph of displacement against time.
   b) Sketch a velocity–time graph for the girl, on the same axes.
   c) What is the phase difference between the two graphs?

4. A mass on a spring bounces up and down at 1.4 Hz. The motion is considered to be SHM with an amplitude of 50 mm.
   a) Calculate the acceleration at displacements of ± 1 cm, ± 2 cm, ± 3 cm, ± 4 cm and ± 5 cm.
   b) Use the values from part (a) to plot a graph of acceleration against displacement.

5.

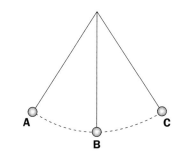

The diagram shows a pendulum oscillating to and fro.
Copy the diagram and add labels showing all the positions where:
a) the acceleration is maximum,
b) the acceleration is zero,
c) the velocity is maximum,
d) the velocity is zero,
e) the kinetic energy is maximum,
f) the potential energy is maximum.

6. A mass of 10 kg bounces up and down on a spring.
The spring constant is 250 N m^{-1}.
Calculate:
a) the time period of the oscillation,
b) the frequency of the bounce.

7. Calculate the time period for a pendulum of length 2.0 m. Take $g = 9.81$ m s^{-2}.

8. Find the length of a pendulum with a time period of 2.00 s. Take $g = 9.81$ m s^{-2}.

9. A baby of mass 9.0 kg bounces with a period of 1.2 s in a baby bouncer.
What is the spring constant for the bouncer?

10. The time periods are measured for a simple pendulum and a mass on a spring on Earth.
The measurements are then repeated with exactly the same apparatus on the Moon. One of the time periods is now different. Which one is it? Explain your answer.

11. The graph shows the motion of a bungee jumper at the end of a jump. You can assume that the motion is simple harmonic.

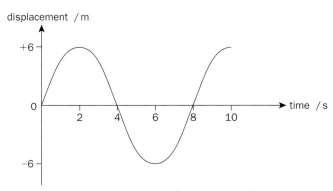

a) What is the amplitude of the motion?
b) What is the time period?
c) Calculate the frequency of the bounces.
d) Use your values to work out the maximum acceleration during this motion.

12. The diagram shows a mass tethered between two springs. It is displaced 10 cm, then released.
The mass oscillates with SHM with a frequency of 0.55 Hz.

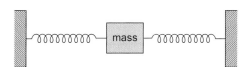

Calculate:
a) the maximum velocity of the mass,
b) the velocity when the mass is 8.0 cm from the equilibrium position.

13. A sewing machine needle moves up and down through a total vertical distance of 2.0 cm.
The frequency of the oscillation is 2.4 Hz.
Assuming the motion is SHM, calculate:
a) the amplitude of the motion,
b) the maximum acceleration of the needle.

14. The table gives values for the time period T against length l for a simple pendulum.

T /s	0.90	1.25	1.80	2.53	3.59
l /m	0.20	0.40	0.80	1.60	3.20
$\sqrt{l}$ /m$^{1/2}$					

a) Copy out the table and fill in the values for $\sqrt{l}$.
b) Use these values to plot a graph with T on the y-axis and $\sqrt{l}$ on the x-axis.
c) Calculate the gradient of your graph.
d) Use the value of the gradient to find the acceleration due to gravity g. (Hint: see page 106)

15. The graph shows how the kinetic energy of a mass on a spring varies with displacement.

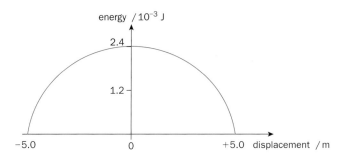

a) What is the amplitude of the motion?
b) Calculate the frequency of the oscillations, given that the time period is 2.0 s.
c) Calculate the maximum velocity of the mass.
d) Calculate the maximum acceleration of the mass.
e) Sketch a curve of total potential energy against displacement for the mass on the spring.

16. A punchbag of mass 0.60 kg is struck so that it oscillates with SHM. The oscillation has a frequency of 2.6 Hz and an amplitude of 0.45 m.
Calculate:
a) the maximum velocity of the bag,
b) the maximum kinetic energy of the bag.
c) What happens to this energy as the oscillations die away?

17. At a certain engine speed a car's rear-view mirror starts to vibrate strongly. Explain why this happens.

Further questions on page 120.

113

Further Questions on Mechanics

▷ Basic Ideas

1. For each of the four concepts listed in the left hand column select the correct example. [4] (Edex)

   A base quantity: mole   length   kilogram
   A base unit: volt   ampere   coulomb
   A scalar quantity: torque   velocity   energy
   A vector quantity: mass   weight   density

2. A student measures the length of an object with a metre rule having 1 mm divisions. Explain which of the following could be a correct measurement of the length. 0.4 m, 0.39 m, 0.392 m, 0.3917 m.

3. Name four different base quantities, stating the unit of each and, by reference to density and force, explain how derived units may be expressed in terms of these base units. [7] (OCR)

4. Which of the following units is equivalent to the SI unit for energy?
   A $kg\,m\,s^{-2}$ B $kW\,h$ C $N\,m^{-1}$ D $W\,s^{-1}$
   [1] (Edex)

5. When a body moves with speed v through a liquid of density ρ, it experiences a force F known as the drag force. Under certain circumstances, this force is related to ρ and v by the expression: $F = k\rho v^2$, where k is a constant. Determine the unit of the constant k in terms of base SI units. [4] (OCR)

6. The figure shows the forces acting on a stage light of weight 120 N held stationary by two separate cables.

   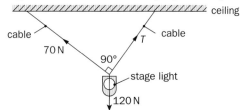

   The angle between the two cables is 90°. One cable has tension 70 N and the other has tension T.
   a) State the magnitude and direction of the resultant of the tensions in the two cables.
   b) Sketch a labelled vector triangle for the forces acting on the stage light. Hence determine the magnitude of the tension T. [6] (OCR)

7. a) Find the resultant of the forces shown below: [2]

   b) Why are perpendicular directions chosen when resolving vectors? [2] (W)

8. Water flows from a nozzle with an initial velocity of $5.8\,m\,s^{-1}$ at an angle of 45° to the horizontal. Show that the horizontal component of the velocity is $4.1\,m\,s^{-1}$. [2] (OCR)

▷ Looking at Forces

9. The free-body force diagram shows the two principal forces acting on a parachutist at the instant of first contact with the ground.

   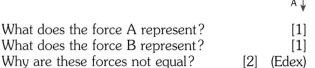

   What does the force A represent? [1]
   What does the force B represent? [1]
   Why are these forces not equal? [2] (Edex)

10. A cylinder of cork of cross-sectional area $5.0 \times 10^{-3}\,m^2$ floats on water with its axis vertical. The length of the cork below the surface of the water is 0.010 m.
    Density of water = $1000\,kg\,m^{-3}$.
    a) Show that the weight of water displaced by the cork is about 0.5 N.
    b) State the weight of the cork and justify your answer. [5] (Edex)

11. Two tugs A and B pull a ship along the direction XO. Tug A exerts a force on the ship of 3.0×10^4 N at an angle of 15° to XO. Tug B pulls with a force of 1.8×10^4 N at an angle θ to XO.

    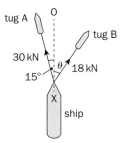

    a) Find the angle θ for the resultant force on the boat to be along XO. [3]
    b) Find the value of the resultant force. [3] (OCR)

12. An athlete with a broken leg is in traction while the bone is healing. The diagram shows a system of pulleys for providing the traction force.
    All the pulleys are frictionless so that the tension in the rope is the same everywhere. $g = 9.8\,N\,kg^{-1}$

    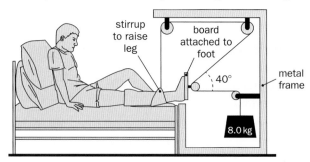

    a) Find the magnitude of the total horizontal force exerted on the leg by the system. [2]
    b) Find the total upward force exerted on the leg by the system. [1]
    c) Explain why the force found in (a) does not move the patient to the bottom of the bed. [2] (AQA)

▷ Turning Effects of Forces

13. A car moves forwards along a straight level road at constant speed. State the direction and origins of
  a) two forces opposing forward motion of the car,
  b) the driving force maintaining the motion of the car against the force opposing motion,
  c) two vertical forces acting on the car. [6] (OCR)

14. An aircraft of weight 1.8×10^6N flies at a constant height. The pressure difference between the top and bottom of the wings is 1.5×10^4Pa. Assuming that the lift is produced only by the wings, calculate the effective area of the wings. [3] (OCR)

15. a) Define the moment of a force.
  b) This figure is of a human arm lifting an object of weight 30N. The lower arm of weight 18N is horizontal with its centre of gravity at a distance 0.150m from the elbow joint. The biceps muscle exerts a vertical force F on the arm at a distance 0.040m from the elbow joint. The arm is in equilibrium.

  i) Define centre of gravity.
  ii) Calculate the total clockwise moment about the elbow joint.
  iii) As the lower arm is moved away from the body, the force F exerted by the biceps muscle acts at an angle θ to the vertical as shown.

The lower arm remains in equilibrium. Describe and explain what happens to each of the following quantities as the angle θ is increased:
1) the anticlockwise moment about the elbow joint;
2) the magnitude of the force, F. [9] (OCR)

16. A beam of length l and weight W is supported horizontally from two spring balances A and B. The reading on balance A is twice that on B.

  a) Find the position of the centre of gravity of the beam. [2]
  b) Explain whether you consider the beam to be uniform. [2] (W)

17. A hanging basket of weight W is supported by three chains of equal length, each at an angle θ to the vertical.
The tension, T, in each chain is given by
A $T = 3W/\cos\theta$ B $T = 3W/\sin\theta$
C $T = W/3\cos\theta$ D $T = W/3\sin\theta$ [1] (Edex)

18. Figure 1 shows a gardener pulling a roller of mass 85kg over a step. The roller has a radius of 0.25m. The handle is attached to an axle through the centre of the roller.

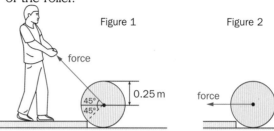

  a) Determine the magnitude of the force that will just move the roller when the force is applied as shown in Figure 1. [2]
  b) Determine the force which would be required if the handle were pulled horizontally (Figure 2). [2]
  c) Without further calculation, draw a sketch graph showing how the magnitude of the force varies with the angle that the handle makes with the horizontal, between 0 and 90°. [2] (AQA)

19. A car wheel nut can be loosened by applying a force of 200N on the end of a bar of length 0.8m as in X. A car mechanic is capable of applying forces of 500N simultaneously in opposite directions on the ends of a wheel wrench as in Y.

What is the minimum length l of the wrench which would be needed for him to loosen the nut?
A 0.16m B 0.32m C 0.48m D 0.64m [1] (AQA)

20. The diagram shows a child's mobile which is supported from the ceiling by a thread of negligible weight. AB is a uniform horizontal rod.

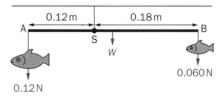

  a) By taking moments about point S, show that the weight W of the rod is 0.12N. [3]
  b) A third fish of weight 0.30N is suspended from the middle of the rod. The thread supporting the mobile is moved along the rod so that the rod remains horizontal. On which side of the centre of the rod is the thread now attached? Explain how you arrived at your answer. [2] (AQA)

Further Questions on Mechanics

▷ Describing Motion

21. A car of mass 800 kg travels along a straight level road. The graph shows its speed against time.

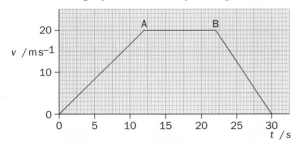

Use the graph to find:
a) the initial acceleration, [1]
b) the total distance travelled, [2]
c) the average speed for the journey. [1] (OCR)

22. The figure shows the variation of velocity v with time t for a small rocket.

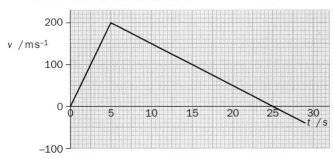

Assume that air resistance has a negligible effect.
a) Without doing any calculations, describe the motion of the rocket from $t = 0$ to $t = 25$ s.
b) Calculate the height reached by the rocket.
c) Sketch a graph of displacement against time for the rocket during the first 25 s. [11] (OCR)

23. A skydiver jumps from an aircraft and initially falls without using a parachute. The graph shows how the speed of the skydiver varies during this part of the drop.

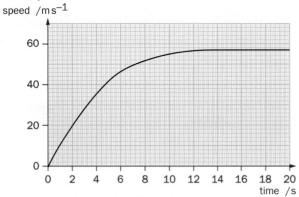

a) Determine the terminal speed. [1]
b) The terminal speed is reached 13 s after leaving the aircraft. Estimate the distance fallen before reaching the terminal speed. [2]
c) Say why the speed becomes constant. [2] (AQA)

24. A ball is thrown vertically upwards and passes a window ledge 0.3 s after being released. It passes the window ledge on its way back down, 1.6 s later. Ignore air resistance.
a) Determine the time of flight of the ball. [1]
b) Calculate the initial velocity of the ball when it is released. [3]
c) Calculate the height of the window ledge above the ground. [2] (W)

25. A stone is projected horizontally from the top of a vertical sea cliff 49 m high, with a speed of $20 \, \mathrm{ms^{-1}}$. Neglecting air resistance, calculate:
a) the time that it takes for the stone to reach the sea, [1]
b) the distance of the point of impact with the sea from the base of the cliff, [1]
c) the velocity of the stone as it hits the sea. [4]
d) If air resistance had been taken into account, explain whether
i) the time of flight,
ii) the final velocity
would increase, decrease or stay the same. [3] (W)

26. An aircraft is at rest at one end of a runway which is 2.2 km long. The aircraft accelerates along the runway with an acceleration of $2.5 \, \mathrm{ms^{-2}}$ until it reaches its take-off speed of $75 \, \mathrm{ms^{-1}}$.

a) Calculate:
i) the time taken to reach take-off speed,
ii) the distance travelled in this time. [4]

b) Just as the aircraft reaches take-off speed, a warning light comes on in the cockpit. The maximum possible deceleration of the aircraft is $4.0 \, \mathrm{ms^{-2}}$ and 2.5 s elapses before the pilot takes any action, during which time the aircraft continues at its take-off speed. Determine whether or not the aircraft can be brought to rest in the remaining length of runway. [4] (AQA)

27. A ball is released from rest at a height of 0.9 m above a horizontal surface.

a) Find its speed as it reaches the surface. [2]

b) The effect of the bounce is to reduce the speed of the ball to two-thirds of the value in part (a). Find:
i) the change in speed in the impact,
ii) the change in velocity in the impact. [3]

c) Plot a graph of the velocity of the ball from the moment of its release until it reaches the maximum height after its first bounce.
The ball takes 0.43 s to reach the surface.
Assume that the bounce takes a negligible time.
Show all your calculations. [5] (OCR)

▷ Newton's Laws and Momentum

28. A jet plane on the deck of an aircraft carrier is accelerated before take-off using a catapult. The mass of the plane is 3.2×10^4 kg and it is accelerated from rest to a velocity of $55\,\text{ms}^{-1}$ in a time of 2.2 s. Calculate
 a) the mean acceleration of the plane, [2]
 b) the distance over which the acceleration takes place, [2]
 c) the mean force producing the acceleration. [2]
 (OCR)

29. The diagram shows a view from above of two air-track gliders, A of mass 0.30 kg and B of mass 0.20 kg, travelling towards each other at the speeds shown. After they collide and separate, A travels with a speed of $0.20\,\text{ms}^{-1}$ from right to left.

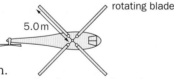

a) Apply the principle of conservation of momentum to this collision, and hence find the velocity of B after the collision.
 b) What general condition must be satisfied for momentum to be conserved in a collision? [5]
 c) Explain the difference between a perfectly elastic collision and an inelastic collision. Show that the collision described above is inelastic. [2] (AQA)

30. The figure shows a hovering helicopter viewed from above. The blades rotate in a circle of radius 5.0 m. The blades propel air vertically downwards with a constant speed of $12\ \text{m s}^{-1}$. Assume that the descending air occupies a uniform cylinder of radius 5.0 m. Density of air is $1.3\ \text{kg m}^{-3}$.
 a) Show that the mass of air propelled downwards in 5.0 s is about 6000 kg.
 b) Calculate
 i) the momentum of this mass of descending air,
 ii) the force provided by the rotating blades to propel this air,
 iii) the mass of the hovering helicopter. [6] (OCR)

31. The Saturn V rockets which launched the Apollo space missions had the following specifications:
 mass at lift-off 3.0×10^6 kg
 velocity of exhaust gases $1.1 \times 10^4\ \text{m s}^{-1}$
 initial rate of fuel consumption $3.0 \times 10^3\ \text{kg s}^{-1}$
 a) Calculate the thrust produced at lift-off, and
 b) resultant force on the rocket at lift-off. [4] (AQA)

32. A parachutist lands with a vertical velocity of $7.0\,\text{ms}^{-1}$ and no horizontal velocity. The parachutist, of mass 85 kg, lands in one single movement, taking 0.25 s to come to rest.
 a) Calculate the average retarding force on the parachutist during the landing. [4]
 b) Explain how the parachutist's loss of momentum on landing is consistent with the principle of conservation of momentum. [2] (AQA)

33. A person is standing on the floor of a lift which is accelerating uniformly upwards.
 a) Draw a free-body diagram showing all the forces acting upon the person. [2]
 b) What are the Newton's Third Law reactions to each of these and on what does each act? [4]
 c) Which, if any, of the forces in (a) changes as the acceleration changes? Give brief reasons. [2] (W)

34. The crumple zone of a car is a hollow structure at the front of the car designed to collapse during a collision. A car of mass 850 kg was driven into a wall at $7.5\,\text{ms}^{-1}$ and was brought to rest in 0.28 s when it hit the wall.

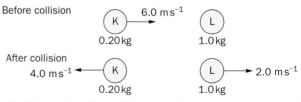

a) Calculate
 i) the deceleration of the car, assumed uniform,
 ii) the average force exerted by the wall on the car.
 b) The crumple zone of the car is designed to absorb 0.45 MJ of energy before any distortion of the passenger cabin occurs. Calculate the maximum safe speed of the car on impact.
 c) In a different test another car of mass 850 kg travelling at a speed of $7.5\,\text{ms}^{-1}$ collides head on with a stationary car of mass 1200 kg. Calculate the speed immediately after impact if both cars move off together at a common speed. [7] (OCR)

35. Which of the following statements about the forces in a Newton's Third Law pair is *not* correct?
 The forces:
 A act along the same line
 B act on the same body
 C are equal in magnitude
 D are of the same type [1] (Edex)

36. The diagram shows a body K of mass 0.20 kg and moving at a speed of $6.0\,\text{ms}^{-1}$, colliding with a stationary body L of mass 1.0 kg. K rebounds with a speed of $4.0\,\text{ms}^{-1}$ and L is driven forwards with a speed of $2.0\,\text{ms}^{-1}$.

Before collision

K 6.0 m s⁻¹ L
0.20 kg 1.0 kg

After collision

4.0 m s⁻¹ K
0.20 kg L → 2.0 m s⁻¹
1.0 kg

a) Show that the collision is elastic. [4]
 b) Calculate the impulse applied to L by K. [2] (Edex)

▷ Work, Energy and Power

37. A person raises and lowers a dumb-bell of mass 2.5 kg through a vertical distance of 0.40 m, 50 times in 60 s.
   a) Show that the useful work performed by the arm muscles during this activity is 490 J. [2]
   b) Estimate the rate of conversion of energy in the muscles for the person during this activity. Assume that the muscles convert energy into useful work with an efficiency of 20%. [3] (OCR)

38. Power can be calculated from the product of force applied and velocity.
   a) Justify this expression, starting from the definition of power. [2]
   b) A cyclist pedalling along a horizontal road provides 200 W of useful power reaching a steady speed of 5.0 ms^{-1}. What is the value of the drag force against which the cyclist is working? [2]
   c) The drag force is proportional to the speed of the bicycle.
      i) Show that the useful power the cyclist must produce at speed v along the flat is proportional to v^2.
      ii) Predict what power the cyclist must produce to reach a speed of 6.0 ms^{-1} along the flat. [4]
   d) What useful power would the cyclist have to develop to maintain a speed of 5.0 ms^{-1} when climbing a hill of 1 in 30? Take the mass of the cyclist plus bicycle to be 100 kg. [4] (OCR)

39. A car of mass 1500 kg is moving at a uniform speed of 12 ms^{-1} down a slope with the engine switched off and its brakes applied.
   The slope makes an angle of 10° with the horizontal.
   a) If the only force opposing the motion is due to the brakes, calculate the magnitude of this force.
   b) Calculate the work done in 2.0 s by the brakes. [5] (AQA)

40. A catapult fires an 80 g stone horizontally. The graph shows how the force on the stone varies with distance through which the stone is being accelerated horizontally from rest.

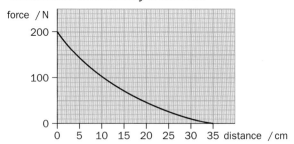

   a) Use the graph to estimate the work done on the stone by the catapult. [4]
   b) Calculate the speed with which the stone leaves the catapult. [2] (Edex)

41. A bullet of mass 3.0×10^{-2} kg is fired at a sheet of plastic of thickness 0.015 m. The bullet enters the plastic with a speed of 200 ms^{-1} and emerges from the other side with a speed of 50 ms^{-1}. Calculate
   a) the loss of kinetic energy of the bullet as it passes through the plastic,
   b) the average frictional force exerted by the plastic on the bullet. [5] (OCR)

42. The figure shows a simplified diagram of a hydroelectric power station.

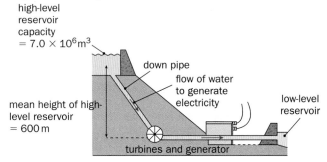

   a) Use the information in the diagram and the density of water = 1000 kg m^{-3} to show that the gravitational potential energy stored in the high-level reservoir is about 4×10^{13} J.
   b) The power plant has six 300 MW (output power) generators. Calculate the time for which the generators could produce full power with the energy stored in the high-level reservoir given that 12% of the stored energy is wasted.
   c) i) Calculate the mean rate of flow of water, in kg s^{-1}, through the turbines when at full power.
      ii) Calculate the wasted energy per second (power lost) during the generation process. [8] (W)

43. A typical take-off speed for a flea of mass 4.5×10^{-7} kg, jumping vertically, is 0.80 ms^{-1}. It takes such a flea 1.2 ms to accelerate to this speed.
   a) Show that the average acceleration is 670 ms^{-2}.
   b) Calculate the force required to produce this acceleration.
   c) Calculate the average power produced during the acceleration. [5] (AQA)

44. a) A car of mass m is at rest. A constant force F acts on the car and it moves a distance x in the direction of the force. The final velocity of the car is v. Write down the equation
      i) for the work done by the force F,
      ii) relating F and acceleration a. [1]
   b) Hence show that the kinetic energy of the car is given by the equation $E_k = \frac{1}{2}mv^2$. [3]
   c) The braking distance of an empty van travelling at a steady speed on a level road is 50 m. The van is now fully loaded with goods and travels at the same speed on the same road. Explain whether or not the braking distance would be the same. Assume the same braking force. [3] (OCR)

▷ Circular Motion

45. A stone of mass 0.50 kg hangs from an inextensible string.
a) Calculate the tension in the string.
The string is now whirled in a vertical circle of radius 0.98 m at a constant speed of $7.0 \, \text{ms}^{-1}$.
b) Calculate
   i) the angular speed of the stone,
   ii) the centripetal acceleration of the stone.
c) i) At which point in the motion is the tension in the string least?
   ii) Calculate the least tension in the string. [8] (W)

46. A model plane, of mass 0.15 kg, flies so that its centre of mass travels in a horizontal circular path of radius 10.0 m.
a) The plane completes two circuits in 21 s. Calculate the angular velocity of the plane. [2]

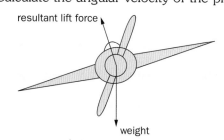

resultant lift force

weight

b) The diagram shows a simple frontal view of the plane showing the forces during flight. It assumes the rudder does not produce any sideways force. Calculate
   i) the total vertical component of the lift force,
   ii) the total horizontal component of the lift force. [3] (W)

47. A boy ties a string to a rubber bung and then whirls it so that it moves in a horizontal circle at constant speed.

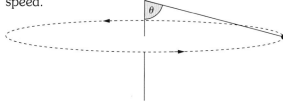

a) i) Copy the diagram. Draw and label arrows representing the forces acting on the bung. Assume that air resistance is negligible.
   ii) Hence explain why the string is not horizontal.
   iii) Give the direction of the resultant force on the bung. State the effect it has on the motion of the bung. [4]
b) The mass of the bung is 0.060 kg, the length of the string from the boy's hand to the bung is 0.40 m and θ is 75°.
   i) Show that the tension in the string is 2.3 N.
   ii) Calculate the resultant force on the bung.
   iii) Find the speed of the bung. [4] (OCR)

▷ Gravitational Forces and Fields

48. A glider is launched with a rope attached to a winch situated on the ground. The diagram shows the forces acting on the glider at one instant during the launch.

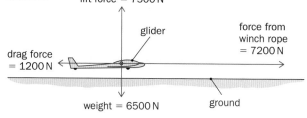

lift force = 7500 N

glider

force from winch rope = 7200 N

drag force = 1200 N

weight = 6500 N   ground

The glider climbs a vertical distance of 600 m in 55 s. The average power input to the winch motor during the launch is 320 kW.
a) Calculate the gain in gravitational potential energy (gpe) of the glider.
b) Calculate the percentage efficiency of the winch system used to launch the glider.
Assume the kinetic energy of the glider after the launch is negligible. [5] (AQA)

49. a) Define gravitational field strength.
b) i) The Moon has a radius of 1.74×10^6 m and a mass of 7.35×10^{22} kg. Calculate the gravitational field strength on the surface of the Moon.
   ii) The Moon moves in a circular orbit of mean radius 3.84×10^8 m around the Earth. The mass of the Earth is 5.98×10^{24} kg. Show that the force on the Moon due to the Earth is 1.99×10^{20} N.
   iii) Calculate the period of the Moon's orbit around the Earth in days.
c) The Earth has many artificial satellites with geostationary orbits. Explain fully what the term 'geostationary' means. [12] (CEA)

50. a) The diagram shows a series of equipotentials around a planet showing values of the gravitational potential.

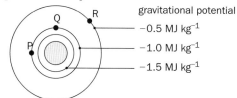

gravitational potential

Q   R
          −0.5 MJ kg⁻¹

P
          −1.0 MJ kg⁻¹

          −1.5 MJ kg⁻¹

A spacecraft of mass 3000 kg orbits the planet. Calculate, showing your reasoning, the changes in the gravitational potential energy of the spacecraft when it moves from
   i) P to Q,   ii) Q to R. [4]
b) With reference to the diagram explain why
   i) the potentials all have a negative sign,
   ii) the equipotential surfaces are spheres centred on the centre of the planet. [2] (OCR)

Further Questions on Mechanics

▷ Simple Harmonic Motion

51. The first graph shows one cycle of the displacement–time graphs for two mass–spring systems X and Y that are performing SHM.

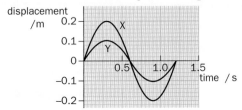

a) i) Determine the frequency of the oscillations. [2]
   ii) The springs used in oscillators X and Y have the same spring constant. Using information from the graph, show that the mass used in oscillator Y is equal to that used in oscillator X.
   iii) Explain how you would use one of the graphs to confirm that the motion is SHM. [4]

b) The second graph shows how the potential energy of oscillator X varies with displacement.

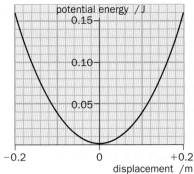

   i) Sketch a copy of the second graph and draw a line to show how the kinetic energy of the mass used in oscillator X varies with its displacement. Label this A. [1]
   ii) Draw a line on your graph to show how the kinetic energy of the mass used in oscillator Y varies with its displacement. Label this B. [2]

c) Use data from the graphs to determine the spring constant of the springs used. [3] (AQA)

52. A friend has a toy which consists of a wooden bird suspended on a long spring, as in the diagram.

a) Describe what measurements you would make to find out whether the vertical oscillations of the bird on the spring approximate to SHM. [6]

b) In an accident, the wings are broken off the bird, reducing its mass and air resistance. The body alone still oscillates in a vertical line. State and explain three changes that would occur in the oscillations when compared with those before the accident. [6] (OCR)

53. a) A mass, m, is attached to a spring and oscillates horizontally with simple harmonic motion on the floor of an ice rink. Its frequency is 0.625 Hz and the spring constant of the spring is 2640 Nm^{-1}.

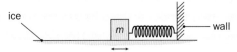

   i) Show the mass m is approximately 170 kg.
   ii) The maximum kinetic energy of the mass is 2.15 kJ. Calculate its maximum speed.
   iii) State the maximum kinetic energy stored in the spring and explain your reasoning.
   iv) Calculate the amplitude of oscillation.
   v) At time $t = 0$, the displacement of the mass is zero. Calculate the acceleration of the mass at time $t = 1.40$ s.

b) Explain why pushing the mass every 1.60 s would result in a large amplitude oscillation. [14] (W)

54. The time period of oscillation of a simple pendulum of length l is the same as the time period of oscillation of a mass M attached to a vertical spring. The length and mass are then changed.
Which row, A to D, in the table would give a simple pendulum with a time period twice that of the spring oscillations? [1] (AQA)

	new pendulum length	new mass on spring
A	2l	2M
B	2l	M/2
C	l/2	2M
D	l/2	M/2

55. A student is asked to do an experiment to find the acceleration due to gravity using a simple pendulum. He is told to vary the length l and determine the time T for one oscillation. He is given the following equation: $T = 2\pi\sqrt{\frac{l}{g}}$ and told to draw a suitable graph.

Which of the following would give a straight-line graph? [1] (Edex)

	y-axis	x-axis
A	T	l
B	T^2	$1/l$
C	$\sqrt{T}$	l
D	T^2	l

56. The cone of a loudspeaker oscillates with simple harmonic motion. It vibrates with a frequency of 2.4 kHz and an amplitude of 1.8 mm.

a) Calculate the maximum acceleration of the cone. [3]

b) The cone experiences a mean damping force of 0.25 N. Calculate the average power needed to be supplied to the cone to keep it oscillating at constant amplitude. [3] (OCR)

▷ Synoptic Questions on Mechanics

57. In a fairground sideshow, a prize (a can of drink) is won by knocking it off a shelf by firing a wooden ball from a spring-loaded gun. The can is made to slide along the shelf by the impact of the ball. The dimensions of the prize and the shelf are shown in the diagram. The can has a mass of 0.40 kg and the ball has a mass of 0.020 kg. $g = 9.8\,\text{ms}^{-2}$

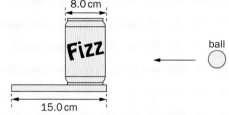

a) When the speed of the can immediately after the collision is $0.90\,\text{ms}^{-1}$, the can just falls off the rear of the shelf. Calculate the kinetic energy of the can immediately after the collision and the average frictional force between the can and the shelf. [3]

b) If the collision were perfectly elastic, the can would just fall off the shelf when the impact is 'head-on' and the speed of the ball is $9.5\,\text{ms}^{-1}$. Determine the velocity of the ball immediately after impact for an elastic collision. [3]

c) The stallholder thinks that there would be less chance of the prize being won if a way could be found of making the ball, travelling at $9.5\,\text{ms}^{-1}$, stick to the can.
Suggest whether this idea is worth following up and briefly justify your answer. [3] (AQA)

58. The diagram shows a smooth wooden board 30 cm long. One end is raised 15 cm above the other. A 100 g mass is placed on the board.
The two forces acting on the 100 g mass are shown on the free-body force diagram.

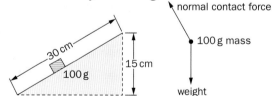

a) Explain why the resultant force on the 100 g mass acts parallel to the board and downwards. [2]

b) Calculate the magnitude of this resultant force. [2]

c) Calculate the kinetic energy gained by the 100 g mass as it slides 20 cm down the slope. [2]

d) The smooth board is replaced by a similar rough board which exerts a frictional force of 0.19 N on the 100 g mass. Calculate the new value for the kinetic energy gained by the 100 g mass as it slides 20 cm down the slope. [2]

e) Explain why the final kinetic energy of the 100 g mass is greater when the board is smooth. [2] (Edex)

59. A car of mass 1200 kg makes the journey shown.

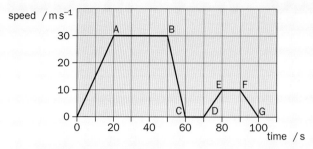

a) Calculate the total distance travelled during the journey of duration 100 s. [3]

b) During the section 0A the road is horizontal. For the section 0A, calculate
   i) the acceleration,
   ii) the force required to accelerate the car,
   iii) the work done in accelerating the car. [5]

c) When the car is moving at $30\,\text{ms}^{-1}$ there is a drag force of 500 N.
   i) Calculate the power required during the section AB to maintain a constant speed on a level road.
   ii) For part of section AB the car travels uphill, rising 1.0 m in every 20 m along the road. Calculate the extra power required when going uphill at a constant speed of $30\,\text{ms}^{-1}$.

d) At B the driver applies the brakes and slows down to stop at C. Describe the energy transfers which occur over the section BC. [2]

e) During section EF of the journey the car travels round a corner of radius 15 m at a constant speed of $10\,\text{ms}^{-1}$.
   i) Calculate the extra force exerted on the car.
   ii) Explain whether this changes the power required to propel the car. [4] (OCR)

60. The speed of a bullet can be estimated by firing it horizontally into a block of wood suspended from a long string, so that the bullet becomes embedded in the centre of the block. The block swings so that the centre of mass rises a vertical distance of 0.15 m. The mass of the bullet is 10 g and the mass of the block is 1.99 kg.

a) Assuming that air resistance can be neglected, calculate the speed of the block + bullet immediately after the impact. [3]

b) Calculate the speed of the bullet before the impact. [2] (AQA)

61. An empty railway truck of mass 10 000 kg is travelling horizontally at a speed of $0.50\,\text{ms}^{-1}$.

a) For this truck, calculate the momentum and the kinetic energy. [2]

b) Sand falls vertically into the truck at a constant rate of $100\,\text{kg s}^{-1}$.
Calculate the additional horizontal force which must be applied to the truck if it is to maintain a steady speed of $0.50\,\text{ms}^{-1}$. [2] (AQA)

10 Wave Motion

During an earthquake, the energy that waves carry is terrifying:

In this chapter you will learn:
- about wavelength, period and amplitude of a wave,
- the difference between transverse and longitudinal waves,
- about phase difference and polarisation.

What is a wave?

Look at the 5 diagrams down this page:

Phiz is holding one end of a long multi-coloured rope that is fastened to a wall at the other end.
The diagrams are *snapshots*. Each snapshot shows what Phiz and the rope look like at one instant of time.
Each diagram is like one 'frame' in a cinema film.

In **snapshot 1**, the rope is straight because Phiz has not begun to move his hand.

1

In **snapshot 2**, Phiz has moved his hand quickly upwards.
Look at the red section of the rope in this snapshot:

2

Can you see that the part held in his hand has moved most, while the part further away is lagging behind?

The sudden pull on the rope has stretched the part near his hand, and this tension is pulling the rope upwards.
The far end of the red part of the rope has not yet moved, so the green section has not yet been pulled upwards.

In **snapshot 3**, he has moved his hand quickly back down to the place it started from. The green section of the rope is being pulled upwards, while his hand is pulling the red section back down.

3

Compare the green section of the rope in snapshot 3 with the red section in snapshot 2. Can you see that it looks exactly the same?

Now look at what has happened in **snapshots 4 and 5**:
Although Phiz has not moved his hand, the up-and-down movement of the rope has continued.
First the green section of rope pulled the blue section upwards, and then the blue section pulled the next one up in its turn.

4

Look at the *shape* that is moving along the rope in these five snapshots. This is a **wave pulse**.
The *energy* put into the rope by Phiz is moving along the rope, being passed on from one section to the next.

5

Although the wave shape and the energy moves right along the rope, each part of the rope is *only moving up and down.*

122

▷ Wavelength λ

On the opposite page you saw how Phiz sent a wave pulse along a rope by moving his hand up and down once.

What if he were to carry on moving his hand up and down in a regular way, oscillating about the central point? The picture shows Phiz doing this with a slinky spring: (A slinky spring is often used to observe waves, because they travel much slower than along a rope.)

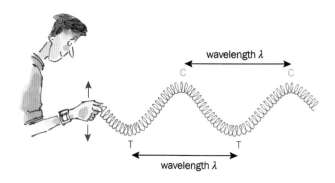

This diagram shows a **continuous wave**, not a pulse. This wave has equally spaced crests (**C**) and troughs (**T**). The wave shape you see here is called a **sine wave**.

The distance between any point on a wave and the next *identical* point on that wave is called the **wavelength**. The symbol for wavelength is λ (the Greek letter lambda).

Wavefronts

You have probably studied waves with a **ripple tank**:

An electric motor is used to vibrate a straight bar or a small dipper which is touching the surface of the water.

When a small dipper is used to make the waves, you get a regular pattern of concentric circles.
(You can use a stroboscope to 'freeze' the motion of the waves, which makes them easier to observe.)

Look at this snapshot of the circular pattern that is produced:

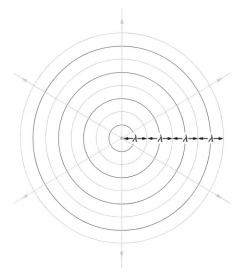

The central blue dot shows the place where the dipper is moving up and down, and the blue circles mark the **crests** of the waves that are produced.
These blue circles are equally spaced, because the distance between each wave crest and the next is one wavelength.

Each blue circle is a **wavefront**. A wavefront is a line joining points at the same position along a wave.

The red circles are the **troughs** of the waves, which are exactly halfway between the wave crests. These red circles are also wavefronts.

Diagrams of wavefronts are useful when you want to see the direction in which waves are moving.
The waves move **at right angles** to the wavefronts, as shown by the green arrows in the diagram.

Example 1
In the scale diagram above, showing circular wavefronts, the diameter of the outermost circle is actually 32 cm. What is the wavelength of the waves?

From the diagram, this distance from one side of the diagram to the other side is 8 wavelengths,

so: $\lambda = \dfrac{32 \text{ cm}}{8}$ $= \underline{4.0 \text{ cm}}$ (2 s.f.)

▷ Period

Look at these snapshots of a wave going along a rope:

The 5 snapshots are equally spaced in time.

The time it took Phiz's hand to go from the centre of the oscillation to the top, between snapshots 1 and 2, is equal to the time for the return trip, between snapshots 2 and 3.

In the same way, moving from the centre of the motion to the lowest point, between snapshots 3 and 4, took the same time as the return trip, between snapshots 4 and 5.

You can measure this regularity in two different ways, using the periodic time or using the frequency.

The *time* it takes an oscillation or a wave to repeat is called the **periodic time** or the **period**. Its symbol is *T*.

This exact repeat of a wave or oscillation is *one cycle*.

After one cycle, every particle along the wave has returned to the same position, and is moving in the same direction as it was at the beginning of that cycle.

Example 2
In the snapshots above, the time between each diagram and the next is 0.20 s. What is the period of the wave?

The rope and Phiz's hand are in exactly the same position in snapshots 1 and 5. The time between these two snapshots is four intervals of 0.20 s.

∴ The time period *T* of the wave is 4×0.20 s = 0.80 s   (2 s.f.)

▷ Frequency

For fast oscillations, the period is too small to measure easily. So instead of counting the number of seconds for 1 oscillation, we count the **number of oscillations (or cycles) in one second.**

This is the **frequency**. It is measured in **hertz** (symbol **Hz**).

The frequency and period are related by this equation:

$$\text{frequency, } f \text{ (Hz)} = \frac{1}{\text{period, } T \text{ (s)}} \quad \text{or} \quad f = \frac{1}{T}$$

See also page 78.

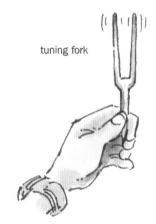

tuning fork

Some oscillations are fast.

Example 3
In Example 2, the period of the wave was 0.80 seconds. What is its frequency?

$$f = \frac{1}{T} = \frac{1}{0.80 \text{ s}} = 1.25 \text{ Hz} = 1.3 \text{ Hz} \quad (2 \text{ s.f.})$$

For the waves we meet in Physics, the periods are often very small, so the frequency is high.
In these cases, the frequency is measured not in hertz but in kilohertz (kHz), megahertz (MHz) or even gigahertz (GHz).

$$1 \text{ kHz} = 1000 \text{ Hz} = 10^3 \text{ Hz}$$
$$1 \text{ MHz} = 10^6 \text{ Hz}$$
$$1 \text{ GHz} = 10^9 \text{ Hz}$$

▷ The wave equation

Look at snapshots 1 and 5 on the opposite page.

The repeat between snapshots 1 and 5 is exactly one cycle.
During this time, the wave has moved one wavelength.
The wave frequency is the number of cycles in one second.

What do you get if you multiply:
(number of waves in 1 second, f) × (length of 1 wave, λ)?

This is the distance moved in one second – so it is the **speed**.

Measuring wave speed.

This shows that we can write this equation:

speed, c = **frequency, f** × **wavelength, λ**
$(m\ s^{-1})$ (Hz)   (m)

or $c = f\lambda$

The usual symbol for wave speed is c, but sometimes v is used.
So you may see: $v = f\lambda$

Example 4
Station Radio 4 on 'long-wave' radio has a frequency of 198 kHz.
What is the wavelength of the waves that arrive at your radio?
The speed of radio waves (electromagnetic radiation) is $3.00 \times 10^8\ m\ s^{-1}$.

In the equation, the frequency must always be in hertz, not kilohertz.
$198\ kHz = 198 \times 10^3\ Hz$

$$c = f\lambda$$
$$3.00 \times 10^8\ m\ s^{-1} = 198 \times 10^3\ Hz \times \lambda$$
$$\therefore \quad \lambda = \frac{3.00 \times 10^8\ m\ s^{-1}}{198 \times 10^3\ Hz} = \underline{1520\ m} \quad (3\ s.f.) \quad (\text{That's nearly a mile!})$$

Can you see why this answer has 3 significant figures?
It is because the precision of your answer should always be the
same as the precision of the data you are given (see page 9).
The frequency and the speed were both quoted to 3 s.f.

Is the wave speed constant?

The speed of electromagnetic waves was given in Example 4
as $3.00 \times 10^8\ m\ s^{-1}$, but this is strictly true only in a vacuum.
The waves slow down in a dense medium like water or glass.

Sometimes when a wave travels through a medium,
different wavelengths travel at different speeds.

This is called **dispersion**, because the waves separate out.

A storm out at sea generates water waves with many different
wavelengths. Hours later, long-wavelength waves arrive at the
shore, with short-wavelength waves following later.
This is because, in deep water, waves with long wavelengths
travel faster than short-wavelength waves.

Surfers know that longer waves arrive first.

▷ Energy transferred by waves

Displacement and amplitude

Look at the diagrams of Phiz sending waves along a rope:

How can you tell these are **not** snapshots of the same wave?

The wavelength of the two waves is identical, but the waves have different displacements at the crests and troughs.

The **displacement** is the distance from the centre of the oscillation to the rope. The vertical arrows show the displacement at different places along each wave. Displacement is a vector, so it is negative in a trough.

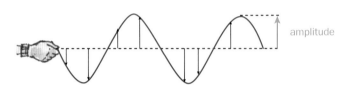

The *maximum* displacement occurs at a crest or a trough of the wave – it is called the **amplitude** of the wave. The diagrams show the amplitude of each wave.

Amplitude and energy

Do small and big amplitude waves have the same energy?

No – it is clear that bigger amplitude waves carry more energy. The larger waves in storms often damage boats and seaside huts.

If you double the amplitude, does that double the energy? No – look at these two diagrams of water waves:

Just like the waves on the rope above, they have the same wavelength, but the lower one has double the amplitude.

To calculate the energy stored in a 1 metre section of each of these waves, we can use the potential energy equation from page 70:

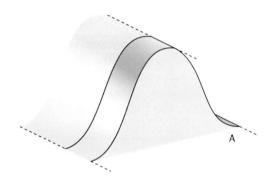

$$\begin{array}{ccccc} \textbf{change in} & & & & \textbf{change in} \\ \textbf{potential energy} & \textbf{= mass} & \times & \textbf{\textit{g}} & \times & \textbf{height} \\ \text{(J)} & \text{(kg)} & & \text{(N kg}^{-1}) & & \text{(m)} \end{array}$$

To calculate the potential energy we need to use the *average* height of the wave.
The average height of wave B is twice that of wave A – so this doubles the energy.
Wave B has twice the mass of A – and this doubles the energy, too.

Together, these factors will increase the energy 4 times.

If we make the amplitude of wave B 3 times that of A, then the average height and the mass both go up 3 times.
This makes the potential energy $3 \times 3 = 9$ times bigger.

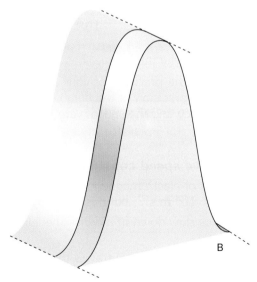

Can you see a pattern in the amplitudes and energies?
$(\text{double})^2 = 4$ times, and $(3 \text{ times})^2 = 9$ times.

Doubling the height quadruples the energy.

The energy is directly proportional to (amplitude)^2

This useful fact is true for all waves and oscillations.

▷ Waves spreading through space

If the light is too dim for reading, you move your lamp closer to your book. You need the source of the waves closer, because the wave energy gets spread more thinly as the waves spread out.

If you go twice as far from the light, will it be half as bright? No – look at this example, which uses aerosol paint to represent the energy in waves.

Move the lamp closer to increase the brightness.

Phiz and his aerosol paint spray

Phiz is holding an aerosol paint spray at 50 cm from a wall:

He squirts the aerosol for one second, and it makes a circular patch of paint, of radius 10 cm, on the wall.

We calculate the area of a circle using $\pi \times$ the radius squared so the area of this patch $= \pi \times (10 \text{ cm})^2 = 314 \text{ cm}^2$.

Now he moves along the wall, and stands **twice** as far from the wall – 100 cm away.
He gives this part of the wall a one-second spray, too.
How big is this patch of paint?

Because he is standing twice as far from the wall, the radius of the patch is doubled, to 20 cm.

The area of this patch is $\pi \times (20 \text{ cm})^2 = 1256 \text{ cm}^2$.

This area is **four times** as big as the area of the first patch.

How does the thickness of paint compare in these patches? The same amount of paint was sprayed on each – a burst lasting one second. But the second patch had **four** times the area, so the paint on it must be **four** times thinner.

At double the distance, a quarter the thickness.

The inverse-square law

The paint thickness and the distance from this spray follow an **inverse-square** relationship, a very common relationship in Physics.

The same relationship is true for the **intensity** of a wave. (The paint thickness represents the intensity of a wave.)

Intensity is the **energy per second per square metre** of surface. Intensity is measured in W m^{-2}.

At a distance r from a source, all the energy is spread out over a sphere of radius r, so the intensity is:

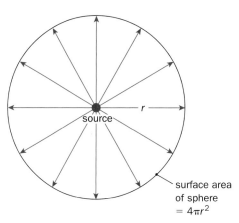

surface area of sphere $= 4\pi r^2$

$\text{Intensity } I = \dfrac{\text{energy/second of the source}}{\text{surface area of a sphere, radius } r} = \dfrac{\textbf{Power of source, } P \text{ (W)}}{4\pi r^2 \text{ (m}^2)}$	or	$I = \dfrac{P}{4\pi r^2}$

Example 5
The intensity of light at a distance 1.3 m from a lamp is 1.0 W m^{-2}.
Calculate: (a) the output power of the lamp, (b) the intensity at 0.7 m from the lamp.

a) $I = \dfrac{P}{4\pi r^2}$ so: $1.0 \text{ W m}^{-2} = \dfrac{P}{4\pi \times (1.3 \text{ m})^2}$ $\therefore P = 1.0 \text{ W m}^{-2} \times 4\pi \times 1.69 \text{ m}^2 = \underline{21 \text{ W}}$ (2 s.f.)

b) $\therefore I = \dfrac{P}{4\pi r^2} = \dfrac{21 \text{ W}}{4\pi \times (0.7 \text{ m})^2} = \dfrac{21 \text{ W}}{4\pi \times 0.49 \text{ m}^2} = \underline{3.4 \text{ W m}^{-2}}$ (2 s.f.)

▷ Phase

Look at these two waves moving from left to right:

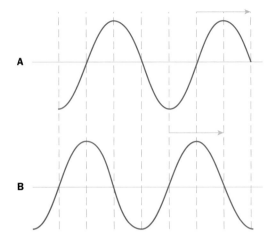

They have the same wavelength, and the same amplitude, but they are not in step with each other.
We say that the two waves are **out of phase.**

The phase is related to the position along the sine curve.
You do not usually need to know the phase of a wave, only the **phase difference** between two waves.

How can you describe this phase difference?
Wave B is a quarter of a wavelength behind wave A.
We say that wave B **lags** wave A with a phase difference of a quarter of a cycle.
Alternatively, we could say that wave A **leads** wave B by a quarter of a cycle.

Phase difference measured as an angle

The mathematics of oscillations is rather like that of circular motion (see also page 78).

One cycle of oscillation is similar to one complete rotation, which in degrees is 360° (or in radian measure it is 2π radians).

A phase difference of **half** a cycle would be referred to as a phase difference of 180° (or π radians).

Example 6
What is the phase difference, in degrees and radians, between the waves in the diagram at the top of this page?

The phase difference is a quarter of a cycle.

One cycle = 360° or 2π radians, so the phase difference $= \dfrac{360°}{4}$ or $\dfrac{2\pi}{4}$ radians $= \underline{90°}$ or $\underline{\dfrac{\pi}{2}}$ radians

---

The most important phase differences are 0 and π radians.

Two waves with a phase difference of 0 are **in phase.**

Two waves with a phase difference of π radians (180°) have exactly opposite phases. They are **in anti-phase.**

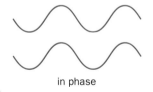

in phase

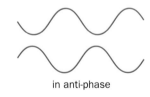

in anti-phase

Every extra complete cycle adds 2π radians to the phase of a wave, but it makes no difference to the appearance of the two waves, so the phase difference will be the same.
So a phase difference of 2π radians (360°) or 4π radians is the same as **no** phase difference. They are in phase.

A phase difference of 3π radians (540°) is the same as a phase difference of π radians (180°).

Example 7
There is a phase difference of 22 radians between two waves. Are they in phase?

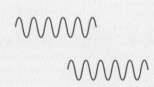

Since $\dfrac{22}{\pi} = 7.0$ (2 s.f.), it means that the phase difference is 7π radians.
7π radians $= 3 \times (2\pi) + \pi$ radians, which is exactly equivalent to just π radians.
So the two waves are completely out of phase – they are in anti-phase.

▷ Transverse and longitudinal

In the snapshots of waves on page 124, each part of the rope oscillated up and down, while the wave went from left to right in each diagram.
That was an example of a **transverse wave.**
The displacements were always *perpendicular* to the wave velocity. The movement of each part of the rope was at right angles to the direction in which the wave moved.

Now look at these snapshots showing Phiz pushing the end of a slinky spring *in* and *out* again and again:
The blue arrows show how Phiz's hand is moving.

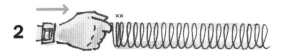

Look at the compressed region of the spring.
Can you see the compression moving steadily to the right?
This is a *compression* wave going along the spring.
As the wave moves along, a *rarefaction* moves along behind it, where the coils of the spring are pulled apart.

Look carefully at the two coils marked 'x' in each snapshot:

The separation of these coils decreases and then increases, as a compression and then a rarefaction passes through them. These coils are oscillating to and fro along the direction in which the slinky spring is lying.

This is an example of a **longitudinal wave.**
The separate displacements of each oscillating part of the spring are *in the same direction* as the wave velocity.

Sound waves travel as longitudinal waves.
Electromagnetic waves, such as radio waves, are transverse.

Using graphs to describe waves

Do the two graphs drawn here look the same?
Look carefully and you will see that they are different.

The first is a snapshot that shows the *displacement* of all of the particles that transmit the wave at one instant in time. The particles are at increasing distances from the wave source.

The second shows how the displacement of just one particle varies as *time* goes by.

We can show the amplitude *A* (the maximum displacement) on both graphs. The first graph can also be used to show the wavelength λ, because the horizontal axis is the distance.

The horizontal axis on the second graph is time.
We can use this graph to show the time period *T*, the time for one cycle of the wave.

Can these graphs be used only for transverse waves?
The graphs show how the displacement of the particles from their rest position varies with distance or with time.
So they can be used for both longitudinal *and* transverse waves.

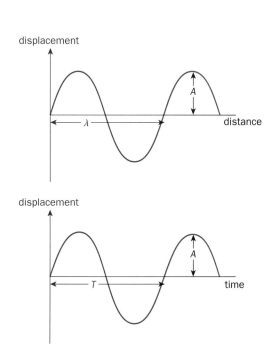

▷ Electromagnetic waves

Do you recognise this well-known experiment?
White light is a mixture of several colours and can be split
by a prism to show a spectrum. Light travels as a transverse
wave and each colour of light has a different wavelength.
Red light has the longest wavelength, violet the shortest.

The visible spectrum is a tiny part of a much wider spectrum,
with wavelengths that range from 10^{-15} m to more than 1 km.

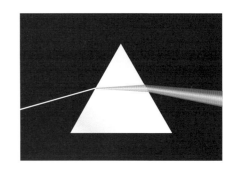

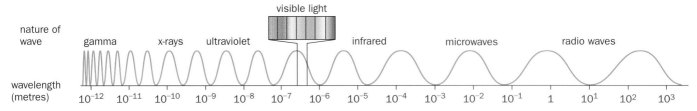

▷ The regions of the electromagnetic spectrum

The diagram above shows how the spectrum is labelled, but
there are no sudden boundaries between the different regions.

Gamma radiation and X-rays

Gamma radiation is emitted by radioactive nuclei (see page 360).
X-rays are produced when high-speed electrons decelerate quickly.
High-energy X-rays have shorter wavelengths than low-energy
gamma rays. They are given different names only because of the
way that they are produced.

Ultraviolet (UV)

Why is the arc welder in the photograph wearing eye protection?
Electric arcs, such as sparks and lightning, produce ultraviolet.
It is also given out by our Sun. Small amounts of UV are good
for us, producing vitamin D in our skin. Large amounts of high-
energy UV can damage living tissue.

Luckily for us the Ozone Layer, high in the Earth's atmosphere,
absorbs UV rays with wavelengths less than 300 nm (3×10^{-7} m).
But its recent thinning increases the risk of skin cancer.

Welders must protect themselves against UV.

Visible light

Human eyes can detect wavelengths from 400 nm (violet light) to
700 nm (red light). Other animals have eyes that are sensitive to
different ranges. Bees, for example, can see ultraviolet radiation.

Infrared (IR)

Every object that has a temperature above absolute zero gives
out infrared waves. Rescue workers can use infrared viewers
to search for survivors trapped below collapsed buildings.
IR was the first invisible part of the spectrum to be discovered
(by the astronomer William Herschel in 1800).

Radio waves

Radio waves range in wavelength from millimetres to tens of
kilometres. Microwaves are the shortest of the radio waves and
they are used for mobile phone and satellite communication.
Longer wavelengths are used for radio transmissions.

Photographs taken with visible light and IR.

130

▷ The nature of electromagnetic radiation

All electromagnetic waves can travel through a vacuum at 3.00×10^8 m s^{-1}. This value is called the 'speed of light'. But what are electromagnetic waves?

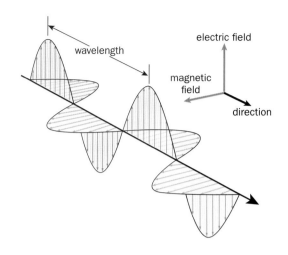

Electromagnetic waves consist of electric and magnetic fields. Can you see that the fields oscillate in phase, and are at 90° to each other and to the direction of travel of the wave? Each field vibrates at the wave frequency.

Who discovered these ideas?
By the early 19th century it was known that an electric current always produces a magnetic field. Michael Faraday then showed that changing a magnetic field produces an electric current.

In 1862 James Clerk Maxwell saw the connection:
If a changing magnetic field produces a changing electric field, the electric field must create a changing magnetic field. The two oscillating fields are linked!

Maxwell predicted that an oscillating electric charge should radiate an electromagnetic wave. He also derived an equation for the speed c of the wave in free space, using the magnetic field constant μ_o and the electric field constant ε_o:

$$c = \frac{1}{\sqrt{\mu_o \varepsilon_o}}$$

Maxwell's calculation for the value of c:

μ_o is the permeability of free space (see page 260)
ε_o is the permittivity of free space (see page 290)

$$c = \frac{1}{\sqrt{4\pi \times 10^{-7} \times 8.85 \times 10^{-12}}}$$

$$c = 3.00 \times 10^8 \text{ m s}^{-1}$$

Using the constants in this equation gives the speed of light! Maxwell had shown that light waves are electromagnetic waves. He thought that light was just one part of a wider spectrum. In 1887, Heinrich Hertz proved that Maxwell's ideas were correct when he discovered radio waves (see page 167).

▷ Measuring the speed of light

Galileo had tried to measure the speed of light in 1638. Can you suggest why he did not succeed?

The first successful direct measurement of the speed of light was achieved by Fizeau in 1849. His apparatus consisted of a rotating toothed wheel with N teeth and a mirror.

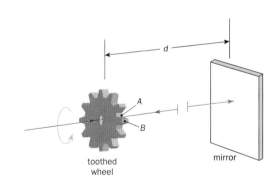

A beam of light passes through a gap A between two teeth. It travels to the distant mirror and is reflected back towards the wheel. The wheel spins, going faster and faster. If light leaving through one gap returns to the wheel as the next tooth B takes the place of the gap, the light is cut off.

We now know the time t for the light to travel to the mirror and back, a distance $2d$.
The wheel spins f times per second and in time t it makes $\frac{1}{2N}$ of a turn while a cog replaces a gap. So $t = \frac{1}{2Nf}$.

This method gave Fizeau a value for c of 313 000 km s^{-1}.

Fizeau's calculation for the value of c:

The wheel had 720 teeth.
The frequency f was 12.6 Hz.
Distance $2d$ was 17.26 km = 1.726×10^4 m.

$$\text{Speed} = \frac{\text{distance}}{\text{time}}, \quad \text{so} \quad c = \frac{2d}{1/(2Nf)} = 2d \times 2Nf$$

$$c = 1.726 \times 10^4 \text{ m} \times 2 \times 720 \times 12.6 \text{ Hz}$$
$$= 3.13 \times 10^8 \text{ m s}^{-1}$$

▷ Polarisation

Phiz has tied a string to a support and is shaking the string
vigorously to produce transverse waves on the string.
Phiz is moving his hand in *all* directions perpendicular to the
direction of the line of the rope.
Can you see that the string will be vibrating in every possible
direction? We say that the waves on the string are **unpolarised.**

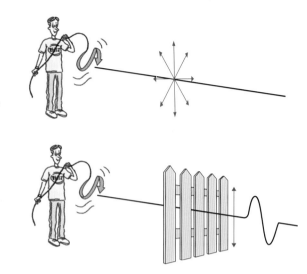

Now Phiz has passed the string through a picket fence.
He shakes the string as before, but what happens?
Only the vertical vibrations get through the slit. The wave
on the far side of the fence is **plane polarised** because the
vibrations are only in a single plane – the vertical plane.

What would happen if Phiz had a second picket fence?
If the second fence is aligned the same way as the first one,
the vertically polarised vibrations will still get through.
But if the second fence is at 90° to the first, the waves will
not get through and the string will stop vibrating.

Do you think sound waves can be polarised?
No, because in longitudinal waves the oscillations are along
the same direction in which the wave is travelling.

▷ Polarisation of radio waves

What do you notice about these television aerials?

Look carefully and you will see that all the small rods on each
aerial are horizontal and that all the aerials are pointing in the
same direction. Why is this necessary?

The aerials are pointing towards the nearest television
transmitter. The transmitter sends out radio waves that carry
the television signal. These radio waves are polarised.

The aerial rods have to be parallel to the plane of polarisation
of the radio waves to receive the signal. They are horizontal,
so it means that the waves from the transmitter must be
horizontally polarised.

Why are the radio waves polarised?

The diagram shows a transmitter radiating radio waves.
An electrical signal is fed to the transmitter and this makes
the electrons in the transmitting aerial oscillate up and down.

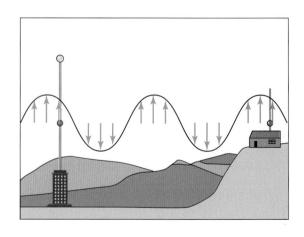

This transmitter produces a *vertically* oscillating electric field.
The magnetic field (not shown) is oscillating horizontally.
By convention, we agree that the direction of polarisation of
the wave is the orientation of the wave's *electric* field. This
transmitter is producing a vertically polarised radio wave.

How would you arrange a transmitting aerial so that it could
produce a horizontally polarised radio wave?

▷ Transverse or longitudinal

How do we know if a wave is transverse or longitudinal?
If a wave can be polarised it must be transverse.
Here is an experiment you can do with a microwave kit.

The transmitter produces vertically polarised microwaves.
When the aerial receiver is vertical, as shown, it absorbs the
microwaves, giving a large output reading on the voltmeter.

What happens when a metal grille is placed between the
transmitter and receiver?
The diagram shows the results:
In which position of the grille is the transmission of the
microwaves stopped? Did you expect the opposite result?

Remember, the microwaves are not mechanical waves.
The metal rods on the grille are acting like aerials.
The oscillating electric field of the microwaves is vertical and
it forces the free electrons in the rods to vibrate up and down.
The energy of the microwaves is absorbed by the rods.

With the rods horizontal, the electrons cannot move far;
little energy is absorbed and the wave passes through the grille.

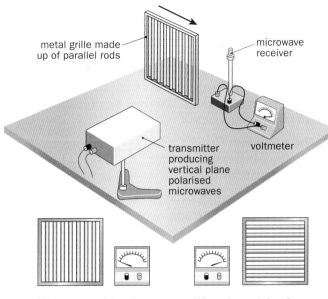

When the metal rods are vertical, the meter output is zero.

When the metal rods are horizontal, the meter output is a maximum.

Polaroid polarises light

Most light sources produce unpolarised light.
When unpolarised light passes through a Polaroid filter, about
half its energy will be absorbed. The light is now polarised
because the filter absorbs all the planes of oscillation but one.

What will happen if you align two filters and then rotate one?
Less and less light gets through. When you have rotated the
second filter through 90°, no light gets through. The filters
are 'crossed' – their transmission paths are at right angles.

What do you think you will find if you carried on rotating the
second filter?

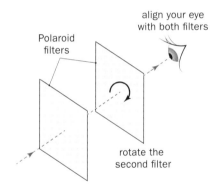

Using Polaroid to reduce reflected glare

Light can also be polarised by reflection. Some of the light reflected from water is polarised horizontally.
(And the light transmitted into the water is mainly vertically polarised.)
Polaroid lenses allow only vertically polarised light through, so light reflected from the water is absorbed.
If you wear Polaroid sunglasses you will not be dazzled by the reflected glare from the water while fishing or skiing!

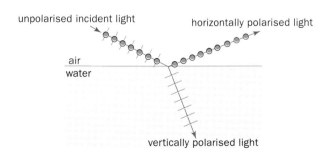

With regular sunglasses   With polaroid sunglasses

▷ The Physics of hearing

How do our ears detect sound?
The longitudinal sound waves cause pressure changes in the
medium through which they are travelling. The air pressure
is higher than normal at the centre of a *compression*. At a
rarefaction, it is lower. Normal air pressure is $\sim 1 \times 10^5$ Pa,
but your ear can detect variations of only 2×10^{-5} Pa!

The *outer ear* acts like a funnel, gathering the sound waves
into the ear canal and passing them on to the *eardrum*.
The fluctuations in air pressure force the eardrum to vibrate.

The vibrations are passed to the *oval window* by the three
bones of the *middle ear*. These bones act as a lever and
magnify the force of the vibrations by a factor of about 1.5.
The oval window has a smaller area than the eardrum, so this
increases the pressure on the oval window by a further factor
of around 15 (see page 26). The pressure on the oval window
is now about 22 times greater than at the eardrum.

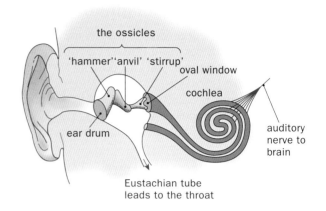

The oval window causes the fluid in the *cochlea* to vibrate.
The basilar membrane stretches along the length of the cochlea,
and the waves in the cochlea force it to resonate.
The lower the wave frequency, the further along the membrane
is the point at which resonance occurs. Hair cells (cilia) at the
point of resonance are forced to move, and as they move they
send electrical impulses along the auditory nerve to the brain.

▷ What is a decibel?

The loudness of a sound wave is related to its intensity.
Intensity was defined on page 127.

The decibel scale is used to compare two sounds of
different intensity. One of these is the intensity of the
sound being measured.
The other is an intensity of 1×10^{-12} W m^{-2}.
This is called the *threshold of hearing*, I_0, because it
is the faintest sound that the human ear can detect.

The range of intensities that the human ear can detect is
large, so the intensity of sound is measured on a scale
based on multiples of 10 (a logarithmic scale).

This is how the **decibel scale (dB)** works:

intensity /W m^{-2}	number of times greater than threshold	decibel rating /dB	description of sound
1×10^{-12}	10^0	0	threshold of hearing
1×10^{-11}	10^1	10	barely audible
1×10^{-10}	10^2	20	whispering
1×10^{-6}	10^6	60	normal conversation
1×10^{-5}	10^7	70	vacuum cleaner
1×10^{-2}	10^{10}	100	pneumatic drill
1×10^0	10^{12}	120	human pain threshold
1×10^3	10^{14}	140	jet take-off at 25 m

> **Relative intensity in dB** $= 10 \times \log_{10} \dfrac{I_1}{I_0}$,   where I_1 = Incident sound intensity (W m^{-2})
> and I_0 = Threshold intensity (W m^{-2})

So, an increase of 10 decibels means the sound has 10 times
more intensity; an increase of 20 dB means it has 100 times
more intensity; and 30 dB has 1000 times more intensity.

Can you show that if the sound intensity doubles the dB level
increases by 3.01?

Have you ever left a rock concert with your ears ringing?
Music at 120 dB has a relative intensity high enough to cause
pain and even permanent damage to your ears!

Example 8
By how much does the decibel level rise if the
sound intensity doubles from I_0 to $2I_0$?

Relative intensity in dB $= 10 \times \log_{10} \dfrac{2I_0}{I_0}$

$= 10 \times \log_{10} 2 = \underline{3.01}$

▷ Human perception of loudness

The greater the intensity of the incoming sound wave, the bigger the amplitude of the vibrations in the cochlea. So do all people agree on what is a loud sound?

We can ask people to judge when they think sounds of different frequencies have the same loudness. The average results from many people have been used to create the chart shown here. The lines on the graph are called *equal loudness curves.* Notice that a logarithmic scale has been used for frequency.

What does the data show? The human ear can detect sounds from 20 Hz to 20 kHz, but it is particularly sensitive to sounds from 2 kHz to 5 kHz. Look at the two blue dots on the chart. Can you see that the relative intensity of sound at 100 Hz needs to be 30 dB greater than at 3 kHz to give the same perception of loudness?

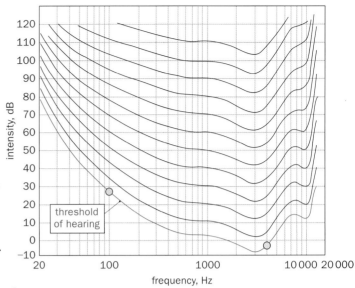

The dBA scale

The chart shows that loudness depends on frequency. So how do we judge if environmental sounds are unacceptably loud? Sound meters can be fitted with an 'A-contour' filter. The filter adjusts the measurement of relative intensity to account for the way in which the ear responds to different frequencies. The unit for this measurement of relative intensity is the dBA.

▷ Defects of hearing

Have you heard of a device called a 'mosquito'? It is designed to stop young people gathering in groups, by emitting a high-pitched sound which they cannot tolerate. It doesn't affect older people because they can't hear it!

The sensitivity of the ear to all frequencies deteriorates with age, but the decline is less for low than for high frequencies. Hearing loss is a maximum at 4 kHz. At the age of 45 years, to detect a sound of frequency 4 kHz you will need 10 dB more relative intensity than at the age of 20. In addition, you will probably be unable to hear frequencies above 12 kHz at all!

Beware those loud sounds! Repeated exposure to loud noises is a common cause of hearing loss. Who do you think may be in danger? People who work with noisy equipment such as pneumatic drills must wear ear protectors. What happens to damage the ear? Aging always causes wear and tear on the cilia in the cochlea, but loud noises may overstimulate and damage these cells. When hair cells for a certain band of frequencies are destroyed, those frequencies can no longer be heard.

How can your hearing be tested? The photograph shows a hearing test. The person sits in a sound-proof booth and hears the sounds though headphones. The data collected is used to produce an audiogram.

Phiz must take care that the music through his headphones is not too loud.

An audiologist conducting a hearing test.

▷ Physics at work: Earthquakes and Tsunamis

Earthquake detection

The crust of the Earth consists of a number of
tectonic plates, rather like the tiles on a bathroom floor.
These plates push against one another, driven by
convection currents in the mantle underneath.
From time to time, the plates suddenly slip under these
pushing forces, and there is an earthquake.

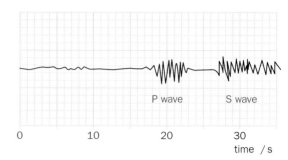

A seismometer recording.

At a distant earthquake detection station, earthquake
waves disturb a delicately-balanced *seismometer*, which
oscillates with the tiny vibrations received from the
earthquake. A trace of the oscillations is produced on a
roll of paper marked with the time the vibrations occurred.

The centre, or *focus*, of the earthquake, is deep
underground. Four sorts of earthquake waves spread out
from the focus. Two of these, called *primary* (**P**) and
secondary (**S**) waves, travel through the Earth, and two
others travel along the surface of the crust.
The P waves and S waves do not travel together: the
P waves are longitudinal compression waves, and they
travel faster than the transverse S waves.

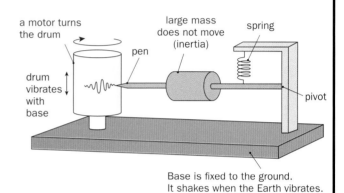

The seismograph trace will show the time delay between
the arrival of P waves and the arrival of S waves.
By knowing the speeds of P waves and S waves,
geologists can use this time delay to calculate the distance
between the earthquake focus and the seismograph.
By working with other seismograph stations, they can
pinpoint the focus of the earthquake exactly.

The principle of a seismometer.

Tsunami!

In many coastal regions in the Pacific Ocean, an earthquake
brings a fresh danger in its wake - a **tsunami**, or tidal wave.

The Hawaii Pacific Tsunami Warning centre monitors data from
numerous sensors around the Pacific Ocean and issues alerts.
The Boxing Day Tsunami of 2004 occurred in the Indian Ocean,
where, at the time, there was no such warning system.
As a result, 275 000 people lost their lives.

If you lift one end of a tray of water a small distance and
then drop it, a wave will be generated which passes along
the water surface. This is a small-scale tsunami.
In 2004, a fault line off the coast of Sumatra suddenly ruptured.
The sea bed moved vertically through 15 m, along a line about
1500 km long. The volume of water displaced was huge.

The huge waves of the tsunami slam into the
shore.

At sea, tsunamis do not look impressive, being only about a
metre high, with a huge wavelength of about 100 km, but
they travel at speeds of about 200 m s^{-1} (over 400 mph!)
As they reach shallow water they slow down, and the tsunami
'tail' catches up with its 'head', giving a dramatic increase in
height. The tremendous amounts of energy in a tsunami can
strip coastal areas of sand, vegetation and houses.

Summary

Oscillations can produce waves, which carry energy.
The energy carried by a wave depends on its amplitude. In fact: energy $\propto$ (amplitude)^2.

The frequency f of a wave is related to its period T by: $\quad f = \dfrac{1}{T}$

The wavelength λ, frequency f and speed v of a wave are related by the equation: $\quad v = f\lambda$

Phase difference, measured in degrees or radians, gives the difference between similar points on two waves.

In *transverse* waves, oscillations are at right angles to the direction in which the wave is moving.
In *longitudinal* waves, oscillations are in the same direction that the wave is moving.

Transverse waves can be polarised by removing all components except those in one particular plane of oscillation, but longitudinal waves cannot be polarised.

As a wave spreads out, its intensity I decreases by an inverse-square law: $\quad I = \dfrac{P}{4\pi r^2}$

Maxwell's equation for the speed of electromagnetic waves in a vacuum states that: $\quad c = \dfrac{1}{\sqrt{\mu_o \varepsilon_o}}$

▷ Questions

1. Can you complete these sentences?
   a) Oscillations often make waves which carry .... away from the place where the waves started.
   b) The time it takes any of the oscillating parts of the wave to repeat is called the ...., while the distance over which it repeats is called the ....
   c) The energy contained in a wave is directly proportional to the square of the ...., which is measured from the .... of oscillations to the highest or lowest value.
   d) .... are lines joining points along the wave with the same phase, and are useful in describing the movement of waves.
   e) When waves do not exactly coincide, they have a phase ...., which is measured in .... or radians.
   f) Waves such as sound, where the oscillations are in the wave direction, are .... waves.
   Waves such as light, where the oscillations are at 90° to the wave direction, are .... waves.
   g) Only .... waves can be polarised.

2. Copy and complete this table:

period T	frequency f
2.0 s	
20 ms	
	440 Hz
	91.5 MHz

3. A ripple tank dipper makes 8 water waves in a time of 2 s. When it is just about to make the 9th wave, the first wave has travelled 48 cm from the dipper.
   a) Calculate the frequency of the waves.
   b) What is the wavelength of the waves?
   c) Use $v = f\lambda$ to find the wave speed.
   d) Check your answer to (c) using the equation speed = distance / time.

4. Primary (longitudinal) earthquake waves travel at 7500 m s^{-1}, while secondary (transverse) waves travel at 4500 m s^{-1}. Calculate the time delay between the two types arriving at a seismographic research station 800 km from the centre of the earthquake.

5. The speed of deep water waves is given by:

$$v = \sqrt{\frac{\lambda g}{2\pi}} \quad \text{where } g = 9.8 \text{ m s}^{-2}$$

   a) How does this equation show that water waves undergo dispersion?
   b) Use the equation to calculate the speed of water waves of wavelength (i) 50 m (ii) 100 m.
   c) Calculate the time delay between 100 m and 50 m waves arriving at a point 20 km away.

6. A water wave has an amplitude of 50 cm. A 100 m wavefront of this wave carries 250 kJ of energy. What would be the energy carried by 100 m of a wave with the same wavelength but an amplitude of 150 cm?

7. Sketch diagrams to show two waves with a phase difference of a) 180°, b) $\pi/2$ radians. Label the phase difference in each case.

8. Radio aerial rods must be in the correct plane, either vertical or horizontal, depending on the transmitter. If you mount the aerial the wrong way, the signal received is weak. Use the terms 'transverse' and 'plane-polarised' to explain these observations.

9. A high-efficiency lamp gives an intensity of 0.4 W m^{-2} on a newspaper 2 m away.
   a) What is the output power of the lamp?
   b) What is the intensity 4 m away from the lamp?

Further questions on page 180.

11 Reflection and Refraction

From ultrasound scans to the Hubble space telescope, from mirages to optical fibres: there is a huge range of uses of reflection and refraction.

In this chapter you will learn:
- how waves are reflected and refracted,
- about critical angle and total internal reflection,
- how images are formed by lenses.

Reflection of waves

How is light reflected? The Dutch scientist Christiaan Huygens was the first to argue that light is a *wave*. (Other scientists at the time, like Newton, argued that light travelled like a bullet.)
Huygens suggested an idea (now called Huygens' construction), to help us explain how waves move.
He said each part of a wavefront is like a point source of tiny circular waves. These 'wavelets' add up to give another wavefront, one wavelength away.

Look at the incident (red) waves in this diagram:

The first wavefront produces sets of circular wavelets that add together to make the next wavefront. This wavefront then does the same thing to make the next wavefront, and so on.

This continues until the waves meet the reflector. Wavelets (coloured blue) are produced at the reflector and add up to make wavefronts (blue) coming back as shown.

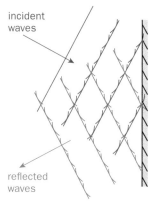

How waves reflect.

It is far easier to follow a **ray diagram** than a wave diagram.
A *ray* is a very narrow band of waves and is drawn as a straight line in the direction of movement of the waves.

Look at this ray diagram of the *same* reflection:

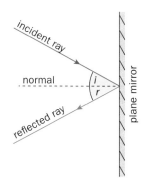

This diagram illustrates the 2 laws of reflection:

1. When light is reflected, the incident ray, the reflected ray and the normal all lie in the same plane.
2. The angle *i* between the incident ray and the normal is the same as the angle *r* between the reflected ray and the normal.

'Normal' means a line drawn at right angles (90°) to the reflector.

Ray diagrams are helpful in complex cases such as the reflection of light by a concave mirror, shown here:

Imagine how difficult it would be to draw this using Huygens' construction! With rays, it is much easier.

The dotted lines, which go to the centre C of the curved mirror, are the normals. So we can easily use the laws of reflection to predict where the rays will be reflected to.

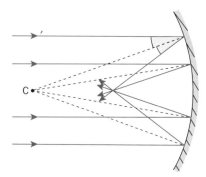

Reflection by a curved mirror.

▷ Refraction and a change in speed

Is the pencil in this photograph really bent?

No, it is the light passing from water to air that is bent. The light is **refracted**. But your brain knows from experience that light normally travels in straight lines, so the pencil looks bent.

Why does refraction occur when light moves from water to air, or from air to water?

Look at this diagram showing wavefronts moving from air into water:

Can you see that the wavefronts are closer together in the water?

This is because **light travels more slowly in water**.
When a wavefront moves into water it slows down, and the one behind catches up with it, until that second wavefront also moves into the water.
It is this slowing down that makes the waves change direction.

This is because the end of the wavefront labelled B has only just entered the water, and so it has had a chance to overtake the end labelled A, which has been travelling more slowly since it entered the water a short time earlier.

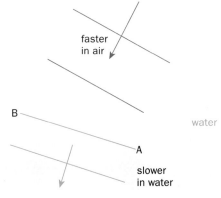

How waves refract.

▷ Snell's Law

The ray diagram shows the refraction more clearly:

The angle of incidence *i* is the angle between the incident ray and the normal. It is larger than the angle of refraction *r* when light moves into a substance where it travels **more slowly**.

A substance in which light travels more slowly is said to be **optically denser**. Water is optically more dense than air.

The young Dutch astronomer Willebrord Snell discovered a mathematical relationship in 1621. It is called Snell's Law:

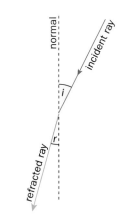

Ray diagram of refraction.

$$\frac{\sin(\text{angle of incidence, } i)}{\sin(\text{angle of refraction, } r)} = \text{constant} \quad \text{or} \quad \frac{\sin i}{\sin r} = n$$

The constant **n** is called the **refractive index**.

The values of **n** in data books are for when light is passing from **air** into the medium. They are often written with a suffix, so the refractive index for light going from air into water is written: n_{water}

As the refractive index is a ratio, it has no units.

Example 1
The diagrams above show light refracting as it travels from air into water.
The angle of incidence $i = 27.0°$, and the angle of refraction $r = 20.0°$.

Calculate the refractive index of water, n_{water}

$$n_{\text{water}} = \frac{\sin i}{\sin r} = \frac{\sin(27.0°)}{\sin(20.0°)} = \frac{0.454}{0.342} = \underline{1.33} \quad (3 \text{ s.f.})$$

▷ Refractive index n and the speed of light

vacuum

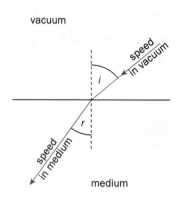

Light is refracted because it slows down or speeds up, as we saw on the previous page.

Because of this, the refractive index n can also be written as a ratio of two speeds:

$$n = \frac{\text{speed of light in vacuum}}{\text{speed of light in the medium}}$$

where the word **medium** means the substance that the light is travelling through.

Example 2
The speed of light in a vacuum, $c = 2.9979 \times 10^8$ m s^{-1}, and the speed of light in air $= 2.9970 \times 10^8$ m s^{-1}.

What is the refractive index of air?

$$n_{\text{air}} = \frac{\text{speed of light in vacuum}}{\text{speed of light in air}}$$
$$= \frac{2.9979 \times 10^8 \text{ m s}^{-1}}{2.9970 \times 10^8 \text{ m s}^{-1}}$$
$$= \underline{1.0003} \text{ (5 s.f.)}$$

Optically, air is almost the same as a vacuum, but in other ways it is very different.

Air and vacuum

Do you agree that the refractive index of air is very close to 1? Light travels in air at virtually the same speed as it does in a vacuum, and so we rarely distinguish between them optically.

The very slight refraction when light goes from the vacuum of space into our atmosphere is normally not noticed, except during a total eclipse of the Moon.

During the total eclipse stage, the Moon is not invisible, but appears very faintly, lit up with sunlight refracted by the Earth's atmosphere:

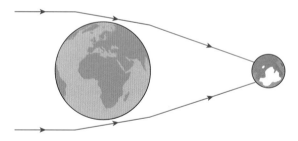

Refraction by the Earth's atmosphere (exaggerated).

Light going from one medium to another

When light goes from one medium to another, the speed of light in both mediums affects the refraction.

Compare this diagram with the one at the top of this page:

Can you see that i and r are now labelled θ_1 and θ_2? Here each θ is the angle between the ray and the normal. Snell's Law now takes this form:

$$\frac{n_2}{n_1} = \frac{\sin \theta_1}{\sin \theta_2}$$

This can be written $\boxed{n_1 \sin \theta_1 = n_2 \sin \theta_2}$

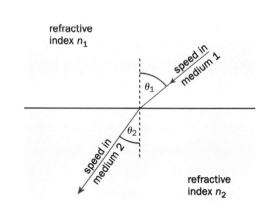

refractive index n_1

refractive index n_2

Moving through several different mediums

This is where the equation $n_1 \sin \theta_1 = n_2 \sin \theta_2$ is very useful.
Look at the diagram of three mediums in parallel:

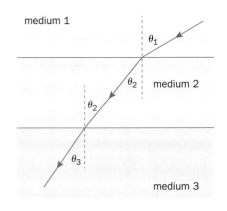

The angle of refraction from medium 1 into medium 2 is the same as the angle of incidence from medium 2 into medium 3.

This means $n_1 \sin \theta_1 = n_2 \sin \theta_2$ and $n_2 \sin \theta_2 = n_3 \sin \theta_3$.

So $n \sin \theta$ is the same all the way through the different layers.

Example 3
In the diagram, $\theta_1 = 60°$ and $\theta_3 = 35°$. Medium 1 is air, medium 2 is water of refractive index 1.33 and medium 3 is glass.
Calculate:
a) the angle θ_2,   b) the refractive index of the glass.

a) $n_1 \sin \theta_1 = n_2 \sin \theta_2$ and medium 1 is air, so $n_1 = 1$
$1 \times \sin 60° = 1.33 \times \sin \theta_2$ $\therefore$ $0.866 = 1.33 \times \sin \theta_2$
$$\sin \theta_2 = \frac{0.866}{1.33} = 0.651 \quad \therefore \ \theta_2 = \sin^{-1} 0.651 = 41° \text{ (2 s.f.)}$$

b) $n_1 \sin \theta_1 = n_3 \sin \theta_3$
$1 \times \sin 60° = n_3 \times \sin 35° = n_3 \times 0.573$
$$n_3 = \frac{0.866}{0.573} = \underline{1.50} \text{ (2 s.f.)}$$

▷ Critical angle and refractive index

Look at the diagram, which shows 3 rays of light passing from water into air:

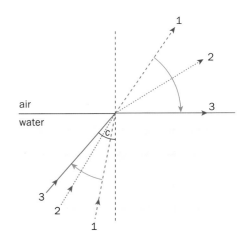

The rays 1, 2, 3 meet the surface at different angles of incidence, and so have different angles of refraction.
As the angle between the incident ray and the normal increases, the refracted ray gets closer and closer to the water surface.
Finally, with ray 3, the light only just escapes from the water.

The angle between ray 3 and the normal in water is called the **critical angle, c**. This is the largest angle at which refraction out of a denser medium is just possible.

How can we calculate the critical angle?
We can apply Snell's Law for light going from air into water along this same path.

$$n_{water} \sin \theta_{water} = n_{air} \sin \theta_{air}$$

$$n_{water} \sin c = n_{air} \sin 90°, \quad \text{but } \sin 90° = 1 \text{ and } n_{air} = 1$$

$$n_{water} \sin c = 1 \quad \therefore \quad \boxed{n_{water} = \frac{1}{\sin c}}$$

Example 4
The critical angle of crown glass is 42°. What is its refractive index?

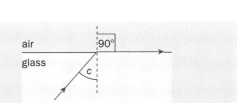

$$n_{glass} = \frac{1}{\sin c} = \frac{1}{\sin 42°} = \frac{1}{0.669} = \underline{1.5} \text{ (2 s.f.)}$$

141

▷ Total internal reflection

Every time that a ray meets a boundary between two mediums, some light is reflected.
Even though glass is transparent, you can see a reflection in a window. Just look at the photograph on page 138.

Now look at the dashed ray in this diagram:

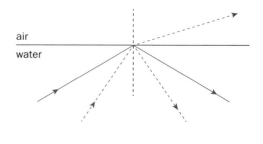

The angle of incidence as it meets the surface is *less* than the critical angle, so this ray is refracted out of the water, but *some* of the light is reflected back into the water.
This is *partial* internal reflection.

Now look at the continuous ray in the diagram:

The angle of incidence here is **greater** than the critical angle, so **no** light is refracted out of the water.
All of the light is reflected back into the water.
This is **total internal reflection**.

▷ Optical fibres

Look at this diagram of an optical fibre:

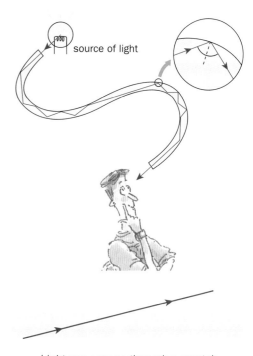

source of light

The angle at which the ray meets the normal to the surface is always much more than the critical angle, so the ray continues down the fibre, with total internal reflection.
At the other end, Phiz can see the light after many total internal reflections.
The thickness of the fibre is greatly exaggerated here, so the angle of incidence in a real fibre is a much bigger angle than in this diagram.

Optical fibres in communications

Fibre-optic cables are now widely used in communications for carrying signals for the internet, television and telephones.

There must be no scratches on the fibre, or light will escape. This is because a ray can meet the surface of the scratch with an angle of incidence less than the critical angle, as shown:

This is easily cured by coating the fibre with a tougher outside layer, made of plastic or glass with a lower refractive index.

See the diagram below.

Any scratch will then occur on the *outside* of the coating layer, and the boundary with the core stays smooth.

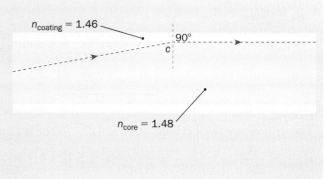

Light can escape through a scratch.

Example 5
The diagram shows an optical fibre with a core of refractive index 1.48 and a coating of refractive index 1.46.
What is the critical angle for this fibre?

$n_{coating} = 1.46$

$90°$
c

$n_{core} = 1.48$

$\boldsymbol{n}\sin\boldsymbol{\theta}$ is constant, so, for light which **just** escapes the core:

$$n_{core}\sin c = n_{coating}\sin 90° = n_{coating} \quad \text{(as } \sin 90° = 1)$$

$$\sin c = \frac{n_{coating}}{n_{core}} = \frac{1.46}{1.48} = 0.9865$$

$$c = \sin^{-1}0.9865 = \underline{80.6°} \text{ (3 s.f.)}$$

Problems with optical fibres

The signals that travel down an optical fibre are usually *digital* signals.

An analogue signal (eg. your voice) is coded into a series of pulses and spaces representing the binary digits 1 and 0:

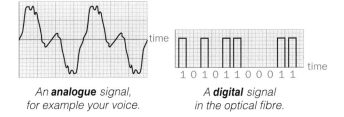

*An **analogue** signal, for example your voice.*

*A **digital** signal in the optical fibre.*

It is important that the digital signal that arrives at the other end of the fibre is clear and not distorted.

The glass in early optical fibres contained contamination which absorbed some of the light. This reduced the signals that they carried, so that every few kilometres a 'repeater' was needed to boost the signal.

Optical fibres are now made from glass which is extremely pure. This means that the signals travel hundreds of kilometres without loss of signal strength. But there are still two problems to solve.

1. Material dispersion

On page 125 you saw that waves can undergo **dispersion** when they travel through a medium, because different wavelengths travel at different speeds.

In a typical glass, violet light travel at 1.95×10^8 m s^{-1} and red light travels at 1.98×10^8 m s^{-1}. If a pulse contains red and violet light, then it will spread out as it travels.
This is like a race where the faster runners are at the front and slower ones at the back, even though they started together.

After travelling a long distance a pulse could overlap with the one in front of it and the one behind it.
This would make a digital signal hard to decode.

This is called *material dispersion* or *spectral dispersion*.

Pulses containing different wavelengths spread as they travel.

To reduce material dispersion, the light source used is an infrared laser. This emits light with a very narrow range of wavelengths so that the pulses do not spread much.

2. Modal dispersion

Look at the diagram:

Can you see that ray 1 is taking a longer route than ray 2, so it takes longer to travel down the fibre?

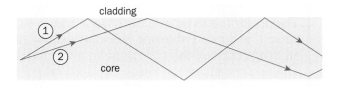

Modern telecommunications uses digital pulses travelling in very quick succession along optical fibres. If the different rays that make up a pulse do not all travel together, the pulse becomes broader. The pulses may overlap and give a faulty signal, just like they did in material dispersion above.

This effect, which causes broader and less-clear pulses, is called *modal dispersion* or *multipath dispersion*.

It can be solved by trapping the light in the very narrow core region in the middle of a *monomode* optical fibre. The core of a monomode fibre is a hundred times thinner than a human hair.

Light travels down the centre of a monomode fibre.

▷ Physics at work: Using refraction and reflection

Coherent and incoherent bundles

For many uses, optical fibres are packed into bundles like the separate strands of wire in an electricity cable.

There are two types of fibre bundle.

A **coherent bundle** has the fibres carefully arranged so that they are in the same order at both ends.

An **incoherent bundle** has the fibres randomly arranged.

Look at the diagram, where different fibres have been labelled with different colours:

Can you see that the fibres in bundle **A** are in the same order at each end, while the fibres in bundle **B** are not?

If you need to observe something down the fibre, you do not want a jumbled image. You need the same pattern of light to come out of the fibre as went in at the other end.

You need a coherent bundle like bundle **A**. A high-resolution image can be obtained by using a bundle containing very many fibres. Each fibre in the bundle connects with one pixel of a CCD detector similar to those found in digital cameras. The digital image is displayed on a monitor.

If you just need to get light into an inaccessible place, an incoherent bundle like bundle **B** will be fine.

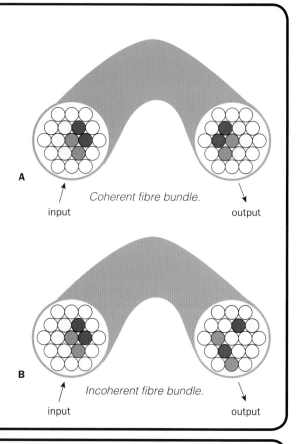

A
Coherent fibre bundle.
input   output

B
Incoherent fibre bundle.
input   output

Endoscopes

An endoscope is often used to examine the digestive tract when a patient develops vomiting or diarrhoea that continues longer than expected.

It is a flexible tube that allows the doctor to look directly at the inside of the digestive tract to make a diagnosis, and possibly even to treat the problem with tiny surgical tools that can be manipulated through the endoscope.

In order to see inside the patient, light has to be taken down the endoscope to light up the region of interest. This is done with an **incoherent** bundle of optical fibres.

The doctor has to be able to view the region inside the patient, and so a **coherent** fibre-optic bundle carries an image back up the endoscope to a CCD detector. The doctor can use the image displayed on a monitor to examine the region inside the patient and identify any problems which need treatment.

To examine regions in the abdomen outside the digestive tract, a rigid endoscope called a **laparoscope** is used. This is inserted through a small incision in the abdomen wall to allow the doctor to examine and treat the condition. A laparoscope often has another bundle of (incoherent) optical fibres, to carry laser light which can be used to cut and seal the tissue.

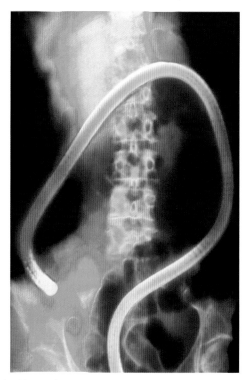

An X-ray of an endoscope inside a patient's colon.

▷ Lenses

On page 138 we saw how a curved mirror can change the direction of rays of light. The rays started out parallel, and they ended up *converging* to a point.
This is exactly how big optical and radio telescopes bring the signals from distant objects to a focus.

Another way to converge light rays is to use a **convex lens**:

The diagram shows some parallel rays passing through a thin glass lens. Each ray is refracted when it meets each surface, following Snell's Law (page 139).
But because the two surfaces are curved, the rays are converged.

Can you see that these *parallel* rays are converged so that they all pass through one point?
This point is called the ***principal focus* F** of the lens.

It lies on the ***principal axis***, the line passing through the centre of the lens and which is perpendicular to the lens.

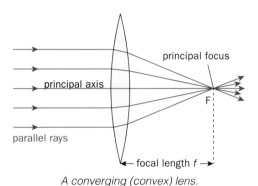
A converging (convex) lens.

Focal length and power

The **focal length *f*** of a lens is the distance from its middle to the principal focus F. The focal length depends on the curvature of the glass surface and on the type of glass that is used.

A more powerful lens is one that bends the light more, and therefore has a *shorter* focal length.
You can use this equation to calculate the power of a lens:

$$\text{Power} = \frac{1}{\text{focal length (m)}} \qquad \text{or} \qquad P = \frac{1}{f}$$

Power is measured in **dioptres (D)**; the focal length must be in metres.

The power *P* of a lens is a useful measure.
If you have several thin lenses in contact, the power of the combination is found by adding up the powers of all the lenses.

Example 6
The optician's prescription for a converging spectacle lens is marked +0.2 D.
What is the focal length of this lens?

$$P = \frac{1}{f}$$

$$\therefore \; 0.2\,\text{D} = \frac{1}{f} \qquad \therefore \; f = \frac{1}{0.2\,\text{D}} = \underline{5\text{ m}} \quad (1 \text{ s.f.})$$

Thick lenses are more powerful.

The lens shown above is a convex or converging lens.
This diagram shows a **concave** or **diverging** lens:

The diverging lens makes parallel light rays spread out or diverge.
They diverge as if they were coming from the one point.
This point is called a **virtual *principal focus***, because the rays do not pass through it but diverge as if they had come from it.

Spectacles or contact lenses that are used to correct short sight use diverging lenses (see page 152).

A diverging lens has a negative power, for example −0.2 D.

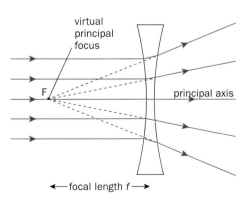
A diverging (concave) lens.

▷ Forming images with lenses

The rays of light from the lorry are converged
by the lens to form an image on the screen.
This image is:
- **inverted** (upside down),
- **real** (because rays of light go through it,
  and the image is shown on the screen),
- **smaller** in size than the object.

convex
lens

ima
rea
in

As the lorry moves nearer to the lens, the image moves away from
the lens and the screen has to be moved to get a clear image.

Drawing diagrams for convex lenses

We draw the lens as a solid line, with the edges of the lens shown
as a dotted outline:

To draw accurate ray diagrams you can use two constructions:

Construction 1: parallel rays of light are refracted through
the principal focus **F** (as in the diagram on page 145).

Construction 2: rays of light passing through the centre of the
lens travel straight on. (This is true for a thin lens because,
at the centre, its sides are parallel.)

Method:

1. Draw the principal axis, the lens and an arrow to represent the object **O**.
2. Mark the two principal foci **F** (equidistant on each side of the lens).
   It is also useful to mark points called '2F' at twice the distance.
3. Use the construction rays to show the paths of two rays from the top of the object.
4. Draw the image **I** between the axis and the point where your construction lines meet,
   to form the image of the top of the object.

The image depends on where the object is placed:

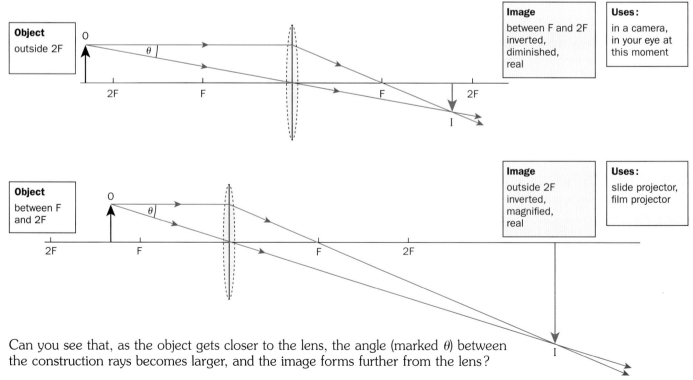

Object outside 2F	Image between F and 2F inverted, diminished, real	Uses: in a camera, in your eye at this moment

Object between F and 2F	Image outside 2F inverted, magnified, real	Uses: slide projector, film projector

Can you see that, as the object gets closer to the lens, the angle (marked θ) between
the construction rays becomes larger, and the image forms further from the lens?

A convex lens cannot form a real image if the object is too close to the lens.
For an object placed at the principal focus, the rays leaving the lens are parallel, as shown here:

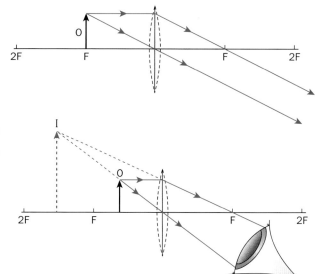

What happens if the object is placed even closer to the lens, inside the principal focus?
Look at the diagram:

Can you see that the rays leaving the lens continue to diverge? The top of the *virtual* image is found by tracing backwards, as shown by the dashed red lines.
This is where the rays of light *seem* to come from.

You can see that the image appears to be upright (right way up), and is bigger than the object. This is a **magnifying glass**.

The image formed by a diverging lens

When light leaves an object, it diverges. A concave lens makes the light diverge even more, so the rays will not meet.
The concave lens produces a *virtual image* behind the lens.

The ray diagram is similar to that of the convex lens, but construction ray 1 bends **outwards**, away from the principal focus, as shown here by the dashed red line:
The image is found where the two rays *seem* to come from.
The image appears to be upright, and is smaller than the object.

The lens equation

We could locate the actual position of an image formed by a lens by drawing a ray diagram carefully to scale.
Or, the lens equation gives us a different method:

$$\frac{1}{\text{object distance, } u} + \frac{1}{\text{image distance, } v} = \frac{1}{\text{focal length, } f} \qquad \text{or} \qquad \frac{1}{u} + \frac{1}{v} = \frac{1}{f}$$

This equation uses a system called *real is positive*, meaning:
- Distances to real objects and real images are positive.
- Distances to virtual images are negative.
- The focal length of a converging lens is positive, because it has a real principal focus; but that of a diverging lens is negative.

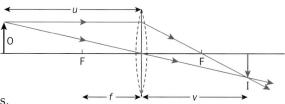

Object and image distances are measured from the centre of the lens.

Example 7
An object is placed 40 cm from a diverging lens of focal length 10 cm.
Find the position of the image.

The lens is diverging (concave), so it has a *virtual* focus. So we write $f = -10$ cm

$$\frac{1}{u} + \frac{1}{v} = \frac{1}{f} \qquad \therefore \frac{1}{40 \text{ cm}} + \frac{1}{v} = \frac{1}{-10 \text{cm}} \qquad \therefore 0.025 + \frac{1}{v} = -0.10$$

$$\therefore \frac{1}{v} = -0.10 - 0.025 = -0.125 \qquad \therefore v = \frac{1}{-0.125} = \underline{-8.0 \text{ cm}} \quad \text{(2 s.f.)}$$

The answer is negative, so the image is virtual and formed 8 cm from the lens, on the same side as the object.

▷ Magnification

Lenses can produce images that are *magnified* (larger than the object) or *diminished* (smaller than the object).
The ratio of the image size to object size is the **magnification**.

Look at the ray diagram:

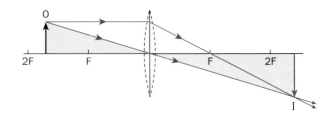

Can you see that the two shaded areas are similar triangles?
The ratio of the sizes is the same as the ratio of distances:

$$\textbf{Magnification} = \frac{\textbf{image size}}{\textbf{object size}} = \frac{\textbf{image distance, } \textit{v}}{\textbf{object distance, } \textit{u}}$$

Looking at distant objects

Look back at the ray diagrams on page 146 and imagine moving the object further and further away from the convex lens.
Do you agree that the angle, marked θ, will become ever smaller?

This means that rays of light from very distant objects are almost parallel by the time they reach us. When these 'parallel' rays of light pass through a convex lens, they meet to form a real image in the focal plane, as shown in this diagram:

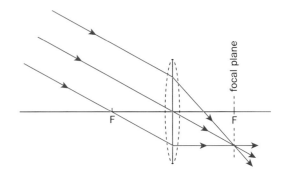

In a cheap, fixed-focus camera, the film (or electronic detector) is always placed at the focal plane of the convex lens.

Angular magnification

Phiz is looking at the full Moon:
The rays from the top and the bottom of the Moon as they enter his *eye* are shown. The angle between these rays is called 'the angle *subtended* at the eye by the Moon'.

When Phiz uses a telescope, the Moon looks larger through the telescope than it does with the unaided eye. This is because the angle *subtended* at his eye by the image in the telescope is greater than the angle subtended at the unaided eye.

But how much bigger does the Moon look?

We cannot use the simple magnification equation.
The telescope's image of the Moon is larger than the unaided eye's image, but it's not bigger than the actual Moon!

Angle subtended at the eye.

We use the *angular magnification* equation:

$$\textbf{Angular magnification} = \frac{\textbf{angle subtended at the eye by the image in the telescope}}{\textbf{angle subtended at the eye by the object with no telescope}}$$

Example 8
The Moon subtends an angle of 0.5° at the unaided eye.
Through a pair of binoculars, the angle between the top and bottom of the image is 4.5°.
Calculate the angular magnification of the binoculars.

$$\textbf{Angular magnification} = \frac{\textbf{angle subtended at the eye with binoculars}}{\textbf{angle subtended at the eye without binoculars}} = \frac{4.5°}{0.5°} = \underline{9\times}$$

▷ The astronomical refracting telescope

The telescope uses two convex lenses:
- a low-power objective lens to form a real image of the astronomical object,
- a high-power eyepiece lens to magnify this real image.

In **normal adjustment**, the eyepiece lens of the telescope is positioned to produce parallel light, and the lenses must be separated by a distance equal to: focal length of objective lens + focal length of eyepiece lens.

In the diagram below, parallel rays of light are entering the telescope from two distant stars, one red and one blue. Three rays are drawn from the red star. One ray is drawn from the blue star and this travels along the principal axis.

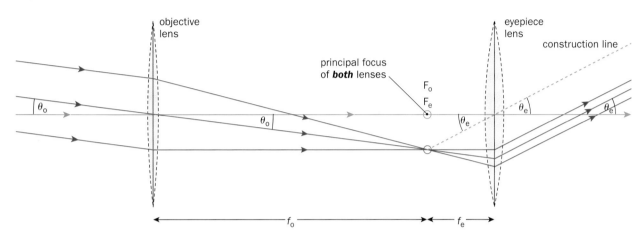

1. The **objective lens** produces a real image of the blue star at F_o. The image position is shown by the blue circle.
2. The objective lens produces a real image of the red star at the point shown by the red circle. The red rays enter at an angle and the red image is in the focal plane, offset from F_o (like the diagram on the facing page).
3. Do you agree that both real images are formed a distance f_o from the objective lens? (f_o is its focal length.)
4. These two real images act like objects for the **eyepiece lens**. Both images are a distance f_e from the eyepiece lens, and so the rays leaving the eyepiece lens will be parallel (as in the diagram at the top of page 147).
5. The blue ray continues along the principal axis. To find the direction of the red rays leaving the eyepiece lens, you need to draw a construction line, shown here in green, from the red image through the centre of the lens.
6. The green construction line is the path of a ray coming from the position of the red image. None of the red rays from the star actually follow this path, but it does show you the angle they all make with the principal axis. So now you can draw the three red rays leaving the eyepiece lens parallel to the green construction line.

Angular magnification of a telescope

Here is a simplified version of the diagram above:
The objective lens forms real images of the blue and red stars inside the telescope, and these images are a distance **x** apart.

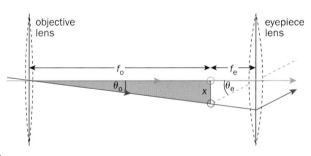

θ_o is the angle subtended at the eye without the telescope, and θ_e is the angle subtended at the eye with the telescope.
Has the telescope produced angular magnification?

Look at the two right-angled triangles, coloured pink and yellow.

Can you see that $\tan \theta_o = \dfrac{x}{f_o}$ and $\tan \theta_e = \dfrac{x}{f_e}$?

These angles are very small, and, for small angles, $\tan \theta \approx \theta$ (measured in radians; see page 475).

$$\text{Angular magnification} = \frac{\boldsymbol{\theta_e}}{\boldsymbol{\theta_o}} = \frac{\tan \theta_e}{\tan \theta_o} = \frac{x}{f_e} \div \frac{x}{f_o}$$

But $\quad \dfrac{x}{f_e} \div \dfrac{x}{f_o} = \dfrac{\cancel{x}}{f_e} \times \dfrac{f_o}{\cancel{x}} = \dfrac{f_o}{f_e} \quad$ so $\quad$ $\boxed{\textbf{angular magnification} = \dfrac{f_o}{f_e}}$

149

▷ Aberrations in lenses

Look at this diagram of parallel rays passing through a lens:
Can you see that rays near the edge of the lens do not
go through the principal focus?
If all the light from a distant object is not focussed at the
same point, the image formed will be blurred!

This defect is called **spherical aberration**.
To avoid spherical aberration in a refracting telescope, it is
important to use only the middle part of the lens.

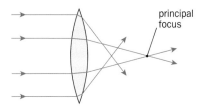

Spherical aberration in a convex lens.

Lenses have another problem.
Look at the diagram showing yellow light being focussed:
The yellow light is a mixture of red and green light.
Can you see that the red and green rays are refracted
differently?

This is called **chromatic aberration**. It results in coloured
blurs (instead of sharp points) when looking at stars.

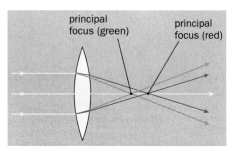

Chromatic aberration in a convex lens.

This problem can be cured by using an **achromatic doublet**.
The diagram shows how this double lens works:

The convex and concave lenses are made from different
types of glass, with different refractive indices. The convex
lens bends the green light inwards more than the red light.
The concave lens bends them both outwards, but by different
amounts. By the time the green and red rays leave the second
lens, they are travelling along the same path.
This means the yellow light is not split up.

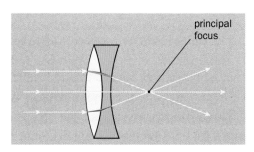

An achromatic doublet.

Lenses used in good-quality optical instruments are always
achromatic doublets. This is true for the lenses in telescopes,
as well as those in microscopes.
High-quality camera lenses are also achromatic combinations.

Reflecting (mirror) telescope

Early astronomers had to make their own lenses.
They found that the images were blurred, with coloured edges,
due to the aberrations described above.

Isaac Newton invented the **reflecting telescope** to overcome
these problems. He used a curved mirror as the objective
instead of a lens. The curved mirror produced a real image of
the distant object, and this was magnified by an eyepiece lens.

His telescope had a small mirror inside it to reflect the light into
an eyepiece on the side.

Almost all of the major telescopes used in astronomy research
are reflecting telescopes.

Most amateur astronomers start by buying a modern Newtonian
reflecting telescope.

(Reflecting telescopes are discussed in more detail on page 154.)

A copy of Newton's reflecting telescope.

▷ The eye

Accommodation of the eye

As light enters the curved cornea of your eye, it is refracted, and then refracted again as it goes into the liquid in front of the lens, then into the lens, and then into the liquid behind the lens.

Most of the bending takes place at the cornea: this is where there is the biggest change in refractive index.

The cornea is convex, so it makes light rays converge towards the sensitive fovea in the middle of the retina. The lens is optically denser than the fluids in the eye, and so it makes the rays from the cornea converge slightly more.

Your eye adjusts the power of the lens until there is a sharp image on the cornea – this action is called **accommodation**.

The lens is elastic, and in its natural state is fat, with greater power. The ciliary muscle is a ring of muscle round the eye, and the lens is attached to it by lots of ligaments.
When the ciliary muscle is relaxed, the lens is pulled thinner by the ligaments and becomes less powerful, so that the eye is focused on distant objects.
When the ciliary muscle is contracted, the ligaments become slack and the lens moves back into its natural fat shape.

What is the role of the iris?
The muscles in the iris contract in bright light, making the pupil smaller and so reducing the amount of light entering the eye. This also reduces spherical aberration (page 150), as only the central part of the lens is used.

Sensitivity of the eye

The retina has two types of sensory cells: rods and cones.

Cone cells detect the colours red, green and blue. All of the colours we see are mixtures of these three colours. Cone cells detect wavelengths from about 380 nm (violet) to 760 nm (red).

The **cone cells** are found near the middle of the retina, around the fovea. The fovea is the most sensitive part of the eye, as there are more sensory cells there. Because the cone cells are close together, the fovea has good spatial resolution. You can look at closely spaced objects and see that they are separate.

Rod cells do not detect colour, but they can detect much lower light intensities than cone cells. Rod cells are limited to detecting wavelengths in the range of 400 nm to 600 nm.

The **rod cells** are found in the region of the retina away from the fovea. As you move further from the fovea, the density of the sensory cells becomes lower, and so the spatial resolution becomes poorer. That is why you turn your head and eyes to look at things: you want to form the image on the fovea.

At low light levels, the cone cells do not respond. That is why everything looks grey outside at night. At night, the resolution of the eye is poorer. It is harder to 'place' things exactly. This is because the rod cells are thinly spread on the retina.

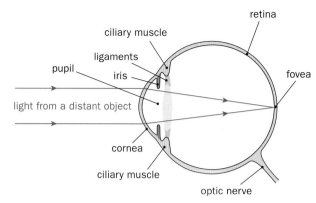

Ciliary muscles relax, so the lens is pulled thin.

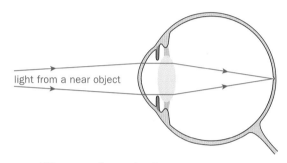

Ciliary muscles contract, so the lens relaxes and gets fatter.

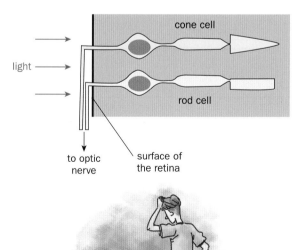

I'm sure I saw something move...

▷ Defects of vision

Myopia (short sight)

In some eyes, the cornea is too sharply curved, and
the rays of light converge too much.
Even with the lens at its thinnest, the light converges
to a point in front of the retina, except for very close
objects. Because the light does not converge on to
the retina, the image is blurred.

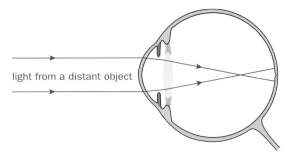

light from a distant object

Light converges in front of the retina.

This condition is called **myopia**, or short sight.
To a normal eye, stars look like sharp points of light. The
distance to the **far point** of the normal eye is infinity.
For a short-sighted person, the far point may be less than a
metre from the eye. Anything further away looks fuzzy.

Correcting short sight

A myopic eye has its *far point* too close to the eye;
it should be infinitely far away. This can be corrected using a
concave (diverging) spectacle lens. This compensates for
the excessive converging power of the cornea.

Look at the diagram:

It shows a diverging spectacle lens in front of the eye.
Light coming from infinity is made to diverge as though it
came from the far point of the short-sighted eye.
Compare this diagram with the diverging-lens diagram at the
bottom of page 145.

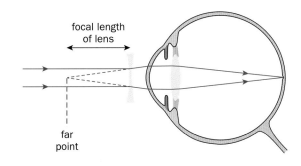

focal length
of lens

far
point

Can you see that a diverging lens makes parallel light diverge
as if it comes from the principal focus of the lens?

This shows that the lens needed to correct myopia is one which
has its principal focus at the far point of the myopic eye.

So, focal length of the diverging lens = distance to the far point

$$\text{Power of corrective lens} = \frac{1}{f} = \frac{1}{\text{distance to far point (m)}}$$

As this is a diverging lens, f is negative.

Hypermetropia (long sight)

If the cornea of your eye is not sharply curved enough,
the rays of light do not meet at all.
Even with the lens at its fattest, the light converges to a
point behind the retina, except for very distant objects.

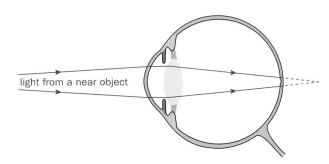

light from a near object

Light converges behind the retina.

This condition is called **hypermetropia**, or long sight.
You can read a book clearly when it is no closer to your eye
than its **near point**. For a normal eye, this is about 25 cm.
For a long-sighted eye, this can be much further away.

Older people often become long-sighted. The eye lens loses its
elasticity with age, and cannot change into its fattest shape
to focus the rays on to the retina.

152

Correcting long sight

A hypermetropic *eye* has its ***near point*** too far from the eye; it should be about 25 cm away. This can be corrected using a convex (converging) spectacle lens.
This compensates for the inadequate converging power of the cornea or the lens.

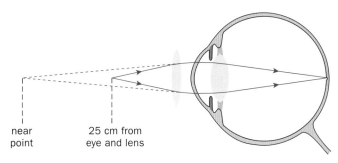

Look at the diagram:

It shows a converging spectacle lens in front of the *eye*.
Light from 25 cm away is made to converge by the spectacle lens, but it is still diverging.
The lens is forming a **virtual image**.
If you trace back the rays to find the position of this virtual image, you can see it is at the near point of this *eye*.

The lens equation can be used to find the focal length *f* and power *P* of the spectacle lens needed, but you must be careful, because the image is virtual.
This means that the image distance, which is the distance to the near point of this hypermetropic *eye*, is negative.

Example 9

A long-sighted eye has a near point at 75 cm from the eye.
What is the power of the correcting lens needed for this eye?
Using the lens equation, $u = 25$ cm, but the image is ***virtual***, so $v = -75$ cm.

$$\frac{1}{f} = \frac{1}{u} + \frac{1}{v} = \frac{1}{25 \text{ cm}} + \frac{1}{-75 \text{ cm}} = \frac{1}{25 \text{ cm}} - \frac{1}{75 \text{ cm}} = \frac{3}{75 \text{ cm}} - \frac{1}{75 \text{ cm}} = \frac{2}{75 \text{ cm}} = \frac{1}{37.5 \text{ cm}}$$

so $f = 37.5$ cm.

The power $P = \dfrac{1}{f}$ when f is in metres; 37.5 cm = 0.375 m $\therefore P = \dfrac{1}{0.375 \text{ m}} = 2.667 \text{ D (dioptres)} = \underline{2.7 \text{ D}}$ (2 s.f.)

Astigmatism

Short sight and long sight are both caused by the cornea having the wrong curvature to form a sharp image on the retina.
Sometimes the cornea has a different curvature in different directions, like a rugby ball. This is called **astigmatism**.

An astigmatic eye needs a **toric** lens to compensate for the different curvatures of the cornea. A toric lens is a lens with different curvatures in different directions.
It is like a slice off a torus (doughnut), as shown here.

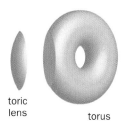

toric lens

torus

Can you see that the horizontal curvature of the toric lens is greater than the vertical curvature?

It has the same effect as adding a cylindrical lens (**B**) to to an ordinary spherical lens (**A**), as shown here:

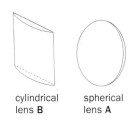

cylindrical lens **B**

spherical lens **A**

The prescription for this correcting lens has three parts:

sph +1.50   cyl +2.00   axis 90

The first two numbers are the powers in D (dioptres) of the spherical lens A and the cylindrical lens B.

The last number tells you the direction of the astigmatism. This is the angle in degrees that the cylinder axis makes with the horizontal.

Horizontal cylindrical lens.

▷ Physics at work: Reflecting telescopes

The reflecting telescope

A reflecting telescope uses a concave mirror to collect the light.
Look at the diagram:

It shows two parallel rays of light incident on a concave mirror.
Because this section is circular, the mirror is described as **spherical**.
It is like a section cut from a hollow ball.
The dashed green lines are normals to the mirror surface.
These normals meet at the centre of curvature of the mirror, **C**.

Look at the angles of incidence and reflection of the two light rays.
Can you see that they obey the laws of reflection (page 138)?
Each ray is reflected back through the principal focus **F**.
The mirror brings parallel light from a distant object to a focus,
and a real image of the distant object is formed at **F**.
The mirror is doing a similar job to the converging lens on page 149.

A spherical concave mirror like this one has one limitation.
Parallel rays must be *close* to the principal axis to pass through
the principal focus. Look at the second diagram:

Can you see that rays that meet the edges of
the spherical mirror do not go through the principal focus?

This defect is called **spherical aberration** (just as it is for lenses).

For this reason, spherical mirrors are not usually used in telescopes.
The mirrors used in telescopes are **parabolic** in section.

Look at the diagram of a parabolic mirror:

Can you see that *all* the parallel rays pass through the principal focus?

Parabolic mirrors are not just found in telescopes.
Satellite TV aerials and car headlamp mirrors are parabolic, too.

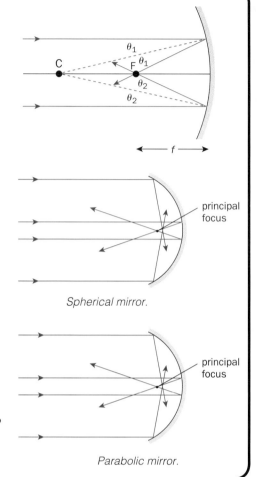

Spherical mirror.

Parabolic mirror.

The Cassegrain reflecting telescope

Just like a refracting lens telescope, the reflecting telescope
uses an eyepiece focussed on the principal focus of the mirror.
The eyepiece lens magnifies the real image formed by the objective.

Another mirror is used to reflect the rays of light out of the telescope.
One type is named after its inventor, the astronomer Cassegrain.

Look at the diagram of the Cassegrain reflecting telescope:

Can you see that the principal focus of the objective is at F_1?
The secondary mirror reflects the rays heading for F_1 and
sends them out through a small hole in the objective mirror.

The secondary mirror is a convex (diverging) mirror. A plane
mirror would be no use here as it would reflect the rays to a
focus point inside the telescope.
Using a diverging mirror makes the rays focus at F_2.

The eyepiece is a magnifying lens, just as in the refracting
telescope. Here it is magnifying the real image formed at F_2.

Often a digital camera is used instead of an eyepiece lens.
It is mounted so that the CCD detector of the camera is at F_2
and records a sharp image of the sky.
You will meet CCD detectors in Chapter 29.

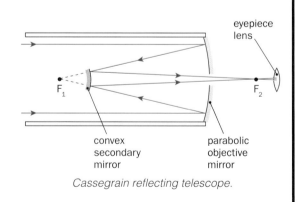

Cassegrain reflecting telescope.

▷ Physics at work: Reflecting telescopes

Refracting or reflecting telescope: which is better?

Advantages of a refracting (lens) telescope:
1. It is less likely that the various parts of the telescope will move out of alignment, so it is more robust.
2. There is no secondary mirror inside the telescope to block any of the light entering the instrument.

Advantages of a reflecting (mirror) telescope:
1. Mirrors do not have chromatic aberration.
2. An objective mirror has only one surface to polish to the correct curvature, while the objective lens in a refractor has two surfaces.
3. The glass in the objective lens in a refractor must be perfectly uniform and transparent, with no blemishes.
4. In a refractor, the lens can be held only at the edges. This limits the size of the objective in a refractor. In a reflector, the mirror can be firmly supported from behind, so it can be much larger.

The Keck telescopes in Hawaii have 10 m diameter mirrors!
There is no way a refracting telescope could be made with a lens this size.
The lens would sag under its own enormous weight!

The reflector gets my vote!

The James Webb Space Telescope (JWST)

The JWST is the successor to the famous Hubble telescope. It will allow scientists to look deeper into space, and so even further back in time. They hope to be able to see how the first stars and galaxies formed.

To achieve this, the instruments on the JWST will collect and analyse radiation in the infrared range of wavelengths, from 0.6 µm (orange light) to 28 µm (deep infrared).

The telescope must be kept at a very low temperature, around 40 K, so that infrared emissions from the telescope itself do not interfere with the observations from space.

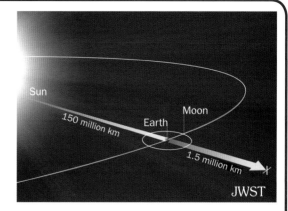

How is the telescope protected from the Earth's radiation? Imagine a straight line from the Sun to the Earth and extending beyond the Earth to a point which is 4 times further away from the Earth than the Moon. This is the position of the JWST. Doesn't this make maintenance visits to the telescope difficult?

Yes, but the position has been carefully chosen. The telescope orbits the Sun, and always maintains the same relative position with respect to the Sun and Earth. This means that the multi-layered radiation shield, which blocks radiation from the Sun, the Earth and the Moon, is *always* positioned between the Sun and the telescope.

How can the telescope orbit the Sun with the same orbital period as the Earth if it is further away from the Sun than the Earth is? Earth's gravity on the satellite provides the extra 'pull' needed to reduce the orbital time of the telescope.

The primary mirror has an area about 7 times larger than that of the Hubble telescope. It consists of 18 hexagonal mirrors, made of beryllium and coated with gold. The mirror segments and the shield unfold after the telescope is launched into position.

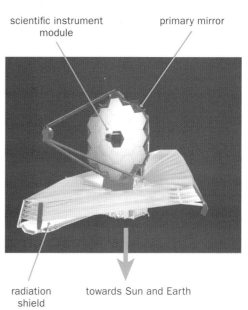

scientific instrument module

primary mirror

radiation shield

towards Sun and Earth

Summary

The laws of reflection: 1. The angle of incidence i = the angle of reflection r.
2. The incident ray, the reflected ray and the normal all lie in the same plane

Snell's Law for refraction: $\dfrac{\sin i}{\sin r} = n$, the refractive index

The refractive index of a medium is $n_{medium} = \dfrac{\textbf{speed of light in vacuum (or air)}}{\textbf{speed of light in the medium}}$

When light moves from medium 1 into medium 2,

$n_1 \sin \theta_1 = n_2 \sin \theta_2$, where θ_1 and θ_2 are the angles made with the normal in those mediums.

The critical angle c for a medium is related to the refractive index n by: $n = \dfrac{1}{\sin c}$

If the angle of incidence, in the optically more dense medium, is greater than the critical angle, then total internal reflection takes place.
Optical fibres carry light signals by total internal reflection.

Converging lenses can form real or virtual images, but diverging lenses form only virtual images.
Ray diagrams can predict the position and size of images in lenses.

The lens equation is: $\dfrac{1}{\text{objective distance}} + \dfrac{1}{\text{image distance}} = \dfrac{1}{\text{focal length}}$ or $\dfrac{1}{u} + \dfrac{1}{v} = \dfrac{1}{f}$
where negative values of v refer to virtual images and negative values of f refer to diverging lenses.

Astronomical telescopes can be made with two converging lenses, or with a concave mirror and a converging lens. In each case a converging lens, acting as a magnifying glass, is the eyepiece, and the other component is the objective.

The angular magnification of a telescope is $\dfrac{f_o}{f_e}$, where f_o is the focal length of the objective lens or mirror and f_e is the focal length of the eyepiece lens.

The eye has a light-sensitive retina containing rod and cone cells, a converging lens and a curved cornea.
Incorrect curvature of the cornea can result in short sight (myopia), long sight (hypermetropia) or astigmatism.

▷ Questions

1. Can you complete these sentences?
 a) When light is reflected, the angles of incidence and reflection are ....
 b) In refraction, the relationship between the angle of incidence and the angle of refraction is given by ....'s Law.
 c) The .... index of a substance is the factor by which the speed of light is slowed down in the substance, compared with the speed in a ....
 d) Light going from a dense medium to a less dense medium speeds up, and the rays bend .... from the normal.
 If the angle with the normal is more than the .... angle, the light cannot escape, and it is totally internally ....
 e) Lenses refract light to form images: a .... lens can form both real and virtual images, but a diverging lens forms only .... images.

 f) The .... length of a lens is the distance from the lens to the .... focus.
 g) The reciprocal of the focal length in metres gives you the .... of the lens, which is measured in .... For converging lenses, the power is positive, while for diverging lenses, the power is ....
 h) A refracting telescope has an objective .... while a reflecting telescope has an objective ....
 Each type also has a converging .... lens.
 i) It is the .... cells of the retina that detect colour. .... cells do not detect colour, but they detect lower light ....
 j) A short-sighted person cannot see .... objects clearly. Their spectacles contain .... lenses.

 A long-sighted person cannot see .... objects clearly. Their spectacles contain .... lenses.
 k) An astigmatic eye needs a .... lens. This is a lens with different .... in different ....

2. Copy and complete this table for refraction of light from air into water, refractive index 1.33.

i /°	30		70
r /°		40	

3. The speed of light in vacuum and two different glasses is given in this table.

Medium	Speed of light /m s^{-1}
vacuum	3.00×10^8
flint glass	1.86×10^8
crown glass	1.97×10^8

a) Calculate the refractive indexes of flint glass and crown glass.
b) Calculate the critical angle for light going from flint glass to crown glass.

4. A stepped-index optical fibre is made of two different glasses. The glass in the central core has a refractive index of 1.53, while that in the outer cladding has a refractive index of 1.49. Calculate the critical angle for light travelling along the core.

5. Copy and complete this table relating the powers of lenses to their focal lengths.

Type of lens	Focal length	Power
converging	0.5 m	
diverging	25 cm	
		+0.2 D
		−0.50 D

6. If you have thin lenses close together, you get the power of the combination by adding the powers of the different lenses together. A combination lens for a camera contains two converging lenses of focal lengths 20 cm and 40 cm, and a diverging lens of focal length 50 cm. Find the power and the focal length of the combination.

7. Use the lens equation to complete this table. A negative focal length means that the lens is a diverging lens.

u /cm	30	45		20	40
v /cm	30		60		
f /cm		15	20	50	−40

8. Use ray diagrams to find the position and size of the virtual image formed when a pound coin, 2.2 cm in diameter, is placed 20 cm from:
a) a diverging lens of focal length 40 cm,
b) a converging lens of focal length 40 cm.

9. The lens in a camera has a focal length that must be the same as the smallest possible distance between the lens and the film.
a) Draw a diagram to show why the focal length must be equal to this distance if the camera is to take a photograph of a very distant object.

The lens in this camera is 5.0 cm from the film when adjusted for a distant object, and 5.5 cm away from the film when adjusted for the closest possible object.
b) What is the focal length of the lens?
c) What is the image distance when the lens is at its furthest from the film?
d) Use the lens equation with the answers to parts (b) and (c) to find the distance to the closest possible object that the camera can photograph.

10. The refractive index of diamond is 2.42.
a) Calculate the speed of light in diamond.
b) Calculate the critical angle for diamond.
c) Diamonds in jewellery are cut to have many flat faces at angles to each other. Explain why they sparkle brilliantly when turned in a beam of light.
d) Diamond is a very dispersive medium, with its refractive index varying from 2.41 for red light to 2.45 for violet light. Describe how this will affect the appearance of a diamond ring in white light, and explain why diamonds are good gemstones.

11. An eyepiece lens of power 20 D can be used with two different telescopes.
Telescope A has an objective lens of power 0.5 D.
Telescope B has an objective mirror of focal length 150 cm.
Calculate the angular magnification produced by each telescope.

12. The optical prescription for a pair of spectacles is:

Eye	Sph	Cyl	Axis
right	−2.00	−1.75	117
left	−3.25	−0.50	57

a) Is this person myopic or hypermetropic?
b) Which is the weaker eye?
c) What do the figures for 'Cyl' and 'Axis' tell you about this person's eyes?

Further questions on page 181.

12 Interference and Diffraction

Isaac Newton did not believe that light could be a wave, but later a physicist called Thomas Young showed that light travels in waves. In fact, the interference and diffraction of light cannot be explained any other way.

In this chapter you will learn:
- what happens when two waves meet and 'superpose',
- how waves diffract through a gap,
- how waves from two or more sources can 'interfere'.

Thomas Young, who showed that waves can bend round corners and add up to give nothing!

The principle of superposition

What happens when 2 waves are *superposed* – are in the same place at the same time?
This principle tells you how to find the result:

> The resultant displacement at any point is found by adding the displacements of each separate wave.

This adding-together of waves is called **interference**.

Look at the diagram. It shows two identical waves (coloured red and blue) which are adding together to give a resultant:

There is no phase difference, so they are *in phase* (see page 128).

The resultant wave is shown in black:

At *point A*, each wave is in the middle of its oscillation, so there is no displacement, and the resultant is zero as well.

At *point B*, the red and blue waves are at their lowest points, at the bottom of a trough. So the resultant has a displacement downwards, of the red amplitude *plus* the blue amplitude.

Point C, like point A, has a zero resultant.
Point D is the same as point B, but the displacements are upwards.

These 2 waves have added together to give a bigger wave.
This is called **constructive interference**.

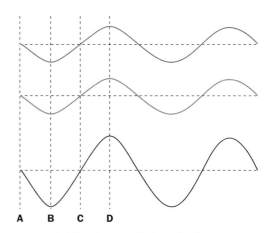

Adding 2 waves that are in phase.

Investigate this with a spreadsheet, page 483.

This second diagram shows 2 waves that are *in anti-phase:*

Look at points W, X, Y and Z.

At each one, the red wave has an equal and *opposite* displacement to the blue wave.
The resultant at every point is therefore zero, and so the waves always cancel each other everywhere, as shown by the black line:

These two waves have cancelled out.
If the 2 waves were sound waves, there would be silence.
If the 2 waves were light waves, there would be darkness.
This is called **destructive interference**.

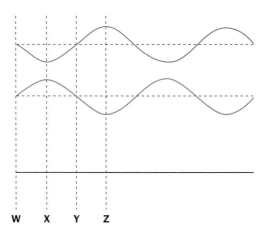

Adding 2 waves that are in anti-phase.

Interference with reflected waves

Have you noticed how the signal on a portable radio, tuned to an FM station, changes when you walk about near it?

This is due to interference – 2 sets of radio waves are getting to the radio, one directly from the transmitter and one reflected off you.

To see how radio waves interfere like this, you can use microwaves. A school microwave transmitter emits electromagnetic waves with a wavelength of about 3 cm (see page 130).

Metal sheets reflect microwaves, just as they reflect the waves inside a microwave oven. They act as a mirror.

Some microwave equipment and a vertical metal sheet are shown in the photograph:

The receiver will receive *two* waves: one directly from the transmitter, and one after it is reflected from the mirror.

By adjusting the position of the sheet, you can get a maximum signal at the detector.
If you move the metal sheet a few millimetres, the signal drops to a minimum.
If you move the sheet a bit more, the signal rises to a maximum again.

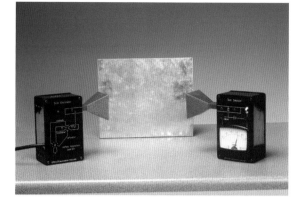

3 cm microwave apparatus.

Phase difference and path difference

Look at diagram 1, which shows the position of the reflector when the signal at the receiver B is a maximum:

It is a maximum because the wave coming directly from A and the wave reflected at X are adding together constructively at B.

Now look at diagram 2, where the reflector has been moved outwards from X. As it moved out, the signal became weaker, and then became a maximum again when the reflector was at Y.

It is a maximum because the waves at B are again adding together constructively. This means that the reflected waves must now be travelling an extra whole wavelength.
It means the **path difference** between path AXB and path AYB is one wavelength (1 λ).

Each extra wavelength in the path difference gives a **phase difference** of 360° (see page 128).

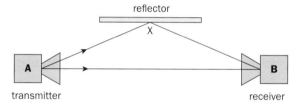

Diagram 1. The waves at B are in phase.

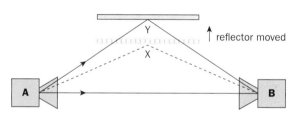

Diagram 2. The waves at B are in phase again.

Path difference of 1 λ = phase difference of 360° (2π rad)

A path difference of a whole number of wavelengths gives constructive interference (a maximum).
If the path difference has a extra half-wavelength it gives destructive interference (a minimum).

Example 1
In the experiment shown above, when the distance AXB is 72.0 cm, the signal at B is very strong (a maximum).
The metal sheet is moved away slowly.
The signal at B weakens and then becomes a maximum again when AYB is 75.1 cm.
Calculate the wavelength of the microwaves.

AYB is exactly one wavelength more than AXB.   ∴ Wavelength λ = 75.1 cm − 72.0 cm = 3.1 cm

▷ Stationary or standing waves

The waves that we have looked at so far are **progressive** waves.
They transfer energy away from the source.
So how can a wave be **stationary**?

Sometimes waves can be trapped in a space, such as the 'twang'
going up and down a guitar string.

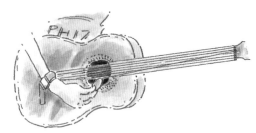

Trapping waves on a string.

This series of 5 snapshots explains how waves can be stationary.
They show a wave, coloured red, moving from left to right.
It meets an identical wave, coloured blue, moving in the **opposite**
direction at the same speed.

The 2 waves superpose to give a resultant wave, shown in black:
Vertical dashed lines in orange and green allow you to align
each snapshot.
Each dashed line is a quarter of a wavelength from its neighbour.

Each snapshot is a quarter of a period after the one before and
so the pattern repeats after 4 snapshots.
Can you see that snapshot 5 is a repeat of snapshot 1?

For snapshots **1**, **3** and **5**, the resultant wave is large.
This is because the red and blue waves are **in phase** and so
they superpose constructively.

For snapshots **2** and **4**, the resultant is zero everywhere.
The red and blue waves are always in **anti-phase**.
So they superpose destructively and cancel each other out.

Look at the waves where they cross the orange and green lines:
Where the resultant crosses the orange dashed lines, the value
is zero for all 5 snapshots. At these points the red and blue
combining waves always cancel.
These points are called **nodes**. There is no oscillation at a node.

Where it crosses the green lines, the resultant wave oscillates
with a large amplitude. It goes from zero to very large positive,
back to zero and to very large negative, to zero … and so on.
These points are the **anti-nodes**, points of maximum oscillation.

Look at any red or blue wave on these diagrams:
Can you see that the distance from a green line to the next green
line is always half a wavelength? And that the same is true for
the orange lines? This means that:

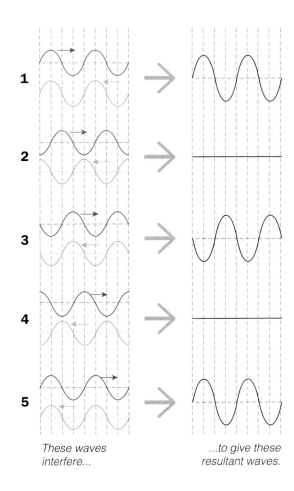

*These waves
interfere...*

*...to give these
resultant waves.*

Distance between nodes = distance between anti-nodes = $\frac{1}{2}\lambda$

We often combine snapshots like those above into one diagram.
Each snapshot, shown here by a different colour, is $\frac{1}{8}$ of a
period later than the previous one. So this diagram represents
one half of a full cycle of the stationary wave.
The letters **N** represent the nodes, where there is no resultant
oscillation. The letters **A** represent the anti-nodes.
These waves are called **stationary waves** or **standing waves**,
because the pattern does not move – it stays in one place.

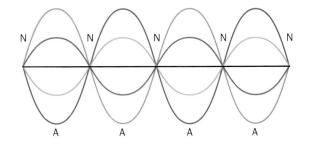

160

Experimenting with standing waves

To produce standing waves we need two identical waves travelling in opposite directions, just like the red and blue waves on the opposite page. How is this achieved?

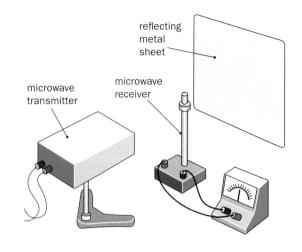

The usual way is to reflect a wave at a boundary. The diagram shows how this is done using microwaves: The transmitter produces microwaves that are reflected from the metal sheet. The reflected wave superposes with the incident wave from the transmitter, and a stationary wave is set up.

If you move the microwave receiver between the transmitter and the metal sheet, it passes through nodes and anti-nodes, and so the output reading on the meter varies.

What is the distance moved by the receiver between two nodes (ie. two points of minimum output)? This is half the wavelength of the microwaves – about 1.5 cm. Why is the meter reading not quite zero at a node?

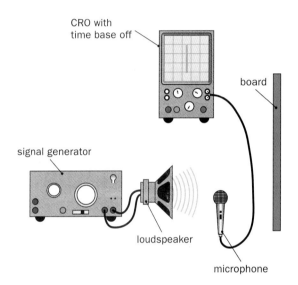

We can do a similar experiment with sound waves: The loudspeaker produces sound waves, which are reflected at the wooden board. As with the microwaves, the reflected wave superposes with the incident wave, and a standing wave is set up.

The microphone detects the sound signal and converts it to an electrical signal. This is displayed on the CRO as a vertical line. As the microphone is moved between the speaker and the board, the length of this line varies. Can you see that it will be a minimum length at a node and a maximum at an anti-node?

Example 2
a) A microphone is moved between the loudspeaker and wooden board, as shown in the diagram above. The distance between two points where the output from the microphone is a minimum is 11.0 cm. What value does this give for the wavelength of the sound waves?

Distance between nodes $= \frac{1}{2}\lambda = 11.0$ cm. So, the wavelength of the sound waves is 2×11.0 cm $= 22.0$ cm.

b) If the frequency of the sound waves is 1.5 kHz, what value does this give for the speed of sound in air? Using $c = f \times \lambda$ (page 125), $c = 1500$ Hz $\times 0.220$ m $= \underline{330 \text{ m s}^{-1}}$.

Comparing progressive and stationary waves

You could be asked to compare the two sorts of waves:

Progressive wave	Stationary wave
A progressive wave transfers energy.	A stationary wave stores energy.
All points on the wave vibrate with the same amplitude.	The amplitude of the wave varies from zero at a node to a maximum at an anti-node.
Every particle oscillates over the same path, but there is a phase lag between each particle and the one before it.	Between two nodes, all of the particles oscillate in phase; on either side of a node they are in anti-phase.

▷ Standing waves and resonance

You can trap waves on a string by attaching a vibration generator to a long cord, fastened firmly at the other end:

The vibrator sends waves along the cord. The waves reflect back from the fixed end, and superpose with the incident waves from the vibration generator, to form a standing wave.

The output of the signal generator can be varied so that the generator forces the cord to vibrate at different frequencies. At certain frequencies, the string vibrates with a large amplitude. These are the **resonant** frequencies of the system.

Resonance occurs when the frequency that is driving the system (from the vibration generator) matches a natural frequency of the system (see also page 110).

The photograph shows what you might see. At a node the cord does not move at all, and you can see the string quite distinctly. At the anti-nodes the string vibrates with a maximum amplitude, and so it appears as a blur.

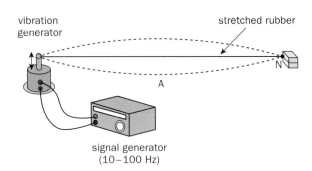

vibration generator

stretched rubber

A

N

signal generator
(10–100 Hz)

Stationary waves in string instruments

The guitar and violin are just two types of string instrument. When one of the strings is plucked or bowed in the middle, transverse waves travel along the string in opposite directions. At the ends of the string the waves are reflected back and a stationary wave, with nodes **N** and anti-nodes **A**, is set up.

The string can only vibrate with certain frequencies. These are the resonant frequencies of the string; we call them **harmonics**. Why are only certain frequencies allowed?

This is because you must have a whole number of stationary wave loops fitting into the length of the string.

The simplest way the string can vibrate is with just one loop. The wavelength is twice the length **L** of the string. Can the stationary wave have a longer wavelength than this? No, one loop produces a sound of the lowest possible frequency. This is the **fundamental frequency** or **first harmonic**, f_1.

In the second diagram, the string vibrates with two loops. The wavelength is now equal to the length of the string. This is half as long as it was for one loop. The frequency f_2 of this **second harmonic** must be twice the frequency of the fundamental frequency.

The third harmonic is at three times the fundamental frequency, because the wavelength is $\frac{1}{3}$ of its value for one loop. Can you predict the pattern for the rest of the harmonics?

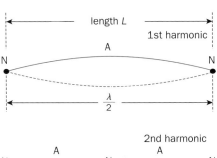

length L

1st harmonic

N   A   N

$\frac{\lambda}{2}$

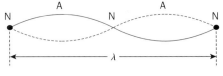

2nd harmonic

N   A   N   A   N

λ

3rd harmonic

N   A   N   A   N   A   N

λ

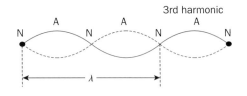

4th harmonic

N   A   N   A   N   A   N

λ

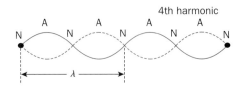

How could you increase the fundamental frequency of a string? You could press down on the string with your finger to shorten its length. Or you could tighten the string, to increase its tension.

Number of loops	1	2	3	4	5
Wavelength	$2L$	$\frac{2L}{2}$	$\frac{2L}{3}$	$\frac{2L}{4}$	$\frac{2L}{5}$
Frequency	f_1	$f_2 = 2 \times f_1$	$f_3 = 3 \times f_1$	$f_4 = 4 \times f_1$	$f_5 = 5 \times f_1$

The first harmonic of a stretched string

Have you noticed that guitar strings have different thicknesses?
Each string has a different *mass per unit length*, μ.

If you tighten any of the strings, increasing its *tension*, T,
why does the frequency of the note from the wire increase?

When you strum the guitar string, waves move in both directions
along the string and superpose to give a stationary wave.
Both the tension *and* the mass per unit length affect the *speed c*
of these waves on the string.

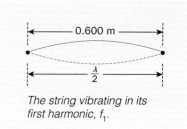

In fact:
$$\text{speed, } c \text{ (m s}^{-1}) = \sqrt{\frac{\text{tension, } T \text{ (N)}}{\text{mass per unit length, } \mu \text{ (kg m}^{-1})}}$$

or

$$c = \sqrt{\frac{T}{\mu}}$$

From the previous page we know that when the string vibrates in
its first harmonic mode, the wavelength λ of the stationary wave
is twice the length L of the string.

For any wave $c = f\lambda$ (page 125)
and so the first harmonic frequency f_1 is given by:

$$f_1 = \frac{c}{\lambda}$$

or

$$f_1 = \frac{1}{2L}\sqrt{\frac{T}{\mu}}$$

Example 3
A guitar string has a mass per unit length of 7.95×10^{-4} kg m^{-1}
The string is 0.600 m long and is under a tension of 78.5 N.
What value is the first harmonic frequency of the guitar string?

$$f_1 = \frac{c}{\lambda} = \frac{1}{2L}\sqrt{\frac{T}{\mu}} = \frac{1}{2 \times 0.600 \text{ m}}\sqrt{\frac{78.5 \text{ N}}{7.95 \times 10^{-4} \text{ kg m}^{-1}}} = \underline{262 \text{ Hz}} \text{ (3 s.f.)}$$

|← 0.600 m →|

*The string vibrating in its
first harmonic, f_1.*

Using a sonometer

The factors that affect the first harmonic frequency of a string
can be investigated using a *sonometer*, as shown:

One experiment is to see how the length L of the wire affects
the frequency of the note it produces. You can adjust the length
by altering the position of the moveable bridge.
Or you can investigate how the tension T in the wire affects
the frequency, by adjusting the weights.

How do you measure the frequency of the sound from the wire?
This is done by adjusting the wire until the sound it produces
matches that of the tuning fork.

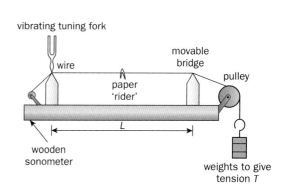

Unless you have a keen ear for music, you may find it difficult to
match the sounds! The paper 'rider' will help.

If the tuning fork is placed as shown, it forces the wire to vibrate.
When the paper rider vibrates vigorously, the wire is resonating.
So then the frequency of the wire equals the frequency of the fork.

▷ Stationary waves in wind instruments

When you blow into a wind instrument you create a sound wave which travels along the air column inside the pipe. The wave is **reflected** at each end of the air column. So there are 2 waves travelling to and fro. These superpose to form a **stationary longitudinal** wave inside the pipe.

As with string instruments, air columns can only vibrate with certain resonant frequencies: the **harmonics**.

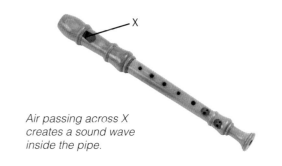

Air passing across X creates a sound wave inside the pipe.

Vibrating air columns can be closed or open.
A **closed pipe** is closed at one end and open at the other.
An **open pipe** is open at both ends.
A node **N** is always formed at a closed end because the air cannot vibrate freely. An open end is always an anti-node **A**.
Here, the air in a closed pipe is vibrating in its simplest way:

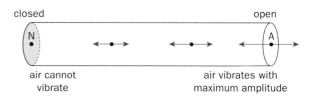

To make it easier to see the stationary waves, we usually draw displacement against distance graphs inside the pipes, so that the diagrams appear to show transverse waves.

Look at these diagrams of the **fundamental frequency** or **first harmonic f_1** in the two pipes. Do you agree that these are the longest waves that can fit into these air columns?

There is just one quarter of a full wave inside the **closed** pipe. So the wavelength of the fundamental frequency is four times as long as the length **L** of the pipe.

There are two quarters or half a wavelength in the **open** pipe, and now the wavelength is two times the length **L**.
The fundamental frequency or first harmonic of an open pipe is twice that of a closed pipe of the same length.

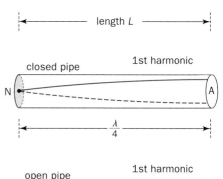

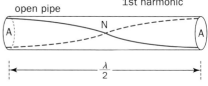

1. The harmonic frequencies of a closed pipe

The diagrams show the next two harmonics for the closed pipe. Can you see that it is possible to get only **odd numbers** of quarter-wavelengths to fit into the pipe?

With three of the quarter-wavelengths fitted into the pipe, the wavelength is $\frac{1}{3}$ of its value for one quarter, and the frequency must be three times the fundamental frequency.

For the next harmonic, with five quarter-wavelengths, the frequency is five times the fundamental frequency.
Do you agree that the closed pipe can only vibrate with the odd-numbered harmonics, as shown in the table below?

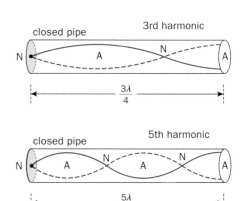

Number of $\frac{1}{4}$ waves	1	3	5	7
Wavelength	$4L$	$\dfrac{4L}{3}$	$\dfrac{4L}{5}$	$\dfrac{4L}{7}$
Frequency	f_1	$f_3 = 3 \times f_1$	$f_5 = 5 \times f_1$	$f_7 = 7 \times f_1$

2. The harmonic frequencies of an open pipe

On the opposite page we saw the diagram for the *first* harmonic vibration of an open pipe. Here are the next two harmonics: Remember, the open ends have to be anti-nodes. Can you see that we now have to fit in *even numbers* of quarter-wavelengths?

With four of the quarter-wavelengths fitted into the pipe, the wavelength is $\frac{1}{2}$ of its value for two quarters, and so the frequency must be twice the fundamental frequency.

Do you agree that the open pipe can vibrate with both odd- and even-numbered harmonics?
The first four are shown in the table below:

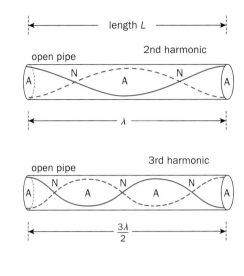

Number of $\frac{1}{4}$ waves	2	4	6	8
Wavelength	$2L$	$\dfrac{2L}{2}$	$\dfrac{2L}{3}$	$\dfrac{2L}{4}$
Frequency	f_1	$f_2 = 2 \times f_1$	$f_3 = 3 \times f_1$	$f_4 = 4 \times f_1$

Measuring the speed of sound in air using a resonance tube

This apparatus is a *closed pipe* of adjustable length:
Beginning with a short air column, strike the fork and hold it over the top of the tube. Sound waves travel down the pipe at the speed of sound in air, c, and are reflected at the water.

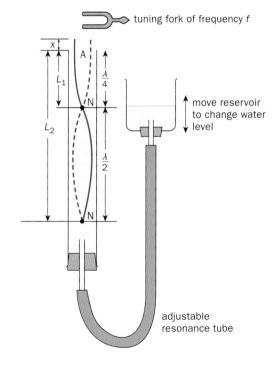

Gradually lengthen the air column until it resonates; you hear a louder sound. Then the natural frequency f of the pipe is the same as the frequency of the fork. Record the length L_1.

A stationary wave, with one quarter of the wavelength λ, has been set up inside the closed pipe.
(In fact the position of the anti-node **A** is just outside the end of the pipe by a distance x, which is called the *end correction*.)

Lengthen the air column again until you find a second point of resonance, at length L_2. The air column is now long enough to fit three of the quarter-wavelengths of the stationary wave.

The distance from the first to the second resonance must be two quarter-wavelengths. Do you agree that $\lambda = 2(L_2 - L_1)$? (This method corrects for the distance x at the end of the pipe.)

Now you know the wavelength λ and the frequency f, you can calculate the velocity of sound, c, using the equation $c = f\lambda$.

Example 4
In the experiment above, loud sounds are heard when the air column is 0.136 m and then 0.468 m long. The tuning fork has a frequency 512 Hz. What value does this give for the speed of sound in air?

The wavelength λ: $\lambda = 2(L_2 - L_1) = 2 \times (0.468\ \text{m} - 0.136\ \text{m}) = 0.664\ \text{m}$

Using: $c = f\lambda$, $c = 512\ \text{Hz} \times 0.664\ \text{m} = \underline{340\ \text{m s}^{-1}}$

▷ Physics at work: Standing waves

Sounding brass

Stationary waves on a string or in a pipe give patterns of frequencies that are mathematically very simple: if the lowest resonant frequency is 100 Hz, then, as you increase the frequency, you get resonance at 200 Hz, 300 Hz, 400 Hz, 500 Hz and so on.

Play these signals on a loudspeaker to a musician, and he or she will recognise them as part of a musical scale – the harmonic series. In fact, these are the very notes that could be played on early trumpets like the one in the photograph:

The lowest notes – 100 Hz, 200 Hz and 300 Hz – are rather widely spaced, and it is hard to play tunes with so many frequencies missing. As you go higher, the notes sound closer together, allowing the trumpeter to play tunes.

The high notes of the natural trumpets used in early music are so difficult to play that only a few very talented players like Crispian Steele-Perkins (above) use them.

In a modern trumpet, valves add extra lengths of tubing when you press keys. If you have a greater length of vibrating air, a new set of lower frequencies is available. These fill in the gaps between the easier low notes, so tunes can be played in the easier low register.

So trumpet playing has now become easy? Not a bit of it! Modern virtuosi like Wynton Marsalis (right) are so skilled that composers just write music that is even more difficult.

The quality of musical sounds

If you listen to the same musical note played on a piano and on a violin, can you distinguish between them? When you strike a tuning fork, it produces a pure note of just one frequency, but other musical instruments produce sounds which are a mixture of the first harmonic frequency f_1 and the other harmonics, $2f_1$, $3f_1$, etc. For example, if you strike the middle A key on a piano, you will hear the first harmonic of 220 Hz, but also 440 Hz and 660 Hz, etc. at the same time. These frequencies add together to give a complicated waveform.

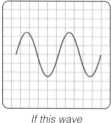   +   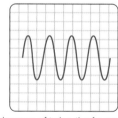   =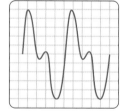

If this wave   *is added to*   *this wave of twice the frequency*   *the result is this waveform*
(the first harmonic)   *(the second harmonic)*

Each instrument produces harmonics at different amplitudes. This produces waveforms of different complexity, to give each instrument its characteristic quality.

The harmonics that are present in a musical note can be identified using **Fourier analysis**:

Graph of displacement against time shows the waveform.

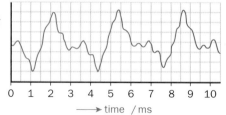

Fourier transform →

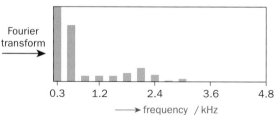

Analysis shows the amplitude of all the harmonic frequencies present in the wave.

▷ Physics at work: Hertz discovers radio waves

In 1862, James Clerk Maxwell predicted that light was part of a wide spectrum of electromagnetic radiation.
It took some time for his ideas to be proved experimentally.

In 1887, Heinrich Hertz generated an electromagnetic wave, sending it from one side of his laboratory to the other.
He had discovered *radio waves*.

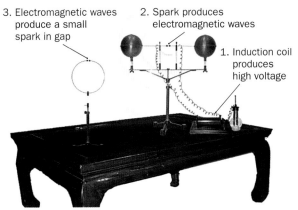

3. Electromagnetic waves produce a small spark in gap

2. Spark produces electromagnetic waves

1. Induction coil produces high voltage

Hertz's apparatus, now in a museum.

Hertz's apparatus is shown here schematically:
An induction coil produces a high voltage which switches on and off rapidly. The high voltage produces an electric field between the two small spheres S_1 and S_2 (see page 292).

The electric field is strong enough to make the air conduct, and sparks oscillate across the gap between the spheres.
The large spheres act as a capacitor (see page 309) and Hertz was able to use them to control the frequency of his sparks.

The sparks set up oscillating electric and magnetic fields, so an electromagnetic wave travels out across the laboratory.
The wave has the same frequency as the sparks, and Hertz showed that this frequency was about 5.0×10^7 Hz.

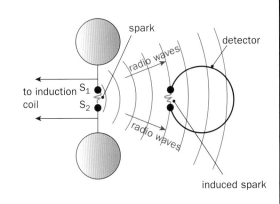

spark

detector

radio waves

to induction coil S_1 S_2

radio waves

induced spark

How did Hertz detect the waves?
The diagram also shows his detector. It is a simple loop of wire with another tiny gap. The high-frequency electromagnetic oscillations induce a voltage in the loop (see page 270).
This causes small sparks to jump between the ends of the loop.

The properties of Hertz's waves

To his surprise, Hertz discovered that, when he took his detector to a different room from the transmitter, he still got sparks.
His waves could travel through insulating materials!
He then went on to show that the waves could be reflected, refracted and diffracted, just like light.

Hertz then found that he could get a stronger beam of waves by placing the spark gap at the focal point of a concave reflector.
With his improved apparatus he was able to set up stationary radio waves and measure their speed.
His value was very close to the speed of light!

The diagram shows the improved transmitter placed several metres away from a large metal sheet which reflected the waves.
As Hertz moved his detector between the transmitter and the reflector the strength of the sparks in the detector varied.
Hertz was moving his detector between nodes and anti-nodes.
(See the experiment with microwaves, described on page 161.)

Hertz found that the distance between nodes was about 3.0 m.
Do you agree that this gives 6.0 m for the wavelength?
Using the wave equation $c = f\lambda$ and the frequency above:

$c = 5.0 \times 10^7$ Hz $\times 6.0$ m $= 3.0 \times 10^8$ m s^{-1}

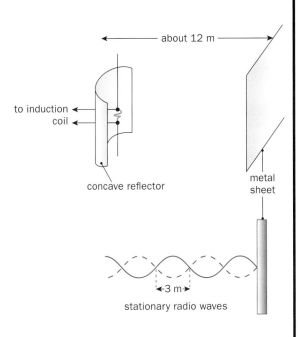

about 12 m

to induction coil

concave reflector

metal sheet

3 m

stationary radio waves

▷ Diffraction

Look at this photograph of water waves entering a harbour:

The wavefronts out at sea are straight, but when they pass through the gap, they curve round into regions you would not expect them to reach.

This bending of waves when they pass through a gap, or curve round edges, is called **diffraction**.

Diffraction and wavelength

How much do waves diffract when they go through a gap? It depends on the *wavelength* of the waves and the *size of the gap* they are passing through.

Look at this first diagram of straight waves in a ripple tank. The waves are passing through a gap between two barriers:

wavelength
λ

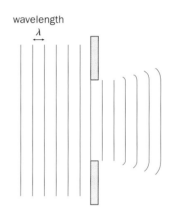

Can you see that the gap is much bigger than the wavelength? When waves pass through a gap much bigger than the wavelength, the wavefronts bend just a little at the edges. There is not much diffraction. There is a clear shadow behind the barrier.

Now look at the second diagram:
Can you see that the gap is about the same size as the wavelength? This time, the wavefronts bend round in circular arcs.

> **When waves pass through a gap that is similar in size to their wavelength, there is a lot of diffraction.**

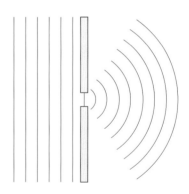

For light waves: the wavelength of light is much, much smaller than the size of a window. That's why sunlight makes sharp shadows in the room, and does not bend round the corners.

Light is seen to be diffracted only if it passes through a *very narrow gap* (see the opposite page). This shows us that light is a wave motion with a very small wavelength.

For sound waves: the wavelength of sound is similar in size to a doorway. Sound waves passing through a doorway are diffracted as in the second diagram. That's why someone outside the room can hear you talking, even when they can't see you!

Example 5
Radio 4 transmits on FM at a frequency of 92.5 MHz.
Calculate the wavelength, and use it to explain why listeners in deep valleys cannot pick up Radio 4 FM.
(Speed of radio waves $c = 3.00 \times 10^8$ m s^{-1})

$$c = f\lambda$$

$$\therefore 3.00 \times 10^8 \text{ m s}^{-1} = 92.5 \times 10^6 \text{ Hz} \times \lambda \qquad \therefore \lambda = \frac{3.00 \times 10^8 \text{ m s}^{-1}}{92.5 \times 10^6 \text{ Hz}} = \underline{3.24 \text{ m}}$$

Because the FM wavelength is small compared with the size of mountains and valleys, it does not diffract much, so it does not bend down into the valleys.

Long wave radio may have a wavelength of 1500 m, so it does diffract round hills and into valleys.

Measuring diffraction

Light has a very small wavelength, so to see its diffraction you need a very narrow slit. Here is what you see on a screen when violet and then red light are passed through a narrow vertical slit:

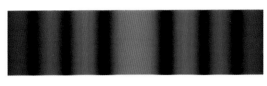

Can you see a wider band of light in the centre of both patterns? This is called the **central maximum**.

There are also **subsidiary maxima**. These are the narrow bright bands that get fainter and fainter as you go out from the centre. These maxima occur where light from some parts of the slit superposes constructively with light from other parts of the slit.

Can you also see the dark bands where there is no light? Each of these is a *minimum*, formed where light from different parts of the slit superposes destructively.

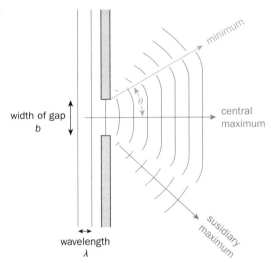

If you use a ripple tank to investigate the diffraction through a gap you will see a pattern like this:

The blue arrow lies along a region with no waves, between the central maximum and the first subsidiary maximum.

If we want to describe the amount of diffraction, we measure the angle between the central maximum and this first minimum. It is labelled θ (theta) on the diagram.

Calculating diffraction angles

Look at any of the diffraction images above.

If you plot *intensity* against *angle* from the centre, you get a graph like this:

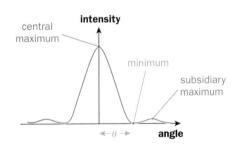

It shows the intensity is greatest in the centre of the pattern. θ is the angle of the first minimum.

You can calculate θ, the angle of the first minimum, by:

Sine of the angle of the first minimum $=$ $\dfrac{\text{wavelength of the waves, } \lambda}{\text{width of the gap, } b}$

or

$$\sin \theta = \frac{\lambda}{b}$$

It does not matter what units you use for the wavelength and for the gap width, but they must be the same units.

Look again at the top of the page.
Why is the red pattern wider than the violet pattern?
Red light has a longer wavelength than violet light, and so the angle θ is greater for red than for violet light.

Example 6
Microwaves of wavelength 3.0 cm pass through a gap of width 5.0 cm.
What is the angle of the first minimum of the diffraction pattern?

$$\sin \theta = \frac{\lambda}{b} = \frac{3.0 \text{ cm}}{5.0 \text{ cm}} = 0.60$$

By using the 'inverse sine' function on the calculator, you will find that $\theta = \sin^{-1}(0.60) = \underline{37°}$

169

▷ Resolution

Imagine using a telescope to look at two stars that are very close together.
Ideally, you want to separate or **resolve** the two stars into separate images.

But diffraction can make waves from different objects overlap and blur.
In a telescope, the gap that causes the waves to diffract is the objective lens.
This a hole, not a slit, and the maxima and minima of the diffraction pattern
become circles instead of bands (see page 169). The images of the stars
become little discs surrounded by the 'haloes' of the subsidiary maxima.

The diagrams on the right show two red stars seen in this way:

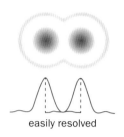

easily resolved

In the first diagram, the images of the two stars can be separated.
Although the diffraction patterns overlap, it is easy to see that there are
two different images. We say that the images **can be resolved**.

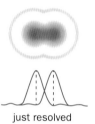

The second diagram shows the images of two stars closer together.
It is **just** possible here to make out that there are two images very close
together. The images here are **just resolved**.

just resolved

The third diagram has the two images even closer together.
The two images blur into one here, and you cannot tell that
there are two different stars. You would think it was one bright star.
The images here are **not resolved**.

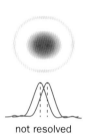

What test can we apply to see if two images will be resolved?

A simple practical test is called **Rayleigh's criterion**.
Look at the graphs underneath the images of the two stars:

not resolved

The critical case is shown in the middle diagram, where the stars are **just**
resolved. It shows that **the centre of one image overlaps the first
minimum of the second image**.

Rayleigh's criterion uses the diffraction equation on page 169 to give
the smallest angular separation of two sources that **can** be resolved:

Sine of the angular separation of two sources $\geq \dfrac{\text{wavelength of the waves, } \lambda}{\text{width of the gap, } b}$ or	$\sin \theta \geq \dfrac{\lambda}{b}$

Example 7
A car's headlights are 1.5 m apart and give out light of average wavelength 500 nm.
You are 5.0 km away, and the diameter of the pupil of your eye is 3.0 mm.
Would you see two separate headlights, or just a patch of light?

When the angles are small like this, we can say:
$\tan \theta \approx \sin \theta \approx \theta$ in radians.

This small-angle approximation lets us write:

$$\sin \theta \approx \theta \approx \frac{1.5 \text{ m}}{5.0 \text{ km}} = \frac{1.5 \text{ m}}{5000 \text{ m}} = 0.000\,30 \text{ radians}$$

Also, $\dfrac{\lambda}{b} = \dfrac{500 \text{ nm}}{3.0 \text{ mm}} = \dfrac{500 \times 10^{-9} \text{ m}}{3.0 \times 10^{-3} \text{ m}} = 0.000\,17$

Since $\sin \theta$ is bigger than $\dfrac{\lambda}{b}$, the sources **will** be resolved by your eye.

You should see two separate headlights (assuming you are not short-sighted).

5.0 km

1.5 m

▷ Physics at work: Diffraction

Radio Astronomy: the Big Dish

The famous 76 m diameter radio telescope at Jodrell Bank, near Manchester, was designed by Sir Bernard Lovell. It is used to investigate radio signals from space. These radio signals give information that you cannot get from optical telescopes.

The Lovell telescope was the first of the huge reflectors built for radio astronomy, and it is still one of the biggest.

The wavelength of these radio signals is about 0.21 m.
A large telescope is needed to resolve radio sources, because the wavelength is large. It is the ratio of wavelength λ to gap width b that determines resolution (see opposite page).

The gap width in this case is the diameter of the reflecting dish. Using Rayleigh's criterion, we can calculate the smallest angle θ that the Lovell telescope can resolve:

The Lovell telescope at Jodrell Bank.

$$\sin \theta = \frac{\lambda}{b} = \frac{0.21 \text{ m}}{76 \text{ m}} = 2.8 \times 10^{-3} \quad \text{so} \quad \theta = 0.16°$$

The Lovell telescope is movable and can be pointed at any part of the sky. But it is a huge dish, and it is not feasible to make a movable dish very much bigger.

One ingenious solution was found in Puerto Rico in the Caribbean: a valley of roughly the right shape was turned into a huge radio telescope 305 m across, four times as big as the Lovell telescope.

Because it is four times bigger, it can resolve an angle that is four times smaller, and because it has 16 times the area, it can detect signals that are 16 times weaker.

The Arecibo telescope in Puerto Rico.

Satellite television

Geostationary communications satellites, which rotate with the Earth and so appear to stay in the same part of the sky, are widely used in communications (see page 90).

They were first suggested by science fiction writer Arthur C Clarke back in 1956, but even he could not have guessed how much use those satellites would get, with telephone, the internet and, above all, television channels.

But how does a satellite send television signals to all of western Europe, for example?

The signal is sent, as a microwave beam, from a dish on Earth to a geostationary satellite over the equator, to the south of Europe.

Because of diffraction at its edges, the transmitting dish must be quite large, to make sure that enough power reaches the satellite.

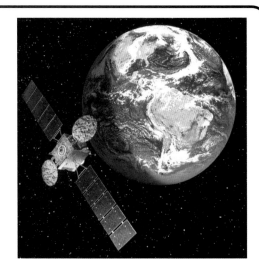

When the satellite re-transmits the TV signal, diffraction is no longer a nuisance, but a help.
The ratio of wavelength to dish size is carefully calculated to ensure that the beam spreads just enough.
Too much spread, and the signal becomes too weak to detect with the satellite dish on your house.
Too little spread, and the beam will not cover the target area (the whole of western Europe).

▷ Two-source interference

We know that when two waves meet they can superpose constructively or destructively.

You can use a ripple tank to show this effect:

Two dippers oscillate up and down together, making two sets of circular waves that overlap. The photograph shows the **interference pattern** that is produced in the water.

Can you see the lines of constructive and destructive interference that are radiating out from the two dippers?

How do we explain this pattern?
The water is not vibrating along the *dark lines* of destructive interference. A crest from one dipper meets with a trough from the other. The two waves cancel each other out.

The water does vibrate along the *bright lines* of constructive interference. A crest from one dipper meets with a crest from the other.
The two waves superpose constructively.

This diagram shows the two sets of overlapping waves:
They are drawn as blue from one dipper and red from the other. The continuous lines are the wave crests and the dashed lines are wave troughs. C is the midpoint of the two dippers.

Look carefully at the points in the diagram marked ✚:
Can you see that these are the places where a crest meets a crest or a trough meets a trough, and constructive interference occurs?

Now look at the points on the diagram marked O:
At all these places, a crest meets a trough and the waves cancel.

Look at the points along the line **CW**:
All of these points are an equal distance from each dipper. We say the **path difference** between the blue and the red waves reaching any of these points is *zero*. So the waves arrive *in phase* to give interference maxima.

Now look at the points along the two lines **CX**.
The path difference between the two waves reaching any of these points is *half* a wavelength, and so the waves arrive in *anti-phase*. The lines **CX** are lines of interference minima.

What about each of the points along the lines **CY**?
Now the path difference is exactly one wavelength, and the waves arrive in phase to give two more lines of constructive interference.

Is the path difference along the lines **CZ** exactly $1\frac{1}{2}$ wavelengths?

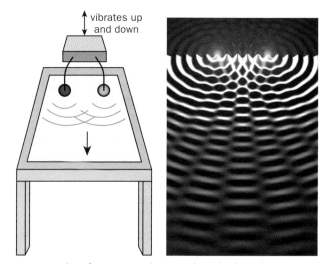

Interference can be seen in a ripple tank.

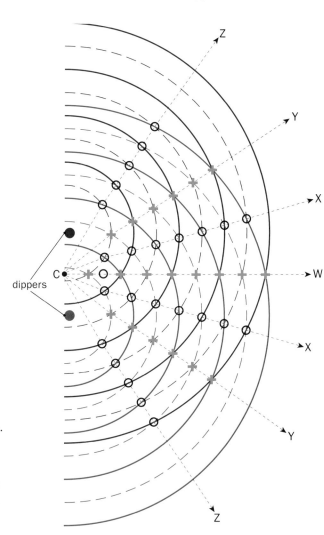

For **interference maxima** the path difference must be whole numbers of wavelengths:

> Path difference = $n\lambda$, where n = 0, 1, 2, 3, ...

For **interference minima** the path difference must be odd numbers of half-wavelengths:

> Path difference = $\left(n + \frac{1}{2}\lambda\right)$, where n = 0, 1, 2, 3, ...

▷ Two-source interference of light

Imagine that our water dippers can be operated independently.
When they vibrate up and down together they produce the
pattern shown on the previous page. What would happen if,
suddenly, one dipper was moving up as the other moved down?
The lines of interference maxima and minima would reverse!
If this happened all the time, we wouldn't see a steady pattern.

To see a **steady** interference pattern we need coherent sources
of waves. **Coherent sources** have the same frequency, and are
always in phase **or** keep the same phase difference all the time.

We can achieve coherent sources of light using a laser.
A laser is monochromatic (the light is of one colour) and
it produces a parallel beam of coherent light.
The light is split into two sources using a double slit as shown:

The light waves diffract at the slits and so overlap to produce
interference maxima (bright light) or minima (darkness).
Can you see that, as the overlapping waves fall on a screen,
you will see a series of stripes or **interference fringes**?
These fringes are parallel to the slits.

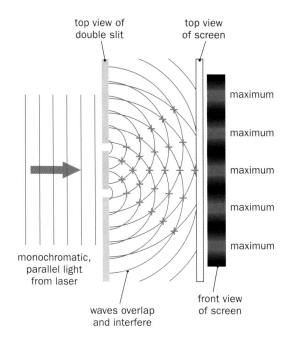

Young's equation

In 1801, Thomas Young showed that light behaved as a wave.
He used a double slit, but without a laser (see next page).

The diagram represents the geometry of his experiment.
Remember the fringes are parallel to the slits:

The spacing between one minimum (dark) and the next is a
constant value w (see the diagram). This is the same as the
distance between the centre of one maximum (bright) and the next.

The diagram is not to scale. The distance s between the two slits
might be $\frac{1}{4}$ millimetre. The distance D might be a metre.

The wavelength of light is so small that the spacing s would be
hundreds of wavelengths, while the distance D would be millions
of wavelengths.

Provided s is very much smaller than D, you can use this equation for
the spacing of maxima or minima for Young's slits:

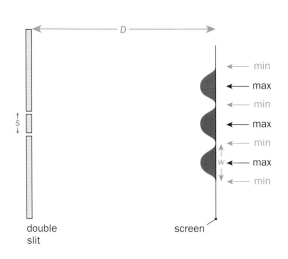

$$\frac{\text{Wavelength of light, } \lambda}{\text{Separation of two slits, } s} = \frac{\text{separation of maxima (or minima), } w}{\text{distance between slits and screen, } D} \quad \text{or} \quad \frac{\lambda}{s} = \frac{w}{D}$$

Example 8
The distance between the slits in a Young's slits experiment is 0.20 mm.
The screen is 84 cm from the slits, and the maxima in the interference pattern are 2.4 mm apart.
What is the wavelength of the light used?

First, convert all lengths to metres: $s = 0.20 \times 10^{-3}$ m, $D = 0.84$ m and $w = 2.4 \times 10^{-3}$ m

$$\frac{\lambda}{s} = \frac{w}{D} \qquad \therefore \quad \frac{\lambda}{0.20 \times 10^{-3} \text{ m}} = \frac{2.4 \times 10^{-3} \text{ m}}{0.84 \text{ m}}$$

$$\therefore \quad \lambda = \frac{0.20 \times 10^{-3} \text{ m} \times 2.4 \times 10^{-3} \text{ m}}{0.84 \text{ m}} = \underline{5.7 \times 10^{-7} \text{ m}} \text{ (2 s.f.)} \quad (= 570 \text{ nm})$$

▷ Young's double-slit experiment

Observing interference fringes using a laser is easy, because the laser produces intense, coherent and parallel light. But how did Young demonstrate the interference of light?

In most light sources, each point in the source is emitting light in very short bursts that last for only about 10^{-9} s. So, to obtain coherent sources, Young first passed the light through a single slit, where the light is diffracted as shown:

He then passed the diffracted light through the double slit. The two light waves coming from the double slit are coherent because they both began together at the first slit.

Can you see the problems with this apparatus? Only a very small amount of light passes through the system. This limits the maximum distance between the double slit and the screen to about 1 m, so the fringes are faint and narrow, and may need to be viewed with a magnifying glass.

The photos here show what you would see if you passed green light and then white light through Young's apparatus: What do you notice about the fringe patterns?

The green fringes are all of equal spacing. White light gives coloured fringes, with a central white band. Can we explain this?

From the previous page we know that the fringe spacing w is given by the equation: $w = \dfrac{\lambda D}{s}$

Red light has a longer wavelength than violet light, and so the fringe spacing is greater for red light than for violet light. Each colour produces its own set of fringes, and these overlap.

The central fringe is white because here the path difference is zero. All the colours produce a bright fringe at this position.

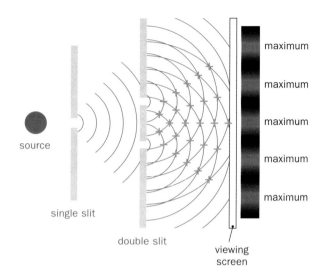

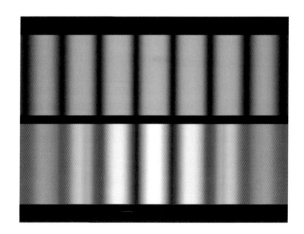

Two-source interference with other waves

It is possible to produce interference fringes with sound waves, and with microwaves. The diagrams below show the experimental arrangements. You can use the equation for λ above to calculate an **approximate** value for the wavelength of the waves. Why is the calculation approximate? Because in both these arrangements s is not very much less than D.

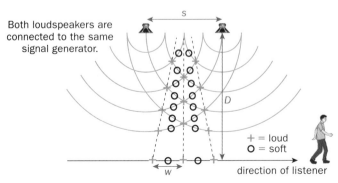

Both loudspeakers are connected to the same signal generator.

+ = loud
O = soft

direction of listener

Interference of sound waves.

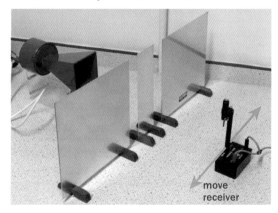

move receiver

▷ Physics at work: Interference

Oil films and soap bubbles

Interference in Young's double-slit experiment happens because the wavefront was divided into two parts that travelled on different routes, and then met and superposed.
But there are other interference effects that work in quite a different way.

Thin-film interference happens when you have a thin layer of transparent material, such as oil on water, or the thin film of soapy water in a bubble.

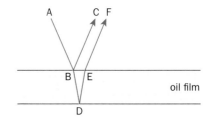

Some waves (ABC in the diagram) reflect off the top of the layer, while some (ABDEF in the diagram) pass through it and reflect off the bottom.
When these 2 waves meet again (at CF), there is a path difference between them, and so they interfere.

Depending on the thickness of the film, they will interfere constructively or destructively.

Colours in a thin soap film.

Why are thin films coloured, as in this photo?

The path difference in the diagram may be a whole number of wavelengths for red light, but not for blue. This will give an interference maximum that looks reddish.

At other places the film is slightly thinner or thicker, and light of a different colour will interfere constructively.

The iridescent colours on the wings of a fly are also due to thin-film interference.

Holography

One of the most spectacular applications of interference is the **hologram**. Light from a laser, which is completely coherent, falls on an object and is reflected in all directions.
Some of the reflected light lands on a photographic plate, where it interferes with light coming directly from the laser.
This interference produces a complex set of fringes of maxima and minima, recorded on the photographic plate.

To see the hologram, light of the same wavelength is allowed to fall on the developed photographic plate.
This produces further interference, allowing you to see a three-dimensional image of the original object.

Holograms are used extensively in scientific measurement and data recording, but their striking three-dimensional images have made them important in art and design: they are on every credit card!

Interference hologram of Rolls Royce blades.
It is used to test for faults in the rotor.

▷ Diffraction grating

Young's double-slit experiment does not give accurate values for the wavelength of light. Can you think why?

The top diagram shows the fringes obtained when green laser light is passed through a double slit. You can see that the fringes have blurred edges, and they are dim. It is difficult to measure the distance between the maxima precisely.

The second diagram shows the pattern obtained on a screen when laser light is passed through a **diffraction grating**.
A diffraction grating has thousands of slits.
Here, there were 80 vertical slits in each mm across the grating!

You can see the results:

● The maxima are very sharp, because constructive superposition occurs only in certain very precise directions.
● The maxima are much brighter, because there are many more slits or *lines* for the light to pass through.

The diffraction grating gives very accurate values for wavelength.

Young's experiment gives faint, blurred 'fringes' of maxima and minima.

A diffraction grating gives bright, sharp maxima.

How does a diffraction grating work?

The diagram below shows part of a diffraction grating:
The slits or lines cause the light to diffract in all directions, but the arrows show the *only* directions in which you can detect light.

The arrows marked **0** show the straight-onwards direction.
These paths are all the same length, so the path differences are zero. There is *no* phase difference between the waves, and they all interfere constructively. This is called the *zero-order maximum*.

The arrows marked **1** show another direction with *no* phase difference. How can this be?

Although the paths are the same length from the straight black line onwards, the distances from the slits to the black line are different.

However, there is no phase difference here, because each path marked **1** is exactly one wavelength longer than the one above it.

Each adjacent path has a phase difference of 360° (2π radians), which is the same as a phase difference of zero (see page 128).

The lines marked **1** give the *first-order maximum*, which is at an angle θ above the zero-order maximum in this diagram.
Another first-order maximum is observed at an angle θ below the zero-order maximum.

In the direction shown by the arrows marked **2**, you find the *second-order maximum*.
For this second-order maximum, the difference between each path and the next is double the difference for the first order – it is *two* wavelengths. So they add constructively again.
At an even bigger angle there can be a third-order maximum.

If the angle is just slightly different from those shown, the diffracted light from all the slits superposes destructively to give darkness. So light is seen only at certain precise angles.

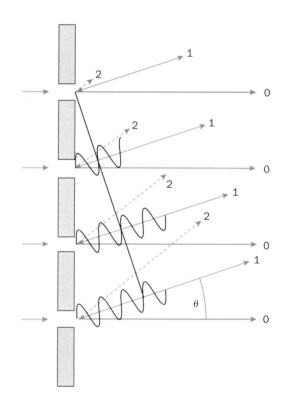

Measuring wavelength with a diffraction grating

Here are just two of the slits, which are a distance d apart:

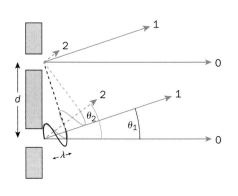

The path difference between the two adjacent paths in direction **1** is one wavelength, λ. The path difference between the two adjacent paths in direction **2** is 2 wavelengths, 2λ.

The angle between the first-order maximum **1** and the zero-order maximum **0** is labelled θ_1 because it is the *first*-order maximum. For the second-order maximum **2**, the corresponding angle is labelled θ_2.

The next diagrams show the triangles made by the dotted lines, the path differences (λ) and the distance d between the slits:

The angles at the top of the triangles are also θ_1 and θ_2. Because this triangle is right-angled, you can write:

$$\sin \theta_1 = \frac{\text{opposite}}{\text{hypotenuse}} = \frac{\lambda}{d} \quad \text{or} \quad \lambda = d \sin \theta_1 \quad \text{for the first order,}$$

and $\quad 2\lambda = d \sin \theta_2 \quad$ for the second order.

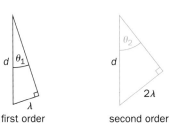

first order   second order

This can be done for a third-order maximum, fourth-order and so on. In fact, one equation can be used for all of them:

$$\boxed{n\lambda = d \sin \theta_n}$$ for the nth-order maximum, where $n = 1, 2, 3, \ldots$

Diffraction grating facts:

d = grating spacing

N = number of lines per mm

d (in mm) $= \dfrac{1}{N}$

Example 9
A student uses the apparatus shown to find the wavelength of red light. His diffraction grating has 300 lines per mm. He measures the second-order maximum at an angle of 25°. What value does he get for the wavelength?

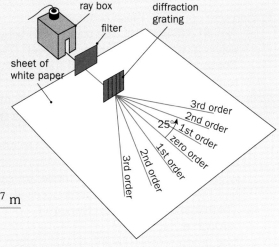

The grating spacing, $d = \dfrac{1 \text{ mm}}{300} = \dfrac{1 \times 10^{-3} \text{ m}}{300} = 3.33 \times 10^{-6}$ m

The order $n = 2$, $\quad n\lambda = d \sin \theta_n$

$$2 \times \lambda = 3.33 \times 10^{-6} \text{ m} \times \sin 25°$$

$$\lambda = \frac{3.33 \times 10^{-6} \text{ m} \times \sin 25°}{2} = \underline{7.04 \times 10^{-7} \text{ m}}$$

Diffraction gratings and spectra

If you pass white light through a diffraction grating you will get a spectrum in each order, as shown in the photograph:
The exception is the zero-order spectrum.
Can you explain these observations?

The equation above tells us that the angle θ depends on the wavelength. Red light has a longer wavelength than violet light and so the red light forms its maxima at larger angles.

The zero-order maximum is white, because in this direction the path difference is zero, and so is not dependent on the wavelength.

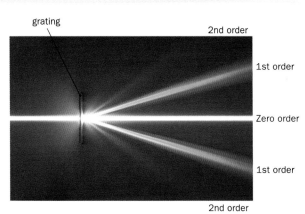

Summary

The principle of superposition states that when two or more waves meet, the resultant displacement at any point is found by adding the displacements of each wave.

Waves interfere constructively when they are in phase, and destructively when they in anti-phase.

When waves reflect forwards and backwards, they superpose and can form stationary (standing) waves. Stationary waves have nodes and anti-nodes. The distance between adjacent nodes is $\frac{1}{2}\lambda$.
A stationary wave system has certain definite frequencies at which it will resonate.

A stretched string or an open pipe can vibrate with all the multiples of the first harmonic frequency f_1.
A closed pipe can vibrate with only the odd-numbered multiples of the first harmonic frequency f_1.

The equation for the first harmonic frequency of a stretched wire is: $f_1 = \dfrac{1}{2L}\sqrt{\dfrac{T}{\mu}}$

When waves pass through a gap, they diffract. Narrow gaps produce more spreading.
The first **minimum** of diffraction through a gap of width b is at an angle θ such that: $\sin\theta = \dfrac{\lambda}{b}$

If 2 sources separated by an angle θ are viewed through a gap of width b, they can be resolved if: $\sin\theta \geq \dfrac{\lambda}{b}$

Young's experiment has two slits a distance s apart which give a pattern of maxima and minima, viewed a distance D away.
If the separation between adjacent maxima (fringes) is w, the equation is: $\dfrac{\lambda}{s} = \dfrac{w}{D}$

A diffraction grating gives sharper, brighter maxima than Young's experiment.
The diffraction grating equation is: $n\lambda = d\sin\theta_n$ for the nth-order **maximum**.

▷ Questions

1. Can you complete these sentences?
   a) A path difference of one wavelength between two waves gives a .... difference of 360° (2π radians). These waves are in .... and they will superpose to give .... interference.
   b) A phase difference of 180° (π radians) means that the waves are in ....–phase, and they give .... interference.
   c) Stationary or .... waves are formed when reflected waves superpose. They form 'loops', where the parts that do not move are called .... and the maximum oscillations occur at ....
   d) There is more diffraction through a .... gap than through a .... gap.
   e) The amount of spreading or .... when waves pass through a gap is measured by the angle between the central maximum and the first.......
   f) ....'s experiment gives interference fringes when waves from two sources of light superpose, but a diffraction .... gives sharper, brighter fringes.

2. Two loudspeakers are connected to the same signal generator. Along the line between the two loudspeakers, there are quiet and loud places.
   a) What are the quiet places called?
   b) The quiet places are 28 cm apart. What is the wavelength of the sound?
   c) Use the speed of sound (340 m s^{-1}) to calculate the frequency of the sound waves.

3. The length of the strings on a guitar is 0.65 m.
   a) What is the wavelength of the lowest note on each string?
   b) When waves travel along the thickest guitar string, they do so at 110 m s^{-1}. Use $c = f\lambda$ to calculate the lowest frequency of that string.
   c) The lowest frequency of the thinnest guitar string is four times that of the thickest string. How fast do waves travel along the thinnest string?

4. A violin string is under a tension of 110 N. The wire is 0.42 m long and has a mass per unit length of 8.2×10^{-4} kg m^{-1}. What is the string's first harmonic frequency?

5. The diagram shows standing waves on a string. The length of the string is 0.60 m, and the frequency of the waves is 25 Hz. Calculate the speed of waves along the string.

0.60 m

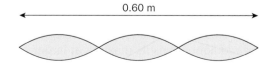

6. a) An open pipe is 0.720 m long. If the speed of sound in air is 340 m s^{-1}, what are the first two harmonic frequencies of the pipe?
   b) What would the first two harmonic frequencies be if the pipe were closed? (Ignore end corrections in your calculations.)

7. In a resonance-tube experiment, a student uses a tuning fork of frequency 440 Hz. She hears loud sounds when the air column is 0.187 m and then 0.573 m long. What value does this give for the speed of sound in air?

8. Sound of frequency 1 kHz and ultrasound of frequency 100 kHz both travel at 340 m s^{-1} through air. Use $c = f\lambda$ to calculate the wavelength of each. Use these wavelengths to explain why the audible waves diffract through a large angle when they go through a doorway, but ultrasound goes straight through without any noticeable bending.

9. The ASTRA satellite transmits radio waves, speed 3.00×10^8 m s^{-1}, at a frequency of 12.5 GHz. (1 GHz = 1×10^9 Hz)
   a) Calculate the wavelength of this signal.
   b) Assuming that the transmitting dish has a diameter of 1.6 m, find the angle θ of the first minimum of diffraction of the radio waves directed towards Earth.

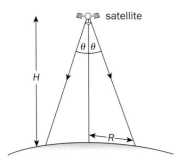

   c) In this diagram, the distance R is the radius of the circular patch on the Earth which receives the radio signal from the ASTRA satellite. The height H of the satellite above the Earth is 3.6×10^7 m. Calculate the radius R, and the area of the Earth that can receive this signal.

10. In a Young's double-slit experiment, laser light of wavelength 630 nm is used to illuminate two slits of separation 0.20 mm.
    a) Calculate the fringe separation on a screen 3.0 m away.
    b) What would happen to the fringes if the screen were moved closer to the slits?

11. A school microwave transmitter sends microwaves of wavelength 2.8 cm to a double-slit arrangement where the gap s is 5.0 cm.
    The microwave receiver is 1.0 m away and is moved along the red arrowed line shown in the diagram, to detect maxima and minima.

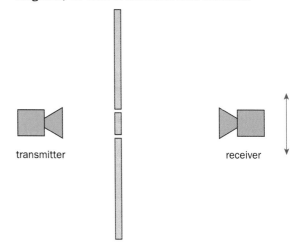

transmitter   receiver

    a) Calculate the distance x between maxima.
    b) Explain why your calculation for part (a) gives only an approximate value for x.

12. An optical telescope has an aperture of 1.5 m. Use Rayleigh's criterion to calculate the angular separation between two stars that can just be resolved.
    Assume that the wavelength of light is 500 nm.

13. A diffraction grating with a grating spacing of 1.000 μm is used to investigate the spectrum of helium. A strong yellow line is observed in the first-order spectrum, at an angle of 36°.
    What is the wavelength of the yellow line?

14. A diffraction grating is marked 500 lines mm^{-1}.
    a) Calculate the grating separation d.
    b) It is used to measure the spectrum of sodium light, which has a strong yellow line of wavelength 589 nm.
       Calculate the angle θ_1 of the first-order maximum of that spectral line.
    c) Calculate the angle θ_2 for the second-order ($n = 2$) maximum for this wavelength.
    d) Calculate the angle θ_3 for the third-order ($n = 3$) maximum for this wavelength.
    e) There is no fourth-order ($n = 4$) or higher spectral line with this grating. Explain why.
       [Hint: is the grating spacing 4 times as big as the wavelength? What is the maximum path difference you can get with this grating?]

Further questions on page 182.

Further Questions on Waves

▷ Wave Motion

1. In a ripple tank a wooden bar touching the surface vibrates with SHM at 8 Hz. The transverse water wave produced is represented below.

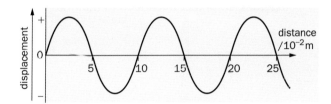

   a) Define displacement and wavelength. [2]
   b) Derive the equation for the speed c of the wave, $c = f\lambda$, where f is the frequency and λ the wavelength. [2]
   c) Calculate the speed of the wave shown in the diagram. [1] (W)

2. The graph shows the displacement of particles in a transverse progressive wave against distance from the source at a particular instant with points labelled A, B, C, D and E.

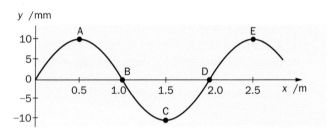

   a) Write down the letters of
      i) all the points at which the speed of the particle is a maximum,
      ii) all the points at which the magnitude of the acceleration of the particle is a maximum,
      iii) two points which are in phase,
      iv) two points which are 90° out of phase. [4]
   b) State the amplitude and wavelength of the wave. [2] (AQA)

3. Write an essay on the topic of waves and wave motion.
   a) Describe the motions of the particles in transverse and longitudinal waves, and explain why polarisation is a phenomenon associated with transverse waves. [6]
   b) Explain the terms *frequency, wavelength, speed, period, displacement* and *amplitude,* and show how these quantities may be obtained from graphical representations of both transverse and longitudinal waves. [10]
   c) Describe an experiment to determine the frequency of a sound wave in air. [5] (OCR)

4. The diagram shows the variation with time t of the displacement x of the cones of two identical loudspeakers A and B placed in air.

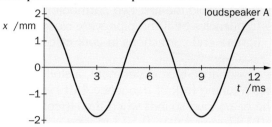

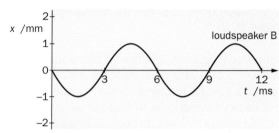

   a) Calculate the frequency of vibration of the loudspeaker cones. [3]
   b) Deduce the phase angle by which the vibrations of cone B lead those of cone A. [2]
   c) State the type of wave produced in the air in front of each loudspeaker. [1]
   d) Suggest, with a reason, which loudspeaker is likely to be producing the louder sound. [2]
   e) The speed of sound waves in air is $340\,\mathrm{m\,s^{-1}}$. Calculate the wavelength of the waves. [2] (OCR)

5. The figure shows part of a transverse progressive wave which is travelling to the right along a string. The horizontal dotted line shows the position of the string when there is no wave present.

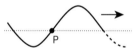

   In which direction is the string at the point P moving at the instant shown?
   A upwards   B downwards
   C to the right   D it is at rest   [1] (OCR)

6. a) Explain what is meant by *unpolarised* light and by *plane polarised* light. [2]

   b) A light source appears bright when viewed through two pieces of Polaroid, as shown in the diagram. Describe what is seen when B is rotated slowly through 180° in its own plane. [2]
   c) State, giving your reasoning, which of the following types of waves can be polarised: *radio, ultrasonic, microwaves, ultraviolet.* [2] (AQA)

▷ Reflection and Refraction

7. The diagram shows a ray of light incident on the face of a cube made of glass of refractive index 1.50.

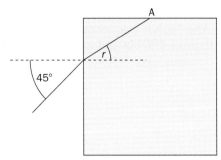

a) Calculate
   i) the angle r,
   ii) the critical angle for the glass–air interface. [3]
b) Use your answer to determine the path of the ray of light through the glass cube.
   Mark the path on a copy of the diagram and show the appropriate angles at the surfaces. [2]
c) When a drop of liquid of refractive index 1.40 is placed on the upper surface of the cube so as to cover point A, light emerges from the block at A. Explain, without calculation, how this occurs.
   [2] (AQA)

8. A postage stamp is examined through a converging lens held close to the observer's eye.
   A virtual magnified image of the stamp is produced with the image being the same way up as the object.
a) Draw a ray diagram showing the situation above. Label the object and image and mark the focal points of the lens. [4]
b) The object is 0.09 m from the lens and the image is formed 0.36 m from the lens.
   Determine the magnification (size of image / size of object) of the stamp produced. [2] (AQA)

9. a) State the laws of reflection of light.
   Draw a diagram to show how a plane mirror can be used to turn a ray of light through 90°. [3]

b) State Snell's Law for the refraction of light. The right-angled prism shown, made of glass of refractive index 1.53, may also be used to turn a ray of light through 90° using *total internal reflection*. Calculate the critical angle for the glass of the prism. Draw the path of a ray of light which is turned through 90° by the prism. [4]
c) Discuss the relative merits of the plane mirror and the right-angled prism as devices to turn a ray of light through 90°. [2] (OCR)

10. a) Describe the structure of a step-index optical fibre outlining the purpose of the core and the cladding. [3]
b) A signal is to be transmitted along an optical fibre of length 1200 m. The signal consists of a square pulse of white light and this is transmitted along the centre of a fibre. The maximum and minimum wavelengths of the light are shown in the table.

Colour	Refractive index of fibre	Wavelength /nm
Blue	1.467	425
Red	1.459	660

Explain how the difference in refractive index results in a change in the pulse of white light by the time it leaves the fibre. [2]
c) Discuss **two** changes that could be made to reduce the effect described in part (b). [2] (AQA)

11. The diagram (not to scale) shows a ray of monochromatic light entering a multimode optical fibre at such an angle that it just undergoes total internal reflection at the boundary between the core and the cladding.

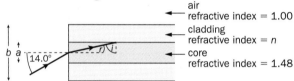

a) Suggest appropriate magnitudes for the diameters labelled a and b. [2]
b) Calculate the angles r and i. [3]
c) Hence estimate the refractive index n of the cladding. [2]
d) On a copy of the diagram, show what would happen to the ray of light if it were incident at an angle slightly greater than 14°. [3] (Edex)

12. The graph shows how refractive index varies across the diameter of a step-index multimode optical fibre.

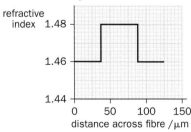

a) Calculate the critical angle for light travelling between the core and the cladding of the fibre. [2]
b) On a copy of the graph, show the variation of refractive index for a step-index monomode optical fibre made from materials of refractive index 1.45 and 1.47. [2]
c) Calculate the time taken for a light pulse to travel 5.00 km along this monomode fibre. [3] (Edex)

▷ Interference and Diffraction

13. The diagram shows an arrangement to produce fringes by Young's two-slits method.

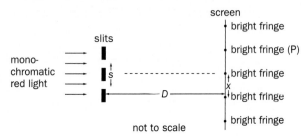

not to scale

a) State suitable values for s and D if clearly observable fringes are to be produced. [2]
b) Explain how the bright fringe labelled P is formed. [2]
c) What would be the effect on the fringe width x of
   i) increasing the slit separation s, and
   ii) illuminating the slits with blue light? [2]
d) To obtain an interference pattern, the light from the two slits must be coherent.
   What is meant by the term *coherent*? [1] (Edex)

14. Two loudspeakers produce identical sounds of frequency 440 Hz which superpose to produce a standing wave. Adjacent nodes are formed 0.75 m apart. Select the correct statement about the waves.
    A The frequency heard is 880 Hz.
    B The speed of the waves is 165 m s^{-1}.
    C The waves are travelling in the same direction.
    D The wavelength of the waves is 1.5 m. [1] (Edex)

15. The diagram shows a loudspeaker which sends a note of constant frequency towards a vertical metal sheet. As the microphone is moved between the loudspeaker and the metal sheet, the amplitude of the vertical trace on the oscilloscope continually changes several times between maximum and minimum values. This shows that a stationary wave has been set up in the space between the loudspeaker and the metal sheet.

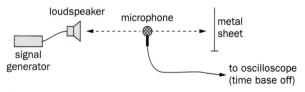

a) How has the stationary wave been produced? [2]
b) State how the stationary wave pattern changes when the frequency of the signal generator is doubled. Explain your answer. [2]
c) What measurements would you take, and how would you use them, to calculate the speed of sound in air? [4]
d) Suggest why the minima detected near to the metal sheet are much smaller than those detected near the loudspeaker. [2] (Edex)

16. The apparatus shown was used to demonstrate a transverse standing wave on a string. Both the weight and the distance between the pin and the pulley were kept constant. At 480 Hz there was a standing wave pattern and each loop was 10 cm long. At a higher frequency there were two more loops than at 480 Hz and each loop was 8 cm long.

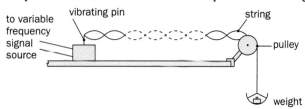

a) Explain why standing waves occur at particular frequencies only. [2]
b) Calculate the speed of the waves in the string. [3]
c) Show that, at 480 Hz, eight loops would be created. [2] (W)

17. Parallel light is incident normally on a diffraction grating.
    a) Light of wavelength 5.9×10^{-7} m gives a first-order image at 20.0° to the normal. Determine the number of lines per metre on the grating. [2]
    b) Light of another wavelength gives a second-order image at 48.9° to the normal. Calculate this wavelength. [2]
    c) What is the highest order in which both these wavelengths will be visible? Justify your answer. [2] (AQA)

18. The diagram shows water waves, produced by two sources A and B, in a ripple tank. The circles represent adjacent crests at a particular instant. The waves are moving with a speed of 80 mm s^{-1}.

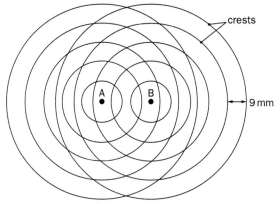

a) From the diagram find the wavelength and frequency of the waves. [4]
b) On a sketch of the diagram draw two lines
   i) joining successive points at which maximum constructive interference occurs (mark this line C),
   ii) joining successive points at which the adjacent destructive interference takes place (mark this line D). [4] (OCR)

▷ **Synoptic Questions on Waves**

19. The diagram shows a woman with long sight looking in a plane mirror. She cannot focus clearly on the image of the end of her nose when it is less than 225 mm from the mirror.

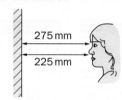

a) Calculate her least distance of distinct vision (near point distance). [2]
b) State what type of lens is needed to correct her near point distance to 250 mm.
Explain how the lens corrects the vision and calculate the power of the lens. [5] (AQA)

20. a) What is meant by the superposition of waves? [2]

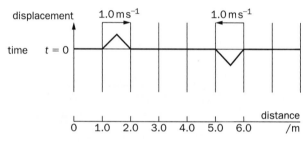

b) The diagram shows two wave pulses moving at $1.0 \, \text{ms}^{-1}$ in opposite directions along a string. They are drawn at time $t = 0$.
On a copy of the diagram, draw the profile of the string at times $t = 1 \, \text{s}, 2 \, \text{s}, 3 \, \text{s}$.
The vertical lines are 1.0 m apart. [5] (OCR)

21. The left diagram shows an arrangement for investigating interference using microwaves. Two vertical slits S_1 and S_2 are at equal distances from the transmitter. A receiver is moved along the line AB. The right graph shows how the amplitude of the received signal varies with position along AB.

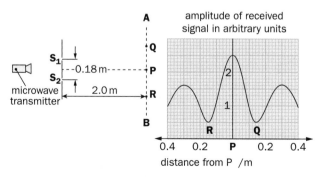

a) Explain why there is a maximum signal at P and a minimum signal at Q. [4]
b) The centres of the slits are 0.18 m apart and P is 2.0 m from the slits. Determine the wavelength and frequency of the microwaves. [4] (AQA)

22. Sound travels by means of longitudinal waves in air and solids. A progressive sound wave of wavelength λ and frequency f passes through a solid from left to right. The diagram X represents the equilibrium positions of a line of atoms in the solid.
Diagram Y represents the positions of the same atoms at time $t = t_0$.

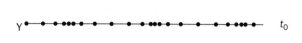

a) Explain why the wave is longitudinal. [1]
b) On a copy of the diagram Y label:
two compressions (C), two rarefactions (R) and the wavelength of the wave. [3]
c) Along the line Z mark the positions of the two compressions and the two rarefactions at a time t given by $t = t_0 + T/4$ (T = period). [2] (Edex)

23. To determine the speed of sound, a student claps her hands at point P, which is a measured distance, d, from a brick wall. A microphone at P is connected to a timing device, arranged so as to record the time, t, between the original clap and its echo. The results are plotted below.

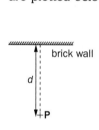

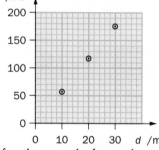

a) Calculate a value for the speed of sound. [2]
b) In another experiment the student sets up two small loudspeakers, S_1 and S_2, connected to the same signal generator, set to 8300 Hz. She moves a microphone along the line AB, and finds maxima of sound at the positions shown by dots, with minima in between.

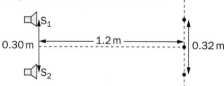

i) Use the equation for Young's double slit to calculate the wavelength. [2]
ii) Hence calculate a value for the speed of sound from this experiment. [1]
iii) When the signal generator is set to 300 Hz the student does not find a succession of maxima and minima as the microphone is moved along the line AB. Explain why this is to be expected. [2] (W)

13 Materials

We are surrounded by a wide range of solid materials, from very light foams to dense metals. Despite their differences, the same Physics is used to describe and analyse their behaviour.

In this chapter, you will learn:
- how to calculate the density of materials,
- how to use Hooke's Law to find extensions and forces,
- how to find the energy stored in a stretched solid,
- how stress and strain lead to the Young modulus.

Aerogel is light, but a good insulator.　　*1 kilogram bars of gold.*

▶ Density

An old riddle asks: *'Which is heavier, a pound of feathers or a pound of lead?'*

People only fall for this if they confuse two different ideas:
- mass – the amount of matter in an object,
- density – the amount of matter in a particular volume.

A pound, like a kilogram, is a measure of **mass**.
The **density** of any substance is given by this equation:

$$\text{Density, } \rho \text{ (kg m}^{-3}\text{)} = \frac{\text{mass, } m \text{ (kg)}}{\text{volume, } V \text{ (m}^3\text{)}} \quad \text{or} \quad \rho = \frac{m}{V}$$

Although kg m^{-3} is the unit always used in Physics, you may meet density in other units, such as g cm^{-3} or g litre^{-1}.

Heavier or denser?

Example 1
The aerogel in the picture at the top of the page measures 1.0 cm × 15 cm × 16 cm and has a mass of 0.46 g. What is its density in kg m^{-3}?

Converting cm to m, volume $V = 0.01 \text{ m} \times 0.15 \text{ m} \times 0.16 \text{ m} = 0.000\,24 \text{ m}^3$

Converting g to kg, mass $m = \dfrac{0.46}{1000} \text{ kg} = 0.000\,46 \text{ kg}$

density $\rho = \dfrac{m}{V} = \dfrac{0.000\,46 \text{ kg}}{0.000\,24 \text{ m}^3} = \underline{1.9 \text{ kg m}^{-3}}$　(2 s.f.)

The gold in the picture has a density of 19 300 kg m^{-3} – about 10 000 times greater than this!

Atoms and crystals

One reason for the high density of gold is that it contains very massive atoms. These atoms are also packed very closely together in a regular arrangement called a **crystalline lattice**.

There are several different three-dimensional lattice structures. The diagram shows the simplest possible lattice:

Each atom is 'sitting' in a small cubical box of width d, where d is the diameter of each atom.
But how big is d?

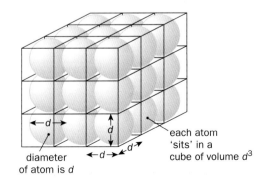

diameter of atom is d

each atom 'sits' in a cube of volume d^3

Measuring the size of atoms

The size of atoms was first measured by Lord Rayleigh in 1890, at a time when many physicists did not believe atoms existed! Rayleigh knew that if oil is spilt on to water, it will spread out until it can spread no further, as it is then as thin as possible. He realised that at this point the film must be just one molecule thick!

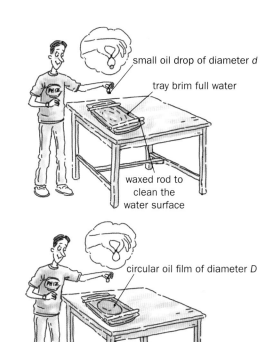

small oil drop of diameter d

tray brim full water

waxed rod to clean the water surface

The volume of oil in the spherical drop must be the same as the volume of oil in the circular film. This allowed Rayleigh to calculate the thickness of the film, as he had measured the diameter of both the tiny drop and the large circular film.

Practical details:

- The water surface must be very clean. Any contamination on the surface can be moved to the edges using waxed rods.
- The water is then lightly dusted with powder to make the oil film visible.
- The oil drop must be very small – less than 1 mm across. It can be picked up and handled using a loop of very thin wire. Its diameter can be measured using a fine scale, viewed through a magnifying glass.
- As soon as the drop touches the water, it spreads into a circular patch. A metre ruler is used to measure the diameter of this patch straight away.

circular oil film of diameter D

Example 2

In an oil film experiment, an olive oil drop of diameter $d = 0.5$ mm spreads into a circular film of diameter $D = 25$ cm.

(a) Calculate the thickness of the oil film.

(b) In the film, the olive oil molecules stand up on end to avoid mixing with the water. If each oil molecule has 12 carbon atoms, calculate the diameter of one atom in nanometres.

(a) Radius of oil drop, $r = 0.25$ mm $= 2.5 \times 10^{-4}$ m. This drop is a sphere, and the volume of a sphere $= \frac{4}{3}\pi r^3$

Volume of oil $= \frac{4}{3}\pi r^3 = \frac{4}{3}\pi \times (2.5 \times 10^{-4}\ m)^3 = 6.54 \times 10^{-11}\ m^3$

Radius of oil film $R = 12.5$ cm $= 0.125$ m, so the area of the film $= \pi R^2 = \pi \times (0.125\ m)^2 = 0.0491\ m^2$
volume of oil in the film = area of the film × its thickness, so

$$\text{thickness} = \frac{\text{volume of oil}}{\text{area of film}} = \frac{6.54 \times 10^{-11}\ m^3}{0.0491\ m^2} = 1.33 \times 10^{-9}\ m = \underline{1.3 \times 10^{-9}\ m}\ \text{(2 s.f.)}$$

(b) One molecule of 12 carbon atoms is 1.33×10^{-9} m long.

$$\text{diameter of one atom} = \frac{1.33 \times 10^{-9}\ m}{12} = 1.1 \times 10^{-10}\ m = \underline{0.11\,nm}\ \text{(2 s.f.)}$$

Modern measurements of atomic size

In 1915, William Bragg and his son Lawrence received the Nobel Prize for using **X-ray diffraction** to determine crystal structure, and this included measurements of the separation of atoms.
The development of **scanning tunnelling microscopes** (STMs) in the 1980s made direct observation of atoms possible.

Look at this STM image:

The magnification of the STM is known, so direct measurement on this image reveals that a copper atom has a diameter of 0.25 nm.

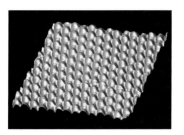

STM image of copper atoms.

▶ Stretching materials

The copper atoms in the STM image on page 185 are packed tightly together, because they attract each other strongly. When you try to bend or stretch a solid, these interatomic forces resist the force you are applying. The more you stretch the solid, the greater the force becomes.

In 1660, the physicist Robert Hooke investigated how metal springs stretched under different forces.

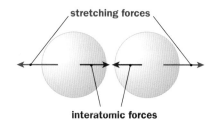

stretching forces

interatomic forces

Hooke's Law

Look at the spring in this diagram:
Can you see what happens when extra weight is added?

The spring extends by equal amounts, because the stretching force is going up in equal steps. The stretching force is directly proportional to the extension. This is called **Hooke's Law**.

You can use this equation to find the force needed:

Stretching force, F = spring constant, k × extension, Δx
(N) $\qquad\qquad$ (N m^{-1}) $\qquad\qquad$ (m)

or $\boxed{F = k\,\Delta x}$

The extension Δx ('delta-x') is sometimes written e or ΔL.
You find the extension from:

Δx = stretched length − original length.

The spring constant k, is sometimes called the **stiffness**.
k has units of newtons per metre (N m^{-1}).

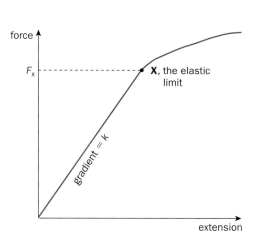

Force–extension graphs

Look at the force–extension graph for a spring:

The line is straight up to the point X, called the **elastic limit**.
Hooke's Law holds true only for this straight-line region.
The **gradient** of this line is the spring constant, k.

For forces less than F_X, the spring behaves **elastically**. If the force is removed, the spring goes back to its original size and shape.

For forces bigger than F_X, the spring does not return to its original size.

Example 3
A spring is 0.38 m long. When it is pulled by a force of 2.0 N, it stretches to 0.42 m.
What is the spring constant? Assume that the spring behaves elastically.

Extension Δx = stretched length − original length $\;=\;$ 0.42 m − 0.38 m $\;=\;$ 0.04 m

$\quad\;\; F = k \times \Delta x$

∴ 2.0 N = k × 0.04 m $\qquad$ so $\quad k = \dfrac{2.0\text{ N}}{0.04\text{ m}} = \underline{50\text{ N m}^{-1}}$

Compressing materials

Elastic materials such as metals may be **compressed** as well as extended. Hooke's Law applies for springs being squeezed as well as for springs being stretched, providing the elastic limit is not exceeded.
Car suspensions use compression springs to make journeys less bumpy.

▶ Elastic potential energy

You have to do work to stretch a spring. This energy is stored in the spring. It is called **strain** potential energy or **elastic** potential energy.

Look at this force–extension graph:
The **blue area** represents the potential energy stored in the spring as it stretches up to point A.

Since the area of this triangle $= \frac{1}{2} \times$ height $\times$ base
this gives us the equation for elastic potential energy:

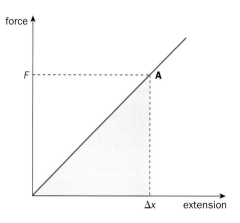

Elastic potential energy = $\frac{1}{2}$ × stretching force × extension (J)$\qquad\qquad$(N)$\qquad\qquad$(m)

or: $\boxed{E_P = \frac{1}{2} F \, \Delta x}$ (provided the elastic limit is not exceeded)

Example 4
A spring is 32 cm long. It is stretched by a force of 6.0 N until it is 54 cm long.
How much energy is stored in the stretched spring?
Hint: make sure the distance measurements are always in metres!

Extension $\Delta x = 0.54 \text{ m} - 0.32 \text{ m} = 0.22 \text{ m}$
Elastic potential energy $= \frac{1}{2} \times F \times \Delta x \quad = \quad \frac{1}{2} \times 6.0 \text{ N} \times 0.22 \text{ m} \quad = \quad \underline{0.66 \text{ J}} \; (2 \text{ s.f.})$

What if the force–extension graph is a curve?

The area under the graph always gives the energy stored, even when the graph is not a straight line.

How can you find the energy if the line is not straight?

You can draw straight-line sections closely fitting the curve, then divide the area under the curve into triangles and rectangles, and add their separate areas together:

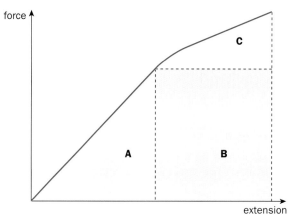

Energy stored = area A + area B + area C

▶ Properties of materials

Materials that are hard to stretch are said to be **stiff**.
Materials that are easy to stretch are described as **flexible**.

Just like the spring on the opposite page, an **elastic** material returns to its original shape when you remove the stretching force. However, if you stretch it past its elastic limit, it will be permanently deformed.

Materials that permanently deform are **plastic**.
If you deform a plastic material, it stays in its new shape.

Plastic materials are also:
- **ductile** – they can be drawn out into wires,
- **malleable** – they can be hammered into sheets.

Malleable materials, like lead, are **tough**. When you try to break lead, it deforms plastically. This makes it give way gradually, and absorb a lot of energy before it snaps.

The opposite of tough is **brittle**.
Brittle materials, like glass, do not deform plastically. They do not absorb much energy before they break. They crack or shatter suddenly.

Lead is malleable and tough.

▶ Stress

Look at the diagram:

It shows 2 copper wires being stretched by the same weight.
The forces are the same, but the extensions are not.
Can you see why?

The wires have different cross-sectional **areas**.
The thinner wire stretches more.

We say that it has a greater **tensile stress**.
('Tensile' means 'under tension' or 'stretched'.)

The stress, σ, is given by this equation:

$$\text{Stress, } \sigma = \frac{\text{force, } \boldsymbol{F}\,(\text{N})}{\text{cross-sectional area, } \boldsymbol{A}\,(\text{m}^2)} \quad \text{or} \quad \sigma = \frac{\boldsymbol{F}}{\boldsymbol{A}}$$

The unit of stress is newton per square metre (N m^{-2}).
Newtons per square metre (N m^{-2}) are called **pascals**.

The pascal (symbol Pa) is also the unit for pressure (see page 26).
Does this mean stress and pressure are the same thing?
No – stress occurs **inside a solid**. Pressure occurs **on a surface**.

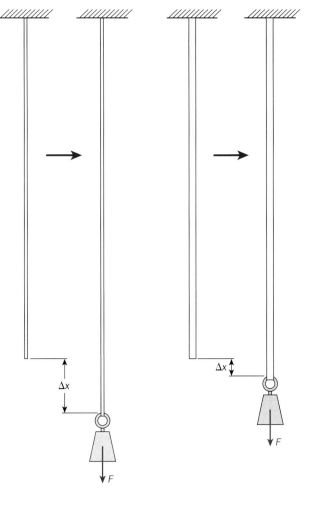

Example 5
A thin wire has a cross-sectional area of $4 \times 10^{-8}\,\text{m}^2$,
and is stretched by a force of 200 N.
What is the tensile stress in the wire?

$$\sigma = \frac{\boldsymbol{F}}{\boldsymbol{A}} = \frac{200\,\text{N}}{4 \times 10^{-8}\,\text{m}^2} = \underline{5 \times 10^9\,\text{Pa}} \quad (5 \times 10^9\,\text{N m}^{-2})$$

Strength

Look at the cartoon:
Is Phiz's rope strong enough to hold him?

It depends on:

- what the rope is made of,
- how thick the rope is – at its **thinnest** point,
- how heavy Phiz is.

The **ultimate tensile stress** is the measure of strength.
This is the tensile stress when the material breaks or yields.

Example 6
Nylon has an ultimate tensile stress of 7×10^7 Pa.
The cross-sectional area of the nylon rope in the diagram
is $3 \times 10^{-5}\,\text{m}^2$.
Phiz weighs 600 N. Will it hold him?

$$\sigma = \frac{\boldsymbol{F}}{\boldsymbol{A}} = \frac{600\,\text{N}}{3 \times 10^{-5}\,\text{m}^2} = \underline{2 \times 10^7\,\text{Pa}} \; (1 \text{ s.f.})$$

This is less than the ultimate tensile stress. He'll be OK.

Phiz goes rock climbing.

▶ Strain

Look at the diagram:

It shows 2 copper wires being stretched by the same weight.
This time, the wires have the **same** cross-sectional area.
The forces are the same, but the extensions are not.
The longer wire stretches twice as much, even though it has
the same tensile stress. Can you see why?

The wire on the right is twice as long to start with.
It stretches by the **same fraction** of its original length.

This fraction is called **tensile strain**.
The strain, ε, is given by this equation:

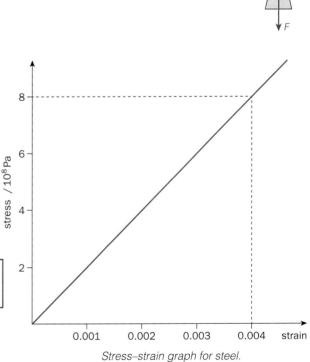

$$\textbf{Strain, } \varepsilon = \frac{\textbf{extension, } \Delta x \text{ (m)}}{\textbf{original length, } L \text{ (m)}} \quad \text{or} \quad \varepsilon = \frac{\Delta x}{L}$$

Strain has no units. It is just a ratio.

Example 7
A steel wire of length 2.3 metres stretches by 1.5 millimetres.
What is the tensile strain?

Extension, $\Delta x = 1.5$ mm $= 1.5 \times 10^{-3}$ m

Strain, $\varepsilon = \dfrac{\Delta x}{L} = \dfrac{1.5 \times 10^{-3} \text{ m}}{2.3 \text{ m}} = \underline{6.5 \times 10^{-4}}$ (2 s.f.)

▶ The Young modulus

When you stretch a wire that obeys Hooke's Law,
the force–extension graph is a straight line.
A graph of stress against strain is also a straight line,
because stress $\propto$ force, and strain $\propto$ extension.

Look at this stress–strain graph for steel:

Notice how small the strain is. A strain of only 0.01 (1%)
would exceed the elastic limit of a metal!

The **gradient** of the stress–strain graph is σ/ε.
This is called the material's **Young modulus of elasticity**.
It is given the symbol E. (Be careful – it is not energy!)

$$\textbf{Young modulus, } E = \frac{\textbf{stress, } \sigma \text{ (Pa)}}{\textbf{strain, } \varepsilon \text{ (no units)}} \quad \text{or} \quad E = \frac{\sigma}{\varepsilon}$$

E is measured in pascals (Pa), the same as stress.
It is a measure of how difficult it is to change the shape of
the material. Rubber has a low value, steel has a high value.

Stress–strain graph for steel.

Example 8
Use the graph shown (for steel) to find the Young modulus of elasticity for steel.

The Young modulus, $E = \dfrac{\text{stress, } \sigma}{\text{strain, } \varepsilon}$ = the gradient of the stress-strain graph $= \dfrac{8 \times 10^8 \text{ Pa}}{0.004} = \underline{2 \times 10^{11} \text{ Pa}}$ (1 s.f.)

▶ Measuring the Young modulus for steel

Metals such as steel do not stretch easily. You need a large stress to stretch the wire. To get such a large stress, you need both a very thin wire and a heavy load.

Metals stretch elastically only for very small strains, and you need a very long wire to get measurable extensions.

Look at this picture of a long wire with a heavy load:

It is hard to measure the stress and the strain, because:

1. The large loads could pull the ceiling attachment downwards, so you would be measuring sag, not stretch.

2. The room temperature may change, making the wire expand (stretch) or contract (shrink).

3. To calculate the strain, you have to measure very small extensions, of less than 1 mm.

4. To calculate the stress, you have to measure the diameter of a very thin wire.

The diagram below shows one laboratory method for measuring the stress and strain in a long, thin steel wire:

The wire to be tested has weights added to the bottom, and a **second** wire is attached to the same support.
This **reference** wire holds the measurement scale, and a weight at the bottom keeps this reference wire taut.

This arrangement deals with the 4 difficulties. In order:

1. The scale for measuring extensions is on the reference wire. If the test wire pulls the ceiling downwards, the reference wire and the scale move with it. So the scale measures only the extension, not the sag of the support.

2. If temperature changes make the test wire expand or contract, the reference wire changes in the same way.

3. A **vernier** is used to measure the tiny extensions accurately. The vernier is a second scale on the test wire itself, with a precision of 0.1 mm. Dividing the extension by the original length gives the strain.

4. You use a **micrometer** to measure the diameter of a thin wire. This has a precision of 0.01 mm. This gives you the cross-sectional area, and you divide the weight of each load (in N) by the area to get the stress.

The test wire may not be uniform, so you should measure the diameter of the test wire at several places and take an average. You can then find the average cross-sectional area of the wire, using the average diameter.

You should measure the extension for several different loads.

You should also remove the loads, one at a time, to check that the length goes back to its original value. This is to check that you didn't exceed the elastic limit of the wire.

With the data provided by this experiment, you can plot a stress–strain graph to find the Young modulus of steel.

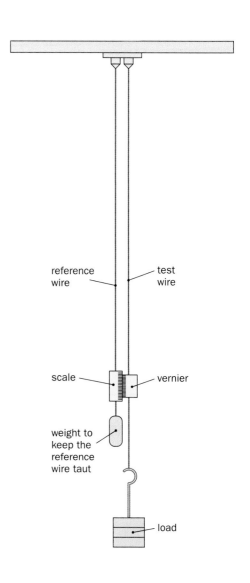

▶ Stress–strain graphs for different metals

Look at these stress–strain graphs for different mixtures of iron with carbon:

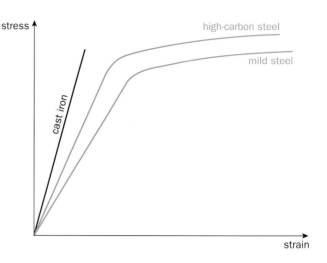

Mixtures of metals are called **alloys**.
Cast iron is a mixture of iron with about 3% carbon.
High-carbon steel contains about 0.3% carbon.
Mild steel has very little carbon – less than about 0.05%.

Notice that the graph for high-carbon steel is steeper than the graph for mild steel.
We say that high-carbon steel is *stiffer* than mild steel, or mild steel is *more flexible* than high-carbon steel.

Can you see that the graph for cast iron stops suddenly, while each of the two steel graphs curve over?
Cast iron is brittle, while most steels are tough.
The steels undergo ductile flow before they break.

Example 9
The Young modulus of elasticity for cast iron is 2.1×10^{11} Pa.
It snaps when the strain reaches 0.000 50.
What is the stress needed to break it? (Assume that it obeys Hooke's Law up to its breaking point.)

$$\text{The Young modulus, } E = \frac{\text{stress, } \sigma}{\text{strain, } \varepsilon}$$

$$2.1 \times 10^{11} \text{ Pa} = \frac{\sigma}{0.000\,50}$$

$$\therefore \quad \text{Stress, } \sigma = 2.1 \times 10^{11} \text{ Pa} \times 0.000\,50 = \underline{1.1 \times 10^8 \text{ Pa}}$$

Area under stress–strain graphs

On page 187, you saw that the *area* underneath the graph of force against extension for a spring gives you the potential energy stored in the stretched spring.

What is the meaning of the area under the graph of stress against strain for a metal wire?

From pages 188–189,

$$\text{Stress} = \frac{\text{force}}{\text{area}} \quad \text{and} \quad \text{strain} = \frac{\text{extension}}{\text{length}}$$

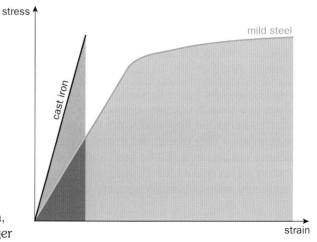

$$
\begin{aligned}
\therefore \text{ the area under the graph} &= \text{average stress} \times \text{strain} \\
&= \frac{\text{average force}}{\text{area}} \times \frac{\text{extension}}{\text{length}} \\
&= \frac{\text{average force} \times \text{extension}}{\text{area} \times \text{length}} \\
&= \frac{\text{work done}}{\text{volume}}
\end{aligned}
$$

since energy stored = work done = force × distance moved

So this area is the *energy stored per unit volume.*

This area is a useful measure of the *toughness* of a metal.
Tougher metals need more energy per unit volume to break them, and so the stress–strain graph up to the breaking strain has a larger area under it.

Mild steel is tougher than cast iron.

▶ Yielding and breaking

Ductile flow

Metals like copper are *ductile* – they can have large plastic deformations without fracturing.

If you stretch a copper wire, it gets thinner by plastic deformation before it snaps. This is shown in the photo:

The copper wire 'necked' before it broke.

It happens because atoms move, as the dislocations in the crystal structure move, to places of lower stress. The metal becomes thinner when the atoms move away from the stressed part. The stress then increases because the cross-sectional area is now less. This increases the ductile flow and so the metal 'yields' or 'gives' and gets thinner and thinner.

Once plastic deformation starts, atoms will continue to flow without any increase in stress. This stretching under a constant load is called **creep**. The thinning of the wire is called **necking**.

Brittle fracture

Brittle materials like glass cannot flow, so they do not 'neck'. The stress cannot be reduced by the movement of atoms.

The photo shows some stressed plastic viewed in polarised light:

The coloured fringes are 'contours' of lines of equal stress.
Can you see where the stress is greatest?
It is concentrated at the tip of a crack in the plastic. The stress is greater if the crack is deeper or sharper.
Tension will make the crack split open further, which increases the stress even more. The brittle solid then shatters.

Compression and tension in brittle materials

Brittle materials can crack under tension as the cracks deepen, but they are strong in compression, because this closes the cracks.

See how Phiz cuts a glass rod:

He cuts a sharp scratch into the rod where he wants it to break. Then he bends the rod with the scratch on the **outside** of the bend. The rod cracks through cleanly.

If he bends the rod with the cut on the **in**side, it would not break at the cut, because the crack is being compressed.

Work hardening

On page 194, you will see that metals can be made harder by heating. Another way to make a metal harder is to bend, stretch or hammer it *repeatedly* – this is called **work hardening**.

When you stretch copper wire past its elastic limit, the metal deforms in ductile flow and starts to 'neck,' due to movement of dislocations. This plastic deformation ends when the dislocations become tangled in a 'traffic jam' and they stop each other from moving. This makes the copper wire harder and more brittle, so it suddenly snaps in brittle fracture.

You can use this work hardening to snap thick copper wire: just bend it backwards and forwards many times.
You will feel it suddenly get stiffer and then snap.

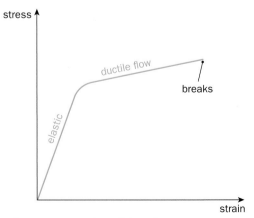
Copper snaps when dislocations cannot move.

▶ Polymers

Rubber and polythene are examples of **polymers**.
Polymer molecules consist of very long chains of atoms.

The diagram shows three polymer chains, tangled together in a random way:

In **rubber**, some chains are joined together by cross-links, shown in black in this diagram:

When you stretch a rubber band and let it go, the cross-links pull the chains back into shape. This is why rubber is elastic.

Rubber is a polymer.

Look at this **stress–strain graph of rubber**:

Why is it such a strange shape? It has two sections:

* Why is it so easy to stretch a rubber band at the start?
  When you pull it, you are straightening the tangled chains.
  This is quite easy to do, so very large strains are possible.

* Why does the rubber band become stiffer at large strains?
  The chain molecules are now as straight as possible.
  To stretch it any more, you must stretch the atomic bonds.
  This is hard to do, and so the rubber is much stiffer.

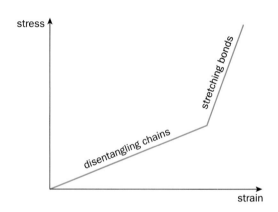

Energy losses in rubber

Look at this **force–extension graph for a rubber band**:

It shows the rubber being stretched (OAB) and then released (BCO). The graph is a similar shape while it is being stretched and being released, but the extensions are greater while it is being released. The straightened chains do not move back easily into place. The 'release' graph does go back to the origin eventually, as the cross-links in the rubber 'remember' the original positions of the chain molecules and pull them back into place.
This delayed-action return of rubber is called **hysteresis**.

Area OABDO shows the work that you do stretching the rubber.
Area OCBDO is the energy given out when the rubber is released.
The difference in energy (OABCO) warms up the rubber.

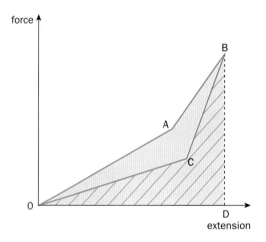

Thermoplastic and thermoset polymers

Artificial polymers, made from oil products, are popular because they are cheap and do not corrode or rot.

Thermoplastic polymers, like polythene and perspex, have no cross-links between molecules.
They soften and deform easily when they are warmed, which makes them ideal for moulding into containers.

Thermoset polymers, such as the melamine used for kitchen work-surfaces and the phenolic resin used for electrical fitments, have many cross-links between the chains, produced chemically when the polymer is cast.
Thermosets are rigid and do not soften with heat.

Thermoplastics and thermosets are part of everyday life.

▶ Faults in crystals

Look at this simplified diagram of a metal crystal. The atoms are mostly in neat rows, and it looks quite easy for them to move. For example, the left-most vertical column could move up or down along the **slip plane** between it and its neighbour.

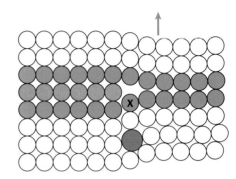

In fact, this slipping cannot happen easily, because you would have to break *all* the bonds joining the slipping atoms to their neighbours. So why is it that some metals, like copper, are ductile?

Look at the odd half-row, coloured orange in the top diagram:

The right-hand end of this part-row of atoms is called an **edge dislocation**. The two neighbouring rows, coloured blue, are very distorted near the dislocation. At the end of the orange part-row there is slightly more space, so the atoms there can move slightly into new positions.

The second diagram shows the crystal after the movement caused by the stress which is indicated by the arrow on the top diagram.

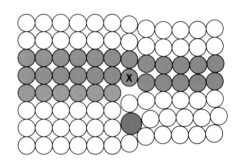

The only atom that has moved much is the one labelled 'X', but because of this the dislocation has moved down one row.
In this way, the dislocation can move down to the edge of the crystal. This is how plastic flow occurs in ductile metals, and it reduces the stress in the metal.

In these diagrams, the dislocation would not be able to move much further. This is because the lattice contains a large green 'foreign' atom that will not easily move into a new position.
This metal is an **alloy**, a mixture of different metals. The foreign atoms hinder the movement of dislocations, making the alloy harder and more brittle.

Polycrystalline and amorphous materials

You may know that a diamond is a crystal, but is copper crystalline? Look at this magnified photograph of a copper surface:

Copper is polycrystalline.

It shows that copper is made of many tiny **crystal grains**. Many solids, such as metals, consist of very tiny crystals, and these materials are described as *polycrystalline*.

Heating metals affects the size of the crystal grains.
If you heat a metal and then cool it *rapidly* – a process called **quenching** – you get very small crystal grains. In a quenched metal, the dislocations cannot move far before they reach the **grain boundary** at the edge of the crystal grain. A dislocation cannot move out of its own crystal, and the absence of plastic flow makes the metal hard and brittle.

In the same way, if you let a hot metal cool *slowly*, large crystal grains are formed. This process is called **annealing**, and it makes the metal malleable and tough.

Some solids, such as glass, are not crystalline at all. Their atoms are not in a lattice, but randomly placed, as in a liquid. These materials are described as **amorphous**. Glass is brittle, because it doesn't have a crystal lattice for dislocations to move in.

Glass is amorphous and brittle.

▶ Tension and compression

Most of this chapter has been about materials in **tension**. However, squashing also produces a change in length. This is not extension but **compression**.

Phiz is painting the ceiling. Look at his plank:

In some places, his weight is stretching the plank. This is tensile stress, and it produces tensile strain.

In some places, his weight is squashing the plank. This is compressive stress, and it produces a compressive strain.

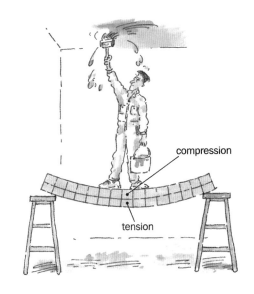

Can you see that his plank is compressed on the top surface where the plank is squashed up? The molecules are closer together.

Can you see that the plank is in tension on the bottom surface where the plank is stretched out? The molecules are farther apart.

If the plank is brittle, it could snap in the middle.
Brittle materials are strong in compression, but they are very weak in tension.

Graphs for tension and compression

Look at the graph showing the effect of tensile (stretching) and compressive (squashing) forces on a solid:

Can you see that the line has the same gradient for small compressions as for small extensions?

The gradient for compression and the gradient for extension are the same because the equation $F = k\,\Delta x$ gives the same spring constant k in both directions.

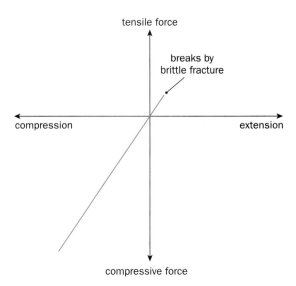

Can you see that the extension graph stops at small extensions, while the compression graph continues much further?

This shows that the solid is **brittle**, and snaps under tension.

Ties and struts

A shelf can be firmly fixed to a wall, but the turning moment produced by a heavy object can make it bend or even give way and snap. To prevent this happening, it needs to be supported at the outside end.

This support can be provided from above by a **tie** (stay), or from underneath by a **strut**.

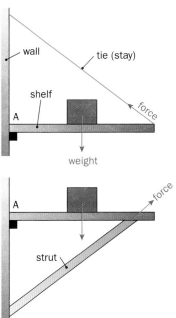

Look at the two ways of supporting the same shelf:

In each case, the shelf is supported by the wall at the edge marked A.

In the first one, a wire tie (coloured blue) provides the vertical component of force needed at the outside end of the shelf. The wire is in tension.

In the second one, a strut (coloured orange) provides the vertical component of force needed at the outside end of the shelf. The strut is in compression.

Of course, you can't use a wire strut – it has to be rigid!

Ceramics

Pottery, tiles and bricks are *ceramic* materials. They are hard and very stiff and they can resist very high temperatures. Glass has similar mechanical properties, but it melts more easily.

The drawback of ceramic materials is that they are very **brittle** and will crack easily if put under tension, even though they are very strong under compression.

Building materials such as stone and concrete have the same properties. The Parthenon is still standing because the stone is kept in compression. The 'span' between pillars must be quite small to prevent the stone from bending enough for the tension at the bottom of span to cause brittle fracture.

The Parthenon in Athens.

Composite materials

A concrete beam is made less brittle by reinforcing it with steel. Steel is very strong under tension, and does not crack. If the reinforced concrete should start to crack underneath, the steel bars will stretch slightly and carry the tensile stress. Reinforced concrete has two components – concrete and steel – and combines the good properties of both. Concrete is cheap and strong under compression. Steel is strong under tension.

Materials with two or more components which combine the good features of each are called **composite materials**.

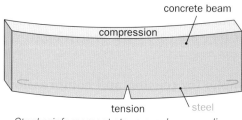
Steel reinforcement stops cracks spreading.

▶ Physics at work: Composite materials

Tennis rackets

Tennis rackets have always been made of **composite materials**. Early tennis rackets were made of wood – a natural composite which consists of stiff, brittle fibres of cellulose embedded in a tough 'background material' or 'matrix' called lignin.

Modern rackets contain strong, stiff fibres of carbon. These fibres are much stiffer than cellulose fibres, but they are brittle and snap easily. So the fibres are embedded in a tough matrix of epoxy resin.

These new rackets are lighter, stronger and much stiffer. They also will not rot or warp like wooden ones.

Climbing ropes

This climber has to trust his rope with his life if he should slip.

Early climbers used hemp ropes. They were light and strong but very stiff, as hemp does not deform plastically. In a fall, the rope does not stretch much. It would stop the climber very quickly and may cause internal injuries.

A rubber rope would have the opposite problem. It would stretch far too much, and the climber could hit the ground.

Modern climbing ropes are composites: a tough woven sheath protects the core.
The nylon core stretches enough to absorb the kinetic energy of the falling climber, without the deceleration force injuring him.

Summary

Density, $\rho = \dfrac{\text{mass, } m}{\text{volume, } V}$

For elastic materials, $F = k\,\Delta x$, where k is the spring constant of the material. In extension or in compression, energy stored in the spring = area between the line and the x-axis = $\frac{1}{2}F\,\Delta x$ if the graph is a straight line.

Stress, $\sigma = \dfrac{\text{force, } F}{\text{cross-sectional area, } A}$ Strain, $\varepsilon = \dfrac{\text{extension, } \Delta x}{\text{original length, } L}$ Young modulus, $E = \dfrac{\text{Stress, } \sigma}{\text{Strain, } \varepsilon}$

The behaviour of solid materials can be explained in terms of their atomic structure. Materials can be classified as crystalline (including polycrystalline), polymeric or amorphous.

▶ Questions

1. Can you complete these sentences?
   a) The density of a material is the .... per unit .... of the material.
   b) The extension of a spring is directly .... to the force extending it, providing that the elastic .... has not been exceeded.
   Calculating the area under the force–extension graph gives you the .... potential energy.
   c) Stress = ... ÷ .... and strain = ... ÷ ....
   For an elastic solid, if you divide stress by strain you get the .... modulus.
   d) Copper is .... under large stresses because energy is absorbed when atoms flow in a plastic way. Materials like glass are .... because atoms cannot flow plastically.
   Polymers like rubber or polythene can extend with large strains because their molecules are long ...., which can be straightened out easily.

2. A force of 5.0 N extends a spring 0.10 m.
   a) Calculate its spring constant.
   b) Calculate the elastic potential energy stored in the spring.

3. An experiment with a spring gave the following data:

Force /N	extension /m
0	0
2.0	0.10
4.0	0.21
6.0	0.29
8.0	0.41
10.0	0.50
12.0	0.64
14.0	0.81

Plot a force–extension graph with this data.
Use it to find:
a) the spring constant k,
b) the energy stored when the spring has been stretched 0.60 metres, and
c) the elastic limit of the spring used.

4. The cross-sectional area of a rope is 1.4×10^{-4} m^2, and it is stretched by a 7000 N force. Calculate the tensile stress.

5. The ultimate tensile stress of aluminium is 1.0×10^8 Pa. What is the maximum tension that is possible in an aluminium wire of cross-sectional area 2.5×10^{-7} m^2?

6. A high-carbon steel wire, circular in cross-section, has a diameter of 0.1 mm. What is the force needed to break it? (The ultimate tensile stress of high-carbon steel is 1.0×10^9 Pa.)

7. A rubber cord of original length 0.2 m can be stretched until it is 1.4 m long.
   a) What is its maximum extension?
   b) Calculate its maximum strain.

8. A copper wire has a strain of 0.000 20 under a tensile stress of 2.6×10^7 Pa. Calculate the Young modulus of copper.

9. In detective films, bank robbers often make off with piles of gold bars.
   The density of gold is 19 300 kg m^{-3}.
   A bar measures 0.2 m × 0.08 m × 0.05 m. What is its mass? Do you think a robber could escape with 5 gold bars?

10. The molar mass of gold is 0.197 kg (see page 214). Use this data and the density in question 9 to show that the size of a gold atom is about 2.6×10^{-10} m. (Avogadro constant $N_A = 6.0 \times 10^{23}$ mol^{-1}.)

11. Canoes are made from fibreglass, which is a composite material made from stiff glass fibres embedded in a tough epoxy resin. Why is this better than glass or resin alone?

Further questions on page 222.

197

14 Thermodynamics

Thermodynamics is the study of energy and its effects on a system.
This branch of Physics started with the men who built the earliest steam engines. They wanted to know how to make their engines more efficient.
This helped to develop our ideas about temperature and energy.

In this chapter you will learn about:
- internal energy, heat and work,
- absolute zero, and temperature changes,
- changes of state between solid, liquid and gas.

James Watt's 1782 steam engine.

Heat and temperature

The words 'hot' and 'heat' can be confusing.
Look at the cartoon of a bath of water and a burning match:

Think about these two questions:
- Which is hotter, the bath water or the burning match?
- Which needed more energy to heat it?

You can use **hot** to mean 'at a high temperature'.
The match is hotter than the bath water. It is at a higher temperature than the bath water.

But the bath water needed more energy to heat it, because there are many more molecules in it. You couldn't heat up a whole bath of water with just one match!
Some people use the name 'heat energy' to mean 'the energy in a hot object', but a more correct name is **internal energy**.

A burning match is hotter than bath water.

Internal energy

Look at the diagram of a hot potato:

The molecules in the hot potato jiggle about vigorously.

They have two types of energy.
The moving molecules have **kinetic energy**, but their movement is disorganised – it changes all the time, randomly.

The **potential energy** of the molecules also changes as they get squeezed closer together and move further apart.

On average, the molecules in a hot potato have more energy than the molecules in a cold potato.

This random molecular energy is called **internal energy**.
The usual symbol for internal energy is **U**.
It is measured in joules (J).

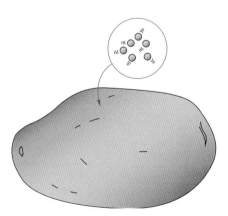

Molecular energy.

Systems and internal energy

In thermodynamics, what is a **system**?
It is the object or group of objects that we are thinking about.
We often need to think about the transfer of energy in and out
of a system.

Look at this diagram of the potato being cooked:

100 joules of energy go into the potato, and 90 joules come
out again.
So there is a change in the internal energy U of the potato,
because more energy went in than came out.
A change in internal energy has the symbol ΔU.
Δ (delta) means 'a change in'.

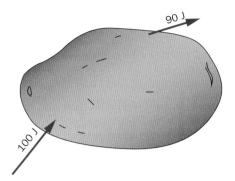

An increase in internal energy.

If you put energy into a system, its internal energy increases,
so ΔU is positive.
If you take energy out of a system, its internal energy drops,
so ΔU is negative.

In the diagram,
change in internal energy, ΔU = energy input − energy output

$$\Delta U = 100 \text{ J} - 90 \text{ J}$$
$$\therefore \quad \Delta U = +10 \text{ J}$$

Heating and working

In the diagram, a cold spoon is put into hot tea:

What are the internal energy changes here?
The tea heats the spoon, so the internal energy of the spoon
increases.
The spoon cools the tea, so the internal energy of the tea
decreases.

Internal energy changes in a cup of tea.

Heating is the name of the process when energy is transferred
from a higher temperature object to a lower temperature object.
In this diagram, the tea is heating the spoon.

Heat is a name for the energy transferred by heating.
The symbol for the energy gained by heating is Q or ΔQ.

Now look at the diagram of Phiz pushing a crate:

The crate and the ground are at the same temperature, but the
crate gets hotter as Phiz pushes it.
The internal energy of the crate goes up, due to friction with the
ground.

Working is the name for a transfer of energy, usually by a force,
when there is no movement of heat.
In this diagram, Phiz is working by pushing the crate against the
friction force exerted by the ground.

Hard work!

Work is the name for the energy transferred by working.
The symbol for the energy gained by work is W or ΔW.

Like heat, it is measured in joules.

The 1st Law of Thermodynamics

This is really just the **law of conservation of energy** (see page 72).

If you put energy into a system, its internal energy goes up.
The equation is:

> **Change in internal energy, ΔU = heat transfer, ΔQ + work done, ΔW**
> (J)   (J)   (J)

or, in symbols:

$$\Delta U = \Delta Q + \Delta W$$

where
ΔQ = heat transferred **into** the system
ΔW = work done **on** the system.

If ΔQ is positive, the system has **gained** energy by heating.

If ΔW is positive, the system has **gained** energy from work done on the system.

A warning:

*Some Examinations and textbooks write the symbol equation differently. They state that ΔW is the work done **by** the system.*

With this rule, if ΔW is positive, the internal energy has dropped. This makes ΔU negative.

The equation is now: $\Delta U = \Delta Q - \Delta W$

Common sense will always tell you what to do. Ask yourself – is the energy going in or out?

Example 1
A hot mug of tea is stirred (very vigorously) so that
5 J of work is done on the tea.
In the same time, 40 joules escape from the mug,
mostly by conduction and convection.
What is the change in internal energy?

$\Delta W = +5\,\text{J}$
It is positive, because the water has **gained** this energy in the stirring.

$\Delta Q = -40\,\text{J}$
It is negative, because the water has **lost** this energy by cooling.

$\therefore \quad \Delta U = \Delta Q + \Delta W = -40\,\text{J} + 5\,\text{J} = -35\,\text{J}$
The internal energy has <u>decreased by 35 joules</u>.

The 0th Law of Thermodynamics

After naming the First Law of Thermodynamics,
physicists realised that there is something more basic,
which they called the zero-th law.

Look at the goldfish in their bowl:

The fish are cold-blooded, so they are at the same
temperature as the water.
Because there is no flow of heat ΔQ between them, we
say that the water and the fish are in **thermal equilibrium**.

This is the 0th (Zero-th) Law:

**If two systems are at the same temperature,
there is no resultant flow of heat between them.**

Thermal equilibrium.

▶ The Celsius temperature scale

You are familiar with the Celsius scale of temperature.
Temperatures on this scale are given the symbol θ (theta).

In melting ice, the temperature $\theta = 0\,°C$.
This is called the **ice point**.
It is a fixed point of the Celsius scale.

The Celsius scale needs a second fixed point.
The steam from boiling water has a temperature $\theta = 100\,°C$.
This is the other fixed point of the Celsius scale.
It is called the **steam point**.

The diagrams show a typical liquid-in-glass thermometer.
How does this thermometer work?
The liquid expands as the temperature rises.

Where is 50 °C on the temperature scale?
It will be halfway between the 0 °C and the 100 °C marks.

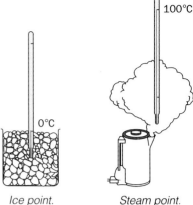

Ice point.   *Steam point.*

▶ Absolute zero and the Kelvin temperature scale

The Celsius scale starts at the freezing point of water.
This is not a fundamental temperature: it was chosen just for
convenience.
There is a more important and fundamental temperature at
−273 °C.
This is the **absolute zero** of temperature. See also page 212.

It is absolute because it is the lowest possible temperature, and
nothing can be cooled below absolute zero.

Absolute zero is the temperature at which the internal energy
of **any** system is at the lowest possible value.
You cannot cool a system that cannot lose internal energy!

260 K is not as hot as it sounds!

The **Kelvin temperature scale** starts at absolute zero,
but it has the same size units as the Celsius scale.
So 0 °C = 273 K   You just add 273.

Temperature in kelvin, T = temperature in degrees Celsius, θ + 273	or	$T = \theta + 273$
(K) (°C)		

The Celsius scale is based on the properties of water, but the Kelvin
scale does not depend on the properties of any particular substance.

- Kelvin temperatures have the symbol T,
  while Celsius temperatures have the symbol θ.
- The symbol for 'degrees kelvin' is K (**not** °K)!

Example 2
Your normal body temperature is 37 °C. What is this on the Kelvin scale?

$$T = \theta + 273 = 37\,°C + 273 = \underline{310\ K}$$

▶ Internal energy and atomic movement

Imagine you have a solid which is being heated steadily by a Bunsen burner.
Here is a graph that shows how its temperature changes:

When atoms in a solid are heated, they vibrate on their interatomic bonds, rather like masses oscillating on springs.

Between the points marked **A** and **B**, the energy is making the temperature of the solid increase, as the atoms oscillate more vigorously. When the atoms of the solid oscillate more, their average *kinetic* energy goes up.

Can you see what happens between **B** and **C**?
The temperature stops increasing at **B**, even though heat is still being supplied. This is where the solid *melts* into a liquid.

As the heating continues, the system still absorbs heat, so its internal energy rises. But because the temperature is constant, the average kinetic energy of the atoms is not increasing. The extra energy is breaking bonds between atoms, and this is increasing the average *potential* energy of the atoms.

Can you see what begins to happen at the point marked **C**?
The temperature begins to increase again, because the entire solid has turned into liquid. The energy supplied is increasing the average kinetic energy of the atoms again.

This carries on until the liquid reaches its boiling point, which is the point marked **D**, and the liquid *boils* into a gas.
The temperature stays constant until all the liquid has turned into gas.

When the internal energy of a system is increased, one of two things can happen:

- it can increase in temperature,
- it can change its state (melt or boil).

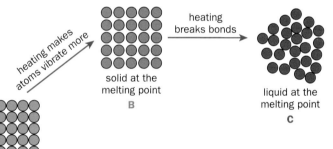

solid at the melting point
B

heating breaks bonds

liquid at the melting point
C

cold solid
A

The heating curve for a larger sample

The graph at the top of the page could be for a 1 kilogram sample of ice, for example.
But would you get the same graph for a 2 kilogram sample, heated at the same rate?

No – doubling the mass would double the number of atoms in the sample, and it would double the number of interatomic bonds.

There will be twice as many atoms to heat up to the melting point, and twice as many bonds to break as the whole sample melts.

Each section of the graph would take twice as long, as shown by the dashed line on this graph:

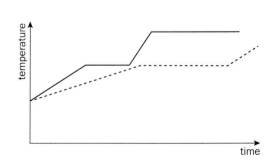

The rise in temperature and change of state both depend on the *mass* of the sample being heated.

▶ Specific latent heat

You saw (opposite) that energy is needed to break bonds when a substance melts or boils. This energy is sometimes called **latent heat**. The temperature is constant during this **change of state**.

To calculate the energy needed for a change of state, we use:

> **Heat transferred, ΔQ** = **mass, m** × **specific latent heat, L**   or $\Delta Q = mL$
> (J)   (kg)   (J kg^{-1})

The specific latent heat L is the energy needed to change the state of 1 kg of the substance (without changing the temperature).

Latent heat of *fusion* refers to the change from solid to liquid (melting).
Latent heat of *vaporisation* refers to the change from liquid to gas (boiling).

Example 3
The specific latent heat of fusion (melting) of ice is 330 000 J kg^{-1}.
What is the energy needed to melt 0.65 kg of ice?

$$\Delta Q = mL$$
$$= 0.65 \text{ kg} \times 330\,000 \text{ J kg}^{-1}$$
$$= 210\,000 \text{ J} \quad (2 \text{ s.f.})$$

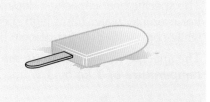

Measuring specific latent heat

Look at the diagram in Example 4 below.
The water is boiling and stays at a constant 100 °C.
Any energy delivered from the heater will turn water into steam.

If you know the power of the heater, then (from page 69) you can use:

> **Energy** (J) = **power** (W) × **time** (s)

Example 4
The power of the immersion heater in the diagram is 60 W.
In 5 minutes, the top pan balance reading falls from 600 g to 592 g.
What is the specific latent heat of vaporisation of water?

Energy ΔQ = power × time = 60 W × (5 × 60) s = 18 000 J

Mass of water evaporated = 600 g − 592 g = 8 g = 8 × 10^{-3} kg

$$\Delta Q = mL$$
$$18\,000 \text{ J} = 8 \times 10^{-3} \text{ kg} \times L$$
$$L = \frac{18\,000 \text{ J}}{8 \times 10^{-3} \text{ kg}} = 2.3 \times 10^6 \text{ J kg}^{-1} \quad (2 \text{ s.f.})$$

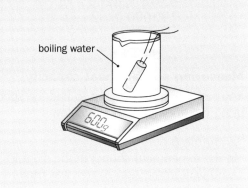

boiling water

This diagram shows a similar method for finding the specific latent heat of fusion (melting):

You fill the funnel with ice at 0 °C, and note the mass of water produced in 5 minutes.

But in a warm room, some ice will melt with energy received from the room, not the heater.
To allow for this you should do the experiment *twice,* for the same length of time: once with the heater *off*, once with it *on*.

The first experiment is a control, which tells you how much ice would have melted by absorbing heat from the room.
The ice actually melted by the heater is the difference between the masses in the two experiments.

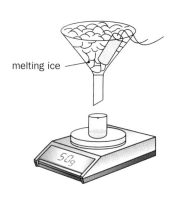

melting ice

▶ Specific heat capacity

For a substance being heated, the rise in temperature depends on the mass of the substance being heated and on how much energy has been put in. It also depends on what the substance is. For example, more energy is needed to give the same temperature rise to 1 kg of water than to 1 kg of copper.
In fact:

Change in internal energy, ΔU = mass, m × specific heat capacity, c × temperature change, ΔT
(J)   (kg)   (J kg^{-1} K^{-1})   (K)

or: $\boxed{\Delta U = mc\Delta T}$

If the energy is supplied as heat, the equation can be written: $\Delta Q = mc\Delta T$

The specific heat capacity c depends on the substance.
For example, water has a high specific heat capacity of 4200 J kg^{-1} K^{-1}, while copper has a value of only 380 J kg^{-1} K^{-1}.

The specific heat capacity is the energy needed to raise the temperature of 1 kg by 1 K (or 1 °C).

Example 5
0.5 kg of water is heated from 10 °C to 100 °C.
How much does its internal energy rise?
The specific heat capacity of water is 4200 J kg^{-1} K^{-1}.

The rise in temperature = 100 °C − 10 °C = 90 °C
Since °C and K are the same size, this is a rise of 90 K.

$\Delta U = mc\Delta T$
     = 0.5 kg × 4200 J kg^{-1} K^{-1} × 90 K
     = 190 000 J   (2 s.f.)

Measuring specific heat capacity

You can do an experiment to find the specific heat capacity using the apparatus shown below:

Example 6
In the experiment shown in the diagram,
a 60 watt immersion heater was used.
The beaker contained a kilogram of water at 21 °C.
After 5 minutes, the heater was switched off.
The temperature of the water went up to 25 °C.
What is the specific heat capacity of water?

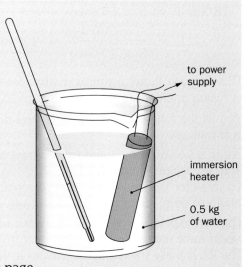

to power supply

immersion heater

0.5 kg of water

Change in internal energy, ΔU = power × time   (see page opposite)
                     = 60 W × (5 × 60) s   = 18 000 J

The rise in temperature = 25 °C − 21 °C = 4 °C = 4 K

$\Delta U = mc\Delta T$
18 000 J = 1 kg × c × 4 K
$\therefore c = \dfrac{18\,000 \text{ J}}{4 \text{ kg K}}$ = 4500 J kg^{-1} K^{-1} (2 s.f.)

This result is too large. Can you see why? This is discussed on the next page.

Energy losses in measuring specific heat capacity

Why was the result for the specific heat capacity in Example 6 too large?

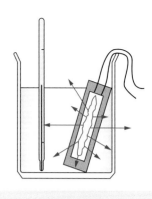

The 18 000 J from electrical heating did not all go to increase the internal energy of the water, because some heat will have escaped to increase the internal energy of the room.
There is also the rest of the apparatus − the heater itself, the beaker and the thermometer. Their internal energies have all gone up.

So what is the increase in internal energy of the water?
It must be *less* than 18 000 J. We can calculate it:

Example 7
What is the increase in internal energy of 1.0 kg of water whose temperature goes up 4.0 K?
The correct specific heat capacity of water is 4200 J kg^{-1} K^{-1}

Increase in internal energy $\Delta U = mc\Delta T$
$$= 1.0 \text{ kg} \times 4200 \text{ J kg}^{-1} \text{ K}^{-1} \times 4.0 \text{ K}$$
$$= \underline{17\,000 \text{ J}} \quad (2\,\text{s.f.})$$

The increase of internal energy of the water is 17 000 J.
This is less than the 18 000 J supplied by the heater in Example 6.
The other 1000 J heated up the apparatus and the surroundings.

Comparing specific heat capacities

This experiment (in Example 6) can also be done with a solid.
The diagram shows a solid block with holes drilled for the heater and the thermometer:

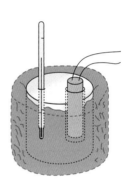

To try to reduce heat losses, the block has been lagged, by covering it with insulating material.

Here are the specific heat capacities for two metals, measured in this way.

1 kg of aluminium.

Substance	Aluminium	Iron
specific heat capacity (in J kg^{-1} K^{-1})	880	470

Can you see that 1 kg of aluminium needs almost twice as much energy as 1 kg of iron for each 1 K rise in temperature. Why is this?

If you have the same number of atoms of aluminium and iron, then the energy needed would be very similar.

But an aluminium atom is about half the mass of an iron atom, so aluminium contains twice as many atoms *per kg* as iron.

What number of atoms should you use for comparison?
The number used in counting atoms is the Avogadro constant, 6.0×10^{23} atoms, which make up one **mole** of the substance. (The mole is discussed in chapter 15.)

The **molar heat capacity** is the energy needed to raise the temperature of *one mole* of the metal by 1 K.

The molar heat capacities of different metals are similar because they contain the same number of atoms.

1 kg of aluminium contains more atoms than 1 kg of iron.

▷ Physics at work

Marathon running

Marathon running takes people to extremes of endurance. What are the risks to the runner?

The human body is only about 20% efficient, and *most* of the energy converted in the muscle raises the internal energy of the body − you get hot!

Unfortunately, the human body cannot tolerate a large rise in temperature. You would die if your body temperature rose by 5°C!

Normally, your body keeps its temperature down to 37°C by perspiration, because the evaporation of water reduces your internal energy and cools you.

To keep his temperature below 40°C, the runner pants as well as perspires, and the evaporation of water from his skin and in his lungs reduces his internal energy and cools him.

The runner loses so much water over the marathon that there is a real danger of becoming dehydrated, so he must drink regularly during the run.

Thermocouples

How do you measure the temperature of a lava flow? This geologist is using a **thermocouple**, which consists of two wires of *different* metals joined at two junctions. If one junction is hot and the other cold, an e.m.f. (voltage) is produced between them, which depends on the temperature difference.

This provides a good way of measuring temperature, as the hot junction can be small, and on long leads. Metals are good conductors of heat, so the thermocouple responds rapidly.

The cold junction is usually at the temperature of the surroundings, and the temperature of the hot junction may be as high as 1500°C − in this case, the lava was at 1160°C.

Heat pumps

Heat normally moves from a higher temperature to a lower one, but it can be made to move the opposite way − this is what a **heat pump** does.

A refrigerator is a heat pump. It works by evaporating and condensing a liquid.

You can use a heat pump to heat your house by having pipes buried in the ground outside. A liquid enters these pipes through an expansion valve. The drop in pressure makes the liquid evaporate, and the latent heat that it needs is absorbed from the ground that surrounds the pipes.

Once back inside the building, the vapour is compressed, which makes it condense into a liquid and give up its energy, so the house is heated.

If the weather is hot, you can reverse the flow of the liquid, and cool your house. This is exactly how refrigerators remove heat from the cold interior − if you put your hand round the back of a fridge, you will find out where the energy goes to.

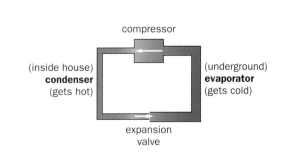

compressor

(inside house)
condenser
(gets hot)

(underground)
evaporator
(gets cold)

expansion
valve

▶ **Physics at work**

Energy movement inside the Sun

Our Sun is an obvious source of infrared as well as visible light. We can feel its warmth on our skin as well as seeing the light. This energy is generated by nuclear fusion in the core.

But how does this heat energy reach us?

Deep within the Sun, the core is surrounded by high-pressure **plasma** (ionised gas). Energy from the core escapes through this, by radiation of high-energy photons.
These photons are constantly absorbed and re-emitted, in different directions, by the densely packed plasma nuclei. Amazingly, it takes tens of thousands of years for the energy to cross this **radiative zone**.

The last part of the journey to the surface is through the **convective zone**. The particles here are no denser than the air in our atmosphere, and heat rises through it in the same way as hot air currents rise on Earth.

At the top of the convective zone, the heat is absorbed by the **photosphere**. The energy is emitted from the photosphere as electromagnetic radiation. This travels through space at the speed of light and reaches us 8 minutes later.

The outermost layers of the sun, the chromosphere and corona, are too thin to absorb or emit much radiation, although you can see them during a total solar eclipse.

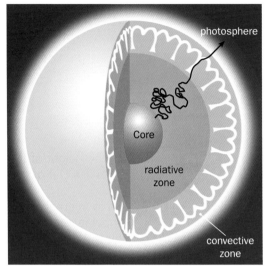

Global warming

The Earth is not an isolated system. It receives radiation from the Sun – about 1.4 kW for every square metre of surface facing the Sun – and it also emits radiation into space. A small amount of heat leaks out of the Earth's core, but the total internal energy U of the Earth remains constant if the incoming radiant energy and the emitted energy balance.

As the diagram suggests, the energy emitted by the Sun (orange) and the energy emitted by the Earth (red) are different. The surface of the Sun is at a temperature of 6000 K. We can use Wien's Law (see page 339) to calculate that the Sun's spectrum has a maximum intensity at a wavelength of 500 nm, which is in the visible region. The Earth's surface is at about 300 K, and the same method shows that the Earth emits radiation with a maximum intensity in the infrared region.

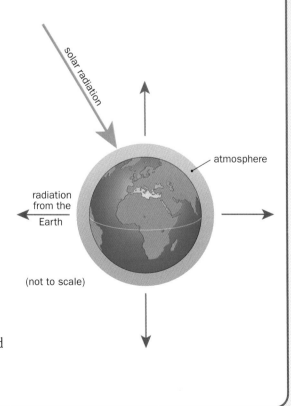

Over the past 200 years, the burning of fossil fuels has greatly increased the percentage of carbon dioxide in the atmosphere. Cutting down forests, which remove carbon dioxide from the atmosphere, makes this situation worse.

Carbon dioxide is a 'greenhouse gas', as it absorbs infrared radiation. This means that some of the Earth's emitted radiation is absorbed by the atmosphere. Although much of this is radiated out into space, some is returned to the Earth.
As a consequence, the internal energy U of the Earth has been slowly rising, which has led to increasing global temperatures.

Summary

A system is the entire collection of material whose energy we are measuring.
Zero-th Law of Thermodynamics: systems are in equilibrium if they are at the same temperature.

A system has internal energy U which can be changed by transferring heat Q or by doing work W.

First Law of Thermodynamics: $\Delta U = \Delta Q + \Delta W$ where Δ means 'a change in'.

Changing temperature: $\Delta U = mc\Delta T$ (or $\Delta Q = mc\Delta T$) where c = specific heat capacity.
Changing state: $\Delta Q = mL$ where L = specific latent heat.

▶ Questions

1. Can you complete these sentences?
   a) The internal .... of a system can be changed by heating or by ....
   b) On average, the molecules in a hot potato have more .... than the .... in a ...... potato.
   c) When you heat a solid, its temperature rises until it reaches its .... point and it starts to .... This is called a change of ....
   d) The energy is then used to break intermolecular .... until it is all liquid.
   e) A similar change happens at the .... point, but it requires more .... because all the intermolecular bonds have to be broken, to become a ....

2. A rubber band is stretched and let go many times. 50 J of work is done on the band in the stretching. While this is happening, 20 J of energy from the band heats the surroundings.
   What are (a) ΔQ, (b) ΔW, (c) ΔU for the system (the rubber band)?

3. The diagram shows a liquid-in-glass thermometer:

   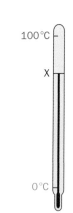

   The ice point (0 °C) and steam point (100 °C) have been marked.

   What is the temperature when the liquid is at the level marked X?

4. Copy and complete this table:

Temperature /°C	−78		6000
Temperature /K		20	

5. A horseshoe of mass 0.8 kg is heated from 20 °C to 900 °C. How much does its internal energy rise? The specific heat capacity of steel is 470 J kg^{-1} K^{-1}.

6. A kilogram of water falls 807 m (the height of the Angel Falls in Venezuela) and gains 7900 J of kinetic energy.
   Assuming that this is all converted to internal energy in the water, calculate the rise in temperature of the water. The specific heat capacity of water is 4200 J kg^{-1} K^{-1}.

7. A kettle contains 1.5 kg of water at 18 °C.
   a) How much heat energy is needed to raise the temperature of the water to 100 °C?
   b) Assuming no energy is lost to the room, calculate how long this will take if the power rating of the kettle is 2000 W.
   c) What else must you assume in part (b)?
   The specific heat capacity of water is 4200 J kg^{-1} K^{-1}.

8. Once the kettle in question 7 has reached 100 °C, the water will boil. Assume the lid has been left off, so the kettle does not switch itself off.
   a) How much energy is needed to boil away 0.5 kg of water?
   The specific latent heat of vaporisation of water, $L = 2.3 \times 10^6$ J kg^{-1}.
   b) Assuming no energy is lost, calculate how long this will take if the power rating of the kettle is 2000 W.

9. The specific latent heat of fusion (melting) of ice is 330 000 J kg^{-1}.
   a) Calculate the energy needed to melt 50 grams of ice which is already at its melting point.
   b) A glass contains 0.4 kg of lemonade at 20 °C, and 50 grams of ice at 0 °C are put in to cool it. Show that the temperature of the lemonade drops to about 10 °C.
   (Assume that the energy needed to melt the ice all comes from the lemonade. Ignore the energy needed to heat the melted ice up to the final temperature. The specific heat capacity of lemonade is 4200 J kg^{-1} K^{-1}.)

Further questions on page 223.

15 Gases and Kinetic Theory

A gas has a very low density, because its molecules are widely separated. But how do balloons keep their shape, when the gas molecules are so far apart? Surely the tight skin of the balloon should pull the molecules together?

In this chapter you will learn:
- about the gas laws and how to use them,
- about the kinetic theory of gases,
- about the Avogadro constant and the Boltzmann constant.

Moving molecules

Steam at atmospheric pressure is about 1000 times *less* dense than liquid water. This is because the molecules are about $10 \times$ further apart in each direction (and $10 \times 10 \times 10 = 1000$).
The molecules are so far apart that intermolecular forces are negligible.

If the molecules are so far apart, why don't they all just fall to the bottom of their container, like liquid molecules?
The molecules in a gas are constantly moving, and their constant movement spreads them out in their container.

If you use a microscope to look at the particles of soot in smoke, you will see evidence for this movement of gas molecules:
The smoke particles look like tiny spots of light as they reflect light into the microscope. These spots of light continually jiggle about in a *random* way. What is happening?

The molecules of the air are crashing into the dust particles in an irregular manner, and these irregular collisions make the smoke particles jiggle about at random.
This random jiggling is called **Brownian motion**.

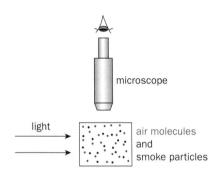

Brownian movement.

Molecules and pressure

What do we mean when we talk of 'gas pressure'?

Look at the diagram of gas molecules in a container:

The molecules all move at high speed and their collisions with the walls of the container are *elastic*. The molecules do not lose energy in elastic collisions, so they travel at the same speed afterwards.

During the rebound, the molecules push on the surface.
The rebound produces a force, and the numerous molecular forces over the area of the container surface produce a pressure.

Gas pressure can be measured only by measuring this force produced by gas molecules rebounding from a surface.
All pressure sensors work by measuring this force.

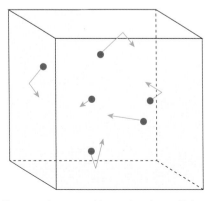

Pressure is caused by molecular collisions.

▶ Boyle's Law

In Chapter 13 we saw how solids stretch and compress. Robert Boyle wondered whether gases compress in the same way or not.

If you want to squash a solid, you must put it under a compressive stress. If you want to squash a gas, you must put it under pressure. When you try to squash the gas in a balloon into a smaller volume, you can feel that its pressure increases.

But what is the relationship between pressure and volume? This diagram shows the apparatus you can use to find out:

The oil traps a fixed mass of gas – air in this case – inside the glass tube. You can change the pressure and measure the volume. You need to wait a minute between readings to let the air return to room temperature, in order to keep the temperature of the gas constant during the experiment.

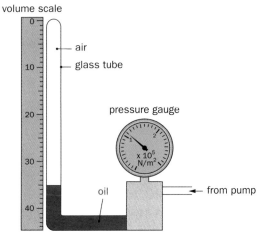

Boyle's Law apparatus.

The graph shows how the volume changes as the pressure changes for a sample of gas at a **constant** temperature:

Can you see that if you double the pressure from 20 kPa to 40 kPa, the volume is halved, from 4.0 m³ to 2.0 m³?
And if you halve the pressure, the volume doubles.

This is **Boyle's Law**:

For a fixed mass of gas, at **constant** temperature,

> **pressure p is inversely proportional to volume V**

Inverse proportion can be written in several different ways, but here are two of the easier ones:

> **pV = constant**   or   **$p_1 V_1 = p_2 V_2$**

p_1 and V_1 are the pressure and volume before the change;
p_2 and V_2 are the pressure and volume after the change.

You can use either of these ways of writing inverse proportion to calculate the changes of pressure or volume, as this example shows:

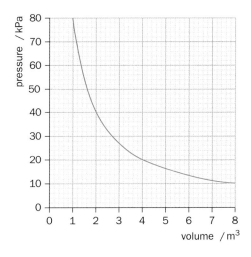

Example 1
A helium balloon has a volume of 200 m³ at an atmospheric pressure of 100 kPa.
It rises into the sky and expands until its volume is 250 m³.
What is atmospheric pressure at this height, assuming that the temperature stays constant?
(You can also assume that the balloon envelope does not exert significant extra pressure on the helium.)

Method 1
Before expanding: p = 100 kPa, V = 200 m³

$$pV = \text{constant}$$

100 kPa × 200 m³ = constant
So the constant = 20 000 kPa m³

After expanding: V = 250 m³ and p = ?

$$pV = \text{constant}$$

p × 250 m³ = 20 000 kPa m³

$$\therefore \ p = \frac{20\,000 \text{ kPa m}^3}{250 \text{ m}^3} = \underline{80 \text{ kPa}} \ (2 \text{ s.f.})$$

Method 2
Before expanding: p_1 = 100 kPa, V_1 = 200 m³
After expanding: V_2 = 250 m³ and p_2 = ?

$$p_1 V_1 = p_2 V_2$$

100 kPa × 200 m³ = p_2 × 250 m³

$$\therefore \ p_2 = \frac{100 \text{ kPa} \times 200 \text{ m}^3}{250 \text{ m}^3}$$

$$= \underline{80 \text{ kPa}} \ (2 \text{ s.f.})$$

Molecules and Boyle's Law

Look at this diagram of molecules in a cylinder:

The cylinder has a movable piston, and the molecules are bouncing off it.

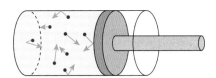

In the second diagram, the piston has been moved halfway along the cylinder, so that the volume has halved.
What happens to the pressure?

The density of molecules in the cylinder has now doubled, so twice as many of them will hit the piston every second.
This doubles the pressure, exactly as Boyle discovered.

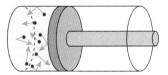

Half the volume, double the pressure.

Energy changes during compression

Look at the diagrams of a gas being compressed:

The work done on the gas, ΔW, is given by the equation

ΔW = force $\times$ distance (from page 68)

but, from page 26: force = pressure $\times$ area ..... (1)

and the distance the piston moves is given by

$$\text{distance} = \frac{\text{change in volume}}{\text{area}} \quad \ldots\ldots (2)$$

If you multiply equation (1) by equation (2), then area will cancel on the right-hand side, giving you:

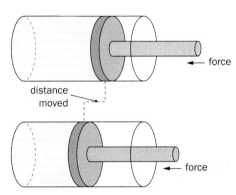

When you compress a gas, you do work on it.

Work done, ΔW = pressure, p $\times$ change in volume, ΔV
(J)   (Pa)   (m^3)

or

$$\Delta W = p\, \Delta V$$

This analysis is only correct if the force stays constant when you move the piston inwards. In practice the pressure, and the force, will change.
The equation is approximately true for **small** changes of pressure, so you can use it in those circumstances.

From the First Law of Thermodynamics (page 200), the internal energy of the gas changes if it gains heat energy or if work is done on it.
If the gas is going to stay at the same temperature, its internal energy must not change.

This means that the gas must **lose** heat when work is done on it during compression if it is to stay at a constant temperature.
That is why you need to wait between measurements in the Boyle's Law experiment opposite.

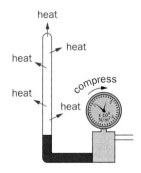

Heat escapes to keep the internal energy constant.

Example 2
A cylinder contains gas at a pressure of 2.0×10^5 Pa.
A piston moves inwards, reducing the volume of the gas by 0.012 m^3.
Calculate the work done.
You can ignore the slight change in pressure during the compression.

$\Delta W = p\, \Delta V$
   = 2.0×10^5 Pa $\times$ 0.012 m^3
   = 2400 J (2 s.f.)

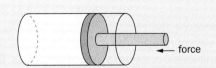

Boyle's Law and thermal equilibrium

Boyle's Law only holds true for a fixed mass of gas at a constant temperature.

- If the **mass** changes, then the number of molecules colliding with the surfaces of the container will change.
  This will change the pressure. Why?
  More molecules give more collisions, so the pressure goes up.
  Fewer molecules give fewer collisions and the pressure drops.

- If you change the **temperature** of the gas, the speed of the gas molecules will change.

  The graph here shows two p–V curves for the same sample of gas:
  The **blue** curve is the same as the one on page 210.
  The **red** curve is the same sample of gas at a higher temperature.

  Why is the red curve higher on the graph?
  This is because the hotter molecules are travelling faster and hitting the surfaces of the container harder, and more often.
  This gives a higher pressure.

  Each of these curves is called an **isotherm**, meaning 'at the same temperature'.

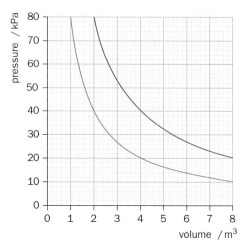

▶ Heating and cooling a gas

Boyle's Law is about the pressure and volume of a gas, if the temperature is kept constant.

Suppose you keep the volume constant and **change the temperature** of that sample of gas. What happens to the **pressure** of the gas?

The diagram shows one way to investigate this:

The sample of gas in the glass bulb is heated in a water bath, and the pressure is measured with a pressure gauge.
You can take readings of the pressure at different temperatures.

This simple method does have a flaw. The gas is not all inside the glass bulb; some is in the tubing and the pressure gauge.
This gas outside the glass bulb is at a different temperature, and you have to allow for this 'dead space' in accurate experiments.

When this is allowed for, the results fit a simple pattern, as the solid line on this graph shows:

If you extend this straight-line graph backwards, it meets the temperature axis at **−273 °C**.
This is **absolute zero** (see also page 201).

Because the pressure changes linearly with the temperature, as shown, apparatus like this can be used as a thermometer.
It is called a **constant volume gas thermometer** and is used to define our temperature scales.
Although it is too big and slow for everyday use, it is used to calibrate other types of thermometer.

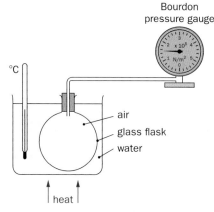

Changing the temperature of a constant volume of gas.

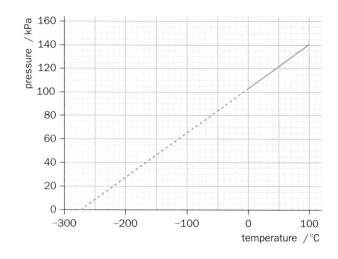

The pressure law

The graph at the bottom of page 212 is even simpler if the x-axis has the **absolute** temperature (in kelvins), as shown here:

This graph shows that the pressure of a fixed mass of gas at constant volume is **directly proportional** to the absolute temperature, T.

This is the pressure law: $p \propto T$ if the volume is constant.

For calculations, we use:

$$\frac{p}{T} = \textbf{constant}$$ or $$\frac{p_1}{T_1} = \frac{p_2}{T_2}$$

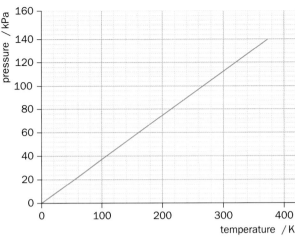

T_1 and T_2 are the **absolute** temperatures, before and after the change.

Example 3
The pressure of a sample of gas kept at constant volume is 101 kPa at 0 °C.
What will it be at 60 °C?

The temperatures T_1 and T_2 must be in kelvin (K) (see page 201), so:
starting temperature $T_1 = 0 + 273 = 273$ K, and final temperature $T_2 = 60 + 273 = 333$ K

Before heating, $p_1 = 101$ kPa, $T_1 = 273$ K, and after heating $T_2 = 333$ K and $p_2 = ?$

$$\frac{p_1}{T_1} = \frac{p_2}{T_2} \quad \text{so:} \quad \frac{101 \text{ kPa}}{273 \text{ K}} = \frac{p_2}{333 \text{ K}} \quad \therefore \ p_2 = \frac{101 \text{ kPa} \times 333 \text{ K}}{273 \text{K}} = \underline{123 \text{ kPa}} \quad \text{(3 s.f.)}$$

▶ Gases expanding at constant pressure

We have seen what happens when you keep the temperature constant (Boyle's Law), or keep the volume constant (the pressure law). What happens if you keep the pressure constant?

If you let the gas expand as it heats, so that the pressure stays constant, then the volume goes up in direct proportion to the absolute temperature. From this it follows that:

$$\frac{V}{T} = \textbf{constant}$$ or $$\frac{V_1}{T_1} = \frac{V_2}{T_2}$$

This is sometimes called **Charles' Law**.

Calculations with Charles' Law are very similar to example 3, but you use volumes V_1 and V_2 instead of pressures p_1 and p_2.

Ideal and real gases

'Ideal' or 'perfect' gases would obey these laws exactly. This is not always true for real gases! At high pressures or low temperatures, real gases behave differently.

Look at this graph of pressure against absolute temperature for a real gas at constant volume:
Compare it with the graph above.

It does not go all the way down to absolute zero, because the molecules start to attract each other as they get closer together. The gas condenses into a liquid.

Boyle's Law also is not always true for real gases. At high pressures a real gas can condense into a liquid.

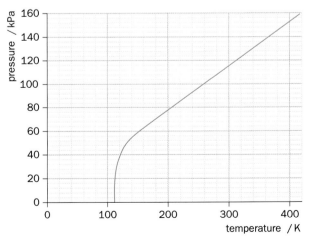

A real gas does not reach absolute zero.

▶ The Avogadro constant and the mole

We use the **Avogadro constant** when counting atoms or molecules.
It defines a quantity of matter called a **mole**.
One mole (abbreviated to mol) contains 6.02×10^{23} particles,
so we say that :

The Avogadro constant N_A = **6.02 × 10²³ particles mol⁻¹**.

Its usual symbol is N_A, but sometimes the symbol L is used.

Atoms and molecules are so very tiny that a real quantity of matter
has a huge number of particles in it.
A mole of water molecules would just fill one tablespoon.
Gas molecules are more spread out, so a mole of air molecules
would fill about 5 balloons.

The mass of one mole of a substance is its **molar mass**.

There are a lot of particles in a mole.

Example 4
The molar mass of oxygen is 32 grams mol⁻¹.
Find the mass of one molecule of oxygen.

One mole of oxygen molecules has a mass of 32 grams = 0.032 kg
So: one mole contains 6.02×10^{23} molecules and has a mass of 0.032 kg

$$\therefore \text{ the mass of one molecule } = \frac{0.032 \text{ kg}}{6.02 \times 10^{23}} = \underline{5.3 \times 10^{-26} \text{ kg}} \text{ (2 s.f.)}$$

When measuring the quantity of gas for calculations, do not use the
mass in kg. Make sure you always use the number of moles, n.

▶ The ideal gas equation

What happens if you change the pressure, the volume and
the temperature of a sample of gas?

If you increase the temperature of the gas, that would make the
volume increase. But if you increase the pressure, Boyle's Law
predicts that the volume would decrease.

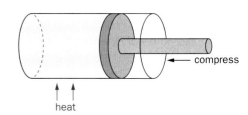

Changing pressure, volume and temperature.

To predict what happens when all three variables change, you need
to use the **ideal gas equation** :

Pressure × **volume** = n × a constant, R × temperature, T	or	$pV = nRT$
(Pa)　　　(m³)　　(mol)　(J K⁻¹ mol⁻¹)　　　(K)		

R, the **gas constant**, is the same for all gases: $R = 8.31$ J K⁻¹ mol⁻¹.　n is the number of moles of the gas.

Example 5
A helium gas cylinder is 0.200 m³ in volume, and contains 50.0 mol of gas at room temperature of 293 K.
What is the pressure in the cylinder? (You can assume that helium is an ideal gas.)

$$pV = nRT$$

$$\therefore \quad p \times 0.200 \text{ m}^3 = 50.0 \text{ mol} \times 8.31 \text{ J K}^{-1} \text{ mol}^{-1} \times 293 \text{ K} = 121\,700 \text{ J}$$

$$\therefore \quad p = \frac{121\,700 \text{ J}}{0.200 \text{ m}^3} = \underline{6.09 \times 10^5 \text{ Pa}} \text{ (3 s.f.)}$$

▶ Kinetic theory

The children's balloons on page 209 are kept inflated by the pressure of the gas inside them. But what are the molecules of the gas *doing* to produce this pressure?

Kinetic theory allows us to answer this question, if we assume the following idealised conditions about the molecules:

- They are perfectly elastic, so they bounce off their container walls and off each other without losing any kinetic energy.
- They exert forces on each other only when they collide.
- They are so tiny that they take up no space at all.

The conditions describe the properties of an **ideal gas**.
Real gases do not meet all these conditions.

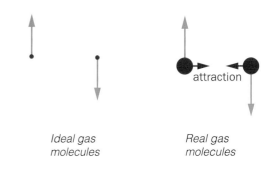

Ideal gas molecules   *Real gas molecules*

Molecular impacts and pressure

The 3 diagrams show a single gas molecule inside a cubical box. The molecule is moving horizontally from side to side at a speed v, bouncing off the opposite faces of the box. We shall be looking at the effects of this on the side of the box coloured yellow.

The three diagrams are shown at equal intervals of time: the time between each drawing and the next is the time for one trip across the box. Note that the molecule goes from left to right with a velocity v, and from right to left with a velocity $-v$.

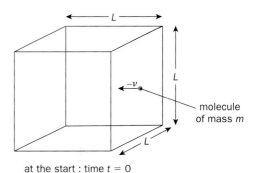

at the start : time $t = 0$

When the molecule bounces off the yellow side of the box, it changes in velocity from $+v$ to $-v$.
This means that it has a momentum change from $+mv$ to $-mv$, so momentum change $= (-mv) - (+mv) = -2mv$.

Momentum is conserved in the collision with the yellow side.
Momentum lost by molecule = momentum gained by the side, so momentum change of the yellow side $= +2mv$.

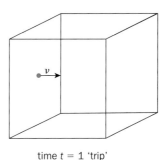

time $t = 1$ 'trip'

One 'trip' later, the molecule bounces off the left-hand side of the box, and heads back to the right-hand side again, where it bounces off two 'trips' later, shown in the third diagram.

Each 'trip' $= L$, so 2 'trips' $= 2L$. At a speed of v,

time for 2 'trips', $t = \dfrac{\text{distance}}{\text{speed}} = \dfrac{2L}{v}$

Force on yellow side $= \dfrac{\text{change in momentum}}{\text{time between collisions}} = \dfrac{2mv}{t}$ (see p. 54)

$$= \dfrac{2mv}{\left(\dfrac{2L}{v}\right)} = \dfrac{2mv \times v}{2L} = \dfrac{mv^2}{L}$$

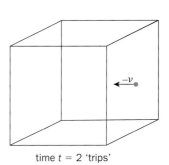

time $t = 2$ 'trips'

On the yellow side, pressure $p = \dfrac{F}{A}$, where $A = L^2$ (see p. 26)

$$\therefore p = F \div A = \dfrac{mv^2}{L} \div L^2 = \dfrac{mv^2}{L^3} = \dfrac{mv^2}{V}$$

because the volume of the cubical box is $V = L^3$.

On the next page, we will see how $p = \dfrac{mv^2}{V}$ can be adapted for many molecules travelling in different directions at different speeds.

The ideal gas equation

On the previous page, you saw that the pressure on the yellow side of the cubical box was $p = \dfrac{mv^2}{V}$ for the single molecule shown in this diagram:

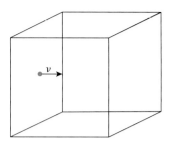

one molecule, one direction

A real gas in the box would be different from this in three ways:
- There would be many molecules.
- They would move in all possible directions.
- They would have a range of speeds, not all the same v.

The next box contains N molecules. If they were all moving in the same direction, this would increase the pressure N times. This is because the molecules all move independently of each other, with each one bombarding the yellow wall, so we add their changes in momentum.

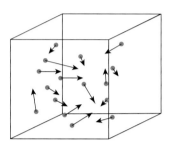

many molecules, many directions

But because they are moving in different directions, we should add the three components of momentum, at right angles to each other, as shown in the next diagram:

The **red** component is along the x-axis, the **blue** along the y-axis and the **purple** component along the z-axis (into the paper).

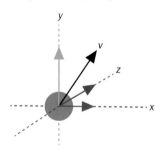

A simple way to average out all the components is to make this assumption: on average, the different components add up in exactly the same way as if one-third of the molecules were moving in the x-direction, one-third were moving in the y-direction and one-third were moving in the z-direction.
This means that the total pressure on the yellow side of the box

is due to $\dfrac{1}{3}N$ molecules, so $p = \dfrac{1}{3}N\dfrac{mv^2}{V} = \dfrac{1}{3}\dfrac{Nmv^2}{V}$.

Multiplying both sides by V, this becomes $pV = \dfrac{1}{3}Nmv^2$.

Different molecular speeds

The analysis above has still has not allowed for the fact that molecules do not all have the same speed v.

Look at this **_Maxwell–Boltzmann distribution_** graph showing how molecular speeds for the same sample of a gas vary at two different temperatures.

Can you see the differences for the two temperatures?

At the higher (**red**) temperature, the average speed is higher, and the spread of values is also higher.

One thing is the same for both graphs: the area under the curve is the same, because that is the total number of gas molecules.

Some molecules in the hotter gas have very high speeds. That is why heating gases makes chemical reactions faster: more of the molecules have enough energy to react.

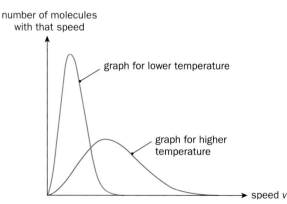

number of molecules with that speed

graph for lower temperature

graph for higher temperature

speed v

Maxwell–Boltzmann distribution.

Using the theoretical ideal gas equation

Because the molecular speeds v vary, as shown opposite, we must use an *average* molecular speed c, giving:

$$pV = \frac{1}{3}Nmc^2$$

Example 6

A mole of air molecules at room temperature and a pressure of 1.0×10^5 Pa has a volume of 0.024 m³.
How fast do these air molecules move? (The average mass of an air molecule is 4.8×10^{-26} kg.)

The Avogadro constant $N_A = 6.0 \times 10^{23}$ molecules mol⁻¹, so $N = 6.0 \times 10^{23}$

$$pV = \frac{1}{3}Nmc^2$$

$$1.0 \times 10^5 \text{Pa} \times 0.024 \text{m}^3 = \frac{6.0 \times 10^{23} \times 4.8 \times 10^{-26} \text{ kg} \times c^2}{3}$$

$$2400 = 0.0096 \times c^2$$

$$c^2 = \frac{2400}{0.0096} = 250\,000$$

so the average molecular speed, $c = \sqrt{250\,000 \text{ m}^2\text{s}^{-2}} = \underline{500 \text{ m s}^{-1}}$ (2 s.f.) (about 1000 mph)

R.m.s. molecular speed

The molecules in a gas do not all travel at the same speed, just as the runners in a race do not all run equally fast.
The equation above refers to 'an average molecular speed'. But what average should be used?

The average value of molecular speed c that we use is not the mean, but a **root-mean-square** (r.m.s.) average. This is the square root of the average (mean) of the squares of the speeds!

This example shows you how to calculate the r.m.s. speed.

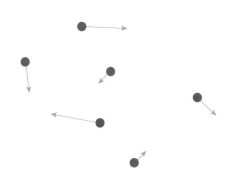

The molecules of a gas have different speeds.

Example 7

Four molecules have speeds of 300, 400, 500 and 600 m s⁻¹.
Calculate (a) the mean speed (average speed), and (b) the r.m.s. speed. Are they the same?

(a) Mean speed $= \dfrac{300 + 400 + 500 + 600}{4} = 450$ m s⁻¹.

(b) To find the r.m.s. speed, you must first find the mean of the squares of the speeds.

$$\text{Mean square speed} = \frac{300^2 + 400^2 + 500^2 + 600^2 \text{ m}^2\text{s}^{-2}}{4} = \frac{90\,000 + 160\,000 + 250\,000 + 360\,000 \text{ m}^2\text{s}^{-2}}{4}$$

$$= \frac{860\,000 \text{ m}^2\text{s}^{-2}}{4} = 215\,000 \text{ m}^2\text{s}^{-2}$$

The r.m.s. speed is the **square root** of the mean square speed, so
r.m.s. speed $= \sqrt{215\,000 \text{ m}^2\text{s}^{-2}} = \underline{460 \text{ m s}^{-1}}$ (2 s.f.) The mean speed is less than the r.m.s. value.

In a real sample of gas, the difference is much bigger.
In Example 6, the average molecular speed
is 500 m s⁻¹ at room temperature. This average is the r.m.s. speed.
In fact, the mean speed of those molecules is much less: 390 m s⁻¹.

Different examination boards write the mean square speed in different ways. For example, $(c_{rms})^2$, $\langle c^2 \rangle$ or $\overline{c^2}$ We have just used c^2 here, to keep the equations simpler.

▶ Kinetic energy and the Boltzmann constant, k

The ideal gas equation on page 214 and the theoretical ideal gas equation on page 217 both have the same subject:

$$pV = nRT \quad \text{and} \quad pV = \frac{1}{3}Nmc^2.$$

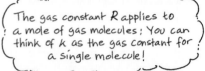

The gas constant R applies to a mole of gas molecules; You can think of k as the gas constant for a single molecule!

For 1 mole of gas, $n = 1$ and N = the Avogadro constant, N_A.

$$pV = RT \quad \text{and} \quad pV = \frac{1}{3}N_A mc^2.$$

You can combine these to give: $\quad mc^2 = \dfrac{3RT}{N_A} = 3kT.$

where $\quad$ | Boltzmann constant, $k = \dfrac{\text{gas constant, } R}{\text{Avogadro constant, } N_A}$

$\frac{1}{2}mc^2$ is the **mean kinetic energy** of a gas molecule, E_k.

so | mean kinetic energy of a gas molecule = $\frac{3}{2}$ × the Boltzmann constant, k × temperature, T | or | $E_k = \dfrac{3}{2}kT$
$\quad$ (J) $\qquad\qquad\qquad\qquad\qquad\qquad$ (J K^{-1}) $\qquad\qquad\qquad$ (K)

so the temperature of a gas allows us to calculate the average kinetic energy of its molecules.

Example 8
Calculate the mean kinetic energy of a gas molecule at 288 K.
The Boltzmann constant k is 1.38×10^{-23} J K^{-1}.

Average kinetic energy $= \frac{3}{2}kT = 1.5 \times 1.38 \times 10^{-23}$ J K$^{-1} \times 288$ K $= \underline{5.96 \times 10^{-21} \text{ J}}$ (3 s.f.)

The internal energy of a gas

In an *ideal* gas, the internal energy is just the sum of all the kinetic energies of the molecules. The gas does not have potential energy, like a spring, as we assume the molecules exert forces on each other only during collisions (page 215).

Look at the two diagrams of an ideal gas at different temperatures:

In an *ideal* gas at absolute zero, the molecules would have no kinetic energy, and so they would just lie on the bottom of the box. Note that a *real* gas would condense into a liquid or a solid!

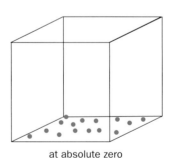

at absolute zero

Above absolute zero, the molecules have mean kinetic energy E_k and exert a pressure p on the sides of the box.
To calculate the internal energy U of the gas, we need to add the kinetic energies of all N gas molecules in the box.

For N molecules with mean kinetic energy E_k,

$$U = N \times E_k = N \times \left(\frac{3}{2}kT\right) = \frac{3}{2}NkT$$

For the special case of 1 mole of gas, $N = N_A$.

$$\therefore \quad U_{\text{mole}} = \frac{3}{2}N_A kT$$

Because $k = \dfrac{R}{N_A}$, $R = N_A k$, and so $\quad \boxed{U_{\text{mole}} = \dfrac{3}{2}RT}$

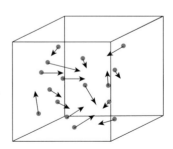

at a temperature T (in kelvin)

R can be replaced by $N_A k$ in any equation, so that
for 1 mole of gas, $\qquad pV = RT = N_A kT$
and for n moles, $\qquad pV = nRT = nN_A kT = NkT$
where $N = nN_A$ is the number of molecules in the gas.

Example 9
Calculate the internal energy of a sample of 50 g of helium at a temperature of 21 °C.
You can assume that the helium is an ideal gas.
The Avogadro constant $N_A = 6.0 \times 10^{23}$ mol^{-1}, the Boltzmann constant $k = 1.4 \times 10^{-23}$ J K^{-1}, and the molar mass of helium = 4.0 g.

50 g helium $= \dfrac{50.0 \text{ g}}{4.0 \text{ g}}$ moles = 12.5 moles

∴ the number of molecules N = 12.5 mol $\times N_A$ = 12.5 mol $\times 6.0 \times 10^{23}$ mol^{-1} = 7.50×10^{24}

$T = 21\,°C = (21 + 273)\,K = 294\,K$

Internal energy $U = \dfrac{3}{2}NkT = 1.5 \times 7.50 \times 10^{24}$ molecules $\times 1.4 \times 10^{-23}$ J K^{-1} $\times$ 294 K = $\underline{46\,000}$ J (2 s.f.)

▶ **Physics at work: Liquefying helium**

If all gases were *ideal*, it would not be possible to liquefy them.

Real gases have forces of attraction between their molecules, and these forces become greater as they get closer together. To liquefy a gas, it is necessary to reduce the kinetic energy of its molecules enough for the molecules to get very close. When the molecules are close enough, the attractive inter-molecular forces bond the molecules together.

Water molecules have strong forces of attraction, so water vapour (steam) does not behave as an ideal gas. It is easy to condense steam. Helium molecules are single atoms, and the forces between them are extremely small. Helium is very difficult to liquefy.

Another factor which makes helium much harder to liquefy than water vapour is the small mass of helium atoms. At the same temperature, helium atoms and water molecules have the same kinetic energy, $\frac{1}{2}mv^2$. Because helium atoms have less than a quarter of the mass of water molecules, they must have a value of v^2 which is more than four times greater to give the same value of $\frac{1}{2}mv^2$. This means that helium atoms move more than twice as fast as water molecules at the same temperature, so helium atoms rarely stay close enough to each other to form interatomic bonds. To liquefy helium, the temperature must be very low indeed: only 4.2 K (-268.9 °C).

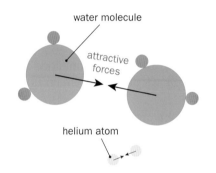
water molecule
attractive forces
helium atom

Helium: the most difficult gas to liquefy.

If helium is so difficult to liquefy, why bother?

The reason is that many, many applications of modern physics will only work well very close to absolute zero, and liquid helium is used to cool them in the same way as water is used to cool a car engine.

Some examples which make use of liquid helium as a coolant are:

* superconducting electromagnets in medical MRI scanners,
* superconducting electromagnets in the LHC (Large Hadron Collider) at CERN,
* sensors in satellites observing infrared emitted by stars.

Testing a superconducting magnet for LHC.

▶ **Physics at work: Gases**

Helium gas in the atmosphere

Saturn and Earth (not to scale).

Why is there hardly any helium in our atmosphere, and none at all on the Moon or the planet Mercury, while the giant planets like Saturn and Jupiter have large proportions of helium?
Although it is thought that the outer planets always had more helium than the inner planets, you would expect to find helium throughout the whole Solar System.

At the higher temperatures of the inner planets, the light helium molecules move very rapidly.
The faster molecules have enough kinetic energy to escape the weaker gravitational pull of the inner planets, and so all the helium gradually leaks away.

The outer planets are much more massive and are very cold. The helium molecules are slower and cannot escape the stronger gravitational fields.

Scuba diving

In normal breathing, you need about 10 to 20 litres of air every minute. Scuba (self-contained underwater breathing apparatus) allows divers to remain under water for many minutes.

A typical scuba gas cylinder holds 10 litres of gas at 200 times atmospheric pressure.
Boyle's Law shows that at ordinary atmospheric pressure this becomes 2000 litres – enough for over an hour of diving.

As the diver goes deeper, the surrounding pressure increases. The air breathed by the diver is at that same higher pressure. At high pressures, the nitrogen dissolves in the blood and can produce nitrogen narcosis, which is a hazard for divers. For this reason, divers often use air tanks with higher proportions of oxygen and less nitrogen than ordinary air.

Divers need a good supply of breathable gas.

Hot-air balloons

Hot air is less dense than cold air.

Hot-air ballooning is a popular sport, but how can a heavy canopy, with basket and balloonists, be lifted by air alone?

A floating balloon, like any floating object, has the same average density as the surrounding fluid.
The hot air in the balloon is at an average temperature of 150 °C or 423 K, which is about 1.5 times the absolute temperature of the surrounding air.
At this higher temperature, the volume of every kilogram of gas will be 1.5 times bigger than the air outside, as the pressure is the same.

This means it has a lower density, and the air in a balloon can weigh about 10 000 N *less* than the same volume of cold air.

This means that the balloon can lift 10 000 N (about 1 tonne) – enough to lift the canopy, basket, burner and balloonists.

Summary

Gas molecules are constantly moving and exert pressure by collisions with the container walls.

For a fixed mass of gas at constant temperature, pV is constant (Boyle's Law) so: $p_1 V_1 = p_2 V_2$

For a small change in volume ΔV, at constant temperature, the work done $\Delta W = p \Delta V$

In all calculations with gases, the absolute (kelvin) temperature scale is used when using

$$\frac{p_1}{T_1} = \frac{p_2}{T_2} \text{ at constant volume} \qquad \frac{V_1}{T_1} = \frac{V_2}{T_2} \text{ at constant pressure} \qquad \text{so:} \qquad \frac{p_1 V_1}{T_1} = \frac{p_2 V_2}{T_2}$$

A mole of any substance contains 6.02×10^{23} particles. For n moles of an ideal gas, $pV = nRT = NkT$

The molecules in a gas have different speeds. The r.m.s. speed c is the value used in calculations.

For N ideal gas molecules of r.m.s. speed c, each of mass m, $\quad pV = \frac{1}{3} Nmc^2$

The mean kinetic energy of a gas molecule, $\frac{1}{2}mc^2 = \frac{3}{2}kT$. $\quad$ Boltzmann constant, $k = \dfrac{R, \text{ the gas constant}}{N_A, \text{ Avogadro constant}}$

Internal energy of an ideal gas, $U = \frac{3}{2}NkT$

▶ Questions

1. Can you complete these sentences?
   a) Gases are much less dense than liquids or .... because in a gas the .... are widely separated. Brownian motion gives us evidence for the movement of .... The collision of gas molecules with the surfaces of the container causes ....
   b) If the temperature does not change, the pressure of the gas is .... proportional to the volume of the container. This is ....'s Law.
   c) If the temperature also changes, you can use the .... gas equation, $pV = nRT$, where R is the .... constant and n is the number of ....
   d) To calculate the average kinetic energy of gas molecules, you need their .... speed. This kinetic energy is directly proportional to the temperature in ....

2. A fixed mass of gas at a constant temperature has a pressure of 2000 Pa and a volume of 0.02 m^3. It is compressed until the volume is 0.005 m^3. What is its new pressure?

3. A gas expands at a constant pressure of 1.0×10^5 Pa. Its volume increases from 0.1 m^3 to 0.15 m^3. How much work is done in the expansion?

4. At a temperature of 200 K, the pressure of air in a flask is 100 kPa. What will the pressure be at a temperature of 300 K? You can assume that the volume of the flask is constant.

5. The molar mass of helium is 0.004 kg mol^{-1}. How many molecules of helium are there in 1 kg? (Avogadro constant $N_A = 6.0 \times 10^{23}$ mol^{-1}.)

6. At a temperature of 200 K, the volume of a sample of gas in a cylinder with a smoothly fitting piston is 0.0024 m^3. The cylinder is heated while allowing the gas to expand at constant pressure. Calculate the volume of the gas at a temperature of 300 K.

7. A room measures 4 m × 5 m × 2.5 m.
   a) Find the volume of air in the room, and calculate its mass. The density of air = 1.2 kg m^{-3}.
   b) The average molar mass of air is 0.030 kg. How many moles of air are there in the room?

8. An air cylinder for a scuba diver contains 150 moles of air at 15 °C. The cylinder has a volume of 0.012 m^3. (Gas constant $R = 8.3$ J K^{-1} mol^{-1}.)
   a) Calculate the pressure in the air cylinder.
   b) Calculate the volume that the gas will have when it is all released at atmospheric pressure of 100 kPa and a temperature of 25 °C.

9. A mole of hydrogen molecules, each of mass 3.3×10^{-27} kg, is contained in a cylinder of volume 0.05 m^3. The molecules have an r.m.s. speed of 800 m s^{-1}. Calculate the pressure of the gas. (Avogadro constant $N_A = 6.0 \times 10^{23}$ mol^{-1}.)

10. Helium at atmospheric pressure of 100 kPa has a density of 0.17 kg m^{-3} at 273 K.
    a) Calculate the r.m.s. speed of the molecules.
    b) At the same temperature and pressure, the density of air is 1.2 kg m^{-3}. What can you say about the speeds of sound in the two gases?

Further questions on pages 224–225.

Further Questions on Matter and Molecules

▶ Materials

1. Materials are often described as being *crystalline*, *polymeric* or *amorphous*.
   Explain the meaning of the terms in italics and name one example of each type of material. [6] (AQA)

2. The table and graph show the properties of TWO materials A and B.

Material	Young modulus / 10^{10} Pa	Ultimate tensile stress / 10^8 Pa	Nature
A			
B	0.34	3.2	brittle

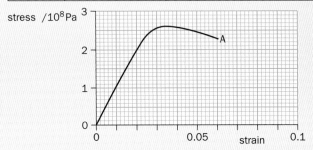

a) Use the graph to complete the table for A. [3]
b) Use the table to draw a graph showing the behaviour of material B. [2] (Edex)

3. a) The Young modulus is defined as the ratio of *tensile stress* to *tensile strain*. Explain what is meant by the two terms in italics. [3]
   b) A long wire is suspended vertically and a load of 10 N is attached to its lower end. The extension of the wire is measured accurately. In order to obtain a value for the Young modulus of the wire's material two more quantities must be measured. State what these are and in each case indicate how an accurate measurement might be made. [4]
   c) Sketch a graph which shows how stress and strain are related for a ductile material. [2] (AQA)

4. The diagram shows the stress–strain graph for a sample of steel up to the point of fracture.

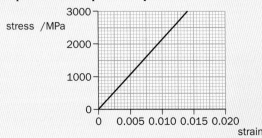

a) Use the graph to find the Young modulus of steel and the ultimate tensile strength of steel. [2]
b) A cable, consisting of 7 strands each of cross-sectional area 2.5×10^{-6} m^2 and length 12 m, is to be made from this steel. Find the force at which the cable is expected to break, the maximum extension of the cable and the elastic strain energy of the cable at maximum extension. [5] (OCR)

5. a) Draw a diagram to illustrate a dislocation in the lattice of a single crystal free from any other defects. By referring to your diagram, explain how the motion of a dislocation can lead to the plastic deformation of such a crystal. [5]
   The variation of load with extension for a specimen may be observed by slowly increasing the extension and recording the corresponding load. The graph illustrates the result of such a test for a steel wire.

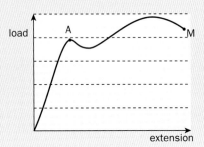

b) On a copy of the graph, draw a line showing what would have happened if the applied load had been gradually reduced to zero after reaching about 90% of the maximum value. [2]
c) Explain why, as the extension continues to increase, there is a reduction in load at A. Name the type of fracture occurring at point M. [3] (AQA)

6. The graph shows the results of an experiment in which stretching force is plotted against extension for a steel cylinder having an original length of 27.2 mm and an initial cross-sectional area of 1.8×10^{-5} m^2. The specimen is extended steadily to **X**, the load reduced to zero at **Y** and then reapplied. The extension is then gradually increased until the specimen breaks.

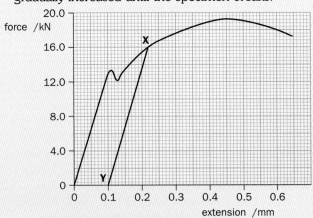

a) State Hooke's Law. Sketch the graph and indicate the limit of proportionality, P. [2]
b) Calculate the Young modulus for the specimen in the region where Hooke's Law is obeyed. [2]
c) Calculate the energy stored in the specimen when the extension is 0.10 mm. [2]
d) Explain in terms of the structure of the material why the specimen does not return to its original length when unloaded along **XY**. [2] (AQA)

7. A load of 3.0 N is attached to a spring of negligible mass and spring constant 15 N m^{-1}.

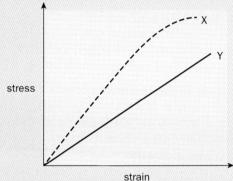

3.0 N

What is the energy stored in the spring?
A 0.3 J   B 0.6 J   C 0.9 J   D 1.2 J   [1]   (AQA)

8. The diagram shows how the stress varies with strain for metal specimens X and Y which are different. Both specimens were stretched until they broke.

stress

X

Y

strain

Which of the following is *incorrect*?
A X is stiffer than Y
B X has a higher value of the Young modulus
C X is more brittle than Y
D Y has a lower maximum tensile stress than X.
[1]   (AQA)

9. a) The diagram shows a length of tape under tension.

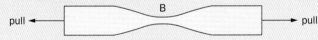

pull ← → pull

B

 i) Explain why the tape is most likely to break at point B.
 ii) Explain what is meant by the statement 'the tape has gone beyond its elastic limit'.   [2]
b) The diagram shows one possible method for determining the Young modulus of a metal in the form of a wire.

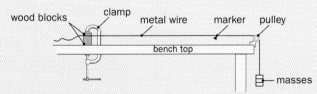

wood blocks   clamp   metal wire   marker   pulley
bench top
masses

Describe how you can use this apparatus to determine the Young modulus of the metal. Describe, with technical terms, spelled correctly, using the following structure:
 i) The measurements to be taken.
 ii) The equipment used to take the measurements.
 iii) How you would determine the Young modulus from your measurements.   [8]   (OCR)

▶ **Thermodynamics and Gases**

10. A student pours 525 g of water into a saucepan of negligible mass, heats it over a steady flame and records the temperature as it heats up. The temperatures are plotted as shown below.

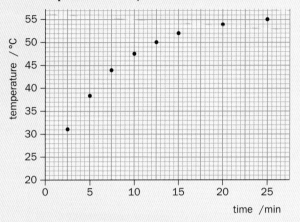

temperature / °C
time /min

a) Calculate the heat capacity of the water and saucepan. (Specific heat capacity of water = 4200 J kg^{-1} K^{-1})   [3]
b) Find the rate of rise of water temperature at the beginning of the heating process.   [2]
c) Find the rate at which energy is supplied to the saucepan and water.   [2]
d) Explain why the rate at which the temperature rises slows down progressively as the heating process continues.   [2]   (Edex)

11. a) A student seals 200 g of ice-cold water in a glass vacuum flask and finds that it warms up by 3.5 K in one hour. The specific heat capacity of water is 4200 J kg^{-1} K^{-1}. Calculate the average rate of heat flow into the flask in watts.   [3]
b) To check this result over a longer period, the student fills the flask with equal amounts of ice and water, all at 0 °C, and leaves it for four hours. The specific latent heat of fusion of ice is 330 kJ kg^{-1}. How much ice would you expect to have melted at the end of four hours?   [3]
c) The student found instead that the glass flask had collapsed into small pieces. Suggest a reason for the pressure inside the glass flask to drop sufficiently for the collapse to occur.   [3]   (Edex)

12. a) Describe an experiment to demonstrate the Brownian motion.   [4]
b) Compare (i) the spacing, (ii) the ordering and (iii) the motion of the molecules in ice at 0 °C and water, also at 0 °C.   [6]   (OCR)

13. Water is boiled in a kettle to produce steam at 100 °C.
a) Describe the motion of a typical water molecule in the steam.   [1]
b) Compare the mean kinetic energy of the water molecules in the steam with those in the boiling water.   [2]   (AQA)

14. 200 kJ of thermal energy is required to vaporise 1.0 kg of liquid oxygen at atmospheric pressure (100 kPa). This vaporisation is accompanied by an increase in volume of 0.23 m³.
    a) For this vaporisation, calculate
       i) the work done on the atmosphere by the expanding oxygen,
       ii) the change in internal energy of the oxygen. [4]
    b) The volume of the oxygen gas is measured as its temperature increases, the pressure remaining constant. At 100 K the volume is 0.25 m³ and at 290 K the volume is 0.79 m³. Show, by means of a calculation, whether oxygen behaves as an ideal gas over this temperature range. [3] (OCR)

15. The table shows data relating to the average power consumption for a person performing various activities.

activity	power consumption /W
sleeping	80
sitting	110
walking	300
swimming	900

    a) In a 24-hour period, a hospital patient spends 15 h sleeping, 7 h sitting up in bed, 1.5 h walking around the hospital gardens and 0.5 h swimming. Calculate
       i) the energy expended by the patient during the 24-hour period,
       ii) the heat energy generated in the patient's body during the 0.5-hour period in which he is swimming. The body converts energy into mechanical work with an efficiency of 30%. [5]
    b) Explain the mechanisms responsible for dissipating excess heat energy from the patient when
       i) he is in the water,
       ii) he leaves the water after his swim. [3] (OCR)

16. The diagram illustrates high-pressure air expelling a disc from a cylinder. The mass of the disc is 7.2 g and the maximum speed is 32 m s⁻¹.

    a) Calculate the energy transferred to the disc. [3]
    b) Use the kinetic theory of gases to account for the cooling of the air in the cylinder during the expulsion of the disc. [3]
    c) Explain why work is done by the air on the disc but no work is done by the disc on the air. [2] (Edex)

▶ Gases and Kinetic Theory

17. With reference to the appropriate physical principles, explain the following in terms of the motion of the gas molecules.
    a) A gas in a container exerts a pressure on the container walls. [4]
    b) The pressure increases if the temperature of a gas is increased, keeping the mass and volume constant. [3]
    c) State what is meant by the root mean square speed of the molecules of a gas. Calculate the r.m.s. speed of four molecules travelling at speeds of 400, 450, 500 and 550 m s⁻¹, respectively. [2]
    d) For a constant mass of gas, explain how the r.m.s. speed of molecules changes, if at all, when the gas expands at constant temperature. [2] (AQA)

18. a) The air pressure inside a constant volume gas thermometer is 1.010×10^5 Pa at the ice point and 1.600×10^5 Pa at the steam point. Determine the Celsius temperature when the pressure of the gas is 1.250×10^5 Pa. [2]
    b) The graph shows the variations of pressure with temperature for two constant volume gas thermometers A and B that contain the same volume of gas.

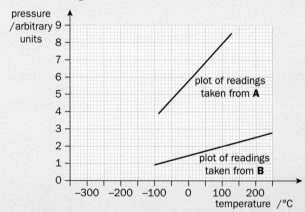

       i) Show that both graphs extrapolate to the same intercept on the temperature axis and determine that temperature. [1]
       ii) A is known to contain 0.024 mol of gas. Determine the number of moles of gas in B. [4] (AQA)

19. A gas cylinder of internal volume 0.050 m³ contains compressed air at 21 °C and pressure 1.2×10^7 Pa. The molar mass of air is 0.029 kg mol⁻¹.
    a) Calculate
       i) the number of moles of air in the cylinder,
       ii) the mass of air in the cylinder. [3]
    b) An additional 1.5 m³ of air at 21 °C and atmospheric pressure, 1.0×10^5 Pa, is pumped into the cylinder. Calculate the new pressure of air in the cylinder, assuming no change in temperature during the process. [4] (OCR)

20. 0.80 mol of an ideal gas is enclosed in a cylinder by a frictionless piston. Conditions are such that the gas expands at constant pressure (A to B); it is then compressed at constant temperature (B to C) until its volume returns to the original value. These changes are shown, with the relevant numerical data, on the graph. (Molar gas constant $= 8.31\,\mathrm{J\,K^{-1}\,mol^{-1}}$)

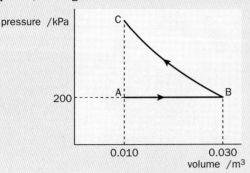

a) Show that the temperature of the gas at A is 300 K. [2]
b) Calculate the temperature of the gas at B. [2]
c) Calculate the pressure of the gas at C. [2]
d) Defining the terms used in your explanations, explain how the first law of thermodynamics applies to the changes from (A to B), and from (B to C). [5]
e) Use the graph to estimate the change in internal energy of the gas when it is returned from C to A at constant volume. Explain whether this is an increase or a decrease. [4] (AQA)

21. A 24 W filament lamp has been switched on for some time. In this situation the first law of thermodynamics, represented by the equation $\Delta U = \Delta Q + \Delta W$, may be applied to the lamp. State and explain the value of each of the terms in the equation during a period of two seconds of the lamp's operation. [6]
Typically, filament lamps have an efficiency of only a few percent. Explain what this means and how it is consistent with the law of conservation of energy. [2] (Edex)

22. a) Explain what is meant by the concept of *work*. Show that the work W done by a gas when it expands by an amount ΔV against a constant pressure p is given by $W = p\,\Delta V$. [5]
b) A fixed mass of oxygen at a pressure of 101 kPa occupies a volume of $1.40\,\mathrm{m^3}$ at 274 K and $1.54\,\mathrm{m^3}$ at 300 K. Calculate the work done to expand the gas against the external pressure. [2]
c) Define *internal energy* and state the *first law of thermodynamics*. State the energy conversion taking place when a gas is heated at constant volume. Explain why, when the temperature of a gas is raised, more heat energy is required when the gas is held at constant pressure than when its volume is kept constant. [7] (OCR)

23. An ideal gas is contained in a hollow cylinder, sealed at one end, with a frictionless piston at the other as in the diagram. The volume of the trapped gas is V_0 when its pressure is p_0 and its temperature T_0.

The system is used as a heat engine carrying out the following cycle of changes ABCDA. The gas is:
from A to B, compressed at constant temperature to half its initial volume V_0,
from B to C, heated at constant volume to 1.5 times its initial temperature T_0,
from C to D, expanded at constant temperature back to its initial volume,
from D to A, cooled at constant volume back to its initial temperature.
a) Find the values of the pressure p of the gas at the end of each of the stages of the cycle in terms of the initial pressure p_0. Give your reasoning. [3]
b) Draw a pV diagram showing the cycle of operation of the engine. Mark the points A, B, C and D on your diagram. [3]
c) Consider the stage A to B of the cycle. Is work W done on or by the gas? Is thermal energy Q supplied to the gas or extracted from it? How are Q and W related? Explain your reasoning. [3] (OCR)

24. The diagram shows a refrigerator running in a room as a thermodynamic system which is effectively insulated from the surroundings.
Within that system there are *two* subsystems: the inside of the refrigerator at a low constant temperature, and the room. The refrigerator cooling fins transfer heat from the refrigerator to the room. Work can be done on the system by means of the electrical energy supplied to the refrigerator.

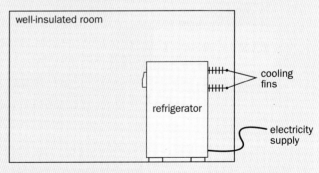

Applying the first law of thermodynamics and justifying your answers, state whether each of the quantities ΔU, ΔQ and ΔW is positive, negative or zero for
a) the refrigerator,
b) the air in the room. [4] (AQA)

225

16 Current and Charge

Electricity is important to all of us, and our lives would be very different without it. The cartoon shows some of the ways that we use electrical energy:

We use electricity to transfer energy from one place to another. How is this energy transferred from the power station to your home?
It is carried by an electric current which flows through cables. These are supported by pylons or buried beneath the ground.

In this chapter you will learn:
* about current as a flow of electric charge,
* about potential difference and resistance,
* how to use the equations for electrical power.

Understanding current electricity

A simple circuit consists of a lamp connected to a battery.
We can use a model to help us understand what is happening.
Our electrical circuit works rather like the water circuit in a central heating system.

Can you see the similarities in the two circuits shown below?

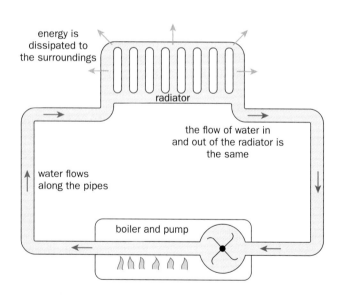

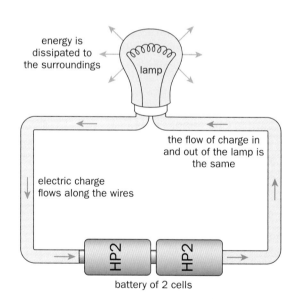

* Water flows around the complete loop of pipes.

* The boiler transfers energy to the water and the pump keeps the water circulating round.

* The radiator transfers energy from the hot water to the surroundings.

* Charge flows around the complete conducting path.

* The battery transfers energy to the charge and keeps it circulating round the circuit.

* The lamp transfers energy to the surroundings as heat and light.

What is an electric current?

Electric current is a flow of electric charges. The lamp lights because charged particles are moving through it.

Like water in the heating system, the charged particles are already in the conductors – but what are they?

A copper wire consists of millions of copper atoms. Most of the electrons are held tightly to their atoms, but each copper atom has one or two electrons which are loosely held.

Since the electrons are negatively charged, an atom that loses an electron is left with a positive charge and is called an **ion**.

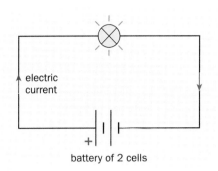

battery of 2 cells

The diagram shows that the copper wire is made up of a lattice of positive ions, surrounded by *'free' electrons*:

The ions can only vibrate about their fixed positions, but the electrons are free to move randomly from one ion to another through the lattice.
All metals have a structure like this.

What happens when a battery is attached to the copper wire?
The free electrons are repelled by the negative terminal and attracted to the positive one.
They still have a random movement, but in addition they all now move slowly in the same direction through the wire with a steady **drift velocity**.
We now have a flow of charge – we have electric current.

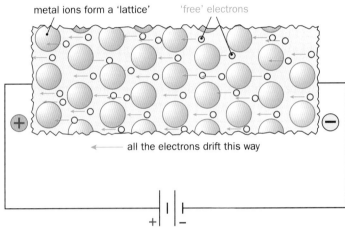

Which way do the electrons move?

At first, scientists thought that a current was made up of positive charges moving from positive to negative.

We now know that electrons really flow the opposite way, but unfortunately the convention has stuck.

Diagrams usually show the direction of 'conventional current' going from positive to negative, but you must remember that the electrons are really flowing the opposite way.

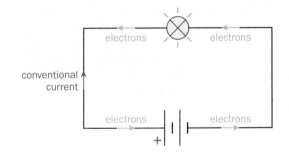

Conductors and insulators

Why are some materials conductors and some insulators?
To answer this question we need to think about the number of free electrons per cubic metre of material: **n**

The table shows some values for **n**:

Copper is a good conductor because in just $1 \, mm^3$ there are about 1×10^{20} or 100 million million million free electrons!

Can you see why an insulator cannot conduct?
In each mm^3 there is only about 1 electron free to move. Almost all of the electrons are firmly fixed to their atoms.

Semiconductors are very important in electronics.
Look at their value for **n**. Can you explain why semiconductors are neither good conductors nor good insulators?

Type of material	Number of free electrons per mm^3	n, number of free electrons per m^3
conductor	$\sim 1 \times 10^{20}$	$\sim 1 \times 10^{29}$
semiconductor	$\sim 1 \times 10^{10}$	$\sim 1 \times 10^{19}$
insulator	~ 1	$\sim 1 \times 10^9$

▷ Electric currents

Current is measured in **amperes** (A) using an ammeter. The ammeter is placed *in* the circuit so that the electrons pass through it. The more electrons that pass through the ammeter in one second, the higher the current reading in amps.

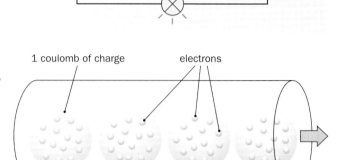

Does the position of the ammeter matter in this simple circuit? No – the number of electrons passing any point per second is constant. The current throughout the circuit is the same.

1 amp is a flow of about 6×10^{18} electrons in each second! The electron is too small to be used as the basic unit of charge, so instead we use a much bigger unit called the **coulomb** (C). The charge on 1 electron is only $\mathbf{1.6 \times 10^{-19}}$ **C**.

In the diagram, we have pictured one coulomb as a group of a large number of electrons. Can you imagine the current as coulombs of charge flowing through the circuit?

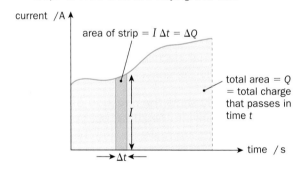

In fact, there are 6×10^{18} electrons in each coulomb!

The more coulombs of charge passing through the ammeter each second, the bigger the current. In fact:

Current = number of coulombs per second.

$$\boxed{\textbf{Current, } I = \frac{\textbf{charge, } Q \text{ (coulombs)}}{\textbf{time, } t \text{ (seconds)}}} \quad \text{or} \quad \boxed{I = \frac{Q}{t}} \quad \text{so} \quad \boxed{Q = I\,t}$$

From this last equation you can see that:

1 coulomb is the amount of charge that passes a point when a current of 1 ampere flows for 1 second.

What if the current in the circuit is changing with time? If we think about a charge ΔQ passing in a small time Δt then we can write:

$$\boxed{I = \frac{\Delta Q}{\Delta t}} \quad \text{or} \quad \boxed{\Delta Q = I\,\Delta t}$$

Graph to show a current varying with time

Can you see that $I\,\Delta t$ is the **area** of the dark strip on this graph? This means that the total charge Q that has passed after time t must be the **total area** under the graph, shown in pale green:

Remember that current is the charge passing per second, or more precisely, **current is the rate of flow of charge**.

Example 1
A current of 0.50 A passes through a lamp for 2.0 minutes.
a) How much charge passes through the lamp?
b) How many electrons pass through the lamp?

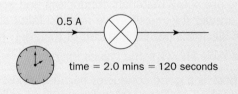

time = 2.0 mins = 120 seconds

a) $\quad \boldsymbol{Q = I\,t}$ (remember that t must be in seconds)
  $\therefore \; Q = 0.50 \text{ A} \times 120 \text{ s} \; = \underline{60 \text{ C}} \;$ (2 s.f.)

b) Each electron has a charge of 1.6×10^{-19} C,
  so 1 coulomb must contain $\dfrac{1}{1.6 \times 10^{-19}}$ electrons.
  This is 6.25×10^{18} electrons
  So in 60 C there are $60 \times 6.25 \times 10^{18}$ electrons
  $\qquad\qquad = \underline{3.8 \times 10^{20} \text{ electrons}} \;$ (2 s.f.)

Hint:
Do you find part (b) difficult?
If so, think about this:
1 penny is equal to £ 0.01
So £1 must contain $\dfrac{1}{0.01} = 100$ pence.

Conduction in liquids and gases

Salt solution is an **electrolyte**. It can conduct electricity.
In order to conduct, the liquid must contain charged particles
that are free to move. What are these moving charges?

Electrolytes contain both positive and negative charged ions.
A positive ion is an atom or group of atoms that is short of
electrons. A negative ion is an atom with extra electrons.

How do the ions move when the electrolyte conducts?
The diagram shows some copper chromate solution after a
power supply has been connected to the electrodes.
The solution contains blue copper ions and yellow chromate ions.
At the start all the solution appeared green, but notice how
the colours have separated.
Which ions are positive and which are negative?

Under normal circumstances, air contains very few ions
and so it does not conduct. Why is this fortunate for us?

During a lightning flash the air becomes highly ionised and
is able to conduct in a spectacular way!

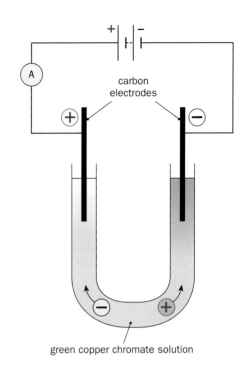

carbon electrodes

green copper chromate solution

Current and drift velocity

What is the relationship between the current I in a wire and
the average drift velocity v of the moving electrons?

The diagram shows part of a wire of cross-sectional area A.
Its length is L and so the volume of the wire is $A\,L$.
If there are n free electrons per metre^3 of the wire, then
the number of electrons in this section of wire is $n\,A\,L$.

The charge on each electron is e.
The total charge Q on the free electrons in the wire is $n\,A\,L\,e$.

If it takes an electron a time t to travel the length of the wire,
after time t all the electrons will have passed through the wire.

Since the current I is the charge passing per second, then:

$$I = \frac{\Delta Q}{\Delta t} = \frac{n\,A\,L\,e}{t}, \text{ but } \frac{L}{t} \text{ is the average drift velocity } v, \text{ so:} \quad \boxed{I = n\,A\,v\,e}$$

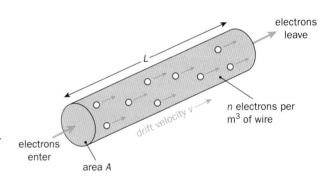

electrons leave

L

drift velocity v

n electrons per m^3 of wire

electrons enter

area A

Example 2
Copper contains 1×10^{29} free electrons per m^3.
What is the drift velocity of electrons in a copper wire of
cross-sectional area 0.25 mm^2 carrying a current of 0.4 A?

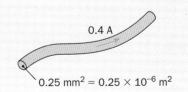

0.4 A

0.25 mm^2 = 0.25 × 10^{-6} m^2

$$I = n\,A\,v\,e$$

$$0.4 \text{ A} = 1 \times 10^{29} \text{ m}^{-3} \times 0.25 \times 10^{-6} \text{ m}^2 \times v \times 1.6 \times 10^{-19} \text{ C}$$

$$0.4 \text{ A} = 4.0 \times 10^3 \text{ C m}^{-1} \times v$$

$$\therefore v = \frac{0.4}{4.0 \times 10^3} = \underline{1 \times 10^{-4} \text{ m s}^{-1}} \text{ or } \underline{0.1 \text{ mm s}^{-1}}$$

Hint:
$$1 \text{ m}^2 = 1000 \text{ mm} \times 1000 \text{ mm} = 1 \times 10^6 \text{ mm}^2$$

$$1 \text{ mm}^2 = \frac{1}{1000} \text{ m} \times \frac{1}{1000} \text{ m} = 1 \times 10^{-6} \text{ m}^2$$

This drift velocity is typical for electrons flowing in metals.
Notice how small it is!

But remember, this is the drift velocity *superimposed* on the
very fast random motion of the free electrons within the metal.

▷ Energy transfer and potential difference

We have seen that an electric current transfers energy from the battery to the lamp. How does this happen?
We can use a model to help us understand.

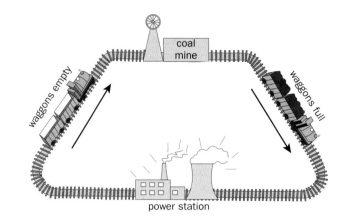

The coal train in the diagram moves at a steady speed and transports energy from the mine to the power station. As soon as it has dropped off one load of coal it returns to the mine to collect another.

In the circuit, each electron gains electrical potential energy as it moves through the battery. Each electron then transports this energy to the lamp.
As they move through the lamp filament, the electrons collide with the filament ions and so transfer energy to them.
The ions vibrate more vigorously and the filament gets hot.

The energy transferred by a single electron is very small and so it is better to think of coulombs moving around the circuit. Can you see that a coulomb is like a wagon on the train?
You can picture it like this:
As each coulomb moves through the battery it 'picks up' a fixed amount of electrical potential energy. The coulomb 'drops off' this energy as it passes through the lamp.
The coulombs then return to the battery to collect more energy.

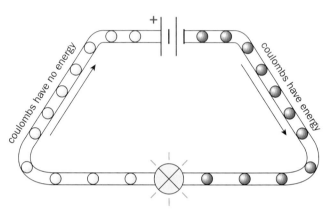

Defining potential difference

The coulombs entering the lamp have electrical potential energy; those leaving have very little potential energy.

There is a **potential difference** (or p.d.) *across* the lamp, because the potential energy of each coulomb has been transferred to heat and light within the lamp.
P.d. is measured in **volts** (V) and is often called voltage.

The p.d. between two points is the electrical potential energy transferred to other forms, per coulomb of charge that passes between the two points.

In the diagram, can you see why the p.d. across the lamp is 3V?

The greater the p.d., the more energy transferred per coulomb.
In fact, if a charge Q transfers energy W, then:

3 joules of energy
transferred by each coulomb

$$\text{p.d., } V \text{ (volts)} = \frac{\text{energy transferred, } W \text{ (joules)}}{\text{charge, } Q \text{ (coulombs)}} \quad \text{or} \quad V = \frac{W}{Q} \quad \text{or} \quad W = Q V$$

If the p.d. is 1 volt, then 1 joule of electrical energy is transferred for each coulomb of charge.

Example 3
In the circuit shown, the p.d. across the lamp is 6.0 V.
If 50 coulombs of charge pass through the lamp,
how much electrical energy is transferred to heat and light?

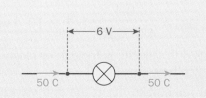

$W = Q V$

∴ $W = 50 \text{ C} \times 6.0 \text{ V} = \underline{300 \text{ J}}$ (2 s.f.)

How do you measure p.d.?

You use a voltmeter which is placed **across** the lamp as shown:
Voltmeter V_1 is measuring the difference in electrical energy
between the coulombs entering the lamp and those leaving it.

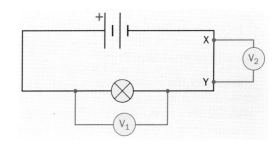

Why is the reading on voltmeter V_2 almost zero?
The moving electrons make far fewer collisions in the thick
copper connecting wire than in the thin tungsten filament of
the lamp, and so they don't lose much energy.
The coulombs at Y have almost the same energy as those at X.

Transferring energy

The p.d. across the car headlamp is 12 volts:
When 1 coulomb passes through the headlamp, 12 joules of
electrical energy is transferred to heat and light.

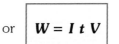

The p.d. across a table lamp is 230 volts. So 230 J of energy
is transferred when 1 coulomb passes through this lamp.

Is the table lamp brighter than the headlamp?
We cannot say, because it depends on the time it takes for
the energy to be transferred.
From the page opposite:

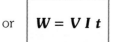

$$\text{Energy transferred, } \mathbf{W} = \text{charge, } \mathbf{Q} \times \text{p.d., } \mathbf{V}$$
$$\text{(joules)} \qquad \text{(coulombs)} \qquad \text{(volts)}$$

and from page 228 we know that: $Q = It$, so:

Energy, W = current, I × time, t × p.d., V		
(joules) (amps) (seconds) (volts)	or $\boxed{W = I\,t\,V}$ or	$\boxed{W = V\,I\,t}$

Example 4
In the circuit shown, the voltmeter reads 10 V and the
ammeter records a current of 0.20 A. How much energy
is transferred to heat and light in the lamp in 5.0 minutes?

   $\mathbf{W = V\,I\,t}$ (remember that t must be in seconds)

5.0 min
(= 300 seconds)

∴ $W = 10\text{ V} \times 0.20\text{ A} \times 300\text{ s} = \underline{600\text{ J}}$ (2 s.f.)

To explain energy transfer we have pictured the electrons
picking up energy at the battery and delivering it to the bulb.
With a drift velocity of 0.1 mm s^{-1}, it would take an electron
about 3000 s to travel along a wire from the battery to a lamp!

Do you have to wait 50 minutes for the lamp to light?
Obviously, our simple model is useful, but far from exact.

How can the energy transfer be immediate?
As soon as the battery is switched on, each electron feels
a force which causes it to accelerate and gain kinetic energy.
The electron soon collides with a metal ion. It transfers its
energy to the vibrating ion and slows down. The electron still
feels the force from the battery and so it accelerates again.

Each electron moves with this stop-start action, but overall
the free electrons appear to flow with a steady drift velocity.

The electron soon collides with a metal ion ...

▷ Resistance

A tungsten filament lamp has a high resistance, but connecting wires have a low resistance. What does this mean?

The greater the resistance of a component, the more difficult it is for charge to flow through it. The electrons make many collisions with the tungsten ions as they move through the filament. But the electrons move more easily through the copper connecting wires because they make fewer collisions with the copper ions.

How is resistance defined?

If there is a p.d. **across** a conductor, a current goes **through** it.
But when you apply the same p.d. across different conductors, the currents are different.
The p.d. across both the kettle and the toaster is 230 V, but the current is 10 A in the kettle and only 5 A in the toaster.
Which has the greater resistance?
The current is smaller in the toaster, so it means the toaster must have more resistance than the kettle.

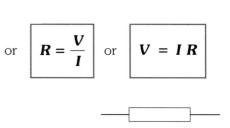

Resistance is measured in **ohms** (Ω) and is defined in the following way:

The resistance of a conductor is the ratio of the p.d. applied across it to the current passing through it.

In fact:

$$\text{Resistance, } R \text{ (ohms)} = \frac{\textbf{p.d. across the conductor, } V \text{ (volts)}}{\textbf{current through the conductor, } I \text{ (amps)}} \quad \text{or} \quad R = \frac{V}{I} \quad \text{or} \quad V = IR$$

Resistors are components that are made to have a certain resistance. They can be made of a length of nichrome wire. The diagram shows the circuit symbol for a resistor:

How is resistance measured?

To find the resistance of a wire you must place it in a circuit, as shown:

The voltmeter tells you the p.d. across the wire, and the ammeter tells you the current through it.
You can calculate the resistance from the two meter readings. The table shows the results that you might record, as you gradually increase the p.d.

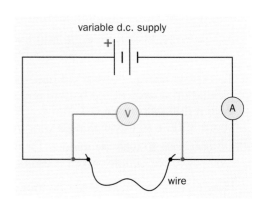

variable d.c. supply

Do you agree that, as the p.d. doubles, the current doubles?
This means that the resistance of the wire is constant.
This important result was discovered by Georg Ohm in 1826, and it is found to be true for all metals at constant temperature:

The current through a metal wire is directly proportional to the p.d. across it (providing the temperature remains constant).
This is Ohm's Law.

Materials that obey Ohm's Law are called **ohmic conductors**.

V /V	I /A	$R = V/I$ /Ω
1.00	0.100	10.0
2.00	0.200	10.0
3.00	0.300	10.0

▷ Resistivity

Are all metals equally good conductors of electricity?
No – for example, nichrome wires have much more resistance
than copper wires.
What other factors affect the resistance that a wire has?
The diagram will give you some clues:

Length – a long wire has more resistance than
a short wire.

Cross-sectional area – a thin wire has more resistance than
a thick wire.

Temperature – in *metals*, a hot wire has more
resistance than a cold wire.

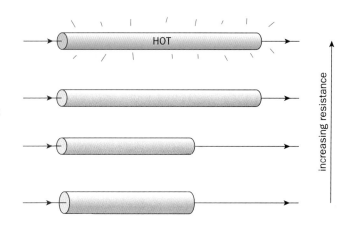

The resistance of a wire at a constant temperature depends
on its dimensions and on the material from which it is made.
Every material has a property called its **resistivity**, ρ. It is
measured in **ohm metres** (Ω m). (ρ is the Greek letter rho.)

For a spreadsheet on this topic, see page 483.

The table shows the resistivities of some materials at 20 °C.
The higher the resistivity, the more difficult it is for charge to
flow through the material.
Can you see why copper is chosen for connecting wires?
What do you notice about the resistivity of insulators?

Material	Type of material	Resistivity /Ω m
copper	conductor	1.7×10^{-8}
nichrome	conductor	110×10^{-8}
silicon	semi conductor	2.3×10^{3}
glass	insulator	1×10^{12}

If we know the resistivity, the cross-sectional area and the
length of a sample of material, we can calculate its resistance.
In fact:

$$\textbf{Resistance, } R \text{ (ohms)} = \frac{\textbf{resistivity, } \rho \text{ (ohm metres)} \times \textbf{length, } l \text{ (metres)}}{\textbf{cross-sectional area, } A \text{ (square metres)}} \quad \text{or} \quad R = \frac{\rho \, l}{A}$$

Notice that the resistance of a wire is:
- proportional to its length, and
- inversely proportional to its cross-sectional area.

Can you design experiments to prove these relationships?

Look carefully at the resistivity equation.
Can you see that, for a sample of length 1 m and area 1 m^2,
its resistance has the same numerical value as the resistivity?
In fact:
***Resistivity is defined as numerically equal to the resistance of a
sample of the material of unit length and cross-sectional area.***

Example 5
A nichrome wire has a length of 0.50 m and a radius of 0.10 mm
a)   What is the cross-sectional area of the wire?
b)   What is its resistance? (See the table for resistivity values)

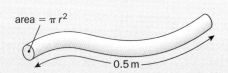

area = πr^2

a)   **Area** = $\pi \, r^2$

∴ Area = $3.14 \times (0.10 \times 10^{-3} \text{ m})^2$ = 3.14×10^{-8} m^2

b) $R = \dfrac{\rho \, l}{A}$ so: $R = \dfrac{110 \times 10^{-8} \, \Omega\text{m} \times 0.50 \text{ m}}{3.14 \times 10^{-8} \text{ m}^2}$ = $\underline{18 \, \Omega}$ (2 s.f.)

Hint: take care with the units:
1 m^2 = $1 \times 10^3 \times 1 \times 10^3$ = 1×10^6 mm^2
1 mm^2 = $1 \times 10^{-3} \times 1 \times 10^{-3}$ = 1×10^{-6} m^2

▷ Current–voltage characteristics

The current in a *metal wire* is proportional to the p.d. across it.
Is this true for other components?
You can use the same circuit (from page 232) to see how
the current I varies with the voltage V, for any component X.

Do you get the same results if the p.d. applied to X is reversed?
Reverse the connections to the supply and repeat the readings.
This time, we say that the V and I values are negative, because
the current is passing through X in the opposite direction.
When you use your results to plot a graph of current against
voltage, you obtain the current–voltage characteristics for X.

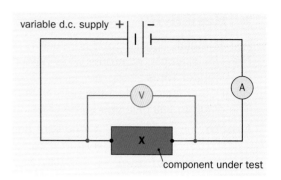

variable d.c. supply

component under test

What do the current–voltage graphs tell us?

When X is a **metal resistance wire**, the graph is a straight line
passing through the origin:
This shows that *I is directly proportional to V*.
If you double the voltage, the current is doubled, and so
the value of V/I is always the same.
Since resistance $R = V/I$, the wire has a constant resistance.

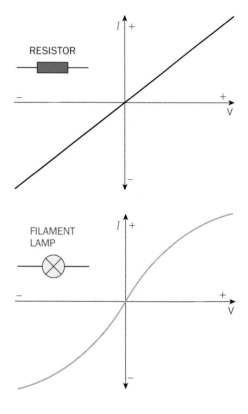

RESISTOR

When X is a **filament lamp**, the graph is a curve, as shown:
Doubling the voltage produces less than double the current.
This means that the value of V/I rises as the current increases.
As the current increases, the metal filament gets hotter and
the resistance of the lamp **rises**.
Do you know why? (This is discussed on the facing page.)

FILAMENT
LAMP

The graphs for the wire and the lamp are symmetrical.
The current–voltage characteristic looks the same, regardless of
the direction of the current.

The semiconductor **diode** behaves differently. There is almost
no current when the voltage is applied in the reverse direction:
the reverse diode has a very high resistance.

In the forward direction, the current increases very rapidly
when the voltage rises above about 0.6 V:
the forward diode has a very low resistance (above 0.6 V).

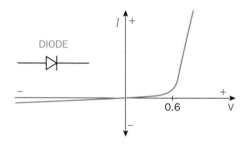

DIODE

The light-emitting diode (LED) behaves in a similar way to the
diode, but it emits light of a certain colour as it conducts.
The forward voltage needed to make an LED conduct is greater
than 0.6 V. In fact, the forward voltage varies with the colour
of the LED. It is higher for a blue LED than for a red LED.
The diagram shows the current–voltage graph for a red LED:

Remember, the diode and the LED conduct **only** when they are
forward-biased, and so they are very useful in electronic circuits.
The arrow on the LED circuit symbol shows the direction of
conventional current when the diode conducts.

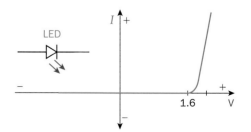

LED

The effect of temperature on resistance

Just like the filament in the lamp, all metals show an increase in resistance as their temperature rises.
Can we explain why?
Remember, metals contain large numbers of free electrons. As these electrons move through the metal lattice they collide with the vibrating metal ions. The collisions oppose the flow of electrons and so the metal has resistance.

What happens to the ions as the metal becomes hotter?
The ions vibrate faster, with greater amplitude, and it is more difficult for the electrons to pass through the lattice.
The resistance of the metal has increased.

Vibrating 'ions' get in the way of an 'electron'.

What about insulators and semiconductors?

In **insulators** at room temperature, there are few free electrons available for conduction.
At high temperatures, some electrons have enough energy to 'escape' from their atoms and the insulator is able to conduct.
So in insulators, resistance **de**creases as the temperature rises.

Have you heard of silicon?
It is one of the best known **semiconductor** materials.
At low temperatures, silicon is a poor conductor.
As its temperature rises, more and more electrons break free from the silicon atoms and so it becomes a better conductor.
However, at about 150 °C, breakdown occurs and the silicon is then permanently damaged.

Like silicon, the resistance of many semiconducting materials **de**creases as their temperature rises. These materials have a **negative temperature coefficient** of resistance (NTC).

The **thermistor** is an important semiconducting component.
On the graph, the resistance of an NTC thermistor is plotted against temperature:
The change in resistance of a metal wire is also shown.

By how much does the resistance of this thermistor decrease as its temperature rises from 20 °C to 100 °C?

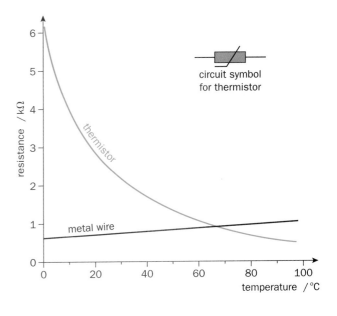

Superconductors

What happens if metals are cooled to very low temperatures?

In 1911, the physicist Heike Kammerlingh-Onnes discovered that, when mercury is cooled to 4.15 K (−269 °C), it loses **all** of its resistance – it becomes a **superconductor**.
Superconducting wires do not become hot, because electrons can flow through them without any transfer of energy.

A material that can 'go superconducting' must be cooled to below its **transition temperature**. For mercury this is 4.15 K.

Some compounds have already been developed with transition temperatures above 100 K (−173 °C).
Scientists are trying to make materials that are superconductors at room temperature.
Why would these be very useful?

Georg Bednorz and Alex Muller won a Nobel Prize for developing new superconductors.

235

▷ Electrical power

The toaster and the kettle transfer electrical energy to heat
but which transfers the energy most rapidly?
You need to know the *power rating* of each of the appliances.

Power is measured in joules per second, called **watts** (W).
See also page 69.
The kettle has a power of 2300 watts.
This means that it transfers 2300 joules of energy in each second.
Can you see that the kettle is twice as powerful as the toaster?

The power of a device is the rate at which it transfers energy.

So:

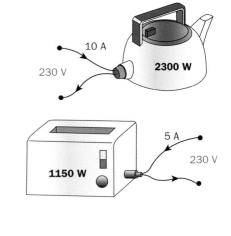

Power, P (watts)	=	**energy transferred, W** (joules) **time, t** (seconds)	or	$P = \dfrac{W}{t}$

How is power related to current and voltage?
The values on the kettle and toaster may give you a clue.
From page 231 we know that:

Energy, **W** = current, **I** × p.d., **V** × time, **t**   or   **W = I V t**
(joules)   (amps)   (volts)   (seconds)

Dividing by time *t* to find the power:

Power, P = **current, I** × **p.d., V** (watts) (amps) (volts)	or	**P = I V**

Power and resistance

For resistors (and lamps) we can combine $P = I \times V$, with
$V = I \times R$ (page 232) to get 3 alternative equations for power:

$$P = I \times V \quad \text{or} \quad P = I^2 \times R \quad \text{or} \quad P = \frac{V^2}{R}$$

The last two equations can be used only when electrical energy
is transferred to heat in a resistance R.
Electric currents often produce wasteful energy transfers,
such as heat in cables. These are called $I^2 R$ heating losses.

*$I^2 R$ heating losses in power cables
should be as small as possible.*

Example 6
a) An isolated cottage uses a generator that gives a
    maximum current of 20 A at a p.d. of 200 V.
    What is the power output of the generator?

   $P = I \times V$

   ∴ $P = 20\ A \times 200\ V = \underline{4000\ W}$ (2 s.f.)

b) The resistance of the cables linking the generator to the cottage is 0.10 Ω.
    How much power is wasted as heat in the cables when the current is 20 A?

   $P = I^2 R$

   ∴ $P = 20\ A \times 20\ A \times 0.10\ \Omega = \underline{40\ W}$ (2 s.f.)

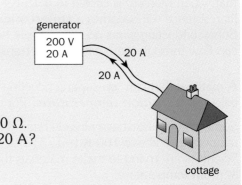

▷ Power in the home

Your home probably has all the electric appliances shown. Usually, the appliance is labelled with its power rating.

In Britain, electricity is supplied to your home at 230 volts. Why 230 volts, rather than a lower voltage such as 23 volts? Imagine using a kettle, a dishwasher, a microwave and a tumble drier. The total power could be about 9200 watts.

Remember, power = current × voltage.

So, to transfer the energy along the cables to the appliances, we could use *either* a) 23 volts and 400 amps (= 9200 W)
or b) 230 volts and 40 amps (= 9200 W)

Even though the copper cables in your home have a very low resistance, the I^2R heating produced by a current of 400 amps would be large enough to melt the cables, and cause fires! 230 V is chosen to keep this heating as low as possible.

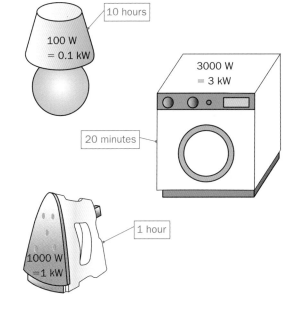

The fused plug

A fuse is a thin wire which melts if the current through it becomes too high. As it melts, it cuts off the current.
Only certain fuse values are available eg. 3 A, 10 A and 13 A.
How can you calculate the correct fuse for an appliance?

Example 7
A vacuum cleaner has a rating of 460 W on the 230 V mains. What fuse should be fitted in the plug?

$$P = I \times V$$

$$460 \text{ W} = I \times 230 \text{ V}$$

$$\therefore I = \frac{460 \text{ W}}{230 \text{ V}} = \underline{2.0 \text{ A}} \quad (2 \text{ s.f.})$$

A 3 A fuse should be fitted, for safety.
A 13 A fuse could allow a damaging current before melting.

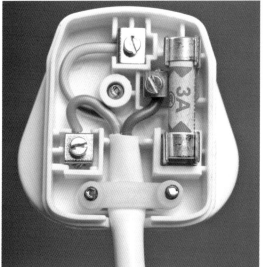

Paying for electricity

You (or your parents!) have to pay for the electrical energy you use. The joule is a very tiny unit of energy and the bill is calculated in a larger unit called the **kilowatt-hour** (kW h)

The kilowatt-hour is the electrical energy transferred by a 1 kW device in 1 hour.

How can we calculate the number of kW h consumed?
From page 236 we know that energy = power × time, so:

Energy transferred	=	**power**	×	**time**
(kW h)		(kW)		(hours)

Since 1 kW = 1000 W and 1 hour = 3600 seconds, it follows that 1 kW h = 1000 W × 3600 s = 3 600 000 joules.

Look at the diagrams at the top of the page.
For how long can each appliance run on 1 kW h of energy?

▷ Physics at work: Using electricity

What is an electric shock?

Put a voltage across the body and a current will pass through it.
This current can override the tiny electrical impulses that make
the nerves and the muscles work correctly, **and** cause burning.
The bigger the current and the longer the time it flows for,
the greater the damage, especially to the heart.
The maximum safe current through the body is only $\sim$ 10 mA.
Why doesn't a fuse protect the body against electric shock?

Is there a maximum safe voltage for an electricity supply?
We need to know the resistance of the body.
This depends on the area of skin in contact with the supply
and whether the skin is damp or dry. A typical value is 10 kΩ.
Since $V = IR$, $V = 10 \times 10^{-3}$ A $\times 10 \times 10^3 \, \Omega = 100$ V,
but even 50 V can kill you!
It is difficult to state a safe voltage, but the greater the p.d.,
the larger the current and so the greater the danger.

15 volts   *50 volts*

Car electrics

New features need more current.

Most cars are fitted with a 12 volt lead-acid battery.
Originally the battery was needed only for the lights and the
indicators, but modern cars have many more electrical features.

This means that the the total current in the wiring can be high
and the heat produced by these large currents can be dangerous.
One solution is to use thick cables.
These carry the current more safely because they have a lower
resistance, and so the $I^2 R$ heating is smaller.

Some designers think that modern cars should be fitted with a
new standard 36 volt battery. Why would this reduce
the currents in the cables to one-third of their present value?

Consider a 72 W headlamp bulb and use the equation $P = IV$:
If $V = 12$ V, $I = $ **6.0** A   (since 6.0 A $\times$ 12 V = 72 W)
If $V = 36$ V, $I = $ **2.0** A   (since 2.0 A $\times$ 36 V = 72 W)
Of course, each component would need to have more resistance
in order to draw the correct current from the 36 V battery.

Electric vehicles

Will we see the battery electric vehicle (BEV) replace
petrol engine vehicles in the 21st century?
The BEV itself produces no atmospheric pollution,
but the electricity it needs to recharge the battery has to
be generated at a power station.

A full petrol tank, with a mass of around 150 kg, can store
about 600 MJ of energy. Lead-acid batteries with a mass
of 500 kg can only supply about 25 MJ of energy!

Could the solution be to use a different type of battery?
One idea is to use the lithium-ion batteries that are used in
laptops and cell phones. These are much lighter and store
more energy. They are also very expensive at the moment!

The electric car of the future may obtain its electrical energy
from a fuel cell (using hydrogen and oxygen).

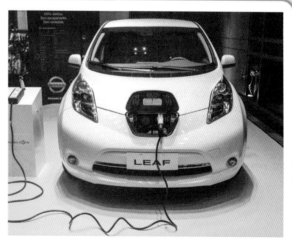

An electric car.

▷ Physics at work: Using electricity

Using superconductors

Our electricity is transmitted by the National Grid (page 285), but at least 5% of this energy is wasted as heat in the cables. With superconducting cables there would be no resistance and no heating losses. Why aren't we using them?

The problem is that miles and miles of cables would have to be cooled to a very low temperature using liquid nitrogen.
We need materials that superconduct at everyday temperatures.

Superconductors are used to produce the strong magnetic fields needed in MRI scanning (page 438) and particle colliders.
In the future, they may also be used in ultra-fast computers.

Superconducting cables going underground.

The light-dependent resistor (LDR)

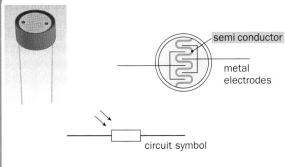

circuit symbol

We have met three semiconducting components in this chapter. The *light-dependent resistor* is another. As its name suggests, its resistance depends on the brightness of light shining on it. The energy of the light shining on the LDR releases extra electrons from the semiconducting material.

As the number of electrons increases, the resistance decreases: from $M\Omega$ in dark conditions to $k\Omega$ in bright light.

Can you think of some uses for an LDR?
How about a circuit which switches the street lights on at dusk?

Semiconductors

Modern electronics depends on semiconducting materials such as silicon and germanium. These pure semiconducting materials are called *intrinsic* semiconductors.
They contain about 1×10^{19} free electrons per cubic metre and are poor conductors, but also poor insulators of electricity.

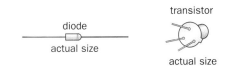

diode
actual size

transistor

actual size

Semiconductors become more useful when they are 'doped' by adding carefully controlled numbers of impurity atoms. These *extrinsic* semiconductors are used in components such as diodes and transistors.
There are two types of extrinsic semiconductor:

In **n-type** semiconductors, each impurity atom donates a free electron, which then takes part in conduction.

In **p-type** semiconductors, each impurity atom is short of one electron. The shortage of electrons creates positive '*holes*'. When a supply is attached, the holes appear to move away from the positive terminal and towards the negative terminal.

In fact what happens is that an electron from one atom jumps across to fill the hole in a neighbouring atom. In jumping, it leaves behind a hole in its own atom. As the process is repeated, the hole appears to drift through the material.

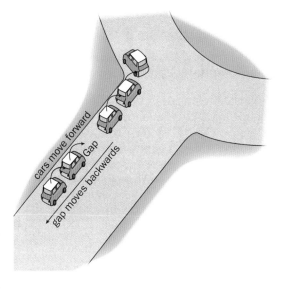

cars move forward
Gap
gap moves backwards

The effect is similar to that seen in a traffic queue. As one car moves forward it leaves a gap. Each car jumps forward to fill the space in front of it and so the gap moves quickly backwards along the queue. The cars themselves hardly move.

▷ Physics at work : Biopotentials

Have you seen a frog's leg twitch when a p.d. is applied to it?
This was first demonstrated by Luigi Galvani in 1786.
We now know that the body generates thousands of electrical
signals as it functions. These signals, called *biopotentials*,
can be measured and used in medical diagnosis.

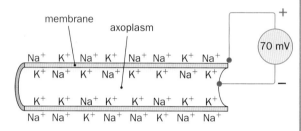

The diagram shows a nerve cell for controlling movement :
The main body of the cell, containing the nucleus, is situated in
the spinal column and messages are sent to the muscle along a
very thin fibre called an **axon**. The axon is insulated from the
surrounding tissue by the **myelin sheath**.

The axon consists of a central core of **axoplasm** surrounded by
a semi-permeable wall or membrane.
The diagram shows the axon in its **resting state** :

A process called the **sodium–potassium pump** transfers
sodium (Na^+) ions out of, and potassium (K^+) ions into,
the axoplasm.

Unequal pumping results in more sodium ions moving out
than potassium ions moving in.
The diagram shows the imbalance :

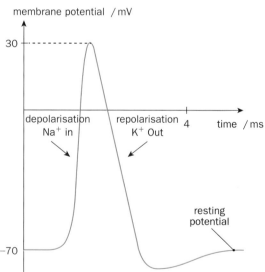

An axon in its resting state.

This imbalance means the axon is *polarised* ; the outside is
positive compared to the inside, and there is a p.d. of about
70 mV across the membrane. This is called the **resting potential**.

When the nerve is stimulated, the membrane suddenly becomes
much more permeable to sodium ions moving into the axoplasm.
As the sodium ions flood back into the axoplasm, the p.d. across
the membrane becomes smaller. This is called **depolarisation**.
The sodium ions continue to move into the axoplasm and the
membrane becomes **reverse polarised** to a peak of about 30 mV.

Now the membrane again becomes less permeable to sodium
ions moving in, but more permeable to potassium ions moving out.
As the potassium ions move out of the axoplasm, the membrane
repolarises and the membrane potential returns to −70 mV.

This rapid rise and fall in the membrane potential is called the
action potential. A typical action potential is shown on the graph :

Depolarisation/polarisation takes only a few milliseconds. Over
a longer period of time (about 50 ms) the sodium–potassium pump
transfers Na^+ and K^+ ions across the membrane, restoring the axon
to its resting state. It is now ready to respond to another stimulus.

So how does the electrical signal pass along the axon?
The depolarised region in the axon acts as a trigger and stimulates
the adjacent region to follow through the same action potential.
Imagine a 'Mexican wave' executed by a crowd of spectators!
In this way, the signal moves along at a speed of around 100 m s^{-1}.

The action potential of a nerve cell in your body.

The 50 ms delay in restoring the membrane to its resting state
ensures that signals can only pass one way along the axon.

▷ Physics at work: The electrocardiogram (ECG)

The heart acts as a double pump, as shown in the diagram:
Can you see how blood from the body flows into the right atrium
and then passes into the right ventricle, where it is pumped to
the lungs to pick up oxygen?
The oxygenated blood returns into the left atrium and passes
into the left ventricle, from where it is pumped around the body.

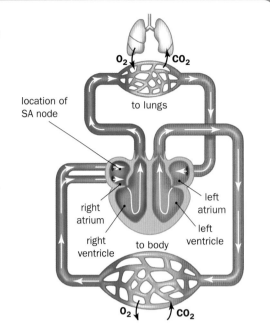

All muscle activity is associated with biopotentials, like that on
the facing page. Your heart is a powerful muscle, which is
controlled by electrical stimuli that originate in the heart itself.
Each beat of the heart is triggered by an electrical pulse from
the **sino-atrial (SA) node**. The SA is also called the **pacemaker**.

The heart muscle consists of millions of cells, each with a resting
potential of about 80 mV. When stimulated, the cells depolarise
and then repolarise, just like the nerve cell on the facing page.

The SA is located in the right atrium. The electrical pulse from
the SA spreads across both of the atria, depolarising the cells
and causing the muscle to contract. The contraction forces
blood out of the atria and into the ventricles.

The cells of the atria repolarise and the muscle relaxes again.
Meanwhile, the electric pulse moves into the ventricles.
The cells of the ventricles depolarise, and the ventricles contract,
forcing blood out of the heart.

The cells of the ventricles repolarise and the muscle relaxes.
This whole sequence is what we call *one beat* of your heart.

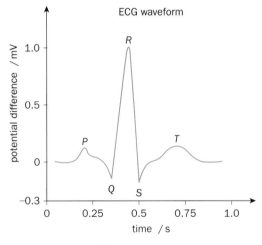

The spread of the electrical pulse across the heart can be detected
at the body surface as fluctuations in potential difference (p.d.).
The diagram shows a typical ECG waveform for a healthy heart:

Can you see the three important features?

The P wave is produced by the depolarisation of the atria.
The QRS wave results from the repolarisation of the atria and
the depolarisation of the ventricles.
The T wave results from the repolarisation of the ventricles.

If the heart is diseased or damaged, the shape of these waves
can give the doctor important clues.

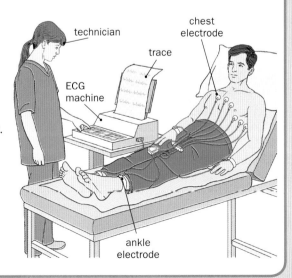

The fluctuations in p.d are detected using metal electrodes.
The electrodes are always connected in pairs.
Good electrical contact is essential, and so the resistance of the
skin–electrode interface must be reduced. This is done by
rubbing the skin with sandpaper, to remove hairs and dead skin.
A conducting gel is applied before the electrode is put into place.

The p.d.s at the body surface are small, and unwanted electrical
signals, called **noise**, can mask the signals from the heart.
The patient needs to be relaxed so that there are no signals
from muscle tremors, and the leads to the electrodes must be
shielded from any nearby electrical equipment.
The signals also have to be amplified. The amplifier used has
to be able to amplify the signal without adding noise to it.

Summary

Electric current I is the rate of flow of charge Q: $\quad I = \dfrac{Q}{t} \quad$ or $\quad I = \dfrac{\Delta Q}{\Delta t} \quad$ and $\quad I = n\,A\,v\,e$
1 amp = 1 coulomb per second.

Potential difference or voltage V is the energy W transferred per coulomb of charge: $\quad V = \dfrac{W}{Q}$
1 volt = 1 joule per coulomb.

The energy W transferred in a time t can be calculated using: $\quad W = V\,I\,t$

Resistance is defined by the equation: $\quad R = \dfrac{V}{I} \quad$ or $\quad V = I\,R$

The resistance of a sample of material depends on its temperature, $\quad R = \dfrac{\rho\,l}{A}$
length l, cross-sectional area A and its resistivity ρ:

The power P of a device is the rate at which it transfers energy W: $\quad P = \dfrac{W}{t}$

Power can be calculated using: $\quad P = I\,V \quad$ and, for resistors and lamps: $\quad P = I^2\,R \quad$ or $\quad P = \dfrac{V^2}{R}$

▶ Questions

1. Can you complete the following sentences?
   a) In metals, current is a flow of .... As the ....
      move round the circuit they transfer ....
   b) The charge that flows (in ....) is equal to the
      current (in ....) multiplied by .... (in ....).
   c) The size of the current passing through a wire
      depends on the .... of the wire and the ....
      applied across it.
   d) Ohm's law states that the .... through a
      conductor is .... to the .... applied across it,
      providing the .... remains constant.
   e) The resistance of a wire increases as the length
      .... , as the temperature .... and as the
      cross-sectional area ....
   f) The power (in ....) of a component is equal
      to the current (in ....) multiplied by the ....
      (in ....).

2. Use the idea of **free electrons** to explain why:
   a) metals are good conductors of electricity,
   b) insulators cannot conduct electricity,
   c) semiconducting materials have higher resistivity
      than conducting materials,
   d) the resistance of a metal wire increases as its
      temperature rises,
   e) the resistance of a thermistor decreases as its
      temperature rises.

3. What charge flows when there is a current of:
   a) 5.0 A for 7.0 s?
   b) 0.2 A for 3.0 min?

4. What is the current when a charge of:
   a) 3.0 C passes through a lamp in 20 s?
   b) 3600 C passes a point in 3.0 minutes?
   c) 4.0 μC flows through a diode in 2.0 ms?

5. A charge of 4000 μC passes each point in a wire in
   50 s. Calculate:
   a) the charge in coulombs,
   b) the current in the wire,
   c) the number of electrons per second passing each
      point in the wire. (Electron charge = 1.6×10^{-19} C.)

6. A cathode-ray tube produces a beam of fast-moving
   electrons, which strike a fluorescent screen. When the
   beam current is 150 μA, how many electrons hit the
   screen in 2.4 s? (Electron charge = 1.6×10^{-19} C.)

7. The current through a wire changes with time as
   shown in the graph. Calculate the total charge that
   passes through the wire.

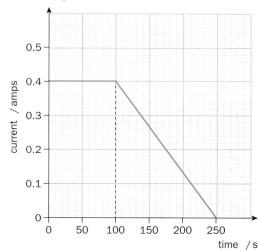

8. A copper wire has a cross-sectional area of 2.5 mm².
   Copper has 1.0×10^{29} free electrons per m³ and the
   charge on each electron is 1.6×10^{-19} C.
   Calculate the drift velocity of the free electrons when
   the wire carries a current of 5.0 A.

9. A lamp has a p.d. of 12 V across it. Calculate how much electrical energy is transferred when:
   a) a charge of 400 C passes through it,
   b) a current of 2.5 A passes through it for 30 s.

10. A 230 V electric heater takes a current of 2.0 A. Calculate the heat produced if it is switched on for 5.0 minutes.

11. Calculate the p.d. across a wire if the energy transferred is:
    a) 600 J when a charge of 50 C passes through it,
    b) 450 J and a steady current of 0.5 A for 20 s.

12. A 230 V kettle transfers 6.9×10^5 joules of energy in 5.0 minutes. What is the current in the kettle?

13. a) What is the p.d. across a wire of resistance $8.0 \ \Omega$ when there is a current of 1.5 A through it?
    b) What is the resistance of a wire if a p.d. of 6.0 V drives a current of 0.25 A through it?
    c) A p.d. of 3.0 V is applied across a wire of resistance $15 \ \Omega$. What is the current in the wire?

14. a) The resistivity of copper is $= 1.7 \times 10^{-8} \ \Omega \text{m}$. Calculate the resistance of 50 m of copper cable with a cross-sectional area of $2.5 \times 10^{-6} \ \text{m}^2$.
    b) What is the voltage drop across this cable when it carries a current of 13 A?

15. Calculate the resistance of a constantan wire of diameter 0.5 mm and length 50 cm if its resistivity is $4.9 \times 10^{-7} \ \Omega \ \text{m}$.

16. The resistivity of nichrome is $1.1 \times 10^{-6} \ \Omega \text{m}$. Calculate the lengths of 0.4 mm diameter nichrome wire needed to make coils with resistances of:
    a) $2.0 \ \Omega$, and b) $5.0 \ \Omega$.

17. The current–voltage characteristics for a metal wire at two temperatures θ_1 and θ_2 are shown below.
    a) Calculate the resistance of the wire at each temperature.
    b) Is θ_1 or θ_2 the higher temperature? Explain.

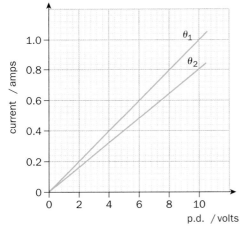

18. The current–voltage characteristics for a diode are shown below. Calculate the resistance of the diode when the p.d. across it is: a) 0.6 V, b) 0.8 V. (Remember that the current is in mA.)

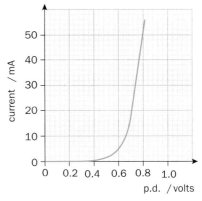

19. A 3000 W electric fire is switched on for $\frac{1}{2}$ hour. Calculate the energy transferred in: a) J, b) kW h.

20. Electricity costs 8 p per kW h. How much does it cost to run a 60 W TV for 30 hours?

21. a) What is the power of a microwave oven that takes a current of 7.5 A from a 230 V supply?
    b) A small 48 W electric heater is connected to a 12 V supply. What current passes through it?

22. A coil of wire has a resistance of $10 \ \Omega$. Calculate the heat produced per second by the coil when the current in it is: a) 2.0 A, b) 4.0 A. What is the heat output when current is doubled?

23. What is the power loss down a copper connecting lead with a resistance of $2.5 \times 10^{-3} \ \Omega$ when it carries a current of 3.5 A?

24. A power station generates electricity at 250 MW and 400 kV. The electricity is transmitted to a distant town along cables with a resistance of $5.0 \ \Omega$.
    a) What is the current in the power lines?
    b) What is the power loss in the cables?

25. When it is hot an electric bar fire has a resistance of $50 \ \Omega$. What is its power rating on a 230 V supply?

26. The filament of a 230 V light bulb is 0.72 m long and has a radius of 6.0×10^{-2} mm. The resistivity of the filament metal is $1.2 \times 10^{-5} \ \Omega \ \text{m}$.
    a) Calculate the resistance of the light bulb.
    b) Calculate its power on a 230 V supply.
    c) The filament of the bulb becomes thinner as the bulb is used. What effect will this have on:
       i) the resistance of the filament?
       ii) its power output? (*Hint*: the p.d. across the bulb remains at 230 V.)

Further questions on page 312.

17 Electric Circuits

A string of Christmas tree lights may consist of 80 lamps, all connected to one 230 V socket.
The lamps could be connected in series, or in parallel.

In this chapter you will learn:
- how to analyse series and parallel circuits,
- about e.m.f. and internal resistance,
- how to use potential dividers.

The series circuit

Here are 2 lamps connected in **series** to a 6 V battery:

The ammeters are placed at different positions in the circuit. Why do they all show the same reading?

Remember that current is the rate of flow of charge.
All the electrons that go through one lamp must also go through the other.
No electrons are lost from the circuit, and so the number of coulombs passing any point each second must be the same.

In a series circuit, the current is the same at all points.

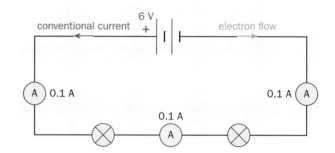

The parallel circuit

Here are 2 lamps connected in **parallel** to a 6 V battery:

You will find that the reading on the ammeter A_1 is the sum of the readings on ammeters A_2 and A_3.
Can you see that the current splits between the two branches?
As the electrons reach point Y, some pass through one lamp and some flow through the other, until they reach X, where their paths join together again.

In a parallel circuit, the current leaving and returning to the supply is the sum of the currents in the separate branches.

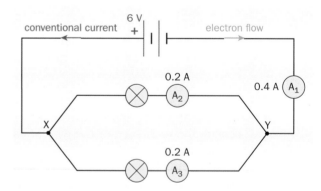

Kirchhoff's First Law:

In both the circuits above, charge is 'conserved'.
In any circuit the charge cannot be created or destroyed, so when charge flows into a point it must flow out again.
This law is usually expressed in terms of current:

The sum of the currents flowing into a point equals the sum of the currents flowing out of that point.

This is Kirchhoff's First Law. It is useful when you apply it to branching circuits like the parallel circuit.
Apply this law to point X.
Is the missing current I equal to 4 amps leaving the junction?

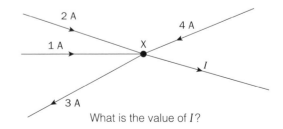

What is the value of I?

▷ Energy transfer in a series circuit

The two lamps in **series** have equal brightness, but each lamp is less bright than if it is connected to the battery on its own.

You can measure the potential differences in this series circuit using voltmeters connected as shown:

What do you notice about the voltmeter readings?
The total p.d. across both lamps is 6 V. This is shared between the two lamps, so that each lamp has a p.d. of 3 V across it.

In a series circuit, the total p.d. across all the lamps is the sum of the p.d.s across the separate lamps.

We can explain this result using our model of energy transfer. (Remember, from page 230, that p.d. is the electrical energy transferred per coulomb of charge).

Each coulomb collects 6 joules of energy from the battery. A single lamp connected to the battery would receive all 6 joules of energy. But in this circuit each coulomb transfers 3 joules at the first lamp and then 3 joules at the second lamp. The energy is shared equally because the lamps are identical.

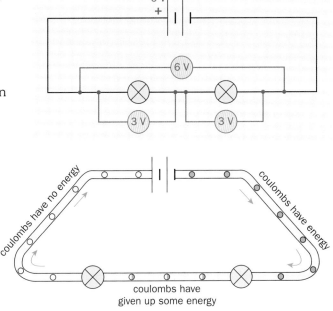

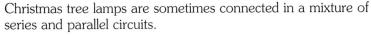

coulombs have given up some energy

▷ Energy transfer in a parallel circuit

In our **parallel** circuit, each lamp is just as bright as if it were connected to the battery on its own.
Can you explain why?
You need to know the potential differences across the lamps.
Look at the voltmeter readings in this diagram:

The p.d. across each bulb is the same and is equal to 6 V.

In a parallel circuit, the p.d. across each branch is the same.

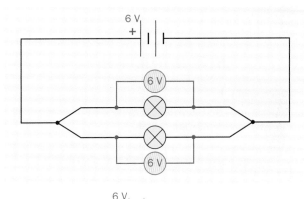

Using our model again, each coulomb of charge transports 6 joules of energy and transfers it all to *one* of the lamps.
Each lamp receives the same energy as if it were connected to the battery on its own.
Twice as many coulombs pass per second through the battery as when a single lamp is connected, and so the battery runs down more quickly.

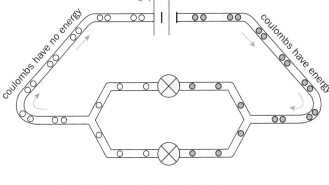

Christmas tree lamps are sometimes connected in a mixture of series and parallel circuits.

In this diagram we have 3 parallel strings of 5 lamps in series:

Can you see why the p.d. across each lamp is 46 V?

If lamp X blows what will happen to the other 14 lamps?
The 4 lamps in the same branch as X will go out, but the other 10 lamps will stay lit at the same brightness.
Can you explain why?

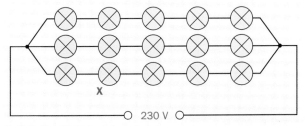

▶ Resistors in series

The diagram shows three resistors connected in **series** :

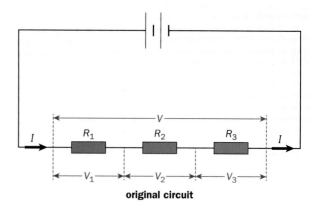

original circuit

There are 3 facts that you should know for a series circuit:

- The current through each resistor in series is the same.
- The total p.d. V across the resistors is the sum of the p.d.s across the separate resistors, so: $V = V_1 + V_2 + V_3$.
- The combined resistance R in the circuit is the sum of the separate resistors :

$$\boxed{R = R_1 + R_2 + R_3}$$

We have seen the first two of these facts before (on pages 244 and 245), but can we prove the third one?

Suppose we replace the 3 resistors with one resistor R that will take the same current I when the same p.d. V is placed across it. This is shown in the diagram. Let's calculate R.

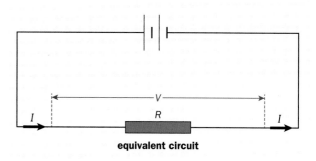

equivalent circuit

We know that for the resistors in series:
$$V = V_1 + V_2 + V_3$$
But, for any resistor, p.d. = current × resistance $(V = I\,R)$.

If we apply this to each of our resistors, and remember that the current through each resistor is the same and equal to I, we get:
$$I\,R = I\,R_1 + I\,R_2 + I\,R_3$$
If we now divide each term in the equation by I, we get:

$$R = R_1 + R_2 + R_3$$

Example 1
A p.d. of 3 V is applied across two resistors (4 Ω and 2 Ω) connected in series, as shown:
Calculate a) the combined resistance,
            b) the current in the circuit,
            c) the p.d. V_1 across the 4 Ω resistor,
            d) the p.d. V_2 across the 2 Ω resistor.

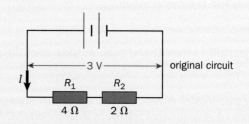

original circuit

a) The combined resistance $R = R_1 + R_2$
$$\therefore \quad R = 4\,\Omega + 2\,\Omega \quad = 6\,\Omega$$

b) Now redraw the circuit as an equivalent circuit:
$$V = I\,R \qquad \text{for the combined resistance}$$
$$3\,V = I \times 6\,\Omega$$
$$\therefore \quad I = 0.5\,A$$

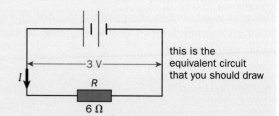

this is the equivalent circuit that you should draw

c) In the original circuit :
$$V_1 = I\,R_1 \qquad \text{for the 4 Ω resistor only}$$
$$V_1 = 0.5\,A \times 4\,\Omega$$
$$\therefore V_1 = 2\,V \qquad \text{across the 4 Ω resistor}$$

d) Also in the original circuit:
$$V_2 = I\,R_2 \qquad \text{for the 2 Ω resistor only}$$
$$V_2 = 0.5\,A \times 2\,\Omega$$
$$\therefore V_2 = 1\,V \qquad \text{across the 2 Ω resistor}$$

Is the p.d. greater across the bigger or the smaller resistor ?
In fact, the p.d. across the 4 Ω resistor is *twice* the p.d. across the 2 Ω resistor.

▷ Resistors in parallel

We now have three resistors connected in **parallel** :

There are 3 facts that you should know for a parallel circuit :

- The p.d. across each resistor in parallel is the same.
- The current in the main circuit is the sum of the currents in each of the parallel branches, so : $I = I_1 + I_2 + I_3$.
- The combined resistance R is calculated from the equation :

$$\boxed{\frac{1}{R} = \frac{1}{R_1} + \frac{1}{R_2} + \frac{1}{R_3}}$$

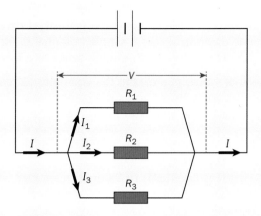

How can we prove this?

Suppose we replace the 3 resistors with one resistor R that takes the same total current I when the same p.d. V is placed across it. This is shown in the diagram. Now let's calculate R.

We know that for the resistors in parallel :

$$I = I_1 + I_2 + I_3$$

But, for any resistor, current = p.d. ÷ resistance ($I = V/R$).

If we apply this to each of our resistors, and remember that the p.d. across each resistor is the same and equal to V, we get :

$$\frac{V}{R} = \frac{V}{R_1} + \frac{V}{R_2} + \frac{V}{R_3}$$

Now we divide each term by V, to get : $\dfrac{1}{R} = \dfrac{1}{R_1} + \dfrac{1}{R_2} + \dfrac{1}{R_3}$

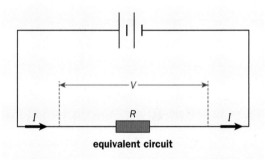

equivalent circuit

You will find that the total resistance R is always *less* than the smallest resistance in the parallel combination.
Check that this is true in the worked example below.

Example 2
A p.d. of 6 V is applied across two resistors (3 Ω and 6 Ω) in parallel. Calculate :
a) the combined resistance,   b) current I in the main circuit.
c) current I_1 in the 3 Ω resistor, d) current I_2 in the 6 Ω resistor.

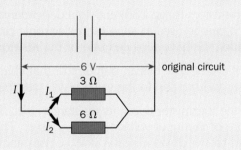

original circuit

a) $\dfrac{1}{R} = \dfrac{1}{R_1} + \dfrac{1}{R_2}$ so : $\dfrac{1}{R} = \dfrac{1}{3} + \dfrac{1}{6} = \dfrac{2+1}{6} = \dfrac{3}{6} = \dfrac{1}{2}$

Since $\dfrac{1}{R} = \dfrac{1}{2}$ then $\dfrac{R}{1} = \dfrac{2}{1}$ so : $\underline{R = 2\,\Omega}$

b) Now redraw the circuit as an equivalent circuit :
$$V = I\,R \qquad \text{for the combined resistance}$$
$$6\,V = I \times 2\,\Omega$$
$$\therefore \quad \underline{I = 3\,A} \qquad \text{in the main circuit.}$$

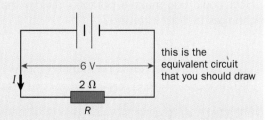

this is the equivalent circuit that you should draw

c) In the original circuit :
$$V = I_1\,R_1 \qquad \text{for the 3 Ω resistor only}$$
$$6\,V = I_1 \times 3\,\Omega$$
$$\therefore \quad \underline{I_1 = 2\,A} \qquad \text{in the 3 Ω resistor.}$$

d) In the original circuit :
$$V = I_2\,R_2 \qquad \text{for the 6 Ω resistor only}$$
$$6\,V = I_2 \times 6\,\Omega$$
$$\therefore \quad \underline{I_2 = 1\,A} \qquad \text{in the 6 Ω resistor.}$$

You can check your answer in the following way : does $I = I_1 + I_2$?
The current is always biggest in the parallel branch with the *least* resistance.

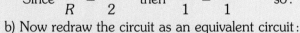

▷ The potential divider (voltage divider)

Sometimes we want to use only part of the voltage provided by a battery. To do this we use a **potential divider** circuit.
The diagram shows a simple potential divider, consisting of two 100 Ω resistors connected in series to a 6 V battery:

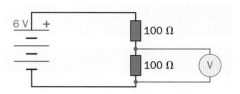

What would you expect the voltmeter to read?
Since the two resistors are identical, they share the applied voltage equally. The voltmeter reads 3 V.

What will the voltmeter read if we replace the lower resistor with a 200 Ω resistor?
200 Ω is two-thirds of the total resistance in this circuit and so the voltage across it is two-thirds of the total voltage. The voltmeter reads 4 V.

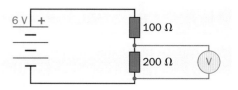

This voltage divider allows us to obtain an output voltage V_{OUT} from a fixed input voltage V_{IN}. V_{IN} is the voltage across both of the resistors. V_{OUT} is the voltage across the resistor R_1 as shown: How can we calculate the value of V_{OUT}?

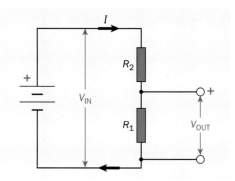

If we apply the equation $V = IR$ to the resistors in this circuit:

$$V_{OUT} = I R_1 \quad \text{for resistor } R_1 \qquad \dots (1)$$
$$V_{IN} = I (R_1 + R_2) \quad \text{for the resistors in series} \quad \dots (2)$$

Dividing equation (1) by equation (2), we get:

$$\boxed{\frac{V_{OUT}}{V_{IN}} = \frac{R_1}{R_1 + R_2}} \quad \text{or} \quad \boxed{V_{OUT} = V_{IN} \times \frac{R_1}{R_1 + R_2}}$$

Using thermistors and LDRs in potential dividers

Thermistors (page 235) and light-dependent resistors (LDRs, page 239) can be used as sensors in electronic circuits. In this potential divider, we have replaced one of the resistors by a **thermistor**:

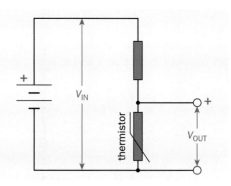

This circuit now has an output p.d. that changes with temperature. What will happen to the value of V_{OUT} as the temperature falls?

As the thermistor cools, its resistance rises. It takes a larger share of the input voltage, so V_{OUT} rises. This circuit could be used to switch on a heater if V_{OUT} rises above a certain value.

How could you make a potential divider that responds to changes in light intensity?

Example 3
The LDR shown has a resistance of 200 Ω (in bright light).
What is the value of V_{OUT}?

$$V_{OUT} = V_{IN} \times \frac{R_1}{R_1 + R_2}$$

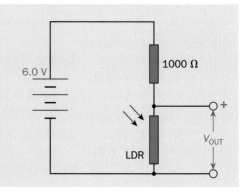

$$\therefore V_{OUT} = 6.0 \text{ V} \times \frac{200 \text{ Ω}}{(200 \text{ Ω} + 1000 \text{ Ω})} = \underline{1.0 \text{ V}}$$

What will happen to the value of V_{OUT} as darkness falls?
Can you think of a use for this circuit?

▷ Variable resistors

Variable resistors can be used in 2 ways in a circuit: to control current, or to control voltage. They consist of a resistance wire and a sliding contact.

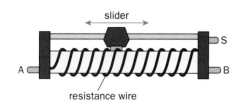

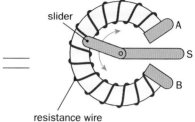

How many connections does a variable resistor have?
Two of these contacts, A and B, are fixed and the third, S, is movable.

The variable resistor as a rheostat

When a variable resistor is used to control current, it is called a **rheostat**. The diagram shows how to connect it into the circuit: We use one of the fixed contacts and the sliding contact S.

Why does the brightness of the bulb change as S is moved? As you move S, you alter the length of resistance wire in the circuit. The longer the wire, the greater its resistance and so the smaller the current.

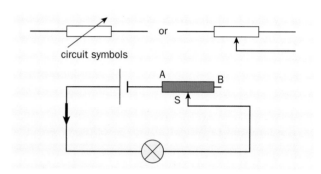

The variable resistor as a potential divider

We can also use a variable resistor as a **potential divider** or **potentiometer** to control voltage.
This needs all three of the contacts, as shown in the diagram:

We apply the supply voltage V_{IN} across the full length of the resistance wire AB. The sliding contact S divides the resistor into two parts, R_1 and R_2.
What will happen to the size of the output voltage as you move the slider from B towards A?
When S is at B, V_{OUT} is zero. If S is at A, V_{OUT} is the full input voltage of 6 V. Where will S be when V_{OUT} is 3 V?

The rotary type of potentiometer is used in radios and other audio equipment as a volume control.

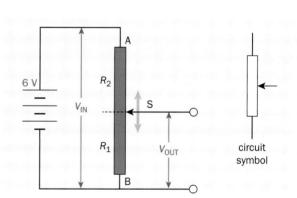

Loading the potentiometer

A potential divider is *loaded* when a resistor is connected across its output terminals. In the diagram, the lamp is the load:

With the sliding contact halfway along the resistor as shown, we might expect the p.d. across the bulb to be 3 V.
In fact, V_{OUT} will be less than 3 V. Can you explain why?
The bulb is in parallel with the resistor R_1 and their combined resistance is *less* than R_1 alone. The bulb and R_1 together take a smaller share of the supply voltage and so V_{OUT} is *less*.

The potential divider equation on the facing page can be used to find V_{OUT}, but only if you use the effective resistance of R_1 and the load in parallel.

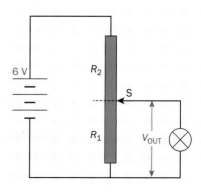

▷ Electromotive force

Resistors and bulbs transfer electrical energy to other forms, but which components **provide** electrical energy?
A dry cell, a dynamo and a solar cell are some examples.

Any component that supplies electrical energy is a source of **electromotive force** or **e.m.f.** It is measured in volts.
The e.m.f. of a dry cell is 1.5 V; that of a car battery is 12 V.

Watches use small button batteries.

Remember our model of energy transfer from page 230.
A battery transfers chemical energy to electrical energy, so that as each coulomb moves through the battery it gains electrical potential energy.
The greater the e.m.f. of a source, the more energy is transferred per coulomb.
In fact:

The e.m.f. of a source is the electrical potential energy transferred from other forms, per coulomb of charge that passes through the source.

Compare this definition with the definition of p.d. (page 230) and make sure you know the difference between them.

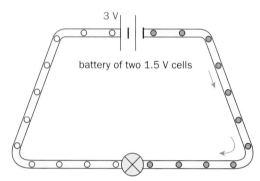

Each coulomb gains 3 J of energy as it moves through the battery of two 1.5 V cells.

If a charge Q moves through a source of e.m.f. **Є**, then:

energy transferred, W = charge, Q × e.m.f., Є
(joules)   (coulombs)   (volts)

or $W = Q\,Є$

The internal resistance of a supply

In the circuit shown, the dry cell has an e.m.f. of 1.5 V:

What happens to the current in the circuit, and the p.d. across the resistance box, as you reduce the value of the resistance R?
The table below shows some typical results.

Did you expect the voltmeter reading to remain at 1.5 V?
In fact, as the current in the circuit increases, the p.d. across the resistance box *falls.*

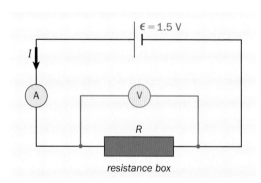

resistance box

The cell gives 1.5 joules of electrical energy to each coulomb that passes through it, but the electrical energy transferred in the resistor is less than 1.5 joules per coulomb and can vary.
The circuit seems to be losing energy – can you think where?

The cell itself has some resistance, its **internal resistance**.
Each coulomb gains energy as it travels through the cell, but some of this energy is wasted or 'lost' as the coulombs move against the resistance of the cell itself.

So, the energy delivered by each coulomb to the circuit is *less* than the energy supplied to each coulomb by the cell.

Value of R / Ω	Current I / A	P.d. across R / V
100	0.015	1.5
20	0.071	1.4
5.0	0.250	1.3
2.0	0.500	1.0

Very often the internal resistance is small and can be ignored.
Dry cells, however, have a significant internal resistance.
This is why a battery can become hot when supplying electric current. The wasted energy is dissipated as heat.

▷ Kirchhoff's Second Law

This is a statement of the conservation of energy in a circuit.
We know that a coulomb gains electrical energy as it moves
through each e.m.f., and loses electrical energy as it moves
through each p.d.
After one loop of the circuit, the energy it has gained must
be equal to the energy it has dissipated. Therefore:

*Around any closed loop in a circuit, the sum of the e.m.f.s
is equal to the sum of the p.d.s.*

▷ The circuit equation

The diagram shows a cell of e.m.f. ϵ together with its internal
resistance r. A border is drawn round ϵ and r to show that they
are really one component. Here, the cell is connected to a circuit
which has a total resistance R.

What does the voltmeter measure?
The voltmeter records the p.d. V across the external circuit,
but notice that V is also the p.d. across the terminals of the cell.
Can you see why V is called the **terminal p.d.**?

The diagram also shows the 'lost volts' v across the internal
resistance.

Since Kirchhoff's Second Law tells us that energy is conserved:

energy **supplied** per coulomb from chemical energy of the cell	=	energy **delivered** per coulomb to the external circuit	+	energy **wasted** per coulomb on the internal resistance

or: e.m.f., ϵ = terminal p.d., V + lost volts, v or $\epsilon = V + v$

We know that the current through the resistors R and r is I.
so, applying voltage = current × resistance to each resistor:

$$\boxed{\epsilon = V + I\,r}$$ or $$\boxed{\epsilon = I\,R + I\,r}$$ or $$\boxed{\epsilon = I\,(R + r)}$$

Example 4
A battery of e.m.f. 6.0 V and internal resistance 1.0 Ω is
connected to two resistors of 4.0 Ω and 7.0 Ω in series:
Calculate a) the total resistance in the external circuit
           b) the current supplied by the battery
           c) the terminal p.d. of the battery

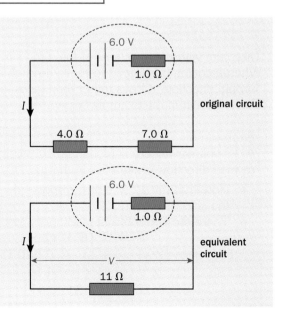

a) $R = R_1 + R_2$ ∴ $R = 4.0\ \Omega + 7.0\ \Omega$ = $\underline{11\ \Omega}$

b) Now draw the equivalent circuit:
  $\epsilon = I\,(R + r)$ for the whole circuit
  6.0 V $= I\,(11\ \Omega + 1.0\ \Omega)$
  ∴ $I = \underline{0.50\ \text{A}}$

c) The p.d. across the resistors in series is the terminal p.d. V:
  $V = I\,R$ where R is the total series
  $V = 0.50\ \text{A} \times 11\ \Omega$ resistance of 11 Ω
  ∴ $V = \underline{5.5\ \text{V}}$

▷ Solving circuit problems

Take care when solving circuit problems using the equation $V = I \times R$.

- Be clear whether the values you put into the equation are for a single resistor, or a group of resistors, or the whole circuit.
- Learn the rules for series and parallel resistors (pages 246–7).
- Try to simplify the circuit by drawing an equivalent circuit.
- Remember to include any internal resistance of the supply when calculating the total resistance of the circuit.
- Remember that the e.m.f. is equal to the sum of all the p.d.s, (including the 'lost volts' due to any internal resistance).

Phiz solves circuit problems.

Example 5

A supply of e.m.f. 4 V and internal resistance 1 Ω is connected to two resistors, as shown in the original circuit diagram.

Calculate
- a) the resistance of the parallel combination,
- b) the current taken from the supply,
- c) the terminal p.d. of the supply,
- d) the current through each parallel branch.

a) Use: $\dfrac{1}{R} = \dfrac{1}{R_1} + \dfrac{1}{R_2}$ for the parallel combination

$$\therefore \quad \frac{1}{R} = \frac{1}{4} + \frac{1}{12} = \frac{3+1}{12} = \frac{4}{12}$$

$$\therefore \quad R = 3\ \Omega \quad \text{where } R \text{ is the total external resistance.}$$

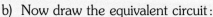
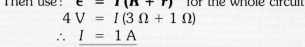

b) Now draw the equivalent circuit:

Then use: $\epsilon = I(R + r)$ for the whole circuit

$$4\ \text{V} = I(3\ \Omega + 1\ \Omega)$$

$$\therefore \quad I = 1\ \text{A}$$

c) The p.d. across both parallel resistors is the terminal p.d. V.

Use: $V = I R$ where R is the combined
$V = 1\ \text{A} \times 3\ \Omega$ parallel resistance of 3 Ω

$$\therefore \quad V = 3\ \text{V}$$

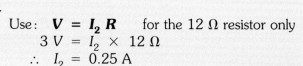

original circuit

equivalent circuit

d) Remember that the p.d. across both parallel resistors is 3 V.

Use: $V = I_1 R$ for the 4 Ω resistor only   Use: $V = I_2 R$ for the 12 Ω resistor only

$3\ \text{V} = I_1 \times 4\ \Omega$ $3\ \text{V} = I_2 \times 12\ \Omega$

$$\therefore \quad I_1 = 0.75\ \text{A} \qquad\qquad\qquad\qquad\qquad \therefore \quad I_2 = 0.25\ \text{A}$$

Take care with the directions of the e.m.f.s in a circuit.
When a coulomb is forced to move through an e.m.f. in the 'wrong' direction, it loses rather than gains electrical energy.
This is shown in the example below.

combinations of 2 V cells

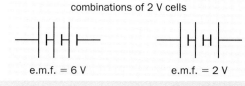

e.m.f. = 6 V   e.m.f. = 2 V

Example 6

A 12 volt car battery is recharged by passing a current through it in the reverse direction using a 14 V charger. Calculate the charging current.

- Note that the 12 V e.m.f. opposes the 14 V e.m.f., so: the sum of the e.m.f.s = 14 V + (−12 V) = 2 V
- There are two internal resistances, and both 'waste' electrical energy.

$$\epsilon = I(R + r)$$

$$14\ \text{V} + (-12\ \text{V}) = I(0 + 0.040\ \Omega + 0.50\ \Omega)$$

$$2\ \text{V} = I \times 0.54\ \Omega$$

$$\therefore \quad I = 3.7\ \text{A} \quad (2\ \text{s.f.})$$

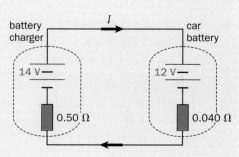

battery charger I car battery

14 V—   12 V—

0.50 Ω   0.040 Ω

▷ Physics at work: Current, voltage, resistance

Ammeters and voltmeters

As you have seen, in order to measure the current,
an ammeter is placed **in series**, *in* the circuit.

What effect might this have on the size of the current?
The *ideal* ammeter has zero resistance, so that placing it
in the circuit does not make the current smaller.
Real ammeters do have very small resistances – around 0.01 Ω.

A voltmeter is connected **in parallel** with a component,
in order to measure the p.d. across it.

Why can this increase the current in the circuit?
Since the voltmeter is in parallel with the component, their
combined resistance is less than the component's resistance.
The *ideal* voltmeter has infinite resistance and takes no current.
Digital voltmeters have very high resistances, around 10 MΩ,
and so they have little effect on the circuit they are placed in.

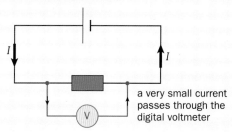

a very small current passes through the digital voltmeter

Can we measure e.m.f directly?

In this diagram we have a cell that is *not* supplying current.
Charge does not move through the cell and so there is no
energy wasted due to the cell's internal resistance.

Remember the circuit equation : $\quad \epsilon = V + I\,r$
Since $I = 0$, $\quad \epsilon = V + 0 \times r \quad$ so : $\quad \epsilon = V$

This means that : ***the e.m.f. of a supply is equal to the
p.d. across its terminals when it is not supplying current.***

An 'ideal' voltmeter takes no current and so we can connect it
across a cell to measure the e.m.f. directly.
Why can a digital voltmeter be placed across a cell, to give a
reading which is very close to the e.m.f ?
The voltmeter takes such a small current that the terminal p.d.
will only be slightly less than the e.m.f.

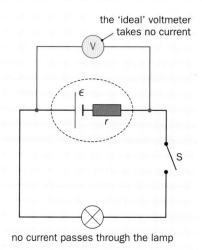

the 'ideal' voltmeter takes no current

no current passes through the lamp

The effects of internal resistance

What is a 'short circuit'? The 1.5 V dry cell in the diagram is
short-circuited by the low resistance copper wire connected to
its terminals. The only significant resistance in this circuit is
0.5 Ω, the internal resistance of the cell itself.

1.5 V
0.5 Ω
thick copper wire, $R \approx 0$

Using $\quad \epsilon = I\,(R + r) \quad$ where $R = 0$, can you show that
the current through the wire must be 3 amps?
This means that 3 A is the *maximum* possible current
that this dry cell can supply.
Some sources, such as the mains supply, have very low internal
resistance, and so they can supply high short-circuit currents.
These current are dangerous because of the heat they produce.

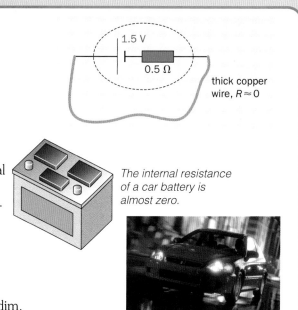

*The internal resistance
of a car battery is
almost zero.*

A 12 V **car battery** must have a very low internal resistance,
because a starter motor needs a current of over 100 A !
What happens if a driver starts a car with the headlamps on?
The current through the battery is so large that the 'lost volts'
are high – even though the battery's internal resistance is low.
The terminal p.d. drops to around 8 V and the headlamps go dim.

Summary

For 3 resistors in series : $R = R_1 + R_2 + R_3$

For 3 resistors in parallel : $\dfrac{1}{R} = \dfrac{1}{R_1} + \dfrac{1}{R_2} + \dfrac{1}{R_3}$

A coulomb of charge **gains** energy as it moves through each source of e.m.f., and **loses** energy as it moves through each resistor (or other component) across which there is a p.d. The unit for both e.m.f. and p.d. is the volt (V).

The e.m.f. ϵ of a supply is the electrical energy supplied per coulomb of charge :

Any source of e.m.f. has some internal resistance r which must be included in circuit calculations (unless you are told to ignore it).

For a potential divider :

$$\dfrac{V_{OUT}}{V_{IN}} = \dfrac{R_1}{R_1 + R_2}$$

1 volt = 1 joule per coulomb

$$\epsilon = \dfrac{W}{Q} \qquad \text{or} \qquad W = Q\,\epsilon$$

$$\epsilon = I\,(R + r)$$

▷ Questions

1. Can you complete these sentences?
   a) In a series circuit, the .... through each component in the circuit is the ....
   b) For resistors in series, the total .... is the .... of the p.d.s across the separate resistors.
   c) In a parallel circuit the .... in the main circuit is equal to the .... of the currents in the branches.
   d) For resistors in parallel, the .... across each .... is the same.
   e) Kirchhoff's Second Law states that around any .... loop in a circuit, the .... of the e.m.f.s is equal to the sum of the ....

2. Explain why:
   a) the combined resistance of two resistors in parallel is always less than either of the separate resistances,
   b) two 1.5 V cells connected together in series can give an e.m.f. of 3.0 V or 0 V,
   c) the terminal p.d. of a dry cell, supplying current to a circuit, is always less than its e.m.f.,
   d) there is an upper limit to the current that a dry cell can deliver to a circuit.

3. Diagrams (a) and (b) show meters that have been incorrectly connected into the circuits.
   For each circuit, (i) state if the lamp is lit or not, and (ii) explain your answer.

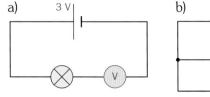

4. Calculate the combined resistance in each of the examples below:

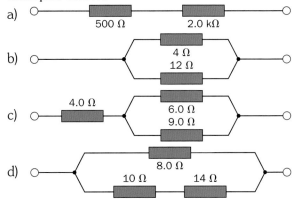

5. a) Calculate the combined resistance of two 6 Ω resistors in parallel.
      What do you notice about the result?
   b) What would be the combined resistance of two 8 Ω resistors in parallel?
   c) Predict the combined resistance of three 9 Ω resistors in parallel.
      Do the calculation to check if you are right.

6. In the circuit below, the p.d. across the 10 Ω resistor is 5.0 V.
   a) What is the current through the 10 Ω resistor?
   b) What is the current through the 8.0 Ω resistor?
   c) What is the p.d. across the 8.0 Ω resistor?

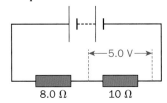

7. In the circuit below:
   a) What is the combined resistance?
   b) What is the p.d. across the combined resistance?
   c) What is the current in the 3 Ω resistor?
   d) What is the current in the 6 Ω resistor?

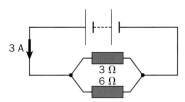

8. In the circuit below, the voltmeter reads 12 V.
   Calculate: a) the total resistance of the circuit
             b) the current I,
             c) the p.d.s V_1 and V_2,
             d) the currents I_1 and I_2.

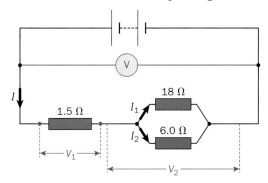

9. How much energy is transferred to 1 coulomb of charge by: a) a battery with an e.m.f. of 2.0 V?
   b) a 5.0 kV high-voltage supply?

10. How many joules of energy does a battery of e.m.f. 3.0 V supply when:
    a) a charge of 5.0 C passes through it?
    b) there is a current of 0.20 A through it for 30 seconds?

11. A battery of e.m.f. 4.0 V and an internal resistance of 1.0 Ω is connected to a 9.0 Ω resistor.
    Draw a circuit diagram and calculate the current that flows around the circuit.

12. For each of the two circuits below, calculate:
    a) the total resistance of the external circuit,
    b) the current supplied by the cell,
    c) the terminal p.d. of the cell,
    d) the energy per coulomb wasted in the cell.

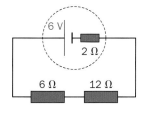
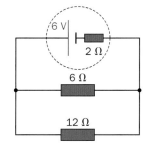

13. Three identical cells, each of e.m.f. 2.0 V and internal resistance 0.50 Ω, are connected in series to a lamp of resistance 2.5 Ω. What current flows?

14. In the circuit below, the high-resistance voltmeter shows a reading of 1.5 V when switch S is open. When switch S is closed the voltmeter reading falls to 1.2 V and the ammeter shows a current of 0.30 A. What is: a) the e.m.f. of the cell?
    b) the internal resistance of the cell?
    c) the resistance of the lamp?

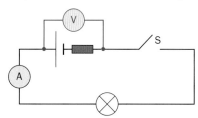

15. A battery charger is used to recharge a cell of e.m.f. 1.5 V and internal resistance 0.50 Ω. The e.m.f. of the battery charger is 6.0 V and its internal resistance is 1.0 Ω.
    Draw a circuit diagram of the arrangement and calculate the recharging current.

16. What is the value of the output voltage V_{OUT} in each of the potential dividers below?

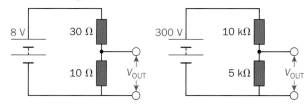

17. In the potential divider shown, what will be the range of values of V_{OUT}, as the variable resistor R is adjusted over its range of 0 Ω to 50 Ω?

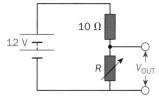

18. In the potential divider shown, the thermistor has a resistance which varies between 200 Ω at 100 °C and 5.0 kΩ at 0 °C. Calculate the value of V_{OUT} at:
    a) 100 °C,   b) 0 °C

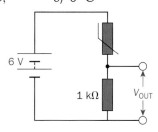

Further questions on page 312.

18 Magnetic Fields

Magnets have fascinated people for centuries. Did you know that the magnetic compass has been used for over 2000 years? The earliest compasses were probably made from lodestone, which is a naturally occurring oxide of iron.

How does a compass show direction?

If you suspend a magnet horizontally from a piece of string, it will always rotate until it points in a north–south direction. The end of the magnet that points north is called the north-seeking pole or **N-pole**. The other end is the south-seeking pole or **S-pole**.

We have come a long way since the invention of the compass! Many everyday devices, from washing machines to computers and swipe-cards, depend on magnetism to work.

In this chapter you will learn:
- about magnetic field patterns and lines of flux,
- how to measure and calculate the strength of magnetic fields,
- how to calculate the force on a current in a magnetic field.

▷ Magnetic fields

What happens if you bring the poles of two magnets together?

You will find that two N-poles repel, and two S-poles repel, but a N-pole and a S-pole attract.
The effect of these forces can be felt even though the magnets are not in contact with each other.
We say that a magnet has a **magnetic field** around it.
The field is a region where the magnetic force can be felt.

How can we 'see the shape' of a magnetic field?
If we place iron filings or plotting compasses in the field of the magnet, they will point along curved lines called **lines of flux**, as shown in the diagram:

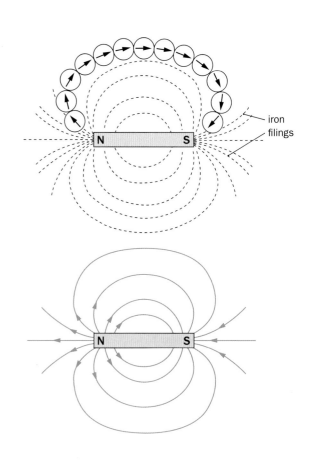

iron filings

The lines of flux show both the **direction** and the **strength** of the magnetic field.
The direction of the line of flux at each point shows the direction of the force that a N-pole would feel if placed at that point in the field.
This means that lines of flux always go from N-poles to S-poles.

Where is the field strongest?

The more closely packed the lines of flux, the stronger the field.
The field is strongest nearest the poles, as shown in the diagram:

More magnetic field patterns

Lines of flux can never cross: if they did it would mean that a compass would point in two directions! If two magnets are placed close to each other, the field produced is the result of the combined effect of both of the magnets:

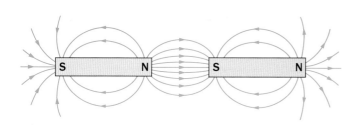 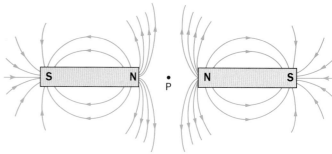

There is an attraction between the N-pole and the S-pole.
Where is the field the strongest?

Can you see the repulsion between the two N-poles? At point P the fields from the two magnets cancel out, and the resultant field strength is zero.
P is called a **neutral point**.

Magnetic flux density

The stronger the magnetic field, the more densely packed the lines of flux. In fact, we describe the strength of the magnetic field by the **magnetic flux density, B**.
This is the quantity of flux passing through unit area, at each point in the field. It is measured in **tesla** (T).

B is a vector quantity, because it has size and direction.
B is discussed in more detail on page 260.

How can we measure magnetic flux density?
A simple way is to use a **Hall probe** (see also page 267).
The diagram shows a Hall probe being used to measure the flux density between the poles of a horseshoe magnet:

The probe must be held in the field so that the lines of flux pass at right angles through its tip. The meter connected to the probe gives a direct reading of the flux density, in tesla.

The diagram shows how the flux passes through the tip of the Hall probe. Can you picture how the flux density will vary as the probe is moved to different positions in the magnetic field?

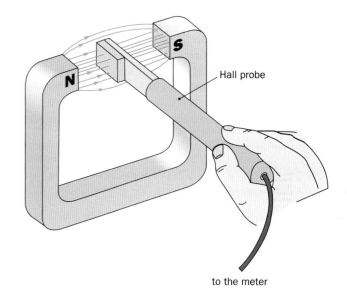

Hall probe

to the meter

The Earth's magnetic field

Why does a compass line up in a north–south direction?
The Earth acts as if there is a huge bar magnet inside it.
In the diagram, notice that the S-pole of the imaginary magnet is in the northern hemisphere so as to attract the N-pole of a compass:

Can you also see from the diagram that the lines of flux act at an **angle** to the Earth's surface, except near the equator?
This angle is called the **angle of dip**. In the UK it is about 70°.

A compass needle held horizontally is only affected by the horizontal component of the magnetic flux. What do you think will happen to a compass needle at the magnetic north pole?

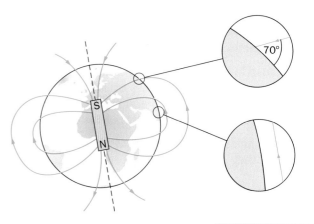

▷ The magnetic effect of a current

In 1819, Hans Christian Oersted noticed that a compass needle was deflected by an electric current in a nearby wire. In this way, he discovered the link between electricity and magnetism.
An electric current is always surrounded by a magnetic field.

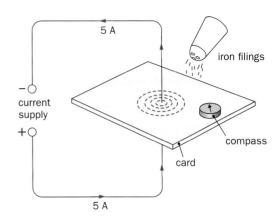

The magnetic field around a long straight wire

The diagram shows a wire carrying a current of about 5 amps.
If you sprinkle some iron filings on to the horizontal card and tap it gently, the iron filings will line up along the lines of flux as shown.
Can you see that the lines of flux are circles around the wire?

You can place a small compass on the card to find the direction of the magnetic field. With the current flowing up the wire, the compass will point anti-clockwise, as shown.
What will happen if you reverse the direction of the current?

The right-hand grip rule gives a simple way to remember the **direction** of the field: imagine gripping the wire so that your **right thumb** points in the direction of the current; your fingers then curl in the direction of the lines of flux:

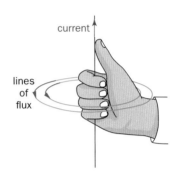

The diagrams show the magnetic field as you look down on the card :
Imagine the current direction as an arrow.
When the arrow moves away from you, into the page, you see the cross (×) of the tail of the arrow.
As the current flows towards you, you see the point of the arrow – the dot in the diagram.

Can you see that the further from the wire the circles are, the more widely separated they become?
What does this tell you?

The flux density is greatest close to the wire.
As you move away from the wire the magnetic field becomes weaker.

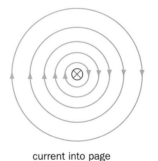

current into page

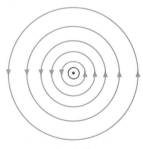

current out of page

The magnetic field of a flat coil

The diagram shows a flat coil carrying electric current :

Again, we can investigate the shape and direction of the magnetic field using iron filings and a compass.

Close to the wire, the lines of flux are circles.
Can you see that the lines of flux run anti-clockwise around the left side of the coil and clockwise around the right side?
What happens at the centre of the coil?

The fields due to the sides of the coil are in the same direction and they combine to give a strong magnetic field.

How would you expect the field to change if the direction of the current flow around the coil was reversed?

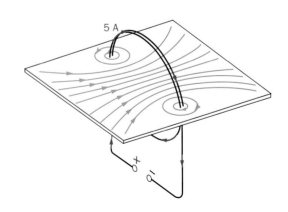

The magnetic field of a solenoid

A **solenoid** is a long coil with a large number of turns of wire. Look at the shape of the field, revealed by the iron filings. Does it look familiar?

The magnetic field **outside** the solenoid has the **same shape as the field around a bar magnet.**

Inside the solenoid the lines of flux are close together, parallel and equally spaced. What does this tell you?
For most of the length of the solenoid the flux density is constant. The field is uniform and strong.

If you reverse the direction of the current flow, will the direction of the magnetic field reverse?

A right-hand grip rule can again be used to remember the direction of the field, **but** this time you must curl the fingers of your **right** hand in the direction of the current as shown:
Your thumb now points along the direction of the lines of flux **inside** the coil – towards the end of the solenoid that behaves like the N-pole of the bar magnet.
This right-hand grip rule can also be used for the flat coil.

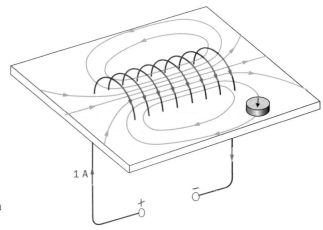

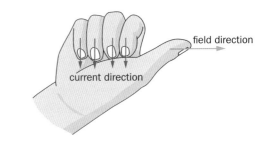

▷ Magnetic materials

We have two different ways of producing magnetic fields: one uses permanent magnets, the other uses electric currents. Is there a link?

Electric current is the movement of charged particles (page 226). In 1864, James Clerk Maxwell proved that all magnetic fields are caused by the movement of charged particles.

Now think about the motion of the electrons in atoms.
As each electron spins, it acts like a tiny electric current and so it produces a very tiny magnetic field.
In some atoms the magnetic effects of all the electrons cancel; in others they do not, so that each atom acts like a tiny magnet.

In **ferromagnetic** materials, these tiny atomic magnets can line up with each other to produce a very strong magnetic field. Iron, cobalt and nickel are well-known ferromagnetic elements.

Using an electromagnet.

Why is the magnetic field of a solenoid greatly increased when a ferromagnetic core is placed inside the coil?
The atomic magnets of the core line up along the lines of flux inside the solenoid, and so the core becomes magnetised.

What happens when the current in the solenoid is turned off?
It depends on the material.

* A **steel** core stays magnetised. The tiny atomic magnets remain lined up, even when the external field is removed.
* An **iron** core quickly demagnetises, because the atomic magnets have enough vibrational energy to turn in random directions. A magnet that can be switched on and off is an **electromagnet**.

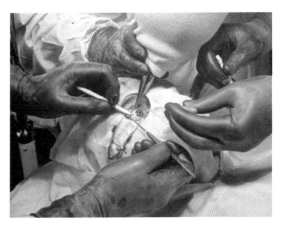

Removing a steel splinter from an eye.

259

▷ Measuring magnetic flux density

We have seen the **shapes** of the magnetic fields around current-carrying conductors, but what do you think affects the **strength** or **magnetic flux density B** of these magnetic fields?

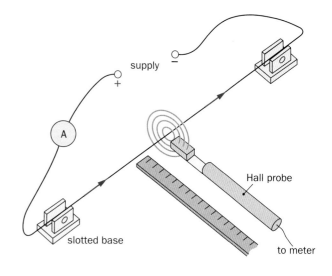

The long straight wire

You can investigate the magnetic flux density around a wire using the apparatus shown in the diagram:

Remember that the lines of flux are circles all along the wire. How can you position the Hall probe so that the lines of flux pass at right angles through its tip?

You can use the apparatus to answer the following questions:
1. When the probe is placed a fixed distance from the wire, how does B vary as the current in the wire is increased?
2. When the current is kept constant, how does B vary as the probe is moved away from the wire?

The graphs show some typical results. Can you see that:
- doubling the current **doubles** the flux density,
- doubling the distance from the wire **halves** the flux density.

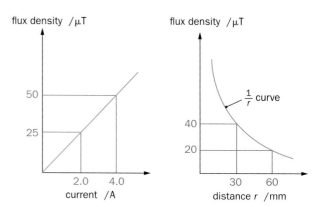

In fact:

Flux density, B $\propto$ $\dfrac{\text{current, } I \text{ (amperes)}}{\text{distance, } r \text{ (metres)}}$ or $B = k\,\dfrac{I}{r}$
(tesla)

k is a constant. It depends on the material around the wire.

If the wire is in a vacuum, we usually write this as:

> **Magnetic flux density, $B = \dfrac{\mu_{\text{o}}\, I}{2\pi\, r}$**

where μ_{o} is a constant called the **permeability of free space**. The permeability μ of a material is a measure of its effect on the strength of the magnetic field.
For a vacuum, $\mu_{\text{o}} = 4\pi \times 10^{-7}$ tesla metre ampere^{-1} (T m A^{-1}).

Our wire is in air, not a vacuum, and so we should really use the permeability of air (μ_{air}) in our equation, instead of μ_{o}.
In fact, air and most other materials have almost the same permeability as a vacuum.
But ferromagnetic materials have much higher permeabilities.
For example, the permeability of iron is about $8\pi \times 10^{-4}$ T m A^{-1}.

Example 1
A vertical wire carries a current of 6.0 A.
What is the flux density at (a) 20 mm and (b) 40 mm from the wire?

a) $B = \dfrac{\mu_{\text{o}}\, I}{2\pi\, r} = \dfrac{4\pi \times 10^{-7}\,\text{T m A}^{-1} \times 6.0\,\text{A}}{2\pi \times 20 \times 10^{-3}\,\text{m}} = \underline{6.0 \times 10^{-5}\,\text{T}}$

b) At twice the distance from the wire, the flux density is halved,

    so: $B = \dfrac{6.0 \times 10^{-5}\,\text{T}}{2} = \underline{3.0 \times 10^{-5}\,\text{T}}$

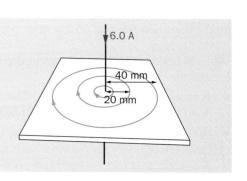

The solenoid

In the diagram, a 'slinky' is used as an adjustable solenoid:

You can use this apparatus to answer the following questions:

1. Is B uniform at all points inside the solenoid?
2. How does the value of B at the centre of the solenoid vary as the current in the solenoid is increased?
3. How does the value of B within the solenoid vary as the solenoid is stretched (so that n, the number of turns per metre, decreases)?

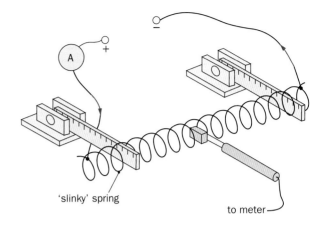

'slinky' spring

to meter

The graphs show some typical results. Can you see that:
- doubling the current **doubles** the flux density,
- doubling the number of turns per metre **doubles** the flux density.

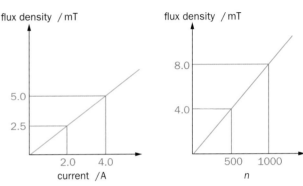

In fact, B is uniform inside the solenoid, and:

Magnetic flux density, B $\propto$ turns per metre, n $\times$ current, I
 (tesla) (amperes)

When the solenoid is in a vacuum (or in air), we can write this as an equation:

> **Magnetic flux density, $B = \mu_o \, n \, I$**

Example 2
A solenoid is 0.40 m long and is made with 800 turns of wire.
a) What is n, the number of turns per metre?
b) What is the flux density at the centre of the solenoid when it carries a current of 2.0 A?

a) $n = \dfrac{\text{the number of turns}}{\text{the length of the solenoid}} = \dfrac{800}{0.40 \text{ m}} = \underline{2000 \text{ m}^{-1}}$

b) $B = \mu_o \, n \, I = 4\pi \times 10^{-7} \text{ T m A}^{-1} \times 2000 \text{ m}^{-1} \times 2.0 \text{ A} = \underline{5.0 \times 10^{-3} \text{ T}}$

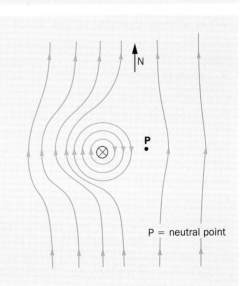

Example 3
Phiz passes a current of 5.0 A down a vertical wire and uses a compass to plot the lines of flux around the wire. He obtains the pattern shown, and finds a neutral point P, 50 mm from the wire:
a) Why has he obtained this pattern?
b) What is the value of the horizontal component of the Earth's magnetic field?

a) This field is the **resultant** magnetic field of both the wire and the Earth in a horizontal plane.

b) At P, the two fields must be **equal**, but in **opposite** directions.

For the wire: $B = \dfrac{\mu_o \, I}{2\pi \, r}$

$\therefore \quad B = \dfrac{4\pi \times 10^{-7} \text{ T m A}^{-1} \times 5.0 \text{ A}}{2\pi \times 50 \times 10^{-3} \text{ m}} = 2.0 \times 10^{-5} \text{ T}$

So the horizontal component of the Earth's field is also $\underline{2.0 \times 10^{-5} \text{ T}}$

P = neutral point

▷ Magnetic force

A wire carrying a current in a magnetic field feels a force.
A simple way to demonstrate this is shown in the diagram:

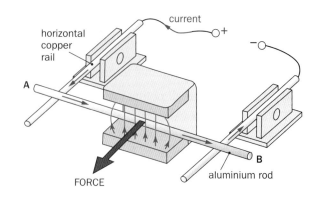

The two strong magnets are attached to an iron yoke with opposite poles facing each other. They produce a strong, almost uniform field in the space between them.

What happens when you switch the current on?
The aluminium rod AB feels a force, and moves along the copper rails as shown.
Notice that the current, the magnetic field and the force are **all at right angles** to each other.

What happens if you reverse the direction of the current flow, **or** turn the magnets so that the magnetic field acts downwards?
In each case the rod moves in the opposite direction.

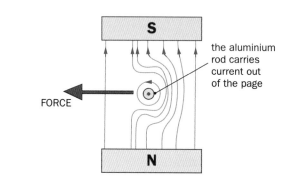

Why does the aluminium rod move?
The magnetic field of the permanent magnets interacts with the magnetic field of the current in the rod.
Imagine looking from end B of the rod. The diagram shows the combined field of the magnet and the rod:

The lines of flux behave a bit like elastic bands.
Can you see that the wire tends to be catapulted to the left?

You can use **Fleming's left-hand rule** to predict the direction of the force.
You need to hold your **left** hand so that the thumb and the first two fingers are at right angles to each other as shown:

If your **F**irst finger points along the **F**ield direction (from N to S), and your se**C**ond finger is the **C**urrent direction (from + to −), then your **Th**umb gives the direction of the **Th**rust (or force).

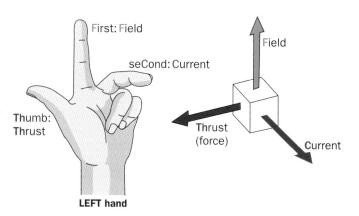

What affects the size of the force?

Look carefully at the apparatus shown. The balance is set to read zero after the magnets have been placed on it.
A current is then passed along the clamped aluminium rod.

Can you use Fleming's left-hand rule to show that the force on the aluminium rod is acting upwards?
If the magnets exert an upward force on the rod, then, by Newton's Third Law, the rod must exert an equal but opposite force on the magnets. This downwards force will cause the reading on the balance to increase.

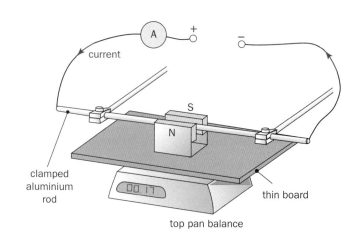

You can use the apparatus to answer the following questions:
1. How does the force vary when the current is increased?
2. How does the length of the rod in the field affect the force?
3. What happens to the force if the strength of the field is increased by adding extra magnets to the yoke?

Some typical results are shown on the next page.

The graph of force against length was obtained by placing a second and a third pair of magnets on the balance. This doubles and then triples the length of the aluminium rod in the field. Look carefully at these graphs. What do they tell us?

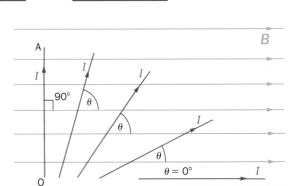

Experiments like this show us that the force F on a conductor in a magnetic field is directly proportional to:

- the magnetic flux density B,
- the current I, and
- the length l of the conductor in the field.

In fact:

> **Force, F = flux density, B × current, I × length, l of wire**
> (newtons)　　　(tesla)　　　(amps)　　　(metres)

or

> **$F = B I l$**

This equation applies when the current is at 90° to the field. Does changing the angle affect the size of the force?

Look at the wire OA in the diagram, at different angles:

When the angle θ is 90° the force has its maximum value. As θ is reduced the force becomes smaller. When the wire is parallel to the field, so that θ is zero, the force is also zero.

In fact, if the current makes an angle θ to the magnetic field the force is given by:

> **$F = B I l \sin \theta$**

Notice that: when $\theta = 90°$, $\sin \theta = 1$ and $F = BIl$ as before.
　　　　　　when $\theta = 0°$, $\sin \theta = 0$ and $F = 0$, as stated above.

The **size** of the force depends on the angle that the wire makes with the magnetic field, but the **direction** of the force does not. The force is **always at 90°** to both the current and the field.

Example 4
The two wires below are in a uniform magnetic field of 0.25 T. Both wires are 0.50 m long and carry a current of 4.0 A. Calculate the size of the force on each wire.

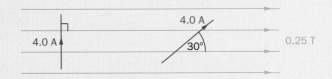

a) The wire is at 90° to the field.
　　　$F = B I l$
　　∴　$F = 0.25$ T × 4.0 A × 0.50 m $=$ __0.50 N__

b) The wire is at 30° to the field.
　　　$F = B I l \sin \theta$
　　∴　$F = 0.25$ T × 4.0 A × 0.50 m × $\sin 30°$ $=$ __0.25 N__

In both cases the force acts at 90° to the wire – into the paper.

Magnetic flux density B and the tesla

We can rearrange the equation $F = BIl$ to give:
$$B = \frac{F}{Il}$$
What is the value of B, when $I = 1$ A and $l = 1$ m? In this case, B has the same numerical value as F.

This gives us the definition of B:
The magnetic flux density B is the force per unit length, acting on a wire carrying unit current, which is perpendicular to the magnetic field.

The unit of B is the **tesla** (T).
Can you see that: 1 T $= 1$ N A^{-1} m^{-1}?

The tesla is defined in the following way:
A magnetic flux density of 1 T produces a force of 1 N on each metre of wire carrying a current of 1 A at 90° to the field.

▷ The magnetic force on a moving charge

The photograph shows the glow of the 'Northern Lights'.
Charged particles from outer space become trapped in the
Earth's magnetic field and spiral from one pole to the other.
As they enter the atmosphere at the poles, they produce this
spectacular glow in the sky.

This phenomenon occurs because a charged particle feels a force
when it moves through a magnetic field.
What factors do you think affect the size of this force?

The force F on the particle is directly proportional to:
- the magnetic flux density B,
- the charge on the particle Q, and
- the velocity v of the particle.

Aurora borealis: the 'Northern Lights'.

When the charged particle is moving at 90° to the field,
the force can be calculated from:

Force, F = flux density, B × charge, Q × velocity, v
(newtons)   (tesla)   (coulombs)   (m s⁻¹)

or $\quad F = B\,Q\,v$

In which direction does the force act?
The force is always at 90° to both the current and the field, and
you use **Fleming's left-hand rule** (page 262) to find its direction.

Try this for the positive particle in the first diagram:
Point your middle finger along the path of the positive particle.
Do you get the result shown?

How can you get the result for the negative particle?
(Note: the left-hand rule applies to conventional current flow.)
A negative charge moving to the right has to be treated as a
positive charge moving to the left.
You must point your middle finger in the opposite direction
to the movement of the negative charge.

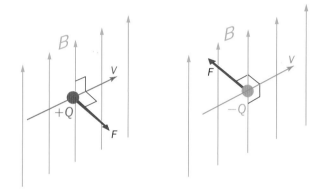

What if the charged particle is moving at an angle θ to the
magnetic field, with speed v?
We can resolve the velocity into two components as shown:
The component $v \sin \theta$ is at 90° to the magnetic field, and so

the force on the particle is given by: $\boxed{F = B\,Q\,v \sin \theta}$

The component $v \cos \theta$ is not affected by the magnetic field,
because it is parallel to it.

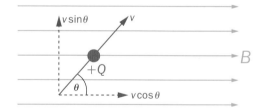

Example 5
An electron moving at 6.0×10^5 m s⁻¹ passes perpendicularly
through a magnetic field of 2.0×10^{-2} T.
The charge of the electron is 1.6×10^{-19} C.
What is the force on the electron?

$F = B\,Q\,v$
$F = 2.0 \times 10^{-2}\text{ T} \times 1.6 \times 10^{-19}\text{ C} \times 6.0 \times 10^5 \text{ m s}^{-1}$
$\underline{F = 1.9 \times 10^{-15}\text{ N}}$ (2 s.f.)

Do the equations $F = BIl$ and $F = BQv$ agree?

Suppose a charge Q moves a distance l, in t seconds.
Current is the rate of flow of charge,
and so (from page 228): $\quad I = \dfrac{Q}{t}$

Substituting for I in $F = BIl$, we get $F = \dfrac{BQl}{t}$.

But $\dfrac{l}{t}$ is the velocity v of the particle, so $F = BQv$.

▷ Circular paths

The movement of charged particles in magnetic fields is very important in atomic physics. In the 'fine beam tube' shown here, electrons move at right angles to a horizontal magnetic field. Can you see the path of the electron beam, a bright blue circle?

What provide the **centripetal force** to make each electron move in a circular path? (see chapter 7)

Look at the diagram. It shows electrons moving in a uniform magnetic field. The flux density B is **into** the paper (as shown by the $\times$ symbols). Each electron in the beam is moving with a speed v, at 90° to the field.

The field exerts a force F on the electron, where $F = BQv$. Can you see that this force is always at 90° to the electron's velocity? Because of this, the magnetic force changes the direction, but not the speed, of the electron. The BQv force provides the centripetal force.

The equation for centripetal force (see page 80) is:

$$F = \frac{mv^2}{r}$$ where m = the mass of the particle, and r = the radius of the circular path.

The centripetal force is provided by the magnetic force, and so:

$$\frac{mv^2}{r} = BQv \qquad \text{or, rearranging:} \qquad r = \frac{mv}{BQ}$$

Can you see from this that the electrons move in a larger circle if their speed is increased, or the magnetic field is made weaker?

What is the time period T for the electron to make one rotation?

Since time = $\dfrac{\text{distance}}{\text{speed}}$, $\quad T = \dfrac{\text{length of circular path}}{\text{speed of electron}} \quad$ or $\quad T = \dfrac{2\pi r}{v}$

but since $r = \dfrac{mv}{BQ}$, we get: $\quad T = \dfrac{2\pi}{v} \times \dfrac{mv}{BQ} \quad$ so: $\quad \boldsymbol{T = \dfrac{2\pi m}{BQ}}$

In a field of constant flux density B, the time period T of the electron does not depend on its speed. A faster electron moves in a circle of larger radius, but it takes exactly the **same time** for one revolution.

What happens when the particle is not moving at 90° to the field? The $v \sin\theta$ component makes it move in a circle. The $v \cos\theta$ component makes it move parallel to the lines of flux (see page 260). This means that the particle spirals around the field lines, as shown:

An electron beam in a fine beam tube.

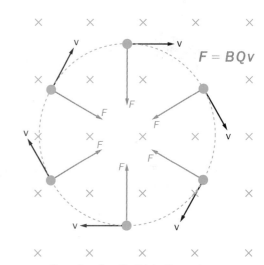
uniform flux density B into the page

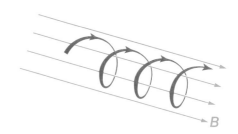
Spiralling particle.

Example 6
A particle of mass 9.1×10^{-31} kg and of charge 1.6×10^{-19} C is moving at 4.5×10^6 ms^{-1}. It enters a uniform magnetic field of flux density 0.15 mT, at 90°, as shown in the diagram: What is the radius of its circular path?

Using: $r = \dfrac{mv}{BQ}$ (as derived above) $\qquad$ 1 mT = 10^{-3} T

$$\therefore r = \frac{9.1 \times 10^{-31} \text{ kg} \times 4.5 \times 10^6 \text{ m s}^{-1}}{0.15 \times 10^{-3} \text{ T} \times 1.6 \times 10^{-19} \text{ C}} = 0.17 \text{ m}$$

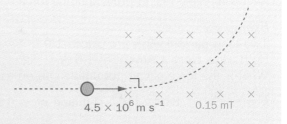

You can use the left-hand rule to show that the particle must be **positively** charged.

▷ Forces between currents

What happens when current is passed along two strips of foil as shown below?
The strips bend as they attract or repel each other.

Two parallel, current-carrying wires exert equal but opposite forces on each other.
Look carefully at these forces, and the resultant magnetic fields around the wires.

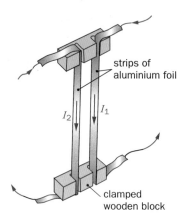

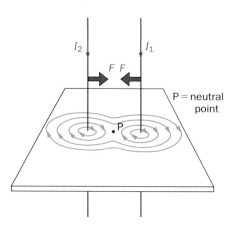

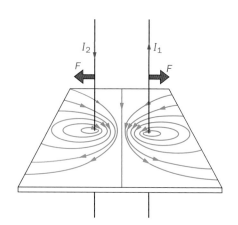

Notice that: currents flowing in the **same** direction **attract**,
currents flowing in **opposite** directions **repel**.

How do these forces arise?
The diagram shows the anti-clockwise field around wire X:
Wire Y is at 90° to this field, and so it experiences a force.
Apply Fleming's left-hand rule (page 262) to wire Y.
Do you find that the force on wire Y is to the left, as shown?

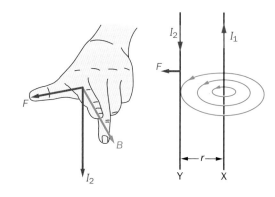

What is the size of the force?
Notice that the wires are a distance r apart.
Wire X carries a current I_1. Wire Y carries a current I_2.
What is the flux density B at wire Y, due to the current I_1 in X?
From page 260 we know that:

$$B = \frac{\mu_o I_1}{2\pi r} \qquad \dots (1)$$

What is the force F on a length l of wire Y?
From page 263 we know that:

$$F = BI_2 l \qquad \dots (2)$$

If we use equation (1) to
replace B in equation (2):

$$\boxed{F = \frac{\mu_o I_1 I_2 l}{2\pi r}} \qquad \dots (3)$$

Similarly, wire X feels a force due to the field of wire Y.
Can you show that equation (3) also gives the force on wire X?
Use the left-hand rule to prove that the force on X is to the right.

The definition of the ampere

The unit of current, the ampere, is defined
in terms of the force between two currents.

As shown in Example 7 below,
when two long wires are parallel
and placed 1 metre apart in air,
and if the current in each wire is 1 ampere,
then the force on each metre of wire is 2×10^{-7} N.

Example 7
Two long parallel wires are placed 1.0 m apart in air as shown.
Both wires carry a current of 1.0 A.
What is the force acting per unit length on each wire?

$F = \dfrac{\mu_o I_1 I_2 l}{2\pi r}$ where $l = 1.0$ m and $\mu_o = 4\pi \times 10^{-7}$ T m A^{-1}

$F = \dfrac{4\pi \times 10^{-7} \text{ T m A}^{-1} \times 1.0\text{A} \times 1.0\text{A} \times 1.0 \text{ m}}{2\pi \times 1.0 \text{ m}} = \underline{2 \times 10^{-7} \text{ N}}$ on each metre

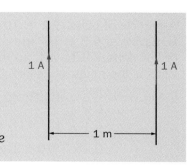

▷ Physics at work: Magnetic fields

The magnetic levitation train

The 'maglev' train in the photograph does not need wheels, because it 'floats' above its track by **magnetic levitation**. The diagram below shows how magnetic levitation works.

The levitation magnets are powerful electromagnets. Can you see that they are attached to the chassis of the train? The magnets are **attracted** upwards towards the levitation rail. This force lifts the train 15 mm above its T-shaped track.

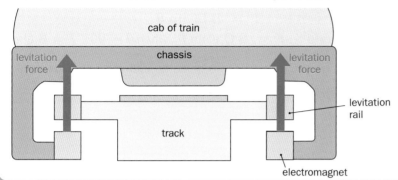

How does the train maintain its exact position above the track? Computers control the current in the electromagnets, and so the strength of the levitation force can be varied.

The Hall effect

How does a Hall probe measure the strength of a magnetic field? The diagram shows the semiconducting slice at the tip of a Hall probe. A small current is passed through the slice as shown: The electrons are moving with a velocity v.

When a magnetic field is applied, the BQv force pushes these electrons towards the upper edge of the slice. This becomes negatively charged. The lower edge, which is short of electrons, becomes positively charged.

If a voltmeter is placed across the top and bottom edges of the slice, it detects a small p.d. The stronger the field, the larger the p.d.

You can calibrate the voltmeter so that it reads in tesla, instead of volts.

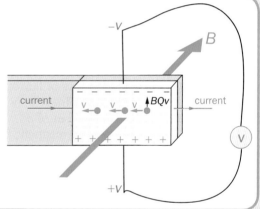

The mass spectrometer

Mass spectrometers are used to identify the different **isotopes** (see page 357) in a sample of material, and measure their relative abundance. The diagram shows the principle of how they work:

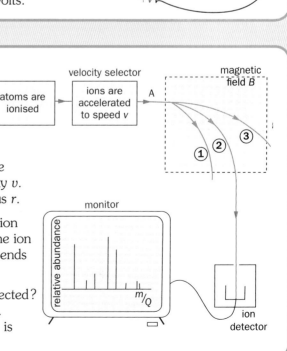

Each isotope in the sample forms ions with a different mass, but the velocity selector ensures that all the ions at A have the same velocity v. The magnetic field B deflects these ions into a circular path of radius r.

From page 265, r is given by: $r = \dfrac{mv}{BQ}$ where m = mass of the ion and Q = charge on the ion

Since B and v are fixed, the radius of the path taken by an ion depends on its value of m/Q. Only the ions following path 2 are detected.

Ions with a smaller value of m/Q take path 1. How are these detected? All the ions are deflected less when the magnetic field B is weaker. The applied magnetic field is gradually increased and each isotope is detected in turn. The screen shows their relative abundance.

Summary

Permanent magnets and electric currents produce magnetic fields. We show fields by drawing lines of flux. The field round a straight wire is in concentric circles. The field inside a solenoid is strong and uniform.

Magnetic flux density B is a vector quantity and is measured in tesla (T). $1\,T = 1\,N\,A^{-1}\,m^{-1}$

Around a long straight wire, $B = \dfrac{\mu_o I}{2\pi r}$ Inside a solenoid, $B = \mu_o n I$

μ_o is called the permeability of free space (a vacuum).

The force on a current in a wire at an angle θ to a magnetic field is: $F = BIl\sin\theta$
The force on a charge moving at an angle θ to a magnetic field is: $F = BQv\sin\theta$; when $\theta = 90°$, $F = BQv$

The direction of the force on the current, and on the moving charged particle, is given by Fleming's **left**-hand rule.
There is a force between two parallel current-carrying wires: this force is used to define the ampere.

▶ Questions

1. Can you complete these sentences?
   a) A magnetic .... can be represented by lines of flux. The more closely packed the lines of flux, the .... the magnetic field.
   b) There is always a .... field around a wire carrying an electric ....
   When the wire is straight and long, the lines of flux are .... around the ....
   c) **Inside** a solenoid, the lines of flux are .... and equally .... This shows that the field is ....
   Outside, the field is like that of a .... magnet.
   d) When a current crosses a magnetic field, it experiences a .... The direction of this .... is at .... to both the current and the .... directions.
   e) The magnetic flux density of a solenoid can be increased by increasing the number of .... per metre, and increasing the ....

2. Explain the following:
   a) A wire clamped firmly at both ends and placed between the poles of a horseshoe magnet, as shown below, vibrates when alternating current is passed through it:

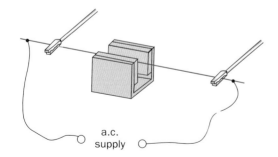

   b) A charged particle moving at 90° to a uniform magnetic field always moves in a circular path.
   c) Two long, parallel wires carrying electric current in opposite directions repel each other.

3. Sketch the magnetic field patterns:
   a) between the poles of a horseshoe magnet,
   b) around a long, straight wire carrying electric current vertically upwards,
   c) around two long parallel wires carrying current in the same direction,
   d) between the poles of the cylindrical magnet below.

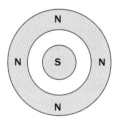

4. The diagram below shows a pair of iron-cored solenoids. The solenoids attract each other. Copy the diagram and add the direction of the current flow in solenoid b.

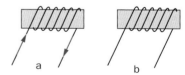

5. Calculate the magnetic flux density at a distance of 5.0 cm from a long, straight wire carrying a current of 4.0 A.

6. The magnetic flux density at a point close to a long wire carrying a current of 2.0 A is 2.0×10^{-5} T. How far is this point from the wire?

7. A solenoid of length 40 cm has 1200 turns. What is the magnetic flux density at the centre of the solenoid when the current through it is 2.5 A?

8. The diagrams below show 4 examples of conductors in magnetic fields.
   a) In which of the diagrams will there be a force acting on the conductor?
   b) If there is a force, in which direction does it act?

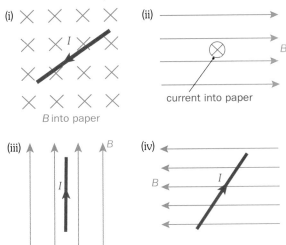

9. A solenoid can carry a maximum current of 10 A. If it is to produce a field of flux density of 0.20 T, how many turns per metre would the solenoid need?

10. Each diagram below shows a **current balance**. This is a rectangular piece of wire supported on pivots so it balances. Which of these current balances will tilt when current flows through the balance as shown?

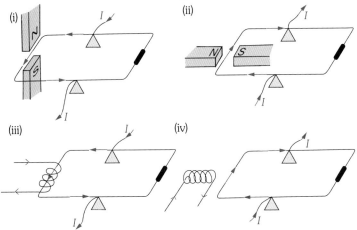

11. Two long, parallel, vertical wires P and Q are 10 mm apart. Both wires carry current upwards. The current in P is 4.0 A; the current in Q is 1.0 A.

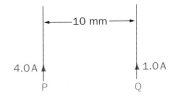

   a) What is the force on each metre length of wire?
   b) There is a neutral point between the two wires. How far is this neutral point from wire P?

12. A wire of length 0.40 m is placed in a field of flux density 0.15 T as shown below.
   The current in the wire is 2.0 A.
   Calculate the size of the force exerted on the wire:
   a) when the wire is at 90° to the magnetic field,
   b) when the wire is at an angle of 60° to the field.

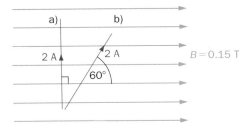

   c) What is the direction of the force in each case?

13. When a current of 4.0 A is passed along the clamped wire shown in the diagram below, the reading on the balance increases by 2.2 g. What is the flux density of the field between the poles of the magnet if the length of the wire in the field is 5.0 cm?
   (Take g to be 10 N kg^{-1}.)

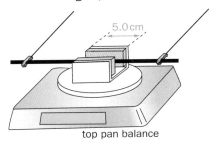

14. A particle X enters a uniform magnetic field.
   a) Is X positively or negatively charged?
   b) What will happen to the path of particle X if:
      i) it is slowed down,
      ii) the strength of the magnetic field is increased,
      iii) the direction of the magnetic field is reversed?

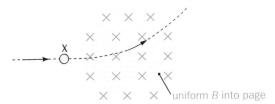

15. An electron moving at 2.0×10^7 ms^{-1} passes at 90° through a magnetic field of flux density 2.5×10^{-2} T. The electron has a mass of 9.1×10^{-31} kg and a charge of -1.6×10^{-19} C.
   a) What force is exerted on the electron by the field?
   b) What is the radius of the electron's circular path in the field?
   c) What happens to the path taken by the electron if it enters at 30° to the magnetic field instead of 90°?

Further questions on page 315.

19 Electromagnetic Induction

The photograph shows a wind farm. Each aerogenerator converts the kinetic energy of the wind into electrical energy. However, its electricity output is small when compared to the 50 gigajoules of energy that Britain's power stations can produce in each second.

How is all this electrical energy produced? Power stations generate electricity by using *electromagnetic induction*. This was discovered in 1831 by Michael Faraday.

In this chapter you will learn:
- about magnetic flux and magnetic flux linkage,
- how to use Faraday's Law and Lenz's Law,
- some of the uses of electromagnetic induction.

What is electromagnetic induction?

The diagram shows a copper rod connected to an ammeter: There is no battery in the circuit.

What happens when you move the copper rod downwards, to cut **across** the horizontal magnetic field?
The pointer on the meter makes a brief 'flick' to the right, showing that an electric current has been **induced**.

What happens when you move the rod upwards?
The meter again gives a 'flick', but this time to the left.
You have now induced a current in the opposite direction.

If you hold the rod stationary, or if you move the rod **along** the field lines, there is no induced current.

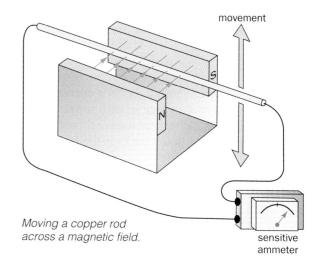

Moving a copper rod across a magnetic field.

Why does electromagnetic induction occur?

When you move the copper rod, its free electrons move with it.
But when a charge moves in a magnetic field it experiences a force on it (the BQv force, see page 264).
You can use Fleming's left-hand rule (page 262) to show that the force on each electron is to the left, as shown in the diagram:
(Remember – an electron moving down has to be treated as a positive charge moving up.)

So electrons accumulate at one end of the rod, making it negative.
This leaves the other end short of electrons and therefore positive.
There is now a **voltage** across the ends of the moving rod.

If the ends of the moving rod are joined to form a complete circuit, the induced voltage causes a current to flow round the circuit, as shown by the 'flick' of the ammeter.
The induced voltage is a source of electrical energy – an **e.m.f.**

When a conductor is moving in a magnetic field like this, **an e.m.f. is induced**, even if there isn't a complete circuit for a current to flow.

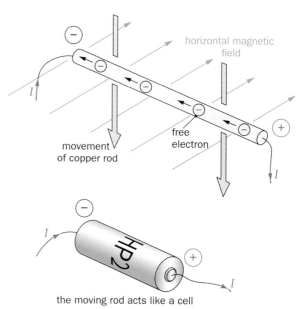

the moving rod acts like a cell

Faraday's model of electromagnetic induction

The diagram shows another way to demonstrate induction:
What happens when you move the magnet **into** the coil?
An e.m.f. is induced across the ends of the coil. This e.m.f.
drives a current round the circuit and the meter gives a 'flick'.

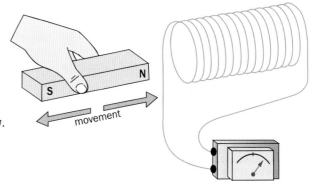

If you move the magnet **out of** the coil, then an e.m.f. is induced
in the reverse direction, and the ammeter 'flicks' the opposite way.

What happens if you move the coil instead of the magnet?
The meter still shows a current. An e.m.f. is induced whenever
there is **relative movement** between the magnet and the coil.

Michael Faraday devised a model to help us visualise what is
happening in electromagnetic induction.
Think about the experiment on the facing page.
As the rod moves downwards, can you visualise it 'cutting'
through the lines of flux between the N-pole and the S-pole?

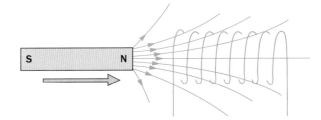

In Faraday's model, this cutting of the lines of magnetic flux
causes an e.m.f. to be induced across the conductor.

The coil cuts through the lines of flux.

Does this idea also work for the magnet and coil?
Look at the diagram. As the magnet is moved towards the coil,
each turn of the coil cuts the lines of flux of the bar magnet.

The more turns on the coil, the greater the e.m.f. This is because
N turns will cut N times more lines of flux than just one turn.

Faraday's flux model can be looked at in a different way.
The diagram shows a small flat coil, called a search coil,
connected to an ammeter:
If the coil is pulled quickly out of the field, the meter gives a flick.
In the diagram, can you see three lines of flux passing through or
linking the coil, when it is in the field?
When it is out of the field the flux linkage of the coil is zero.
This **change in flux linkage** induces an e.m.f. across the coil.

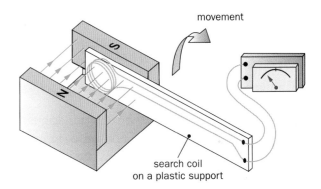

search coil
on a plastic support

So Faraday's model tells us that an e.m.f. is induced when
- a conductor like the copper rod cuts through lines of flux, or
- when there is a change in the flux linkage of a coil.

Flux cutting and changes in flux linkage are just two ways
of looking at the same thing. Look at the lines on the diagram:

When the coil is moved from A to B we can say that the number
of lines of flux linking it changes from 5 to 9 – a change of **4**.
Or we can say that the coil has cut through **4** lines of flux.
These are two ways of describing the same change.

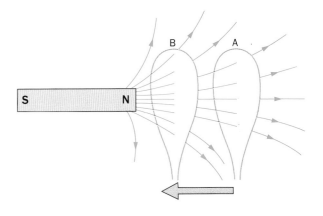

You can increase the size of the e.m.f. induced across a coil by:
- using a stronger magnet,
- using a coil with more turns of wire,
- using a coil with a greater cross-sectional area,
- moving the magnet faster.

This is because the larger and the faster the change in flux linkage
of the coil, the greater the induced e.m.f.

▷ Electromagnetic induction using two coils

Faraday was not using a magnet when he first discovered electromagnetic induction. He was using two coils, wrapped around an iron ring. What did he see?

The diagram shows two coils placed side by side:
Coil C is connected to a battery and a switch.
Coil D is connected to a sensitive ammeter.
Think about these 4 questions:

1. What do you think happens when the switch is closed?
The meter gives a flick to the right, and then goes back to zero. (You would see the same effect if you were to move the N-pole of a bar magnet quickly into coil D.)

2. Why is a current induced in coil D?
When you switch on the current, a magnetic field 'grows' around coil C. The 'growing' flux lines link the turns of coil D. As its *flux linkage changes*, an e.m.f. is induced across coil D and this induced e.m.f. drives a current through the ammeter.

3. Why does the ammeter reading go back to zero?
When the current in coil C reaches its final steady value, the magnetic field around C is steady, so the flux linkage of D is *not changing*, and so there is no induced e.m.f. and no current.

4. What do you think will happen when the switch is opened?
As the current in C falls, its magnetic field collapses.
As the flux linkage of D *changes*, an e.m.f. is induced again. The meter gives a flick again, but this time to the left, because the induced e.m.f. and the current are in the opposite direction.

Can you explain why the e.m.f. induced across coil D can be increased by:
• using an iron core, as shown in this diagram?
• using coils with a greater number of turns?
• passing a larger current around coil C?
• making the current in C change more quickly?
Remember: the larger and the faster the change in flux linkage of coil D, the greater the e.m.f. induced across it.

If the current in coil C is switched on and off repeatedly, what happens in coil D?
You would see the pointer on the meter move to and fro as the current in coil D constantly changes direction.
This is alternating current or a.c.

Now suppose you pass alternating current through coil C.
The field around C will grow and decay in one direction, and then grow and decay in the opposite direction.
The magnetic flux linkage of coil D will always be changing.
Passing a.c. through coil C induces a.c. in coil D, as shown:

This apparatus is a simple **transformer** (see also page 284).
Coil C is called the *primary* coil; coil D is the *secondary* coil.

Michael Faraday.

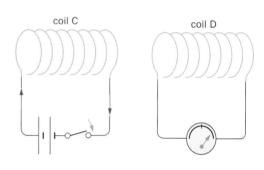

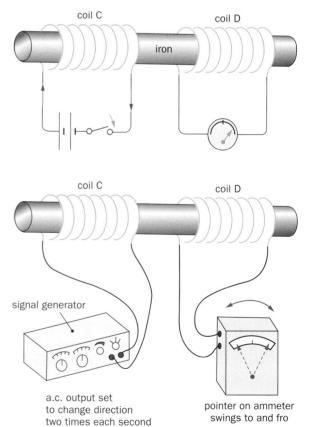

signal generator

a.c. output set
to change direction
two times each second

pointer on ammeter
swings to and fro

▷ Magnetic flux, Φ

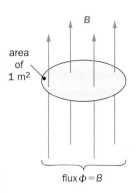

We draw lines of magnetic flux so that the density of the lines shows the strength of the field. So the number of lines passing at 90° through unit area represents the magnetic flux density B (see page 257).

Magnetic flux has the symbol Φ (the Greek letter phi). It represents the *total number* of lines of flux that 'link', or pass at 90° through, a given area. Magnetic flux is measured in webers (Wb).

If B is the flux per unit area, the total flux linking the area A can be calculated from:

Magnetic flux, Φ = magnetic flux density, B × area, A
(weber)   (tesla)   (m²)

or | $\Phi = B\,A$ |

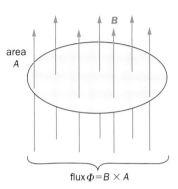

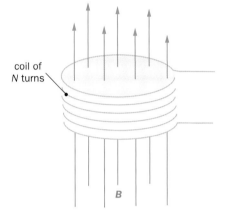

Here, the perpendicular to the area A makes an angle θ with the flux density B. We can resolve B into two components:

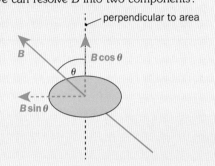

What is the flux linkage of the **coil** of cross-sectional area A and N turns? Each **turn** of the coil has flux Φ linking it, where $\Phi = BA$, and so:

Total flux linkage of the coil $= N\,\Phi = N\,B\,A$

$B\sin\theta$ **does not link** the area. The flux linkage is only by the $B\cos\theta$ component. So $\Phi = B\cos\theta \times A$, or:

Magnetic flux, $\Phi = B\,A\cos\theta$

A *change* in flux linkage is given the symbol $\Delta\Phi$. The flux linking a coil may change for two reasons: either (1) the flux density B changes, or (2) the area A linked by the flux changes.

(1) If a single turn coil of area A is placed at 90° to a magnetic flux density that is changing, then:

$\Delta\Phi = \Delta B\,A$

where ΔB is the *change in flux density*

(2) If a single turn coil moves in a flux density B so that the area linked by the flux changes, then:

$\Delta\Phi = B\,\Delta A$

where ΔA is the *change in area*

Example 1
How much flux is linking a 200 turn coil of area 0.10 m², when it is placed at 30° to a magnetic field $B = 2.5 \times 10^{-3}$ T?

$$\text{Total flux linkage} = N\Phi = NBA\cos\theta$$
$$= 200 \times 2.5 \times 10^{-3}\,\text{T} \times 0.10\,\text{m}^2 \times \cos 30° = \underline{0.043\,\text{Wb}}$$

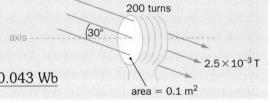

Example 2
A coil of 100 turns and area 0.20 m² is at 90° to a flux density which decreases from 0.50 T to 0.20 T.
What is the change in flux linkage of the coil?

$$\text{Change in flux linkage} = N\Delta\Phi = N\Delta BA$$
$$= 100 \times (0.50 - 0.20)\text{T} \times 0.20\,\text{m}^2$$
$$= 100 \times 0.30\text{T} \times 0.2\,\text{m}^2 = \underline{6.0\,\text{Wb}}$$

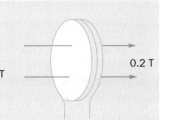

▶ Faraday's Law

We know that an e.m.f. is induced when there is a change in the flux linking a conductor.

Think again about the experiments on pages 270–272.
Do you agree that they all show that the larger and the faster the change in flux linkage, the greater the induced e.m.f.?

Faraday's Law makes the connection between the size of the induced e.m.f. and the *rate* at which the flux is changing.
It states that:

> *the magnitude of the induced e.m.f. is equal to the rate of change of flux linkage.*

Phiz makes the link with Faraday's law.

If a change in flux $\Delta\Phi$ occurs in a time Δt, then:

Induced e.m.f., ϵ (volts) = $\dfrac{\text{change in flux linkage}, \Delta\Phi \text{ (weber)}}{\text{time taken}, \Delta t \text{ (s)}}$

or

$$\epsilon = \frac{\Delta\Phi}{\Delta t}$$

For a coil with N turns, the e.m.f. will be N times as great, and so:

$$\epsilon = N\frac{\Delta\Phi}{\Delta t}$$

Remember (from page 273): $\Delta\Phi$ can be calculated using
either $\Delta\Phi = \Delta BA$ **or** $\Delta\Phi = B\Delta A$

Dropping a bar magnet

The diagram shows a bar magnet falling through a horizontal coil:

The coil is connected to a data-logger, which records the induced e.m.f. every millisecond.
The graph below shows how the induced e.m.f. varies with time:

Section AB shows us that as the magnet enters the coil, the flux linkage of the coil increases and an e.m.f. is induced.

Why is the induced e.m.f. negative over section BC?
As the magnet leaves the coil, the flux linkage decreases and the e.m.f. is induced in the opposite direction.

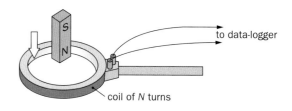

Rearranging Faraday's equation we get: $\epsilon\Delta t = N\Delta\Phi$

So $\epsilon\Delta t$ is equal to the flux change of the coil in the time Δt.
Can you see the area $\epsilon\Delta t$: the dark pink strip on the graph?
This small area tells us the change in flux linkage over time Δt.
Therefore, the total area under the graph for part AB (shaded pink) is the total change in flux linkage as the magnet enters the coil.

The area under the graph for the section AB (pink) is equal to the area under the graph for the section BC (shaded grey).
What does this tell us about the total change in flux linkage as the magnet leaves the coil and as the magnet enters the coil?

Why is the peak BC narrower but taller than the peak AB?
Remember the magnet is accelerating as it falls.

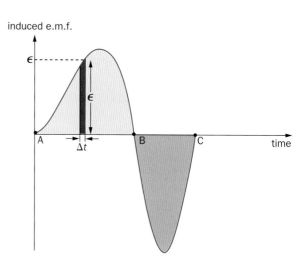

induced e.m.f.

Solving problems using Faraday's Law

Example 3

A search coil has 5000 turns and an area of 1.0×10^{-4} m². It is placed in a long current-carrying coil as shown, so that its face is at 90° to the lines of magnetic flux inside the coil. What e.m.f is induced across the coil when the flux density inside it changes from 2.5×10^{-3} T to 1.3×10^{-3} T in 0.40 s?

Note that it is the **flux density** linking the coil that changes.

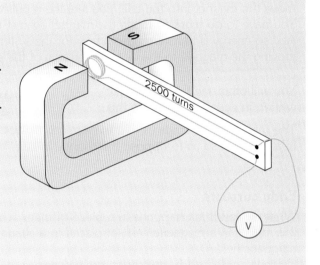
current
5000 turns
V

$$\text{e.m.f., } \epsilon = N \frac{\Delta \Phi}{\Delta t} \qquad \text{where} \quad \Delta \Phi = \Delta B\, A$$

$$= N \frac{\Delta B A}{\Delta t}$$

$$= \frac{5000 \times (2.5 \times 10^{-3} - 1.3 \times 10^{-3})\,\text{T} \times 1.0 \times 10^{-4}\,\text{m}^2}{0.40\,\text{s}}$$

$$= \underline{1.5 \times 10^{-3}\,\text{V}} \quad (2 \text{ s.f.})$$

Example 4

A search coil has 2500 turns and an area of 1.5×10^{-4} m². It is placed between the poles of a horseshoe magnet, as shown. As the coil is rapidly pulled out of the field in 0.30 seconds, the average value of the e.m.f. induced across the coil is 0.75 V. Calculate the flux density between the poles of the magnet.

Note that as the coil moves, the **area** linked by the flux changes.

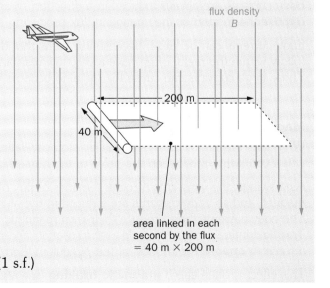
S
N
2500 turns
V

$$\text{e.m.f., } \epsilon = N \frac{\Delta \Phi}{\Delta t} \qquad \text{where} \quad \Delta \Phi = B\, \Delta A$$

$$= N \frac{B \Delta A}{\Delta t}$$

$$0.75\,\text{V} = \frac{2500 \times B \times (1.5 \times 10^{-4} - 0)\,\text{m}^2}{0.30\,\text{s}}$$

$$\therefore \quad B = \frac{0.75\,\text{V} \times 0.30\,\text{s}}{2500 \times 1.5 \times 10^{-4}\,\text{m}^2} = \underline{0.60\,\text{T}} \quad (2 \text{ s.f.})$$

Example 5

An aeroplane with a wingspan of 40 m moves at 200 m s^{-1}. It flies horizontally through an area where the vertical component of the Earth's magnetic flux density is 5×10^{-5} T. What is the e.m.f. induced across the wing-tips of the plane?

We can treat the wingspan as a straight conductor, moving through the magnetic flux as shown in the diagram:

In each second the conductor moves 200 m and 'sweeps out' an area ΔA through the flux, where $\Delta A = 40$ m $\times$ 200 m.

flux density
B
200 m
40 m
area linked in each second by the flux
= 40 m × 200 m

$$\text{e.m.f., } \epsilon = N \frac{\Delta \Phi}{\Delta t} \qquad \text{where } N = 1 \text{ and } \Delta \Phi = B\, \Delta A$$

$$= \frac{B \Delta A}{\Delta t}$$

$$= \frac{5 \times 10^{-5}\,\text{T} \times (40\,\text{m} \times 200\,\text{m})}{1\,\text{s}} = \underline{0.4\,\text{V}} \quad (1 \text{ s.f.})$$

▷ Lenz's Law

Faraday's Law tells us the size of the induced e.m.f.,
but we can find its direction using **Lenz's Law**:

> ***The direction of the induced e.m.f. is such that it will try
> to oppose the change in flux that is producing it.***

Lenz's Law is illustrated in the diagrams:

As you move the N-pole *into* the coil, an e.m.f is induced
which drives a current round the circuit as shown.
Now use the right-hand grip rule (from page 259).
Can you see that the current produces a magnetic field
with an N-pole at the end of the coil nearest to the magnet?

So the coil *repels* the incoming magnet, and in this way
the induced current *opposes* the change in flux.

Why is the current reversed as you move the N-pole *out*?
By Lenz's Law, the coil needs to attract the receding N-pole.

Lenz's Law is a result of the conservation of energy. If you
move the magnet into the coil, you feel the repulsive force.
You have to **do work** to move the magnet against this force.
And so energy is transferred from you (or the system that is
moving the magnet) to the electrical energy of the current.

You will often see Faraday's Law
written in equation form like this:
$$\epsilon = -N \frac{\Delta \Phi}{\Delta t}$$

The minus sign reminds us that the e.m.f. is always induced
in a direction so as to **oppose** the change in flux.

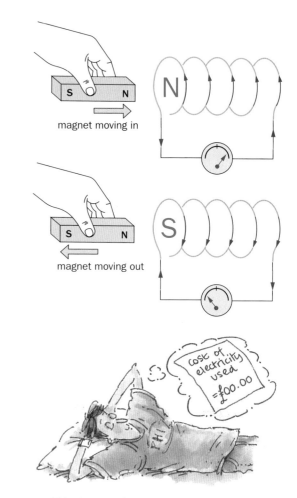

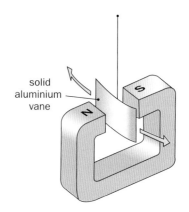

Phiz dreams of creating energy without work.

Eddy currents

What do you think happens in a piece of metal when
it is moved in a magnetic field, or placed in a changing
magnetic field?
Currents, called **eddy currents**, are induced in the metal.
The diagram shows a simple example:

Eddy currents are always induced in a direction so as to
oppose the motion or the change in flux producing them.

Eddy currents can be large, because the paths that they
follow within the metal often have a very low resistance.
These large currents can cause the metal to become very hot.

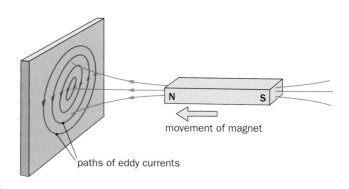

Magnetic braking

Why does this aluminium vane quickly stop oscillating
when you swing it between the poles of the horseshoe magnet?

As it swings, the vane cuts through the lines of flux of the
magnet, and so eddy currents are induced in the aluminium.
The eddy currents then produce their own magnetic field.
This field interacts with the field of the magnet, and produces
a force to **oppose** the motion of the pendulum (Lenz's Law).

This effect is often called **magnetic braking** (see also page 278).

▷ The a.c. generator

A generator converts *kinetic energy into electrical energy*.
It consists of a rectangular coil that rotates in a magnetic field:

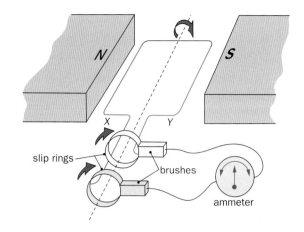

Each end of the coil is connected to a *slip-ring*. The slip-rings
rotate with the coil and press against stationary carbon brushes.
Each side of the coil always makes contact with the same brush.
This is shown by the blue and green shading on the diagram.

What happens if you rotate the coil at a steady rate?
As it turns the flux linkage of the coil constantly changes
and the pointer on the ammeter swings from side to side,
showing that alternating current (a.c.) is induced.

The graph shows how the size of the induced e.m.f. changes
as the coil rotates once in the magnetic field:
One complete revolution of the coil gives one cycle of a.c.

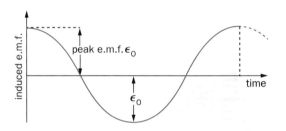

How can we increase the size of the peak e.m.f. ϵ_0?
We must increase the rate of change of flux linkage of the
coil as it spins in the magnetic field, by:
- using a coil with more turns N,
- using a coil with a larger cross-sectional area A,
- increasing the strength of the magnetic field B, or
- increasing the frequency of rotation of the coil, f.

Why is the graph of induced e.m.f. against time sinusoidal?
The diagram shows the coil end-on. It is rotating about
its horizontal axis with angular speed ω (see page 79).
The flux linkage of the coil must be: $N\Phi = NBA\cos\theta$,
where N is the number of turns and A is the area of the coil.

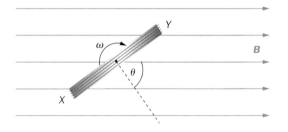

We start the time when $\theta = 0$; then after a time t the coil
will have turned through an angle θ, where $\theta = \omega t$.
This means that the flux linkage of the coil is: $NBA\cos\omega t$.

Applying Faraday's Law (page 274) to the flux linkage equation, we get:

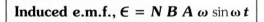

Induced e.m.f., $\epsilon = N B A \omega \sin\omega t$	or	$\epsilon = B A N \omega \sin\omega t$

The graph shows how the flux Φ and the e.m.f. ϵ vary as the
coil rotates. Notice that:
- the induced e.m.f. is zero when the flux linkage is a maximum,
- the e.m.f. is a maximum when the flux linkage is zero.

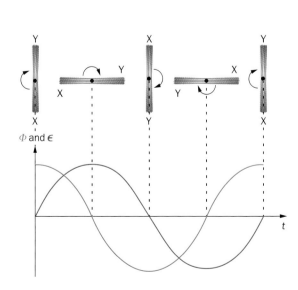

Can we explain these results?
Faraday's Law tells us that the magnitude of the induced e.m.f.
is equal to the rate of change of flux linkage.

When Φ is a maximum, the gradient of the Φ curve is zero.
There is no rate of change of flux linkage and no induced e.m.f.

When Φ is zero, the gradient of the Φ curve is a maximum. The
rate of change of flux linkage is a maximum and so is the e.m.f.

What is the maximum value ϵ_0 of the induced e.m.f.?
The maximum value that the sine function can have is 1, and so:

$$\epsilon_0 = B A N \omega \quad \text{or} \quad \epsilon_0 = B A N \times 2\pi f \quad (\text{because } \omega = 2\pi f)$$

▷ Physics at work: Using electromagnetic induction

The metal detector
How does Phiz's metal detector work?
A.c. flows through the search coil in the search head, and this produces a constantly changing magnetic field around the coil.

What happens if there is a metal object in this changing field?
Eddy currents are induced in the metal object. This means that energy is transferred from the search coil to the metal object, and the current in the search coil changes. This is detected by the control box, which produces an audible signal.

How effective is the metal detector?
This depends on the size and the orientation of the metal object and how far below the ground it is buried.

Traffic lights are triggered in a similar way.
There is an induction loop in the road surface, and this creates an alternating magnetic field (like that of the search coil).
A car passing over the loop is detected, because it changes the current in the loop – just like a metal object in the metal detector.

Induction heating
A ceramic hob is a type of electric cooker. It consists of three or four flat coils embedded in a ceramic surface. Each coil is supplied with high-frequency a.c. – about 18 kHz.

What happens when a metal pan is placed over one of the coils? Eddy currents are induced in the base and the sides of the pan. As these currents flow through the metal pan, it becomes hot and so the food in the pan is cooked.
Can you explain why the ceramic surface stays relatively cool?

Electromagnetic brakes
Eddy-current brakes are often fitted to coaches and lorries.
Two discs (called rotors) are fixed to the wheel-shaft, so that they spin as the wheels turn.
Can you see the stator, fixed in place between the rotors?
The stator is an electromagnet made up of a number of coils.

What do you notice about the way the coils are wound?
When the driver activates the brake, a current is passed through the coils so that adjacent coils produce reversed magnetic fields.
The spinning rotors cut through the magnetic flux, and so eddy currents are induced in the discs.
The eddy currents flow so as to *oppose* the change producing them; hence the braking effect on the rotors and the wheels.

Why are the brakes not effective at slow speeds?
The discs spin slowly and the rate of flux cutting is reduced.
The smaller eddy currents produce a smaller braking effect.
Contact brakes (see page 30) are still needed for the final braking at slow speed.

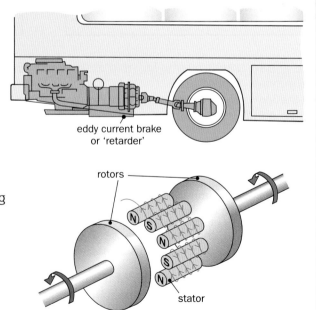

eddy current brake or 'retarder'

rotors

stator

278

Summary

The magnetic flux Φ linking an area A when the perpendicular to the area makes an angle θ to a field of flux density B is: $\Phi = BA\cos\theta$

For a coil with N turns, total flux linkage $= NBA\cos\theta$

An e.m.f. is induced when the magnetic flux linkage changes, because either the flux density changes or the area linked by the flux changes.

Faraday's Law gives us the **size** of the induced e.m.f. It states that the e.m.f. is equal to the **rate** of change of flux linkage:

$$\epsilon = N\frac{\Delta\Phi}{\Delta t}$$

Lenz's Law gives us the **direction** of the induced e.m.f. It states that the e.m.f. is induced so as to try to **oppose** the change that is causing it.

The e.m.f. in a coil rotating in a uniform field B is $\epsilon = BAN\omega\sin\omega t$

▷ Questions

1. Can you complete these sentences?
   a) When the magnetic .... linking a coil changes, an e.m.f. is ....
   b) Faraday's Law states that the .... of the induced .... is equal to the .... of change of .... linkage.
   c) Lenz's Law states that the .... of the induced .... is always so as to .... the change causing it.
   d) When a magnet moves towards a coil, the size of the induced e.m.f depends on the .... of the movement, the .... of the magnet, and the cross-sectional .... and number of .... on the coil.

2. Explain the following:
   a) No e.m.f. is induced when a straight wire is moved parallel to lines of magnetic flux.
   b) A brass ring set to swing in a magnetic field quickly comes to rest; a plastic ring swings for much longer.
   c) When a magnet is dropped through a plastic pipe it accelerates towards the ground; if dropped down an identical sized copper pipe it falls more slowly.

3. In the U.K. the vertical component of the Earth's magnetic field is 5.0×10^{-5} T. If the area of a village is 2.2×10^{6} m², how much flux passes through it?

4. A rectangular coil measures 5.0 cm by 8.0 cm. It is placed at 90° to a magnetic field of flux density 1.5 T. What is the flux linkage of the coil if it has 250 turns?

5. A search coil has 3400 turns and an area of 1.0 cm². It is placed between the poles of a magnet where the flux density is 0.40 T.
   What is the average e.m.f. induced across the coil when it is snatched out of the field in 0.20 s?

6. A search coil with 2000 turns and an area of 1.5 cm² is used to measure the flux density between the poles of a horseshoe magnet. The coil is pulled out of the field in 0.30 s and the induced e.m.f. is 0.50 V. What is the flux density of the magnet?

7. A wire of length 10.0 cm is moved downwards at a speed of 4.0 cm s^{-1} through a horizontal field of flux density 2.0 T. What is the e.m.f. induced in the wire?

8. An aircraft of wingspan 40 m flies horizontally at a speed of 250 m s^{-1}. An e.m.f. of 0.2 V is induced across the tips of the wings. What is the vertical component of the Earth's magnetic flux density?

9. The diagram shows a coil being turned in a horizontal magnetic field of flux density 50 mT.
   a) In the position shown, does the induced current flow in the direction ABCD or DCBA?
   b) The coil has 300 turns, area 15 cm² and rotates 10 times per second. What is the peak e.m.f.?

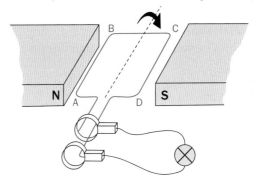

10. A search coil with 2500 turns and area 1.5 cm² is connected to a data-logger, placed between the poles of a magnet and quickly withdrawn.
    a) Use the graph shown below to estimate the total change in flux linkage of the coil. (You can treat the graph as a triangle, as shown.)
    b) Calculate the flux density of the magnet.

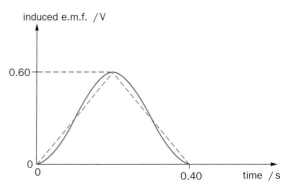

Further questions on page 316.

20 Alternating Current

Have you ever experienced a power cut?
If so, you will know how much we rely on our electricity supply.
Mains electricity has only been available for about 100 years, and initially each area had its own different supply.
Now we all receive the same standardised a.c. supply, and this is delivered to our homes by the National Grid system.

In this chapter you will learn:
- the meaning of the root-mean-square (r.m.s.) value,
- how to use an oscilloscope to measure time and p.d.,
- about the transformer.

The control room in a nuclear power station.

▷ What is a.c.?

The diagram shows a bicycle 'dynamo' connected to a lamp:
The dynamo is a small a.c. generator.
What will happen as you turn the generator slowly by hand?

The generator produces an *alternating* e.m.f. This causes current to flow one way and then the other way round the circuit.
The free electrons release energy as they flow to and fro through the lamp filament, but as they stop to change direction there is no current and so the lamp flashes on and off.

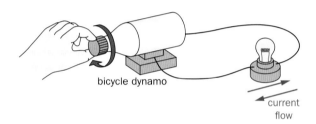

The lamp has an alternating p.d. across it. How does the current I through the bulb change as the p.d. V varies?
We can investigate this using the apparatus in the diagram:
The a.c. supply is a signal generator set at a low frequency.

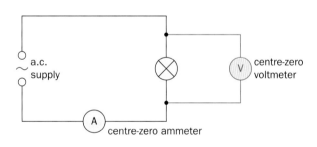

The graph shows the values of V and I plotted against time:
Can you see that the graphs for both V and I are **sine** curves?
They both vary *sinusoidally* with time.
Can you see that the p.d. and the current rise and fall *together*?
We say that V and I are **in phase** (see also page 102).

The **time period T** of an alternating p.d. or current is the time for one complete cycle. This is shown on the graph:

The **frequency f** of an alternating p.d. or current is the number of cycles in one second.

The **peak values V_0** and **I_0** of the alternating p.d. and current are also shown on the graph:

Why don't we see our lights flickering as they flash on and off?
The frequency of the mains supply is 50 Hz and so each cycle lasts for only $\frac{1}{50}$ second, or 20 milliseconds.
At this frequency the filament has no time to cool down as the current through it varies, and so we see constant illumination.

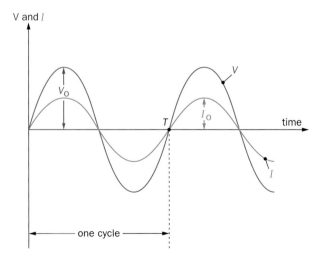

▶ Root-mean-square (r.m.s.) values

How do we measure the size of an alternating p.d. (or current) when its value changes from one instant to the next?

We could use the peak value, but this occurs only for a moment.
What about the average value?
This is zero over a complete cycle and so is not very helpful!

In fact, we use the **root-mean-square (r.m.s.)** value.
This is also called the **effective value**.
The r.m.s. value is chosen because it is the value which is *equivalent to a steady direct current*.

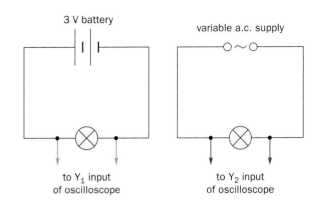

You can investigate this using the apparatus in the diagram:
Place two identical lamps side by side.
Connect one lamp to a battery, the other to an a.c. supply.
The p.d. across each lamp must be displayed on the screen of a double-beam oscilloscope.

Adjust the a.c. supply, so that both lamps are equally bright.
The graph shows a typical trace from the oscilloscope:
We can use it to compare the voltage across each lamp.

Since both lamps are equally bright, the d.c. and a.c. supplies are transferring energy to the bulbs at the same rate.
Therefore, the d.c. voltage is equivalent to the a.c. voltage.
The d.c. voltage equals the r.m.s. value of the a.c. voltage.
Notice that the r.m.s. value is about 70% of the peak value.

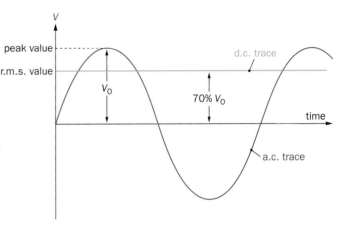

In fact:

$$\text{r.m.s. value} = \frac{\text{peak value}}{\sqrt{2}}$$
or
$$\text{peak value} = \sqrt{2} \times \text{r.m.s. value}$$

These equations apply to both the current and the p.d. and so:

$$I_{rms} = \frac{I_o}{\sqrt{2}}$$
or
$$V_{rms} = \frac{V_o}{\sqrt{2}}$$

Example 1
The r.m.s. value of the mains in Britain is 230 V.
What is its peak value?

$$V_o = \sqrt{2} \times 230\text{ V} = 325\text{ V} \quad (3\text{ s.f.})$$

The mains supply in Britain is 230 V, 50 Hz, but the electricity companies are allowed to vary the supply between +10% and −6% of 230 V.
This means that the r.m.s. value of the voltage ranges from 216 V to 253 V.
Have you ever noticed this fluctuation?

Why $\sqrt{2}$?
The power dissipated in a lamp varies as the p.d. across it, and the current passing through it, alternate.
Remember (page 236): **power, P = current, I × p.d., V**

If we multiply the values of I and V at any instant, we get the power at that moment in time, as the graph shows:

The power varies between $I_o V_o$ and zero.

$$\therefore \text{average power} = \frac{I_o \times V_o}{2} = \frac{I_o}{\sqrt{2}} \times \frac{V_o}{\sqrt{2}} = I_{rms} \times V_{rms}$$

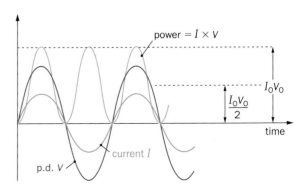

281

▶ The cathode-ray oscilloscope (CRO)

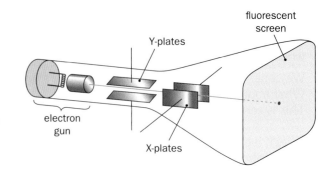

The oscilloscope is an important measuring instrument.
It contains a cathode-ray tube, as shown in the diagram:
An electron gun produces a fine beam of fast-moving electrons.
(There are more details on pages 320 and 321.)

The electrons move along the length of the tube and hit the
fluorescent screen. Can you see that a spot of light is produced?

Two sets of parallel plates are used to deflect the electron beam
and vary the position of the spot on the screen.
The X-plates move the spot horizontally, the Y-plates vertically.

The diagrams show how applying a p.d. across each set of plates
can deflect the spot to any position on the screen:

How could the spot be deflected to the upper left-hand corner?

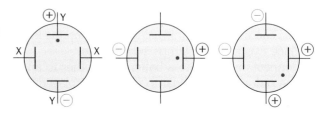

Built into the oscilloscope is a circuit called the **time-base**.
The time-base applies a p.d. across the **X**-plates, so that the spot
moves at a constant speed across the screen, from left to right,
and then jumps back very quickly to start again.

You can vary the speed of the spot. At high speeds it appears as
a line, due to your 'persistence of vision'.

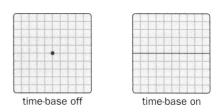

time-base off   time-base on

The controls

The picture shows a simple oscilloscope:
It has several controls:

1. The **brilliance control** changes the brightness of the spot
by varying the number of electrons that strike the screen
each second.

2. The **focus control** is adjusted to give a well-defined spot.

3. The voltage signal from an external circuit is connected to
the **input terminals** as shown.
This signal is amplified and then applied to the Y-plates.
In which direction does the input signal move the spot?

4. The **Y-gain control** varies the amplification of the input p.d.
What is the setting of the Y-gain control in the picture?
5 V/cm means that the spot moves 1 cm up (or down)
for every 5 V applied across the input terminals.

5. The **time-base control** is used to vary the speed of the spot.
In the picture the time-base is set to 10 ms/cm.
This means that it takes 10 milliseconds for the spot to move
1 cm across the screen.

6. You can use the **X-shift** and **Y-shift** controls to move the spot
in the X or Y directions, if necessary.

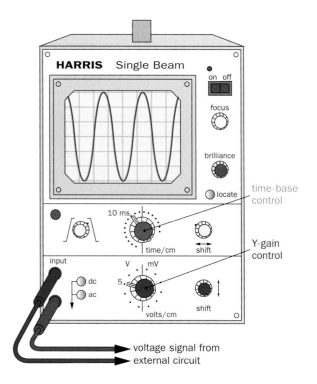

▶ Using the oscilloscope

The oscilloscope has 3 main uses:

1. Measuring d.c. and a.c. voltages

First move the spot or line to the centre of the screen.
Then connect the p.d. to be measured to the input terminals.

A **d.c. voltage** makes the spot move up (or down) as shown.
The deflection **d** in cm of the spot must be measured.
How can you convert this into volts?
Check the Y-gain setting in volts per cm. You then multiply **d** by
the Y-gain to find the d.c. voltage, eg. 2 cm × 5 V/cm = 10 V.

no input
time-base off

d.c. input
time-base off

d.c. input
time-base on

An **a.c. input** makes the spot move up and down repeatedly.
If the time-base is on, it moves the spot horizontally at the same
time. You can then see the shape of the voltage waveform.

Notice that when you multiply the length **y** by the Y-gain,
you are measuring the **peak-to-peak voltage** of the a.c. input.
Remember, this is *twice* the value of the peak voltage V_o.

no input
time-base off

a.c. input
time-base off

a.c. input
time-base on

2. Displaying waveforms

An oscilloscope can be used to display any varying p.d., as long
as the time-base is switched on.

A microphone converts sound signals into electrical signals.
If you connect a microphone to the oscilloscope input terminals
you can see the waveforms that you make as you sing!

In hospitals, sensors are used to convert a heartbeat into an
electrical signal, which is then displayed on an oscilloscope.

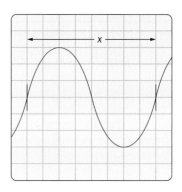

3. Measuring time and frequency

How can you measure the time period T for an alternating p.d.?
Measure the horizontal distance for one cycle of the a.c.
This is **x** cm in the diagram.
Now check the time-base setting.
This tells you that each centimetre represents a certain time.
If you multiply **x** by the time-base setting, you get the time T.

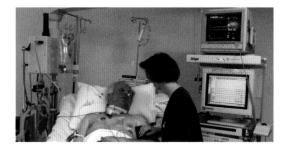

Remember (from page 78): **frequency, $f = \dfrac{1}{\text{time period, } T}$**

You can now use the time period to find the frequency of the a.c.

Example 2
The CRO trace shows the output from a laboratory power supply:
The time-base is set at 5 ms/cm, and the Y-gain is 2 V/cm.
Calculate: a) the r.m.s. value of the a.c. voltage,
  b) the time period T and the frequency f of the a.c.

a) Peak-to-peak voltage = 9.0 cm × 2 V/cm = 18.0 V

$$\therefore \ V_o = \frac{18.0 \text{ V}}{2} = 9.0 \text{ V} \quad \text{and so:} \quad V_{rms} = \frac{9.0 \text{ V}}{\sqrt{2}} = \underline{6.4 \text{ V}}$$

b) T = 4.0 cm × 5 ms/cm = 20 ms or <u>0.02 s</u>

$$\therefore \ f = \frac{1}{T} = \frac{1}{0.02} = \underline{50 \text{ Hz}}$$

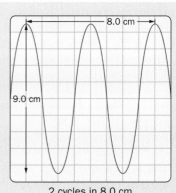

2 cycles in 8.0 cm

▶ The transformer

A transformer changes the value of an **alternating** voltage.
It consists of two coils, wound around a soft-iron core, as shown:
In this transformer, when an input p.d. of 2 V is applied to the
primary coil, the output p.d. of the secondary coil is 8 V.

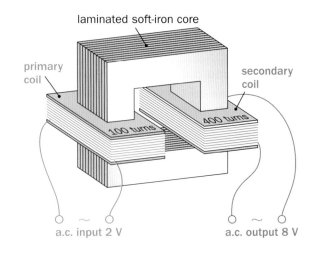

How does the transformer work?
An alternating current flows in the **primary** coil. This produces
an alternating magnetic field in the soft-iron core. This means
that the flux linkage of the **secondary** coil is constantly changing
and so an alternating voltage is induced across it.

A transformer cannot work on d.c. Can you explain why?

Look at the diagram to find the connection between the primary
and secondary voltages and the number of turns on each coil.
For an ideal transformer we can write:

Secondary voltage	=	number of turns on the secondary coil		or	$\dfrac{V_S}{V_P} = \dfrac{N_S}{N_P}$
Primary voltage		number of turns on the primary coil			

A **step-up** transformer increases the a.c. voltage, because
the secondary coil has more turns than the primary coil.

In a **step-down** transformer the voltage is reduced, because
the secondary coil has fewer turns than the primary coil.

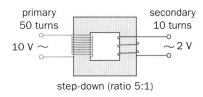

primary 50 turns · secondary 200 turns · 10 V ~ · ~ 40 V
step-up (ratio 1:4)

In an ideal transformer no energy is lost and so we can write:

Power supplied	=	power delivered		or	$V_P\,I_P = V_S\,I_S$
to primary coil		to secondary coil			(from page 236)

primary 50 turns · secondary 10 turns · 10 V ~ · ~ 2 V
step-down (ratio 5:1)

We can combine the two equations on this page to give us:

$$\frac{V_S}{V_P} = \frac{I_P}{I_S} = \frac{N_S}{N_P}$$

Can you see that this means that if the voltage is stepped **up**,
the current in the secondary is stepped **down** by the **same** ratio?

Note:
- In the transformer equations, the voltages and
  currents that you use must all be peak values
  or all r.m.s. values. Do not mix the two.

- Strictly, the equations apply only to an ideal
  transformer, which is 100% efficient.

Example 3
A step-up transformer has a primary coil with 100 turns.
It transforms the mains voltage of 230 V a.c. to 11 500 V a.c.
a) How many turns must there be on the secondary coil?
b) When the current in the secondary coil is 0.10 A,
    what is the current in the primary coil?

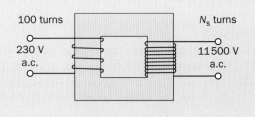

100 turns · N_S turns · 230 V a.c. · 11 500 V a.c.

a) $\dfrac{V_S}{V_P} = \dfrac{N_S}{N_P}$

∴ $\dfrac{11\,500\ \text{V}}{230\ \text{V}} = \dfrac{N_S}{100}$ ∴ $N_S = \dfrac{11\,500\ \text{V}}{230\ \text{V}} \times 100 = 50 \times 100 = \underline{5000\ \text{turns}}$

b) The secondary voltage is 50 times the primary voltage and so
    the secondary current must be $\frac{1}{50}$ of the primary current.
    Therefore, the primary current is: $50 \times 0.10\ \text{A} = \underline{5.0\ \text{A}}$

Energy losses in a transformer

In practice the efficiency of a transformer is never 100%, although it may be as high as 99%.

One possible cause of energy loss is reduced by the design of the transformer.
As the magnetic field alternates, **eddy currents** (see page 276) are induced in the soft-iron core. These currents could cause the core to become very hot.
To reduce this energy loss, the core is made of thin sheets of iron, called laminations, separated by insulating material.
This makes it much harder for the eddy currents to flow.

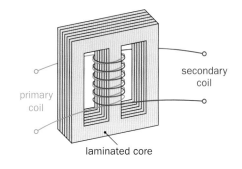

primary coil

secondary coil

laminated core

circuit symbol:

Transmission of electrical energy

The National Grid transmits large amounts of electrical energy each second, from the power stations to the consumers, often over large distances.
Since power = current × voltage, we could use:
either a) a low voltage and a high current,
   or b) a high voltage and a low current.

Why does the National Grid always use method (b)?
Remember that a current always produces heat in a resistor.
If the cables have resistance R and carry a current I,
the energy converted to heat each second is I^2R (see page 236).

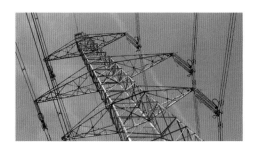

This means that in method (a) the high current produces a lot of heat in the cables, and little of the energy from the power station gets to the consumer.

Method (b) is used because the low current minimises the power loss.
Transformers at each end of the system step the voltage up and then down:

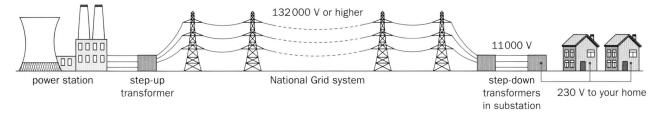

132 000 V or higher

11 000 V

power station   step-up transformer   National Grid system   step-down transformers in substation   230 V to your home

---

Example 4
A small power station generates 2 MW of electricity.
How much power is lost in cables of resistance 2 Ω
if the electricity is transmitted at (a) 40 kV, (b) 4 kV?

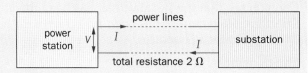

power station   power lines   substation
V I I
total resistance 2 Ω

First, we need to calculate the current in the cables using $P = I\,V$:

In (a) $\quad I = \dfrac{P}{V} = \dfrac{2 \times 10^6 \text{ W}}{40 \times 10^3 \text{ V}} = 50 \text{ A}$ In (b) $\quad I = \dfrac{P}{V} = \dfrac{2 \times 10^6 \text{ W}}{4 \times 10^3 \text{ V}} = 500 \text{ A}$

Now we know the current, we can calculate the I^2R heating losses in the cables:
In (a) $P = I^2R = (50 \text{ A})^2 \times 2 \text{ Ω}$ In (b) $P = I^2R = (500 \text{ A})^2 \times 2 \text{ Ω}$
$\therefore P = 5000 \text{ W} = \underline{5 \text{ kW}}$ $\therefore P = 500\,000 \text{ W} = \underline{500 \text{ kW}}$

When the current is 10 times bigger the power loss is 100 times bigger.

Hint: Notice that V is not the p.d. across the cables – it is the p.d. across the terminals at the power station.
As shown in the circuit diagram, when $V = 40$ kV, the p.d. across the 2 Ω resistance of the power cables is, in total, 100 V.

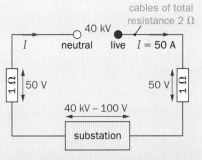

cables of total resistance 2 Ω

40 kV
I neutral   live $I = 50$ A
1 Ω   50 V   50 V   1 Ω
40 kV – 100 V
substation

equivalent circuit for transmission at 40 kV

▶ Rectification

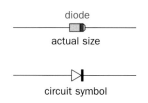

diode

actual size

circuit symbol

Although we transmit our electrical energy as an alternating
current, many electronic devices need a d.c. power supply.
How is a.c. converted to d.c.?
One way is to use the semiconducting **diode** (see page 234).
Remember, the diode allows a current to flow *only one way*.

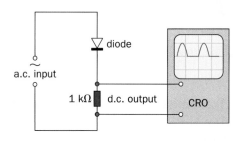

diode

a.c. input

1 kΩ   d.c. output

CRO

The diagram shows a diode connected in series with a 1000 Ω
resistor and a low-voltage a.c. supply:
You can use an oscilloscope (CRO) to look at the waveform.
First connect the CRO across the a.c. supply (the input) and
look at the a.c. waveform. Then connect the CRO across
the 1000 Ω load resistor and look at the output waveform.

Why is only half the a.c. waveform still present?
The diode conducts for only half of the a.c. cycle – only when
it is forward-biased. The alternating current has been **rectified**
to give an uneven direct current.

Full-wave rectification

The half-wave rectification shown above is wasteful.
Is it possible to convert the whole of the input cycle
of a.c. to d.c.?

This can be done using the **bridge rectifier** shown:
As you can see, this is made up of 4 diodes.

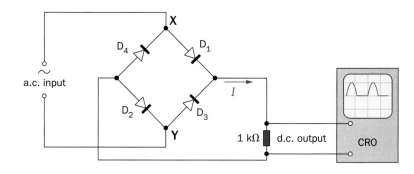

X

D_4 D_1

a.c. input

D_2 D_3

Y

I

1 kΩ   d.c. output

CRO

Can you see how this works?
When **X** is positive, conventional current flows from
X through diode D_1, *downwards* through the load
resistor, through D_2 to Y and back to the input.

When **Y** is positive, conventional current flows from
Y through diode D_3, *downwards* through the load
resistor, through D_4 to X and back to the input.
Current always flows in the same direction through
the resistor.

You can see the waveform on the oscilloscope.
Both halves of the waveform are present.
Can you see that the negative half has been made positive?
The alternating current has been *fully rectified*, but the
direct current is still uneven.

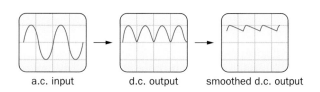

a.c. input   d.c. output   smoothed d.c. output

The output can be *smoothed*.
This is done by placing a capacitor (see Chapter 22) in parallel
with the load resistor (across the output terminals).
The capacitor stores some energy when the voltage is high,
and then releases it when the voltage falls.
The diagram shows the smoothed d.c output:

d.c to a.c.

The diagram shows a panel of photovoltaic cells (PV cells):
You may have noticed these on house roofs.
PV cells convert sunlight directly into electricity, but the cells
produce d.c. In order to feed the electricity into the household
supply, the d.c must be converted into a.c.
This is done using an **inverter**. The inverter essentially switches
the d.c. supply repeatedly on and off at a very fast rate.

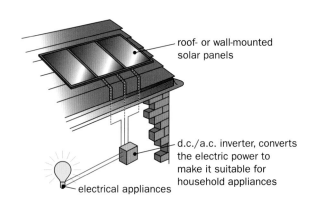

roof- or wall-mounted
solar panels

d.c./a.c. inverter, converts
the electric power to
make it suitable for
household appliances

electrical appliances

Summary

An a.c. supply causes charge to flow alternately one way and then the other way around the circuit.
A diode allows current to flow one way only, and can be used to rectify a.c. to d.c.

The r.m.s. value of an alternating current or p.d. $= \dfrac{\text{peak value}}{\sqrt{2}}$ For a transformer: $\dfrac{V_S}{V_P} = \dfrac{N_S}{N_P}$

When the voltage is stepped down the current is stepped up, in almost the same ratio.
In the National Grid, electricity is transmitted at high voltage and therefore low current, so less energy is lost.

▶ Questions

1. Can you complete these sentences?
   a) The peak value of a.c. is ...... times its ...... value.
   b) In a transformer, when the voltage is stepped up the current is .... in the same .... (providing the transformer is .... % efficient)
   c) In a cathode-ray oscilloscope the vertical deflection of the spot is proportional to the .... applied across the Y-plates. The horizontal movement of the spot is controlled by the ....

2. Explain why:
   a) the core of a transformer is laminated,
   b) a transformer will not work with direct current,
   c) in the National Grid system,
      i) electricity is transmitted at high voltage,
      ii) a.c. is used rather than d.c.

3. The r.m.s. value of an alternating voltage is 12 V. What is the value of the peak voltage?

4. What is (a) the time period, (b) the frequency, (c) the peak value and (d) the r.m.s. value of the alternating p.d. shown in the graph below?

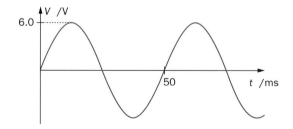

5. 3.0 MW of power is transmitted at a p.d. of 100 kV along cables of total resistance 5.0 Ω.
   a) What current flows in the cables?
   b) What is the power wasted in the cables?

6. Complete the following table of transformers:

7. A CRO screen with 1 cm squares displays this trace:

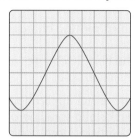

   The Y-gain is set at 0.5 V/cm and the time base is set at 5 ms/cm. Calculate:
   a) the peak to peak voltage, (b) the r.m.s. voltage,
   c) the time period and (d) the frequency of the signal.

8. The CRO trace below shows a patient's heartbeat obtained with the time-base set at 200 ms/cm. Calculate the number of heartbeats per minute.

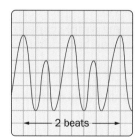

2 beats

9. A transformer has a primary coil of 5000 turns. When the output p.d. is 12 V r.m.s., the input p.d. is 240 V r.m.s.
   a) How many turns are there on the secondary coil?
   b) The output from the secondary coil is connected across a 20 Ω lamp.
      i) What is the r.m.s. current in the secondary coil?
      ii) What is the r.m.s. current in the primary coil?
      (Assume the efficiency of the transformer is 100%.)

Further questions on page 319.

Primary p.d.	Secondary p.d.	Primary turns	Secondary turns	Step-up or step-down
230 V r.m.s.		500	50	
230 V r.m.s.	1150 V r.m.s.		2000	
11 000 V r.m.s.	132 000 V r.m.s.	1000		

21 Electric Fields

Static electricity can often be a nuisance in everyday life. Have you ever had an electric shock when getting out of a car?

Lightning is a spectacular display of static electricity. In 1752, in a famous but highly dangerous experiment, Benjamin Franklin flew a kite during a thunderstorm and proved that lightning is a form of electricity. Some people were killed as they tried to repeat Franklin's work!

You have already studied moving charges or electric current in Chapter 16.
In *static* electricity, the electric charges are either at rest or only flow for a very short time.

Electrostatics has many useful applications. These include the electrostatic spraying of paints and powders and removing smoke particles from power station chimneys.

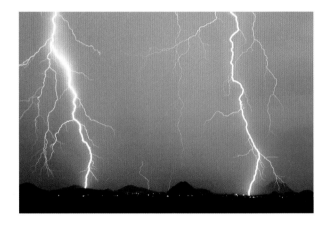

In this chapter you will learn:
- how to calculate the electric force between two charges,
- about electric fields,
- about electric field strength and electric potential.

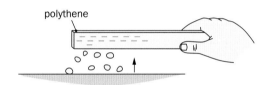

The diagrams show some simple experiments in electrostatics that you may have done before.

Do you remember the following facts?

- Objects are usually uncharged and so are electrically neutral.

- Objects can become charged by friction, when one material is rubbed against another material. This is discussed on the opposite page.

- A charged object can attract an uncharged object, for example, small pieces of paper.

- There are two types of electric charge. We call them positive and negative.

- A rubbed polythene strip gains a negative charge. A cellulose acetate strip gains a positive charge.

- When two charged objects are brought together, we find that:
    Like electric charges repel.
    Opposite electric charges attract.

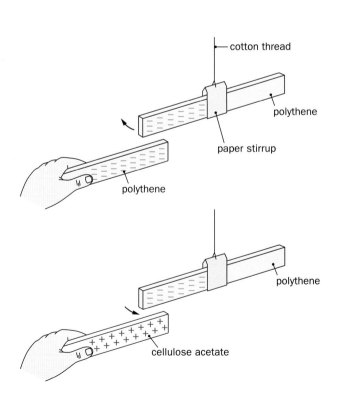

How do objects become electrically charged?

Normally, objects are uncharged because each atom has the same number of positive protons as negative electrons. Charging occurs when *electrons are transferred* from one material to another.

Friction can transfer electrons, as shown in the diagram:

The polythene has gained extra electrons and so it is negatively charged. The duster is left short of electrons. What is its charge?

Polythene (like all insulating materials) has few free electrons (see Chapter 16) and so the negative charge does not flow away from the rubbed region – it is a *static charge*.

Can you put a metal rod in your hand and charge it by rubbing? Any charge transferred to the rod will flow along the rod and through your body to Earth, because your body and the Earth are poor insulators of electricity.
A conductor can only be charged if it is insulated from Earth.

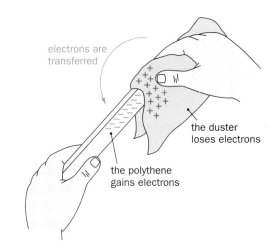

Charging by electrostatic induction

A charged conductor will share its charge with an uncharged conductor that is placed in contact with it. Electrons can flow between the two conductors because they are touching.

A charged strip can also be used to *induce* a charge in an uncharged conductor *without* touching it.
The diagrams show how two conducting spheres can be oppositely charged by **induction**:

1. When a negative strip is brought close to 2 metal spheres, electrons from sphere A are repelled to sphere B.

2. The spheres are separated with the strip still held nearby.

3. Then the strip is removed. The free electrons in each sphere now spread out, so that the charge is distributed evenly over the surface of each sphere.

Why does a balloon, charged by rubbing, stick to a wall? The negative charge on the balloon repels some of the electrons in the wall away from the surface. This leaves the wall surface positively charged, and so the balloon is attracted to the wall.

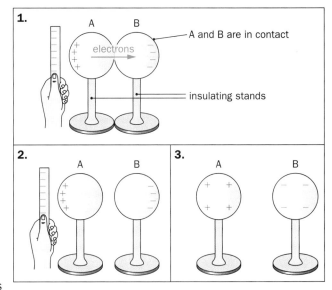

Measuring charge

One way of measuring charge is to use a **coulombmeter**: Some of the charge on the polythene strip is transferred to the coulombmeter when the strip is scraped across its metal cap. The coulombmeter then measures the charge placed on its cap.

Why is this coulombmeter giving a negative reading? This confirms that the polythene strip is negatively charged.

The size of this charge is 32 nC or 32×10^{-9} coulombs. Remember, the coulomb is a very large unit (see page 228). In fact, the coulombmeter has been given 200 billion electrons from the polythene strip!

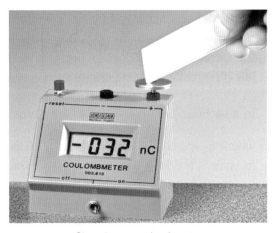

Charging a coulombmeter.

▷ Forces between charges

What affects the force that charged objects exert on each other?
You can investigate this using the apparatus in the photograph:

Place a charged polythene strip on top of an insulating material
on an electronic balance and set the balance to read zero.
What happens as you bring up a second charged polythene strip?

The strip in your hand repels the strip on the balance and
this downwards force causes the balance reading to increase.

Does the force increase as the strips are brought closer together?
Does the force increase if you increase the charge on each strip?

Coulomb's Law

In 1785, the French physicist Charles Coulomb carried out
a series of experiments and discovered that the force F
between two charges was:

- directly proportional to each of the charges Q_1 and Q_2,
- inversely proportional to the **square** of their separation r.

Coulomb's Law is written:

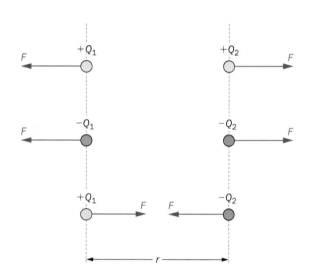

$$F \propto \frac{Q_1 Q_2}{r^2} \quad \text{or} \quad \boxed{F = \frac{k\, Q_1\, Q_2}{r^2}}$$

where: F is the force, in newtons
Q_1 and Q_2 are the charges, in coulombs
r is the distance apart, in metres
k is the constant of proportionality

Strictly, Coulomb's equation applies only to point charges, but
a sphere of charge behaves as if all its charge is concentrated at
its centre (providing it is not placed too close to another charge).

The value of the constant k

When the two charges are separated by a vacuum,
the value of the constant k is $9.0 \times 10^9 \text{ N m}^2 \text{ C}^{-2}$.
When they are separated by air, it is effectively the same.

In fact, this is the largest value that k can have because the size of
the force is always reduced when an insulating material, other than
air, separates the two charges.

In a vacuum (or air) the value of k can be written as: $k = \dfrac{1}{4\pi\varepsilon_0}$

where ε_0 is a constant called the **permittivity of free space**.
Its value is $8.85 \times 10^{-12} \text{ F m}^{-1}$ (The farad, F, is explained in Chapter 22.)

So $k = \dfrac{1}{4\pi\varepsilon_0} = \dfrac{1}{4\pi \times 8.85 \times 10^{-12} \text{ F m}^{-1}} = 9.0 \times 10^9 \text{ N m}^2 \text{ C}^{-2}$

Every insulating material has a permittivity ε which is greater than ε_0.

▷ Another inverse-square law

Can you see that the two equations below are very similar?

Coulomb's law:

$$F = \frac{k\, Q_1\, Q_2}{r^2}$$

Newton's law of gravitation (page 86):

$$F = \frac{G\, m_1\, m_2}{r^2}$$

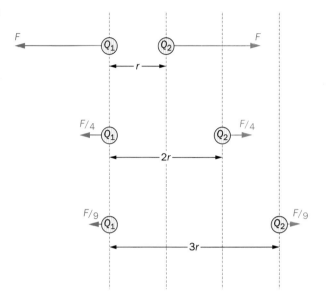

Coulomb's law is similar to Newton's law, except that for gravity the force can only attract, whereas the force between charges can attract **or** repel. The diagram shows 2 charges repelling:

Like Newton's law, Coulomb's law is an inverse-square law. This means that if we double the distance between the charges, the force decreases to $(\frac{1}{2})^2$ or $\frac{1}{4}$ of its previous value.

Look at the diagram:
Why does the force drop to $F/9$ when the separation is tripled?
The force decreases to $(1/3)^2$ or $1/9$ of its previous value.

We can make other comparisons between these two laws.
Can you find the similarities and the differences in the table?

Newton's law of gravitation	Coulomb's law of charge
The force is 'felt' by objects with mass.	The force is 'felt' by objects with charge.
The force is proportional to the size of the masses.	The force is proportional to the size of the charges.
The force is inversely proportional to the square of the separation of the masses.	The force is inversely proportional to the square of the separation of the charges.
There is a gravitational field round a mass.	There is an electric field round a charge (see page 292).
There is only one type of mass.	There are two types of charge – positive and negative.
The force is always attractive.	The force can be attractive or repulsive.
The constant of proportionality is G. G always has the value 6.67×10^{-11} N m^{-2} kg^{-2}.	The constant is k, and its value depends on the material. In air, $k = 1/4\pi\varepsilon_0 = 9.0 \times 10^9$ N m^{-2} C^{-2}.
The gravitational force is very weak, unless one of the masses is large.	The electric force is very strong, but not normally noticed as the charges cancel out within a neutral atom.

Example 1
Two small conducting spheres are placed 0.10 m apart in air, and are given positive charges of 12 nC and 15 nC, as shown:
Calculate the size of the repulsive force between them.
(The permittivity ε_0 of air is 8.85×10^{-12} F m^{-1}.)

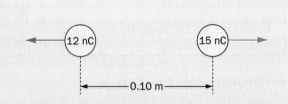

$$F = \frac{k\, Q_1\, Q_2}{r^2} \qquad \text{where } k = \frac{1}{4\pi\varepsilon_0}$$

$$\therefore F = \frac{Q_1 Q_2}{4\pi\varepsilon_0 r^2} = \frac{12 \times 10^{-9}\,\text{C} \times 15 \times 10^{-9}\,\text{C}}{4\pi \times 8.85 \times 10^{-12}\,\text{F m}^{-1} \times (0.10\,\text{m})^2} = \underline{1.6 \times 10^{-4}\,\text{N}} \quad (2\ \text{s.f.})$$

▷ Electric fields

The charged polythene strip is attracting Phiz's hair towards it.
We say that there is an **electric field** around the strip.
An electric field is a region where a charge 'feels' a force, just
as a gravitational field is a region where a mass 'feels' a force.

Is it possible to *see* the shape of an electric field?
One way is to use semolina grains sprinkled on castor oil.
Two electrodes in the oil produce the electric field, as shown.
The semolina grains line up along the electric field lines :

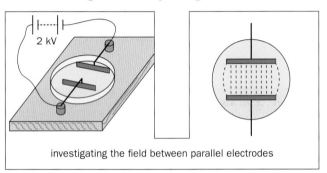

investigating the field between parallel electrodes

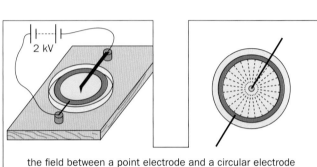

the field between a point electrode and a circular electrode

The arrows on the field lines tell us the **direction** of the field.
This is defined as the direction of the force on a **positive** charge
and so the arrows on the lines go from positive to negative.

The spacing of the lines tells us about the **strength** of the field.
The more closely packed the field lines, the stronger the field.

Look at the electric field caused by the parallel electrodes.
The field lines are parallel to each other and equally spaced.
This field is **uniform**. It has the same strength at all points.

The field of the point electrode is called a **radial** field.
The diagrams show the radial field of an isolated point charge.
How can you tell that the field is strongest close to the charge?

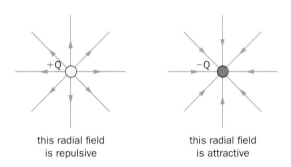

this radial field
is repulsive

this radial field
is attractive

Electric field strength, *E*

On page 88 we defined gravitational field strength *g* as the
force acting per kilogram of mass.
Electric field strength *E* is defined in a similar way:

***The electric field strength at a point is the force per coulomb
exerted on a positive charge placed at that point in the field.***

The diagram shows a charge $+Q$ placed in a field of strength E :
If E is the force per coulomb (in newtons per coulomb), then:

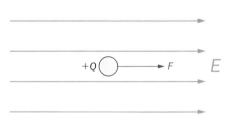

*The charge $+Q$ 'feels' a force F
in the same direction as E.*

$$\textbf{Electric field strength, } E = \frac{\text{force, } F \text{ (N)}}{\text{charge, } Q \text{ (C)}} \quad \text{or} \quad E = \frac{F}{Q}$$
(N C^{-1})

It follows that : $\boxed{F = Q\,E}$

- Electric field strength E is a vector quantity. The direction
  of the field is the direction of the force on a positive charge.

- The unit here for E is N C^{-1}, but an alternative unit is
  volts per metre (V m^{-1}). You will see why on the next page.

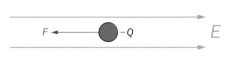

*The force on the negative charge $-Q$
is in the opposite direction to E.*

The uniform electric field

In the diagram, there is a p.d. of about 1000 V across the parallel metal plates, so that they become charged as shown:
A *uniform* electric field is produced between the plates.

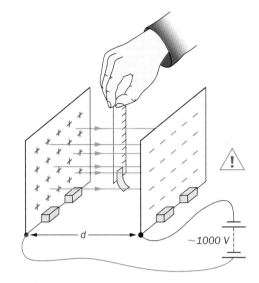

You can test the electric field using a home-made **electroscope**.
This consists of a strip of metal foil attached to a plastic ruler.
You charge the metal foil and hold the electroscope in the field.
The stronger the field, the greater the force on the charged foil, and the more it deflects from the vertical position.

How can you make the field stronger? You can:
- increase the p.d. across the plates, or
- move the plates closer together.

In fact, the field between the plates can be calculated by:

Electric field strength, E = $\dfrac{\text{p.d. across the plates, } V \text{ (V)}}{\text{separation of plates } d \text{ (m)}}$

(V m^{-1})

or $E = \dfrac{V}{d}$

Note that the unit here for E is V m^{-1}. (1 V m^{-1} = 1 N C^{-1})

The radial electric field

What is the electric field strength at any point in a *radial* field?
In the diagram, a point charge $+Q$ produces the radial field, and a *small* positive test charge $+q$ is placed in the field as shown:

Using Coulomb's Law the force on the test charge $+q$ is given by: $F = \dfrac{kQq}{r^2}$ where $k = \dfrac{1}{4\pi\varepsilon_0}$

The electric field strength is the force per unit charge. Therefore: $E = \dfrac{F}{q} = \dfrac{kQ\cancel{q}}{\cancel{q}r^2}$

So the field strength E a distance r away from a point charge $+Q$ is:

$$E = \dfrac{kQ}{r^2}$$

The field strength E a distance r away from a negative point charge $-Q$ is: $E = -\dfrac{kQ}{r^2}$
The $-$ sign tells us the field is inwards.

Example 2
Two plates are placed 10 cm apart and connected to a 5.0 kV supply.
Calculate: a) the strength of the uniform field between the plates,
   b) the force on a $+4.0$ nC charge placed in the field.

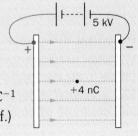

a) $E = \dfrac{V}{d}$

  $E = \dfrac{5.0 \times 10^3 \text{ V}}{0.10 \text{ m}} = \underline{5.0 \times 10^4 \text{ V m}^{-1}}$

b) $F = QE$
  $F = 4.0 \times 10^{-9} \text{ C} \times 5.0 \times 10^4 \text{ N C}^{-1}$
  $F = \underline{2.0 \times 10^{-4} \text{ N, to the right}}$ (2 s.f.)

Example 3
The charge on the dome of a Van de Graaff generator is 15 μC.
The dome radius is 20 cm. What is the field strength at its surface?
Remember, a sphere acts as if all its charge is at its centre.

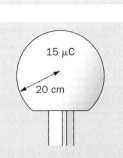

$E = \dfrac{kQ}{r^2}$ where $k = \dfrac{1}{4\pi\varepsilon_0}$ and ε_0 is 8.85×10^{-12} F m^{-1}

$E = \dfrac{Q}{4\pi\varepsilon_0 r^2} = \dfrac{15 \times 10^{-6} \text{ C}}{4\pi \times 8.85 \times 10^{-12} \text{ F m}^{-1} \times (0.20 \text{ m})^2} = \underline{3.4 \times 10^6 \text{ V m}^{-1}}$ (2 s.f.)

▷ Electric potential

If Phiz wants to move his charge closer to the charged sphere he has to push against the repulsive force:
Phiz **does work** and his charge **gains** electric potential energy.

If Phiz lets go of the charge it will move away from the sphere, losing electric potential energy, but gaining kinetic energy.

When you move a charge in an electric field, its potential energy changes. This is like moving a mass in a gravitational field.

Gravitational potential was discussed in Chapter 8.
Electric potential is a similar quantity:

The **electric potential V** at any point in an electric field is the *potential energy that each coulomb of positive charge would have if placed at that point in the field.*

The unit for electric potential is the joule per coulomb (J C^{-1}), or the **volt** (V). Like gravitational potential it is a **scalar** quantity.

Earthing

In electrostatics, all the points on a conducting surface are at the same potential. If not, electrons would flow until the potential of all of the conductor becomes the same.

Where is the zero of electric potential?
Strictly, electric potential is defined to be zero at infinity, just like it was for gravitational potential on page 92.
In practice, it is convenient to say that Earth is at zero potential.
Why must a conductor connected to Earth be at zero potential?

These spheres lose their charge when they are earthed:
Can you see which way the electrons flow in each case?

A conductor can retain a charge even when it is earthed!
Look at the parallel plates:
The lower plate is at zero potential (earthed), but even so it has a negative charge which is attracted by the nearby positive plate.

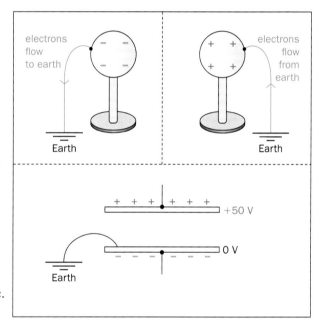

Potential difference

We often need to know the difference in potential between two points in an electric field:

The potential difference or p.d. is the energy transferred when one coulomb of charge passes from one point to the other point.

The diagram shows some values of the electric potential at points in the electric field of a positively charged sphere:
What is the p.d. between points A and B in the diagram?

When one coulomb moves from A to B it gains 15 J of energy.
If 2 C move from A to B then 30 J of energy is transferred. In fact:

Energy transferred, ΔW = charge, Q × p.d., ΔV (joules) (coulombs) (volts) or	$\Delta W = Q \times \Delta V$

Do you recognise this equation from Chapter 16?
P.d. has the same meaning in static and current electricity.

Equipotential surfaces

All the points that have the *same* potential in an electric field are said to lie on an **equipotential** surface.
The diagram shows how you can investigate equipotentials:

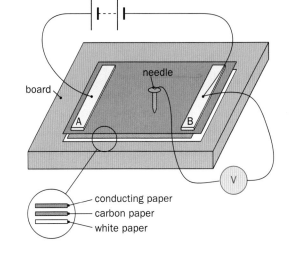

The 2 electrodes A and B are clamped to the conducting paper. You can trace the equipotential lines using the voltmeter. It measures the p.d. between the contact needle and electrode B.

You can place the needle on electrode A and adjust the d.c. supply until the voltmeter reads 4.0 V.
Then move the needle across the paper, so that you draw a line through all the points where the voltmeter reads 3.0 V. Then draw lines for points where the potential is 2.0 V and 1.0 V.

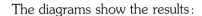

You can draw the equipotentials in a radial field by repeating this experiment using a point electrode and circular electrode.

The diagrams show the results:

Look at the equipotentials between the **parallel** plates:
They are evenly spaced because the electric field is *uniform*.

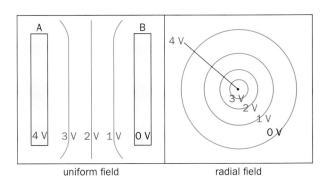

uniform field   radial field

In a **radial** field the equipotentials are concentric spheres:

Why are they closer together near the point charge?
Imagine moving a small positive test charge in this field.
As you get closer to the point charge the field gets stronger, the repulsive force increases and you have to do more and more work to move the test charge the same distance.

When a charge moves *along* an equipotential, *no work* is done because its potential energy does not change.
This means that the equipotentials must cross electric field lines at 90°, just like in gravitational fields (see page 92).

Lightning

An electric field of about 3×10^6 V m^{-1} will ionise dry air. This means that there must be a p.d. of around 3 million volts across each metre of air for it to conduct electricity.

Look at the equipotentials between the cloud and the church:
Can you see how the pointed spire distorts these lines?
The field is strong where the equipotentials are close together.

Why is the lightning most likely to strike the spire?
The field around the spire is strong enough to ionise the air, so that charge can flow between the cloud and the spire.

Example 4
The diagram shows the equipotentials between two plates.
How much energy is transferred when a charge of 2.0 nC
moves a) from point A to B,   b) from B to C?

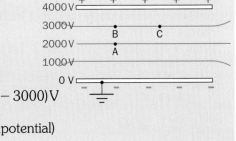

$\Delta W = Q \Delta V$ where V is the p.d. between the points

a) $\Delta W = 2.0 \times 10^{-9}$ C $\times (3000 - 2000)$ V   b) $\Delta W = 2.0 \times 10^{-9}$ C $\times (3000 - 3000)$ V
$\quad = 2.0 \times 10^{-9}$ C $\times 1000$ V $\quad = 2.0 \times 10^{-9}$ C $\times 0$ V
$\quad = 2.0 \times 10^{-6}$ J (2 s.f.) $\quad = 0$ J (on the same equipotential)

Electric potential in a radial field

The electric potential V at a point in the field of a point charge depends on the charge Q and the distance r from the point charge.

For a positive point charge: $\boxed{V = \dfrac{kQ}{r}}$ and for a negative point charge: $\boxed{V = -\dfrac{kQ}{r}}$ where $k = \dfrac{1}{4\pi\varepsilon_0}$

(This value of k applies when the charge is in a vacuum or in air; see page 290.)

Can you see that as r becomes larger, the potential gets smaller?
The absolute electric potential at a point in an electric field is defined as:

the work done per unit positive charge, when the charge is moved from infinity to that point.

So $V = 0$ at infinity (as we said on page 294).

Why is the potential positive for charge $+Q$ and negative for $-Q$?

Suppose you move a small positive test charge **towards** the charge $+Q$.
You have to **do work against** the repulsive force, and so the test charge gains energy.
But for the charge $-Q$ we do work against the attractive force as the positive test charge moves further **away**. V must have negative values so the energy of the test charge can increase and reach zero at infinity.

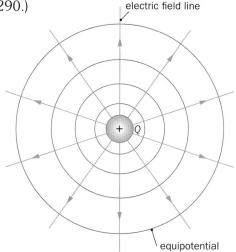
The electric field of a positive point charge.

Linking electric potential V and field strength E

When a charge $+Q$ is placed on an isolated conducting sphere, the charge spreads uniformly over the surface of the sphere.
Here the radius of the sphere is r_s.

The upper graph shows how the electric **potential** V varies with the distance r from the centre of the sphere:
All of the conducting sphere has the same potential V, where $V = \dfrac{kQ}{r_s}$ (the same value as if all the charge were at its centre).
Notice how, beyond r_s, the potential V decreases as $1/r$.

The lower graph shows how the electric **field strength** E varies with the distance r from the centre of the sphere:
At any point **outside** the sphere, E decreases as $1/r^2$ and has the same value as if all the charge is at the centre of the sphere.

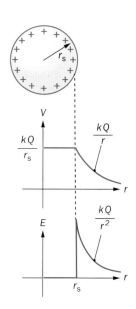

In fact, the negative gradient of the V graph at any point gives the value of E, the electric field strength, at that distance: $\boxed{E = -\dfrac{\Delta V}{\Delta r}}$

For $r \le r_s$, the gradient of the V graph is zero, and so inside the conducting sphere the electric field strength E is zero.

Example 5
An isolated conducting sphere of radius 12 cm is given a charge of 0.16 nC.
What is the electric potential of the sphere?

$$V = \frac{Q}{4\pi\varepsilon_0 r} = \frac{0.16 \times 10^{-9}\ \text{C}}{4\pi \times 8.85 \times 10^{-12}\ \text{F m}^{-1} \times 0.12\ \text{m}}$$

$$= \underline{12\ \text{V}} \quad (2\ \text{s.f.})$$

$E = -\dfrac{\Delta V}{\Delta r}$

$\therefore \Delta V = E\Delta r$

The area of the shaded strip is $E \times \Delta r$.
So adding up the total area under the curve would give us the total change in potential, ΔV.

▷ Physics at work: Using static electricity

Electrostatic crop spraying

Most crop pests live on the underside of leaves. Electrostatics can be used to target these 'hard to get at' places.
The boom of the spray is connected to a high-voltage supply.
This produces positively charged pesticide droplets, and also
induces an opposite charge in the plants and in the ground.

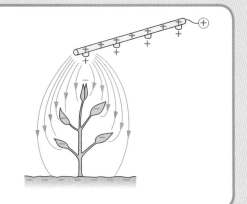

The droplets repel each other to form an even cloud of spray,
and are attracted to all parts of the oppositely charged plant.
Can you see how the spray follows the electric field lines and
reaches the underside of the leaves?

Bicycles and cars are painted using a similar technique.

The photocopier

The diagrams show four of the stages in Xerox photocopying.

In stage **1**, a high-voltage wire charges the insulating surface
of the drum. The drum is now positively charged.
An intense light will allow charge to escape from this drum.

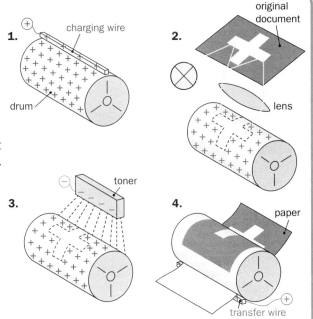

In stage **2**, intense light is flashed across the original document.
The reflected light is then projected on to the drum surface.
Which parts of the drum remain charged?
The black parts of the original document do not reflect the light
and so the corresponding regions of the drum remain charged.
The drum surface now holds an image of the original.

In stage **3**, negatively charged toner particles are attracted
to the parts of the drum which are still positively charged.

In stage **4**, a sheet of paper is passed over the drum surface.
There is a positively charged transfer wire below the drum.
This attracts the toner from the drum on to the paper.

Finally, the paper is passed through heated rollers, which melt
the toner and fuse it to the paper to make a permanent copy.

The ink-jet printer

One type of ink-jet printer is shown in the diagram:
The ink gun produces 10 000 tiny droplets of ink every second.
The droplets are positively charged by the charge electrode,
which surrounds the ink gun nozzle.

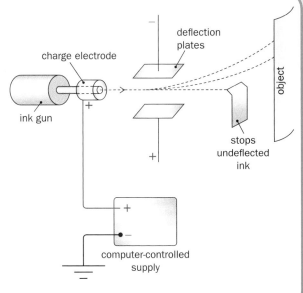

The ink droplets are deflected by a vertical electric field.
In this printer, the p.d across the deflector plates is constant.

The voltage of the charge electrode is computer-controlled,
so that each ink droplet is given a different size charge.
The greater the charge on the droplet, the more it is deflected
as it passes between the deflector plates.
So the computer directs each droplet to a precise position.

How are the spaces between the words produced?
The charge electrode does not charge all the droplets.
Any uncharged droplets are not deflected.
They are stopped, as shown, and the ink is returned to the gun.
'Sell-by' dates are printed on cans using this technique.

Summary

The force F between two charges is given by Coulomb's law: $\quad F = \dfrac{k\,Q_1\,Q_2}{r^2} \quad$ where $\quad k = \dfrac{1}{4\pi\,\varepsilon_0}$

ε_0 is the permittivity of a vacuum or air.

The electric field strength E is the force acting per unit positive charge: $\quad E = \dfrac{F}{Q}$

E is a vector quantity. It is measured in $N\,C^{-1}$ or $V\,m^{-1}$.

In a radial field: $\quad E = \dfrac{k\,Q}{r^2} \quad$ In the uniform field between two parallel plates: $E = \dfrac{V}{d}$

In a radial field: $V = \dfrac{k\,Q}{r}$ for a positive charge, and $V = -\dfrac{k\,Q}{r}$ for a negative charge.

The electric potential V gives the potential energy per coulomb of positive charge at a point in an electric field. It is defined as the work done per unit positive charge when the charge is moved from infinity to that point.

V is a scalar quantity. It is measured in volts. Energy transferred is $\Delta W = Q\,\Delta V$.

▷ Questions

You *will need the following data*:
Permittivity of free space ε_0 is 8.85×10^{-12} F m^{-1}.
Charge on the electron is 1.6×10^{-19} C.
Mass of an electron is 9.1×10^{-31} kg.

1. Can you complete these questions?
   a) The electric force between two charges depends on their ....and their ....
   b) Charge is measured in ....
      One .... is the charge on 6.25×10^{18} ....
   c) An electric field is a region in which a .... feels a .... Electric field lines can indicate the .... and the .... of a field.
   d) In a uniform field the electric field lines are .... and ....
   e) A surface joining all the points with the same potential is called an .... surface.
   f) To find the energy change when a charge moves from one point in a field to another, we multiply the .... difference by the .... on the object.

2. The diagram below shows an electrostatic paint spray.
   a) Explain how the paint spray works.
   b) Why does this technique waste little paint?

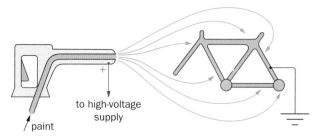

3. Describe two similarities and one difference between gravitational and electric fields.

4. Explain:
   a) how two insulated conductors can be oppositely charged by induction,
   b) why no work is done when a charge moves along an equipotential surface,
   c) why the charged conducting ball (in the diagram below) is attracted to the earthed metal plate.

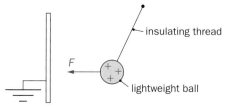

5. The two spheres A and B below are both given a charge $+Q$. The repulsive force between them is F.

   What is the size of the repulsive force when:
   a) the charge on A is doubled?
   b) the charge on both spheres is doubled?
   c) the charge on both the spheres is $+Q$, but the separation of the spheres is doubled to $2r$?

6. Calculate the electrostatic force between the proton and the electron in the hydrogen atom if the separation of the two particles is 1.0×10^{-10} m.

7. Two small, equally charged spheres are 10 cm apart. The repulsive force between them is 3.6×10^{-6} N.
   a) What is the charge on each sphere?
   b) The spheres are negatively charged. How many excess electrons are there on each sphere?

8. Calculate the electric field strength at distances of
   a) 2.0 cm and b) 4.0 cm from a point charge of 16 nC. Comment on your answers.

9. Two point charges of $+4.0$ nC and $+2.0$ nC are placed 6.0 cm apart, as shown below.

   a) What is the resultant field strength at point P, which is midway between the two charges?
   b) What would be the field strength at P if the $+2$ nC charge was replaced with a -2.0 nC charge? (*Hint*: Find the field due to each charge and then add them together. Remember E is a vector.)

10. The diagram below shows three parallel, conducting plates, each connected to a different voltage.

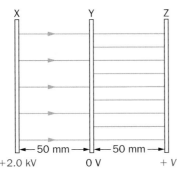

   a) What is the strength of the uniform electric field between plates X and Y?
   b) What is the direction of the electric field between plates Y and Z?
   c) What is the strength of the uniform electric field between plates Y and Z?
   d) What is the potential V of plate Z?

11. A spark can pass through dry air when the electric field strength is 3.0×10^6 V m^{-1} or greater.
   a) The gap between the electrodes in the spark plug of a car is 0.80 mm. What p.d. is needed across the electrodes to produce a spark?
   b) An overhead electricity cable reaches a maximum voltage of 1.86×10^5 V. How close could you get to the cable before receiving an electric shock?

12. Phiz is foolishly standing in an open field during a thunderstorm. Draw the equipotentials between the cloud and the ground and use them to explain why Phiz is in danger of being struck by lightning.

13. What is the force on a charge of 2.4 nC placed in an electric field of strength 4.0×10^3 N C^{-1}?

14. The diagram shows two metal plates, A and B. Plate B is earthed; plate A is at 1200 V.

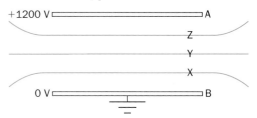

   a) What is the potential of each of the equipotential lines X, Y and Z?
   b) A charge of -3.2×10^{-16} C is moved from X to Y. What is its change in electrical potential energy?
   c) Has the system lost or gained electrical potential energy as the charge moves from X to Y?

15. A tiny negatively charged oil drop is held stationary in the electric field between two horizontal parallel plates, as shown below. Its mass is 4.0×10^{-15} kg.

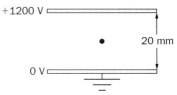

   a) What are the two forces acting on the drop and in which direction do they act?
   b) Use the fact that the 2 forces balance to calculate the charge on the oil drop. ($g = 10$ N kg^{-1}.)

16. The dome of a Van de Graaff generator has a radius of 15 cm.
   a) What is the charge on the dome when the electric field strength at its surface is 3.0×10^6 V m^{-1}? (Remember, the sphere acts as if the charge is concentrated at its centre.)
   b) What is the potential of the dome?

17. How much work is done when a charge of 3.6 pC is moved from 0.50 m to 0.75 m away from a point charge of -4.0 nC?

18. An electron beam passes between deflecting plates.

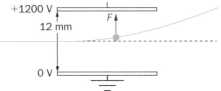

   Calculate:
   a) the electric field strength between the plates,
   b) the deflecting force F on each electron.

Further questions on page 317.

22 Capacitors

The photograph shows some **capacitors**.
They are important in both electric and electronic circuits.
Have you seen any of them before?
You will find capacitors in radios, TV sets and computers.
They are used to control the timing in car and burglar alarms.
The flash unit of a camera also uses a capacitor.

Although there are many different types of capacitors,
they all work in the same way.
Capacitors can store and then release electrical energy.
The value of a capacitor is often measured in microfarads (μF).

In this chapter you will learn:
- how capacitors work in simple circuits,
- about series and parallel combinations of capacitors,
- how the charge on a capacitor varies with time.

Can you see that all the capacitors have two leads?
The leads connect to two plates, which become charged.
The amount of energy stored by the capacitor depends on how much charge is moved on to the plates.
The conducting plates are separated by an insulating material, which is called the **dielectric**.

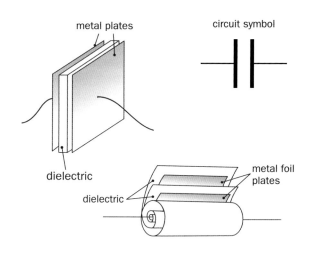

The plates need to have quite a large area.
How is this achieved in the capacitor shown in the diagram?
The two long strips of metal foil are separated by an insulator and then rolled up like a 'Swiss roll'.

▶ Charging and discharging a capacitor

In order to charge up a capacitor, we can connect it to a battery:

What happens when the switch S is closed?
Both the ammeters show equal size 'flicks', and then go back to zero. (Note: the zero is in the centre of the scale here.)
There is a surge of current (a flow of electrons) in the circuit, but *only* while the capacitor charges up.

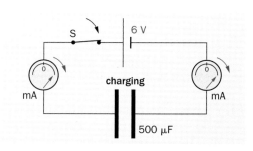

If the battery is removed, we can discharge the capacitor by connecting its leads together:

Now what happens when the switch S is closed?
Both meters again give equal flicks, and then return to zero.
Again, there is a brief current (a flow of charge) in the circuit.

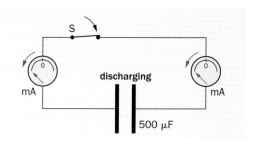

What do you notice about the direction of the flicks now?
This current is in the *opposite* direction to the charging current.

How does a capacitor charge up?

The capacitor plates are separated by an insulator. So how can charge flow?
This is explained in the diagrams below.
When the switch is closed, electrons flow from the negative terminal of
the battery on to plate B of the capacitor. At the same time, electrons
are repelled away from plate A back to the positive terminal of the battery.
This brief flow of electrons causes the flicks on the ammeters.

A negative charge builds up on B as the electrons collect on it. An equal
but opposite positive charge builds up on A as the electrons move off it.
Notice that as the charge builds up, the p.d. across the plates rises.

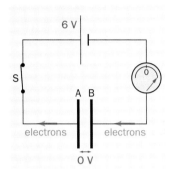

*The initial charging current
is high.*

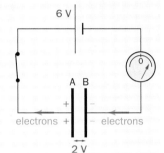

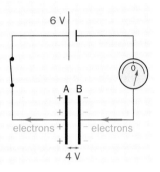

*As the charge builds up on the
capacitor plates, the current falls.*

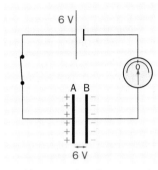

*No more charging current.
It is fully charged.*

Eventually, the build-up of charge prevents any more charge arriving
(because like charges repel). So the movement of the electrons stops.

Now the p.d. across the capacitor is 6 V, the same as the supply voltage.
During this charging process, **no** charge has crossed the insulating strip.

How does a capacitor discharge?

The charged capacitor acts rather like a very small battery. Look at the
circuit below. When the switch is closed, the electrons on plate B are
driven round the circuit to neutralise the positive charge on plate A.
The charge on both of the plates falls, as shown in the diagrams.
What happens to the p.d. across the plates as the capacitor discharges?

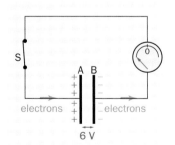

*The initial discharge current
is high.*

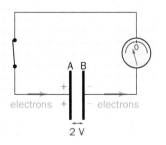

*As the charge left on the plates falls,
the current falls.*

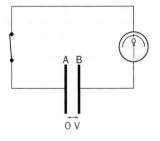

*No more discharge current.
It is discharged.*

The charging and discharging processes happen very quickly.
Can we slow them down?
Remember, a resistor makes it more difficult for charge to flow.
If we place a resistor in series with the capacitor, as shown here,
the capacitor charges and discharges more slowly.

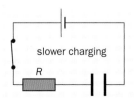

slower charging

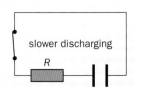

slower discharging

▷ Capacitance

The ability of a capacitor to 'store' charge on its plates is
called its **capacitance**. Capacitance is measured in **farads** (F).
The microfarad (μF) is often used, where $1 \ \mu F = 10^{-6}$ F.

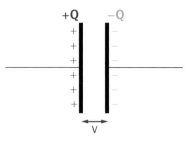

The p.d. across the plates of the capacitor in the diagram is V:
One of its plates has a charge of $+Q$; the other has a charge $-Q$.
We say that the charge stored by the capacitor is Q.

The larger the capacitance, the more charge the capacitor stores
for each volt applied across its plates. In fact:

$$\text{Capacitance, } C \text{ (farads)} = \frac{\text{charge, } Q \text{ (coulombs)}}{\text{p.d., } V \text{ (volts)}} \quad \text{or} \quad C = \frac{Q}{V}$$

The 100 μF capacitor stores 10 times more charge per volt than the 10 μF capacitor.

Can you see that capacitance is the charge stored per volt?

How can we measure the charge on the capacitor plates?

One way is to measure the current as the capacitor charges up.
On the previous page we saw that, as the capacitor charges up,
the current in the charging circuit falls.
The graph shows how the charging current varies with time:

From page 228 we know that, when a current I flows for a
short time Δt, the charge ΔQ that passes is given by: $\Delta Q = I \Delta t$
Can you see that this is the area of the dark strip on the graph?

The total charge Q that flows on to the capacitor after t seconds
is the **total area** under the graph.

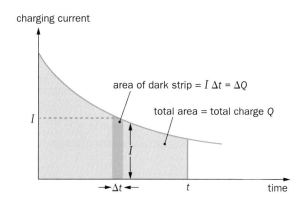

Measuring capacitance

To measure the capacitance C of a capacitor we must know the
charge Q on its plates and the corresponding p.d. V across it.
We can then calculate the capacitance from the formula above.

Is there an easy way to measure the charge?
If you could keep the charging current constant, you could
calculate the charge on the capacitor plates after time t using:

   $Q = I \, t$ where I is the **constant** charging current.

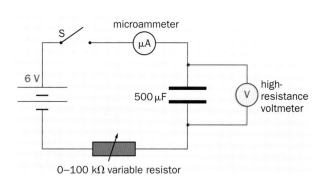

Can you see how this is done in the circuit shown here?

As the capacitor charges up, you can reduce the resistance
of the variable resistor to keep the current at a steady value.
You may need a few attempts to master the technique!

You can set the resistor at its maximum value and close switch S.
Adjust the resistor to keep the current constant at, say, 60 μA,
and record the p.d. across the capacitor plates every 10 seconds.

The table shows some typical results:
As the p.d. doubles, does the charge on the capacitor double?

The charge Q is proportional to the p.d. V: $Q = C V$

t /s	I /μA	$\therefore Q = It$ /μC	V /V	$C = \dfrac{Q}{V}$ /μF
0	60	0	0	
10	60	600	1.2	500
20	60	1200	2.4	500
30	60	1800	3.6	500
40	60	2400	4.8	500

302

▷ The energy of a charged capacitor

Look at the diagram shown here:
When the switch is in position A, the capacitor charges up.
What happens if you move the switch to B?
The lamp flashes briefly as the capacitor discharges.
The capacitor has released its stored electrical energy.

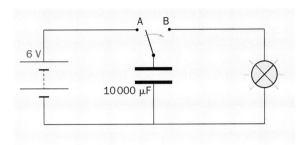

Where did this energy come from?
The energy was transferred to the capacitor by the supply.
The capacitor behaves a bit like a small rechargeable battery,
but it stores its energy in the electric field between its plates.

The p.d. across the capacitor and the size of its capacitance
both affect the energy stored. In fact:

Energy stored, $W = \frac{1}{2} \times$ capacitance, $C \times$ p.d., V^2
(joules) (farads) (volts)^2

or

$$W = \frac{1}{2} C V^2$$

How can we derive this energy equation?

The charge on a capacitor is proportional to the p.d. across it.
A graph of p.d. against charge is a straight line through (0,0).
The graph shows a capacitor charged to a p.d. V and charge Q.

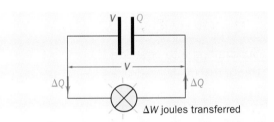

Now we allow the capacitor to discharge and release its energy.
At the start, the p.d. across the capacitor is V as shown:
A small charge ΔQ flows across this p.d. V and transfers
some energy ΔW.
Since p.d. is the energy transferred per coulomb (page 230),
the energy transferred, ΔW, is given by:

$$\Delta W = \Delta Q \times V$$

Can you see this is the area of the shaded strip on the graph?

When the capacitor discharges so that V and Q fall to zero,
the **total** energy transferred = the total area of all the strips
$\qquad\qquad\qquad\qquad\quad$ = the area of the triangle
$\qquad\qquad\qquad\qquad\quad$ = $\frac{1}{2}QV$

This must be the energy W originally stored in the capacitor.
Therefore:

$W = \frac{1}{2} Q V$

or, since $Q = CV$,

$W = \frac{1}{2} C V^2$

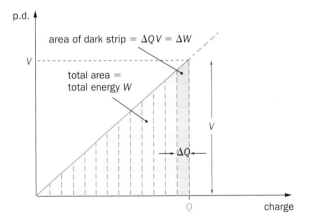

Example 1
A capacitor is charged using a steady current of 20.0 μA,
from a 5.0 V supply. After 55 s it is fully charged.
Calculate   a) the charge on the capacitor, b) its capacitance,
$\qquad\qquad$ c) the energy stored in the charged capacitor.

a) $\;\; Q = It$

$\qquad = 20.0 \times 10^{-6}$ A $\times$ 55 s

$\therefore\; Q = 1.1 \times 10^{-3}$ C (2 s.f.)

b) $\; C = \dfrac{Q}{V} = \dfrac{1.10 \times 10^{-3} \text{ C}}{5.0 \text{ V}}$

$\therefore\; C = 220 \times 10^{-6}$ F (= 220 μF)

c) $\; W = \frac{1}{2} C V^2$

$\qquad = \frac{1}{2} \times 220 \times 10^{-6}$ F $\times (5.0$ V$)^2$

$\therefore\; W = 2.8 \times 10^{-3}$ J (2 s.f.)

d) How much energy is stored if the charging p.d. is doubled?
$\qquad W = \frac{1}{2} CV^2$. C is constant. V doubles, so W = 4 times as much = $4 \times 2.8 \times 10^{-3}$ J = 1.1×10^{-2} J (2 s.f.)

▷ Capacitors in parallel

Circuits often contain combinations of two or more capacitors. Like resistors, they can be connected in series or in parallel.

The diagram shows two capacitors connected in **parallel**: Here are 3 facts that you should know for this circuit.

- The p.d. across each capacitor is the same, and equal to V.

- The total charge Q is the sum $Q = Q_1 + Q_2$
  of the charges on each capacitor:

- The combined capacitance C
  is the **sum** of the two capacitances: $\boxed{C = C_1 + C_2}$

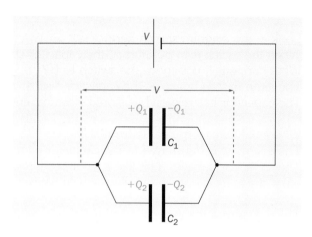

How can we prove this formula for capacitors in parallel?
We wish to replace C_1 and C_2 with a single capacitor C,
so that Q is the charge on C when the p.d. across it is V.
We know that: $Q = Q_1 + Q_2$ . . . . . (1)

We now apply the equation $Q = CV$ to each capacitor, to get:

$Q = CV$ $Q_1 = C_1 V$ $Q_2 = C_2 V$

Substituting into equation (1): $C\cancel{V} = C_1\cancel{V} + C_2\cancel{V}$
When the Vs are cancelled we get the equation shown above.

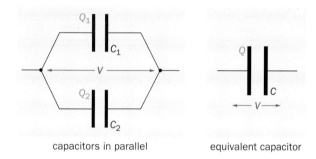

capacitors in parallel   equivalent capacitor

Example 2
Calculate the charge on each capacitor in this diagram:

Using $\mathbf{Q = CV}$ $Q_1 = C_1 V = 400\ \mu\text{F} \times 6.0\ \text{V} = \underline{2400\ \mu\text{C}}$

$Q_2 = C_2 V = 200\ \mu\text{F} \times 6.0\ \text{V} = \underline{1200\ \mu\text{C}}$

Notice that the 400 μF capacitor has twice as much charge as the 200 μF capacitor

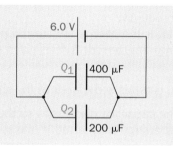

Example 3
A 20 μF capacitor is charged to 9.0 V, then disconnected from the supply, and then connected across an uncharged 10 μF capacitor.

Calculate: a) the initial charge on the 20 μF capacitor,
b) the capacitance of the parallel combination,
c) the p.d. across the parallel combination.

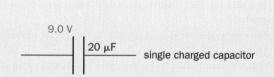

single charged capacitor

a) $\mathbf{Q = CV}$
$Q = 20 \times 10^{-6}\ \text{F} \times 9.0\ \text{V} = 180 \times 10^{-6}\ \text{C} = \underline{180\ \mu\text{C}}$

b) $\mathbf{C = C_1 + C_2}$ and leaving the values in μF
$C = 20\ \mu\text{F} + 10\ \mu\text{F} = \underline{30\ \mu\text{F}}$

c) The two capacitors now share the total charge $= 180\ \mu\text{C}$.
Charge (as electrons) flows between the two capacitors until the p.d. across each capacitor is the same.

$V = \dfrac{Q}{C} = \dfrac{180\ \mu\text{C}}{30\ \mu\text{F}} = \dfrac{180 \times 10^{-6}\ \text{C}}{30 \times 10^{-6}\ \text{F}} = \underline{6.0\ \text{V}}$ (2 s.f.)

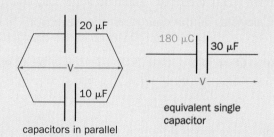

capacitors in parallel   equivalent single capacitor

▷ Capacitors in series

The diagram shows two capacitors connected in **series**.
Here are 3 facts that you should know for this circuit:

- The charge on each capacitor is the same and equal to Q, and the total charge stored by both capacitors together is Q.

- The supply p.d. V is shared between the two capacitors so that:

$$V = V_1 + V_2$$

- The combined capacitance C of the series combination is given by:

$$\frac{1}{C} = \frac{1}{C_1} + \frac{1}{C_2}$$

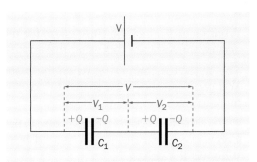

Did you expect that the total charge stored would be $2Q$?
The cartoon shows you why this is not the case:
During the charging process, a charge Q (not $2Q$) flows between the positive and negative terminals of the supply.

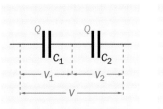

How can we prove this formula for capacitors in series?
We wish to replace C_1 and C_2 with a single capacitor C, so that Q is the charge on C when the p.d. across it is V.
We know that: $V = V_1 + V_2$ . . . . . (1)

We can apply the equation $V = \dfrac{Q}{C}$ to each capacitor, to get:

$$V = \frac{Q}{C} \qquad V_1 = \frac{Q}{C_1} \qquad V_2 = \frac{Q}{C_2}$$

Substituting into equation (1): $\dfrac{\cancel{Q}}{C} = \dfrac{\cancel{Q}}{C_1} + \dfrac{\cancel{Q}}{C_2}$

When the Qs are cancelled we get the equation shown above.

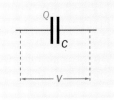

capacitors in series   equivalent capacitor

Example 4
In the circuit shown, what is   a) the combined capacitance,
b) the charge on each capacitor,   c) the p.d. across each capacitor?

a) $\dfrac{1}{C} = \dfrac{1}{C_1} + \dfrac{1}{C_2}$ and leaving the values in μF

$$\frac{1}{C} = \frac{1}{200\ \mu F} + \frac{1}{300\ \mu F} = \frac{3 + 2}{600\ \mu F} = \frac{5}{600\ \mu F}$$

$$\therefore C = \frac{600\ \mu F}{5} = \underline{120\ \mu F}$$

b) Now we can draw the equivalent circuit:
   $\mathbf{Q = C\,V}$ for the combined capacitor
       $= 120\ \mu F \times 6.0\ V$
$\therefore Q = \underline{720\ \mu C}$

c) We now know that the charge on each capacitor is 720 μC.

   Using $\mathbf{V = \dfrac{Q}{C}}$ for the 200 μF capacitor, $V = \dfrac{720\ \mu C}{200\ \mu F} = \underline{3.6\ V}$

               for the 300 μF capacitor, $V = \dfrac{720\ \mu C}{300\ \mu F} = \underline{2.4\ V}$

Notice that the *larger* capacitor in the series combination has the *smaller* p.d. across it.

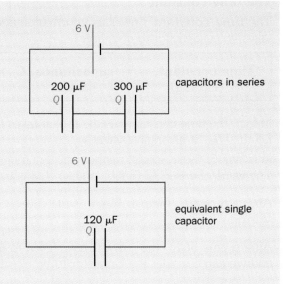

capacitors in series

equivalent single capacitor

Note that the combined capacitance of capacitors in series is *smaller* than their individual capacitances.
Compare this to resistors in parallel.

▶ Discharging a capacitor

What factors affect the **time** taken for a capacitor to discharge?
You can investigate this using the apparatus shown:

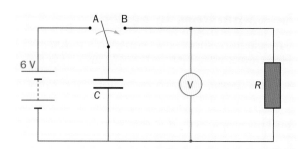

The capacitor charges up when the switch is in position A.
You can then move the switch to position B and discharge
the capacitor through the resistor.

The voltmeter records the p.d. across the capacitor.
Can you see that this is also the p.d. across the resistor?
Why will the voltmeter read 6 V at the start of the discharge?

You can record the p.d. every 10 s as the capacitor discharges.
The graph shows the results that you would get using
different values for the capacitance C and the resistance R:

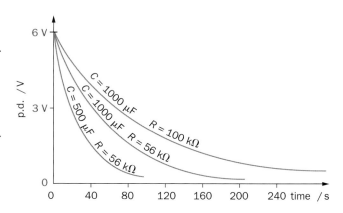

Why does the discharge take longer as C and R increase?
The larger the resistance, the more it resists the flow of charge.
The charge moves more slowly around the circuit.

The greater the capacitance, the greater the charge stored.
It takes longer for the charge to flow off the capacitor plates.

Why does the p.d. fall more and more slowly with time?
Look at the circuit diagrams:
Notice that the smaller the p.d., the smaller the current:
in other words, the smaller the rate of flow of charge.

As the p.d. falls, the charge flows off the plates more slowly
and the time for the p.d. to drop takes longer and longer.

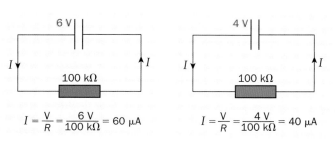

$$I = \frac{V}{R} = \frac{6\ V}{100\ k\Omega} = 60\ \mu A \qquad I = \frac{V}{R} = \frac{4\ V}{100\ k\Omega} = 40\ \mu A$$

The time constant

The **time constant** τ gives us information about the time it
takes for a capacitor to discharge:

> **Time constant, τ = capacitance, C × resistance, R**
> (seconds)   (farads)   (ohms)

In fact, the time constant τ is the time it takes for the p.d. to
fall to $1/e$ of its original value V_0.

(**e** is one of those special mathematical numbers, rather like π).
The value of e is 2.72 (3 s.f.) and so $1/e$ equals **0.37** (2 s.f.).

After a time CR, the p.d. V across the capacitor equals $0.37 \times V_0$.
V is now 37% (very roughly $1/3$) of its original value.

After a time $5CR$, the p.d. will have fallen to about 0.7% of V_0
and we can consider the capacitor to be effectively discharged.

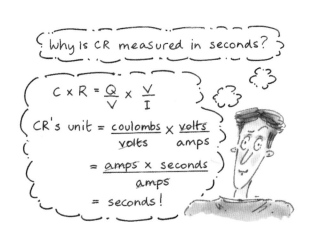

Why is CR measured in seconds?

$$C \times R = \frac{Q}{V} \times \frac{V}{I}$$

CR's unit = $\frac{\text{coulombs}}{\text{volts}} \times \frac{\text{volts}}{\text{amps}}$

$= \frac{\text{amps} \times \text{seconds}}{\text{amps}}$

= seconds!

> *Example 5*
> What is the time constant for the circuit shown?
>
>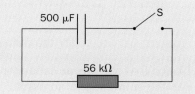
>
> **Time constant, τ = capacitance, C × resistance, R**
> $$= 500 \times 10^{-6}\ F \times 56 \times 10^3\ \Omega$$
> $$\tau = \underline{28\ \text{seconds}}$$

Exponential discharge

How does the charge on the capacitor plates vary as it discharges?

The diagram shows our discharge circuit again:

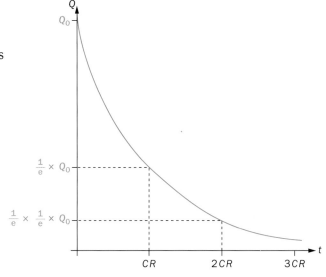

At this instant, the charge on the capacitor is Q, the discharge current is I and the p.d. across the resistor and the capacitor is V.

In a short time Δt, a small charge ΔQ flows off the capacitor plates. From page 228 we know that: $\Delta Q = I\Delta t$

But for the resistor, $I = \dfrac{V}{R}$, and for the capacitor, $V = \dfrac{Q}{C}$, so $I = \dfrac{Q}{CR}$.

Substituting for I in the equation for ΔQ: $\Delta Q = \dfrac{Q}{CR}\,\Delta t$ or $\boxed{\dfrac{\Delta Q}{\Delta t} = -\dfrac{Q}{CR}}$

(The minus sign tells us that Q decreases as time passes.)

This equation tells us that the rate of flow of charge off the capacitor plates, $\Delta Q/\Delta t$, depends on the time constant CR and on the charge Q on the capacitor plates. As time passes, Q falls and the capacitor discharges more and more slowly.

Let's look in detail at the graph of charge Q against time t:
After a time CR, the charge has fallen to $1/e$ (about $1/3$) of Q_0.

After a further time CR, the charge has again fallen by $1/e$. Now the charge is $1/e \times 1/e$ (roughly $1/9$) of its original value.

A further CR seconds and the charge is $1/e \times 1/e \times 1/e$ of Q_0.

The charge continues to fall by $1/e$ every CR seconds.
We say that the charge falls **exponentially with time**, and

$$\boxed{Q_t = Q_0\, e^{-t/CR}}$$

Q_0 is the initial charge
Q_t is the charge after time t
CR is the time constant

$Q = CV$, and so V is proportional to Q.
∴ The p.d. V must also fall exponentially.

$V = IR$, and so I is proportional to V.
∴ The current I must fall exponentially.

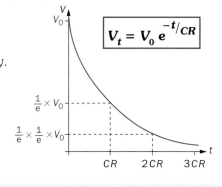

$$\boxed{V_t = V_0\, e^{-t/CR}}$$

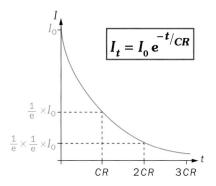

$$\boxed{I_t = I_0\, e^{-t/CR}}$$

Example 6
In the circuit shown, the capacitor is charged to 6.0 V.
The switch is now closed. Calculate:
a) the time constant,
b) the p.d. across the capacitor after 20 s,
c) the charge on the capacitor after 20 s.

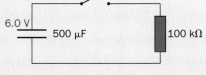

a) $\tau = C \times R$
   $= 500 \times 10^{-6}\,\text{F} \times 100 \times 10^{3}\,\Omega$
   ∴ $\tau = \underline{50\,\text{s}}$ (2 s.f.)

b) $V_t = V_0\, e^{-t/CR}$
   $= 6.0 \times e^{-20\,\text{s}/50\,\text{s}}$
   $= 6.0 \times e^{-0.4}$
   ∴ $V_t = \underline{4.0\,\text{V}}$ (2 s.f.)

c) $Q_t = C\,V_t$
   $= 500 \times 10^{-6}\,\text{F} \times 4.0\,\text{V}$
   ∴ $Q_t = \underline{2.0 \times 10^{-3}\,\text{C}}$ (2 s.f.)

▷ Charging a capacitor: Exponential growth

What factors affect the time taken for a capacitor to charge up?
You can investigate this using the apparatus shown:

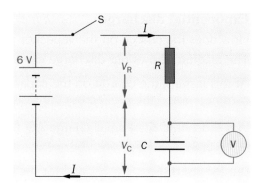

Here we are using a 6 V supply.
The voltmeter is recording the p.d. across the capacitor.
Is this also the p.d. across the resistor?
No, because the capacitor and resistor **share** the supply p.d.

When the switch S is closed, the capacitor starts to charge up,
and you can record the p.d. across the capacitor every 10 s.
The graph shows the results that you would get using
different values for the capacitance C and resistance R:

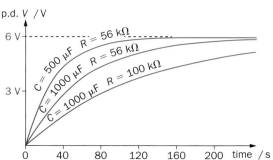

Notice that, as the capacitor charges, the p.d. increases ever
more slowly, eventually reaching the supply p.d. of 6.0 V.

Can you see that charging takes longer as C and R increase?
The time constant τ affects the charging process.

Since the charge on the capacitor is proportional to the p.d.
across its plates, the graph of charge Q against time must
be the same shape as the graph of p.d. against time.

Let's look in more detail at the graph of charge against time:
Note that the charge on the fully charged capacitor is $\mathbf{Q_0}$.

After a time CR, the charge has risen to $(1 - 1/e)$ of Q_0.
This is about 2/3 of the final charge Q_0.

After time $2 \times CR$, the charge has risen to $(1 - (1/e \times 1/e))$
or $(1 - 1/e^2)$ of Q_0. This is about 8/9 of Q_0.

Can you see a pattern?

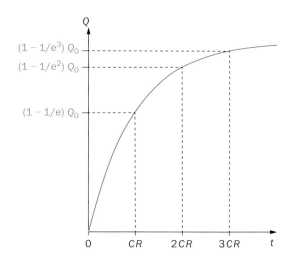

We say the charge **increases exponentially** with time, and:

$$Q_t = Q_0 (1 - e^{-t/CR})$$

Q_0 is the final charge
Q_t is the charge after time t
CR is the time constant

We can write a similar equation for the p.d. across the
capacitor as it charges up, but what about the current?

Look at the circuit diagram again.
V_C is the p.d. across the capacitor, V_R the p.d. across the resistor.
Can you see that $V_C + V_R$ must **always** equal the supply voltage?

So as V_C rises exponentially, V_R must fall exponentially.
The current is proportional to the p.d. V_R across the resistor, so
the charging current **falls** exponentially as the capacitor charges.

$$V_t = V_0 (1 - e^{-t/CR})$$

$$I_t = I_0 e^{-t/CR}$$

Example 7
The switch is closed in the circuit shown. Calculate:
a) the time constant,
b) the p.d across the capacitor after 40 s,
c) the charging current after 40 s.

250 μF

6.0 V

100 kΩ

a) $\tau = C \times R$

$\quad = 250 \times 10^{-6}\text{F} \times 100 \times 10^3\,\Omega$

$\quad \tau = \underline{25\text{ s}}$ (2 s.f.)

b) $V_t = V_0 (1 - e^{-t/CR})$

$\quad = 6.0 \times (1 - e^{-40/25\text{ s}})$

$\quad = 6.0 \times (1 - e^{-1.6})$

$\quad V_t = \underline{4.8\text{ V}}$ (2 s.f.)

c) The p.d. across $R = 6.0\text{V} - 4.8\text{V} = 1.2\text{V}$

Using $V = I \times R = 1.2$ V:

$\quad 1.2\text{ V} = I \times 100 \times 10^3\,\Omega$

$\quad I = \underline{1.2 \times 10^{-5}\text{ A}}$ (2 s.f.)

▷ Physics at work: Capacitors

The capacitance of charged spheres

Have you seen this 'experiment' being carried out?

The student stands on a block of polystyrene and touches the dome of the Van de Graaff generator. When the Van de Graaff is switched on, the dome and the student become charged. Her hair demonstrates that like charges repel!

You need to take care when using a Van de Graaff generator. As the dome becomes highly charged, an electric field is created at its surface. The electric field strength can be so large that the insulation of the air breaks down, and sparks pass from the dome to any nearby object that is connected to Earth (see page 294). You may get an unexpected electric shock!

The dome of the Van de Graaff is storing electric charge. It is acting as a capacitor. Can we calculate its capacitance C?

The diagram shows an isolated sphere of radius r_s, in a vacuum: The dome of the Van de Graaff is (almost) a sphere.

The charge on the sphere is $+Q$. From page 296 we know that the electric potential V at the surface of the sphere is:

$$V = \frac{kQ}{r_s}, \text{ where } k = \frac{1}{4\pi\varepsilon_0}$$

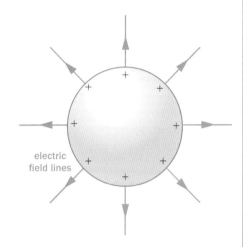

electric field lines

For any capacitor $C = \dfrac{Q}{V}$, and so: $\quad C = Q \div \dfrac{Q}{4\pi\varepsilon_0 r_s}$

$$C = \cancel{Q} \times \frac{4\pi\varepsilon_0 r_s}{\cancel{Q}}, \qquad \therefore \quad \boxed{C = 4\pi\varepsilon_0 r_s}$$

Example 8
Treating the Earth as an isolated sphere, what is its capacitance? The radius of the Earth is 6.4×10^6 m.

(From page 290, the value of ε_0 is 8.85×10^{-12} F m^{-1}.)

$C = 4\pi\varepsilon_0 r_s$
C $= 4\pi \times 8.85 \times 10^{-12}$ F m^{-1} $\times$ 6.4×10^6 m $= 7.1 \times 10^{-4}$ F, or $\underline{710\ \mu\text{F}}$ (2 s.f.)

▷ Physics at work: Earthing

When a charged capacitor is connected to an uncharged capacitor (as in Example 3), electrons flow between the capacitors until the p.d. across both capacitors is the same. The capacitors share the charge, but not equally.

If the uncharged capacitor has a much larger capacitance than the charged capacitor, the charged capacitor loses almost all of its charge to the uncharged capacitor, and it is effectively discharged. Example 8 above shows that the Earth has a large capacitance. So when a charged conductor is connected to the Earth, it is (effectively) discharged.

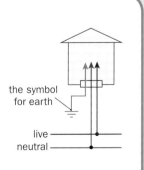

the symbol for earth

live
neutral

The Earth has such a large capacitance that any loss or gain of charge causes a very small change in its potential. Therefore, we can treat the Earth as a convenient zero of electrical potential energy, as we saw in Chapter 21.

▷ Physics at work: Designing practical capacitors

The parallel-plate capacitor

How does the construction of a capacitor affect its capacitance?

The capacitance C of a parallel plate capacitor can be calculated using:

$$C = \frac{\varepsilon A}{d}$$

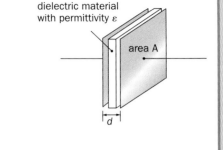

dielectric material with permittivity ε

area A

Can you see that: increasing the area A of the plates,
decreasing the separation d of the plates,
increasing the permittivity ε of the dielectric
all increase the capacitance of the capacitor?

In the capacitors below, how is A made large and d made small?

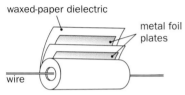

waxed-paper dielectric

metal foil plates

wire

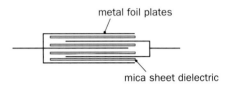

metal foil plates

mica sheet dielectric

1. A waxed-paper capacitor consists of two long strips of metal foil separated by strips of waxed paper. It is rolled up to form a cylinder. Typical value is 0.1 μF.

2. An electrolytic capacitor has a similar 'Swiss roll' construction. Its very thin dielectric is formed chemically. The plate marked + must never be charged negatively. Typical value is 100 μF.

3. The mica capacitor consists of layers of metal foil separated by thin sheets of mica. Typical value is less than 0.01 μF, but they have high stability − they keep their value with age.

Action of the dielectric

In an electric field, the nuclei in each molecule of the dielectric are pushed in the direction of the electric field; the electron cloud is pulled the opposite way. As the equilibrium positions of the positive and negative charges move, electric dipoles form: The molecules of some materials (such as water) are already polarised. These **rotate** to align with an electric field.

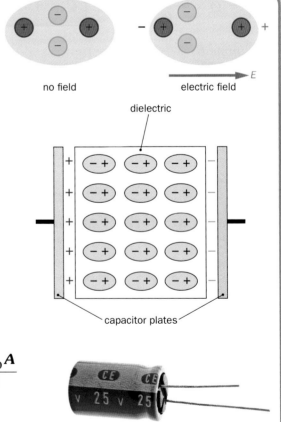

no field

electric field

E

dielectric

capacitor plates

The diagram shows polarised molecules in a capacitor: Can you see that the charges at the surfaces of the dielectric are opposite to the charges on the adjacent capacitor plates?

The electric field due to the charges on the capacitor plates is to the right, but each dipole produces a tiny field to the left. The resultant electric field E is made smaller by the dipoles.

The p.d. V across the plates is given by $V = E d$ (see page 293). Since E is smaller, V is less than it would be with no dielectric. More charge can be added to the plates before the p.d. across them equals the applied p.d. So, the capacitance is increased.

Why is the maximum working voltage marked on a capacitor? If the p.d. applied to the plates is greater than this, the dipoles will be 'stretched' so much that the dielectric will break down.

The parallel plate capacitor equation can be written:

$$C = \frac{\varepsilon_r \varepsilon_0 A}{d}$$

where ε_r is the relative permittivity, the ratio of the material's permittivity to the permittivity of free space: $\varepsilon_r = \dfrac{\varepsilon}{\varepsilon_0}$

Summary

A capacitor of capacitance C stores a charge Q when there is a p.d. V across it: $\quad C = \dfrac{Q}{V} \quad$ or $\quad Q = CV$

Energy W is stored in the electric field between the plates of a capacitor: $\quad W = \frac{1}{2}QV = \frac{1}{2}CV^2 = \frac{1}{2}\dfrac{Q^2}{C}$

The combined capacitance C of two capacitors *in parallel* is given by: $\quad C = C_1 + C_2$

The combined capacitance C of two capacitors *in series* is given by: $\quad \dfrac{1}{C} = \dfrac{1}{C_1} + \dfrac{1}{C_2}$

The rate at which a capacitor discharges depends on the time constant τ: $\quad \tau = CR$

The charge on a capacitor, its p.d. and the discharge current all *fall* exponentially with time. The exponential equation for charge is: $\quad Q_t = Q_0 e^{-t/CR}$

The charge on a capacitor and the p.d. across its plates both *rise* exponentially with time. The exponential equation for charge is: $\quad Q_t = Q_0 (1 - e^{-t/CR})$

▷ Questions

1. Can you complete these sentences?
   a) As a capacitor charges, a positive .... builds up on one plate and an equal .... charge builds up on the other plate.
   b) The charging process stops when the .... across the capacitor plates equals the p.d. of the ....
   c) If the p.d. across a capacitor is doubled, the .... that is stored is quadrupled.
   d) The p.d. is the same across each capacitor in ....
   e) The charge is the same on each capacitor in ....

2. Explain:
   a) How can there be a current in a circuit containing a capacitor, since no charge is able to pass through the insulator between the capacitor plates?
   b) Why does the current in a capacitor discharge circuit decrease with time?

3. When a capacitor is charged from a 12 V supply, 6.0 μC moves on to its plates. What is its capacitance?

4. Some computer chips use a capacitor to provide energy for a short time if there is a power failure. A 100 μF capacitor is charged to 9.0 V.
   a) What is the charge on the capacitor? b) How long would it be able to supply a steady current of 50 μA?

5. A capacitor stores a charge of 60 μC when charged from a 12 V supply. How much energy does it store?

6. What is the combined capacitance of each of the examples shown below?

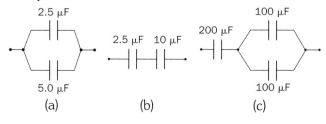

    (a)   (b)   (c)

7. Some capacitors are marked 10 μF, 25 V. How could you combine these capacitors to make:
   a) a 20 μF capacitor with a working voltage of 25 V?
   b) a 5 μF capacitor with a working voltage of 50 V?
   c) a 10 μF capacitor with a working voltage of 50 V?

8. A 5.0 μF capacitor is charged to a p.d. of 12 V and then connected across an uncharged 20 μF capacitor. Calculate: a) the initial charge on the 5 μF capacitor, b) the total capacitance of the combination, c) the p.d. across the capacitors in parallel.

9. A 6.0 V battery is used to charge a 100 μF capacitor in series with a 25 μF capacitor. Calculate:
   a) the capacitance of the combination,
   b) the charge on the capacitors,
   c) the p.d. across each capacitor.

10. A 4.0 μF capacitor is charged to 15 V and then discharged through a 0.80 MΩ resistor. What is:
    a) the initial charge on the capacitor?
    b) the time constant of the discharge circuit?
    c) the charge 5.0 s after the discharge starts?

11. A 10 000 μF capacitor and a 20 kΩ resistor are connected in series to a 6.0 V supply. Calculate:
    a) the initial charging current,
    b) the p.d. across the plates after 100 s.

12. A timer uses a 1000 μF capacitor to switch off a security light after a pre-set time. The capacitor is charged to a p.d. of 9.0 V and allowed to discharge through a resistor of 330 kΩ. The switch is triggered when the p.d. falls to 2.5 V. Calculate the time for which the light is on.

Further questions on page 318.

▷ Current and Charge

1. An electric heater consists of three heating elements connected to a 240 V supply as in the diagram.

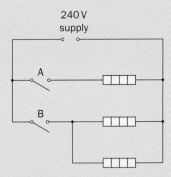

240 V
supply

A

B

Each heating element is rated at 240 V, 960 W.
a) Calculate the output power of the heater for the indicated positions of switches A and B. [3]

switch A	switch B	output power /W
closed	open	
closed	closed	
open	closed	

b) For one heating element at its normal operating temperature, calculate the current in and the resistance of the element. [4]
c) Calculate the total resistance of the circuit when all three heating elements are in use. [2] (OCR)

2. In the following circuits the battery has negligible internal resistance and the lamps are identical.

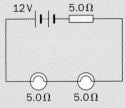

12 V   5.0 Ω

5.0 Ω   5.0 Ω

Figure 1

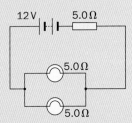

12 V   5.0 Ω

5.0 Ω

5.0 Ω

Figure 2

a) For the circuit shown in Figure 1 calculate the current in each lamp and the power dissipated in each lamp. [2]
b) For the circuit shown in Figure 2 calculate the current in each lamp. [2]
c) Explain how the brightness of the lamps in the circuit in Figure 1 compares with the brightness of the lamps in the circuit in Figure 2.
Explain why the battery would last longer in the circuit shown in Figure 1. [3]
d) One of the lamps in the circuit in Figure 2 is faulty and no longer conducts.
Describe and explain what happens to the brightness of the other lamp. [2] (AQA)

3. The graph shows the current–voltage plot for a filament lamp.

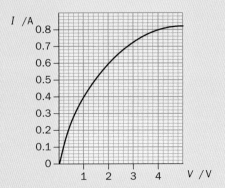

a) Explain why the graph is curved.
Use the graph to calculate the resistance of the lamp filament when connected across 4.0 V. [3]

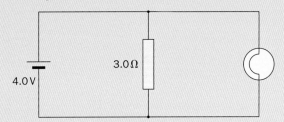

4.0 V   3.0 Ω

b) The lamp is connected in parallel with a 3.0 Ω resistor to a 4.0 V supply of negligible internal resistance.
Calculate the total current drawn from the supply and the power dissipated in the circuit. [3]
c) The lamp filament is made of 0.015 m of tungsten wire, which has a resistivity of $3.9 \times 10^{-7}\,\Omega\,\text{m}$ in these conditions.
Calculate the diameter of the wire. [3] (OCR)

4. In the circuit shown each resistor has a resistance of 10 Ω; the battery has an e.m.f. of 12.0 V and is of negligible internal resistance.

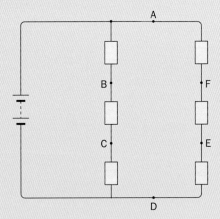

A

B   F

C   E

D

a) Calculate the power delivered by the battery.
b) Write down the p.d. between C and D, and between F and D, and hence calculate the p.d. between C and F. [5] (AQA)

5. a) Name the charge carriers responsible for electric current (i) in a metal and (ii) in an electrolyte. [1]

b) A copper rod, cross-sectional area $3.0 \times 10^{-4}\,m^2$, is used to transmit large currents.
A charge of $650\,C$ passes along the rod every $5.0\,s$. Calculate

   i) the current I in the rod, [1]

   ii) the total number of electrons passing any point in the rod per second, [1]

   iii) the mean drift velocity of the electrons in the rod given that the number density of free electrons is $1.0 \times 10^{29}\,m^{-3}$. [2]

c) The copper rod X in (b) is connected to a longer thinner copper rod Y.

   i) State why the current in Y must also be I. [1]

   ii) Rod Y has half the cross-sectional area of rod X. Calculate the mean drift velocity of electrons in Y. [1] (OCR)

6. The circuit shown is used to produce a current–voltage graph for a 12 V, 24 W lamp.

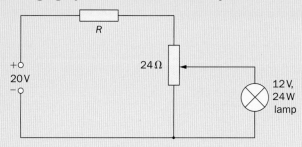

a) Show on a copy of the diagram the correct positions for a voltmeter and an ammeter. [2]

b) Calculate the resistance of the lamp in normal operation. [3]

c) Calculate the value of R which would enable the voltage across the lamp to be varied between 0 V and 12 V. [4] (Edex)

7. The e.m.f. of the electricity supply to a remote cottage is 230 V. The resistance of the cables to the cottage may be considered as the internal resistance of the supply. When an electric cooker is used in the cottage, the measured voltage across the cooker is 210 V. The resistance of the cooker is 35 Ω.

a) Calculate

   i) the current to the cooker,

   ii) the power of the cooker,

   iii) the resistance of the cables to the cottage. [4]

b) Explain why the voltage measured at the cooker is less than the supply voltage when the cooker is in use. [2]

c) Suggest two disadvantages of this power supply. [2] (OCR)

8. A designer uses the circuit shown for the sidelights of a car. The lamps are each rated at 6 V, 12 W. The battery used is of negligible resistance.

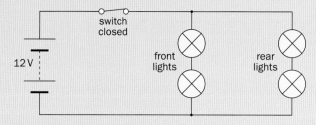

a) Calculate the current in the battery. [3]

b) A fully charged battery is able to supply a total charge of $1.2 \times 10^5\,C$.
How long could the lamps operate when connected to a fully charged battery? [2]

c) A user replaces one of the front lamps with one rated at 12 V, 24 W. State and explain whether

   i) this change has any effect on the brightness of the rear sidelights,

   ii) the front sidelights will operate at their normal rated power.

d) The original circuit is not normally used. In practice, four lamps, each rated at 12 V, are connected in parallel as shown below. Give one reason why this is preferred. [1] (AQA)

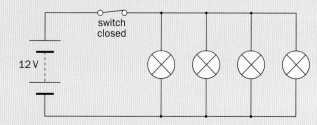

9. a) Derive a formula for the resistance of three resistors R_1, R_2 and R_3 in series. [2]

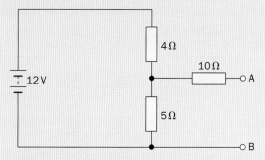

b) Calculate the p.d. between A and B in the above circuit when the 12 V battery has negligible internal resistance. [2]

c) What would the p.d. between A and B become if a 10 Ω resistor were connected between the two points? [2]

d) Calculate the current in the 10 Ω resistor connected to A and B. [2] (AQA)

10. a) Distinguish between the e.m.f. and the terminal p.d. of a cell. [2]

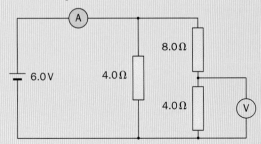

b) Find values for the ammeter and voltmeter readings in the circuit shown. Assume that the ammeter and cell have negligible resistance and the voltmeter has a very large resistance. [4]

c) In fact the cell does have an internal resistance. A student finds that the voltmeter reads 1.6 V.
   i) Show that the terminal p.d. of the cell is 4.8 V.
   ii) If the ammeter reads 1.6 A, find the value of the internal resistance of the cell. [4] (OCR)

11. The graph shows how the resistance of a thermistor varies with temperature.

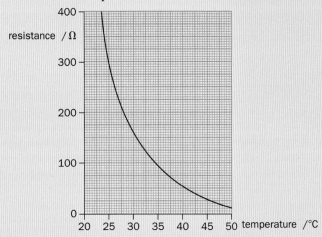

a) Determine the resistance of the thermistor at a temperature of 30 °C. [1]

b) The circuit diagram shows the thermistor connected in series with a 300 Ω resistor and a 6.0 V battery of negligible internal resistance.

   i) Explain, without calculation, how you would expect the p.d. across the thermistor to change as its temperature is increased. [3]
   ii) Calculate the p.d. across the resistor when the temperature of the thermistor is 30 °C. [2] (AQA)

12. A student uses the circuit shown to determine the resistivity of a metal in the form of a wire.

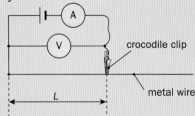

The length L of the wire is changed with the help of a crocodile clip. The current in the wire is I, the p.d. across the wire is V and the wire has resistance R. The table shows the results recorded by the student:

L /m	V /V	I /A	R /Ω
0.050	0.40	0.160	2.50
0.200	0.40	0.140	2.86
0.400	0.40	0.072	
0.800	0.40	0.036	11.1
1.000	0.40	0.029	13.8

a) Complete the table by calculating the resistance of the wire of length 0.400 m. [1]
   Plot a graph of R against L for these results.

b) The student observed that the wire was significantly hotter when the shortest length $L = 0.050$ m was used. The cross-sectional area of the wire is 8.0×10^{-8} m². Use your graph to determine the resistivity of the metal. [3]

c) The voltmeter used in the experiment had a zero error. The potential difference recorded in the experiment was smaller than it should have been. Discuss how the actual value of the resistivity of the metal would differ from the value calculated in (b). [3] (OCR)

13. The graph shows the I–V characteristic of a semiconductor diode.
Which statement about the resistance of the diode can be deduced from the characteristic?

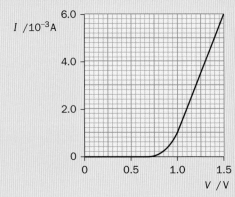

A It is zero between 0 V and 0.70 V.
B It is constant between 1.0 V and 1.5 V.
C It is 0.4 Ω at 1.2 V.
D It decreases between 0.70 V and 1.0 V. (OCR)

Magnetic Fields

14. The diagram shows a horizontal cross-section through two long vertical wires P and Q. P carries a current into the plane of the paper. The circles on the diagram represent the magnetic field of this current. Q carries a current of 3.0 A in the opposite direction to the current in P.

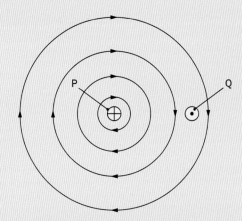

a) The magnetic flux density of the field at Q caused by the current in P is 2.0×10^{-5} T. Calculate the force experienced by unit length (1.0 m) of Q.
b) Sketch a diagram and on it draw an arrow to show the direction of this force.
Explain the rule you used to obtain the direction of the force. [6] (OCR)

15. The long solenoid shown in the diagram has 1200 turns per m and carries a current of 2.0 A.

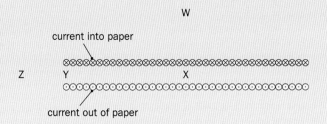

a) Draw an arrow on the diagram at X to show the direction of the field at X.
b) Describe the magnetic field inside the solenoid.
c) Arrange the letters W, X, Y and Z in the order of size of magnetic field strengths (weakest to strongest) at the points which correspond to these letters in the diagram.
d) The magnetic flux density at a point well inside and on the axis of a long solenoid is given by $B = \mu_0 NI/l$, where N is the number of turns on the solenoid, l is the length of the solenoid, and μ_0 and I have their usual meanings.
Show that the flux density of the magnetic field at X is 3.02×10^{-3} T. [4] (AQA)

16. A horseshoe magnet, made from two magnets and a steel yoke, is placed on a top pan balance as shown in the upper diagram. A current is passed through a long wire which is fixed. The wire is between, and parallel to, the poles of the magnet, as shown in the lower diagram.

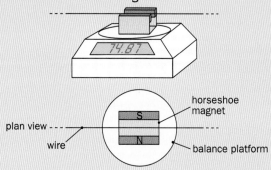

The current is varied. The following values of balance reading m and current I are recorded.

I /A	0	1.1	2.2	3.0	4.3	−1.8
m /g	74.87	75.02	75.17	75.29	75.46	74.63

a) Write down the relationship between the force F on the wire, the magnetic field B acting on length l of the wire, and the current I. [1]
b) Explain why the force on the wire is equal in magnitude to the force on the magnet on the top pan balance. [2]
c) Use the data to plot a graph to test the relationship between the current and the force it exerts on the magnet. [2]
d) What happens when the current is reversed and why? [2] (OCR)

17. The magnetic force F that acts on a current-carrying conductor in a magnetic field is given by the equation $F = BIl$.
a) State the condition under which this equation applies. [1]
b) The unit for magnetic flux density B is the tesla. Express the tesla in base units. [2] (Edex)

18. Charged particles are travelling at a speed v, at right angles to a magnetic field of flux density B. Each particle has a mass m and a charge Q. Which of the following changes would cause a decrease in the radius of the circular path of the particles?
A an increase in B B an increase in m
C an increase in v D a decrease in Q [1]   (Edex)

19. An α particle travelling with speed 8.8×10^6 m s^{-1} passes through a point where the magnetic field strength is $2.4\,\mu$T. Explain briefly in which direction the α particle is travelling when it experiences
(a) no force (b) a force of 3.38×10^{-18} N.
Charge on α particle $= +3.2 \times 10^{-19}$ C [2] (W)

Further Questions on Electricity

▷ Electromagnetic Induction

20. a) Explain why you would expect an e.m.f. to be developed between the wing tips of an aircraft flying above the Earth. [1]

b) An aircraft, whose wing tips are 20 m apart, flies horizontally at a speed of 600 km h^{-1} in a region where the vertical component of the Earth's magnetic flux density is 4.0×10^{-5} T. Calculate:
  i) the area swept out by the aircraft's wings in 1 s,
  ii) the e.m.f. between the wing tips. [4] (AQA)

21. The diagrams show two identical coils aligned on a common axis XX. Coil P is connected to a battery and a switch. Coil Q is connected to a sensitive meter M.

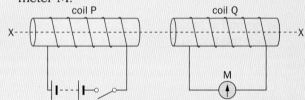

a) Before the switch is closed, the reading on M is zero. On closing the switch, M records a pulse of current. The reading on M then returns to zero and remains at that value. Explain these observations by reference to Faraday's law of electromagnetic induction. [5]

b) As the switch is closed, the current in coil P creates a magnetic field B in the region between the coils. The direction of this field along the axis XX is shown in the lower diagram. The pulse of current in coil Q also produces a magnetic field along the axis XX.

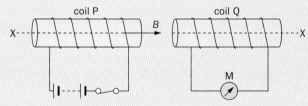

  i) Sketch the diagram and draw an arrow labelled C to show the direction of this field.
  ii) How is this explained by Lenz's Law? [3](OCR)

22. The magnetic flux through a coil of 5 turns changes uniformly from 15×10^{-3} Wb to 7.0×10^{-3} Wb in 0.50 s. What is the magnitude of the e.m.f induced in the coil due to this change in flux?
A 14 mV   B 16 mV   C 30 mV   D 80 mV   [1] (AQA)

23 A horizontal straight wire of length 0.30 m carries a current of 2.0 A perpendicular to a horizontal uniform magnetic field of flux density 5.0×10^{-2} T. The wire 'floats' in equilibrium in the field. What is the mass of the wire?
A 8.0×10^{-4} kg   B 3.1×10^{-3} kg
C 3.0×10^{-2} kg   D 8.2×10^{-1} kg   [1] (AQA)

24. The diagram shows a magnet being dropped through the centre of a narrow coil. The e.m.f. across the coil is monitored on an oscilloscope.

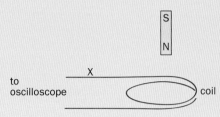

a) What is the induced polarity on the top side of the coil as the magnet falls towards the coil? Sketch the diagram and draw an arrow at position X to indicate the direction of current flow as the magnet falls towards the coil. [2]

b) The second diagram shows the variation of e.m.f. with time as the magnet falls.

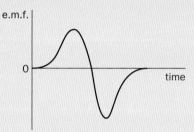

State, and explain, two ways in which the graph would change if the magnet were dropped from a greater height than that used to produce the graph in the diagram. [2]   (AQA)

25. Faraday's and Lenz's Laws may be summarised in a formula as

$$\varepsilon = -\frac{d(N\phi)}{dt} \quad \text{or} \quad \varepsilon = -\frac{\Delta(N\phi)}{\Delta t}$$

a) State the meaning of the term $N\phi$. [2]
b) Explain the significance of the minus sign. [3] (Edex)

26. An electromagnet is used to produce a uniform magnetic field in the gap between its poles. The field outside the gap is negligible. The coil C encloses all the flux of the magnetic field. The magnetic flux density in the gap is 0.030 T and coil C has 70 turns.

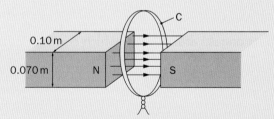

a) Calculate the flux linkage through the coil. [2]
b) The current in the electromagnet is reduced to zero at a uniform rate over a time of 2.5 s. Find the average e.m.f. induced in the coil. [2] (OCR)

▷ Electric Fields

27. Define *electric field strength* at a point in space.   [1]
Two point charges, $+1.6 \times 10^{-19}$C and
-6.4×10^{-19}C, are held at a distance 8.0×10^{-10}m
apart at points A and B, as shown in the diagram.

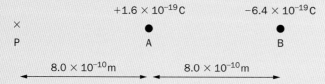

a) Find the magnitude and direction of the electric
field at the midpoint between the two charges.   [2]
b) Find the electric field at point P in the diagram.   [2]
c) What can you say about the direction of electric
field on either side of P along AB?   [2]   (OCR)

28. Two identical table tennis balls, A and B, each of
mass 1.5 g, are attached to non-conducting threads.
The balls are charged to the same positive value.
When the threads are fastened to a point P, the
balls hang as shown in the diagram.

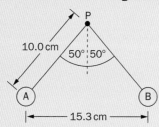

a) Draw a labelled free-body force diagram for A.   [3]
b) Calculate the tension in one of the threads.   [3]
c) Show that the electrostatic force between the
two balls is 1.8×10^{-2}N.   [1]
d) Calculate the charge on each ball.   [3]   (Edex)

29. Two parallel plates are set a distance of 12 mm apart
in a vacuum as shown below. The top plate is at a
potential of $+300$ V and the bottom plate is at a
potential of -300 V.

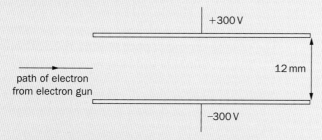

a) Sketch the diagram and on it draw lines to show
the electric field lines between the plates.   [3]
b) At a point midway between the plates the field is
uniform. Calculate the magnitudes of
i) the electric field strength at this point,
ii) the force on an electron at this point.
[5]   (OCR)

30. The diagram shows part of the region around a
small positive charge.

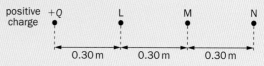

a) The electric potential at point L due to this
charge is $+3.0$ V. Calculate the magnitude Q of
the charge.
b) Show that the electric potential at point N, due to
the charge, is $+1.0$ V.
c) Show that the electric field strength at point M
is 2.5 Vm^{-1}.   [5]   (AQA)

31. Two charges, each of $+0.8$ nC,
are 40 mm apart.
Point P is 40 mm from
each of the charges.

What is the electric
potential at P?

A zero   B 180 V   C 360 V   D 4500 V [1] (AQA)

32. The diagram shows two charged parallel conducting
plates.

a) Copy the diagram, then add solid lines to show
the electric field in the space between and just
beyond the edges of the plates.   [2]
b) Add to your diagram dotted lines to show three
equipotentials in the same regions.   [2]
c) Define electric potential at a point. Is electric
potential a vector or a scalar quantity?   [3]   (Edex)

33. Two parallel metal plates P_1 and P_2 are placed 10 cm
apart in air. P_1 is maintained at -200 kV whilst
P_2 is connected to Earth through a microammeter.
Suspended between the plates by an insulating
thread is a light plastic sphere coated with
conducting paint so that it will store charge.

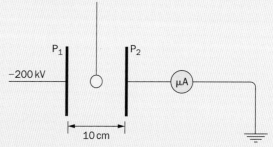

a) Calculate the electric field strength between the
plates.   [1]
b) Explain why the ball shuttles back and forth
between the plates, contacting each plate
alternately.   [3]   (AQA)

Further Questions on Electricity

▷ Capacitors

34. a) Sketch a graph to show the variation with charge Q of the p.d. V across the plates of a capacitor. [2]

b) Use your graph to show that the energy E stored in a capacitor of capacitance C with a p.d. V across its plates is given by $E = \frac{1}{2}CV^2$. [2] (OCR)

35. A $47\,\mu\text{F}$ capacitor is used to power the flashgun of a camera. The average power output of the flashgun is $4.0\,\text{kW}$ for the duration of flash, which is $2.0\,\text{ms}$.

a) Show that the energy output of the flashgun per flash is $8.0\,\text{J}$. [2]

b) Calculate the p.d. between the terminals of the capacitor immediately before the flash. Assume that all the energy stored by the capacitor is used to provide the output power of the flashgun. [2]

c) Calculate the maximum charge stored by the capacitor. [2]

d) Calculate the average current provided by the capacitor during the flash. [1] (AQA)

36. The circuit shows a $0.47\,\mu\text{F}$ capacitor which can be charged through a $3000\,\Omega$ resistor when switch S is closed.

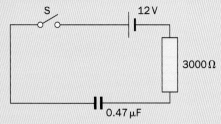

The capacitor is initially uncharged. Find:

a) the initial charging current,

b) the charging current after a long time,

c) the maximum charge stored on the capacitor,

d) the maximum energy stored on the capacitor.

[5] (OCR)

37. The circuit shown is used to investigate the discharge of a capacitor. With the switch in position S_1 the capacitor is charged. The switch is then moved to S_2 and readings of current and time are taken as the capacitor discharges through the resistor. The results are plotted on the graph below.

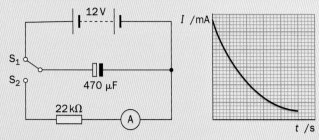

a) Calculate the maximum charge stored on the capacitor. [1]

b) Make suitable calculations to enable you to add scales to both axes of the graph. [4] (Edex)

38. An engineer required a $20\,\mu\text{F}$ capacitor which operated safely using a voltage of $6.0\,\text{kV}$. Ten similar capacitors connected in series were found to provide a suitable arrangement when operating at their maximum safe voltage.

a) State the maximum safe working voltage of each capacitor. [1]

b) Determine the capacitance of each capacitor. [1]

c) Calculate the total energy stored by the combination of capacitors when fully charged to $6.0\,\text{kV}$. [2] (OCR)

39. A radio tuning circuit uses a variable capacitor with a range of $50–500\,\text{pF}$. A designer decides to change the range using a $500\,\text{pF}$ capacitor.

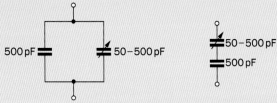

Determine the range of capacitance available when the $500\,\text{pF}$ capacitor is connected

a) in parallel with the variable capacitor (above left),

b) in series with the variable capacitor (right-hand diagram). [4] (AQA)

40. A $0.1\,\text{F}$ capacitor is charged at a constant rate with a **steady current** of $40\,\text{mA}$ for a time of $60\,\text{s}$. Calculate the final

a) charge stored by the capacitor, [2]

b) energy stored by the capacitor. [2] (OCR)

41. The diagram shows a capacitor of capacitance $370\,\text{pF}$. It consists of two parallel metal plates of area $250\,\text{cm}^2$. A sheet of polythene that has a relative permittivity 2.3 completely fills the gap between the plates.

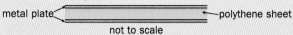

metal plate — polythene sheet

not to scale

a) Calculate the thickness of the polythene sheet. [2]

b) The capacitor is charged so that there is a potential difference of $35\,\text{V}$ between the plates. The charge on the capacitor is then $13\,\text{nC}$ and the energy stored is $0.23\,\mu\text{J}$. The supply is now disconnected and the polythene sheet is pulled out from between the plates without discharging or altering the separation of the plates.

i) Show that the potential difference between the plates increases to about $80\,\text{V}$. [2]

ii) Calculate the energy that is now stored by the capacitor. [2]

iii) Explain why there is an increase in the energy stored by the capacitor when the polythene sheet is pulled out from between the plates. [2] (AQA)

▷ The CRO and a.c.

42. An alternating current (a.c.) source is connected to a resistor to form a complete circuit.
The trace obtained on an oscilloscope connected across the resistor is shown in the diagram.

The oscilloscope settings are: Y gain 5.0 V per division, time-base 2.0 ms per division.

a) Calculate the peak voltage of the a.c. source. [1]
b) Calculate the r.m.s. voltage. [1]
c) Calculate the time period of the a.c. signal. [1]
d) Calculate the frequency of the a.c. signal. [2] (AQA)

43. In each of the circuits below, the p.d. across the d.c. Y-input terminals of an oscilloscope is displayed on the oscilloscope screen. The Y-input voltage sensitivity is set at $1.0\,\mathrm{V\,cm^{-1}}$ and the time-base at $5.0\,\mathrm{ms\,cm^{-1}}$. The oscilloscope is adjusted so that an input of 0 V corresponds to a horizontal straight-line trace across the centre of the grid. In each case sketch the trace you would expect to see. [8] (AQA)

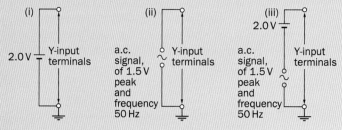

44. High-voltage power lines and distribution cables may consist of four conductors. Three of these conductors supply the power and the fourth conductor is a common return conductor called the neutral. The p.d.'s between each of the three supply conductors and Earth vary as shown in graphs 1, 2 and 3 in the diagram below. The peak voltage of each supply conductor relative to Earth is 325 kV.

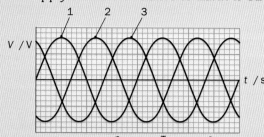

a) Estimate the maximum p.d. between any two supply conductors. Sketch the graph and show on it how you obtained this value. [2]
b) Show that the sum of the voltages in the three conductors at any instant is zero. [2] (Edex)

▷ Synoptic Questions on Electricity

45. The circuit in the diagram contains an ideal diode.

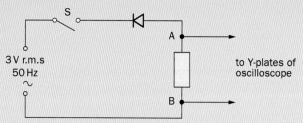

a) The power supply produces a sinusoidal alternating voltage with a peak value of 5.0 V and a frequency of 50 Hz.
For this supply calculate the period. [2]
b) i) The Y-plate sensitivity and the time-base of the oscilloscope are set to $2.0\,\mathrm{V\,cm^{-1}}$ and $5.0\,\mathrm{ms\,cm^{-1}}$, respectively. The switch is now closed. Sketch a graph showing the trace displayed on the oscilloscope screen.
ii) Draw an arrow to indicate the direction of the conventional current in the resistor when the diode is conducting. [4]
c) The trace you have sketched is rectified a.c. and not d.c. Draw on your graph the trace displayed on an oscilloscope screen when a 9.0 V d.c. supply is connected to the CRO. [4] (OCR)

46. Below is a diagram of a laboratory power supply which has an e.m.f. of 3200 V and internal resistance r. The capacitor C ensures a steady output voltage between X and Y when the switch S is in the 'on' position. The resistor R allows the capacitor to discharge when S is in the 'off' position.

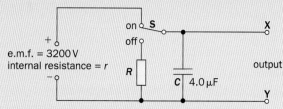

a) For safety reasons, the power supply is designed with an internal resistance r so that if X and Y are short-circuited the current does not exceed 2.0 mA. Calculate the value of r. [1]
b) For safety reasons, when the supply is switched off the capacitor must discharge from 3200 V to 50 V in 5 minutes.
i) The time constant of the circuit is 70 s. Show that the capacitor discharges from 3200 V to 50 V in less than 5.0 minutes. [3]
ii) Calculate the value of the resistor R. [2]
iii) Explain briefly whether or not increasing the value of R would make the supply safer. [1]
c) Calculate the charge on the capacitor when the p.d. across it is 3200 V. Sketch a graph to show the variation of charge on the capacitor with time as is discharges through the resistor. [3] (AQA)

23 Electrons and Photons

Just over a hundred years ago, everyone thought that Physics was nearly finished. They thought everything important had been discovered, and nothing was left except to tidy up some loose ends, and make some more accurate measurements.

How wrong that was! Within a few years, everyone's understanding of Physics had to be radically changed.

In this chapter you will learn:
- how metals can emit electrons,
- about photons – particles of light,
- how electrons can behave as waves.

Electrons and photons at work.

The discovery of the electron

In the 19th century, physicists tried to see if gases would conduct by connecting two metal electrodes in a sealed tube.
They *do* conduct – you can see this in neon lights – but very high voltages are needed.

When experiments were done with gases that were at low pressures, unusual things were observed. The **cathode** (the negative electrode) gave off strange invisible 'rays', and these were called '*cathode rays*'.

In 1897, J J Thomson measured the bending of these 'rays' in electric and magnetic fields (see page 322). This shows that they are not a form of radiation, but are streams of negatively charged particles, which he called **electrons**. Soon everyone accepted that all matter contained these tiny negative particles.

J J Thomson: his discovery won him the 1906 Nobel Prize.

Heating metals to drive off electrons

Electrons can be pulled out of metals by very high voltages, but an easier method is used in X-ray tubes, and in the cathode-ray oscilloscope (or CRO, see page 282).

If a metal **cathode** (a negative electrode) is *heated*, electrons leave the surface without needing a high voltage.
This is called **thermionic emission**.

This heating is done with a hot wire like an electric lamp filament, and sometimes the hot wire itself is used as the cathode.

If a positive electrode (an **anode**) is nearby, then the electric field between the cathode and the anode exerts a force on each free electron, as shown in the diagram:

The negative electrons are attracted to the positive anode. The force makes the electrons accelerate towards the anode, and they gain kinetic energy.

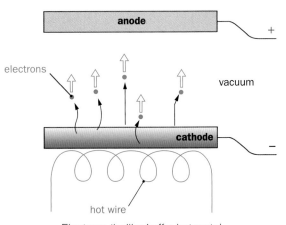

Electrons 'boiling' off a hot metal.

▶ Electron beams

Oscilloscopes, X-ray machines and some older televisions and computer monitors produce fast-moving beams of electrons. This Maltese Cross tube shows the principle of the **electron gun**:

The low-voltage supply heats the cathode, so that electrons can escape from its surface by thermionic emission.
The positive anode attracts the electrons. Some pass through the anode and shoot across the tube to the fluorescent screen.

How can we calculate the speed v of the electrons in the beam? As the electrons accelerate in the field between the anode and the cathode, they lose potential energy, but gain kinetic energy. For electrons of mass m and charge Q moving through a p.d. V:

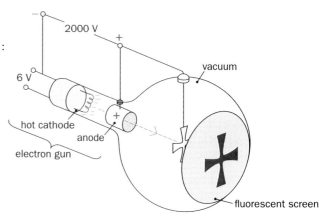

gain in kinetic energy = loss in electrical potential energy

$$\tfrac{1}{2}\, m\, v^2 \;=\; Q\, V$$

▶ Electron kinetic energy and the electron-volt (eV)

The kinetic energy E_k gained when a charge accelerates across a potential difference is given by:

Energy transferred, E_k	**=**	**charge, Q**	**×**	**p.d., V**
(joules)		(coulombs)		(volts)

or $\boxed{E_k = Q\, V}$

(See also page 230, but note that here we are using E, not W, as the symbol for energy.)

For electrons (and other subatomic particles), a smaller unit called the **electron-volt**, symbol **eV**, is often used:

Energy transferred, E_k	**=**	**charge**	**×**	**p.d., V**
(electron-volts)		(electrons)		(volts)

The charge on the electron is 1.6×10^{-19} C. It follows that: **$1\ eV = 1.6 \times 10^{-19}\ J$**

Example 1
An electron is accelerated by a p.d. of 2000 V. The electronic charge is 1.6×10^{-19} C. Calculate the gain in kinetic energy a) in joules, b) in electron-volts.

a) $E_k = Q\, V = 1.6 \times 10^{-19}\,\text{C} \times 2000\,\text{V} = \underline{3.2 \times 10^{-16}\,\text{J}}$ (2 s.f.)

b) $E_k = \text{charge} \times \text{p.d.}, V = 1\text{e} \times 2000\,\text{V} = \underline{2000\,\text{eV}}$

Deflecting electron beams

Look at the diagrams of an electron in electric and magnetic fields:

Electrons are negatively charged. If an electron is in an electric field, a force will act on it (page 292). The force on the electron is in the opposite direction to the electric field E, as the electron is negatively charged.

A beam of electrons – moving charges – is an electric current. If this beam is in a magnetic field, a force will act on it (page 264).

Because electrons are negatively charged, a beam of electrons moving in the direction of v in the diagram (left to right) is an electric current going from right to left.

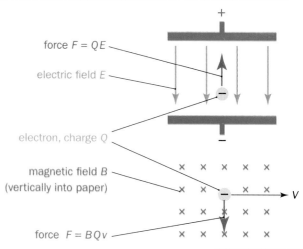

▶ Thomson's measurement of $\frac{Q}{m}$ for electron

In 1897 J J Thomson found a way to measure the ratio of charge Q to mass m for the electron.

The diagram shows an electron beam moving through a vacuum tube called an **electron deflection tube**.

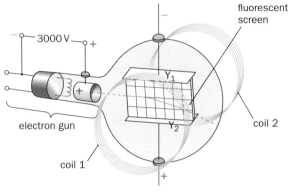

The beam can be deflected in 2 ways:
- by an *electric* field between the plates Y_1 and Y_2,
- by a *magnetic* field produced by the 2 coils carrying a current.

An electric field between Y_1 and Y_2 can push electrons *downwards* by a constant force $Q\,E$ (see page 292), as shown in the diagram:
The value of the electric field E is found by measuring the p.d. V between the plates Y_1 and Y_2 and the separation d between those plates, as we know that $E = \dfrac{V}{d}$ (see page 293).

A current in coils 1 and 2 provides a magnetic field B across the electron deflection tube at right angles to the electric field. The coils are connected in series so they have the same current. The magnetic field gives an **upwards** force $B\,Q\,v$ on the electron beam (see page 264).

The current through the coils is adjusted until the beam of electrons is a horizontal line. At this point the upward magnetic force balances the downward electric force. The current is noted.

The electron deflection tube is removed, and the magnetic field B midway between the two coils is measured using a sensor such as a Hall probe (see page 257) while the coils carry the same current as before.

force $F = BQv$

V

force $F = QE$

Balanced forces on the moving electron.

Combining the electric and magnetic forces

We don't know the electron velocity v, so we need to eliminate it from our calculations. The kinetic energy gained by the electron from the electron gun is $Q\,V_{gun}$ (see page 321).

so $\dfrac{1}{2}\,m\,v^2 = Q\,V_{gun}$ $\therefore v^2 = \dfrac{2QV_{gun}}{m}$

But $B\,Q\,v = Q\,E$ $\therefore v = \dfrac{E}{B}$ and so $v^2 = \dfrac{E^2}{B^2}$

$\therefore \dfrac{2QV_{gun}}{m} = \dfrac{E^2}{B^2}$ and dividing both sides by $2\,V_{gun}$

$$\boxed{\dfrac{Q}{m} = \dfrac{E^2}{2\,V_{gun}\,B^2}}$$

We know that $BQv = QE$, and we now know both B and E, but what is v?

Example 2
In this experiment, the electrons are accelerated by a p.d. of 2900 V. The same p.d. V_{gun} is applied across the plates Y_1 and Y_2, which are 10 cm apart. The magnetic field B needed to balance the forces on the electron beam is 9.1×10^{-4} tesla.

Calculate the value of $\dfrac{Q}{m}$ given by this experiment.

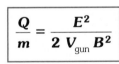

$E = \dfrac{V_{gun}}{d} = \dfrac{2900\text{ V}}{0.10\text{ m}} = 29\,000$ V m^{-1} (or 29 000 N C^{-1}). Note that you must use d in metres!

$\dfrac{Q}{m} = \dfrac{E^2}{2\,V_{gun}B^2} = \dfrac{(29\,000\text{ V m}^{-1})^2}{2 \times 2900\text{ V} \times (9.1 \times 10^{-4}\text{ T})^2} = \underline{1.8 \times 10^{11}\text{ C kg}^{-1}}$ (2 s.f.)

▶ Physics at work: Particle accelerators

Particle physicists can create new particles by colliding other particles together. To do this, the colliding particles need to be given large amounts of energy by an **accelerator.** Accelerators use electric fields to accelerate particles such as protons to energies much greater than the 2 000 eV of the electron gun on page 321.

Linear accelerator or 'linac'

This accelerates particles in a straight line:
The cylindrical electrodes are connected to an alternating supply so that they are alternately positive and negative. The frequency of the p.d. is set so that as the particles emerge from each electrode they are accelerated across the next gap.
Why do the electrodes get progressively longer? To keep in step with the alternating p.d., the particles must take the same time to travel through each electrode. As the particles get faster the tubes must get longer. The accelerator may be 3 km long!
A linear accelerator can accelerate electrons to about 50 GeV.

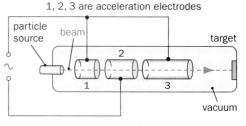

A linear accelerator.

Cyclotron

A cyclotron consists of two semicircular metal 'dees' separated by a small gap. When a charged particle enters the cyclotron, a perpendicular magnetic field makes it move along a circular path at a steady speed. Each time the particle reaches the gap between the dees, an alternating p.d. accelerates it across.

The force on a moving charged particle in a magnetic field is given by $F = B Q v$ (see page 264). Since this provides the centripetal force ($F = m v^2/r$, see page 80), we can write:

$$\frac{m v^2}{r} = BQv \qquad \therefore \quad v = \frac{BQr}{m}$$

This shows that the velocity is proportional to the radius, so as the particles get faster they spiral outwards. The time spent in each dee stays the same (see the coloured box):

So the alternating p.d. must reverse every $\dfrac{\pi m}{BQ}$ seconds.

Cyclotrons are used to accelerate heavy particles such as protons and α-particles. These can reach energies of about 25 MeV.

How long does a particle take to travel along its semicircular path within each dee?

$$\text{time} = \frac{\text{distance}}{\text{speed}} = \frac{\pi r}{BQr/m} = \frac{\pi m}{BQ}$$

This is independent of the speed and radius.
See also page 265.

Synchrotron

In modern accelerators, particles travel at speeds close to the speed of light. At these very high speeds, Einstein's theory of **special relativity** (see Chapter 28) is needed to explain the way in which particles move. As particles approach the speed of light, a constant force does not produce a constant acceleration – you cannot use $E_k = \frac{1}{2}m v^2$ in special relativity!

Synchrotrons contain electromagnets which keep the particles moving in a circle. Regions of electric field at various points around the loop give the particles extra energy, in 'pushes' synchronised with the arrival of pulses of particles.

As the particles gain energy, the magnetic field is increased to keep them moving in a circle of constant radius.

Synchrotrons can accelerate particles such as protons to energies of more than 1000 GeV.

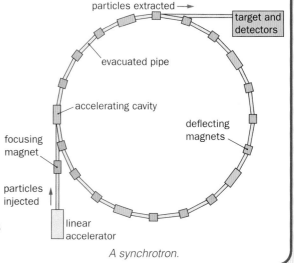

A synchrotron.

▶ Millikan's experiment to measure electron charge

In 1913, just 16 years after Thomson's experiment,
Robert Millikan measured the charge of an electron.
The diagram shows the arrangement he used:

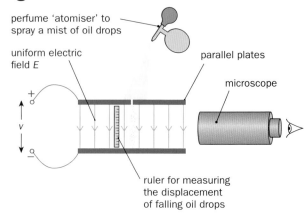

Tiny droplets of oil, squirted from a perfume 'atomiser', enter
a region of uniform electric field though a hole in the top plate.
The drops are so tiny that they quickly reach their terminal
velocities and fall at different steady speeds.

The act of spraying the oil through the atomiser nozzle gives the
drops electrical charge by friction (see page 289). So when
a p.d. V is applied to the plates, some drops are pulled upwards
and some downwards.

Millikan focussed on one drop, and then adjusted the voltage
to change the magnitude of the upwards force until the drop was
stationary between the two plates.

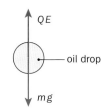

For this stationary drop, its weight was equal to the electric force,

$$m\,g = Q\,E = Q\,\frac{V}{d}$$

where d is the separation of the parallel plates (see page 293).

So, if he knew the mass m of the drop, Millikan could calculate Q.

Millikan knew that falling oil drops obey Stokes' Law (see page 23):
weight of the oil drop = viscous drag forces, at terminal velocity
and the tiny oil drops reach terminal velocity very quickly.

So $m\,g = 6\,\pi\,\eta\,r\,v$, where η is the viscosity of air, which is known.

v was found by timing the drop falling in front of a small ruler
when the electric field was switched off.

$m\,g = \rho \times V \times g = \rho \times \dfrac{4}{3}\pi\,r^3 \times g$ for the spherical oil drops,

so
$$\rho \times \frac{4}{3}\pi\,r^3 \times g = 6\,\pi\,\eta\,r\,v$$

You do not have to solve $\rho \times \frac{4}{3}\pi r^3 \times g = 6\pi\eta r v$!

Just remember that, if you know all of the values except r, it is possible to calculate r and then m = volume of drop $\times$ its density.

Because the density ρ of the oil was easily measured, this equation
allowed Millikan to calculate r for each drop, which then gave m,
so then he could calculate Q from the first equation above.

Millikan was also able to change the charge Q on an oil drop,
using ionising radiation. This enabled him to do several different
measurements on each oil drop.

Example 3
An oil drop of mass 2.6×10^{-17} kg was held stationary in Millikan's apparatus when the p.d. between the plates
was 19 V. The plates were separated by a distance $d = 1.2$ cm.
Calculate the charge on the oil drop. ($g = 9.8$ N kg^{-1})

$$m\,g = Q\,E = Q\,\frac{V}{d}$$

$(2.6 \times 10^{-17} \text{ kg}) \times 9.8 \text{ N kg}^{-1} = Q\dfrac{19 \text{ V}}{1.2 \text{ cm}} = Q\dfrac{19 \text{ V}}{0.012 \text{ m}} = Q \times 1580 \text{ V m}^{-1} \text{ (or N C}^{-1})$

$2.5 \times 10^{-16} \text{ N} = Q \times 1580 \text{ N C}^{-1} \qquad \therefore\; Q = \dfrac{2.5 \times 10^{-16} \text{ N}}{1580 \text{ N C}^{-1}} = \underline{1.6 \times 10^{-19} \text{ C}} \quad \text{(2 s.f.)}$

Deducing the value of the electronic charge

Suppose that you repeat Millikan's experiment five times and obtain the values of charge Q on the oil drops in this table:

Experiment number	1	2	3	4	5
charge $Q/10^{-19}$ C	−3.2	−6.4	−1.6	−3.2	−4.8

What do you notice about the values of Q?

Can you see that they are all multiples of the same value?

All of Millikan's oil-drop measurements gave charges which were multiples of -1.6×10^{-19}C, and never gave a smaller value. This smallest possible charge is the charge of a **single** electron. Because this charge is so important – it is also the size of the charge of a proton – it is given a special symbol: e.

there are only five charges in this table. Millikan had thousands to look at!

What is the mass of an electron, m_e?

Thomson's experiment showed that $\dfrac{e}{m_e} = 1.75 \times 10^{11}$ C kg^{-1}.

Millikan's value for e allows us to find the mass of an electron.

$$\frac{e}{m_e} = \frac{1.6 \times 10^{-19} \text{ C}}{m_e} = 1.75 \times 10^{11} \text{ C kg}^{-1}$$

$$\therefore m_e = \frac{1.6 \times 10^{-19} \text{ C}}{1.75 \times 10^{11} \text{ C kg}^{-1}} = 9.1 \times 10^{-31} \text{ kg}$$

Comparing the electron with the proton

Thomson found $\dfrac{e}{m_e}$ for the electron, but $\dfrac{e}{m_p}$ for the proton can be easily found using the experiment shown in this diagram:

The hydrogen ions H$^+$ and oxygen ions O^{2-} move through the dilute acid, completing the circuit. At the cathode, each H$^+$ ion receives an electron, and hydrogen gas is given off.

The experiment is allowed to run for a measured time t. By measuring the volume of hydrogen gas, the mass of hydrogen produced in the time t can be calculated, as its density is known.

The current flowing is measured, giving the charge $Q = I\,t$.

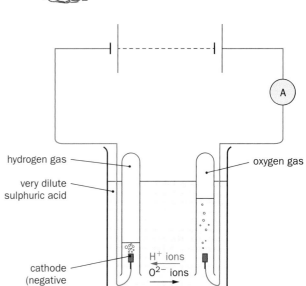

hydrogen gas

oxygen gas

very dilute sulphuric acid

H$^+$ ions

O^{2-} ions

cathode (negative electrode)

Example 4
In the experiment, hydrogen ions carry a current of 2.5 A through dilute acid for 1 hour 4 minutes (3840 seconds). At the negative electrode, hydrogen gas was given off, and 0.10 g of hydrogen was collected. Electrons are so small you can assume that a hydrogen ion H$^+$ and a hydrogen atom H have the same mass.

a) Calculate $\dfrac{e}{m_p}$ for hydrogen ions,   b) find the mass of the hydrogen ion (a proton).

a)   Charge flowing $= I\,t = 2.5$ A $\times$ 3840 s $= 9600$ C

$$m = 0.10 \text{ g} = 1.0 \times 10^{-4} \text{ kg} \quad \therefore \frac{Q}{m} = \frac{\text{total charge flowing}}{\text{total mass discharged}} = \frac{9600 \text{ C}}{1.0 \times 10^{-4} \text{ kg}} = \underline{9.6 \times 10^7 \text{ C kg}^{-1}} \quad \text{(2 s.f.)}$$

b)   The charge Q of the proton is the same size as that of the electron e: 1.6×10^{-19} C

$$\frac{e}{m_p} = \frac{1.6 \times 10^{-19} \text{ C}}{m_p} = 9.6 \times 10^7 \text{ C kg}^{-1} \quad \therefore m_p = \frac{1.6 \times 10^{-19} \text{ C}}{9.6 \times 10^7 \text{ C kg}^{-1}} = \underline{1.7 \times 10^{-27} \text{ kg}} \quad \text{(2 s.f.)}$$

Using the electron mass m_e found above, $\quad \dfrac{m_p}{m_e} = \dfrac{1.7 \times 10^{-27} \text{ kg}}{9.1 \times 10^{-31} \text{ kg}} = 1900 \quad$ (2 s.f.)

The proton is nearly 2000 times more massive than the electron!

▶ Using light to free electrons

One way to drive out electrons is by heating (see page 320).
This diagram shows another way to emit free electrons:

A negatively charged, clean zinc plate will keep its charge in dry air for a long time.
However, if you shine ultraviolet light on it, it loses its charge very rapidly. The ultraviolet light gives the electrons the energy that they need to escape the metal.
The coulombmeter shows the negative charge decreasing.

This effect can also be demonstrated with a negatively charged zinc plate on a gold-leaf electroscope instead of a coulombmeter. The deflection of the gold leaf shows that the electroscope is charged. When the ultraviolet light falls on the zinc plate, the gold leaf falls, showing that it has lost charge.

If this experiment is repeated – using either a coulombmeter or a gold-leaf electroscope – with a sheet of glass between the ultraviolet lamp and the zinc plate, the negative charge does not decrease as fast. This is because the glass absorbs ultraviolet light.

This is the **photoelectric effect**.
Electrons given off in this way are often called *photoelectrons*.

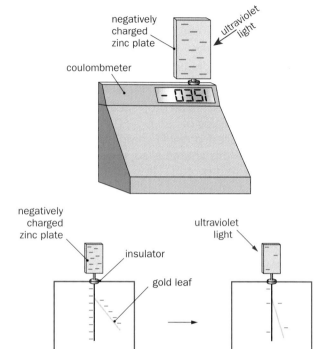

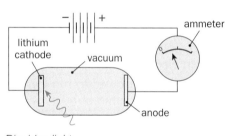

The gold leaf falls, showing the electroscope is discharging.

▶ The photoelectric effect

When you heat the metal cathode in an electron tube, like the one on the previous page, it gives off electrons. If you heat the metal more, it liberates more electrons, because the increased internal energy excites the free electrons more.

So what happens when you shine light on a metal?

In Chapter 10 on Waves you learnt that waves with bigger amplitude carry more energy. For light, bigger amplitude means brighter light, so a brighter light will always give off more photoelectrons.
Is this right? No, surprisingly, it is *wrong!*

Look at these diagrams, which show what happens when light shines on a clean surface of the metal lithium:

- A dim blue light will make the lithium give off a few electrons, so there will be a small, measurable current.

- A **brighter blue** light will give a bigger photoelectric current, because there are more photoelectrons.

    This seems exactly what you would expect . . . *BUT*

- A **red** light, just as bright as the bright blue light, gives off *no* electrons at all, and the current is zero!

This is impossible to explain – *if* light is a wave.

If light was a wave, then a stronger red light would be a wave of greater amplitude, and so have enough energy to liberate electrons. But this does not happen! Even a red laser is unable to free electrons from lithium.

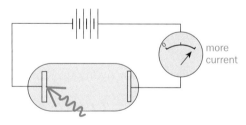

Dim blue light

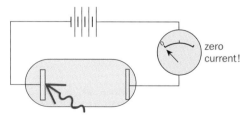

Bright blue light

Bright red light

Light: particles and not waves?

The German physicist Max Planck suggested that electromagnetic *energy comes only in 'lumps'*, called *quanta*.
Although he suggested this to solve a quite different problem (in thermodynamics), Albert Einstein used the idea to explain this photoelectric effect.

Einstein suggested that *each* lump, or *quantum*, of light energy can provide the energy for *one* electron to escape the metal.
If the quantum is too small, the electron cannot escape from the metal surface.

If light is made of tiny particles like this, then it makes sense to give them a particle name. Modern particle names usually end in '–on', so a light particle is called a **photon**.

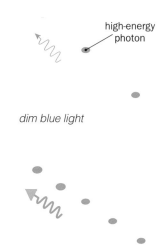

dim blue light

Now we can explain the 3 results listed on the opposite page:

- The photons of blue light contain enough energy to eject electrons from lithium. A dim blue light has few photons, so few electrons are liberated. This gives a small photoelectric current.

- A bright blue light has more of these high-energy photons, so more electrons are liberated per second, giving a larger photoelectric current.

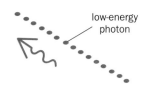

bright blue light (more photons)

- But red light consists of photons that do *not* have enough energy to emit electrons from lithium.
  Even though a bright red light has very many of these photons, not one of them has enough energy to eject an electron.

bright red light
(photons have less energy)

▶ Physics at work: Corpuscles and waves

When Einstein showed that light can behave like a particle, he was following in the steps of his most famous predecessor: Isaac Newton. Newton had puzzled long over the nature of light, and decided that a beam of light was a stream of particles, which he called **'corpuscles'**. This allowed him to explain why light *travels in straight lines* and how it *reflects* off a mirror. He thought he could explain how light is *refracted* when it goes into glass: he suggested that the light was pulled inwards by forces acting at the surface of the glass, making the stream of particles bend towards the normal.

Isaac Newton investigating light.

Christiaan Huygens.

The Dutch physicist Christiaan Huygens had a different suggestion: he thought light was a **wave**, and used wave ideas to explain how light is emitted, travels in straight lines and is reflected (see page 138). Waves, like corpuscles, can explain refraction, but in Huygens' explanation light moves *more slowly* in glass than in air: Newton's model had it travelling *faster*. At the time this was impossible to check, but modern measurements confirm Huygens was right.

Newton was such a great physicist that it was assumed that he would be right in all matters of science. Furthermore, England was at war with Holland at the time, which was another reason not to believe Huygens, a foreign scientist!

Not until Thomas Young's famous experiment (see page 173) was the argument settled – in favour of waves. It is impossible to explain **interference** in terms of 'corpuscles'. And that settled the question of what light must be – until the 20th century, and Albert Einstein!

▶ Planck's equation and photon energy

In his work on thermodynamics, Planck produced this equation:

Energy of a photon, E	**=**	**the Planck constant, h**	**×**	**frequency, f**
(joules)		(joules-second)		(hertz)

or

$$E = hf$$

The Planck constant $= 6.6 \times 10^{-34}$ J s.

This is an astonishing equation!
On the left-hand side, you have the energy of a *particle*.
On the right-hand side, you have the frequency of a *wave*.
For light, you need to use *both* particle and wave ideas.

Example 5
Blue light has a frequency of 7.7×10^{14} Hz, while red light has a frequency of 4.3×10^{14} Hz.
Calculate the energy of a photon of each. The Planck constant $h = 6.6 \times 10^{-34}$ J s.

Blue light: $E = hf = 6.6 \times 10^{-34}$ J s $\times 7.7 \times 10^{14}$ Hz $= \underline{5.1 \times 10^{-19}$ J}$ (2 s.f.)

Red light: $E = hf = 6.6 \times 10^{-34}$ J s $\times 4.3 \times 10^{14}$ Hz $= \underline{2.8 \times 10^{-19}$ J}$ (2 s.f.)

These figures agree with the observations on the opposite page.
The energy of a photon of blue light is much more than
the energy of a photon of red light.

The energies in Example 5 seem incredibly small.

It is often clearer to write energies in electron-volts,
and Example 1 on page 321 shows that there is a
'scaling factor' of 1.6×10^{-19} J per eV:

Lower-energy photons   *Higher-energy photons*

Energy in joules = energy in eV $\times 1.6 \times 10^{-19}$ J eV^{-1}

Example 6
Convert the energy of the blue photon in Example 5 from J into eV.

$$\text{Energy of photon} = 5.1 \times 10^{-19} \text{ J} = \frac{5.1 \times 10^{-19} \text{ J}}{1.6 \times 10^{-19} \text{ J eV}^{-1}} = \underline{3.2 \text{ eV}} \quad (2 \text{ s.f.})$$

2 eV

The energy of a typical visible photon.

Photons and the electromagnetic spectrum

Photons of visible light have energies between 1.7 eV and 3.2 eV.
But what about the invisible parts of the electromagnetic spectrum?

Infrared radiation has a smaller frequency than red light, so its
photons have *less* energy. Microwaves and radio waves have
photons with even less energy.

Going along the electromagnetic spectrum in the other direction,
ultraviolet radiation has higher frequency than blue light, so its
photons are *more* energetic. That explains why UV light is
hazardous.

X-rays and gamma-rays have even higher frequency, and so their
photons have even more energy (and so are even more dangerous).

Ultraviolet photons are dangerous.

▶ Escaping from a metal: The work function

We've seen that all photons carry energy, but red light cannot liberate electrons from lithium metal, even though blue light can.

Look at the graph shown here:

It shows the **energy** of the photo-electrons that are emitted from lithium metal for different **frequencies** of light.
Light of frequency less than 5.6×10^{14} Hz does not give any photoelectrons. This is called the cut-off or **threshold frequency**, f_0.

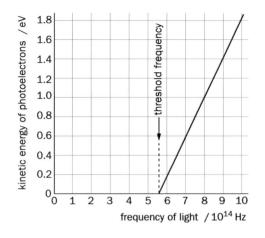

The graph goes up as a straight line – as the frequency of light goes up, the energy of the emitted electrons goes up as well.
The gradient of this line is the Planck constant, **h**.
This is true for any metal, not just lithium.

Look at the graph above:

Can you see that, at the threshold frequency f_0, any photoelectron that is emitted will have zero kinetic energy?

This means that **all** of the energy of the photon goes to removing the electron from the metal. The amount of energy needed to remove the electron is called the **work function**.

The work function is given the symbol ϕ (the Greek letter phi).

$$\boxed{\text{work function, } \phi = h \, f_0}$$

▶ Einstein's photoelectric equation

When a photoelectron has absorbed a photon and escaped from the metal, it has only kinetic energy.
Conservation of energy shows us that:

  kinetic energy of the electron = energy of photon − energy needed to escape from the metal

The minimum energy needed to escape from the metal is the work function ϕ. This gives us **Einstein's equation**:

Maximum kinetic energy of electron, E_k = the Planck constant, h × frequency, f − work function, ϕ

So: $\boxed{E_k = h \, f - \phi}$ or $\boxed{\tfrac{1}{2} m \, v^2 = h \, f - \phi}$

You can use the last form of the equation if you need to calculate the speed of the electrons.

Example 7
Light of frequency 6.70×10^{14} Hz shines on to clean caesium metal.
What is the maximum kinetic energy of the electrons emitted?

The work function of caesium is 3.43×10^{-19} J and
the Planck constant $h = 6.63 \times 10^{-34}$ J s.

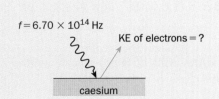

$$E_k = h \, f - \phi$$
$$= 6.63 \times 10^{-34} \text{ J s} \times 6.70 \times 10^{14} \text{ Hz} - 3.43 \times 10^{-19} \text{ J}$$
$$= 4.44 \times 10^{-19} \text{ J} - 3.43 \times 10^{-19} \text{ J}$$
$$= \underline{1.01 \times 10^{-19} \text{ J}} \quad \text{(3 s.f.)}$$

▶ Measuring the Planck constant

Light-emitting diodes (LEDs) emit light when a current flows.
Using LEDs, we can find the Planck constant **h**.

Look at the circuit diagram:

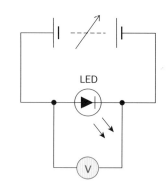

A light-emitting diode is connected to a variable power supply.
The LEDs used in this experiment each give out light of a **single**
wavelength, so that all photons have the same energy. As the p.d.
is increased from 0, each electron passing though the LED gives
more energy to the LED. The LED will not emit a photon until
the electron has enough energy to release a photon.

In the experiment, the p.d. across the LED is increased until the
LED **just** emits light. The smallest possible voltage which can
emit photons is best judged by 'squinting' down a black paper
tube with the LED at its bottom while the p.d. is slowly increased
until a glimmer of light is seen. This 'turn-on' voltage allows us
to calculate the photon energy **E**, in electron-volts (eV).

This is repeated with LEDs of different colours.

Clear LEDs must be used for this experiment as 'cloudy' ones
have diffusing screens which absorb light and make it difficult to
find the smallest p.d. which emits light, thus making the voltage
and energy readings too high.

The frequency of each LED must be found. This can be done
by measuring each wavelength using a diffraction grating
(see page 177) and using **c = f λ**.

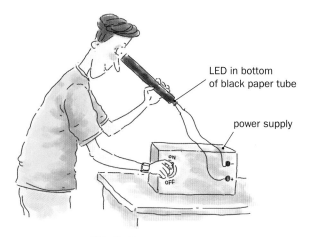

Phiz finding the 'turn-on' p.d.

Data obtained in this way gives the graph shown:

Since **E = h f**, the graph is a straight line through the origin.
The Planck constant **h** must be the gradient of the graph.

It is possible to convert all the p.d.s into energies in joules
using **E = e × V** (see page 321) before plotting the graph,
but it is easier to use energies in eV and then to convert
the final answer into the correct unit, as $1 \text{ eV} = 1.6 \times 10^{-19} \text{ J}$.

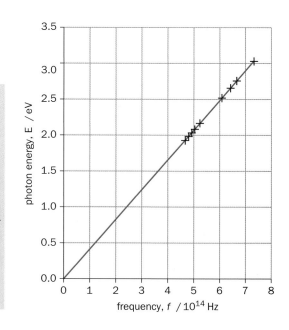

> *Example 8*
> Use the data in the graph to calculate the Planck constant **h**.
> Accurate plotting shows that the straight line passes through
> $f = 6.0 \times 10^{14}$ Hz and **E** = 2.47 eV, and the origin.
>
> the gradient $\mathbf{h} = \dfrac{2.47 \text{ eV}}{6.0 \times 10^{14} \text{ Hz}} = 4.12 \times 10^{-15} \text{ eV Hz}^{-1}$.
>
> $1 \text{ eV} = 1.6 \times 10^{-19} \text{ J}$
> so $\mathbf{h} = (1.6 \times 10^{-19}) \times (4.12 \times 10^{-15}) \text{ J Hz}^{-1} = 6.6 \times 10^{-34} \text{ J Hz}^{-1}$
>
> $\text{Hz} = \text{'per second'} = \left(\dfrac{1}{s}\right)$, so $\text{Hz}^{-1} = \left(\dfrac{1}{s}\right)^{-1} = s$,
>
> and the units of **h** are usually written as J s rather than J Hz^{-1}.
> This gives $\underline{\mathbf{h} = 6.6 \times 10^{-34} \text{ J s}}$ (2 s.f.)

The Planck constant is the smallest of the fundamental constants:
imagine trying to write it out without using standard form!

▶ Physics at work: Photons and electrons

Photocells and energy production

The photoelectric effect was first studied with reactive metals such as caesium. Nowadays there is far more interest in semi-conductors, because some of these can also convert photon energy into electrical energy.
That is, they convert light energy directly into electrical energy. These **photovoltaic** photocells, made from silicon, have been used for over 40 years.

Early uses were in places where other electric sources were inconvenient. The orbiting satellite Vanguard 1 was fitted with photovoltaic solar panels as long ago as 1958.
Devices with low current demands, such as calculators, often use photovoltaic cells.

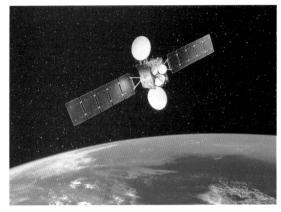
Photovoltaic cells are used to power satellites.

Solar power station in California.

By the end of 2013, there were 1600 photovoltaic power stations ('solar farms') around the world with a total generating capacity of 22 GW. Many of these are small local power stations with capacities of a few MW, but their output is increasing: at the time of writing, power stations of capacity over 500 MW were being constructed.

As photovoltaic panels continue to come down in price, and governments try to reduce the production of greenhouse gases by burning less fossil fuel, more and more solar farms will be built. Although solar farms cannot replace all other types of power stations, every MW h produced by photovoltaic panels is one less produced by burning fossil fuels. This reduces the carbon dioxide released into the atmosphere.

Photovoltaic units are not just found in large solar farms: helped by government grants, many householders are putting them on their roofs to help reduce energy bills. Panels cannot be put on north-facing roofs in the UK, as the Sun is to the south, but roofs facing east or west can be used. At 250 W per panel, most UK houses could install several panels on the roof and so reduce their energy demands.

Even though the UK is often cloudy, and dark at night, a 4-panel system rated at 1 kW would still produce 850 kW h over a year.

And on days when your solar panels are harvesting solar energy which you are not using, you can sell it to the National Grid!

Solar panels on a house in the UK.

Example 9
A house in London was fitted with 8 solar panels on a south-facing roof at a cost of £ 4000.
Assuming that one kW h costs 13 p, calculate the time taken for the saving in electricity bills to cover the cost.

Using the information in Physics at work above, 8 panels will produce 2×850 kW h = 1700 kW h per year on average.
After N years, savings = $N \times 1700$ kW h $\times$ 13p kW h^{-1} = $N \times$ 22100 p = $N \times$ £ 221
When the system 'breaks even', $N \times$ £ 221 = £ 4000

so the break-even time = N years = $\dfrac{£ 4000}{£ 221}$ years = 18 years (2 s.f.)

▶ Electrons as waves

Electron beams produce a very strange effect if they strike thin layers of graphite carbon, as this diagram shows:

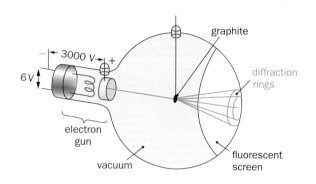

Most of the electrons pass straight through the very thin graphite, but others pass through at certain angles only, giving rings on the fluorescent screen.

These rings are like the interference maxima you get when light waves pass through a diffraction grating (see page 176).

But where is the diffraction grating in this case?
And what are the waves that are diffracting?

De Broglie's idea

In 1924, nearly 20 years after Einstein explained the photoelectric effect, a young French student named Louis de Broglie (pronounced *Broy*) made a bold suggestion.

We thought that light was a wave, he said. And now we see it can be a particle as well. Is it possible that electrons, which we thought were particles, can also be waves?

In Chapter 12, you saw the two key facts that convinced physicists that light was a wave:

- light can be diffracted by a gap, and
- light can superpose to give interference patterns.

The electron-tube experiment above shows exactly that! Electrons are diffracted by the gaps between atoms, and give maxima on the fluorescent screen where they are in phase.

Waves or particles? Wavicles?

De Broglie's theory suggested that the wavelength of electron waves was very small, about the size of an atom. So the separation of the slits in a diffraction grating for electrons would have to be small as well – about the size of an atom.

Look at the diagram of electrons striking atoms in graphite:

The atoms are in regular columns, shown by the green lines. The gaps between these columns of atoms are regularly spaced, just like a diffraction grating. There are interference maxima on each side of the central maximum, just as in an ordinary diffraction grating exposed to light (see page 176).

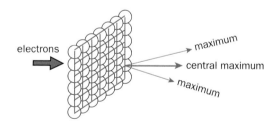

Graphite atoms diffracting electrons.

The atoms are also in regular rows, shown by the red lines. These give interference maxima above and below the central maximum, but they are not shown on the diagram.

In a real sample of graphite, there are many layers of atoms:

These are not parallel but jumbled, rather like a pile of loose papers on a desk.
The diffraction maxima from these different layers are all at the same angle to the central maximum, but in all possible directions. These add together to give the rings seen on the fluorescent screen.

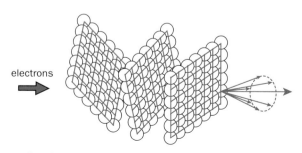

Graphite has different layers arranged at random.

▶ De Broglie's equation

Planck's equation for the energy of a photon is $E = hf$.

De Broglie suggested a similar equation for the electron.
In his equation, the left-hand side is not energy, but another
important measure of a moving particle – its momentum.

(Momentum is mass multiplied by velocity, mv; see page 54.)

In a similar way, the right-hand side has the other important
measure of a wave – its wavelength.
De Broglie's equation is:

Momentum = mass × velocity.

$$\text{Momentum of the particle, } mv \atop (\text{kg m s}^{-1}) = \frac{\text{the Planck constant, } h \text{ (J s)}}{\text{wavelength of the wave, } \lambda \text{ (m)}} \quad \text{or} \quad mv = \frac{h}{\lambda} \quad \text{so} \quad \lambda = \frac{h}{mv}$$

Remember:

- For photons, always use Planck's equation: $E = hf$
- For the wave behaviour of subatomic 'particles'
  like electrons, always use de Broglie's equation: $\lambda = \dfrac{h}{mv}$

Example 10
Electrons are travelling at a speed 2.0×10^6 m s^{-1}.
Calculate the de Broglie wavelength of the electrons.

The mass of the electron is 9.1×10^{-31} kg.
The Planck constant $h = 6.6 \times 10^{-34}$ J s.

Momentum = mass × velocity $= 9.1 \times 10^{-31}$ kg $\times 2.0 \times 10^6$ m s$^{-1} = 1.8 \times 10^{-24}$ kg m s^{-1}

$$\lambda = \frac{h}{mv} = \frac{6.6 \times 10^{-34} \text{ J s}}{1.8 \times 10^{-24} \text{ kg m s}^{-1}} = \underline{3.6 \times 10^{-10} \text{ m}} \quad (2 \text{ s.f.})$$

Electron wavelength λ and the electron gun voltage V

As de Broglie's equation above shows, the wavelength is inversely

proportional to the momentum: $\lambda \propto \dfrac{1}{mv}$ This means that the

way in which kinetic energy, momentum and wavelength change
with p.d. V is: larger V $\Rightarrow$ larger E_k $\Rightarrow$ larger mv $\Rightarrow$ smaller λ.

But how can we calculate the wavelength produced by a given p.d.?

From page 322, $eV = \dfrac{1}{2} mv^2$ $\therefore v^2 = \dfrac{2eV}{m}$ Multiplying by m^2, $m^2v^2 = m^2 \dfrac{2eV}{m} = 2meV$

From above, $\lambda = \dfrac{h}{mv} = \dfrac{h}{\sqrt{2meV}}$ $\boxed{\lambda = \dfrac{h}{\sqrt{2\,m\,eV}}}$

Example 11
Calculate the wavelength of electrons accelerated through 3000 V.
(electron mass $m = 9.1 \times 10^{-31}$ kg, electronic charge $e = 1.6 \times 10^{-19}$ C and the Planck constant $h = 6.6 \times 10^{-34}$ J s)

$$\lambda = \frac{h}{\sqrt{2\,m\,eV}} = \frac{6.6 \times 10^{-34} \text{ J s}}{\sqrt{2 \times 9.1 \times 10^{-31} \text{ kg} \times 1.6 \times 10^{-19} \text{ C} \times 3000 \text{ V}}} = \frac{6.6 \times 10^{-34} \text{ J s}}{\sqrt{8.74 \times 10^{-46} \text{ kg C V}}} = \frac{6.6 \times 10^{-34} \text{ J s}}{2.96 \times 10^{-23} \text{ N s}}$$

$= \underline{2.2 \times 10^{-11} \text{ m}} \quad (2 \text{ s.f.})$

▷ Electron waves and the atom

Are electrons waves? Electron diffraction shows this is true, but there is other evidence as well.

The model of the atom that you studied at GCSE was first suggested by Ernest Rutherford (see page 355).
It has electrons orbiting a central positive nucleus, like the planets around the Sun.

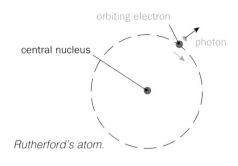

Rutherford's atom.

It was quickly realised that this model is physically impossible.
If the electrons, which are negatively charged, moved in circles, electromagnetic theory shows they would radiate energy.
But this would make them slow down and spiral into the nucleus.
Luckily, this does not happen, or matter could not exist!

But what keeps electrons in their orbits?

The answer comes from the fact that electrons behave as waves. Electrons are trapped within the atom in the same way as stationary waves are trapped on a rope (see page 162). The diagram shows a very simplified view of an electron trapped in this way:

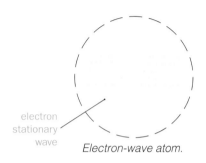

Electron-wave atom.

Theory based on stationary-wave ideas can predict the spectra of atoms accurately.

Electrons and the nucleus

On page 354, you will see that Rutherford investigated how alpha-particles are scattered from atomic nuclei. Later experiments to probe the nucleus used very high-energy beams of electrons. De Broglie's equation (page 333) shows that high-energy electrons, which have a high momentum $m\,v$, have very small wavelengths.

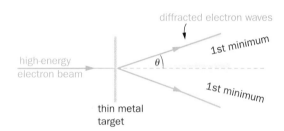

If the electron wavelength is similar to the diameter of the nucleus, the electrons will be diffracted in the same way that light is diffracted by a single slit (see page 168).
Measuring the angle of the first minimum in the diffraction pattern allows calculation of the size of the nucleus.

To a good approximation, the single-slit diffraction equation

$\sin\theta = \dfrac{\lambda}{b}$, where b is the diameter of the nucleus, can be

used to calculate the diameter of the nucleus.

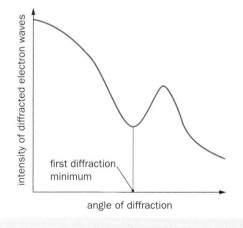

Example 12
A beam of high-speed electrons, each of momentum $m\,v = 5.3 \times 10^{-19}$ kg m s^{-1}, strikes a thin metal target.
They are diffracted as shown in the graph, with a first diffraction maximum of 42°.
Calculate the size of the nucleus of the metal atoms in the target. The Planck constant $h = 6.6 \times 10^{-34}$ J s.

$$\lambda = \frac{h}{m\,v} = \frac{6.6 \times 10^{-34}\ \text{J s}}{5.3 \times 10^{-19}\ \text{kg m s}^{-1}} = 1.24 \times 10^{-15}\ \text{m}$$

$$\sin\theta = \frac{\lambda}{b} \quad \therefore\ b = \frac{\lambda}{\sin\theta} = \frac{1.24 \times 10^{-15}\ \text{m}}{\sin 42°} = \frac{1.24 \times 10^{-15}\ \text{m}}{0.669} = \underline{1.9 \times 10^{-15}\ \text{m}}\ \ (2\ \text{s.f.})$$

A more accurate value can be obtained by using the equation for the diffraction of waves by a round hole, instead of a slit:
this gives an answer which is 1.22 times greater (2.3×10^{-15} m), but it is of the same order of magnitude.

▶ Physics at work: Modern microscopes

Electron microscopes

A problem with ordinary optical microscopes occurs at high magnifications. When the object you are looking at is about the same size as the wavelength of light – for example, a chromosome in a cell nucleus – then diffraction occurs and confuses the image. A solution to this is to use a beam of electrons, which can have much smaller wavelengths if they are accelerated through a high enough voltage.

Optical microscopes use lenses to produce a magnified image of the specimen, rather like the telescope on page 149. There are no lenses for electron beams, but remember that electrons are particles as well as waves, and they are negatively charged. By passing the electron beam through the centre of an electromagnet coil, the magnetic force on the moving charges 'bends' the beam like a lens.

Look at the three electromagnet coils in this diagram of a **transmission electron microscope** (TEM):

In each coil, there is a small gap in the soft-iron casing in the centre. This allows the magnetic field lines to 'leak out' so that they interact with the electron beam and focus it on the fluorescent screen.

One disadvantage of the TEM is that the specimen being examined must be very thin indeed, but it does give excellent resolution.

A different design, the **scanning electron microscope** (SEM), has poorer resolution than a TEM, but it does not need such thin specimens. SEM is usually used for examining surfaces.

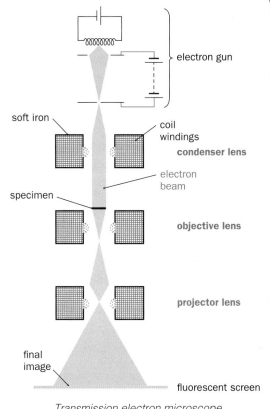

Transmission electron microscope.

Scanning tunnelling microscope

Electron microscopes use one quantum property of electrons: they are waves so they can produce images, just like light. **Scanning tunnelling microscopes** (STMs) use a different quantum property, called **tunnelling**.

Electrons in the STM get across a tiny gap by 'borrowing' energy to perform the process, then 'paying it back' afterwards. It is like a rolling ball which does not have enough kinetic energy to roll over a hill but mysteriously makes a tunnel through the hill:

The diagram shows the probe in an STM:

Piezoelectric transducers move the probe along the surface and can change the height of the probe. These transducers can control movement to a precision of 1 picometre (1 pm $= 10^{-12}$ m). The electron tunnelling current varies rapidly with the height of the probe above the atom beneath, and the probe can be moved vertically to keep it exactly the same height above the atoms below.

As the probe moves along, its vertical movements give a record of the shape of that part of the surface, to the nearest 1 pm. After passing across the surface, another transducer (acting perpendicular to the diagram) moves the probe over slightly, so that it can scan the next line of atoms. Moving in this way, the STM can build up an image of the entire surface.

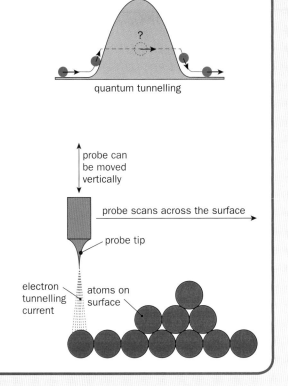

Summary

Electrons can be driven out of metals by heating (thermionic emission) or by shining light upon them (photoelectric effect). Electron energies are often measured in electron-volts (eV), where $1\,eV = 1.6 \times 10^{-19}\,J$.
The photoelectric effect shows that light can be thought of as particles, called photons.

Planck's equation: The energy of a photon, $E = h\,f$. A packet of energy is called a quantum (plural: quanta).

To remove electrons from a metal requires a minimum amount of energy called the work function, ϕ.
Einstein's photoelectric equation: Maximum kinetic energy of a photoelectron, $E_k = h\,f - \phi$.
The work function $\phi = h\,f_0$, where f_0 is the threshold frequency.

Beams of electrons can be diffracted and interfere, showing that electrons can act like waves.

De Broglie's equation for 'particle' waves: momentum $m\,v = \dfrac{h}{\lambda}$,   or $\lambda = \dfrac{h}{m\,v}$

▶ Questions

1. Can you complete these sentences?
   a) Negative particles called .... were first discovered by J J Thomson.
   These can be driven out of a metal by heating, (called .... emission) or by shining .... on it (called the .... effect).
   b) A convenient small unit of energy for use in atomic calculations is the ....
   The conversion factor between energy in joules and energy in electron-volts is: $1\ eV = $ .... J.
   c) When light of high enough frequency shines on a metal, photoelectrons are ....
   The lowest frequency is called the .... frequency. Blue light has a .... frequency than red light.
   d) The energy arrives in quanta or 'lumps' whose size depends on the .... of the light.
   A particle of light is called a ....
   To drive out a photoelectron, the photon must have more energy than the .... function of the metal.
   e) Light can behave as a .... as well as a wave. Electrons are normally thought of as particles, but they can also behave as ....

2. Calculate the energy (i) in joules and (ii) in electron-volts when an electron is accelerated through a p.d. of
   a) 10 V,
   b) 10 kV.
   (Charge on electron, $e = 1.6 \times 10^{-19}$ C)

3. Calculate the energy (i) in joules and (ii) in electron-volts of a photon of:
   a) radio waves of frequency 100 MHz,
   b) microwaves of wavelength 10 cm,
   c) infrared of wavelength 1 mm,
   d) ultraviolet light of wavelength 100 nm,
   e) X-rays of wavelength 10 nm.
   (Speed of light, $c = 3.0 \times 10^8$ m s^{-1})
   (The Planck constant, $h = 6.6 \times 10^{-34}$ J s)

4. The table gives the kinetic energy of photoelectrons emitted from a metal surface for different frequencies of light.

Electron energy /10^{-19} J	Frequency /10^{14} Hz
0.22	5.00
0.88	6.00
1.54	7.00
2.20	8.00
2.87	9.00

   a) Plot a graph of electron energy (vertically) against frequency (horizontally)
   b) Use the graph to find the work function of the metal.
   c) Find the gradient of the graph. Can you recognise what this is?

5. A photocell cathode is made of a metal with a work function of 2.8 eV.
   a) Light of wavelength 400 nm shines on the cathode. Will photoelectrons be emitted?
   b) The cathode is now illuminated with ultraviolet radiation, and electrons are emitted with energy 1.4 eV. Calculate the wavelength of the ultraviolet radiation.
   (Speed of light, $c = 3.0 \times 10^8$ m s^{-1})
   (Charge of electron, $e = 1.6 \times 10^{-19}$ C)
   (The Planck constant, $h = 6.6 \times 10^{-34}$ J s)

6. Electrons are accelerated to a kinetic energy of 500 eV. Calculate their
   a) velocity,
   b) momentum,
   c) de Broglie wavelength.
   (Mass of electron, $m = 9.1 \times 10^{-31}$ kg)
   (Values of e and h as in Question 5)

Further questions on page 352.

24 Spectra and Energy Levels

From the time that our ancestors first gazed at rainbows, spectra have been a source of both wonder and new understanding.

From gamma rays to radio waves, spectra have given us new understanding, ranging from the structure of atoms to the nature of distant quasars.

In this chapter you will learn:
- about the electromagnetic spectrum,
- about absorption and emission spectra,
- how spectra depend upon energy jumps.

Natural spectral analysis.

The electromagnetic spectrum

The diagram shows the entire electromagnetic spectrum:

What do you notice about the numbers on the scales?
The scales are both **logarithmic** – they do not go up in equal steps, but in equal **ratios**. Every number is ten times bigger than the adjacent smaller one.

Why is it not possible to have a regular, linear scale?
The smallest division here is about 10^{-12} m. If you try to draw this diagram using a linear scale with just one millimetre of scale representing 10^{-12} m, the diagram would have to be very big to show the longest radio wavelength – it would stretch out beyond the dwarf planet Pluto!
Logarithmic scales have their uses!

Don't forget, when reading these scales, that $10^0 = 1$, and that you must key 10^8 (for example) into a calculator as 1×10^8 not as 10×10^8! (See also page 473.)

Photon energies and the electromagnetic spectrum

In Chapter 23, you calculated the energy of photons, both in joules and in a useful smaller unit, electron-volts (eV).
The diagram shows the regions of the electromagnetic spectrum, described by **wavelength** and by **photon energy** (in eV):

- Wavelength is used when you need to consider the **wave properties** of the radiation.
  You needed to do this when we looked at diffraction and interference (in Chapter 12).

- Photon energy is used when you need to consider the **particle properties** of the radiation.
  You needed to do this for the photoelectric effect, and you will need it in this chapter too.

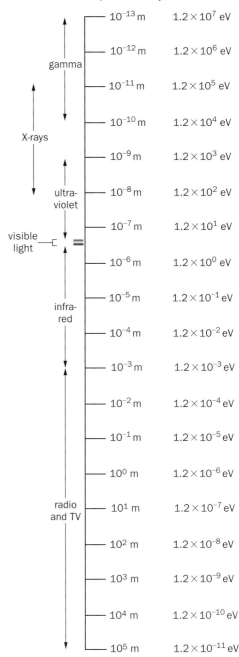

gamma	10^{-13} m — 1.2×10^7 eV
	10^{-12} m — 1.2×10^6 eV
	10^{-11} m — 1.2×10^5 eV
X-rays	10^{-10} m — 1.2×10^4 eV
	10^{-9} m — 1.2×10^3 eV
ultra-violet	10^{-8} m — 1.2×10^2 eV
	10^{-7} m — 1.2×10^1 eV
visible light	10^{-6} m — 1.2×10^0 eV
	10^{-5} m — 1.2×10^{-1} eV
infra-red	10^{-4} m — 1.2×10^{-2} eV
	10^{-3} m — 1.2×10^{-3} eV
	10^{-2} m — 1.2×10^{-4} eV
	10^{-1} m — 1.2×10^{-5} eV
	10^0 m — 1.2×10^{-6} eV
radio and TV	10^1 m — 1.2×10^{-7} eV
	10^2 m — 1.2×10^{-8} eV
	10^3 m — 1.2×10^{-9} eV
	10^4 m — 1.2×10^{-10} eV
	10^5 m — 1.2×10^{-11} eV

Electromagnetic wavelengths and energies.

337

▶ Heat radiation

This thermal image shows hot and cool regions of a face:

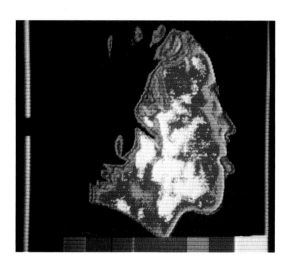

The hottest parts are white, and the coolest are black and blue. The photograph is not taken with visible light, but with **infra-red radiation**. This is also called **heat radiation**, and it is given off by all hot objects.

Look at the picture:

Can you see that the cheeks are giving out most heat, and that the man is poking out his tongue, which is hot?
And that hair reduces radiation from the back of the head?

Absorption and emission of infra-red radiation

Infrared radiation is part of the electromagnetic spectrum. The radiation travels in small packets called photons (see page 327).

In infrared radiation, the photons have energies of about the same size as the energies needed to excite atoms in matter. When an infrared photon strikes matter, it can be *absorbed*:

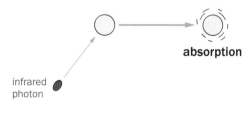

The energy of an absorbed photon is transferred into random energy in an atom of the matter – the internal energy of the matter goes up. It gets hotter.

In exactly the opposite process, an excited atom can *emit* an infrared photon and become less excited:
So the internal energy of the matter goes down, and it gets cooler.

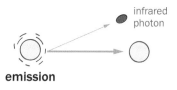

Infrared and light radiation are similar to each other, because they are adjacent in the electromagnetic spectrum.

That is why matt black objects, which are good absorbers and emitters of visible light, are also good absorbers and emitters of infrared radiation.

However, infrared and visible light are not identical: many matt white paints actually absorb infrared quite well!

Example 1
Greenhouse glass is transparent to visible light, but opaque to infrared radiation. How does this explain why greenhouses get hot in sunlight?

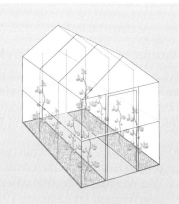

Visible light passes through the glass into the greenhouse, and some is absorbed by the soil and plants inside. They get warmer.
As the internal energy of these materials goes up, the excited atoms give off infrared radiation.
This infrared radiation has a longer wavelength than visible light.
It does not pass through the glass, but is reflected or absorbed by it.
The greenhouse gets warmer.

The **greenhouse effect** in our atmosphere is similar.
The 'glass' in this case consists of certain gases, particularly carbon dioxide and methane. These 'greenhouse gases' transmit both the visible light and short-wavelength infrared radiation that we get from the Sun, but they absorb the long-wavelength infrared given off by the warm Earth.
This is gradually making our atmosphere hotter and hotter.

▶ Black bodies

The usual meaning of 'black' is 'absorbs visible light'.
But in Physics, a **black body** is one that absorbs all the radiation
that falls on it, at **all** wavelengths. It is a perfect absorber.

A black body is also a perfect emitter or **perfect radiator**.
It emits energy in **all** regions of the electromagnetic spectrum.

One familiar 'black body' looks far from black: our Sun!

Black bodies and temperature

Every black body gives out a different amount of energy at
different wavelengths, depending on its temperature.

The graph shows three different spectra:

Each curve shows the energy per second given out per nm of
wavelength by each m² of surface, at different wavelengths.

The **blue** line is the measured spectrum for a black body at
8000 K. This is the temperature of the blue-white star Procyon.

The **red** line is the measured spectrum for a black body at
6000 K. This is the temperature of our Sun.

Why is the blue line so far above the red one?
The blue-white star Procyon is much hotter than our Sun,
so it gives out more energy/second from each m² of surface.

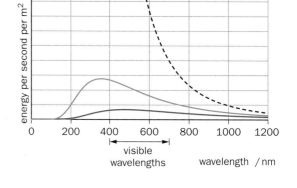

The dashed black line is the **predicted** behaviour of a black
body at 6000 K, based on classical thermodynamic **theory**
developed at the end of the 19th century.

Can you see that the predicted energy gets greater and greater
as the wavelength becomes smaller?
In fact, the theory predicts that the energy becomes infinite as
the wavelength approaches zero. This is clearly not true!

The way in which classical theory goes so very wrong for shorter
wavelengths was called the **ultraviolet catastrophe**.
It led Max Planck to develop **quantum theory** (page 327).

Colour and temperature

Look again at the **blue** and **red** curves on the graph.

Can you see that the peak of the blue curve, for the star Procyon,
occurs at a **shorter** wavelength than the red curve, our Sun?

Max Planck's colleague, Wilhelm Wien, found a relationship
between the **peak wavelength** of the spectrum and the
temperature of the black body. This is **Wien's Law**:

Peak wavelength, λ_{max} × **temperature, T = 0.0029** (m) (K) (m K)	or $\lambda_{max}\, T = 0.0029$ m K

For our Sun, this peak wavelength is close to the middle of
the visible spectrum, while for Procyon it is near the blue end.
So the Sun looks yellow-white, while Procyon looks blue.

▶ The luminosity of stars

In Chapter 29 you will see that stars are classified by their
magnitudes, based on how bright they appear to the eye.
In this chapter we are interested in their **luminosity**, *L*.
This is the energy emitted as electromagnetic radiation every
second, measured in joules per second. As energy per second
is power, luminosity is often given the symbol *P*.

The diagram shows part of the surface of a star. It has area *A*:
The luminosity of this area, shown by the red arrows, depends on how
large *A* is, and how hot it is.

The equation linking these factors for a *black body*, like a star,

is **Stefan's Law**: the luminosity of the black body $\boxed{L = \sigma A T^4}$

where *T* is the absolute temperature (in kelvins), the area *A* is
measured in m² and σ is the Stefan constant.

If *A* is the *total* surface area of a star (which is $4\pi\, r^2$, where
r is the radius of the star), then *L* is the luminosity of the star.

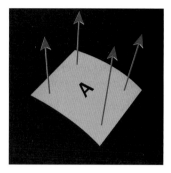

*Energy emitted per second depends
on temperature and surface area.*

Remember: the surface area of a sphere is
$$A = 4\pi\, r^2$$

Example 2
The surface area of the Sun is 6.1×10^{18} m. Its surface temperature, to 2 s.f., is 5800 K.
Calculate the energy emitted by the Sun each second. Stefan constant $\sigma = 5.7 \times 10^{-8}$ W m^{-2} K^{-4}

$L = \sigma A T^4 = 5.7 \times 10^{-8}$ W m^{-2} K$^{-4} \times 6.1 \times 10^{18}$ m² $\times (5800\text{ K})^4$
$= 5.7 \times 10^{-8}$ W m^{-2} K$^{-4} \times 6.1 \times 10^{18}$ m² $\times 1.1 \times 10^{14}$ K$^4 = 3.9 \times 10^{26}$ W $= \underline{3.9 \times 10^{26}\text{ J s}^{-1}}$ (2 s.f.)

Stefan? Boltzmann?

Joseph Stefan, an Austrian physicist, deduced his law by
analysing the results of laboratory experiments on black
bodies. Another Austrian physicist, Ludwig Boltzmann, was
able to prove the equation using thermodynamic theory, so
Stefan's Law is often called the Stefan–Boltzmann Law.

Be aware that:
- $L = \sigma A T^4$ may be written $P = \sigma A T^4$
- σ is often called the Stefan–Boltzmann constant

Estimating the radius of a star

Look at the equation of Stefan's Law in the yellow box above.

Can you see that the luminosity depends on two variables,
the *area* of the surface and its *temperature*?

Wien's Law (see page 339) allows us to calculate the temperature
of a star from its spectrum. Combining this with Stefan's Law
allows us to calculate the radius *r* of a star of known luminosity:

Example 3
The luminosity of the star Betelgeuse is 5.4×10^{31} W and the peak wavelength in its spectrum is 8.5×10^{-7} m.

Calculate the radius *r* of Betelgeuse.   Stefan constant $\sigma = 5.7 \times 10^{-8}$ W m^{-2} K^{-4}

Using Wien's Law, $\lambda_{\text{max}} T = 0.0029$ m K $\therefore T = \dfrac{0.0029 \text{ m K}}{\lambda_{\text{max}}} = \dfrac{0.0029 \text{ m K}}{8.5 \times 10^{-7} \text{ m}} = 3410$ K

Using Stefan's Law, $L = \sigma A T^4$ $\therefore A = \dfrac{L}{\sigma T^4} = \dfrac{5.4 \times 10^{31}\text{ W}}{5.7 \times 10^{-8}\text{ W m}^{-2}\text{ K}^{-4} \times (3410)^4} = 7.01 \times 10^{24}$ m²

Surface area of a sphere $A = 4\pi\, r^2$ $\therefore r = \sqrt{\dfrac{A}{4\pi}} = \sqrt{\dfrac{7.01 \times 10^{24}\text{ m}^2}{4\pi}} = \sqrt{5.58 \times 10^{23}\text{ m}^2} = \underline{7.5 \times 10^{11}\text{ m}}$ (2 s.f.)

▶ Physics at work: Electromagnetic waves

Synchrotron radiation

In the past, X-rays were always produced by high-speed electrons colliding with a heavy metal anode. Now researchers can use a different source – a synchrotron. Synchrotrons were developed as particle accelerators, and their working is explained on page 323.

Synchrotrons had one big disadvantage: when charged particles accelerate, they lose energy by radiation. Anything which moves in a circular path is accelerating all the time, so the charged particles moving in circles lose energy (as electromagnetic radiation) all the time. Depending on the kinetic energy of the charged particles, this radiation can have different wavelengths, from infrared to gamma-rays.

Synchrotron radiation.

The disadvantage is now turned into an advantage: at certain energies the synchrotron radiation turns out to give a beautiful beam of X-rays. Synchrotron radiation is used to study many aspects of the structure of matter at the atomic and molecular scale, from semiconductors to protein molecules.

Colour printing

The photographs and diagrams in this book contain a great range of colours, but they are all printed using just a few colours. The standard printing technique uses three coloured pigments: cyan, magenta and yellow, as shown in the diagram:

The cone cells in the human eye (see page 151) detect only the primary colours red, green and blue and combinations of them.

But how are these colours produced from cyan, magenta and yellow?

Each of the pigments in the printing ink absorbs some colours and reflects others from the white light falling on them. Cyan absorbs all colours except blue and green, while yellow absorbs all colours except green and red. When cyan and yellow are printed on to white paper at the same place, the only colour which is not absorbed is **green**, so that is the colour that we see in the yellow–cyan overlap region in the diagram.

In a similar way, magenta absorbs all colours except blue and red, so printing magenta and yellow pigments at the same place will reflect **red**, the only colour not absorbed by magenta or by yellow.

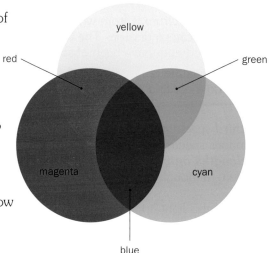

Colour printing with yellow, cyan and magenta produces a range of colours.

Mixing magenta and cyan pigments will reflect **blue**, the last primary colour. If all three pigments are combined in one place, no colours are reflected, and that part of the print will be **black**.

In this way, the light reflected from the colour print can be made any combination of red, green and blue. Different shades are produced by using different proportions of cyan, magenta and yellow.

To save ink, and to give denser colours using less ink, the colour black is not made by mixing all three pigments but by using black as a fourth pigment or 'key'.

This four-colour process is often called **CMYK printing**.

All of the these subtle colours are printed using CMYK.

▶ Coloured spectra

Look at the two kinds of yellow lights in the photographs:

They both look yellow, because they both stimulate the same cells in the retina of your eye. The only real difference that you can see is that the street light is more powerful, so it seems brighter.

However, if you examine their spectra, you get a quite different picture.

Continuous and line spectra

Look at these three diagrams of spectra. They show the wavelengths present in visible light from 3 different sources.

The *first* spectrum is from an ordinary light bulb like the one inside the car's indicator. This is a **continuous spectrum**.
All visible wavelengths are present in the light, and the eye sees the light as white.

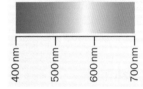
Continuous spectrum: white light.

The *second* spectrum is from white light filtered by the yellow plastic of the car's indicator. The blue end of the spectrum, and part of the red end, have been *absorbed* by the yellow plastic.
The brain sees the remaining range of wavelengths as yellow. This is also a continuous spectrum, but over a limited range.

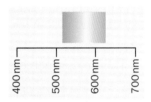
Filtered spectrum: yellow light.

The *third* spectrum is from the sodium street-light. It is not a continuous spectrum at all – it is a **line spectrum**. It has a *few wavelengths only*. There are about 90 different lines in this spectrum, although only the 5 brightest ones are shown here:

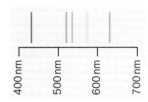
Sodium line spectrum.

Why does this light appear yellow, when the spectral lines are evenly spread across the visible range?
The sodium line spectrum appears yellow because most of the lines are relatively faint. Over 98% of the energy is given out by just two spectral lines, of wavelengths 589.0 nm and 589.6 nm.
In this diagram they are too close together to resolve separately, and so they appear as the slightly thicker yellow line.

Sodium light is often referred to as *monochromatic* (= one colour) because almost all the light has the same wavelength.
Lasers (see page 347) produce pure monochromatic light.

What causes line spectra?

You always get line spectra from atoms that have been excited in some way, either by heating or by an electrical discharge.
In the atoms, the energy has been given to the electrons, which then release it as light.

Line spectra are caused by changes in the energy of the electrons. Large, complicated atoms like neon give very complex line spectra, so physicists first investigated the line spectrum of the simplest possible atom, hydrogen, which has only one electron.

This is discussed on the next page.

Heating sodium atoms produces the same line spectrum

▶ The hydrogen spectrum

It is not only sodium that gives a line spectrum. All elements have a line spectrum, which you can see if you pass an electric current through the element when it is vaporised.

If you look at a hydrogen discharge tube through a diffraction grating or a prism, you will see a spectrum with just 4 sharp lines. The 4 wavelengths are:

656 nm, 486 nm, 434 nm and 410 nm

Why only these four wavelengths? We need to look further.

Physicists soon found there were many more spectral lines in the invisible ultraviolet and infrared regions of the spectrum, and found that they also are grouped in a similar series of lines.

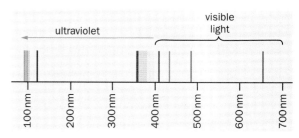

The hydrogen spectrum.

Some lines in the hydrogen spectrum are shown in the diagram:

Can you see two different groups or **series** of lines, one series near 100 nm and the other series between 350 nm and 650 nm?

Each is named after the physicist who investigated it: the **Lyman** series in the ultraviolet near 100 nm, and the **Balmer** series between 350 nm and 650 nm.

Photon energies in the hydrogen spectrum

To see what causes these series of spectral lines, we must look at the photon **energies** instead of the wavelengths.

Example 4
Calculate the photon energy of the 656 nm line in the hydrogen spectrum.

The speed of light $c = 3.00 \times 10^8$ m s^{-1}, and the Planck constant $h = 6.63 \times 10^{-34}$ J s.

From page 125: $c = f \lambda$ so: 3.00×10^8 m s^{-1} = $f \times 656 \times 10^{-9}$ m

$$\therefore f = \frac{3.00 \times 10^8 \text{ m s}^{-1}}{656 \times 10^{-9} \text{ m}} = 4.57 \times 10^{14} \text{ Hz}$$

From page 321: $E = hf = 6.63 \times 10^{-34}$ J s $\times 4.57 \times 10^{14}$ Hz $= \underline{3.03 \times 10^{-19} \text{ J}}$ (3 s.f.)   This is 1.9 eV.

Repeating this calculation for other spectral lines shows that the photon energies (in eV) of the first few lines in each series are:

Lyman energies /eV: 10.2   12.1   12.8

Balmer energies /eV: 1.9   2.6

Can you spot a pattern in these numbers?

Each of the Balmer energies is the **difference** between two of the Lyman energies.
This is true of all the Balmer lines, not just the two given here.

The explanation of the spectrum of hydrogen is a story of **energy differences**. This is discussed on the next page.

► Energy levels and quanta

Planck and Einstein's quantum theory of light (page 327) gives us the key to understanding the regular patterns in line spectra.

The photons in these line spectra have certain energy values only, so the electrons in those atoms can only have certain energy values. This energy level diagram shows a very simple case. It is for an atom in which there are only two possible energy levels:

The electron, shown by the blue dot, has the most potential energy when it is on the upper level, or *excited state*. When the electron is on the lower level, or *ground state*, it has the least potential energy.

The diagram shows an electron in an excited atom dropping from the excited state to the ground state. This energy jump, or *transition*, has to be done as one jump. It cannot be done in stages. This transition is the smallest amount of energy that this atom can lose and is called a **quantum** (plural = quanta).

The potential energy that the electron has lost is given out as a photon. From $E = h f$, this energy jump corresponds to a specific frequency (or wavelength), giving a specific line in the line spectrum.

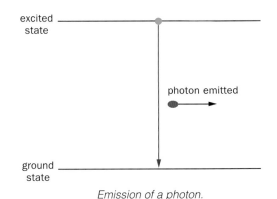

Emission of a photon.

Example 5
The diagram shows an atom with three electron energy levels. What are the photon energies, in eV, that this atom can emit?

There are *three* energy transitions:
- the electron can drop from excited state 1 to the ground state, emitting a 10 eV photon, or
- from excited state 2 to excited state 1, emitting a 5 eV photon, or
- from excited state 2 to the ground state, emitting a photon of energy (10 eV + 5 eV) = 15 eV.

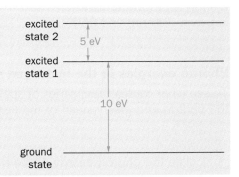

A potential well

In this diagram, Phiz has fallen into a steep-sided pit:

He slipped down 3 metres, and lost 2000 joules of potential energy. We often calculate potential energy from ground level, which means that his potential energy was zero on the level soil outside the pit. At the bottom of the pit, Phiz's potential energy is 2000 J less than zero: it is −2000 J.

When Phiz jumps with all his might, he gains 1300 J of kinetic energy. Can he jump out of the pit?
No, because his kinetic energy is not enough to make up the potential energy he lost by falling down the pit.
His total energy is 1300 J + (−2000 J) = −700 J.
If the sum of (kinetic energy + potential energy) is negative, we say that the system is **bound**. Phiz is stuck in the pit.

This situation is described as a **potential well**.

Now think about a similar situation for an atom.
Suppose you want to ionise the atom by removing an electron.
To remove the electron completely from the attraction of the positive nucleus, you must provide enough energy for the electron to jump from the ground state to the very top of the potential well.
This is called the **ionisation energy**.

Phiz in a 'potential well'.

▶ The energy levels of hydrogen

The Danish physicist Niels Bohr found that the hydrogen spectrum could be explained by the set of energy levels shown in the diagram:

The lowest level is the ground state, and all the other levels are excited states.
The ground state is a long way below the excited states, and the excited states get closer together as you go upwards.

Look at the energy values of each level:

Just like Phiz in his pit, the electron is bound in the atom, and does not have enough energy to get out.
To get the electron out of the hydrogen atom, you have to give it extra energy.

At the bottom of his pit, Phiz had a potential energy of -2000 J. In the same way, the potential energy of all these levels is negative. Zero potential energy occurs at the very top, when the electron escapes, and leaves an ionised atom.

$E = 0$
$E = -0.85$ eV
$E = -1.51$ eV ——————— second excited state

$E = -3.40$ eV ——————— first excited state

$E = -13.61$ eV ——————— ground state

Energy levels of the hydrogen atom.

Example 6
Use the diagram to find the ionisation energy of hydrogen.

The ionisation energy is the energy needed to raise the electron from the ground state to the highest possible energy level, when the electron will be free.
The highest possible state is always defined as zero potential energy.

Ionisation energy = energy of highest level − energy of ground state
$$= 0 \text{ eV} - (-13.61 \text{ eV})$$
$$= \underline{13.61 \text{ eV}} \quad (4 \text{ s.f.})$$

The hydrogen emission spectrum

The simple two-level diagram at the top of the opposite page has only one possible energy jump down from an excited state.
Example 5 has three energy levels and three possible energy jumps.

In the hydrogen atom, with all those excited states, there are many possible transitions.

Look at this diagram:
The arrows all show *down*ward energy transitions, so each would give *out* a photon. This is an **emission** line spectrum.
(We will look at an absorption spectrum on the next page.)

The transitions on the left, going down to the ground state, are all large. This is the Lyman series, giving out energetic photons in the *ultraviolet* region of the spectrum.

The smaller transitions on the right, going down to the $E = -1.51$ eV excited state, give out less energetic *infrared* photons.

Between these two sets of emissions is the Balmer series of lines, going to the first excited state. This series includes the four *visible* lines in the spectrum, coloured in this diagram.

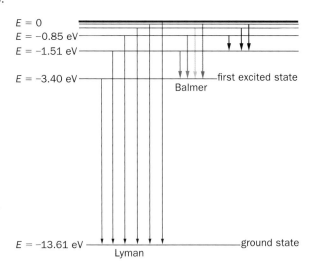

Jumps between the energy levels of the hydrogen atom.

▶ Absorption spectra

The spectra on the previous pages are *emission* spectra, because the electron started in an excited state and dropped *down*.
But where did the electron get this energy from in the first place?
Absorption of a photon is one way.

This diagram shows **absorption** in a simple two-level atom:

It is the exact opposite of the diagram at the top of page 344.
The electron starts in the lower state and absorbs a photon, which raises it to the excited state.

This photon must *exactly* match the energy jump. If it is too big or too small, it is not absorbed.

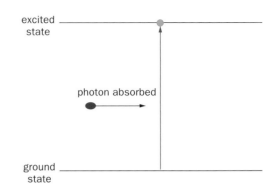

Absorption of a photon.

Example 7
A hydrogen atom has its electron in the energy level at -1.51 eV.
It absorbs a photon, which promotes the electron to the -0.85 eV level.
What is the wavelength of this photon?
The electronic charge $e = 1.6 \times 10^{-19}$ C, the speed of light $c = 3.0 \times 10^8$ m s^{-1} and the Planck constant $h = 6.6 \times 10^{-34}$ J s.

Energy of the photon = energy jump $= -0.85$ eV $- (-1.51$ eV$) = 0.66$ eV

1 eV = 1.6 × 10^{-19} J (from page 321) so: 0.66 eV $= 0.66 \times 1.6 \times 10^{-19}$ J $= 1.1 \times 10^{-19}$ J

E = h f (page 328) so: 1.1×10^{-19} J $= 6.6 \times 10^{-34}$ J s $\times f$ $\quad \therefore f = \dfrac{1.1 \times 10^{-19} \text{ J}}{6.6 \times 10^{-34} \text{ J s}} = 1.6 \times 10^{14}$ Hz

c = f λ (page 125) so: 3.0×10^8 m s$^{-1} = 1.6 \times 10^{14}$ Hz $\times \lambda$

$$\therefore \lambda = \frac{3.0 \times 10^8 \text{ m s}^{-1}}{1.6 \times 10^{14} \text{ Hz}} = \underline{1.9 \times 10^{-6} \text{ m}} \quad (2 \text{ s.f.}) \text{ (infrared)}$$

The Sun's spectrum

One of the first places that an absorption spectrum was observed was in sunlight. The continuous spectrum from the Sun is covered with vertical *dark* lines:
These were systematically measured and classified by the Bavarian instrument maker Joseph Fraunhofer.

These dark lines are due to cooler gases in the outer layers of the Sun. As light from the hot photosphere passes out from the Sun, some light is absorbed by these cooler atoms, promoting their electrons to excited states. The absorbed photons must match the energy jumps exactly, so only certain wavelengths are absorbed.

The absorbed photons are later re-emitted, but in all possible directions, so fewer photons end up going directly outwards.
The spectrum of this light is dimmer now at these wavelengths, because fewer photons reach us, giving the dark Fraunhofer lines.

Absorption spectra are very useful for astronomers.
The absorption lines in the spectrum of a star or galaxy give us a 'fingerprint' of the elements present. If the Doppler effect (page 415) shifts this 'fingerprint' to a longer wavelength, we can calculate how fast the star or galaxy is moving away from us.

Absorption lines in the Sun's spectrum.

These galaxies are moving away from us at 1200 km s^{-1}.

▶ Stimulated emission

In his analysis of quantum theory, Albert Einstein realised that emission and absorption were not the only possible ways to cause energy jumps.

An atom that is already in an excited state can be 'persuaded' to emit a photon, by a passing photon of exactly the right energy.

There will then be two identical photons – the original one and the one created by the downward transition of the electron.

The first photon stimulated the atom into emitting the second photon, so this is called **stimulated emission**.

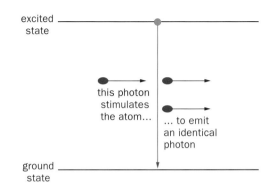

Stimulated emission of photons.

When a beam of light contains identical photons like this, the light is monochromatic – its spectrum has only one wavelength.
The light is also coherent (page 173), as the phase is constant across the beam, so interference effects can be achieved much more easily.

This way of producing extremely regular, uniform radiation was first done with microwaves, but a far more famous application uses photons in or near the optical range, and is called a **laser**.

Light Amplification by Stimulated Emission of Radiation

This is such a long-winded title that it made sense to shorten it to its initial letters – LASER.

Since their invention in 1958, lasers have become very common – they are in every CD player and DVD player, for example.

Infrared lasers can cut through metal.

Everyone knows that laser light is a narrow, parallel beam which is very intense, but the laser's scientific usefulness is due to two facts:

- the light is *monochromatic* – one wavelength only,
- the light is *coherent* – all the waves are in step.

Laser action ('lasing') can take place in solids, liquids or gases.

What is necessary for the medium to 'lase'?
Before stimulated emission can happen, there must be more atoms with their electrons in higher excited states than in lower levels.
Under normal circumstances, the number of atoms with electrons in an excited state is always less than the number in the state below.

This means that electrons must be 'pumped' up to the excited state, often using an electric field, as in the helium–neon gas laser shown in this diagram:

To begin with, one of the excited atoms emits a photon, at random. This photon quickly stimulates another emission. The two photons then stimulate another two emissions, which rapidly becomes an avalanche of identical photons.

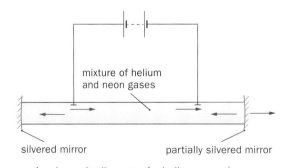

A schematic diagram of a helium–neon laser.

Mirrors at each end of the laser reflect the light, making the photons pass to and fro along the laser.
One of the two mirrors is only partially silvered, so a small percentage of the photons can continually escape.

347

▷ Physics at work: Fluorescence

A fluorescent lamp.

Fluorescent lamps

The long fluorescent tubes in office and classroom ceilings are so familiar we hardly notice them. But how do they work? They are certainly more efficient than ordinary filament lamps.

These tubes are in fact gas discharge tubes, rather like the neon lamps used in advertising. The gas inside is mercury vapour, and it has a line spectrum, like hydrogen and sodium.

Mercury atoms are much larger than sodium or hydrogen atoms. Because the nucleus has a large positive change, the energy levels are separated by bigger gaps. Although some of the emission lines are visible, most energy is emitted in the ultraviolet region of the spectrum, as the lowest energy jumps are large.

This is not a problem if you actually want ultraviolet light, and mercury discharge tubes are used as ultraviolet lamps.

But you don't want your workplace flooded with UV, or your health would be seriously damaged. The high-energy photons of ultraviolet light can cause conjunctivitis and even skin cancer.

The answer lies in **fluorescence**. The inside of the glass tube is coated with an opaque white material called a *phosphor*. The phosphor absorbs the ultraviolet radiation, and its electrons are raised to excited states, as shown in the diagram:

The electrons in the phosphor fall down from one excited state to a lower one, emitting photons each time, until they are back in the ground state. As these transitions are all smaller than the original jump, the photons emitted are not ultraviolet, but visible.

By a careful choice of phosphor materials, the mixture of photons that are emitted blend to give the appearance of white light.

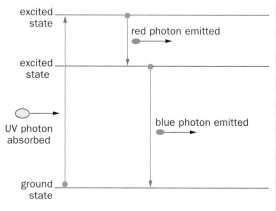

Fluorescence.

Liquid scintillation counters

Liquid scintillation counting was developed to tackle the difficulty of detecting the low energies produced by β (beta) emission from radioactive isotopes such as hydrogen-3 and carbon-14.

It works by **fluorescence** of a solute (a dissolved chemical). This fluorescent chemical is excited by collision with a beta particle, not by a photon.

This is how it works. The kinetic energy of the beta particle is absorbed by an electron in a molecule of the solvent, which then jumps up to a higher energy level.

This electron does not easily drop down from this level, so the solvent molecule stays in an excited state until it collides with a dissolved molecule of the solute. The solvent molecule passes on its energy to the solute molecule. This energy makes an electron jump to a higher level and then drop down immediately, emitting a photon.

This photon is detected by the photoelectric effect in a photomultiplier tube. This is a device for magnifying the effect of a single photon, so that it is easily detected.

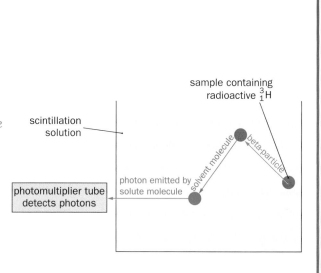

▶ Physics at work: Emission and absorption in Chemistry

Flame tests and fireworks

Chemists use the brightly coloured flames produced by certain reactive metals to identify their presence in unknown compounds.
In flame tests you see yellow for sodium, lilac for potassium and brick-red for calcium.

In a more attractive way, these same metals give fireworks their bright colours. But surely this is Chemistry, not Physics?

The whiz and bang of the fireworks are definitely Chemistry, but the colours given by the fireworks are most certainly Physics.

The reactive metals calcium, strontium and barium are often used in fireworks because they give distinctive colours in flames.

These particular colours are emitted because there are only a few energy transitions possible in those metals when they are ionised, so only a few types of photon are emitted.
(Atoms with a greater range of energy transitions give off a wider range of photons, and the light they emit looks white.)

In fireworks, the reactive metals are used in compounds that vaporise in the heat of the flame, releasing the ionised metal atoms.
The atoms are excited by the heat, and give out photons – just like a sodium streetlight, but far more impressive!

Electrons making energy jumps.

Infrared spectroscopy

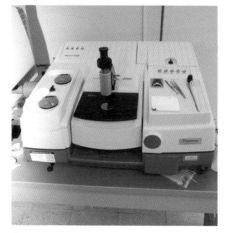

An infrared spectrometer used for chemical analysis.

The picture shows an infrared spectrometer. It is an instrument that is designed to obtain an infrared spectrum of a substance.

The spectrum is obtained by first irradiating a sample with a source of infrared radiation, usually a hot filament lamp.
The infrared passes through the sample, which can be in solution or contained within a gel, and then on to a detector.

Molecules absorb electromagnetic energy in the infrared region of the electromagnetic spectrum, because infrared photons have the right energy to make molecular bonds vibrate.
The energy levels for the vibration of molecular bonds are closer together than the electron energy levels you saw earlier in this chapter.

Bonds between different atoms will absorb different infrared photons, so absorption spectroscopy can be used to identify the different chemical bonds in a sample.

The infrared spectrum is analysed by examining at which wavelengths these absorptions occur:

Each wavelength corresponds to a particular resonance, such as that of the chemical bond joining a carbon atom to an oxygen atom.
Each chemical compound, with its own combination of bonds, has its own infrared spectrum, so the spectrum provides a method of identifying compounds.

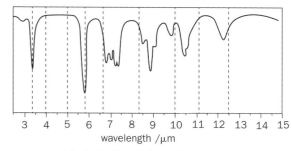

A typical infrared absorption spectrum.

Summary

The electromagnetic spectrum ranges from high-energy gamma radiation to long wavelength radio waves. Different parts of the electromagnetic spectrum are produced in different ways, but there is no clear distinction between adjacent regions.

A continuous spectrum has a complete range of wavelengths.
An emission line spectrum has only a few exactly defined wavelengths.
An absorption line spectrum is a continuous spectrum with a few exactly defined wavelengths removed.

Black bodies, such as stars, are perfect radiators and absorbers of electromagnetic radiation.
The shape of the spectrum emitted from a black body depends on its temperature.
The maximum of the black-body spectrum occurs at a wavelength given by Wien's Law: $\lambda_{max} T = 0.0029$ m K

Stefan's Law: the luminosity L of a star of radius r and surface area A is given by: $L = \sigma A T^4$, where $A = 4\pi r^2$

An atom has definite electron energy levels, and transitions between these levels produce a line spectrum.
The lowest energy level is the ground state, and all the other levels are excited states.
The highest possible level occurs when the atom is ionised.
The energy level at ionisation is usually counted as 0 eV; all other levels have negative potential energies.

A photon is emitted when an electron drops from a higher level to a lower level.
The frequency can be found from $E = h f$ and the wavelength from $c = f \lambda$.

For a photon to be absorbed, it must have exactly the right energy to raise an electron from a lower level to a higher level. Stimulated emission (laser action) occurs when a photon of the right size makes an excited atom emit an identical second photon.

▶ Questions

1. Can you complete these sentences?
   a) The electromagnetic spectrum ranges from high-energy .... radiation to long .... radio waves. Gamma rays come from the nucleus of .... atoms, while .... are generated by collisions of high-energy electrons.
   A well-defined part of the electromagnetic spectrum, in the wavelength range 400 nm to 700 nm, is the .... part.
   b) Wave effects such as diffraction and .... are most noticeable in the long wavelength region of .... Particle effects are most noticeable for .... rays.
   c) A continuous spectrum has .... the wavelengths in a range.
   An emission .... spectrum has only a few wavelengths, while the .... spectrum for the same element is a continuous spectrum with exactly those same wavelengths missing.
   d) When an atom is excited, an electron is raised to an .... state. A line spectrum is caused by jumps between energy .... in the atom.
   e) In an emission spectrum, an electron moves from a .... level to a .... level, while in an .... spectrum the reverse happens.
   Absorption often starts with the electron in the lowest level, known as the .... state.
   f) If the atom is given enough energy for the electron to escape, it is said to be .... and the energy needed for this is called the .... energy.

2. Copy and complete the table below.
   The electronic charge, $e = 1.6 \times 10^{-19}$ C,
   the speed of light, $c = 3.0 \times 10^8$ m s^{-1},
   the Planck constant, $h = 6.6 \times 10^{-34}$ J s.

Region of the electromagnetic spectrum	Wavelength	Energy
	500 nm	
		50 eV
		5.0×10^{-21} J

3. A star has a luminosity of 2.1×10^{29} W, and its spectrum has its greatest intensity at a wavelength of 450 nm.
   The Stefan constant $\sigma = 5.7 \times 10^{-8}$ W m^2 K^{-4}.
   a) Calculate the surface temperature of the star.
   b) Calculate the radius r_{star} of the star.

4. The diagram shows the lowest two energy levels of an atom.

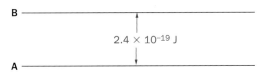

Use the constants from Question 2 to calculate the wavelength of the photon emitted when an electron drops from level B to level A.

5. Here is the simplified energy level structure of the electrons in an atom.

0 eV ——————————————

−0.4 eV ——————————————

−0.9 eV ——————————————

−3.6 eV ——————————————

a) How many excited states are shown in this diagram?
b) Explain why the potential energy of the lowest three states is negative.
c) What is the ionisation energy of this atom (in eV)?
d) What is the largest photon energy (in eV) that can be absorbed by this atom?
e) What is the smallest photon energy (in eV) that can be absorbed by this atom?

6. Use the constants given in Question 2 opposite to calculate the wavelength of the radiation given by the following energy level transitions in the diagram above, in Question 5.
a) The transition from the first excited state to the ground state. Show that it is in the visible part of the spectrum.
b) The transition from the second excited state to the ground state. Show that it is in the ultraviolet part of the spectrum.
c) The transition from the second excited state to the first excited state. Show that it is in the infrared part of the spectrum.

7. The diagram shows three energy levels of neon.

−4.03 eV ——————————————
−5.99 eV ——————————————

−21.57 eV ——————————————

A helium–neon laser emits light of wavelength 633 nm.
a) Use the constants from Question 2 to show that this light emitted by the helium–neon laser has a photon energy of 1.96 eV.
b) On a copy of the diagram, use an arrow to show the electron transition in the neon atom that emits photons of this energy.

8. The *nuclear* energy levels of an atom are similar to the electron energy levels that you have met in this chapter. The diagram shows the lowest two energy levels for a heavy nucleus.

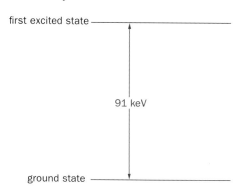

a) Compare the diagram with those in Questions 5 and 7, and explain how this diagram shows that nuclear energies are much greater than electron energies.
b) Use the constants from Question 2 to calculate the wavelength of the radiation emitted in a nucleus change that takes it from the first excited state to the ground state.
c) To what part of the electromagnetic spectrum does this photon belong?

9. The diagram shows three of the electron energy levels of mercury.

a) What is the ionisation energy of mercury in eV?
b) Calculate the three possible energy transitions in eV between these energy levels.
c) Use the constants from Question 2 to calculate the wavelengths of the radiation given off by the three transitions in (b), and state in which part of the electromagnetic spectrum each is to be found.
d) Ultraviolet light emitted by a mercury-vapour lamp is absorbed by cold mercury vapour, in which most of the atoms are in the ground state. Use the energy level diagram to explain why cold mercury vapour absorbs ultraviolet but does not absorb visible light.

Further questions on page 353.

Further Questions on Quantum Physics

▶ Electrons and Photons

Where necessary in these questions, use values for the Planck constant, and the mass and charge of an electron.

1. a) What is the *photoelectric effect*? [1]
    b) What is meant by the *threshold frequency* of the incident light? How is this quantity related to the *work function* of the metal surface? [3]
    c) Explain how the photoelectric effect provides important experimental evidence for the photon model of light. Do this by describing two experimental observations of the effect which cannot be explained using a continuous wave model but can be explained using a photon model of light. [6]

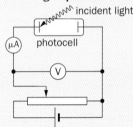
incident light
photocell
μA
V

    d) In an experiment using the circuit shown, blue light of wavelength 450 nm is incident on a potassium metal cathode. The current in the circuit falls to zero for a voltage of 0.76 V across the photocell. [This is called the **stopping potential**.] Find the work function of potassium and the threshold frequency of the light for this surface. [4] (OCR)

2. a) Explain the difference between *thermionic emission* & *photoelectric emission* of electrons. [4]
    b) Write down Einstein's photoelectric equation relating the maximum kinetic energy E_{max} of the photoelectrons with the frequency f of the incident radiation and the work function ϕ for the emitting surface. [4]

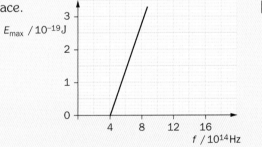

    c) The graph shows how E_{max} varies with f for a particular surface. Use the graph to find the value of ϕ and the threshold wavelength of the radiation. [5]
    d) Describe and explain the effect of increasing the intensity of the incident radiation. [3] (W)

3. In a demonstration, ultraviolet light is incident on a zinc plate and electrons are emitted. The intensity of the ultraviolet light is increased. Explain the following:
    a) the number of electrons emitted per second increases,
    b) the maximum kinetic energy of an electron does not change. [4] (Edex)

4. A spherical charged droplet of unknown mass is observed between two horizontal parallel metal plates in Millikan's apparatus.

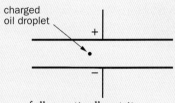
charged oil droplet

    a) The droplet, of radius r, falls vertically at its terminal speed, v, when the potential difference (p.d.) between the plates is zero. Derive an expression for r in terms of v, η, ρ and g where η is the viscosity of air and ρ is the density of the oil droplet.
    b) Explain how the mass of the oil droplet can be calculated from its radius and other relevant data.
    c) The p.d. is adjusted to 1560 V at which point the droplet is held stationary. The metal plates are 15 mm apart and the mass of the droplet is 3.4×10^{-15} kg. Calculate the charge on the droplet. [5] (AQA)

5. In order for the unaided eye to detect a distant source the rate of energy arriving at the eye from this source must be 1.0×10^{-16} W. The energy of each photon arriving at the eye is 3.6×10^{-19} J and the light emitted by the source has a wavelength of 550 nm.
    a) Calculate the number of photons arriving at the eye per second if the eye can just detect the source. [1]
    b) Give one reason why fewer photons reach the retina than are incident on the pupil of the eye. [1] (AQA)

6. Electrons are accelerated through a p.d. of 1500 V in a vacuum. They collide with a thin film of graphite.
    a) Show that the speed of the electrons before impact is 2.3×10^7 m s^{-1}. [2]
    b) Calculate the wavelength associated with the electrons in (a). [2]
    c) Explain why the electrons would be diffracted through an appreciable angle by the graphite. [2]
    d) Electron diffraction can be used to measure nuclear radii. Explain why the electrons used in such measurements would need to have much greater kinetic energies than those in the question above. [2] (AQA)

7. Electrons are accelerated by a potential difference into a narrow beam and enter a region of space occupied by both uniform electric and magnetic fields.

path of electrons
E
S
region of uniform magnetic and electric fields

    The electric field is E and the magnetic flux density is B, which is perpendicular to E and directed into the plane of the paper. B is increased until all the electrons pass through the slit **S** at a speed v. Derive an expression for v in terms of E and B. [2] (OCR)

▶ Spectra and Energy Levels

8. a) Describe the physical appearance of
   i) a line emission spectrum,
   ii) a line absorption spectrum. [2]
   b) Describe the relationship between the two spectra named in (a). [1]
   c) The highest frequency light emitted from a mercury discharge lamp is 2.5×10^{15} Hz. In which part of the spectrum is this frequency? Calculate the ionisation energy of a mercury atom in eV. [3] (W)

9. The radiation emitted from an asteroid is monitored and the following spectrum obtained. State the wavelength at which the peak radiation flux from the asteroid occurs and use it to estimate the temperature of the asteroid. [3] (Edex)

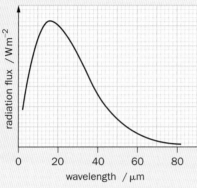

10. The star Rigel in the constellation Orion has a surface temperature that is approximately twice that of the Sun, and the radius of Rigel is 70 times the radius of the Sun. Use Stefan's Law to estimate the ratio

$$\frac{\text{total power of electromagnetic radiation emitted by Rigel}}{\text{total power of electromagnetic radiation emitted by the Sun}}$$

[3] (W)

11. Delta Cephei is a star whose surface has a maximum temperature of 6700 K. The star's luminosity (total power emitted as em-radiation) at its maximum temperature is 1.46×10^{30} W. Calculate its diameter. [4] (W)

12. The diagram shows some of the energy levels for a hydrogen atom. The level $n = 1$ is the ground state.

Assume $c = 3.0 \times 10^8 \, \text{m s}^{-1}$ and $h = 6.6 \times 10^{-34} \, \text{J s}$.
a) State what is meant by the *ground state*. [1]
b) With reference to the energy levels, explain how an electrical discharge through hydrogen gas gives rise to a visible line spectrum. [4]
c) Show that the shortest wavelength of em-radiation that can be emitted by a hydrogen atom is approximately 90 nm. [3] (AQA)

13. Describe evidence you have seen in a school laboratory which shows that different elements have different characteristic optical spectra. [3]

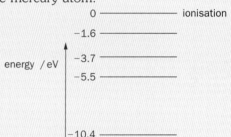

a) The diagram is a simplified energy level diagram for atomic hydrogen. A free electron with kinetic energy 12 eV collides with an atom of hydrogen and causes it to be raised to its first excited state. Calculate the kinetic energy (in eV) of the free electron after the collision. [2]
b) Find the wavelength of the photon emitted when the atom returns to its ground state. [2] (Edex)

14. The diagram shows some of the outer energy levels of the mercury atom.

$$
\begin{array}{ll}
0 & \text{ionisation} \\
-1.6 & \\
-3.7 & \\
\text{energy} / \text{eV} \quad -5.5 & \\
-10.4 & \\
\end{array}
$$

a) Calculate the ionisation energy in J for an electron in the -10.4 eV level. [2]
b) An electron has been excited to the -1.6 eV level. On a copy of the diagram, show all the possible ways it can return to the -10.4 eV level. [3]
c) Which change in energy levels will give rise to a yellowish line ($\lambda \sim 600$ nm) in the mercury spectrum? [4] (Edex)

15. The light from a hydrogen discharge tube contains em-radiation of wavelengths 6.6×10^{-7} m (red light) and 4.9×10^{-7} m (blue light). The light can be separated to form a line spectrum by diffraction or by refraction.
a) Explain how an electrical discharge through hydrogen gas in a discharge tube gives rise to a spectrum which consists only of certain frequencies. [3]
b) Determine the angular separation of the red and blue light of hydrogen for:
   i) the second-order diffraction pattern produced by a diffraction grating which has 5.0×10^4 lines per metre; [3]
   ii) the refraction of a beam of hydrogen light passing from air to glass when the angle of incidence is 80°.
   (The refractive index from air to glass for red light is 1.65, and for blue light is 1.67.) [3] (AQA)

25 Radioactivity

How many different materials can you think of?
Your list could include thousands of substances, but in fact
they are all composed from around 100 elements.
Just as we can make words from the 26 letters of our alphabet,
the elements combine together chemically to form compounds.

So what are elements made up of?
The Greek philosopher Democritus first suggested that all matter
is made up of tiny particles, that he called atoms.
His ideas were rejected, and for 2000 years scientists believed
that every substance was made up of earth, fire, air and water!

Our modern atomic theory was first put forward by John Dalton
in 1803. He imagined atoms to be tiny, indivisible particles.

In this chapter you will learn about:
- the structure of the atom,
- the properties of ionising radiations,
- the process of radioactive decay.

▶ Alpha particle scattering

John Dalton said that atoms were tiny indivisible spheres, but
in 1897 J J Thomson discovered that all matter contains tiny
negatively charged particles (see page 320).
He showed that these particles are *smaller* than an atom.
He had found the first *subatomic* particle – the electron.

Scientists then set out to find the structure of the atom.

Thomson thought that the atom was a positive sphere of matter
and the negative electrons were embedded in it as shown here:
This 'model' was called the 'plum-pudding' model of the atom.

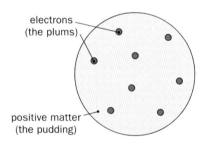

Thomson's plum-pudding model

Ernest Rutherford decided to probe the atom using fast moving
alpha (α) particles. He fired the positively charged α-particles
at very thin gold foil and observed how they were scattered.

The diagram shows his apparatus:

A fine beam of α-particles fell on a very thin foil of gold.
A tiny flash of light was seen through the microscope
each time a scattered α-particle struck the zinc sulphide screen.

The microscope could be rotated around the foil, and so
set at different angles to the path of the incident beam of
α-particles.
The number of α-particles hitting the screen at each angle
could then be counted.

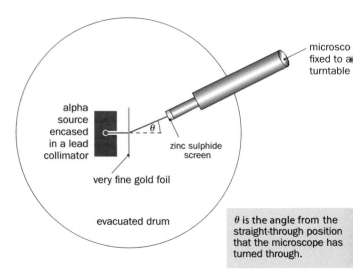

θ is the angle from the
straight-through position
that the microscope has
turned through.

The diagram summarises Rutherford's results:

Most of the α-particles passed straight through the foil, but to his surprise a few were scattered back towards the source. Rutherford said that this was rather like firing a gun at tissue paper and finding that some bullets bounce back towards you!

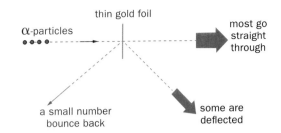

The nuclear model of the atom

Rutherford soon realised that the positive charge in the atom must be highly concentrated to repel the positive α-particles in this way.

The diagram shows a simple analogy:

The ball is rolled towards the hill and represents the α-particle. The steeper the 'hill', the more highly concentrated the charge. The closer the approach of the steel ball to the hill, the greater its angle of deflection.

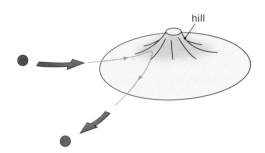

In 1911 Rutherford described his **nuclear model** of the atom. He said that:

- All of an atom's positive charge and most of its mass is concentrated in a tiny core. Rutherford called this the **nucleus**.
- The electrons surround the nucleus, but they are at relatively large distances from it. The atom is mainly empty space!

Can we use this model to explain the α-particle scattering?

The concentrated positive charge produces an electric field which is very strong, close to the nucleus.
The closer the path of the α-particle to the nucleus, the greater the electrostatic repulsion and the greater the deflection.

Most α-particles are hardly deflected because they are far away from the nucleus and the field is too weak to repel them much. The electrons do not deflect the α-particles because the effect of their negative charge is spread thinly throughout the atom.

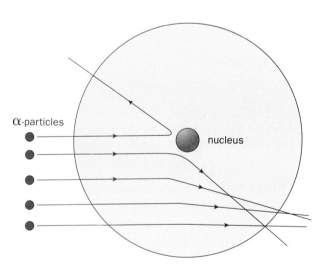

Using this model Rutherford calculated that the diameter of the gold nucleus could not be larger than 10^{-14} m.

We now know that the nuclei of atoms have diameters of a few femtometres (10^{-15} m). Compare this with the diameters of atoms (around 10^{-10} m).

This diagram is not to scale. With a 1 mm diameter nucleus, the diameter of the atom would have to be 100 000 mm or 100 m! The nucleus is like a full stop at the centre of a football pitch.

Probing the structure of matter

Rutherford's α-particle scattering experiment was very important. Not only did it reveal the nuclear structure of the atom, but it also showed that energetic particles could be used as probes to investigate the nature of matter.

This technique has led to many important discoveries in Physics. Rutherford went on to show that the nucleus was in fact made up of two particles, the proton and the neutron.

▶ The structure of the atom

We know now that the nucleus is made up of two particles, the **proton** and the **neutron**. Protons and neutrons are both referred to as **nucleons** because they are found in the nucleus. All atoms (except for the hydrogen atom) are made up of three kinds of particles. The table shows their properties:

Since the masses of these particles are so small it is useful to measure them in a very small unit – the **atomic mass unit (u)**.

$$1 \text{ u} = 1.6605 \times 10^{-27} \text{ kg}$$

particle	location	mass	charge
proton	in the nucleus	1.00728 u	$+ e$
neutron	in the nucleus	1.00867 u	0
electron	orbiting the nucleus	0.00055 u	$- e$

$e = 1.6 \times 10^{-19}$ coulombs

Notice that:
- the mass of the proton and the neutron are each approximately 1 u,
- the electron has a much smaller mass – about $1/1800$ u,
- the charge on the proton is equal and opposite to the electron charge.

The neutrons and protons make up the tiny positive nucleus. The electrons spin round the nucleus, giving the effect of a thin cloud of negative charge.

The diagrams below show some examples.
Which two particles make up the simplest atom, the hydrogen atom?

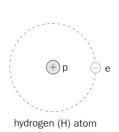

hydrogen (H) atom

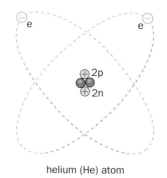

helium (He) atom

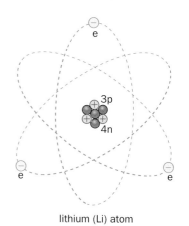

lithium (Li) atom

What do you notice about the number of protons and electrons in each atom?
The atom is *neutral* – unless it loses or gains an electron to become an **ion**.

Proton number and nucleon number

There are over 100 elements listed in the periodic table.
Some of these are not found naturally, but they can be made.

Each element has atoms with a *specific* number of protons.
For example, a lithium atom always has 3 protons.
A carbon atom always has 6 protons.
The number of protons in an atom is its **proton number, Z**.
This is sometimes called the atomic number.

Another important number is the **nucleon number, A**.
This is the total number of nucleons (= protons + neutrons).
It is sometimes called the mass number.
The nucleon number of lithium is 7 (= 3 protons + 4 neutrons).

The word **nuclide** is often used for a particular combination of protons and neutrons in a nucleus.
The panel shows the shorthand way of describing a nuclide:

Can you see that a beryllium nucleus consists of 4 protons and 5 neutrons?

All nuclides can be described using this format:

nucleon number A
                    chemical symbol of element
proton number Z

Eg.
$$^{7}_{3}\text{Li}$$

number of nucleons (protons + neutrons)

number of protons

Eg. $^{9}_{4}\text{Be}$ Be = beryllium

Isotopes

Every atom of oxygen has a proton number of 8. That is, it has 8 protons (and so 8 electrons to make it a neutral atom).

Most oxygen atoms have a nucleon number of 16. This means that these atoms also have 8 neutrons. This is $^{16}_{8}\text{O}$

Some oxygen atoms have a nucleon number of 17. These atoms have 9 neutrons (but still 8 protons). This is $^{17}_{8}\text{O}$.
$^{16}_{8}\text{O}$ and $^{17}_{8}\text{O}$ are both oxygen atoms.
They are called **isotopes** of oxygen.

There is a third isotope of oxygen: $^{18}_{8}\text{O}$.
How many neutrons are there in the nucleus of an $^{18}_{8}\text{O}$ atom?

> **Isotopes are atoms with the same proton number, but different nucleon numbers.**

Since the isotopes of an element have the same number of electrons, they must have the same *chemical* properties. The atoms have different masses, however, and so their *physical* properties are different.

All the elements have more than one isotope. Hydrogen has three, as shown in the diagram: These are so important that each has its own name.

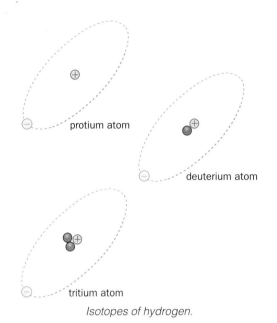

Isotopes of hydrogen.

The table shows some elements.

Proton number	Element + symbol		Commonest isotope
1	Hydrogen	H	$^{1}_{1}\text{H}$
2	Helium	He	$^{4}_{2}\text{He}$
3	Lithium	Li	$^{7}_{3}\text{Li}$
4	Beryllium	Be	$^{9}_{4}\text{Be}$
5	Boron	B	$^{11}_{5}\text{B}$
6	Carbon	C	$^{12}_{6}\text{C}$
8	Oxygen	O	$^{16}_{8}\text{O}$
79	Gold	Au	$^{197}_{79}\text{Au}$
92	Uranium	U	$^{238}_{92}\text{U}$

Nuclear diameters

The periodic table lists the elements in order of increasing proton number. Hydrogen is the first element in the table. Uranium is the last of the naturally occurring elements.

Would you expect the uranium nucleus, with 238 nucleons, to have the same radius as the hydrogen nucleus?

Suppose each nucleon is a sphere with a constant volume v.
Then $v = \frac{4}{3}\pi r_o^3$, where r_o represents the radius of a nucleon.

The volume V of a nucleus with A nucleons will be roughly Av.
What value does this give for the radius of the nucleus r?

Assuming that the nucleus is a sphere, then $V = \frac{4}{3}\pi r^3$

Since $V = Av$, then: $\qquad \frac{4}{3}\pi r^3 = A \times \frac{4}{3}\pi r_o^3$

Cancelling and taking the cube root:

$$r = r_o A^{1/3}$$

r_o is the radius of a nucleon and is found by experiment to be 1.2×10^{-15} m.

This equation gives nuclear radii in the range 2×10^{-15} m to 8×10^{-15} m. These values have in fact been confirmed by experiments using high-energy electrons.

Example 1
What is the radius of a lead nucleus of nucleon number 206?

$r = r_o A^{1/3}$
$r = 1.2 \times 10^{-15}\text{m} \times 206^{1/3}$ *(Use x^y button on your calculator)*
$r = 1.2 \times 10^{-15}\text{m} \times 5.91 = \underline{7.1 \times 10^{-15} \text{ m}}$ (2 s.f.)

Nuclear density

The proton and the neutron have almost the same mass. Suppose each nucleon has a mass m and a volume v. The mass of a nucleus with A nucleons is therefore Am and the volume of the nucleus is about Av.

What is the density of the nucleus?

$density = \dfrac{mass}{volume}$ and so density of the nucleus is $\dfrac{Am}{Av}$

A cancels to give: nuclear density $= \dfrac{m}{v}$

This means that all nuclei have about the same density. This result is supported by experimental evidence.

Estimating the density using data on this double page:

$\rho = \dfrac{m}{v} = \dfrac{1\,u}{\frac{4}{3}\pi r^3} = \dfrac{1 \times 1.7 \times 10^{-27} \text{ kg}}{\frac{4}{3} \times \pi \times (1.2 \times 10^{-15})^3 \text{ m}^3}$

nuclear density $\rho = 2.3 \times 10^{17}$ kg m^{-3}
The density of everyday materials is about 10^3 kg m^{-3}.

▶ Radioactivity

In 1896, Henri Becquerel discovered, almost by accident, that uranium can blacken a photographic plate, even in the dark. Uranium emits very energetic radiation – it is **radioactive**.

Then Marie and Pierre Curie discovered more radioactive elements, including polonium and radium. Scientists soon realised that there were three different types of radiation. These were called alpha (α), beta (β) and gamma (γ) rays from the first three letters of the Greek alphabet.

α, β and γ-rays cannot be seen, so how can we detect them? As the rays pass through matter, energy is transferred from the radiation to the atoms of the material.
This transfer of energy produces effects that can be observed.

Radioactive materials are dangerous.
You must always handle radioactive sources with tongs and point them away from your body.

Marie Curie.

▶ Detecting radioactivity

1. Photographic film

The photograph shows how some film has been blackened by radioactivity, except in the shadow of a key:

People who work in radioactive environments such as a nuclear power plant or a hospital X-ray department wear badges of film. Each month the film is developed in order to monitor the amount of radiation that the workers have been exposed to.

2. The leaf electroscope

A leaf electroscope can be used to detect electric charges.
It consists of a metal cap connected to a metal rod and a 'leaf' made of a very thin strip of metal, as shown in the diagram:
The cap, the rod and the leaf are insulated from the metal case.

The electroscope in the diagram is negatively charged.
Why is the hinged leaf deflected away from the metal rod?
The negative leaf is repelled by the negatively charged rod.
When given a charge, the leaf will stay up for several minutes.

When you bring a radium source close to the cap, the leaf falls. Can you explain why?
The radiation from the radium source is highly energetic and it uses this energy to remove electrons from the air molecules – it *ionises* the air. It produces electrons and positive air ions.

The negative electroscope attracts the positive air ions to it, and so it is discharged.

α, β and γ-rays are called *ionising* radiations.
The ionising effect of radiation is used in many other detectors. In fact, the photographic film above darkens because the radiation ionises the photographic chemicals in the film (as does light).

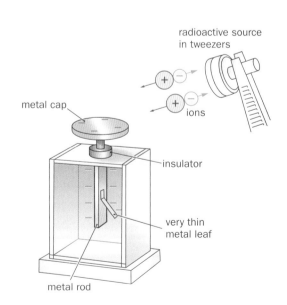

3. The cloud chamber and the bubble chamber

Have you looked at the sky and seen a cloud-trail behind a high-flying aircraft?
Water vapour in the air condenses on the ionised exhaust gases from the engine
to form droplets that reveal the path of the plane.

A **cloud chamber** produces a similar effect using alcohol vapour.
Radiation from a radioactive source ionises the cold air inside the chamber.
Alcohol condenses on the ions of air to form a trail of tiny white droplets along the
path of the radiation. The diagrams below show some typical tracks :

α-radiation

high energy β-radiation

γ-radiation

The α-radiation produces dense straight tracks showing intense ionisation.
Notice that all the tracks are similar in length.

The high energy β-ray tracks are thinner and less intense. The tracks vary
in length amd most of the tracks are much longer than the α-ray tracks.

The γ-rays do not produce continuous tracks.

A **bubble chamber** also shows the tracks of ionising radiation.
The radiation leaves a trail of vapour bubbles in a liquid (often liquid hydrogen).

4. The Geiger–Müller tube (G–M tube)

The G–M tube is capable of detecting a single ionising event.
It consists of a metal tube, the cathode, containing argon gas
at low pressure. The thin wire down the centre is the anode (+).
The p.d. between the anode and the cathode is about 400 V.
This creates a strong electric field around the central anode.

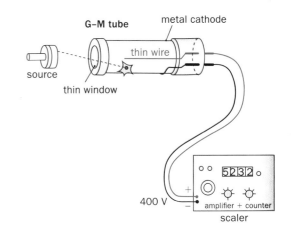

Radiation enters the tube through the thin mica window and
may ionise an argon atom to give a positive ion and an electron.
The free electron (−) accelerates rapidly towards the anode (+)
and gains enough energy to ionise many more argon atoms. The
sudden flow of charge in the tube gives a pulse of electric current.

This is amplified and passed to either :

   a **scaler**, which counts the pulses and shows the total number,

   or a **ratemeter**, which records the counts per second.

When a loudspeaker is connected, each pulse gives a 'click'.
Rapid, random clicking is produced when a G–M tube is held
near to a source of radioactive material.

Background radiation

The photograph shows a G–M tube connected to a ratemeter :

Why is a count registered, even if there is no source present?

The count is the result of background radiation. It is always
present and is mainly caused by radioactivity from the Earth's
rocks and by cosmic rays from the Sun. But the air and even
living things are slightly radioactive.

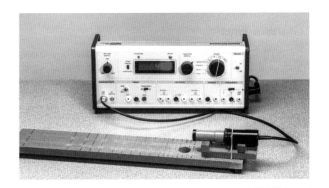

▶ Alpha, beta and gamma radiation

Scientists soon identified the nature of α, β and γ-rays.
You can repeat some of their experiments using sealed sources.

Remember, the count recorded on a G–M tube will be caused
by the radiation from the source **and** the background radiation.

First measure the background count and then **deduct** this value
from your other counts to find the count caused by just the source.
This is called the **corrected count rate**.

Alpha-rays

Place an americium-241 source, which emits α-rays, close
to the window of a G–M tube, as shown in the diagram:

What happens to the count rate when you place some
paper between the source and the tube?
The count rate drops to zero, so the paper must have
stopped the α-radiation.

Now remove the paper and record the count rate as you move
the α-source further away from the tube.
You will find that α-rays can only penetrate about 5 cm of air.

It is also possible to show that α-radiation can be deflected by
very strong electric and magnetic fields.

We now know that **α-radiation consists of positively-charged
particles moving at high speed.**
In fact, an α-particle consists of **2 protons and 2 neutrons**.
Do you recognise this as the **nucleus of a helium atom**?

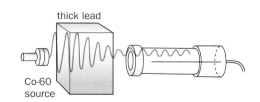

to counter

Am-241 source
in holder

wooden supports

Beta-rays

You can investigate β-rays using a strontium-90 source.
You will find that paper does not stop these rays, but
what happens if you use different thicknesses of aluminium?

β-radiation is stopped by about 3 mm of aluminium.
The β-rays from this source can penetrate about 50 cm of air.

What happens if you place a powerful magnet between the
source and the tube as shown in the diagram?
The count rate falls, but rises again when the G–M tube is moved.

If you use Fleming's left-hand rule (page 262) you can show that
β-rays must be negatively-charged particles.
In fact **β-particles are electrons moving at very high speeds**.

Sr-90 source

N

S

no magnet

with magnet

Gamma-rays

Repeat the experiments using γ-rays from a cobalt-60 source.
You will find that γ-radiation is unaffected by a magnetic field
and is very penetrating. Although it is reduced, it is not stopped
completely, even by thick pieces of lead.
γ-rays are electromagnetic waves of very short wavelength.

It is often useful to think of γ-rays as a stream of γ-ray photons
rather than waves (see chapter 23). Remember, the shorter the
γ-wave wavelength, the higher the energy of the γ-ray photons.

thick lead

Co-60
source

▶ How is radiation absorbed?

α-particles, β-particles and γ-ray photons are all very energetic particles.
We often measure their energy in electron-volts (eV) rather than joules.
Typically the kinetic energy of an α-particle is about 6 million eV (6 MeV).

We know that radiation ionises molecules by 'knocking' electrons off them.
As it does so, energy is transferred from the radiation to the material.
The diagrams below show what happens to an α-particle:

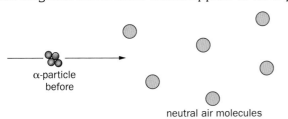
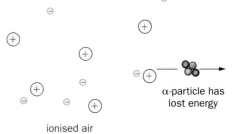

It takes about 30 eV to produce a positive ion and a free electron – an ion pair –
from an air molecule. How many ion pairs could a 6 MeV α-particle produce?
After producing about 2×10^5 ion pairs, the α-particle has given up all its energy
and has been stopped or absorbed.

Why do the 3 types of radiation have different penetrations?

Since the **α-particle** is a heavy, relatively slow-moving particle,
with a charge of $+2e$, it interacts strongly with matter.
It produces about 1×10^5 ion pairs per cm of its path in air.
After passing through just a few cm of air it has lost its energy.

The **β-particle** is a much lighter particle than the α-particle
and it travels much faster. Since it spends just a short time in
the vicinity of each air molecule and has a charge of only $-1e$,
it causes less intense ionisation than the α-particle.
The β-particle produces about 1×10^3 ion pairs per cm in air,
and so it travels about 1 m before it is absorbed.

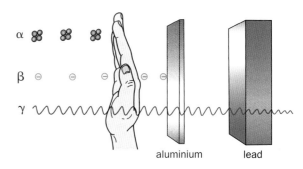

Absorption and penetration of α, β, γ rays.

A **γ-ray photon** interacts weakly with matter because it is
uncharged and therefore it is difficult to stop.
A γ-ray photon often loses all its energy in one event.
However, the chance of such an event is small and on average
a γ-ray photon travels a long way before it is absorbed.

Here is a summary:

Property	Alpha-particle	Beta-particle	Gamma-ray
Nature:	2 protons + 2 neutrons (a helium nucleus)	an electron	an electromagnetic wave of very short wavelength
Symbol:	^4_2He or $^4_2\alpha$	$^0_{-1}e$ or $^0_{-1}\beta$	γ
Mass:	~ 4 u	~ 0.000 55 u	0
Charge:	+2	−1	0
Ionising ability:	very strong	strong	weak
Speed:	~ 5% the speed of light	up to 98% the speed of light	speed of light
Penetration:	stopped by paper, skin or a few cm of air	stopped by 3 mm of aluminium or about 1 m of air	reduced significantly by several cm of lead
Affected by electric and magnetic fields?	yes	yes	no

▶ Why is radiation dangerous?

α, β and γ-rays are dangerous because of their energy.
They can ionise the molecules in living cells and in this way
release energy into the cells and damage them.

Once the radiation has been stopped by 'colliding' with matter,
it is no longer dangerous. Compare this to a bullet: you would
be wise to avoid a fast-moving bullet from a gun, but once the
bullet has been stopped, it is no longer a threat.

The release of energy into body cells can cause burning.
Ionisation can also disrupt the chemistry of the cells and cause
radiation sickness and hair loss.
Cells may become genetically changed by radiation. This can
lead to illnesses such as cancer and leukaemia, which may
occur years after the body was exposed to the radiation.
Damage to the genetic make-up of sex cells can also be passed
on and can cause abnormalities in the person's children.

Which type of radiation is the most dangerous? This depends on
whether the source of radiation is inside or outside the body:

- An α-source outside your body is safe. The α-particles cannot
  pass through your outer skin to get to the internal organs.
  But if you swallow an α-source or breathe in an α-emitting
  gas, that is very dangerous. It causes a lot of ionisation,
  releasing a lot of energy into the surrounding cells.
- A γ-source outside your body can be dangerous because the
  radiation is very penetrating and can affect cells deep inside.

Workers must be protected from the radiation emitted by highly radioactive materials.

▶ The inverse-square law for gamma radiation

Place a gamma source close to the window of a G–M tube
and record the count rate. Now change the distance.
What happens to the count rate as you increase the distance
between the source and the tube?

The γ-rays spread out from the source and so fewer and fewer
rays enter the tube as it is moved away from the source.

In fact, if you double the distance between the source and the
tube the count rate falls to $\frac{1}{4}$.

At three times the distance the count rate is only $\frac{1}{9}$.

The intensity of light from a lamp changes in the same way.
Like all electromagnetic radiation, the intensity of γ-rays
obeys an **inverse-square** relationship (see also page 127).

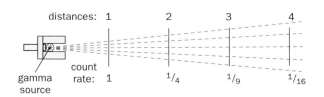

Can we explain this inverse-square law? Look at the diagram:

The radiation passes through an area A at a distance r from the
source. At a distance $2r$ the radiation has spread out and now
passes through an area $4A$.
So the intensity of the radiation at $2r$ is $\frac{1}{4}$ of that at distance r.

The inverse-square law applies to α and β-rays in a vacuum,
but not in air. Why?
Unlike γ-radiation, the air absorbs both α and β-radiation.

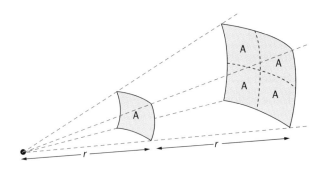

Investigating the inverse-square law

You can investigate the inverse-square law using the apparatus in the diagram. **Before** placing the gamma source in position you need to measure the background count rate.

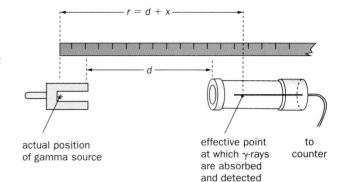

actual position of gamma source

effective point at which γ-rays are absorbed and detected

to counter

Start with the γ-ray source close to the G–M tube and measure the count rate. Remember to subtract the background count to find the **corrected** count rate.

Gradually increase the distance r between the source and the tube and record the corrected count rate C at each position.

If the inverse-square law is true, then you should find that:

$$C \propto \frac{1}{r^2} \quad \text{or} \quad \boxed{C = \frac{k}{r^2}} \quad \text{where } k \text{ is a constant.}$$

But there is a problem in measuring r. Can you see what this is?

The γ-source is in a sealed container and so you do not know its exact position. Also you do not know the exact point where the γ-rays are detected within the G–M tube.

How can we solve this problem?
You can measure the distance d from the end of the source to the tube as shown, so that $r = d + x$, where x is unknown.

What graph can you then plot to prove the law, without knowing x?

Since $C = \dfrac{k}{r^2}$ and $r = d + x$, then: $C = \dfrac{k}{(d+x)^2}$

Rearranging: $(d+x)^2 = \dfrac{k}{C}$ and so: $d + x = \dfrac{\sqrt{k}}{\sqrt{C}}$

Therefore: $d = \sqrt{k} \times \dfrac{1}{\sqrt{C}} - x$

which is like: $y = m \quad x + c$ (see page 478).

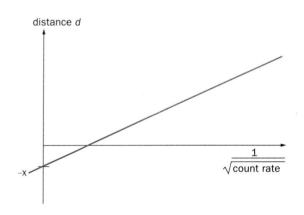

distance d

$\dfrac{1}{\sqrt{\text{count rate}}}$

$-x$

Can you see that if γ-rays do obey the inverse-square law, the graph of d against $1/\sqrt{C}$ should be a straight line as shown:

Notice that this method overcomes the problem of not knowing exactly where the source is, or the effective point where the γ-rays are detected in the G–M tube.

Example 2
A G–M tube is placed 200 mm from a point source of γ-rays and registers a corrected count rate of 60 s^{-1}.
a) What will be the corrected count rate at 1000 mm from the source?
b) How far is the tube from the source when the count rate is 240 s^{-1}?

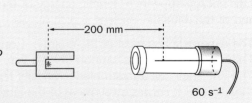

200 mm

60 s^{-1}

a) The separation of the source and the tube is 5 times as great, and so the count rate must fall to $1/5^2$ or $\frac{1}{25}$ of its original value.
   Therefore the new count rate is $\frac{1}{25} \times 60$ s^{-1} = 2.4 s^{-1}

b) The count rate has increased from 60 s^{-1} (at 200 mm) to 240 s^{-1}.
   The count rate is now 4 times as great, so the distance must have halved.
   The separation of the source and G–M tube is $\frac{1}{2} \times 200$ mm = 100 mm

▶ Radioactive decay

Most of the atoms that we find in our world have stable nuclei. Yet all nuclei (except hydrogen) contain at least two protons, and the positively charged protons *repel* each other.
In fact, using Coulomb's law (page 290) you can show that the force between two protons 2×10^{-15} m apart is about 50 N!
So why doesn't the nucleus blow apart?

The answer is that there is another force between nucleons. This is the **strong nuclear force**.
It is about 100 times as strong as the electric repulsive force, but acts only over very short distances, as shown on the graph: The strong nuclear force is *attractive* (at distances greater than about 0.5×10^{-15} m) and so it holds the nucleons together.

In stable nuclei the effects of these forces are balanced, but an imbalance makes the nucleus unstable.

The atoms of radioactive materials have *un*stable nuclei.
In radioactive decay an unstable nucleus attempts to become more stable by emitting α, β or γ radiation.
When an unstable **parent nucleus** emits an α or β-particle, it is left with a *different* number of protons and neutrons. This new nucleus is called the **daughter nucleus**.

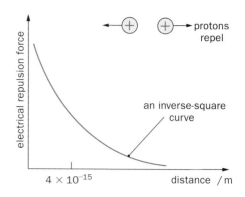

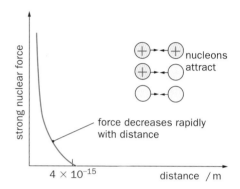

Alpha-decay

An α-particle is a helium nucleus and is written ^4_2He or $^4_2\alpha$.
It consists of 2 protons and 2 neutrons.

When an unstable nucleus decays by emitting an α-particle. it loses 4 nucleons and so *its nucleon number decreases by 4*.
Also, since it loses 2 protons, *its proton number decreases by 2*.

The nuclear equation is: $\quad ^A_Z\text{X} \quad \Longrightarrow \quad ^{A-4}_{Z-2}\text{Y} \quad + \quad ^4_2\text{He}$

Note that the top numbers balance on each side of the equation. So do the bottom numbers.

For example, polonium-208 emits an α-particle and becomes an isotope of lead (Pb):
$$^{208}_{84}\text{Po} \quad \Longrightarrow \quad ^{204}_{82}\text{Pb} \quad + \quad ^4_2\text{He}$$

This decay is shown in the diagram:

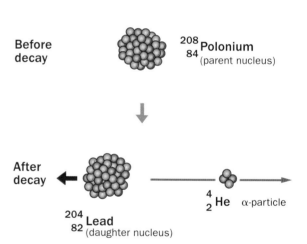

Every decay of a Po-208 nucleus releases 5.1 MeV of energy. This energy appears as kinetic energy of the ejected α-particle and of the daughter nucleus, which recoils.

The α-particle has most of the energy. Can you explain why?
Think about a massive cannon firing a very small shell.
The tiny shell has a much higher velocity than the recoiling cannon and it carries away most of the kinetic energy.

Each α-particle from Po-208 has about the *same* kinetic energy – almost 5.1 MeV. In fact, nearly all α-emitting nuclides eject most of their α-particles with a single energy value, and this energy value is characteristic of the nuclide.

The photo shows some α-particle tracks in a cloud chamber:

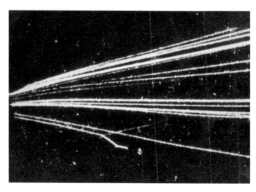

α-particle tracks

Beta-decay

Many radioactive nuclides (radionuclides) decay by β-emission.
This is the emission of an electron *from the nucleus.*
But there are no electrons in the nucleus!
What happens is this: one of the neutrons changes into a proton
(which stays in the nucleus) *and* an electron (which is emitted
as a β-particle).
This means that *the proton number increases by 1*, while the
total nucleon number remains the *same*.

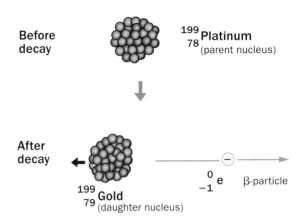

Before
decay

$^{199}_{78}$ Platinum
(parent nucleus)

After
decay

$^{199}_{79}$ Gold
(daughter nucleus)

$^{0}_{-1}$ e   β-particle

The nuclear equation is: $^{A}_{Z}X \implies \,^{A}_{Z+1}Y \,+\, ^{0}_{-1}e$

Notice again, the top numbers balance, as do the bottom ones.

For example, platinum-199 changes to gold-199 by β-decay,
like this:

$$^{199}_{78}Pt \implies \,^{199}_{79}Au \,+\, ^{0}_{-1}e$$

This decay is shown in the diagram.

Every decay of a Pt-199 nucleus releases 1.8 MeV of energy.
As with α-decay, this energy appears as kinetic energy.
You might expect all the β-particles from the decay of Pt-199
to be emitted with almost 1.8 MeV of kinetic energy.
In fact this does not happen!
The graph shows that the β-particles from Pt-199 are emitted
with a *range* of kinetic energies up to a maximum of 1.8 MeV.

How can we explain this spread of energies?
In 1930, Wolfgang Pauli suggested that as the nucleus decays
another particle is emitted in addition to the electron.
The energy released in the nuclear decay is then shared
randomly between the electron and the second particle.
This second particle must have no charge and almost no mass.
It is called an **antineutrino** and given the symbol $\overline{\nu}$ ('nu-bar').

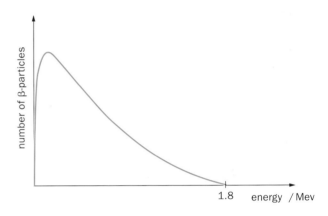

number of β-particles

1.8   energy / Mev

If we include the antineutrino in the equation for β-decay,
we get: $^{199}_{78}Pt \implies \,^{199}_{79}Au \,+\, ^{0}_{-1}e \,+\, ^{0}_{0}\overline{\nu}$

Notice that the numbers on the top, and the bottom, still balance.

All β-emitters produce β-particles with a range of energies up
to a maximum value. This value is characteristic of the nuclide.

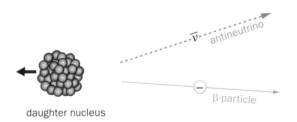

$\overline{\nu}$ antineutrino

β-particle

daughter nucleus

Gamma emission

Gamma emission does not change the structure of the nucleus,
but it does make the nucleus more stable, because it reduces
the energy of the nucleus.

This is what happens:
A nucleus that has emitted an α or β-particle is often left in an
excited state. It loses its surplus energy by emitting a γ-ray.

In the example, aluminium-29 changes to silicon-29 by first
emitting a β-particle and then a γ-photon of energy 1.4 MeV:

The energy, and therefore the wavelength, of the γ-rays emitted
by a radionuclide are characteristic of that nuclide (see page 344).

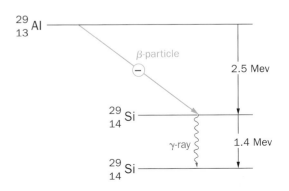

$^{29}_{13}$ Al

β-particle

2.5 Mev

$^{29}_{14}$ Si

γ-ray   1.4 Mev

$^{29}_{14}$ Si

▶ Nuclear stability

If you plot the neutron number N against the proton number Z for all the known nuclides, you get the diagram shown here:

Can you see that the stable nuclides of the lighter elements have approximately equal numbers of protons and neutrons? However, as Z increases the 'stability line' curves upwards. Heavier nuclei need more and more neutrons to be stable.

It is the strong nuclear force that holds the nucleons together, but this is almost zero above 3×10^{-15} m. The repulsive electric force between the protons is a longer-range force. So in a large nucleus all the protons repel each other, but each nucleon attracts only its nearest neighbours. More neutrons are needed to hold the nucleus together (although adding too many neutrons can also cause instability).

There are no stable nuclides with Z higher than 82!

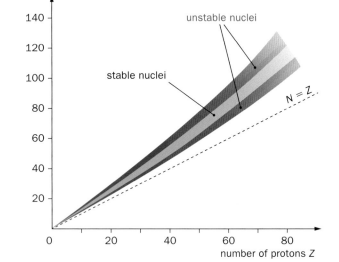

Beta⁺-decay

A radionuclide **above** the stability line decays by β-emission. Because it loses a neutron and gains a proton, it moves diagonally **towards** the stability line, as shown on this graph:

But how does a radionuclide **below** the line become more stable? There is another process called **positron-decay** or β⁺-decay.

The positron is the **antiparticle** of the electron (see Chapter 27). It has the same mass as the electron, but the opposite charge. Its symbol is $^{0}_{+1}e$ or $^{0}_{+1}\beta$.

In β⁺-decay, one of the protons in the unstable nucleus changes into a neutron and a positron, which is emitted as a β⁺-particle. The nucleus loses a proton but gains a neutron and so moves towards the stability line, in the opposite direction to β⁻-decay.

A second particle, the neutrino (ν), is emitted with the positron. (The neutrino ν and antineutrino $\bar{\nu}$ are also antiparticles.)

The nuclear equation is: $\quad ^{A}_{Z}X \Rightarrow ^{A}_{Z-1}Y + ^{0}_{+1}e + ^{0}_{0}\nu$

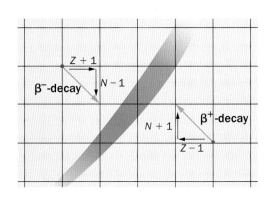

Compare these two decays:

β⁻ -decay: $^{14}_{16}C \Rightarrow ^{14}_{7}N + ^{0}_{-1}e + ^{0}_{0}\bar{\nu}$

β⁺ -decay: $^{15}_{8}O \Rightarrow ^{15}_{7}N + ^{0}_{+1}e + ^{0}_{0}\nu$

Electron capture

A radionuclide below the stability line may also decay by **electron capture**. In this process, the unstable nucleus captures an orbital electron from one of its lowest energy levels. A proton in the nucleus then changes to a neutron, and a neutrino is emitted. So the nucleus loses a proton, but gains a neutron.

The nuclear equation is: $\quad ^{A}_{Z}X + ^{0}_{-1}e \Rightarrow ^{A}_{Z-1}X + ^{0}_{0}\nu$

Can you see that this decay process has the same effect on the structure of a nucleus as β⁺-decay?

Potassium-40 can decay by all three types of β decay!

The electron capture equation is: $\quad ^{40}_{19}K + ^{0}_{-1}e \Rightarrow ^{40}_{18}Ar + ^{0}_{0}\nu$

After electron capture, the atom is short of an electron and so is in an excited state. The atom can return to the ground state by emitting an X-ray. The nucleus may also be in an excited state. It loses its surplus energy by emitting a γ-ray.

$^{40}_{19}K$ nucleus captures an orbital electron.

▶ The random nature of radioactive decay

Suppose you have a sample of 100 identical nuclei.
All the nuclei are equally likely to decay, but you can never
predict which individual nucleus will be the next to decay.
The decay process is completely **random**.

Also, there is nothing you can do to 'persuade' one nucleus to
decay at a certain time. The decay process is **spontaneous**.

You can observe the random nature of radioactive decay by
using a G–M tube to detect the radiation from a weak source.
The counts do not occur regularly. They occur randomly.

Does this mean that we can never know the rate of decay?
No, because for any particular radionuclide there is a certain
probability that an individual nucleus will decay. This means
that if we start with a large number of identical nuclei we can
predict how many will decay in a certain time interval.

Which 6 numbers will be drawn at random tonight ?

Radioactive decay using dice!

You can produce a good model of radioactive decay using
a large number of dice, with one face marked as shown:
Think of each of the dice as an 'atom' with an unstable nucleus,
so that the group of dice represent the identical atoms of a
particular radionuclide.
If you throw one of the dice and it lands with its marked face
upwards, then we say that this dice-atom has 'decayed'.

Can you predict when one individual dice-atom will decay?
No, but each dice-atom has a $\frac{1}{6}$ chance of decaying each time
it is thrown. If you throw a large number of dice-atoms at
the same time, you can predict that $\frac{1}{6}$ of them will decay.

> number of dice-atoms = the probability × number of dice-atoms
>   that decay   of decay   in the sample

Count and record the number of dice-atoms in your sample.
Throw the dice-atoms and remove those that have decayed.
Now count the number of dice-atoms that have survived.

Throw the surviving dice-atoms again and repeat the process
until only a few dice-atoms remain undecayed.

Plot a graph of the number of dice-atoms surviving against
the number of throws. It will look like the one drawn here:

Does it take about 3.8 throws for the number of dice-atoms to
fall from 200 to 100? And does it take another 3.8 throws
for the number of dice-atoms to halve again to 50?
The number of dice-atoms **always halves** in 3.8 throws.

The graph is an **exponential decay curve** because the number
of dice-atoms falls by the same fraction ($\frac{1}{6}$) after each throw.

Notice that the results are unreliable after about 8 throws.
When the sample is too small it becomes impossible to predict
accurately the number that will decay in each throw.

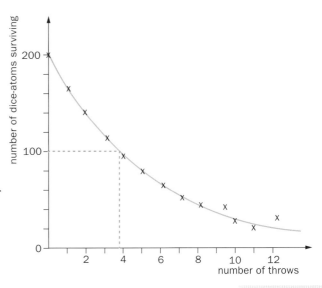

▶ The decay constant, λ

A dice-atom has a $\frac{1}{6}$ chance of decay per throw. Similarly:

The decay constant λ of a radionuclide is the probability that an individual nucleus will decay within a unit of time.

Since λ is the chance per unit time, its unit is s^{-1}, h^{-1}, day^{-1}, etc.

If the value of λ for a particular nuclide is $0.2\ s^{-1}$, then in one second each nucleus has a 0.2 or $\frac{1}{5}$ chance of decay.
In a sample of 1000 nuclei, 200 will have decayed after 1 second.

The value of λ is a constant for any particular nuclide.
Can you see that the value of λ is zero for a stable nuclide?

Phiz imagines dice-atoms with different decay probabilities.

Activity, A

The activity A of a source is the number of its nuclei that decay in unit time.

Since each decay produces ionising radiation, the activity of a source also tells you the number of ionising particles that it emits in unit time.

Activity can be measured in decays per hour, per day etc., but the SI unit of activity is the **becquerel (Bq)**.
1 Bq is an activity of 1 decay per second: $1\ Bq = 1\ s^{-1}$.

The activity of a sample must depend on the decay constant λ of the nuclide.
The greater the probability of decay, the more nuclei that decay in unit time.
Also, the activity will increase if there are more undecayed nuclei in the sample.
In fact:

Activity, A = decay constant, λ × number of undecayed nuclei, N
(s^{-1}) (s^{-1})

or $\boxed{A = \lambda N}$

You will see this equation written like this: $\boxed{\dfrac{\Delta N}{\Delta t} = -\lambda N}$

ΔN is the number of nuclei that decay in time Δt,

and so $\dfrac{\Delta N}{\Delta t}$ is the activity A.

The − sign tells us that N decreases as time passes.

> **Example 3**
> A sample of a radionuclide initially contains 200 000 nuclei. Its decay constant is $0.30\ s^{-1}$. What is the initial activity?
> $$A = \lambda N$$
> $$A = 0.30\ s^{-1} \times 200\,000 = \underline{60\,000\ s^{-1}}$$

Measuring the decay of protactinium-234

The protactinium emits β-particles which pass through the walls of the plastic bottle. They are then detected by the G–M tube.
You should record the count rate every 10 s for several minutes.
(Remember to deduct the background count from each result.)

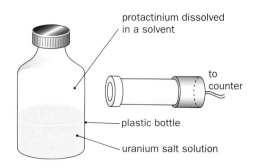

protactinium dissolved in a solvent

to counter

plastic bottle

uranium salt solution

Does this method record the activity of the protactinium?
No, not all the β-particles emitted by the protactinium are detected and so the count rate is lower than the activity of the source.
If we assume that we always detect the same percentage of the emitted radiation, the count rate is proportional to the activity.

You will find that the count rate falls, because as time goes by there are fewer and fewer protactinium nuclei left to decay.
It takes 72 s for the activity to fall to half. After another 72 s it has halved again. And after another 72 s it halves yet again.

▶ Exponential decay

Experiments like that on the facing page show that the number of nuclei in a sample always decreases in the same way.
The graph shows this decrease:
Do you recognise the shape of the graph?
It has the same form as the one from the dice-atom experiment.
It takes a time $t_{\frac{1}{2}}$ for the number of nuclei to fall from N_0 to $N_0/2$.

Can you see it takes the same time for the number of undecayed nuclei to fall from $N_0/2$ to $N_0/4$, and from $N_0/4$ to $N_0/8$?

The time $t_{\frac{1}{2}}$ is called the **half-life** of the radioactive nuclide.
It is the time taken for half the radioactive nuclei to decay.
The half-lives of different radionuclides vary widely, from fractions of a second up to millions of years.

This graph is an **exponential decay curve** because the number of nuclei always falls by the same fraction in the same time.
The equation for this curve is:

$$N = N_0 \, e^{-\lambda t}$$

where: N = the number of nuclei at time t
N_0 = the original number of nuclei
λ = the decay constant
See also page 484.

We can also write:

$$A = A_0 \, e^{-\lambda t} \quad \text{and} \quad C = C_0 \, e^{-\lambda t}$$

For spreadsheet modelling of this, see pages 482 and 486.

since the activity A is proportional to the number of nuclei present ($A = \lambda N$) and the recorded count rate C is proportional to the activity (see facing page).

The decay constant and the half life

How does the decay constant of a nuclide affect its half-life?
The higher the probability of decay λ, the more rapidly the nuclide decays, and so the shorter its half-life. In fact:

$$t_{\frac{1}{2}} = \frac{0.693}{\lambda}$$

More details are given in the box:

- Take care with the units. If λ is in s^{-1}, t and $t_{\frac{1}{2}}$ must be in s.
- The exponential equation no longer applies when only a few nuclei remain, because of the random nature of the decay.

Deriving the link between λ and $t_{\frac{1}{2}}$:

Rearranging $N = N_0 \, e^{-\lambda t}$ gives: $\dfrac{N_0}{N} = e^{\lambda t}$

When $t = t_{\frac{1}{2}}$, $N = N_0/2$ so: $\dfrac{N_0}{N_0/2} = e^{\lambda t_{\frac{1}{2}}}$

Therefore: $2 = e^{\lambda t_{\frac{1}{2}}}$
taking logs to base e: $\ln 2 = \lambda t_{\frac{1}{2}}$

So: $t_{\frac{1}{2}} = \dfrac{\ln 2}{\lambda} = \dfrac{0.693}{\lambda}$

Example 4
A radionuclide has a half-life of 55.0 s.
Initially a sample of the nuclide contains 5000 nuclei.
a) What is the decay constant for this nuclide?
b) How many nuclei remain undecayed after 200 s?

a) $t_{\frac{1}{2}} = \dfrac{0.693}{\lambda}$ so: $55.0 = \dfrac{0.693}{\lambda}$

$\therefore \lambda = \dfrac{0.693}{55.0 \text{ s}} = \underline{1.26 \times 10^{-2} \text{ s}^{-1}}$

b) $N = N_0 \, e^{-\lambda t}$
$N = 5000 \times e^{-1.26 \times 10^{-2} \times 200}$
$N = 5000 \times e^{-2.52}$
$N = \underline{400 \text{ nuclei}}$

calculator hint:
you want the value of e^x when $x = -2.52$

▶ Physics at work: Investigating the atom

Specific charge

How did J J Thomson discover the electron?
Thomson was studying cathode rays. These are produced when gases at very low pressure conduct electricity. He discovered that cathode rays are streams of negatively charged particles.
By deflecting the particles in electric and magnetic fields, Thomson was able to measure their **specific charge**. He found that he always obtained the same value, regardless of the gas he was using.
The particles were a universal constituent of all atoms: electrons.
The specific charge is often called the charge-to-mass ratio:

Specific charge, Q/m $=$ **total charge on the particle, Q** (C)
(C kg^{-1})   **total mass of the particle, m** (kg)

The charge of an ion is due to its protons and electrons.
Do you agree that for a neutral atom the total charge will be zero?
The total mass of an ion depends on the number
of protons, neutrons and electrons.

Example 5
An atom of magnesium, $^{25}_{12}$Mg, is ionised by removing two electrons.
Assuming each nucleon has a mass of 1 u, what is the specific charge of the ion?
(The mass of an electron is 9.11×10^{-31} kg. The value of u is 1.66×10^{-27} kg)

The ion has 12 protons, 10 electrons (because 2 have been removed) and 13 neutrons.

$$Q/m = \frac{2 \times \text{charge on electron, } e}{\text{mass of 25 nucleons } + \text{ mass of 10 electrons}}$$

$$Q/m = \frac{2 \times 1.6 \times 10^{-19} \text{ C}}{(25 \times 1.66 \times 10^{-27} + 10 \times 9.11 \times 10^{-31}) \text{kg}}$$

$$Q/m = \underline{7.7 \times 10^6 \text{ C kg}^{-1}}$$

Estimating the diameter of the nucleus

How did Rutherford arrive at an upper limit for the diameter of a nucleus?
He used ideas about energy and electric fields.
Here, an α-particle is heading directly towards the nucleus:

The α-particle is repelled by the nucleus, and as it approaches it loses kinetic energy and gains electric potential energy.

When all its initial kinetic energy E_k, is transferred to potential energy, the α-particle is at its closest point to the nucleus.
This is the distance r on the diagram. The α-particle stops, momentarily, before moving back along its original path.
If the nucleus has atomic number Z, then its charge will be Ze

The electric potential at distance r from the nucleus is: $\dfrac{Ze}{4\pi\varepsilon_0 r}$
(See page 296.)
The α-particle has charge $2e$ and so its electric potential energy
at its closest approach (distance r) is: $2e \times \dfrac{Ze}{4\pi\varepsilon_0 r} = \dfrac{2Ze^2}{4\pi\varepsilon_0 r}$

Kinetic energy transferred to potential energy is: $E_k = \dfrac{2Ze^2}{4\pi\varepsilon_0 r}$

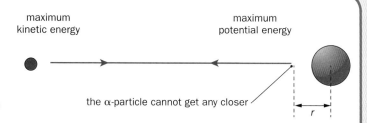

maximum kinetic energy   maximum potential energy

the α-particle cannot get any closer

Example 6
An α-particle has 5.0 MeV of kinetic energy.
What value does this give for the upper limit of a gold nucleus? Z for gold is 79.
(ε_0 has the value 8.85×10^{-12} F m^{-1})

$$5.0 \times 10^6 \times 1.6 \times 10^{-19} \text{ J} = \frac{2 \times 79 \times (1.6 \times 10^{-19})^2}{4\pi \times 8.85 \times 10^{-12} \times r}$$

$$\therefore \quad r = \frac{2 \times 79 \times (1.6 \times 10^{-19})^2}{4\pi \times 8.85 \times 10^{-12} \times 5.0 \times 10^6 \times 1.6 \times 10^{-19}}$$

$$r = \underline{4.5 \times 10^{-14} \text{ m}}$$

(The radius of the nucleus cannot be bigger than this.)

Decay chains

A radionuclide often produces an unstable daughter nuclide.
The daughter will also decay, and the process will continue until finally a stable nuclide is formed.
This is called a **decay chain** or a decay series.
One decay chain is shown here:
The stable nuclide at the very end of the chain is lead-206.

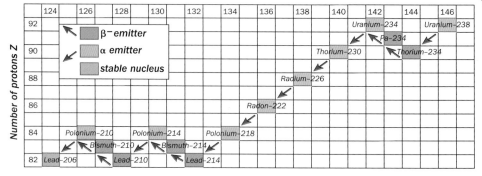

▶ Physics at work: Radioactive dating

The age of rocks

Radioactivity can be used to find the age of a rock. For example, uranium-238 (^{238}U) has a very long half-life of 4500 million years. It changes slowly by a decay chain into a stable nuclide, lead-206 (^{206}Pb).

After one half-life, half of it is unchanged and the other half has changed into lead. After two half-lives, $\frac{1}{4}$ of the ^{238}U is left and $\frac{3}{4}$ is lead.

By measuring how much of the ^{238}U in a rock has changed to ^{206}Pb, it is possible to calculate the age of the rock.

But what about the other nuclides in the decay chain? They can be ignored, because the half-life of ^{238}U is more than 20 000 times longer than any other half-life in its decay chain.

A rock from the Moon. Radioactive dating shows it is 4500 million years old, the same age as the Earth.

Example 7
In a rock sample, the proportion of ^{238}U atoms to ^{206}Pb atoms was found to be 4:1. How old is the rock?

This means that, on average, for every 5 atoms of ^{238}U when the rock was formed, 1 atom has decayed and 4 atoms have not decayed yet.
That is, 4/5 of the atoms are still radioactive ^{238}U.

So the activity of the ^{238}U in the rock is 80% of its initial value.

From the half-life graph for ^{238}U we find that this has taken <u>about 1500 million years</u>:

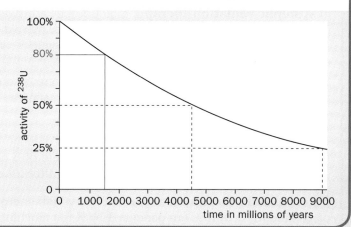

Carbon dating

There are three naturally occurring isotopes of carbon. One, carbon-14 (^{14}C), decays with a half-life of 5700 years. ^{14}C is produced in the upper atmosphere by cosmic rays.

All living things take in ^{14}C, as well as the more usual ^{12}C, as a result of photosynthesis. When living things die, they stop taking in carbon. The ^{14}C slowly decays, and so the percentage of ^{14}C in organic matter slowly decreases with time. This can be used to date bones, wood, paper and cloth.

The Dead Sea Scrolls: ^{14}C dating shows they are 1900 years old.

Example 8
The ratio of ^{14}C atoms to ^{12}C atoms in living material is $1.0 : 10^{12}$.
In an Egyptian mummy, the ratio ^{14}C : ^{12}C is $0.60 : 10^{12}$. How old is it?

The relative number of ^{14}C atoms in the mummy has fallen from 1.0 to 0.60.

Step 1:
$$t_{\frac{1}{2}} = \frac{0.693}{\lambda}$$
$$\lambda = \frac{0.693}{5700\,y}$$
$$\lambda = 1.2 \times 10^{-4}\,y^{-1}$$

Step 2:
$$N = N_0\,e^{-\lambda t}$$
$$0.60 = 1 \times e^{-1.2 \times 10^{-4}\,t}$$
logs to base e: $\ln 0.6 = -1.2 \times 10^{-4}\,t$
$$-0.5 = -1.2 \times 10^{-4}\,t$$
$$\underline{t = 4200\,y} \quad (2\text{ s.f.})$$

The Shroud of Turin: ^{14}C dating shows that it is 600 years old.

Alternatively you could read the time from a decay graph, as in Example 7. Carbon dating is not exact. The count rates involved are very small and the percentage of ^{14}C in our atmosphere has not always been constant, as we have assumed here. Modern methods try to allow for this variation.

▶ Physics at work: Radiotherapy

Radiotherapy uses high-energy radiation to kill cancer cells.
The patient can be exposed to the radiation in different ways.

In **brachytherapy**, radioactive isotopes are sealed in tiny
titanium pellets called 'seeds'. The seeds are then placed in,
or very close to, the tumour.

What isotopes are used for this therapy?
Palladium-103 and iodine-125 are used for prostate cancer.
Both decay by electron capture and subsequently emit *low-
energy* gamma rays, which are absorbed within the tumour.
Eye tumours can be treated with ruthenium-106. It emits
beta-particles which penetrate about 5 mm into the tumour.

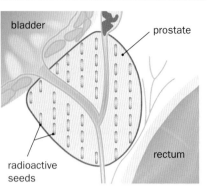

*Bracytherapy seeds, about the size of a
grain of rice, in the prostate gland.*

In **systemic radiation therapy**, radioisotopes are swallowed, or
injected into the patient. Iodine-131 is used for thyroid cancer.
The thyroid gland absorbs iodine and the energetic β-radiation
emitted by the isotope destroys the cancer cells.

Strontium-89, which is strongly absorbed by bone, also emits
β-radiation, and may be injected to treat bone cancers.

Isn't it dangerous to ingest radioactive material?
The low penetration of β-radiation causes little damage to
the tissues around the tumour. ^{131}I has a half-life of 8 days,
so the activity of the iodine falls to about 10% after 24 days.

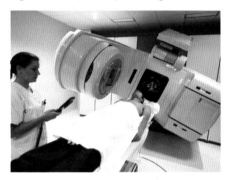

A radiation source is rotated about the patient.

External radiotherapy uses beams of very high-energy X-rays
or gamma-rays. X-rays are generated by X-ray machines.
Gamma radiation usually comes from the decay of cobalt-60.

To limit the damage to healthy cells, short doses of radiation
are usually made to enter the body at several different points,
but each dose is focussed on the tumour. This may be done by
by rotating the source about the patient's body as shown:

Electron or proton beams can also be used for radiation therapy.
Can you see from the graph that electrons have a short range?
They can be used to treat skin cancers.
Protons are strongly absorbed at a certain tissue depth, as shown:
So in proton beam therapy, the protons can be targeted at the
tumour, minimising the dose to the surrounding healthy tissue.

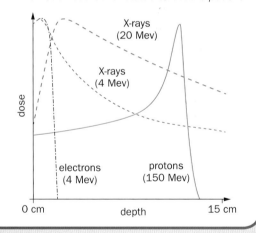

Summary

Rutherford's α-scattering experiment shows us that an atom has a small positive nucleus.
It is surrounded by a cloud of negative electrons.

In radioactive decay, a nucleus attempts to become more stable by emitting radiation.
There are 3 types of ionising radiation, α, β and γ, each with different properties (see page 361).

Nuclear decay is a spontaneous and random process. It is predictable only for large numbers of nuclei.

λ is the decay constant: the probability that a nucleus will decay in a certain time. $\qquad t_{\frac{1}{2}} = \dfrac{0.693}{\lambda}$
$t_{\frac{1}{2}}$ is the half-life: the time taken for half the nuclei in a sample to decay.

The radioactive decay process can be described by an exponential equation: $\qquad N = N_0\, e^{-\lambda t}$
N is the number of undecayed nuclei, N_0 is the value of N at $t = 0$.
A similar equation gives the activity A, or the count rate C, after a time t.

Radioactive decay and binding energy

In radioactive decay, an unstable nucleus emits radiation and becomes more stable.
The daughter nucleus always has a *higher* binding energy per nucleon than the parent.

Energy is given out when the nucleus decays. Where does this energy come from?
The total mass of the products is *less* than the mass of the parent nucleus.
The mass difference is released as energy. Look at these examples:

α-decay

Thorium-228 decays by α-emission:

$$^{228}_{90}\text{Th} \implies {}^{224}_{88}\text{Ra} + {}^{4}_{2}\alpha$$

Mass of thorium-228 nucleus = 227.979 29 u

Mass of radium-224 nucleus + α-particle
$\quad$ = 223.971 89 u + 4.001 51 u = 227.973 40 u

∴ Mass difference = 227.979 29 u − 227.973 40 u
$\qquad\qquad\qquad$ = 0.005 89 u
$\qquad\qquad\qquad$ = 5.49 MeV (as 1 u = 931.5 MeV)

What happens to this energy? It appears mostly as the kinetic energy of the α-particle. The radium nucleus also recoils slightly (and so momentum is conserved).

β-decay

Aluminium-29 decays by β-emission:

$$^{29}_{13}\text{Al} \implies {}^{29}_{14}\text{Si} + {}^{0}_{-1}\beta + {}^{0}_{0}\bar{\nu}$$

Mass of aluminium-29 nucleus = 28.973 30 u

Mass of silicon-29 nucleus + β-particle + antineutrino
$\quad$ = 28.968 80 u + 0.000 549 u + 0 = 28.969 349 u

∴ Mass difference = 28.973 30 u − 28.969 349 u
$\qquad\qquad\qquad$ = 0.003 951 u
$\qquad\qquad\qquad$ = 3.68 MeV (as 1 u = 931.5 MeV)

Some of this energy is carried away by a γ-ray (see page 365). The rest becomes the kinetic energy of the decay products.

Transmutation and energy

Transmutation is the conversion of one element into another.
Radioactive decay is a *spontaneous* transmutation.
It releases energy.

Some elements can also be made to change by firing particles at them. The particles need to be moving fast enough to penetrate the nucleus. This is called **artificial transmutation**.
Energy must be supplied for artificial transmutation to occur.

The first successful artificial transmutation was carried out by Rutherford, Marsden and Chadwick in 1919.
They converted nitrogen into oxygen by bombarding it with α-particles:

$$^{14}_{7}\text{N} + {}^{4}_{2}\alpha \implies {}^{17}_{8}\text{O} + {}^{1}_{1}\text{H}$$

Here the mass of the products is *greater* than the total mass of the nitrogen nucleus plus the α-particle.
Energy must be supplied to make this reaction happen, and balance the equation. The energy comes from the kinetic energy of the α-particles.

A modern use of artificial transmutation is the production of medical radioisotopes. For example, iodine-131 is a gamma-emitter used for the treatment of thyroid problems.
It is made by bombarding an element called tellurium (Te) with neutrons:

$$^{130}_{52}\text{Te} + {}^{1}_{0}\text{n} \implies {}^{131}_{52}\text{Te} \implies {}^{131}_{53}\text{I} + {}^{0}_{-1}\beta + {}^{0}_{0}\bar{\nu}$$

Energy, in the form of kinetic energy of the neutrons, is needed to make this reaction happen.

Ernest Rutherford

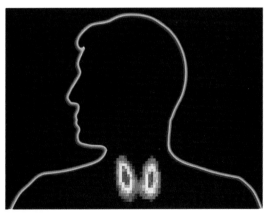

A gamma camera scan of a person injected with a radioisotope that is taken up by the thyroid.

▶ Fission

Fission means 'splitting up'.
In a fission reaction a large nucleus ($A > 200$) splits in two.

Look again at the curve of binding energy per nucleon:

If a nucleus with $A > 200$ splits in half, the two fragments have
a higher binding energy per nucleon than the parent.
This means that the fragments are more stable than the parent.
The *excess energy is released* by the reaction.

Spontaneous fission is very rare. Uranium is the largest nucleus
found on Earth. Its isotopes will sometimes fission naturally.

Bombarding the nucleus with neutrons can trigger a fission
reaction. For example:

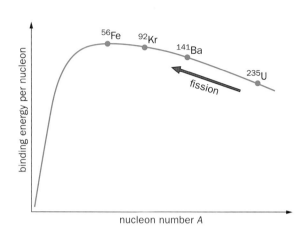

$$^{235}_{92}\text{U} + {}^{1}_{0}\text{n} \ \Rrightarrow\ ^{236}_{92}\text{U} \ \Rrightarrow\ ^{141}_{56}\text{Ba} + {}^{92}_{36}\text{Kr} + 3\,{}^{1}_{0}\text{n} + \ \text{energy}$$

How does the neutron make the nucleus less stable?
The diagram shows what happens:

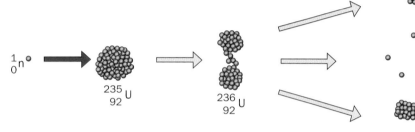

The strong forces that hold the nucleus together only act
over a very short distance (see page 364).
When a uranium nucleus absorbs a neutron, it knocks the
nucleus out of shape. If the nucleus deforms enough,
the electrostatic repulsion between the protons in each half
becomes greater than the strong force. It then splits in two.

The nucleus splits randomly. In the diagram, the fission
fragments are shown as isotopes of barium and krypton.
This is just one of the many possible combinations.

Fission of a uranium nucleus gives out about 200 MeV
of energy.

*Lise Meitner explained nuclear fission.
She worked with Otto Hahn on the first fission
experiments in 1938.*

When the uranium nucleus splits, a number of neutrons
are also ejected. If each ejected neutron causes another
uranium nucleus to undergo fission, we get a **chain reaction**:
The number of fissions increases rapidly and a huge amount
of energy is released.

Uncontrolled chain reactions are used in nuclear bombs.
The energy they unleash is devastating.

Nuclear power stations use the heat released in carefully
controlled fission reactions to generate electricity.
We will look in detail at how these operate on page 380.

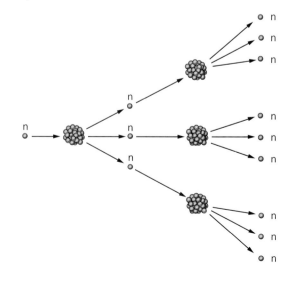

▶ Fusion

Fusion means 'joining together'.
In a fusion reaction, two light nuclei join together to make a heavier nucleus. Fusion gives out more energy per kilogram of fuel than fission. Can you see why from the graph?

The increases in binding energy per nucleon are much larger for fusion than for fission reactions, because the graph increases more steeply for light nuclei. So fusion gives out more energy *per nucleon* involved in the reaction than fission does.

The stars are powered by fusion reactions. In our Sun more than 560 million tonnes of hydrogen fuse together to make helium every second. One series of reactions for this is shown here:

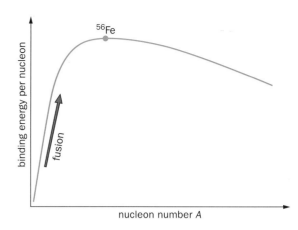

$$^{1}_{1}H + {}^{1}_{1}H \Rightarrow {}^{2}_{1}H + {}^{0}_{1}\beta^{+} + {}^{0}_{0}\nu + \boxed{\text{energy}}$$

$$^{2}_{1}H + {}^{1}_{1}H \Rightarrow {}^{3}_{2}He + {}^{0}_{0}\gamma + \boxed{\text{energy}}$$

$$^{3}_{2}He + {}^{3}_{2}He \Rightarrow {}^{4}_{2}He + 2\,{}^{1}_{1}H + \boxed{\text{energy}}$$

The energy released is radiated by the Sun at a rate of 3.90×10^{20} MW. This is the power output of a million million million large power stations!

Not surprisingly scientists are keen to develop fusion as a source of power.
One possible reaction is the fusion of deuterium and tritium. These are isotopes of hydrogen (see also page 357).

$$^{2}_{1}H + {}^{3}_{1}H \Rightarrow {}^{4}_{2}He + {}^{1}_{0}n + \boxed{\text{energy}}$$

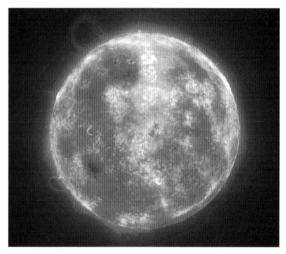

The Sun's enormous energy output comes from fusion.

Fusion has a number of advantages over fission:
- The power output per kilogram is greater.
- The raw materials are cheap and readily available.
- No radioactive elements are produced directly.
- Irradiation by the neutrons leads to radioactivity in the reactor materials, but these have relatively short half-lives and only need to be stored safely for a short time.

So why don't we use fusion in nuclear power stations?
The JET (Joint European Torus) project was set up to carry out research into fusion power. It has yet to generate a self-sustaining fusion reaction.
The main problem is getting two nuclei close enough for long enough for them to fuse. The enormous temperatures and pressures in the Sun's core provide the right conditions.
On Earth, temperatures of over 100 million kelvin are needed. At this temperature all matter exists as an ionised gas or *plasma*.

Another problem is containment. What can you use to hold something this hot? JET uses magnetic fields in a doughnut-shaped chamber called a torus, to keep the plasma away from the container walls.
Unfortunately, generating high temperatures and strong magnetic fields uses up more energy than the fusion reaction produces! We are still some years away from a fusion power station.

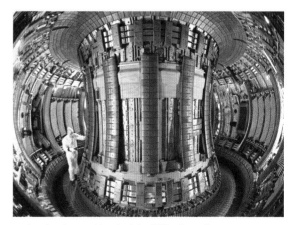

A scientist working inside JET's 6 m-diameter torus.

▶ Nuclear power stations

How is electricity generated in a nuclear power station?
In exactly the same way as in a conventional power station.
Water is heated and turned to steam. The steam drives turbines,
and the turbines power the generators:
The rotating electromagnets generate the electricity.

The only difference between a nuclear and a conventional
power station is the source of energy to heat the water.
In a conventional power station, the water is heated by burning
fossil fuels. Nuclear power stations use the energy released in
fission reactions.
Nuclear fuel is a concentrated energy source.
Fission of 1 kg of uranium-235 provides more energy than
burning 2 million kilograms of coal!

There are five key components in a nuclear reactor.
Let's take a look at them in more detail:

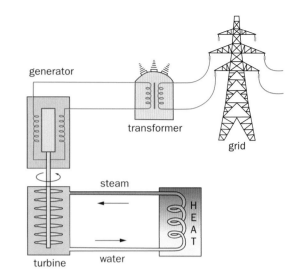

The fuel
Nuclear power stations in the UK are fuelled
by fission of uranium-235:

$$^1_0n \; + \; ^{235}_{92}U \; \longrightarrow \; ^{141}_{56}Ba \; + \; ^{92}_{36}Kr \; + \; 3\,^1_0n \; + \; \text{energy}$$

Uranium occurs naturally in the Earth's crust.
Unfortunately it contains only 0.7% uranium-235.
Natural uranium contains mostly uranium-238.
This absorbs neutrons without undergoing fission.
Uranium for use in reactors is usually artificially enriched
to increase the uranium-235 content to a few percent.

The fuel is contained in cylindrical canisters
made of magnesium alloy or stainless steel.
These **fuel rods** are inserted into tubes in the
reactor core. The rods can be easily removed
and replaced once the uranium is used up.

The moderator
Fission is triggered by the absorption of a *slow-moving*
neutron.
The neutrons produced in the reaction are very fast-moving.
They need to be slowed down before they can cause further
fission reactions.
This is the job of the **moderator**.

The neutrons collide elastically with the nuclei of the
moderator material. Energy is transferred to the moderator
and the neutrons slow down.
After repeated collisions, the neutrons' kinetic energy falls
to the average for material at that temperature.
They are then known as **thermal neutrons**.

For collisions to transfer energy efficiently, the moderator
material must have small nuclei.
Also, it must not absorb too many neutrons.
The moderator in most reactors is graphite or water.

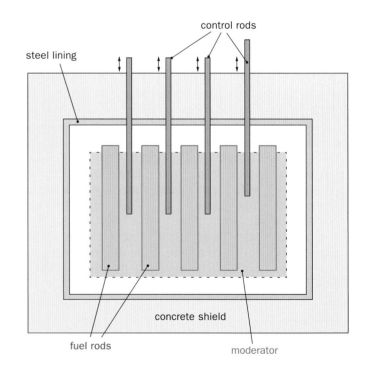

Control rods

There is only one allowable reaction rate in the reactor.
On average, *one* neutron released in each fission must cause
a further fission reaction. Any more than this and the reaction will
quickly grow out of control. Any less and the reaction will stop.

A minimum mass of uranium is needed to set up a chain reaction.
This is called the **critical mass**.
If the mass of uranium present is too small, all the neutrons
will escape from the surface before causing further fission.

Individual fuel rods are below critical mass. Their narrow shape
allows the neutrons to escape into the moderator material.
They can then be slowed down, ready to cause fission in
the next fuel rod.

The control rods control the number of neutrons available
to cause fission. The rods are usually cadmium or boron.
These are materials that absorb neutrons.
The control rods can be raised or lowered into the reactor core.
Lowering the rods causes more neutrons to be absorbed.
This reduces the reaction rate.
Raising the rods has the opposite effect.
Lowering the rods fully will shut down the reactor.

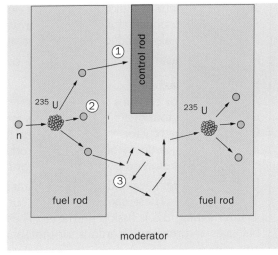

The fate of the neutrons.

○ = neutron

① Some neutrons are absorbed by the control rods.

② Some neutrons are absorbed by uranium-238 or the fission products in the fuel rods.

③ Some neutrons are slowed by the moderator and go on to cause further fission.

The coolant

A coolant extracts the heat generated in the reactor.
The coolant is pumped through the reactor core. It then passes
to a **heat exchanger**, where the heat energy of the coolant
is passed on to water to create steam.
Coolants need to be fluids that are stable at high temperatures
and non-corrosive. Common coolants are carbon dioxide and water.
Water needs to be pressurised in order to remain liquid at the
core temperatures of several hundred degrees Celsius.

The coolant becomes radioactive as it passes through the core,
so it must be fully contained.
Loss of coolant can be very dangerous.
Even if the reactor is shut down quickly, the core temperature
can continue to climb due to radioactive decay of the fission
products. Eventually **meltdown** may occur, in which
the core overheats and melts.
Emergency cooling systems need to be in place to prevent this.

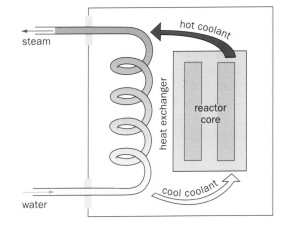

Shielding and safety

The reactor core sits inside a steel pressure vessel.
This contains the high-pressure coolant.
The reactor is then surrounded by a 5 m-thick concrete shield.
This stops neutrons and radiation escaping.

The whole thing is then housed in a steel and concrete
containment building. This is designed to contain any radiation
even in the event of meltdown of the core.

*Reaction rate, temperatures and pressures
in the reactor are constantly monitored
to ensure they are within safety limits.*

▶ Physics at work: Nuclear issues

Nuclear waste

The management of radioactive waste is one of the most controversial issues surrounding nuclear power. Although disposal and storage of nuclear waste is a problem, the quantities of waste produced are relatively low. Nuclear power also produces none of the atmospheric pollution caused by burning fossil fuels.

Radioactive waste falls into one of three categories:

High-level waste

A typical reactor contains 200 fuel rods. These need to be replaced every four years. The fuel rods are highly radioactive. Once out of the reactor, the rods continue to generate heat due to the radioactive decay of the fission products. The rods are placed in water storage pools for a year or more to cool down.

Used rods contain approximately 96% uranium isotopes and 1% plutonium. These can be separated out and reused. The remaining 3% is fission products. These have long half-lives and remain hazardous for thousands of years.

This high-level waste is sealed inside glass blocks in a process called vitrification. The blocks are placed in sealed containers and then buried underground.
The technology for packaging waste safely is straightforward. What isn't certain is how long the storage can remain intact.

The Thermal Oxide Reprocessing Plant (THORP) at Sellafield recovers valuable fuel.

Intermediate-level waste

Empty fuel rods, reactor components and chemical sludges used in treating nuclear fuel are classed as intermediate waste. This is 100 times less radioactive than the spent fuel.
This waste is encased in cement inside stainless steel drums. The drums are then stored in concrete vaults or underground.

Low-level waste

Solid low-level waste includes protective clothing, packaging and laboratory instruments. It is only slightly radioactive. Low-level waste is compacted and placed in steel containers. The containers are then stored in concrete-lined vaults.

Liquid wastes, such as cooling pond water, are cleaned and then discharged into the sea.

BNFL's low-level waste disposal site in Cumbria

Nuclear accidents

Despite stringent safety precautions, accidents do still happen.

The world's worst nuclear disaster occurred at Chernobyl in the Ukraine in 1986. Technicians had disabled vital safety systems while testing the core! An uncontrolled chain reaction led to chemical explosions that blew a hole in the reactor.
Large amounts of radiation were released into the atmosphere and carried across Europe by the wind.
More than 30 people died in the accident and over 100 000 people were evacuated from the region. The final death toll due to radiation-related illness is not yet known.

In 2011, three reactors at Fukushima in Japan went into meltdown after the coolant systems were damaged by a tsunami. Substantial amounts of radioactive material were released, leading to the evacuation of 300 000 people. Cleaning up will take decades.

Damaged reactors at Fukushima

Summary

An atomic nucleus has less mass than the separate protons and neutrons it contains.
The difference is known as the **mass difference** or **mass defect**.

Einstein's equation $E = mc^2$ gives the energy equivalent E (in J) of a mass m (kg). c is the speed of light (m s^{-1}).

The energy equivalent of the mass difference for a nucleus is called its **binding energy**.
The greater the binding energy per nucleon, the more stable the nucleus.

Fission is the splitting up of a large nucleus into two smaller nuclei, releasing energy.
Fusion is the joining together of two light nuclei to make a heavier nucleus, releasing energy.
In both fission and fusion the binding energy per nucleon *increases*. This means that the products are *more* stable.

$$1 \text{ u} = 1.66 \times 10^{-27} \text{ kg} \qquad 1 \text{ u} = 931.5 \text{ MeV}$$
$$1 \text{ MeV}/c^2 = 1.78 \times 10^{-30} \text{ kg} \qquad 1 \text{ eV} = 1.60 \times 10^{-19} \text{ J}$$

Note: For simplicity, **nuclear** masses have been used in calculations throughout the chapter.
If you are provided with atomic masses, you must remember to subtract the mass of the Z electrons ($m_e = 0.000549$ u).

▶ Questions

You will need the following data:

proton mass m_p = 1.007 28 u
neutron mass m_n = 1.008 67 u
mass of α-particle = 4.001 51 u
velocity of light c = 3.00×10^8 m s^{-1}

mass of $^2_1$H = 3.3425×10^{-27} kg
mass of $^4_2$He = 6.6425×10^{-27} kg

nuclear masses: radium-226 = 225.977 1 u
radon-222 = 221.970 3 u
uranium-235 = 234.993 4 u
lanthanum-146 = 145.868 4 u
bromine-87 = 86.902 8 u
nickel-60 = 59.915 3 u
zinc-63 = 62.920 5 u

1. Can you complete these sentences?
 a) A nucleus has .... mass than the protons and neutrons it contains. The missing mass is known as the mass .... or mass ....
 b) The energy equivalent of the mass difference is known as the .... energy.
 c) The binding energy per .... is a useful indication of stability.
   This is the average energy needed to remove each nucleon from the ....
 d) In spontaneous nuclear decay, the products weigh .... than the parent nucleus.
   The difference is released as ....
 e) In a .... reaction, a large nucleus splits in two.
   In a .... reaction, two light nuclei join together.
   In both reactions, the binding energy per nucleon .... and the products are more stable.

2. Calculate the mass difference in u and in kilograms for the following nuclei:
 a) lithium ($^7_3$Li), nuclear mass = 7.014 353 u
 b) silicon ($^{28}_{14}$Si), nuclear mass = 27.969 24 u
 c) copper ($^{63}_{29}$Cu), nuclear mass = 62.913 67 u
 d) gold ($^{197}_{79}$Au), nuclear mass = 196.923 18 u

3. For each of the nuclei in question 2, calculate:
 a) the total binding energy in eV,
 b) the binding energy per nucleon in eV.

4. The equation shows the decay of radium into radon:
$$^{226}_{88}\text{Ra} \rightarrow {}^{222}_{86}\text{Rn} + {}^4_2\alpha$$
 a) Can this reaction happen naturally? Justify your answer.
 b) Calculate the energy released in eV.

5. Calculate the energy in eV released by the following reaction:
$$^2_1\text{H} + {}^2_1\text{H} \rightarrow {}^4_2\text{He} + \text{energy}$$

6. A possible reaction for the fission of uranium is:
$$^{235}_{92}\text{U} + {}^1_0\text{n} \rightarrow {}^{146}_{57}\text{La} + {}^{87}_{35}\text{Br} + 3\,{}^1_0\text{n}$$
Calculate the energy released in the reaction.

7. Calculate the minimum energy needed to make the following reaction happen:
$$^{60}_{28}\text{Ni} + {}^4_2\alpha \rightarrow {}^{63}_{30}\text{Zn} + {}^1_0\text{n}$$

Further questions on page 396 and 397.

27 Particle Physics

What is everything made of? What holds everything together? Particle physicists are continuing to search for answers to these questions. This chapter introduces you to some of the key ideas.

In this chapter you will learn:
- what the building blocks of all matter are,
- what antimatter is and how it is created,
- about the fundamental forces that hold everything in the Universe together.

Paul Dirac predicted the existence of antimatter in 1928.

▶ Matter and antimatter

Antimatter is not just science fiction! Every particle has an equivalent antiparticle – for example, antiprotons and antineutrons. These are real particles.

Antimatter is a bit like a mirror image. A particle and its antiparticle have the same mass. They carry equal but opposite charge and they spin in opposite directions.

Some antiparticles have special names and symbols, but most are represented by a bar over the particle symbol. For example, $\bar{p}$ ('p–bar') represents an antiproton.

The existence of antimatter was predicted mathematically by British physicist Paul Dirac in 1928. Dirac's equation links the complex theories of special relativity and quantum mechanics. The equation describes both negative electrons and an equivalent *positive* particle. These **anti-electrons** or **positrons** (e^+ or β^+) were discovered experimentally by Carl Anderson in 1932.

Antiprotons and antineutrons were first observed in accelerator experiments in the mid-1950s. Since then antiparticles have been observed or detected for all known particles.
Many antiparticles occur naturally. They are created by high-energy collisions of cosmic rays with the molecules in our atmosphere. Positrons are also created in β^+-decay (page 366). This reaction is routinely used in hospitals for medical imaging by positron emission tomography (PET) (see page 432).

Can antiparticles join together to make anti-atoms and real antimatter? In theory, yes. In 1995, scientists at the European Laboratory for Particle Physics (CERN) created their first atoms of antihydrogen by joining positrons with antiprotons. Only 9 atoms were made in total. They lasted just 10^{-10} s!

By 2011, several hundred antihydrogen atoms could be made and stored, but only for around 17 minutes.
At current production rates it would take us 100 billion years to produce just 1 g of antihydrogen!

mirror image

particle   anti-particle

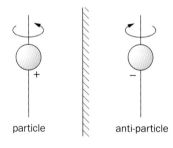

antiproton + positron ⟹ antihydrogen

$\bar{p}$ + e^+ ⟹ $\bar{H}$

Annihilation

Why does antimatter not last long? As soon as an antiparticle meets its particle, the two destroy each other. Their mass is converted to energy. This is called **annihilation**.

For example, when an electron and a positron collide, they annihilate, producing two γ-ray photons of energy:

$$_{-1}^{0}e^{-} \ + \ _{+1}^{0}e^{+} \ \Longrightarrow \ 2\ _{0}^{0}\gamma$$

Why are two photons produced? One photon could conserve charge and mass/energy. But *momentum* must also be conserved, and for that to happen, two photons are needed.
If an electron and positron with the same speed collide head-on, the total momentum before the collision is zero.
For zero momentum after the collision, we must have two identical photons moving in opposite directions.

When sufficient energy is available, annihilation can produce short-lived particles. The energy is converted back into *matter*. This is how new particles are created in accelerator experiments.

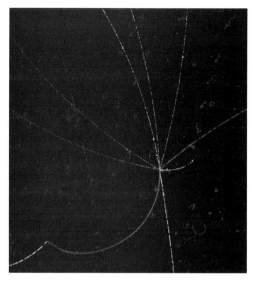

An antiproton (blue) collides with a proton in the bubble chamber liquid. The annihilation creates 4 positive (red) and 4 negative (green) particles.

Example 1
An electron and positron with negligible kinetic energy annihilate and produce two identical γ-ray photons.
Calculate a) the energy released and b) the frequency of the γ-photons.
Rest mass of an electron = 9.11×10^{-31} kg $h = 6.63 \times 10^{-34}$ J s $c = 3.00 \times 10^{8}$ m s^{-1}

(a) Using Einstein's equation (see page 375),
$E = mc^{2} = (2 \times 9.11 \times 10^{-31} \text{ kg}) \times (3.00 \times 10^{8} \text{ m s}^{-1})^{2} = \underline{1.64 \times 10^{-13} \text{ J}}$

(b) Using Planck's equation $E = hf$ (see page 328),
Energy of each photon $= \frac{1}{2} \times 1.64 \times 10^{-13}$ J $\therefore$ $f = \dfrac{E}{h} = \dfrac{\frac{1}{2} \times 1.64 \times 10^{-13} \text{ J}}{6.63 \times 10^{-34} \text{ J s}} = \underline{1.24 \times 10^{20} \text{ Hz}}$

Pair production

Pair production is the opposite of annihilation. High-energy photons can vanish, creating particle–antiparticle pairs in their place.
For example, a γ-ray can produce an electron–positron pair:

$$_{0}^{0}\gamma \ \Longrightarrow \ _{-1}^{0}e^{-} \ + \ _{+1}^{0}e^{+}$$

A third particle such as an atomic nucleus or electron is often involved indirectly. This recoils, carrying away some of the photon energy.

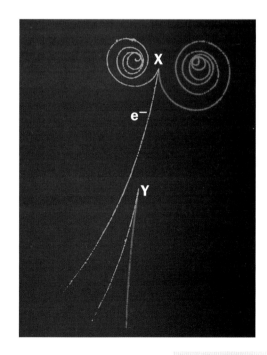

The photograph shows the creation of electron–positron pairs at X and Y in a bubble chamber (see page 359). The tracks are curved due to a magnetic field. They are shown in green (electrons) and red (positrons). Notice that the γ-ray photons leave no tracks as they are uncharged.

In event X, a third particle is involved. (This is an electron ejected from the bubble chamber liquid during the interaction.)

Why are the tracks at X more curved than at Y?
In event X the ejected electron carries off some of the energy.
This leaves less energy for the electron–positron pair.
The lower the energy (and speed) of the particles produced, the more they are deflected in the magnetic field. This is why their paths curve more.

▶ Fundamental particles

So, what is everything made of? Scientists have long searched for order in the millions of materials we see around us.

In 1869, Dmitri Mendeleev grouped elements based on their properties, to produce the first periodic table. He used the patterns he'd spotted to successfully predict the existence of new undiscovered elements. Elements were thought to consist of different indivisible atoms until 1897 when J J Thomson found the first subatomic particle, the electron. Rutherford's experiments (see page 354) then showed that atoms are made of protons, neutrons and electrons. It seemed the properties of all known matter could be explained in terms of combinations of these three subatomic particles.

Then, in the 1930s, the study of cosmic rays led to the discovery of mysterious new particles. Cosmic rays are high-energy particles (mostly protons and atomic nuclei) from beyond our Solar System. They interact with atoms in the upper atmosphere to produce showers of secondary particles. These include pions, kaons and muons (more on these later). The properties of these particles could not be explained in terms of protons, neutrons and electrons.
To aid the study of these new particles, scientists constructed particle accelerators to replicate cosmic-ray interactions in the laboratory. This allowed scientists to study the new particles produced when high-energy particles collide.
Soon a '**particle zoo**' of hundreds of new particles had been found, and physicists started to run out of symbols to label them!

Enter the quark

In 1963 the American physicist Murray Gell-Mann came up with an idea that simplified things. He spotted patterns or 'symmetries' in groups of particles (including protons and neutrons) based on their mass, charge, spin and lifetime.
He realised that this could be explained if the observed particles were made up of different combinations of smaller particles.
He called these smaller particles **quarks**.

Quarks are never found in isolation and they cannot be observed directly, but experiments have confirmed their existence.
We cannot see subatomic particles with visible light, as the wavelength is too large.
Rutherford used α-particles with de Broglie wavelengths (see page 333) of 10^{-15} m to study the properties of the nucleus.
To 'see' quarks requires an even smaller wavelength, achieved using high-energy electrons ($\sim$10 GeV).
When these electrons are fired at protons and neutrons, some pass almost straight through, while others are deflected through large angles. This is known as **deep inelastic scattering** (*inelastic* because the electrons lose kinetic energy, which is converted to short-lived particles in the collision).
This scattering can only be explained if protons and neutrons contain charge concentrated at three points. These are the three quarks that make up each proton and neutron.

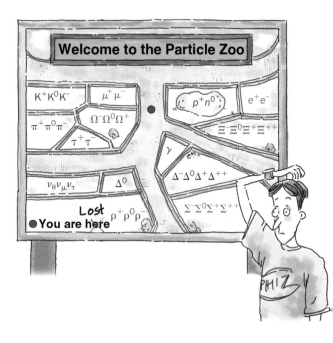

Murray Gell-Mann received the 1969 Nobel Prize for Physics for his work on fundamental particles.

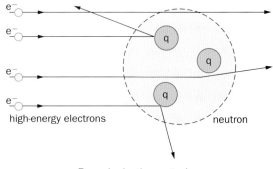

Deep inelastic scattering

The quark family

The original model proposed just three quarks called *up*, *down* and *strange*.

As accelerator technology developed, the higher-energy collisions allowed physicists to create heavier and heavier particles. These are made of more massive quarks.

We now know that there are **6** types or **flavours** of quark, arranged in three groups called **generations**:

Each of these also has an antiquark.

1st generation	increasing mass ↓	up (u)	down (d)
2nd generation		charm (c)	strange (s)
3rd generation		top (t)	bottom (b)

The quarks

Evidence for the final quark, the top quark, was found in 1995, after its existence was predicted by the symmetry of the model. This is the heaviest subatomic particle ever observed.

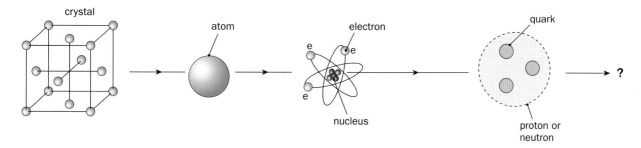

The leptons

Not everything is made of quarks. Particles can be split into two groups: **hadrons** and **leptons**.
(See also the particle family tree in the Summary on page 393.)

Hadrons are made up of quarks. These are held together by strong nuclear forces. Protons and neutrons are hadrons.

Leptons are fundamental particles. As far as we know they have **no** internal structure. They are not made up of smaller particles. Leptons do not feel strong nuclear forces.

There are **6** leptons, in three generations, to match the quarks:
Each lepton also has an antiparticle.

The muon and tau are heavier versions of the electron. They carry the same charge $(1.6 \times 10^{-19} \text{ C})$.
Charge is often given relative to this number.
The **relative charge** of the electron, muon and tau is -1. Neutrinos are uncharged.

The leptons

1st generation	increasing mass * ↓	electron e^-	electron-neutrino ν or ν_e
2nd generation		muon μ^-	muon-neutrino ν_μ
3rd generation		tau τ^-	tau-neutrino ν_τ

* neutrinos are thought to have zero rest mass

Only the electron and the electron-neutrino are stable. The other leptons rapidly decay into these.

All ordinary matter is made up of first-generation particles: up and down quarks, electrons and electron-neutrinos.
Second and third-generation particles are routinely created in accelerators and the hot centres of stars, but they are unstable. Most decay in less than a millionth of a second.

So what is everything made of? The answer appears to be leptons and quarks. This idea is known as the **Standard Model**.

Inside the hadrons

Particles that are made up of quarks are called **hadrons**.
These are split into two types:

1. **Baryons.** These contain *three quarks*.
   Protons and neutrons are baryons.
   Protons contain 2 up (u) and 1 down (d) quark.
   Neutrons contain 1 up (u) and 2 down (d) quarks.

   proton p = (u u d)   antiproton $\overline{\text{p}}$ = ($\overline{\text{u}}$ $\overline{\text{u}}$ $\overline{\text{d}}$)

   neutron n = (u d d)   antineutron $\overline{\text{n}}$ = ($\overline{\text{u}}$ $\overline{\text{d}}$ $\overline{\text{d}}$)

   The only stable baryon is the proton. All other baryons decay
   into protons. Even neutrons are unstable outside the nucleus.
   They decay with a half-life of 11 minutes:

   $$\text{n} \implies \text{p} + e^- + \overline{\nu}_e$$

2. **Mesons.** These contain *a quark and an antiquark*.
   There are no stable mesons. They rapidly decay into leptons
   and photons (energy). The only mesons that last long enough
   to leave tracks in a detector such as a bubble chamber are
   pions and kaons:

   pions: π^0 = (u $\overline{\text{u}}$) or (d $\overline{\text{d}}$) π^+ = (u $\overline{\text{d}}$) π^- = ($\overline{\text{u}}$ d)

   kaons: K^0 = (d $\overline{\text{s}}$) or ($\overline{\text{d}}$ s) K^+ = (u $\overline{\text{s}}$) K^- = ($\overline{\text{u}}$ s)

Confused by all these new names? Have a look at the particle
family tree in the Summary on page 393. This shows all the
particles you need to know at A-level.

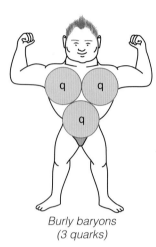

*Burly baryons
(3 quarks)*

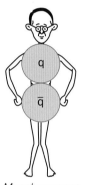

*Measly mesons
(quark + anti-quark)*

Quark properties and conservation laws

The table shows the properties of the six quarks
and their antiquarks:

Charge
This is a property that you are familiar with.

Individual quarks have fractional charges.
They combine together to make particles
with a total relative charge of 0 or ±1.

For example (using the table),

proton (u u d)
total charge $= +\frac{2}{3} +\frac{2}{3} -\frac{1}{3} = +1$

neutron (u d d)
total charge $= +\frac{2}{3} -\frac{1}{3} -\frac{1}{3} = 0$

K^- ($\overline{\text{u}}$ s)
total charge $= -\frac{2}{3} -\frac{1}{3} = -1$

In any particle interaction, **total charge
is conserved** – it stays the same.

		Symbol	Relative charge*	Baryon number	Strangeness
quarks	up	u	$+\frac{2}{3}$	$+\frac{1}{3}$	0
	down	d	$-\frac{1}{3}$	$+\frac{1}{3}$	0
	charm	c	$+\frac{2}{3}$	$+\frac{1}{3}$	0
	strange	s	$-\frac{1}{3}$	$+\frac{1}{3}$	-1
	top	t	$+\frac{2}{3}$	$+\frac{1}{3}$	0
	bottom	b	$-\frac{1}{3}$	$+\frac{1}{3}$	0
antiquarks	anti-up	$\overline{\text{u}}$	$-\frac{2}{3}$	$-\frac{1}{3}$	0
	antidown	$\overline{\text{d}}$	$+\frac{1}{3}$	$-\frac{1}{3}$	0
	anticharm	$\overline{\text{c}}$	$-\frac{2}{3}$	$-\frac{1}{3}$	0
	antistrange	$\overline{\text{s}}$	$+\frac{1}{3}$	$-\frac{1}{3}$	$+1$
	anti-top	$\overline{\text{t}}$	$-\frac{2}{3}$	$-\frac{1}{3}$	0
	antibottom	$\overline{\text{b}}$	$+\frac{1}{3}$	$-\frac{1}{3}$	0

* relative to a charge of 1.6×10^{-19} C, the charge on an electron

Baryon number

Some particle interactions never happen even though charge and mass/energy can be conserved.
For example, a proton will **not** decay into a positron and pion:

$$p \not\rightarrow e^+ + \pi^0$$

To explain this, scientists came up with another property that must be conserved, called **baryon number**.
All quarks carry a baryon number of $+\frac{1}{3}$. For antiquarks it is $-\frac{1}{3}$.

What is the total baryon number for a proton containing 3 quarks? And for a meson containing a quark and an antiquark?

Baryon numbers for different types of particles are given in the table:

Particle	Baryon number
All baryons (eg. p, n)	+1
All antibaryons	−1
Mesons (eg. π, K) and leptons	0

In the interaction shown above, the total baryon number is +1 on the left and 0 on the right. This is why the reaction cannot happen.
Baryon number is always conserved.

Lepton number

In a similar way leptons are allocated a property called **lepton number**. **Lepton number is always conserved.**
This explains why leptons are always formed from non-leptons in pairs: one lepton and one antilepton.

Particle	Lepton number
All leptons	+1
All antileptons	−1
Hadrons (baryons & mesons)	0

eg.

$$\pi^+ \rightarrow \mu^+ + \nu_\mu$$

lepton number: $\quad 0 \quad \rightarrow \quad -1 + 1$

An antilepton is created with a neutrino of the **same** generation.

Strangeness

Particles that contain strange quarks are called strange particles. Strange particles are unusually long-lived. A particle can have a strangeness number from +3 to −3 depending on the number of strange or antistrange quarks it contains.

Given its name it is not surprising that the conservation law is a bit odd too! Strangeness is conserved in interactions involving the strong nuclear force. In weak interactions, strangeness can be conserved or it can change by ± 1.

eg.

$$K^+ \rightarrow \mu^+ + \nu_\mu$$

strangeness: $\quad 1 \quad \rightarrow \quad 0 + 0$

This is allowed, as kaon decay is a weak interaction.

- Strange particles are **produced** in pairs via the strong interaction.
  Strangeness must be conserved.

- Strange particles **decay** via the weak interaction.
  Strangeness does *not* need to be conserved.

Example 2
Use the conservation laws to identify the unknown particles X and Y in the equations for electron (β^-) decay, and positron (β^+) decay.

β^-– decay.
A neutron changes to a proton and an electron:

$$n \rightarrow p + e^- + X$$

charge:	0 →	1	−1	?
baryon no.:	1 →	1	0	?
lepton no.:	0 →	0	1	?

To balance the equation, X must have charge = 0, baryon number = 0 and lepton number −1. This is an uncharged antilepton. So X is an antineutrino ($\bar{\nu}_e$).

β^+– decay.
A proton changes to a neutron and a positron:

$$p \rightarrow n + e^+ + Y$$

charge:	1 →	0	1	?
baryon no.:	1 →	1	0	?
lepton no.:	0 →	0	−1	?

To balance the equation, Y must have charge = 0, baryon number = 0 and lepton number +1. This is an uncharged lepton. So Y is a neutrino (ν_e).

Electron-neutrinos and anti-neutrinos are needed here since the decays involve first-generation leptons.
Remember that a neutrino is needed in the equation to explain the energy distribution of the particles (see page 365).

▶ Fundamental forces and exchange particles

We now know what everything is made of: quarks and leptons.
But what holds these together to make everything you see around you?
The Standard Model of particle physics describes not only the fundamental particles but also the **four fundamental forces** that control all their interactions.
You met these briefly in Chapter 2:

- **Gravitational force** (see also page 86)
  The weakest of the four forces, but it acts over an infinite distance. It pulls objects towards the Earth and holds stars and galaxies together.
- **Electromagnetic force** (see also page 290)
  This also has an infinite range. It holds atoms and molecules together. It is responsible for the chemical, mechanical and electrical properties of matter.
- **Weak nuclear force**
  Its range does not extend beyond the nucleus.
  It is responsible for β-decay and fusion reactions in stars.
- **Strong nuclear force** (see also page 364)
  This is the strongest force but it has a very short range.
  It acts only between neighbouring nucleons.
  It binds quarks and antiquarks to hold nucleons together.

We know that these forces exist. We can feel them and measure them. But what actually causes the force?
The theory is that individual particles interact by exchanging particles called **exchange particles**.

There are different exchange particles for each type of force.
For example, the exchange particle for electromagnetic force is a 'virtual photon' ('virtual' as they are not directly observable).
So two electrons repel each other by exchanging virtual photons.

To get a very simple idea of how an exchange can lead to a force, think about the ice skaters in the diagrams:
By throwing a heavy ball to each other, they are pushed further apart. The exchange produces a repulsive force.
Standing back-to-back and exchanging a boomerang will push them closer together, like an attractive force.

Exchange particles can transfer force, energy, momentum and charge between interacting particles.
The exchange particles for each force are given in the table below.
All have now been discovered, apart from the graviton.

Force	Acts on	Relative strength	Range	Exchange particle	Rest mass
strong nuclear	quarks	1	10^{-15} m	gluon (g)	0
electromagnetic	charged particles	10^{-2}	∞	virtual photon (γ)	0
weak nuclear	quarks and leptons	10^{-5}	10^{-17} m	Z^0, W^+, W^- particles	non-zero
gravity	everything with mass	10^{-40}	∞	graviton	0

▶ Feynman diagrams

The photograph shows American physicist Richard Feynman (1918–1988). Regarded as both a brilliant scientist and gifted communicator, Feynman suggested a simple way of representing particle interactions in a diagram.

Feynman diagrams consist of:

- straight lines (with one free end) to represent the physical particles *before* and *after* the interaction,

- wavy lines connecting the straight lines and representing the particle exchange. Charge must be conserved at each junction.

Here are the Feynman diagrams for 8 common interactions. Notice the 'before' and 'after' labels. Time advances from the bottom to the top of each diagram.

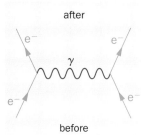

Electrostatic repulsion between 2 electrons

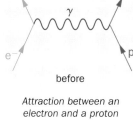

Attraction between an electron and a proton

Electromagnetic interactions

Notice that the particle lines point in the same direction for attraction and repulsion:
The direction of the lines is *not* significant.
They do *not* show the direction of the particles.
The 'virtual photon' γ that is exchanged is very short-lived and never directly observed.

Weak interactions

In β⁻-decay, a neutron changes into a proton. For this to happen, a down (d) quark changes to an up (u) quark, emitting a W⁻ particle. This decays into an electron and an antineutrino:

Feynman diagrams can be used to show changes in quark structure. Individual quarks can be represented by their own lines on the diagram (far right):
Notice that only the quark that changes is connected to the exchange-particle line.

β⁻-decay

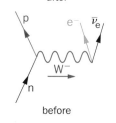

β⁻-decay showing the quarks

Electron capture and β⁺-decay are other common weak interactions:
Again these could be redrawn (as with β⁻-decay) to show the changes to individual quarks.

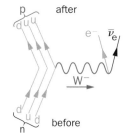

β⁺-decay

Electron capture

Strong interactions

Gluons interact with each other as well as with quarks. Gluons hold quarks together to make the hadrons. (Remember that leptons do not feel the strong nuclear force):

The strong force between protons and neutrons in the nucleus is often shown as a pion exchange. This actually results from gluon exchange between all the quarks that the protons, neutrons and pions are made of.

Gluon exchange is sometimes represented by curly rather than wavy lines.

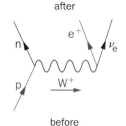

Gluons hold quarks together

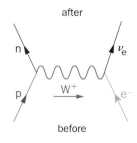

Pion exhange holds protons and neutrons together to form a nucleus

▶ Physics at work: Particle physics

World Wide Web

The internet has changed the way we work, learn, shop and
even socialise. It is an essential part of modern life.
But did you know that the way you surf the net is an offshoot
of particle physics research?

The amount of information available on-line is staggering.
The World Wide Web (www) helps to make this information
more accessible by cross-referencing related web pages.
The web was developed in 1989 by British scientist
Tim Berners-Lee. He created a new data format called
HyperText Transfer Protocol (http). It allows you to browse
through connected sites by clicking on hyperlinks that take you
directly to other pages. This was designed to allow particle
physics groups around the world to quickly share data held on
their different computers, via the internet.

*Tim Berners-Lee developed the internet
while working for CERN.*

Identifying particles

Fundamental particles are too small to see. So how do we
identify the particles created in high-energy collisions? Simple
detectors such as bubble chambers (page 359) can show the
tracks created by the particles. Analysis of these tracks can provide
all the information needed to identify the particles:

- **The greater the ionisation, the thicker the tracks.**
  Slow, heavy, charged particles leave thick tracks.
  Fast, light particles such as electrons leave thin, irregular tracks.

- **Uncharged particles leave no tracks.**
  You can guess they are there from the gaps between other tracks.

- **Positive and negative particles curve in opposite directions**
  when the detector is operated in a strong magnetic field.
  The directions are given by Fleming's left-hand rule (page 262).

- **The greater the momentum, the less curved the tracks.**
  Particles spiral inwards as they lose energy through collisions.

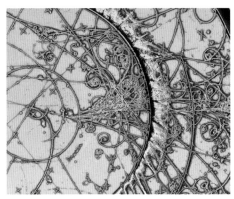

*Particle tracks in a bubble chamber.
Modern detectors use complex charge
detection networks and powerful computer
analysis but still rely on the same basic ideas.*

The LHC and the hunt for the Higgs

In July 2012 CERN announced the probable discovery of the
final piece of the Standard Model: the Higgs boson.
After a 40-year search, a new particle with mass 125–127 GeV $/c^2$
was detected. The particle existed for only a billionth of a billionth
of a billionth of a second, but by the following year tests confirmed
that this was consistent with the Higgs boson that had been
predicted by British physicist Peter Higgs. The Higgs is important
because the field it produces is thought to give particles their mass.

Detecting the Higgs required the construction of the world's largest
and most powerful particle accelerator. In the Large Hadron Collider
(LHC) at CERN, two beams of particles are accelerated along a 27 km
ring before colliding at close to the speed of light.
Superconducting magnets cooled to $-271.3\,°C$ direct the particles along their circular paths through
ultra-high-vacuum tubes. Six layers of detectors operated by international teams of scientists
analyze the particle tracks resulting from the collisions. This is physics and engineering on a grand scale.

*The LHC is huge. Can you spot the
scientist working on it in this photo?*

Summary

A particle family tree:

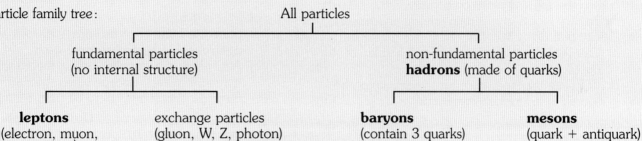

In any interaction, some properties are *always* conserved: charge, mass/energy, baryon number and lepton number. Strangeness is conserved in strong nuclear interactions. (Strangeness can change by 0 or ± 1 in weak interactions.)

Particle interactions can be represented using Feynman diagrams.

All particles have an antimatter equivalent with the same mass and equal but opposite charge and spin.

▶ Questions

1. Can you complete these sentences?
   a) A particle and its antiparticle have the same ...., and equal but opposite .... and ....
   b) When a particle and its antiparticle meet they .... Their mass is converted to ....
   c) The opposite process to annihilation is called .... production.
   d) The six types or flavours of quark are ....
   e) The six leptons are ....
   f) The .... nuclear force is felt by hadrons but not by leptons.
   g) Particles that are made up of three quarks are called ....
      Particles that contain a quark and an antiquark are called ....
   h) .... and .... are examples of baryons.
      .... and .... are examples of mesons.
   i) In any interaction, the charge, the mass/energy, the .... number and the .... number are always conserved.
   j) The exchange particle for the strong nuclear force is the ....
      Z and W particles carry the .... force.
      Electromagnetic forces are carried by ....

2. What combination of quarks do the following particles contain?
   a) proton
   b) neutron
   c) π^0, π^+, π^-
   d) K^0, K^+, K^-

3. Using the data in the table on page 388, calculate the charge, baryon number and strangeness for each of the particles in Question 2.

4. Each particle in Question 2 has an antiparticle. Use your answer to Question 2 to work out the quark combinations for these antiparticles. What do you notice about π^0 and its antiparticle?

5. What is the minimum photon energy needed to create a proton–antiproton pair?
   ($m_p = 1.67 \times 10^{-27}$ kg, $c = 3.00 \times 10^8$ m s^{-1})

6. Use the conservation laws and the data tables on pages 388–9 to work out whether the following *weak* interactions can happen:
   a) muon decay: $\mu^- \xrightarrow{?} e^- + \bar{\nu}_e + \nu_\mu$
   b) electron capture: $p + e^- \xrightarrow{?} n + \nu_e$
   c) kaon decay: $K^+ \xrightarrow{?} \pi^+ + \pi^0$

7. Sketch Feynman diagrams showing:
   a) electron capture (showing the quark changes),
   b) electrostatic repulsion between two protons,
   c) strong interaction between two protons,
   d) β^+-decay (showing the quark changes).

8. Use the table of quark properties on page 388 to explain why mesons and antimesons can only have strangeness values of 0 or ± 1.

9. Protons are injected into a synchrotron with an energy of 50 MeV. The protons are accelerated by a p.d. of 6.0 kV at 12 points around the ring. Calculate:
   a) the energy gained by the protons in one lap,
   b) the number of laps needed to reach 50 GeV.

Further questions on page 397.

▶ Radioactivity

1. A source of ionising radiation contains 3.87×10^{24} radioactive atoms of a certain nuclide and has an activity of 4.05×10^{13} Bq.
   a) Calculate the decay constant and half-life of the nuclide. [4]
   b) Calculate the time taken for the activity of the source to fall to 1.26×10^{12} Bq. [2] (OCR)

2. Boron has two isotopes represented by the nuclide symbols $^{11}_{5}$B and $^{12}_{5}$B.
   a) In what way do the nuclei Boron-11 and Boron-12 differ? [1]
   b) Calculate the number of atoms in a sample of Boron-12 of mass 6.0 kg. [2] (OCR)

3. A student carries out an experiment to determine the half-life of a radioactive isotope M. After subtracting the background count from the readings, the student plots the smooth curve shown in the diagram.

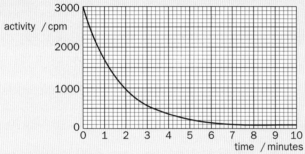

   a) From this graph he concludes that the isotope M is not pure, but contains a small proportion of another isotope C with a relatively long half-life. State a feature of the graph which supports this conclusion and estimate the activity of isotope C. [2]
   b) Determine the half-life of isotope M. Show clearly how you obtained your answer. [3] (Edex)

4. A certain radioactive source contains several radioactive isotopes and is known to emit α, β and γ radiation. The source is to be used to produce a beam of β radiation.
   a) How would you remove α radiation from the beam without significantly removing β or γ radiation? [1]
   b) The diagram shows a proposed method of separating the remaining two types of radiation.

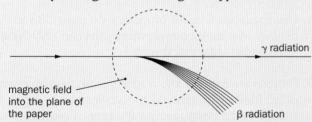

   i) Why does the magnetic field leave the γ radiation undeflected? [1]
   ii) Why does the magnetic field cause the β radiation to spread out? [3] (AQA)

5. $^{235}_{92}$U and $^{238}_{92}$U are isotopes of uranium. The half-life of $^{235}_{92}$U is 7.0×10^{8} y and that of $^{238}_{92}$U is 4.5×10^{9} y.
   a) State the number of protons, neutrons and electrons in a neutral atom of each. [2]
   b) According to a popular theory, there were equal numbers of each isotope when the Earth was formed. Assuming this to be true, estimate the ratio $R = $ (number of $^{235}_{92}$U)/(number of $^{238}_{92}$U) at a time 4.5×10^{9} years after formation. [3]
   c) The present day value of R is found to be about 0.007. What does the application of this theory suggest about the age of the Earth? [2] (OCR)

6. Radioactive sources can often be very useful, yet it is known that they are hazardous. The table lists some sources and their properties.

type of source	useful radiation	half-life
Manganese-52	γ	5.6 day
Cobalt-60	γ	5.3 year
Strontium-90	β	28 year
Thallium-210	β	1.3 minute
Radon-220	α	52 second
Americium-241	α	460 year

   a) Explain why the radiation from such sources is hazardous. [5]
   b) Radioactive sources are to be used
   i) to trace the path of an underground stream,
   ii) to kill cells in a tumour deep inside a human body, using a radiotherapy machine,
   iii) to act as a smoke detector.
   For each situation choose an appropriate source from the table, explain your choice and explain any safety precautions to be taken. [10] (OCR)

7. The fluorine isotope $^{18}_{9}$F can be produced in the process represented by
   $$X + {}^{1}_{1}p \ \Rightarrow \ {}^{18}_{9}F + {}^{1}_{0}n$$
   a) Determine the number of protons and neutrons in the nucleus X. [2]
   b) Determine the charge/mass ratio for the $^{18}_{9}$F nucleus, in C kg^{-1}. [2]
   c) Show that the charge/mass ratio for the $^{18}_{9}$F nucleus is larger than that for nucleus X. [1] (AQA)

8. Excavations near Welshpool, Powys, have revealed the remains of a large wooden monument two-thirds the size of Stonehenge. Archaeological finds suggest that the site is about 4000 y old. If carbon-14 dating were used to check this estimate, what activity would you expect to find for 1 g of the wood of the monument? (Half-life of carbon-14 is 5700 y, activity of living wood is 19 counts min^{-1} for 1 g.) [7] (W)

9. Radiocarbon dating is possible because of the presence of radioactive carbon-14 ($^{14}_{6}C$) caused by collisions of neutrons with nitrogen-14 ($^{14}_{7}N$) in the upper atmosphere. The equation for the reaction is:

$$^{14}_{7}N + ^{1}_{0}n = ^{14}_{7}C + X$$

The half-life of carbon-14 is 5700 y.

a) Identify the particle X. [2]

b) The mass of carbon-14 produced by this reaction in one year is 7.5 kg. 14 g of carbon-14 contains 6.0×10^{23} atoms. Show that the number of carbon-14 atoms produced each year is approximately 3×10^{26}. [1]

c) Calculate the decay constant of carbon-14 in y^{-1}. [2]

d) Assuming that the number of carbon-14 atoms in the Earth and its atmosphere is constant, then 3×10^{26} carbon-14 atoms must decay each year. Use this fact to calculate the number of carbon-14 atoms in the Earth and its atmosphere. [1]

e) A sample of wood containing carbon-14 from a tree which had recently been chopped down had an activity of 0.80 Bq. A sample of the same size from an ancient boat had an activity of 0.30 Bq. Sketch a graph to show how the activity of the sample, having an initial activity of 0.80 Bq, will vary with time over a period of three half-lives. Use the graph to estimate the age of the boat. [3]

f) Explain why an activity of 0.80 Bq would be difficult to measure in a school laboratory. [2] (AQA)

10. A space probe to one of the outer planets is to be powered by a radioactive source containing 2.0 kg of plutonium-238 of half-life 87 y. ($1\,y \approx 3.2 \times 10^7\,s$)

a) Calculate the approximate number of plutonium atoms initially in the source. [2]

b) Calculate the decay constant, in s^{-1}, of plutonium-238. [2]

c) Hence determine the initial activity of the source. [2]

d) Unfortunately the rocket carrying the space probe explodes on take-off and the probe's radioactive fuel is released into the Earth's atmosphere. How long will it take for the activity due to this release to fall to 10% of its initial value? [2] (Edex)

11. The activity of a sample of a radioactive isotope of iodine is 3.7×10^4 Bq. 48 hours later the activity is found to be 3.1×10^4 Bq.

a) Show that the decay constant is about $1.0 \times 10^{-6}\,s^{-1}$. [2]

b) Find the half-life of the iodine isotope. [2] (AQA)

The Nuclear Model of the Atom

12. a) Describe the principal features of the *nuclear model of the atom* suggested by Rutherford. [4]

b) When gold foil is bombarded by alpha particles it is found that most of the alpha particles pass through the foil without significant change in direction or loss of energy. A few particles are deviated from their original direction by more than 90°. Explain, in terms of the nuclear model of the atom and by considering the nature of the forces acting, why:

i) some alpha particles are deflected through large angles,

ii) most of the alpha particles pass through the foil without any significant change in direction or significant loss in energy. [5] (AQA)

13. The 3.4 MeV (5.4×10^{-13} J) α particle shown in the diagram approaches a gold nucleus. Its trajectory is such that, at its distance of closest approach, it has lost half of its kinetic energy.

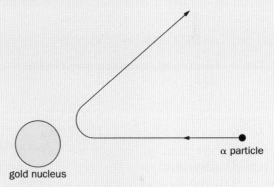

gold nucleus

α particle

a) Explain why the α particle loses kinetic energy as it approaches the gold nucleus. [2]

b) State the value of the potential energy of the α particle at the distance of closest approach. [1]

c) Sketch, on a copy of the diagram, the path of another 3.4 MeV α particle for which all the kinetic energy is converted into potential energy at its position of closest approach to the nucleus. [3] (AQA)

14. When α particles are projected at a thin metal foil in an evacuated enclosure they are scattered at various angles.

a) In which direction will the maximum number of α particles coming from the foil be detected? [1]

b) Describe the angular distribution of the scattered α particles around the foil. [2]

c) What do the results suggest about the structure of the metal atoms? [2]

d) Explain why the foil used must be thin. [1]

e) Explain why the incident beam of α particles should be parallel and narrow. [2] (AQA)

▶ **Nuclear Energy**

15. A neutral atom of a nuclide of tin has the nuclear symbol $^{120}_{50}$Sn. The total mass of the atom is not equal to the mass of all its constituent protons, neutrons and electrons. State which is larger, the mass of the atom or the mass of all its constituent particles, and explain why there is this small but important difference. [3] (OCR)

16. Describe the relevance of binding energy to nuclei. In your essay you should make reference to, and, where appropriate, include sketches, graphs, equations and examples, illustrating the following:
a) the meaning of the term binding energy, [4]
b) the relation between mass and energy, [4]
c) the variation of binding energy per nucleon with nucleon number, [4]
d) fission and fusion reactions. [4] (OCR)

17. When a high-speed α particle strikes a stationary nitrogen nucleus, it may cause a nuclear reaction in which an oxygen nucleus and a proton are formed as shown by the following equation:

$$^{14}_{7}N + {}^{4}_{2}He \rightarrow {}^{17}_{8}O + {}^{1}_{1}H$$

The masses of the nuclei are given below.

Nitrogen-14	13.999 22 u
Helium-4	4.001 50 u
Oxygen-17	16.994 73 u
Hydrogen-1	1.007 28 u

Explain how mass and energy are balanced in such a nuclear equation. [7] (OCR)

18. A nuclear reaction is represented by the equation

$$^{235}_{92}U + {}^{1}_{0}n \rightarrow {}^{143}_{54}Xe + {}^{90}_{38}Sr + 3\,{}^{1}_{0}n + E$$

a) Name the physical process represented by this equation and describe what takes place in the reaction. What does E represent? [4]
b) The mass of an atom of each of the nuclides in the equation is given below.

Uranium-235	235.043 u
Xenon-143	142.908 u
Strontium-90	89.907 u
mass of neutron	1.009 u

Calculate the difference between the total mass on the left-hand side of the equation and the total mass on the right-hand side. Express this mass difference in terms of the unified atomic mass constant u and in kg. Calculate the energy equivalent to this mass difference. [6] (OCR)

19. Describe what happens to a Uranium-235 nucleus when it is struck by a thermal neutron in a nuclear reactor. How is a working nuclear reactor shut down in an emergency? [5] (AQA)

20. The moderator, the control rods and the coolant are essential components of a thermal nuclear reactor which is supplying power. For each component, explain its function in the reactor, suggest one suitable material, and indicate one essential physical property the material must possess, apart from a tolerance of high temperatures. [11] (AQA)

21. The figure shows an enlarged portion of a graph indicating how the binding energy per nucleon of various nuclides varies with their nucleon number.

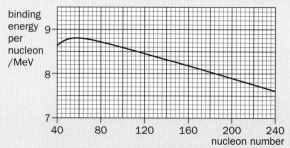

a) State the value of the nucleon number for the nuclides that are most likely to be stable. Give your reasoning.
b) When fission of Uranium-235 takes place, the nucleus splits into two roughly equal parts and approximately 200 MeV of energy is released. Use the graph to justify this figure, explaining how you arrive at your answer. [5] (AQA)

22. Calculate the energy released by the following reaction: $^{2}_{1}H + {}^{2}_{1}H \rightarrow {}^{3}_{2}He + {}^{1}_{0}n$

mass of $^{2}_{1}H = 3.344 \times 10^{-27}$ kg
mass of $^{3}_{2}He = 5.008 \times 10^{-27}$ kg
mass of $^{1}_{0}n = 1.675 \times 10^{-27}$ kg [3] (Edex)

23. a) Sketch a graph showing the relationship between binding energy per nucleon and nucleon (mass) number. Label the approximate positions of hydrogen, iron and uranium. [4]
b) State the difference between nuclear fission and nuclear fusion.
With reference to your graph, explain why both processes release energy. [4] (Edex)

24. The fusion of deuterium nuclei can be represented by the equation $^{2}_{1}H + {}^{2}_{1}H = {}^{4}_{2}He$
a) Calculate the energy released by this reaction.
Mass of $^{2}_{1}H = 2.014\,19$ u

Mass of $^{4}_{2}He = 4.002\,77$ u
1 u is equivalent to 1.49×10^{-10} J [2]
b) 2 kg (1000 mol) of deuterium is caused to fuse and it is proposed that the energy released by the fusion is used to generate electricity in a power station.
If the efficiency of the process were 52% and the electrical output of the station is to be 5.0 MW, how long would the deuterium fuel last? [5] (W)

▶ **Particle Physics**

25. a) A bubble chamber is used to investigate a reaction between sub-atomic particles. Explain how visible tracks are produced. [4]

   b) A charged elementary particle produced as a result of a collision in a bubble chamber has a lifetime of 2.9×10^{-10} s.
   By means of a suitable calculation, determine the length of its track in the bubble chamber if the particle is travelling at $0.8 \times$ the speed of light. You may ignore the effects of relativity. [2] (AQA)

26. The Feynman diagram represents the β^+ decay process.

   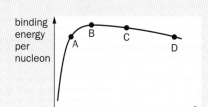

   a) What quantity changes continuously in moving from the bottom to the top of the diagram?
   b) Name the particles represented by letters A to D.
   c) What type of interaction is responsible for β^+? [5] (AQA)

27. a) A gamma ray photon can interact with a proton as represented in the following equation:
   $$p + \gamma \rightarrow n + e^+ + x$$
   Identify x, giving your reasoning.
   b) Another possible interaction is:
   $$p + \gamma \rightarrow n + y$$
   Identify y, giving your reasoning.
   c) Discuss whether the weak force is involved in each of the interactions in (a) and (b). [7] (W)

28. A graph of binding energy per nucleon against nucleon number is shown.

   Which nucleus, A, B, C or D, on the graph has the largest magnitude of binding energy? [1] (OCR)

29. This question is about hadrons.
   a) Describe three characteristics of hadrons. [3]
   b) Distinguish between mesons and baryons. [2]
   c) Determine the charge Q, the baryon number B and strangeness S of the particles X, Y and Z produced in the following strong interaction processes:
   i) $p + \pi^- \rightarrow n + X$
   ii) $p + p \rightarrow \Delta^{++} + Y$
   iii) $p + \pi^- \rightarrow K^+ + Z$
   (see pages 388–389, and given that:
   Δ^{++} has $Q = +2$, $B = +1$, $S = 0$, and
   K^+ has $Q = +1$, $B = 0$, $S = +1$) [6] (OCR)

▶ **Further Questions on Nuclear Physics**

30. A small proportion of the hydrogen in the air is the isotope tritium $^3_1$H. This is continually being formed in the upper atmosphere by cosmic radiation so that the tritium content of air is constant.
   Tritium is a beta-emitter with a half-life of 12.3 y.
   a) Write down the nuclear equation that represents the decay of tritium using the symbol X for the daughter nucleus. [2]
   b) Calculate the decay constant for tritium in y^{-1}. [1]
   c) When wine is sealed in a bottle no new tritium forms and the activity of the tritium content of the wine gradually decreases with time. At one time the activity of the tritium in an old bottle of wine is found to be $12\frac{1}{2}\%$ of that in a new bottle.
   Calculate the approximate age of the old wine. [3]
   d) Calculate the mass change, in kg, when a tritium nucleus is formed from its component parts, and the binding energy, in J, of a tritium nucleus.

mass of tritium nucleus	3.016050 u
mass of proton	1.007277 u
mass of neutron	1.008665 u
atomic mass constant u	1.660566×10^{-27} kg

   [4] (AQA)

31. The diagram represents a photograph of two events, labelled X and Y, from a bubble chamber with a magnetic field directed out of the plane of the photograph.
   In the events, pair production has occurred as a result of a gamma ray photon creating an electron and a positron.
   In event X an electron has also been ejected from an atom. The track created by this electron is labelled 'atomic electron'. The tracks created by pair production are labelled A, B, C and D.

   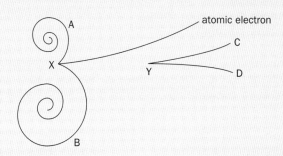

   a) Which two of the tracks were produced by positrons? [2]
   b) Explain why gamma ray photons do not leave visible tracks. [2]
   c) Explain why the tracks created by pair production in event X are much more curved than those produced in event Y. [2] (AQA)

28 Special Relativity

Special relativity is a theory first proposed by Albert Einstein in 1905. The key idea is that **the speed of light is invariant**. This means that the speed of light is completely independent of the speed of the observer or the speed of the source. This simple statement leads to some remarkable consequences.

In this chapter you will learn:
- why moving clocks run slow due to 'time dilation',
- how the length of a moving object gets shorter,
- about the dependence of mass on velocity.

Albert Einstein in California in 1933.

▷ The Michelson–Morley experiment

Can you ever be truly stationary?
Even when sitting still, you are moving at hundreds of kilometres per second as the Earth hurtles through space.

In the late 19th century it was thought that the stars and planets moved against a fixed background. A massless, non-viscous substance called 'ether' permeated the whole of space. Everything was thought to move relative to the ether.

If ether existed, the Earth's motion would create an 'ether wind'. The direction of the ether's relative motion would change depending on the month and time of day.

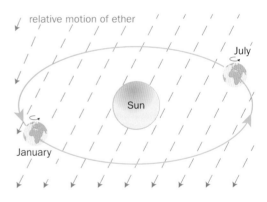

At this time it was still thought that light needed a medium through which to propagate. The ether provided this medium. It was expected that the speed of light would be affected by the ether wind, in the same way that the resultant velocity of a swimmer in a stream changes when he swims with, against or across the water current.

By finding the change in the speed of light measured in different directions and at different times of the day and year, it was thought possible to calculate the speed of the Earth relative to the ether.

In 1887 Albert Michelson and Edward Morley designed an experiment to test this out. The diagram shows the basic set-up:

The experiment uses an *interferometer* – a device that splits a ray of light into two perpendicular beams.
This is achieved using a partially reflecting mirror or 'beam splitter'. Half the incident light is reflected (beam A) and half is transmitted (beam B). Each beam travels an equal distance to a mirror and back.

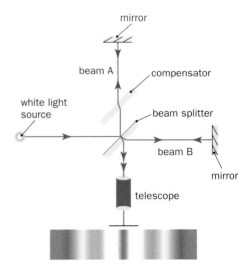

The two beams superpose when they meet up again, creating interference fringes (see page 173), viewed using the telescope. The additional plane glass block is known as a **compensator**. This ensures that both beams have the same optical path length as they travel through equal amounts of glass and air before superposing.

Each beam is moving in a different direction relative to the theoretical ether.
Rotating the apparatus through 90° should affect the time taken for each beam to reach the telescope.
This was expected to produce a shift in the interference pattern.

The experiment was repeated many times, at different times of the day and year, but *no change* in the interference pattern was ever seen. The time taken for the light to travel between the mirrors was unaffected by rotating the apparatus.

Michelson and Morley's original apparatus. It was mounted on a large stone slab floating on a pool of mercury. This helped to reduce vibrations and allowed the apparatus to rotate easily.

This **null result** led to the following important conclusions:
- **It is impossible to detect absolute motion.**
  There is no fixed background. All motion is relative.
- **The speed of light is invariant.**
  It is independent of the speed of the light source or of the observer.

Frames of reference

Since all motion is relative, to describe motion we need to use a *frame of reference* that the objects are moving relative to.

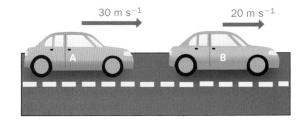

For example, the cars in the diagram are moving to the right at 30 m s^{-1} and 20 m s^{-1} in the frame of reference of the road.

If we take car B as our frame of reference, however, everything in car B is stationary relative to the frame. In this frame, car A appears to be approaching B at 10 m s^{-1} to the right, while the road ahead moves towards car B at 20 m s^{-1} to the left.

There are two types of frames of reference:
- In an **inertial frame** of reference, Newton's First Law is obeyed.
  A train carriage that is either stationary or moving at steady speed would be an inertial frame.
  A marble placed on a table in the carriage stays where it is placed.

- In an **accelerating frame** of reference Newton's First Law is not obeyed.
  If the train accelerates, a person sitting in the carriage would see the marble roll back across the table. It appears to start moving without the application of a resultant force.

[Note:
An observer outside the accelerating frame of the carriage would see that the marble's motion is actually unchanged. It only accelerates with the train once a force is applied, for example when it meets the raised edge of the table.]

▷ Einstein's postulates

Albert Einstein introduced his theory of special relativity in 1905, based on just two statements or 'postulates':

- The laws of physics are the same for all observers in any inertial frame.
- The speed of light in a vacuum is invariant.
  It is independent of the speed of the light source or the speed of the observer.

These postulates have some remarkable consequences that we will now look at in more detail.

Albert Einstein (1879–1955)

▷ Time dilation

Did you know that you can slow down time by moving faster? So a moving clock will actually run slower than a stationary one! This effect is called time dilation.

To see why time dilation must be true if the speed of light is fixed for all observers, consider this simple example:

Imagine that a person inside a fast-moving rocket fires a pulse of light from a torch towards a screen. He records that the light travels a distance d_0 to the screen, in time t_0.

An observer outside the rocket watches this happen. In the time between the pulse leaving the torch and arriving at the screen, the rocket moves forward. So, to the outside observer, the light travels a greater distance d.

We know that **speed = distance ÷ time**, but the speed of light c must be the same for both observers.
The outside observer measures a greater distance travelled by the light, so they must also find that they measure a longer time taken, in order to get the same answer for the speed. This longer time depends on the speed of the rocket.

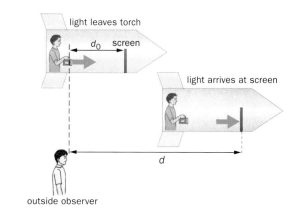

The longer time t recorded by an observer outside the rest frame of the rocket is given by:

$$t = \frac{t_0}{\sqrt{1 - v^2/c^2}}$$

t_0 = 'proper time' (s) (see below)
c = speed of light (3.00×10^8 m s^{-1})
v = relative velocity (m s^{-1}) between the outside observer and the frame in which the events occur

The **proper time t_0** is the time as measured within the inertial frame where the event occurs. In our rocket example, this is the time recorded by the person inside the rocket. Somewhat confusingly, this person is referred to as the **stationary observer**, as they are *stationary relative to the frame in which the event occurs*.

The expression $\dfrac{1}{\sqrt{1 - v^2/c^2}}$ is sometimes referred to as the

Lorentz factor or **relativistic factor**, γ.

The effects of special relativity are only really apparent at speeds approaching the speed of light.

Look at the expression for the Lorentz factor again.
At speeds much lower than the speed of light, v^2/c^2 becomes very small, and so the relativistic factor is approximately equal to 1. Therefore we can generally ignore time dilation.
The graph shows how the relativistic factor becomes increasingly significant, however, as an object approaches the speed of light, c.

The following example is clearly not very realistic, but it does illustrate how to apply the equation.

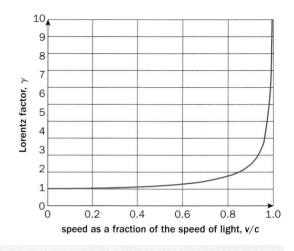

Example 1
Phiz is on a (very!) high-speed train, travelling at $0.90c$. He raises his hand for 2.0 s, timed with his stopwatch. His friend Alex times this same event from the station platform. How long was Phiz's hand raised, according to Alex?

Phiz is stationary relative to the event, so his time is the proper time, ie. $t_0 = 2.0$ s.

Time recorded by Alex: $t = \dfrac{t_0}{\sqrt{1 - v^2/c^2}} = \dfrac{2.0}{\sqrt{1 - \dfrac{(0.90c)^2}{c^2}}} = \dfrac{2.0}{\sqrt{1 - 0.90^2}}$

$$= \dfrac{2.0}{\sqrt{0.19}} = \underline{4.6 \text{ s}}$$

So, according to Alex, Phiz's stopwatch must be running slow.

Time dilation can actually be demonstrated experimentally.
An atomic clock flown around the world on a jet aircraft will show a tiny but measurable time difference when compared to an identical clock that remained on the ground. GPS satellites also need to have their time signals corrected to account for time dilation, as the satellite clocks are orbiting the Earth at about $14\,000$ km hr^{-1}.

You may have heard of the **Twins Paradox**. This is a 'thought experiment' that relies on time dilation. Suppose one twin leaves Earth on a fast-moving rocket. Time on the moving rocket will pass more slowly. The twin on the rocket will not notice anything out of the ordinary, but what seems like one second to the travelling twin will appear much longer to the twin on Earth. When our traveller returns many years later, special relativity predicts that he will be younger than his twin who remained on Earth! The paradox is that surely, in the rest frame of the rocket, the Earth appears to move away at high speed. So doesn't this mean that the twin on Earth appears to age more slowly according to the twin on the rocket?

The solution to this paradox is that **Einstein's equations of special relativity only apply in an inertial frame of reference**.
The Twins Paradox conveniently ignores the fact that, as the rocket leaves Earth, and when it turns around to come home, it would be an accelerating frame. We generally treat the Earth as an inertial frame, so only the twin on Earth can apply the equation, so he will age faster than the twin on the rocket.

▷ Muon decay

Some good experimental evidence for time dilation comes from observation of muon decay. Muons are created a few kilometres above the Earth when cosmic rays strike our atmosphere. They quickly decay into electrons, with an average lifetime of just 2.2×10^{-6} s.

This should mean that, in the time taken by the muons to travel to the Earth's surface, nearly all of them decay away.

In fact, muons are detected on Earth in much higher numbers than expected. So how does relativistic time dilation explain this?

A muon travels at about 98% of the speed of light. The muon's internal clock is therefore running slow.

This means its half-life as recorded by an observer on Earth is much longer than if the muon was at rest relative to the observer. This longer half-life gives the muon time to reach the ground.

This idea forms the basis of many exam questions. To see how this works, look at the following worked example:

Example 2
Suppose a million (1.00×10^6) muons are created 15.0 km above the Earth. If they are moving at an average speed of $0.980c$, calculate how many on average should reach the Earth's surface,
   a) if you ignore relativistic effects, and
   b) allowing for the effect of time dilation.

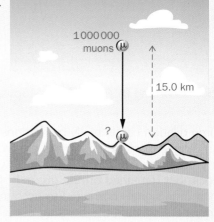

The muons' rest half-life is 1.56×10^{-6} s.

a) Time to travel 15 km $= \dfrac{\text{distance}}{\text{speed}} = \dfrac{15\,000 \text{ m}}{0.980 \times 3.00 \times 10^8 \text{ m s}^{-1}}$
$= 5.10 \times 10^{-5}$ s

The number of muons remaining after this time can then be found using the exponential decay equation (see page 369):

$$N = N_0 e^{-\lambda T} = N_0\, e^{-\left(\frac{\ln 2}{T_{1/2}} \times t\right)} = 1.00 \times 10^6 \times e^{-\left(\frac{\ln 2}{1.56 \times 10^{-6}} \times 5.10 \times 10^{-5}\right)} = \underline{1.44 \times 10^{-4} \text{ muons}}$$

(far fewer than actually detected)

b) From the frame of reference of the Earth, the muons' half-life will be dilated, as the muons are moving at high speed relative to the observer. Their half-life is effectively a ticking clock that appears to an observer on Earth to run slow.

Applying the time-dilation formula to the half-life:

Half-life as viewed from Earth: $t = \dfrac{t_0}{\sqrt{1 - {v^2}/{c^2}}} = \dfrac{1.56 \times 10^{-6}}{\sqrt{1 - \dfrac{0.980^2\, c^2}{c^2}}} = 7.84 \times 10^{-6}$ s

Now find the number remaining using the dilated half-life:

$$N = N_0 e^{-\lambda T} = N_0\, e^{-\left(\frac{\ln 2}{T_{1/2}} \times t\right)} = 1.00 \times 10^6 \times e^{-\left(\frac{\ln 2}{7.84 \times 10^{-6}} \times 5.10 \times 10^{-5}\right)}$$
$$= \underline{1.10 \times 10^4 \text{ muons}}$$

So, allowing for relativity, about 1.1% reach the ground, which is the proportion actually detected.

▶ Length contraction

Another strange consequence of special relativity is that **an object moving parallel to its length is shorter than an identical stationary object**. Again the effect is only noticeable close to the speed of light.

Length contraction is really just another way of looking at time dilation. In the muon example on page 402, in the reference frame of an observer on Earth, a muon's lifetime is dilated (increased) because its moving clock is running more slowly.

In the frame of reference of a muon, however, its half-life is still at the rest value. The Earth appears to be moving towards it.
To a muon, the distance between the top of the atmosphere and the Earth's surface would appear to contract.
The muon can reach the Earth before decaying, since from its viewpoint it decays at the normal rate but has less distance to cover.

The formula for length contraction is similar to the one for time dilation. The length l of an object measured by an observer moving at a velocity v relative to the object being measured is given by:

$$l = l_0 \sqrt{1 - v^2/c^2}$$

l_0 = proper length (m), ie. the length measured in the object's rest frame
c = speed of light (3.00×10^8 m s^{-1})
v = relative velocity (m s^{-1}) between the outside observer and the moving object

Time is dilated (increased) but length is contracted (decreased).
If your calculation is correct, you should always find that $l < l_0$.

Example 3
A spaceship of length 60.0 m passes the Earth at a speed of $0.965c$.
What is the length of the spaceship as seen by an observer on Earth?

$$l = l_0 \sqrt{1 - v^2/c^2} = 60.0 \times \sqrt{1 - \frac{0.965^2 \, c^2}{c^2}} = 15.7 \text{ m}$$

Example 4
Let's repeat the calculation of Example 2 but from the point of view of the muon.
Again let's assume one million muons are created 15.0 km above the Earth's surface (as measured in our rest frame).
How many muons on average would reach the ground if they travel at $0.980c$?
Muon rest half-life $= 1.56 \times 10^{-6}$ s.

Distance to Earth in the muon's rest frame: $l = l_0 \sqrt{1 - v^2/c^2} = 15.0 \times 10^3 \times \sqrt{1 - \frac{0.980^2 \, c^2}{c^2}} = 2985$ m

Time to reach Earth $= \dfrac{\text{distance}}{\text{speed}} = \dfrac{2985 \text{ m}}{0.980 \times 3.00 \times 10^8 \text{ m s}^{-1}} = 1.015 \times 10^{-5}$ s

So, the number of muons remaining after this time:

$$N = N_0 e^{-\lambda t} = N_0 e^{-\left(\frac{\ln 2}{T_{1/2}} \times t\right)} = 1.00 \times 10^6 \times e^{-\left(\frac{\ln 2}{1.56 \times 10^{-6}} \times 1.015 \times 10^{-5}\right)} = \underline{1.10 \times 10^4 \text{ muons}}$$

This is the same answer we found using the time-dilation formula applied from the point of view of an observer on Earth.

▶ Relativistic mass

Einstein also showed that **the mass of an object increases as its speed increases**.
Again this change is only noticeable at close to the speed of light.
The mass m of a moving object is given by:

$$m = \frac{m_0}{\sqrt{1 - \frac{v^2}{c^2}}}$$

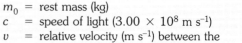

m_0 = rest mass (kg)
c = speed of light (3.00×10^8 m s^{-1})
v = relative velocity (m s^{-1}) between the outside observer and the moving object

What happens to the mass of an object as its speed approaches the speed of light, c?
The equation tells us that the mass approaches infinity.
This means that it is impossible to accelerate a mass to the speed of light, because this would require an infinite amount of energy.

The increase in mass of an object due to its velocity may seem improbable, but it is easily demonstrated in modern particle accelerators. These use powerful electric fields to accelerate charged particles such as protons and electrons.
It is found that these particles become heavier and heavier as the speed of light is approached, and greater and greater forces are needed for further acceleration.

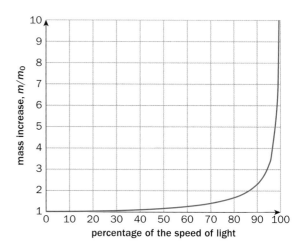

The explanation for the relativistic mass increase comes from the **principle of conservation of momentum** (see page 62), which still applies with special relativity.
To help you understand this, think about two identical rockets moving towards each other with equal but opposite velocities close to the speed of light. The rockets are close to, but on opposite sides of, an imaginary line in space, so that they collide with a slight glancing blow:

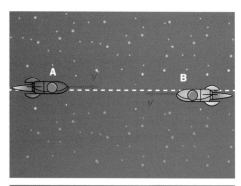

After impact the rockets must move off with equal but opposite velocities in both the x- and y-directions, in order to conserve momentum. Observers on each rocket will see themselves moving away from the line in the y-direction at the same rate.
But how fast does rocket A appear to move to an observer on rocket B, and vice versa?

We do not need to worry about length contraction in the y-direction as the rockets are only moving close to the speed of light in the x-direction, but there will be time dilation.

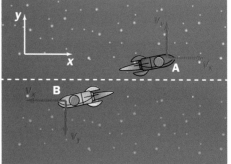

Suppose an observer on A times how long they take to move 10 m in the y-direction. An observer on B will agree with their measurement of the 10 m, but the clock on rocket A will appear to run slow so the time taken will appear longer. So to an observer on B, the rocket A has a lower velocity in the y-direction than his own rocket. We still need to conserve momentum, however, so if the velocity is lower, the mass of the rocket must have increased.

The observer on rocket A has exactly the same experience when looking at rocket B. They see the velocity of B as lower, so the mass of rocket B must be higher.

▶ Mass and energy

We came across Einstein's mass–energy equation, $E = mc^2$, on page 375. This tells us that mass and energy are essentially the same thing.

As an object speeds up, its kinetic energy increases and so does its mass. Substituting our equation for the relativistic mass increase into $E = mc^2$ gives us an equation for the **total energy** E of the object:

$$E = \frac{m_0 c^2}{\sqrt{1 - v^2/c^2}}$$

This includes the kinetic energy and takes account of the relativistic mass increase.

If we need to find the kinetic energy of an object moving at relativistic speeds, we **cannot use** $\frac{1}{2}mv^2$. That was derived assuming mass was constant, and so it is no longer valid at high speeds. Kinetic energy is the energy gained due to an object's motion, so, at relativistic speeds:

William Bertozzi, Professor of Physics at MIT

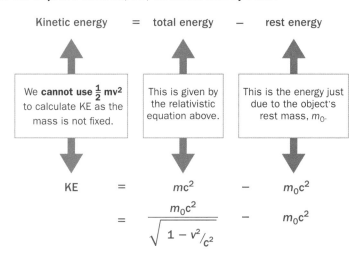

Bertozzi experiment

In 1964 a young researcher at MIT, William Bertozzi, carried out the first experiment to directly test the effect of speed on kinetic energy.

Bertozzi used a particle accelerator to accelerate electrons to high speeds. The electrons then travelled across an 8.4 m gap before colliding with an aluminium disc. The speed of the electrons was found by measuring their time of flight across the gap. The basic set-up is shown in the diagram:

The electrons were absorbed by the aluminium, which caused the disc to heat up. We saw on page 204 that the energy ΔU supplied to a material can be linked to the resulting temperature change ΔT by the specific heat capacity equation: $\Delta U = mc\Delta T$ By measuring the temperature change of the aluminium disc, the energy supplied by the electrons was found. Since the heat energy gained must equal the kinetic energy lost, this gave a measurement of the electrons' kinetic energy.

Bertozzi repeated his experiment at five different speeds. The results, shown on the graph, were in excellent agreement with the values predicted by special relativity.

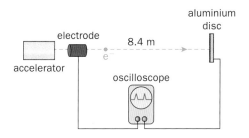

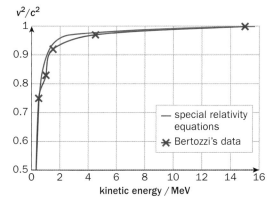

Example 5
The rest mass of an electron is 9.11×10^{-31} kg. What is the mass of an electron travelling at $0.998c$?

$$m = \frac{m_0}{\sqrt{1 - v^2/c^2}} = \frac{9.11 \times 10^{-31}}{\sqrt{1 - \frac{0.998^2 \, c^2}{c^2}}} = \frac{9.11 \times 10^{-31}}{\sqrt{3.996 \times 10^{-3}}}$$

$$= \underline{1.44 \times 10^{-29} \text{ kg}} \quad \text{(3 s.f.)}$$

$$\text{(ie. } 15.8 \times \text{ the rest mass)}$$

Example 6
A muon has a rest mass of 1.88×10^{-28} kg. What would be the kinetic energy (in MeV) of a muon travelling at a speed of $0.996c$? (Remember, $1 \text{ eV} = 1.60 \times 10^{-19}$ J)

Rest energy $= m_0 c^2 = 1.88 \times 10^{-28} \times (3.00 \times 10^8)^2 = 1.692 \times 10^{-11}$ J

Total energy: $E = \dfrac{m_0 c^2}{\sqrt{1 - v^2/c^2}} = \dfrac{1.88 \times 10^{-28} \times (3.00 \times 10^8)^2}{\sqrt{1 - \frac{0.996^2 \, c^2}{c^2}}}$

$$= 1.894 \times 10^{-10} \text{ J}$$

KE = total energy − rest energy $= 1.894 \times 10^{-10} - 1.692 \times 10^{-11} = 1.725 \times 10^{-10}$ J

$$= \frac{1.725 \times 10^{-10} \times 10^{-6}}{1.60 \times 10^{-19}} \text{ MeV} = \underline{1080 \text{ MeV}}$$
$$\text{(3 s.f.)}$$

Example 7
Taking account of the relativistic increase in mass, calculate the speed of an electron that has been accelerated from rest through a p.d. of 2.0×10^6 V.
Electron rest mass $m_0 = 9.11 \times 10^{-31}$ kg

KE gained $=$ work done by the accelerating p.d. $= e \times V = 1.60 \times 10^{-19} \times 2.0 \times 10^6 = 3.2 \times 10^{-13}$ J

Rest energy $= m_0 c^2 = 9.11 \times 10^{-31} \times (3.00 \times 10^8)^2 = 8.20 \times 10^{-14}$ J

Total energy $=$ kinetic energy + rest energy $= 3.2 \times 10^{-13} + 8.20 \times 10^{-14} = 4.02 \times 10^{-13}$ J

Total energy $= \dfrac{m_0 c^2}{\sqrt{1 - v^2/c^2}}$ Substituting into this equation gives:

$$4.02 \times 10^{-13} = \frac{9.11 \times 10^{-31} \times (3.00 \times 10^8)^2}{\sqrt{1 - v^2/c^2}}$$

Rearranging, $\dfrac{9.11 \times 10^{-31} \times (3.00 \times 10^8)^2}{4.02 \times 10^{-13}} = \sqrt{1 - v^2/c^2}$

$$\therefore \quad 1/4.903 = \sqrt{1 - v^2/c^2}$$

$$\therefore \quad 1/4.903^2 = 1 - v^2/c^2$$

$$\therefore \quad v^2/c^2 = 1 - 1/4.903^2 = 0.958$$

$$\therefore \quad v^2 = 0.958c^2$$

$$\therefore \quad v = \sqrt{0.958} \times 3.00 \times 10^8 = \underline{2.94 \times 10^8 \text{ m s}^{-1}} \quad \text{(3 s.f.)}$$

Summary

Einstein's two postulates:

- The laws of physics are the same for all observers in any inertial frame.
- The speed of light in a vacuum is invariant. It is independent of the speed of the light source or the speed of the observer.

All motion is relative. There is no absolutely fixed background (as shown by the Michelson–Morley experiment).

Time dilation, length contraction and increased mass are consequences of the invariance of the speed of light.

$$t = \frac{t_0}{\sqrt{1 - v^2/c^2}} \qquad l = l_0\sqrt{1 - v^2/c^2} \qquad m = \frac{m_0}{\sqrt{1 - v^2/c^2}}$$

The expression $\dfrac{1}{\sqrt{1 - v^2/c^2}}$ is sometimes called the **Lorentz factor** or **relativistic factor**, γ.

For an object moving at relativistic speeds:

Total energy E = kinetic energy + rest energy $(m_0 c^2)$ = $\dfrac{m_0 c^2}{\sqrt{1 - v^2/c^2}}$

t_0 = proper time, ie. the time measured in the inertial frame in which the event occurs
l_0 = proper length, ie. the length measured in the object's rest frame
m_0 = rest mass
c = speed of light (3.00×10^8 m s^{-1})
v = relative velocity between the outside observer and the moving object

Questions

1. Can you complete these sentences?
   a) The speed of light is independent of the speed of the .... or the speed of the ....
   b) In an .... frame of reference, Newton's First Law is obeyed. Einstein's equations of special relativity only apply in an .... frame of reference.
   c) A moving clock will run more .... than an identical stationary clock. The .... the speed, the greater the time difference. This effect is known as time ....
   d) Length contraction only affects lengths .... to the direction of motion.
   e) As the speed of an object increases, its mass ....
   f) The relativistic expression for the total energy of an object accounts for both its .... energy and its .... energy.

2. a) A person on a train moving at 20 m s^{-1} throws a ball in the direction of the train's motion at 4 m s^{-1}. What is the speed of the ball as seen by an observer on the station platform?
   b) A second person on the same train fires a light pulse from a laser. What is the speed of the light pulse according to (i) the person on the train, and (ii) the observer on the platform?

3. a) Why would a twin who travels away into space be younger than her non-travelling twin when she returned to Earth?
   b) What assumption about the Earth does this rely on?

4. An observer on a platform sees a 5.0 s flash of light from a torch moving past on a high-speed train. If the train is moving at 85% the speed of light, how long did the person on the train actually shine the torch for?

5. Will the measured half-life be *increased*, *decreased* or *unchanged* compared to its rest value in each of the following cases?
   a) You run past a radioactive source while measuring its half-life.
   b) A person runs past you carrying the source while you measure the half-life.
   c) You run along carrying the source while measuring the half-life.

6. An observer measures the length of a metre rule to be 80 cm. What is the speed of the rule relative to the observer?

7. Using a high p.d., a proton is accelerated from rest to a speed of $0.950c$. (Proton rest mass = 1.67×10^{-27} kg)
   a) What is the mass of the proton at this speed?
   b) Calculate the total energy of the moving proton.
   c) What is the rest energy of a proton?
   d) Find the final kinetic energy of the proton.
   e) What p.d. was needed to accelerate the proton?

8. Why can the kinetic energy of an object be increased without limit, even though its speed cannot exceed the speed of light?

Further questions on page 442.

29 Astrophysics

Astronomers have made huge strides in observing the Universe. Today we use all ranges of the electromagnetic spectrum to give us insights into the processes in distant stars and galaxies.

In this chapter you will learn:
- how reflecting and refracting telescopes are used,
- how stars are classified, and how they change,
- how Doppler shift and cosmological red-shift are used.

▷ Observing the sky

Naked-eye astronomy

Early astronomers observed the sky with the naked eye. They saw that the fixed stars appeared to rotate around the Earth, taking a little less than a day to do this. They also saw things which did not move with the stars: the Sun, the Moon and the planets Mercury, Venus, Mars, Jupiter and Saturn.
They saw other mysteries: comets which appeared from nowhere, and cloudy blurs called nebulas which moved with the stars.

Halley's comet was seen just before the Norman Invasion in 1066.

The Solar System

The use of telescopes gave us a much better understanding of the Solar System. We now know that the Sun is a star, the Moon is a planetary satellite and the Earth is one of the family of planets, dwarf planets and asteroids orbiting the Sun.
The orbits are all in the same direction because they all condensed from the same rotating cloud of interstellar dust and gas.

Comets are icy, dusty masses from outer parts of the Solar System. If disturbed by the gravity of a nearby star, they may fall inwards towards the Sun and flash past in long elliptical orbits with their tails of evaporated ice streaming away from the Sun.

Nebulas were a mystery until well into the 20th century, when Edwin Hubble showed that many are distant galaxies like our Milky Way. Other nebulas are clouds of gas and dust.

The eye and CCDs

In telescopes the image may be observed directly with the eye. But modern telescopes use the detectors found in digital cameras: CCDs.

A CCD (charge-coupled device) is a silicon chip divided into many tiny, sensitive areas (pixels). Photons hitting the chip liberate electrons and the charge on each pixel creates the image.

Look at the graph of quantum efficiency:

It shows the percentage of incoming photons that can be detected by a CCD and by the eye. The eye is very good, but a CCD is more sensitive and it can detect a much wider range of wavelengths. CCDs are now made for each part of the electromagnetic spectrum.

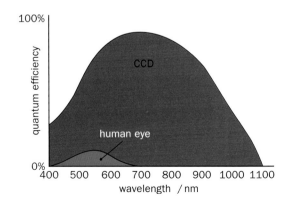

The atmosphere and the electromagnetic spectrum

Look at the diagram:

It shows the **absorbance** of different parts of the electromagnetic spectrum by the Earth's atmosphere.

Can you see which parts get through to the Earth's surface? The diagram shows that visible light, microwaves and some radio waves are transmitted by the atmosphere, together with some infrared and very small amounts of ultraviolet. Telescopes on Earth observe these regions of the electromagnetic spectrum.

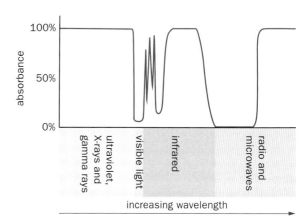

Regions of the electromagnetic spectrum

Telescopes and observatories

As you saw on page 150, astronomical telescopes are usually reflectors, and the best reflectors have large objective mirrors. There are **two** reasons why large mirrors are used:

1. Brighter images
A larger objective mirror has a larger area, so it will collect more light and the images will be less faint.

2. Better resolution
When you look through a telescope, the objective mirror or lens is the hole you are looking through. As you saw on page 168, light passing through a bigger hole spreads out much less.

small objective mirror   large objective mirror

View of the same region of the sky with two telescopes at the same angular magnification.

Example 1
The 10 m diameter Keck reflecting telescope observes infrared radiation of wavelength 2.4 μm.
What is the smallest angle θ that the telescope can resolve at this wavelength?

Rayleigh's criterion (on page 170) states that $\sin \theta \geq \dfrac{\lambda}{b}$, where b = diameter of the objective mirror.
λ = 2.4 μm = 2.4×10^{-6} m and b = 10 m

$$\sin \theta \geq \frac{2.4 \times 10^{-6} \text{ m}}{10 \text{ m}} = 2.4 \times 10^{-7} \qquad \therefore \theta \geq 0.000\,014° = \underline{0.050 \text{ seconds of arc}} \text{ (2 s.f.)}$$

(This is the angle subtended at your eye by a £1 coin placed about 100 km away!)

Radio telescopes are also reflectors, but the wavelengths that they detect are nearly half a million times bigger. This means that they need to be huge to be able to resolve even as well as the naked eye. A bigger reflector also increases the collecting power of the radio telescope, allowing astronomers to study very faint radio sources.

Where should observatories be sited?

Modern optical observatories on Earth are placed where the sky is clear and where there is little air pollution. Most are on high mountains, such as the Chilean Andes or Mauna Kea in Hawaii.

Putting observatories into Earth orbit is very expensive, but it avoids problems caused by the atmosphere. Besides detecting visible and infrared light without any atmospheric distortion, these orbiting telescopes detect other regions of the electromagnetic spectrum.

Orbiting telescopes which can 'see' ultraviolet, X-rays and gamma rays have led to much of our current knowledge of the Universe. These telescopes use specially designed CCD detectors which respond to these high-energy electromagnetic photons.

The Compton gamma-ray telescope.

▶ Measuring distances

1. Astronomical units

Distances in the solar system are often measured in **astronomical units (AU)**. The AU is the mean radius of the Earth's orbit, which is 1.5×10^{11} m. Even though this distance is huge, it's too small for measuring the distances to stars and galaxies.

For these we need *much* bigger units.

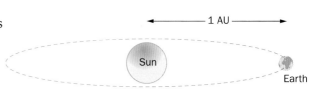

2. Light-years

A **light-year** is the distance that light travels in a year.
This is the unit usually used in popular science writing.

> *Example 2*
> What is a light-year in metres?
> One year = $365 \times 24 \times 60 \times 60$ seconds = $31\,536\,000$ s = 3.15×10^7 s
> Speed of light, $c = 3.00 \times 10^8$ m s^{-1}
> Distance = speed × time = 3.00×10^8 m s^{-1} × 3.15×10^7 s = $\underline{9.45 \times 10^{15} \text{ m}}$ (3 s.f.)

3. Parallax and the parsec

If you measure the position in the sky of a nearby star, it seems to shift slightly from January to June and then back again.
This effect is called **parallax**. You can see parallax easily by lining up your thumb on a distant part of the room with one eye closed. If you then open that eye and close the other, you will see that your thumb appears to shift from side to side.

In the sky, a nearby star (your thumb) seems to move through a very tiny angle over six months when compared with distant stars. This angle is measured in seconds of arc, and it gives us a new unit of distance: the **parsec (pc)**.

A star is one parsec away from us if it seems to move through an angle of one second (1″) when observed from two positions separated by a distance of one astronomical unit (AU).

Demonstrating parallax

> *Example 3*
> What is a parsec in metres?
> The triangle shows a star 1 parsec away being observed from two points of the Earth's orbit, A and B, 1 AU apart.
>
> $\tan(1'') = \dfrac{\text{opposite}}{\text{adjacent}} = \dfrac{1 \text{ AU}}{1 \text{ pc}}$
>
> $1'' = \dfrac{1°}{60 \times 60} = 0.000\,278°$ $\therefore 1 \text{ pc} = \dfrac{1 \text{ AU}}{\tan(0.000\,278°)} = \dfrac{1.5 \times 10^{11} \text{ m}}{0.000\,004\,85} = \underline{3.09 \times 10^{16} \text{ m}}$ (3 s.f.)

The parsec is the unit used by professional astronomers.
A star that is further away than 1 pc will have a parallax angle less than 1 second of arc. Parallax angles are so small that the small-angle approximation (see page 475) gives:

$$\text{distance in pc} = \frac{1}{\text{parallax angle in seconds of arc}}$$

1 parsec = 3.26 light-years

Even though a parsec is a huge distance, for most astrophysics the distances being measured are much, much bigger and are measured in kiloparsecs (kpc) or megaparsecs (Mpc).

▷ Classifying stars: how bright are they?

Hipparchus and the brightness of stars

The first catalogue of stars was made over 2000 years ago by the Greek astronomer Hipparchus. He sorted stars into six classes of brightness. The brightest he called **1st magnitude**. Stars that are just visible in a dark sky he called **6th magnitude**.

Apparent magnitude, m

The modern scale of magnitudes is based on measuring the light intensity received from stars. This allows each star to be given a value which is not just a whole number.

The two brightest stars in the constellation of Orion are Rigel and Betelgeuse. In Hipparchus's old system they would both be 1st magnitude. Using modern measurements of light intensity, the apparent magnitude of Betelgeuse is $m = 0.50$. Rigel is slightly brighter and has an apparent magnitude of $m = 0.12$. Note that a star which appears **brighter** has a smaller apparent magnitude, and magnitudes can be less than 0. The brightest star in the night sky is Sirius. It has an apparent magnitude $m = -1.46$.

These magnitudes are called **apparent** magnitudes because it is how they appear to us on Earth. The Sun (magnitude -27) is not really brighter than Sirius. It is just very much closer to us than Sirius.

A logarithmic scale

When the light intensities from stars of different magnitudes are compared, it is found that each decrease of 1 in magnitude corresponds to 2.51 times more light. This is a logarithmic scale, just like the decibel scale for hearing (page 134).

This relationship is shown on the graph:

Each magnitude decrease means that the star is $2.51 \times$ brighter. From magnitude 6 to magnitude 1, a decrease of 5 magnitudes, the star is $2.51 \times 2.51 \times 2.51 \times 2.51 \times 2.51 = 100$ times brighter.

Absolute magnitude, M

To make a proper comparison between different stars, we must allow for the fact that they are at different distances from us. We need to use the **absolute** magnitude, M.
This is defined as the magnitude a star would appear to have if it were at a distance of 10 pc from us.

The inverse-square law (page 127) allows us to calculate how the intensity of light from a star would change if we could 'move' it to a position 10 pc from the Earth.

From the inverse-square law, we get this equation:

$$\boxed{m - M = 5 \log \frac{d}{10}}$$ where d = distance in pc to the star and

m is its apparent magnitude. You must use base-10 logs here!

The brightest star in the daytime sky has magnitude -27.

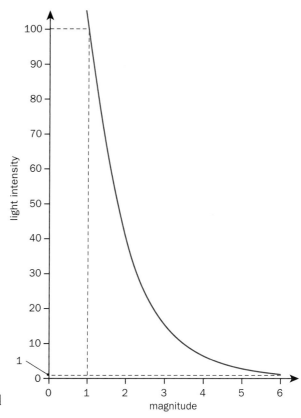

light intensity (y-axis) vs magnitude (x-axis)

Example 4

The star Betelgeuse has an apparent magnitude of 0.50. It is 430 pc from us. What is its absolute magnitude?

$m - M = 5\log\dfrac{d}{10}$ $\therefore 0.50 - M = 5 \times \log\dfrac{430}{10} = 5 \times \log 43 = 5 \times 1.633 = 8.165$

$M = 0.50 - 8.165 = -0.7665 = -7.7$ (2 s.f.)

▶ Classifying stars: how hot are they?

Colour and temperature

On page 339 you saw how Wien's Law can be used to find the temperature of a star from the peak wavelength it emits: its colour. Red stars are cooler than blue stars, so measuring the peak wavelength allows us to classify stars by their surface temperature.

Spectral classes

Stars were originally classified by the appearance of their spectra, and each class was given a letter. Each class corresponds to a range of temperatures, as shown in the table:

Spectral class	Colour	Temperature /K
O	blue	25000–50000
B	blue	11000–25000
A	blue-white	7500–11000
F	white	6000–7500
G	yellow-white	5000–6000
K	orange	3500–5000
M	red	<3500

Oh, Be A Fine Girl: Kiss Me!

One way to remember the spectral classes.

The temperature is related to the absorption spectral lines of hydrogen, particularly the Balmer series where atoms drop from the second excited state (see pages 343 and 344).

Each class is divided into a number of sub-classes. As an example, our Sun is G2, while Sirius is class A1.

Absorption lines in the spectra of stars

On page 346 you saw that Joseph Fraunhofer observed dark absorption lines in the Sun's spectrum. These lines allow us to identify the elements in the Sun's outer layers. The same is true for other stars. The important absorption lines are shown in the table.

Spectral class	Temperature /K	Absorption lines
O	25000–50000	ionised helium, helium and hydrogen
B	11000–25000	helium and hydrogen
A	7500–11000	hydrogen (strongest), ionised metals
F	6000–7500	ionised metals
G	5000–6000	ionised and neutral metals
K	3500–5000	neutral metals
M	<3500	neutral atoms, titanium oxide

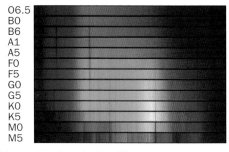

Absorption lines in the different spectral classes

As explained on page 342, absorption lines are due to atoms or ions in the outer layer of the Sun absorbing radiation coming from further inside the Sun, resulting in wavelengths 'missing' from the continuous spectrum.

Spectral class O stars do not have any neutral atoms in their outer layers because the very high temperature has ionised all of them. Class M stars, however, are not hot enough to ionise atoms. Absorption in these stars is due to neutral atoms and molecules.

▶ The Hertzsprung–Russell (HR) diagram

In 1910 the Dane, Ejnar Hertzsprung, and the American, Henry Norris Russell, plotted hundreds of different stars on a scattergram, and found that the stars were all in a few places on the diagram.

Look at the *y*-axis of the HR diagram:

Can you see that the brighter stars are near the top? Remember: a negative magnitude means that a star is very bright.

Look at the *x*-axis of the HR diagram:

Can you see that the temperature seems to be plotted in the wrong direction? This is done because the first HR diagrams were arranged by spectral class, not temperature.

Most stars, including our Sun and the bright star Sirius, are found in the diagonal band called the **main sequence**. Bigger stars are both brighter and hotter, and will be found at the upper left.

Smaller, cooler stars are at the lower end of the main sequence.

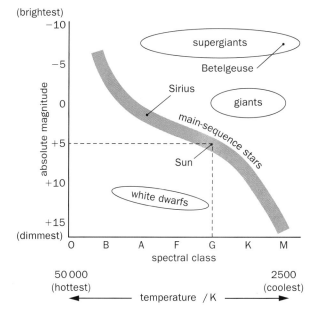

Example 5
Use the HR diagram to identify the colours of a) Betelgeuse, b) Sirius and c) the Sun.

a) Betelgeuse is class M, so it is red.   b) Sirius is class A, so it is blue-white.
c) The Sun is class G, so it is yellow-white. (You probably already knew this.)

The life cycle of our Sun

Our Sun was created when a cloud of interstellar dust and gas collapsed due to the gravitational attraction between the different parts. This collapse caused the gas in the core of the new star to heat up. When the core reached a temperature of 10 000 000 K, nuclear fusion of hydrogen into helium started (page 379).

This is when our Sun took its present place in the main sequence, about 5 billion years ago. The Sun has been stable since that time, as the inward gravitational pull is balanced by the internal pressure and radiation pressure from photons leaving the core.

When the hydrogen in the Sun's core gets used up, these balanced forces will change. The Sun will become a **red giant**. As it does this, it may eject its outer layer to form a cloud of dust and gas called a **planetary nebula**. (These are nothing to do with planets, but they looked like planets in early telescopes, and the name stuck.)

Red giant stars fuse helium into more massive nuclei such as oxygen and carbon. Radiation from the core will make the Sun expand greatly, swallowing up the planets Mercury and Venus. This bigger Sun will emit more energy, even though it is cooler. This will be the end of life on Earth (in about 5 billion years).

When the red giant Sun has used up its helium, it will start to cool. Gravity will make the dead star collapse to a density about 100 000 times greater than lead. The collapse will heat up the dying Sun; it will become a **white dwarf**, then slowly cool as it emits radiation.

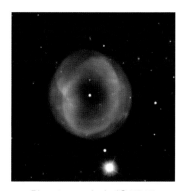

Planetary nebula IC 5148

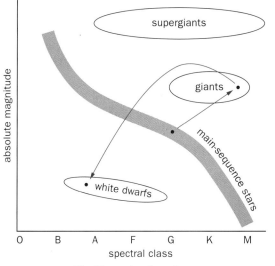

The life cycle of our Sun.

▶ The Doppler effect

Have you ever stood by the side of the road while a police car or a fire engine raced past with its siren blaring?
Did you hear the change in pitch (frequency) as it passed?

This change in frequency is called the **Doppler effect**.
It applies to electromagnetic radiation as well as sound.

Why do the frequency and wavelength change?

Look at the first two snapshots of a rocket emitting a radio signal:

In these first two snapshots, the rocket is not moving.
The first snapshot shows a **wavefront**, in blue, just leaving the rocket. The second snapshot is one period later, so the rocket is emitting a second wavefront. The first wavefront has moved forwards, and the distance between the two wavefronts is one wavelength, λ.

Stationary rocket giving out radio waves

Now look at the second pair of snapshots of the rocket:

This time the rocket has a constant velocity v, so the moving rocket has given the second wavefront a 'start' in its race after the first.

Can you see that the new wavelength, λ', is less than λ?

The change in wavelength $\Delta\lambda = \lambda' - \lambda$, so $\Delta\lambda$ is negative.

We can use this equation to calculate the speed v from $\Delta\lambda$:

Moving rocket giving out radio waves

$$\frac{\text{Change in wavelength, } \Delta\lambda\,(\text{m})}{\text{Original wavelength, } \lambda\,(\text{m})} = -\frac{\text{speed of the moving source of waves, } v\,(\text{m s}^{-1})}{\text{speed of waves, } c\,(\text{m s}^{-1})} \quad \text{or} \quad \frac{\Delta\lambda}{\lambda} = -\frac{v}{c}$$

Alternatively, the equation can be written in terms of frequency, f: $\boxed{\dfrac{\Delta f}{f} = \dfrac{v}{c}}$

Example 6

A **binary star** consists of two similar stars orbiting each other.

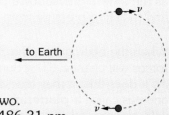

This orbit is seen 'edge on' from the Earth, as indicated in the diagram:
The hydrogen Balmer absorption line, which has a wavelength 486.13 nm when measured in a laboratory, can be observed in the spectrum of the binary star.

As the two stars move away from and towards the Earth, this spectral line splits into two.
At the position in the diagram, the wavelengths of the split lines are 485.95 nm and 486.31 nm.
a)   How can you prove that the two stars are moving at the same speed v?
b)   Calculate v.   (Speed of light, $c = 3.0 \times 10^8$ m s^{-1})

a)   The two Doppler shifts are 486.31 nm − 486.13 nm = 0.18 nm, and 486.13 nm − 485.95 nm = 0.18 nm
     As the Doppler shifts are the same, they are moving at the same speed (in opposite directions).

b)   For the star moving away from us,

$$\frac{\Delta\lambda}{\lambda} = -\frac{v}{c} \qquad \therefore \quad \frac{0.18 \text{ nm}}{486.13 \text{ nm}} = -\frac{v}{3.0 \times 10^8 \text{ m s}^{-1}} \qquad \therefore \quad 0.00037 = -\frac{v}{3.0 \times 10^8 \text{ m s}^{-1}}$$

$$v = -0.00037 \times 3.0 \times 10^8 \text{ m s}^{-1} = \underline{-1.1 \times 10^5 \text{ m s}^{-1}} \quad \text{(2 s.f.)}$$

The minus sign shows that the star is moving away from us, in the opposite direction to the light.

Red-shift and blue-shift

In Example 6, one star was moving away from us, and the absorption line in its spectrum had been moved to a longer wavelength. This shifted it towards the red end of the spectrum, and is called **red-shift**.
The other star was coming towards us, so the absorption line is shifted towards the blue end of the spectrum – this is **blue-shift**.

Red-shift always refers to the wavelength getting longer.
This is true even if the absorption line has been moved out of the red part of the spectrum into the infrared!

Cosmological red-shift

Studies of distant galaxies reveal that they *all* show red-shift, and the further away the galaxy is, the greater the red-shift is.
This led to the Big Bang theory: the Universe all started at a point and is expanding outwards.

Look at the diagram:

It is a model of the entire Universe!

As you blow up the balloon, the galaxies marked on it all move apart.
The rubber in the balloon expands, carrying the galaxies with it.
Space is modelled by the rubber of the balloon, not the air inside it.

The galaxies were not thrown out in a gigantic explosion – space itself has expanded, carrying the galaxies with it.

A model of the expanding Universe.

Light in transit from these galaxies is also stretched, so that its wavelength increases.
This is called **cosmological red-shift**. It is given the symbol **z**,
where $z = \dfrac{\Delta\lambda}{\lambda}$.

Although cosmological red-shift is not Doppler shift, the same equation can be used to link the speed of the moving galaxy and the red-shift.

Cosmological red-shift, $z = \dfrac{\Delta\lambda}{\lambda} = \dfrac{v}{c}$

Example 7
A gamma-ray burst picked up by the Swift satellite identified a new distant galaxy. The spectrum of infrared radiation emitted by this galaxy has an absorption line at a wavelength of 1100 nm.
This was identified as a hydrogen absorption line, which has a wavelength of 120 nm when observed in a laboratory.

What is the red-shift of this galaxy?

True λ = 120 nm = 120×10^{-9} m
observed λ = 1100 nm = 1100×10^{-9} m

$\Delta\lambda = 1100 \times 10^{-9}$ m $- 120 \times 10^{-9}$ m $= 980 \times 10^{-9}$ m

$z = \dfrac{\Delta\lambda}{\lambda} = \dfrac{980 \times 10^{-9} \text{ m}}{120 \times 10^{-9} \text{ m}} = \underline{8.2}$ (2 s.f.)

Note: we cannot use $z = 8.2 = \dfrac{v}{c}$ to find the speed of this galaxy, as it is travelling so fast that special relativity (Chapter 28) applies.

First discovery of the distant galaxy GRB 090423 by the Swift satellite

415

▶ Hubble's Law

Measuring the distances to galaxies

In 1920 Edwin Hubble estimated the distance to nebula M31 (a fuzzy object in the night sky) and showed that it is far outside our galaxy. It is a galaxy like our Milky Way.

Hubble made his measurement by finding the apparent magnitude of a type of star of known absolute magnitude inside nebula M31. He went on to observe many more of these nebulas, which are now called galaxies. He estimated their distances by assuming they all had the same absolute magnitude. He also looked at their spectra and found their red-shifts.

From his observations, Hubble produced the equation we now

know as Hubble's Law: $\boxed{v = H_0\,d}$

v is the velocity at which the galaxy is moving away from us and d is the distance to that galaxy. H_0 is the Hubble constant.

Because astronomers always use megaparsecs for intergalactic distances and kilometres per second for speeds, the Hubble constant is in very odd units: $\text{km s}^{-1}\,\text{Mpc}^{-1}$.

Edwin Hubble ...

... and the space telescope named in his honour.

Example 8
A galaxy at a distance of 280 Mpc is observed to have a red-shift $z = 0.065$.
What is the value of the Hubble constant?

$$z = \frac{v}{c} \quad \therefore 0.065 = \frac{v}{3.0 \times 10^8 \text{ m s}^{-1}} \qquad \therefore v = 0.065 \times 3.0 \times 10^5 \text{ km s}^{-1} = 19\,500 \text{ km s}^{-1}$$

$$v = H_0\,d \quad \therefore H_0 = \frac{v}{d} = \frac{19\,500 \text{ km s}^{-1}}{280 \text{ Mpc}} = \underline{70 \text{ km s}^{-1} \text{ Mpc}^{-1}} \quad \text{(2 s.f.)}$$

The age of the Universe

The Hubble constant allows us to work out how old the Universe is! First, we must convert $\text{km s}^{-1}\,\text{Mpc}^{-1}$ into a proper SI unit.

On page 410 you saw that that $1 \text{ pc} = 3.09 \times 10^{16} \text{ m}$.

$$1 \text{ km s}^{-1} \text{ Mpc}^{-1} = \frac{1 \text{ km}}{s} \times \frac{1}{\text{Mpc}} = \frac{1000 \text{ m}}{s} \times \frac{1}{10^6 \text{ pc}}$$

$$= \frac{1000 \text{ m}}{s} \times \frac{1}{10^6 \times 3.09 \times 10^{16} \text{ m}} = 3.2 \times 10^{-20} \text{ s}^{-1}$$

So, from Example 8, the Hubble constant is:

$$H_0 = 70 \times (3.2 \times 10^{-20} \text{ s}^{-1}) = 2.3 \times 10^{-18} \text{ s}^{-1}$$

Assuming that the Hubble constant has always had the same value, the time taken for a galaxy moving at a speed $v = \dfrac{d}{t}$ to travel

a distance d is given by time $t = \dfrac{d}{v}$. As $v = H_0 d$, $t = \dfrac{d}{H_0 d} = \dfrac{1}{H_0}$.

This is the **Hubble time**.

This is the time taken for any galaxy to reach its present position, starting at the Big Bang: it is the age of the Universe.

$\dfrac{1}{H_0} = \dfrac{1}{2.3 \times 10^{-18} \text{ s}^{-1}} = 4.3 \times 10^{17} \text{ s}$

$= \dfrac{4.3 \times 10^{17}}{365 \times 24 \times 60 \times 60}$ years

$= 14\,000\,000\,000$ years

Phiz calculates the age of the Universe.

416

Supernovas and distance measurement

Hubble's measurements of distances were not accurate, because he assumed that all galaxies were of the same brightness. Supernovas, the brightest objects in the Universe, have now corrected this error.

Look at the diagram:

This is a light curve for a Type 1a supernova over several months.

Can you see that it rises quickly to a maximum, then dies away?

All Type 1a supernovas are produced in the same way, and they all have the same peak brightness: an *absolute* magnitude of -19.3. Astronomers recording a supernova in a distant galaxy can easily identify a Type 1a supernova from the light curve. If they have measured the *apparent* magnitude of the supernova at its peak, they can calculate the distance to the galaxy.

This method allowed astronomers to make accurate measurements of the distance to distant galaxies. This confirmed Hubble's Law and gave an accurate value of the Hubble constant H_0.

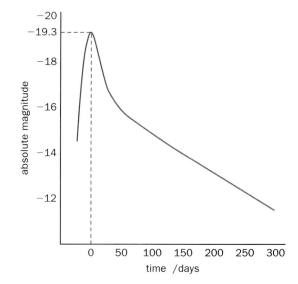

Example 9

A Type 1a supernova in a distant galaxy has a maximum *apparent* magnitude of 18.1.
Calculate the distance to this galaxy.

$m - M = 5 \log \dfrac{d}{10}$ and we know that the **absolute** magnitude $M = -19.3$.

$18.1 - (-19.3) = 5 \log \dfrac{d}{10}$ $\therefore 18.1 + 19.3 = 37.4 = 5 \log \dfrac{d}{10}$ $\therefore \log \dfrac{d}{10} = \dfrac{37.4}{5} = 7.48$ $\therefore \dfrac{d}{10} =$ inverse log of 7.48

$\therefore \dfrac{d}{10} = 10^{7.48} = 30\,000\,000$ $\therefore d = 10 \times 30\,000\,000$ pc $= 300\,000\,000$ pc $= \underline{300 \text{ Mpc}}$ (2 s.f.)

▷ Physics at work: Dark matter and dark energy

In the 1960s, astronomer Vera Rubin measured the rotation of galaxies and found that they did not seem to obey Newton's Laws. The only explanation seemed to be that galaxies were surrounded by vast clouds of invisible **dark matter** which had gravitational effects on the stars, but could not be detected in any other way. The existence of dark matter is now accepted by most physicists, although it is not agreed what it actually is.
The most popular theory is that it consists of weakly interacting massive particles (WIMPs) left over from the Big Bang.

In the 1990s, more problems cropped up. Astronomers used Type 1a supernovas to measure the distances to extremely distant galaxies. They expected to find the galaxies were slowing down as gravity pulled back against the outward motion of the expanding Universe. They found the opposite. The galaxies were all dimmer than was expected. This suggested that the expansion of the Universe is speeding up, not slowing down.

To explain this, scientists say that **dark energy** is pushing everything outwards. No one yet knows what dark energy is, but it is agreed that 'normal' matter like you, me and the rest of our Solar System makes up only about 5% of the Universe. The other 95% is about one-third dark matter and two-thirds dark energy.

Astronomer Vera Rubin

417

▷ Physics at work: Supermassive black holes

At the end of its life, a star the size of our Sun will finish its life cycle as a white dwarf. A star which is between 1.3 and 3 times the mass of the Sun will eventually become a neutron star. Stars larger than this will become black holes, and these may be 50 or 60 times more massive than the Sun. Many black holes have now been observed.

They cannot be observed directly, as they emit no light, but their strong gravitational effects on other stars can produce radiation. Some sources of strong X-ray emissions are due to matter being 'sucked in' to a black hole from a nearby star.

Studies of the rotation of our Milky Way galaxy shows that it has an extremely large amount of mass concentrated at its centre. The Chandra X-ray satellite measured strong X-ray emissions from a region in the constellation of Sagittarius. The movement of stars in this region shows that they are orbiting an object smaller than our Solar System, and that the object they are orbiting has a mass about 4 million times greater than the mass of our Sun!

This object is a **supermassive black hole** at the centre of our galaxy. It is thought that most, if not all, galaxies have supermassive black holes at their centres. It is thought that these have been formed by black holes pulling in and merging with nearby stars and with smaller black holes.

X-ray image of the centre of our galaxy

▷ Physics at work: Quasars

Quasi-stellar radio sources – **quasars** for short – were discovered in the 1960s by radio-astronomers Thomas Matthews and Allan Sandage. 'Quasi-stellar' means 'apparently, but not really, stars'. Quasars were given the name because they emit radio waves like galaxies, but appear small, like stars.

Quasars have small angular sizes because they are vast distances from us, and are the furthest known objects in our Universe. For them to be such strong radio sources, yet be so far away, they must be emitting enormous amounts of energy.

Quasar 3C 273 is emitting about 100 times the energy of the entire Milky Way. In the 1960s, there was no known way to explain how this energy was generated.

Quasars are found at the centres of certain **active galaxies**. These are galaxies which emit much more energy than a normal galaxy. The supermassive black hole at the centre of an active galaxy creates energy by pulling matter from nearby stars into **accretion discs**. These accretion discs orbit rapidly as they fall inwards, emitting jets of radiation along their axis of rotation.

When we observe very distant galaxies, we are looking at radiation which left them billions of years ago. This means that the galaxies containing quasars were young galaxies formed in the early Universe.

It has been suggested that quasars can form when two galaxies merge. This could happen to the Milky Way and the Andromeda galaxy M31, which are moving towards each other. But don't worry: this will not happen for at least 3 billion years.

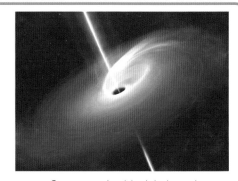

Supermassive black hole and accretion

The Andromeda galaxy: on a collision course with us!

▷ Physics at work: Neutron stars

The chair Phiz is sitting on is nearly all empty space. Why does it not collapse under his weight?

The explanation from quantum theory is that it is not possible to squeeze matter closer because electrons would be forced to be in the same quantum state, and this is impossible. This resistance to compression is called **electron degeneracy pressure**. This is what stops white dwarf stars from contracting more.

For more massive stars, the end of the red giant stage is dramatic. If the star has a mass greater than 1.4 times the mass of the Sun – this is called the *Chandrasekhar limit* – then nuclear fusion in the core produces nuclei as massive as iron. Iron nuclei cannot undergo nuclear fusion, so all nuclear processes stop. Gravitational collapse compresses the star, which then explodes in a violent change called a **supernova**. Supernovas emit more light than all of the rest of the stars in their galaxies, so they are easy to detect.

Supernovas are often accompanied by **gamma-ray bursts**. These are narrow beams of gamma rays emitted along the axis of the collapsing star.

During a supernova, the remains of the core form a super-dense object called a **neutron star**. All of the protons and electrons have been combined into neutrons.
A neutron star is far denser than a white dwarf: a teaspoonful of neutron star matter would have a mass of 5000 million tonnes!

Electron degeneracy pressure in action.

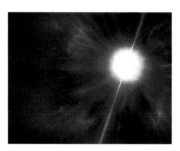

A gamma-ray burst

▷ Physics at work: Black holes

Electron degeneracy pressure stops atoms in a white dwarf from collapsing into neutrons. In the same way, **neutron degeneracy pressure** stops a neutron star from further collapse.

But what would happen if the mass of the star is so big it can overcome this pressure?
In this case, the gravitational attraction makes the star continue to contract. There are no other forces which can resist the collapse, so the star will contract down until it has no size at all, called a **singularity**. This totally collapsed star is a **black hole**.

The gravitational field near a black hole is immense. A spacecraft at a distance *r* from the singularity would need great kinetic energy to escape from the gravitational potential well (see page 96).

$\frac{1}{2}mv^2 > \frac{GMm}{r}$. Dividing both sides by $\frac{1}{2}m$, we get $v^2 > \frac{2GM}{r}$
where *v* is the escape velocity.

The spacecraft in the diagram can escape, but as you get closer to the singularity, *r* gets smaller and so *v* gets bigger.
At the **event horizon**, a distance R_s from the singularity, $v = c$.

R_s is called the **Schwarzschild radius**, which is given by equation $\boxed{R_s = \frac{2GM}{c^2}}$.

From any point closer to the singularity than this, not even light can escape! That is why it is called a black hole.

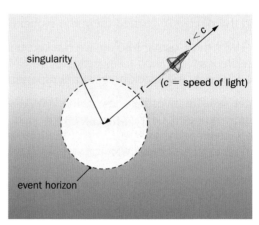

A black hole.

▷ Support for the Big Bang theory

1. The hydrogen:helium ratio

According to the Big Bang theory, in the early Universe protons and neutrons were formed in equal numbers. Like electrons, free protons are stable. However, free neutrons are not stable. They decay with a half-life of about 10 minutes. As the Universe continued to cool, most neutrons decayed into protons and electrons. According to this theory, by the time the Universe was cool enough for stable nuclei to form, protons outnumbered neutrons by about 7 to 1. The nuclei formed would then be about one helium-4 nucleus for every 12 hydrogen-1 nuclei, with tiny proportions of other light nuclei.

Measurements of the spectra of stars and galaxies have confirmed that the ratio of hydrogen to helium throughout the Universe is very close to this value, supporting the Big Bang prediction.

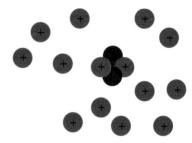

Hydrogen nuclei and helium nuclei are found throughout the Universe in the proportion predicted by the Big Bang theory.

2. Cosmic microwave background radiation

As the Universe continued to expand and cool, it reached a temperature at which electrons and nuclei combined into atoms. Matter changed from a plasma into a gas, and became transparent. This early Universe was then a black body full of radiation.

As the Universe expanded, all the radiation in it 'stretched' as well; the wavelengths increased. Big Bang theory predicts that it should now be at a temperature of a few K. Wien's Law (page 339) tells us that the peak wavelength of a black body at this temperature would be in the microwave region of the electromagnetic spectrum.

In 1964, Arno Penzias and Robert Wilson detected microwaves coming from all directions in space. This is called the **cosmic microwave background radiation (CMBR)**, and the discovery earned them the Nobel Prize in 1978.

Wilson and Penzias with their microwave detector.

In 1989 the Cosmic Microwave Background Explorer (COBE) satellite was launched to investigate the CMBR. It confirmed that the Universe is a black body at a temperature of 2.7 K.

Later satellite investigations by the WMAP and Planck satellites greatly refined COBE's measurements and confirmed the age of the Universe, the temperature of the Universe and the amounts of dark energy and dark matter (page 417) with high precision.

The hydrogen:helium ratio and the presence of the CMBR are examples of the **Cosmological Principle**: we believe that the laws of Physics are the same everywhere in the Universe. Everywhere we look, we observe the same physics in action.

WMAP microwave map of the Universe, showing tiny variations in temperature

Example 10
Calculate the peak wavelength of the cosmic microwave background radiation, given that the temperature of the Universe is 2.7 K.

Wien's Law states that $\lambda_{max} \times T = 0.0029 \ \text{m K}$

$$\lambda_{max} \times 2.7 \ \text{K} = 0.0029 \ \text{m K}$$

$$\lambda_{max} = \frac{0.0029 \ \text{m K}}{2.7 \ \text{K}} = 0.00107 \ \text{m}$$

$$= 1.07 \ \text{mm} = \underline{1.1 \ \text{mm}} \qquad \text{(2 s.f.)}$$

▶ Physics at work : Exoplanets

The possibility of life outside Earth has fascinated people for many years.
Probes have not yet found any traces of life elsewhere in the Solar System, but the idea that planets like ours could be found around other stars made astronomers look further away.

The Cosmological Principle states that the same laws of Physics apply everywhere in the Universe, so if life can exist here (and it does), then why not on a planet around a distant star?

It is thought that life needs liquid water to develop, so the search has been for planets in the 'Goldilocks' zone around their star – not too hot and not too cold – where water will be a liquid and not just ice or steam. In those conditions, life could possibly develop.

Finding **exoplanets** – planets around other stars – is very difficult.
The stars are so far away that the angular separation between them and their planets is very small. Any light reflected off those planets is swamped by the light from their star, so direct imaging of these planets is extremely hard to do, although some have been seen.

The first method used to detect exoplanets was to observe changes in the Doppler shift in the light emitted by a star. A star shifts to and fro as its planet orbits. This is just like the way the hammer-thrower on page 76 moves as his hammer 'orbits' him. The regular to-and-fro tug on the star by its planet alternately increases and decreases its velocity relative to us. This gives a regular change in the frequencies of light detected on Earth.

The other method used is based on careful measurement of the intensity of the light emitted by the star.
While a planet is in **transit**, passing between us and its star, the intensity of light from the star drops slightly.
This method has been used by NASA's Kepler mission to identify nearly a thousand exoplanets.

The Kepler mission scientists were helped in this search by 'crowd-sourcing'. They used the internet to enlist hundreds of volunteers who used their computers to look at light curves for possible transits, finding over 500 exoplanets so far!

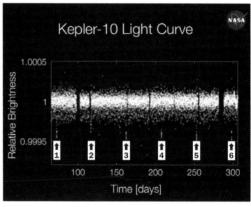

Light curve of Kepler 10b

The six drops in brightness each show a transit of the planet.
The black bands are where the telescope was turned off.

Summary

Astronomical distances can be measured in AU, light-years or parsecs.

Astronomical refracting telescopes have angular magnification $= \dfrac{f_o}{f_e}$ (see Chapter 11, pages 148–149).

Distance in parsecs (pc) $= \dfrac{1}{\text{parallax angle in seconds}}$

Absolute magnitude M, apparent magnitude m and distance d in parsecs are linked by $m - M = 5\log\dfrac{d}{10}$.
Small stars like our Sun will become red giants and end their lifetimes as white dwarfs.
Larger stars like Sirius will become red supergiants and end their lifetimes, after supernova explosions, as neutrons stars or black holes.

For a black hole, the Schwartzschild radius $R_s = \dfrac{2GM}{c^2}$ is the limit inside which light cannot escape.

Doppler effect for nearby planets, stars and galaxies: $\dfrac{\Delta\lambda}{\lambda} = -\dfrac{v}{c}$

Cosmological red-shift for distant galaxies: $z = \dfrac{\Delta\lambda}{\lambda} = -\dfrac{v}{c}$
Hubble's Law for distant galaxies: $v = H_0\,d$

▶ Questions

1. Can you complete these sentences?
   a) The atmosphere absorbs all electromagnetic radiation except ...... and ...., some ...... and a little ......
   b) The detector used in modern telescopes to detect light is a ....
   c) Modern telescopes have big objective mirrors because they collect more ...... and because they can ...... stars which are closer together.
   d) Astronomers measure distances to stars in ......
   e) The brightness of stars is based on the ...... scale. On this scale, a larger number means the star is ...... (dimmer/brighter).
   f) The ...... diagram is used to show the brightness and temperature of different stars.
   g) A star moving towards us has wavelengths in the light it emits which are ......-shifted according to the ...... effect.
   h) Hubble's Law shows that the Universe is ......

2. The Hubble space telescope has an objective mirror of diameter 2.4 m. It can observe ultraviolet light of wavelength 320 nm. Calculate the smallest angle θ that the telescope can resolve.

3. Put these distances in increasing order of size: astronomical unit, kiloparsec, light-year, megametre.

4. A star has apparent magnitude $m = 3.4$. It is a distance $d = 50$ pc from us.
   Calculate its absolute magnitude M.

5. Use data from the tables on page 412 to write down the spectral class of each of these stars:
   a) a blue-white star,
   b) a star with surface temperature 12 000 K,
   c) a star with titanium oxide absorption lines in its spectrum.

6. Sketch a labelled diagram to show the main regions of the Hertzsprung–Russell diagram, showing the position of a class B main sequence star and where that star will be when it starts fusion of helium. Explain what will happen to that star after all nuclear fusion has finished.

7. Explain why red giant stars are often surrounded by clouds of dust and gas called planetary nebulas.

8. Describe the differences between white dwarfs, neutron stars and black holes, and explain why different stars end their life cycles as one of these.

9. A Type 1a supernova in a distant galaxy had a maximum apparent magnitude of +12.5.
   a) Calculate the distance d to this galaxy.
   b) Explain the way in which the active galaxies called quasars produce powerful beams of energy.

10. Observations of the spectrum of different parts of the Sun show that the sodium absorption line, normally of wavelength 589 nm, is blue-shifted at the eastern edge by 0.0078 nm and red-shifted by the same amount at the right-hand edge.
    Explain what this shows about the Sun's motion.

11. Use the galaxy data below to plot a velocity–distance graph and find the Hubble constant H_0.

Distance /Mpc	7.8	9.1	11.5	14.6	20.8
Velocity /km s^{-1}	515	640	705	1100	1380

12. The Lovell radio telescope at Jodrell Bank, Manchester, has a mirror of diameter 76 m. The radio telescope at Arecibo in Puerto Rico has a mirror of diameter 300 m. Compare these two telescopes in terms of:
    a) their resolving powers,   b) their collecting powers.

13. The graph below is an absorption spectrum from a distant galaxy showing two absorption lines.

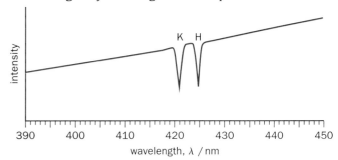

The unshifted values of these two lines are K: 393 nm and H: 397 nm. Use both values of wavelength from the graph to find the red-shift z of this galaxy.

14. It is thought that baryonic matter – the ordinary matter of which we are made – is only a small fraction of the Universe. What does the rest of the Universe consist of?

15. Calculate the Schwarzschild radius R_S of a black hole 100 times more massive than the Sun.
    (Mass of Sun = 2.0×10^{30} kg)

16. The Big Bang theory was supported by the discovery of the cosmic microwave background radiation. Explain how the theory accounts for this radiation.

17. Explain why observing exoplanets is so difficult, and describe the two methods used to find them.

Further questions on pages 443, 445.

30 Medical Imaging Techniques

This is an ultrasound image of a 12-week foetus:
Ultrasound scanning allows a doctor to 'see' the developing foetus inside the womb.
100 years ago this would have been done using X-rays.
Why are X-rays no longer used for foetal imaging?

To fully answer this question you need to understand how these imaging techniques work.

In this chapter you will learn:

- how Physics is used in medical measurements and imaging,
- about advantages and disadvantages of different techniques, X-rays, radioactive tracers, PET scans, ultrasound, MRI,
- about the benefits and risks of exposure to radiation.

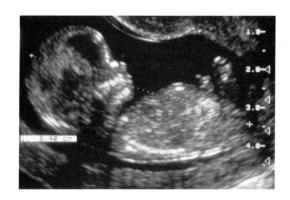

▷ X-ray imaging

Have you ever had an X-ray?

Before 1895 the only way a doctor could see inside the body was by doing surgery. This changed when Wilhelm Röntgen discovered X-rays and found that they could show the bones in his wife's hand!

What are X-rays?

X-rays are electromagnetic waves, with wavelengths from about 0.01 nm to 10 nm. They can penetrate matter, but penetration is less for materials of high density.
Like light, X-rays also affect photographic film, causing it to 'fog'.

When the film is developed, the areas exposed to X-rays are black. The areas of the film not exposed to X-rays remain white.

Think about the shadow of your body on a sunny day. The light is blocked by your body and casts a shadow on the ground. The most common method of obtaining an X-ray works in a similar way.

X-rays pass through soft tissue, but are absorbed by the bones. The bones cast a shadow on the photographic plate, so they appear white against the black background.

We now know that exposure to X-rays can be harmful. They cause ionisation of materials through which they pass, and this can damage living tissue and affect metabolic functions. Cells can die and may also be genetically altered.

Today, special measures are taken to protect both the patient and the radiographer who administers the X-ray examination.

Wilhelm Röntgen (1845–1923)

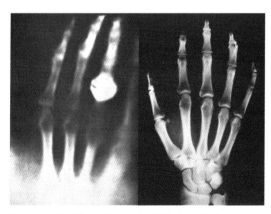

The very first X-ray (taken by Wilhelm Röntgen of his wife's hand and ring) and a modern X-ray.

▶ The X-ray machine

When very fast-moving electrons smash into a solid target,
some of their kinetic energy is converted into X-ray photons.

The diagram shows a typical X-ray tube:
Can you see the evacuated glass tube surrounded by lead?
The window in the bottom of the lead shield allows some X-rays
to exit and pass through the patient's body.

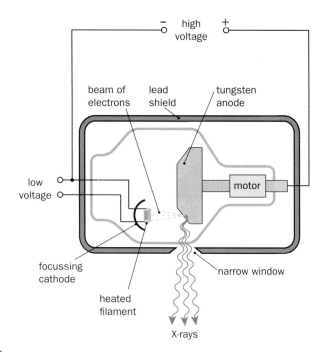

The electrons escape from a heated filament (see page 320).

The filament is connected to the negative cathode, and the
electrons are accelerated through the high potential difference
between the cathode and the metal anode.
The concave shape of the cathode means that the electrons
are focussed on to just a small area of the bevelled anode.

Tungsten is the usual target material, and typical tube voltages
range from 25 kV to 120 kV for diagnostic X-ray tubes.
(In tubes used for radiotherapy, the voltage can be 10 MV.)

Less than 1% of the electrons' kinetic energy is converted
into X-rays. Most becomes internal energy (heat) in the anode.

One of the reasons that tungsten is used is that it has a
high melting point, but, to avoid overheating:
● the anode is rotated to distribute the heat more evenly,
● the tungsten is mounted on copper to conduct the heat.

How are the X-rays produced?

There are two processes that produce X-rays from the target.
1. In **bremsstrahlung**, a fast-moving electron interacts with
the nucleus of a target atom and is slowed down. The energy
lost by the electron is released as an X-ray photon.

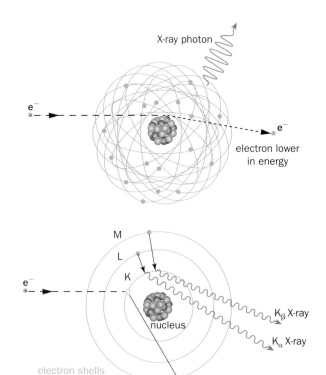

An electron may lose any fraction of its energy, so any photon
energy and hence any wavelength of X-rays can be produced.

In fact, most electrons lose their energy too gradually for X-rays
to be emitted, so they just increase the target's internal energy.

2. Incoming electrons can also **ionise** the atom by 'kicking'
an orbital electron right out of one of the deepest energy
levels in the target atom, referred to as the K and L shells.
An electron then falls from a higher energy level to fill this
gap, and the energy it loses is released as an X-ray photon.

Do you remember studying visible line spectra in Chapter 24?
This process is similar, but the energies involved here are
much higher because the target atoms have many electrons.
Only certain energy levels exist, and so the X-ray photons
produced by this mechanism have precise frequencies or
wavelengths. These are characteristic of the target material.

424

▶ The X-ray spectrum

Here is an X-ray spectrum for a tungsten anode:

It shows the intensity of the X-rays plotted against the wavelength. In fact there are two spectra, here shown by the red and the blue lines. These two spectra are produced by different tube voltages.

Each spectrum consists of two parts:
- a *line* spectrum, which is superimposed on...
- a *continuous* spectrum. Can you explain why?

The line spectrum is produced by ionisation, and the continuous spectrum is produced by bremsstrahlung. The spectrum has some important features. Look at the diagram labels to see what these are:

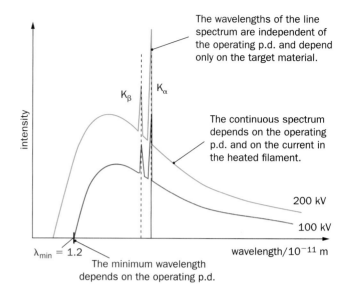

The wavelengths of the line spectrum are independent of the operating p.d. and depend only on the target material.

The continuous spectrum depends on the operating p.d. and on the current in the heated filament.

The minimum wavelength depends on the operating p.d.

Why is there a **minimum wavelength**?
The maximum energy of an X-ray photon is the energy of the electron colliding with the target. So this gives a maximum frequency and a minimum wavelength λ_{min}. This is produced when an electron loses *all* its energy at once.

When an electron of charge e accelerates across a p.d. V, the kinetic energy E_k it gains is given by: $E_k = eV$

All the kinetic energy of the electron is given to the X-ray photon, so the maximum energy E_{max} of the photon is given by: $E_{max} = eV$

Since the photon has a maximum energy, it must have a minimum wavelength. Using Planck's equation:

$$E_{max} = \frac{hc}{\lambda_{min}} = eV; \qquad \text{rearranging:} \qquad \boxed{\lambda_{min} = \frac{hc}{eV}}$$

The shorter the wavelength, the more penetrating the X-rays. The **quality** of the X-ray beam is its *penetrating* ability. Can you see that this will increase as the tube voltage gets higher?

Intensity is the energy per second per square metre of surface. Can you see two ways of increasing the intensity of the X-rays? Raising the tube voltage gives the electrons more kinetic energy, so this affects the quality of the beam as well as its intensity.

The intensity can also be increased by increasing the current to the heated filament. This liberates more electrons so that more electrons collide with the target each second.

Some revision points to help you:

1. From page 321, we know that when a charge Q accelerates across a potential difference (p.d.) V, the gain in kinetic energy is $E_k = QV$.

2. From page 328, Planck's equation tells us that the energy E of a photon is given by $E = hf = hc/\lambda$, so the higher the energy, the shorter the wavelength.

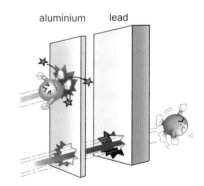

*Low-penetration X-rays are called **soft** X-rays; high-penetration X-rays are **hard** X-rays.*

Example 1
The graph above shows that the minimum X-ray wavelength is 1.2×10^{-11} m when the tube voltage is 100 kV. Can you show this is correct?

$$\lambda_{min} = \frac{hc}{eV} = \frac{6.6 \times 10^{-34} \text{ J s} \times 3.0 \times 10^8 \text{ m s}^{-1}}{1.6 \times 10^{-19} \text{ J} \times 100 \times 10^3 \text{ V}} = \underline{1.2 \times 10^{-11} \text{ m}} \quad \text{(2 s.f.)}$$

▶ Interaction of X-rays with matter

An X-ray tube can be designed to produce an almost parallel or **collimated** beam of X-rays, as shown in the diagram:

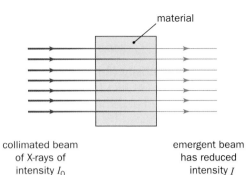

material

collimated beam
of X-rays of
intensity I_0

emergent beam
has reduced
intensity I

When X-rays pass through matter, such as a patient's body, the intensity of the beam is reduced, because the X-rays are **absorbed** and **scattered** by the atoms of the matter.

This reduction in intensity is called **attenuation**.
The thicker the matter, the greater the attenuation.

Look at the graph:
It shows how the intensity I of a collimated beam of X-rays attenuates with the thickness x of the material.
Do you recognise the shape of the curve?

It is an **exponential decay curve**, and you met this before when studying radioactive decay and capacitor discharge.
The equation for this curve is:

$$I = I_0\, e^{-\mu x}$$

where: I_0 is the initial intensity,
I is the intensity after thickness x,
μ is the attenuation coefficient.

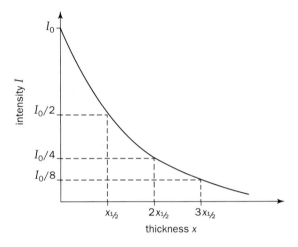

The **attenuation coefficient** μ is a constant for a given material, and its unit is m^{-1}.

Look at the graph again:
Can you see that there is a value of x at which the intensity falls to $I_0/2$, half of its original value?

This value of x is called the **half-value thickness** $x_{1/2}$.
At twice the half-value thickness the intensity has halved again to $I_0/4$. At three times $x_{1/2}$, the intensity is $I_0/8$.

The higher the value of the coefficient, the smaller the thickness of material needed to reduce the intensity by half.

In fact: $$x_{1/2} = \frac{\ln 2}{\mu} \quad \text{(where } \ln 2 = 0.693\text{)}$$

The attenuation coefficient of a material can vary depending on the wavelength of the X-rays. This is because the wavelength affects the attenuation processes, as described on the next page. However, typical values of the coefficient for medical X-rays are:
$\mu = 50\ \text{m}^{-1}$ for soft tissue, and $\mu = 250\ \text{m}^{-1}$ for bone.

> The density ρ of the attenuating material is one of the factors that affects the value of the attenuation coefficient.
>
> Data books often quote the value of the **mass absorption coefficient**, μ_m.
>
> μ_m and μ are related by: $\mu_m = \dfrac{\mu}{\rho}$
>
> The unit for μ_m is $\text{m}^2\ \text{kg}^{-1}$.

Example 2
Using the value of $250\ \text{m}^{-1}$ for the attenuation coefficient of bone, calculate:
a) the fraction of the original intensity of X-rays **remaining** after passing through 0.60 cm of bone,
b) the half-value thickness for bone.

a) $\mu x = 250\ \text{m}^{-1} \times 0.60 \times 10^{-2}\ \text{m} = 1.5$

$I = I_0\, e^{-\mu x}$ so: $\dfrac{I}{I_0} = e^{-\mu x} = e^{-1.5} = \underline{0.22}$ (2 s.f.)

b) $x_{1/2} = \dfrac{\ln 2}{\mu} = \dfrac{0.693}{250\ \text{m}^{-1}}$

$= \underline{2.8 \times 10^{-3}\ \text{m or 2.8 mm}}$ (2 s.f.)

▷ Attenuation processes

There are a number of ways X-rays can be attenuated to:

1. Scattering occurs when an X-ray photon interacts with an atom and changes direction, without transferring energy to the atom. The photon has not lost energy, but the change in its path reduces the intensity of the X-ray beam in the original direction.

2. Do you remember the **photoelectric effect** (page 326)? In this process, X-ray photons of *lower energy* are absorbed. A photon collides with an electron and loses *all* of its energy. The electron escapes from its atom with a lot of kinetic energy and leaves a gap in one of the energy levels. Another electron fills this gap, emitting a photon as it drops down.

3. Look at the diagram, which shows **Compton scattering**: Can you see that this is a bit like a snooker-ball collision?

A *high-energy* X-ray photon collides with an electron, transfers *some* of its energy to it and 'kicks' it out of the atom.
The photon has lost energy and is deflected, changing direction.
The larger the angle of deflection is, the more energy is lost.
Do you agree that, if the photon loses energy, then the scattered X-ray must have a longer wavelength.
So a 'hard' X-ray photon has produced a 'soft' X-ray photon.

4. In **pair production**, a *very high-energy* X-ray photon interacts with the *nucleus* of an atom. The photon is absorbed and its energy produces an electron and a positron.

The voltage of the X-ray tube needs to be above 1 million volts to produce photons with enough energy for pair production. Diagnostic X-ray tubes do not normally work at such a high p.d.

X-rays can be absorbed and scattered in different ways.

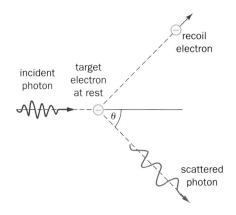

Contrast media

Can you see that X-ray attenuation is mainly due to collisions between photons and the electrons in the attenuating medium?

Attenuation is greater in materials of high atomic number and high density, because there are more electrons per unit volume.

Dense materials like bone attenuate X-rays more than tissue such as flesh, and so good shadow pictures can be obtained. But what if the tissues have similar attenuation coefficients?

A contrast medium has to be used.
For example, swallowing a **barium meal** allows the digestive tract to be studied. Barium has a high attenuation coefficient, and so the tract shows up clearly against the surrounding soft tissue.

In a similar way, iodine can be introduced into the bloodstream in order to show the blood vessels more clearly.

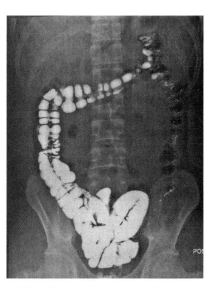

The digestive tract can be imaged using a contrast medium.

▶ Radiography

Our exposure to X-rays must be limited.
They eject electrons as they are scattered and absorbed, and
the electrons cause ionisation in living cells, damaging them.

Medical X-rays have to be a compromise between producing
sharp, clear images and keeping the radiation dose low.
What techniques are adopted to achieve this?

1. Look at your shadow in sunlight and under a streetlight.
Do you notice a difference?
Both have 'fuzzy' edges, but the shadow edges are sharper in
sunlight. A **point source** of radiation gives you sharp shadows.

To achieve a small X-ray source, the cathode focusses the
electrons on to a small area of the bevelled anode. This edge
of the anode is inclined at a sharp angle to the path of the electrons,
and so a narrow beam of X-rays leaves the tube (see page 424).

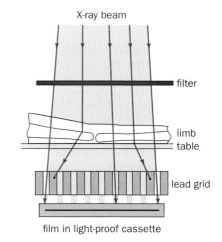

2. Low-energy photons are absorbed by soft tissue *and* bone,
so very few soft X-rays emerge from the far side of the body.
Can you see that soft X-rays increase the patient's dose but
do not contribute to the contrast between different tissue types?

The X-ray beam is **filtered** by passing it through a metal plate,
so the soft X-rays are removed before they reach the patient.

3. X-ray photons are scattered as they pass through a patient.
Would the scattered X-rays reduce the sharpness of the shadow?
Look at the diagram:

The beam passes first through the patient and then through a
lead grid, which absorbs the scattered X-rays.
This stops them reaching the film and reducing the contrast.
(The grid oscillates rapidly so that its shadow is not recorded.)

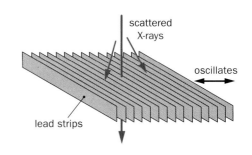

4. The photographic film is double-sided, and is placed between
two **intensifier screens**, as shown in the diagram:

Each screen consists of a layer of zinc sulfide crystals.
The crystals fluoresce: they absorb X-ray photons and re-emit
some of the energy as light photons. One X-ray photon causes
many light photons to be emitted. Photographic film is more
sensitive to light than to X-rays, and so there is extra fogging of
the film. The exposure time and radiation dose are reduced.
Why do you think there is a metal backing plate?

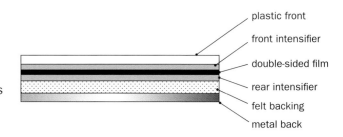

5. The film and screens are contained in a light-tight box called
a **cassette**. The cassette is placed immediately below the
patient as shown in the diagram above.

428

▶ Advances in X-ray imaging

Fluoroscopic image intensification

How does a radiographer observe **movement** inside
the patient's body? The diagram shows the equipment:
In the place of film is a fluorescent screen, A. This glows when
irradiated by the X-rays, and gives a visual image of the patient.

You could look at this screen directly, but high-intensity
X-rays would be needed to produce a clear image.
The **image intensifier** is used to avoid a large X-ray dose.

Light falls on the photocathode, causing it to emit electrons.
The electrons are accelerated onto fluorescent screen B
and produce an image at least 1000 times brighter than at A.
The image at B can be viewed directly or recorded.

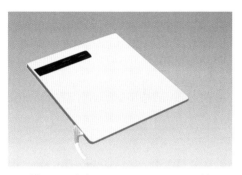

Flat-panel detectors (FPD)

Many radiography departments now use digital methods to
capture X-ray images. The FPD is one of these:

X-rays from the patient strike a layer of scintillating material.
This converts the X-rays into light photons.
The light strikes an array of millions of photodiode pixels,
similar to the pixels used in a digital camera or video.

The light activates the pixels, and the electrical signals they
produce are amplified and encoded to give an accurate
digital representation of the X-ray image. This can be
displayed on a computer monitor or sent all over the world.

*Flat-panel detectors are more sensitive
and faster than film and so allow a lower
radiation dose for the patient.*

Computerised axial tomography (CAT)

Traditional X-rays use shadows, and so an image of the chest
is difficult to analyse due to the overlap of all the structures.
How can we get **3D images**?

A CAT scanner produces a thin fan-shaped beam of X-rays.
Can you see this in the diagram?

A ring of detectors is positioned directly opposite the source,
and these register the X-rays that pass through the patient.

The X-ray source is rotated around the patient, irradiating a
very thin 'slice' of the body. In the time the source has made
one revolution, the motorised table has moved about 1 cm.
This means that on the next revolution the machine 'looks' at
the next slice of the body. Can you see the spiralling effect?

Complex computer analysis is used to assemble all the data
and give either 2D slice images or a 3D image of an organ.
The doctor can rotate the 3D image to view it from any angle.

CAT scans are more expensive than traditional 2D X-rays,
and the patient also receives an increased radiation dose.
But they have higher resolution and reveal much more detail.

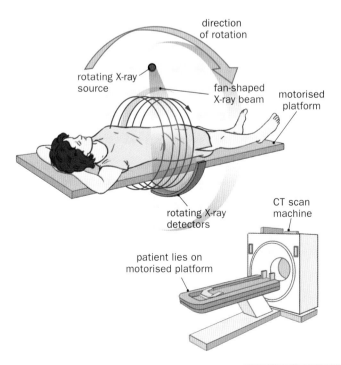

▶ Radioactive medical tracers

Medical tracers allow us to see both the structure and *function* of the organs. How do these *radioactive* tracers work?

A tracer is swallowed, or injected into a patient. It is chemically bound or *tagged* to a substance that will transport it to a certain location within the body. The radioisotope emits radiation which is detected by a *gamma camera* placed outside the body.

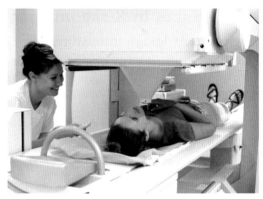

Imaging the thyroid gland using a gamma camera

Technetium-99m (^{99m}Tc) is often used as a radioactive tracer. What makes this isotope so useful?

- It emits *only* gamma radiation, of energy about 140 keV. This travels easily through matter, causing little ionisation within the body, but it is readily detected by the camera.

- It has a half-life for gamma emission of about 6 hours. This is long enough for the investigation, but short enough so that the concentration of ^{99m}Tc in the body falls quickly.

- It decays to form an isotope with a very long half-life, so its decay product does not contribute to the radiation dose.

- Its chemical properties enable it to be tagged to a *range* of compounds, each with an affinity for a particular organ.

Physical, biological and effective half-lives

Do we know how long the tracer remains active in the patient? This depends on both the physical properties of the tracer and the metabolic processes of the body:

The **biological half-life** T_B is the time for a biological system to eliminate half of the amount of a substance that has entered it.

The **physical half-life** T_P is the time taken for half the unstable nuclei in the sample to decay (see page 369).

The **effective half-life** T_E combines the two factors above. It is the time for the activity of a tracer to be reduced to half of its initial value, by a combination of radioactive decay and biological elimination processes. In fact:

$$\frac{1}{T_E} = \frac{1}{T_B} + \frac{1}{T_P}$$

| | Half-life (hours) | | |
Isotope	T_P	T_B	T_E
^{131}I	192	3300	181
^{123}I	13	3300	13
^{111}In	67	6.0	5.5

Can you see that, providing either T_P or T_B is short, the radioactive tracer will clear quickly from the body after the scan, keeping the patient's radiation exposure as low as possible?

Example 3
^{99m}Tc has a physical half-life of 6.0 hours and a biological half-life of 24 hours. Calculate the effective half-life.

$$\frac{1}{T_E} = \frac{1}{T_B} + \frac{1}{T_P} \qquad \text{so:} \qquad \frac{1}{T_E} = \frac{1}{24} + \frac{1}{6.0} = \frac{1+4}{24} = \frac{5}{24}$$

Since $\dfrac{1}{T_E} = \dfrac{5}{24}$ then $\dfrac{T_E}{1} = \dfrac{24}{5}$ so: $T_E = \underline{4.8 \text{ hours}}$

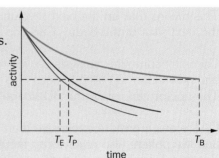

▶ The gamma camera

The diagram shows a gamma camera:
Can you see that the camera consists of 5 main parts?

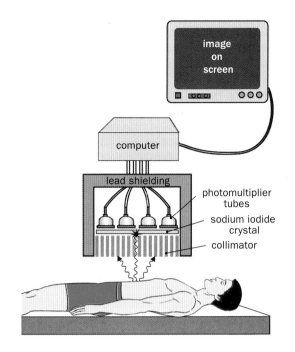

Lead shield

This stops radiation from other sources entering the camera.

Lead collimator

This is a piece of lead with thousands of parallel *holes*.
Only γ-rays travelling parallel to the holes can pass through;
all others will be absorbed by the lead. To reach the crystal,
any γ-ray emerging from a hole must have been emitted
from a point in the patient directly in line with that hole.

Large sodium iodide crystal

This may be 400 mm in diameter and 10 mm thick.
The γ-rays cause the crystal to scintillate: the crystal emits
a flash of light at the exact point where it absorbs a γ-ray.

Photomultiplier tubes

Depending on the camera size, there may be 37 to 91 tubes,
all arranged in a hexagonal pattern. These photomultiplier
tubes convert the tiny flashes of light from the crystal into
pulses of electricity, and these are fed to the computer.
Each tube gives an output for *every* flash of light, but the
greater the distance of a tube from the point in the crystal
where the flash occurs, the weaker its output pulse.

Computer

The position from which each γ-ray was emitted is computed
by comparing the strength of each of the electrical pulses.
The computer uses all the data to process an image.

Diagnosis using medical tracers

Medical tracers can be used to study a range of conditions,
including those of the brain, kidneys, liver and lungs.
In each case the radioactive tracer is chosen carefully, so as
to match the medical procedure.

For example, iodine is naturally absorbed by the thyroid gland.
Iodine-123 is a γ-emitter with an *effective* half-life of 13 hours.
These properties make this isotope ideal for imaging the uptake
of iodine by the thyroid gland.
Can you see how the scans vary for different thyroid conditions?

Indium-111 (^{111}In) is another gamma emitter. It has a half-life
of 5.5 hours, and is used as a tracer in white blood cell scans.
These scans look for infection or inflammation in the body.

White blood cells are taken from a sample of the patient's
blood and tagged with the indium. The tagged blood cells are
returned to the body and gather in any sites of infection.
These are identified by the high intensity of γ-ray emission.

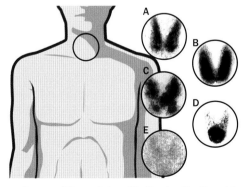

*Scans of the uptake of iodine by the thyroid
gland. Only scan A is normal.*

431

▷ Obtaining radioactive tracers

Technetium-99m is the most commonly used radioactive tracer, but its physical half-life is only 6 hours, making it impossible to store. So, where does a hospital get its supplies of ^{99m}Tc?

It is the parent nuclide molybdenum-99 that is transported to the hospital, in a **molybdenum-technetium generator**. Once at the hospital, the two isotopes are separated chemically. The generator is designed so that chemical separation is simple, and it also provides shielding from the nuclear radiation emitted.

The parent nuclide, ^{99}Mo, undergoes β^--decay with a half-life of 66 hours. After 132 hours, three-quarters of the original number of ^{99}Mo nuclei in the generator will have decayed to form technetium-99m. Do you agree that the generator will need to be replaced after about one week?

Many useful isotopes are produced by the fission of uranium in nuclear reactors. Iodine-131 is one such product. Unfortunately, it decays by beta- and gamma-emission, with the beta-particles accounting for most of the energy released. This beta-radiation does not contribute to an image, because it is absorbed by body tissue and damages it. So iodine-123 is now the preferred tracer for thyroid imaging, although this isotope is far more expensive to produce.

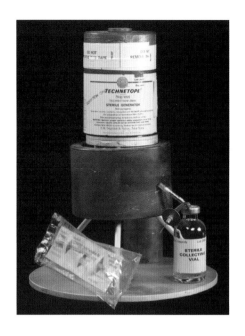

▷ Positron emission tomography (PET)

This is an imaging technique that uses the annihilation of antimatter in a patient's body to measure *metabolic activity*.

The patient is injected with glucose tagged with fluorine-18. Fluorine-18 decays by positron emission (see page 366). The positrons soon collide with electrons in the body, causing the positron and the electron to *annihilate*. This produces two γ-rays, which fly off in opposite directions from the site of the annihilation.

A ring of detectors surrounds the body. Each detector is a scintillation crystal connected to a photomultiplier tube, and so is similar in operation to the gamma camera.

Two detectors pick up the gamma rays as shown: Can you see that the site of the annihilation must be along the line joining the detectors?

The computer builds up an image using the signals from all the detectors. This image shows **metabolic activity** because the greater the activity of the cells, the greater the uptake of the tagged glucose, and so the more γ-rays emitted.

This diagram shows two brain images taken by a PET scanner: Which do you think is the scan for a patient with Alzheimer's?

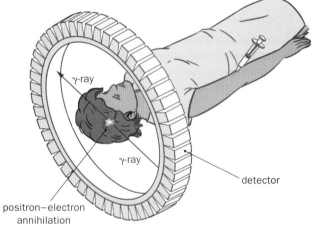

The principle of the PET scanner.

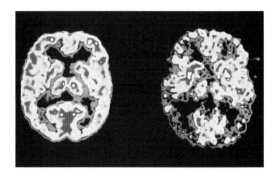

▶ Ultrasound imaging

Most people have heard of the use of ultrasound for foetal scans, but ultrasound is also used for imaging other organs, including the heart and the prostate.
Ultrasound imaging is very different from X-ray imaging.
X-rays must pass through your body, but ultrasound uses *echoes*.

Do you know what ultrasound is?
Ultrasound waves are sound waves with frequencies higher than the human ear can detect. This means above 20 kHz, although scanning uses very high frequencies, between 1 and 5 MHz.

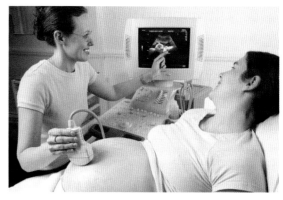

A sonographer conducts an ultrasound scan.

Look at the photograph:
Can you see the scan head in the sonographer's hand?
This contains the *transducers* that produce pulses of ultrasound.
The transducers also detect the echoes: the reflected ultrasound.

These are the principles of ultrasound scanning:

- The ultrasound transducer or *probe* transmits *short pulses* of the ultrasound waves into the body.

- The sound waves meet a boundary between different tissue types. Some of the sound energy is reflected back and some is transmitted across the boundary.

- The transmitted part of the wave travels on towards further boundaries, where reflection and transmission occur again.

- The reflected wave pulses arrive back at the transducer to be detected, and are relayed to the ultrasound machine. This uses the echoes to build up the display on a monitor.

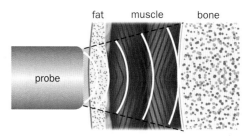

Ultrasound waves travel across tissue boundaries. The reflected waves are not shown here.

What data does the ultrasound machine record?
The *intensity* of each of the reflected pulses is measured.
The *time* it takes for each echo to return to the transducer is also recorded, and this allows the distance from the probe to the boundary to be calculated.

Echo sounding uses a similar method. Look at the cartoon:
Suppose the submarine sends out a sound wave and receives an echo after 0.50 s. How far away is the rock?

It must take the sound 0.25 s to reach the rock.
The speed of sound in water is 1500 m s⁻¹.
So, using distance = speed × time,
  distance = 1500 × 0.25 s = 375 m.

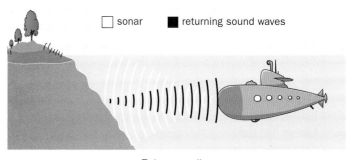

Echo sounding

Why must the probe produce short, sharp pulses of sound?
The pulses must be short, just a few ms long, so that the timing is accurate. After the pulse is transmitted, there must be a pause so that the reflected waves have time to return and be detected.

▶ Ultrasound: acoustic impedance

An ultrasound scan depends on pulses of sound being partially reflected and partially transmitted at the tissue boundaries. Ideally, there will be little absorption of the ultrasound.

In order to determine the fraction of ultrasound reflected, we need to know the **acoustic impedance Z** of each of the materials.

In fact: $\boxed{Z = \rho\, c}$ where: ρ is the density of the material,
c is the speed of sound in the material.

Z has a unit of $kg\, m^{-2}\, s^{-1}$. Can you show why?
(The proof is in the teal box below the table.)

The table shows the acoustic impedances for some materials:
Do you agree that many of the acoustic impedances are similar?

Material	ρ /kg m^{-3}	c /m s^{-1}	Z /kg m^{-2} s^{-1}
Air	1.23	340	4.18×10^2
Fat	952	1450	1.38×10^6
Water	1000	1480	1.48×10^6
Brain	1030	1550	1.60×10^6
Heart	1050	1570	1.65×10^6
Muscle	1065	1580	1.68×10^6
Liver	1060	1590	1.69×10^6
Skin	1150	1730	1.99×10^6
Bone	1910	2800	5.35×10^6

The unit for Z

The unit for ρ is: $kg\, m^{-3}$
The unit for c is: $m\, s^{-1}$
So, the unit for Z is: $kg\, m^{-3} \times m\, s^{-1}$
$\qquad\qquad\qquad\qquad = kg\, m^{-2}\, s^{-1}$

Calculating the reflected wave intensity

The diagram shows an ultrasound wave of intensity I_0 incident on a boundary between materials with impedances Z_1 and Z_2:

The intensity of the reflected wave is I_r, and the intensity of the transmitted wave is I_t. The diagram has been drawn so that the incident and reflected waves are easy for you to see, but in fact the waves need to meet a boundary at near normal incidence in order for the reflected wave to return to the transducer.

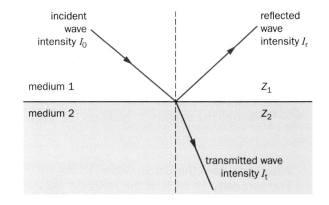

The fraction of the wave intensity that is **reflected** is I_r/I_0, and this is calculated using the equation:

$$\boxed{\dfrac{I_r}{I_0} = \dfrac{(Z_2 - Z_1)^2}{(Z_2 + Z_1)^2}}$$

Look carefully at this equation.

What happens if the two materials have the same impedance?
The term $(Z_2 - Z_1)$ is zero, and so there will be no reflection.
The tissue boundary will be 'invisible'.

What do you think happens if there is a *large* difference in the acoustic impedances of the two materials?
The terms $(Z_2 - Z_1)$ and $(Z_2 + Z_1)$ will have similar values, and so most of the sound energy will be reflected, as shown in the example below.

Example 4
Ultrasound is incident on an air-to-skin boundary.
Using the values for acoustic impedance given in the table above, calculate the fraction of ultrasound intensity reflected at the boundary.

$$\dfrac{I_r}{I_0} = \dfrac{(Z_2 - Z_1)^2}{(Z_2 + Z_1)^2} = \dfrac{(1.99 \times 10^6 - 4.18 \times 10^2)^2}{(1.99 \times 10^6 + 4.18 \times 10^2)^2} = \underline{0.999} \quad \text{(3 s.f.)}$$

At an air-to-skin boundary, 99.9% of the energy is reflected!

Impedance matching

How can a scan be obtained if almost all the ultrasound is reflected at the skin?
The photograph shows the answer:

A special gel is smeared onto the skin. This displaces the air, so that the transducers are transmitting the sound *into the gel*. The gel has an impedance close to that of the skin and so the ultrasound is transmitted across the gel-to-skin boundary. The gel is a **coupling medium** used for **impedance matching**.

Air anywhere in the path of the sound causes problems. An expectant mother has to have a full bladder for a foetal scan!

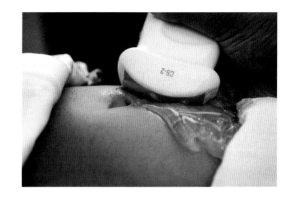

Amplitude scan (A-scan)

A-scans are used when distance measurements are needed, for example, to measure the length of the eyeball.

- A *single* transducer sends ultrasound pulses into the eye.
- The spot on a CRO starts to move across the screen.
- The wave is reflected and transmitted at each boundary.
- The reflected pulses are shown as **vertical deflections** on the CRO screen and are used to calculate the distances.

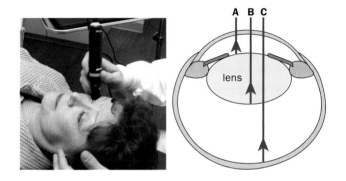

Example 5
This is the output for the A-scan of the eye above.
The pulse reflected at the back of the eye arrives back at the probe after 30 μs.
Sound travels at 1600 m s^{-1} inside the eye.
What is the length of the eyeball?

It must take the sound 15 μs to reach the back of the eye.
Using: distance = speed × time
length of eyeball = $1600 \times 15 \times 10^{-6}$ s = 0.024 m, or 24 mm

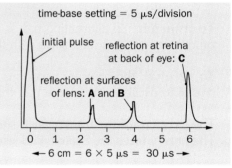

time-base setting = 5 μs/division

initial pulse   reflection at retina at back of eye: **C**

reflection at surfaces of lens: **A** and **B**

0   1   2   3   4   5   6
← 6 cm = 6 × 5 μs = 30 μs →

Brightness scan (B-scan)

The B-scan is widely used in foetal scanning.
The scan head is made up of an array of transducers, so that many ultrasound pulses are sent into the body simultaneously. The scan head is also rocked about while it is moved through the gel, so that the foetus is scanned from a range of angles.

The CRO beam moves across *and down* the monitor screen, and the amplitude of each of the many reflected pulses is displayed as the **brightness of the spot** on the screen.
The dots build up to form a 2D image, just like the ink dots that form a newspaper picture.
Huge amounts of data are computer-processed to build the image.

Real-time B-scans take a series of pictures in rapid succession and can show the movement of the foetus in the womb.
The latest scanners can actually build up 3D images.

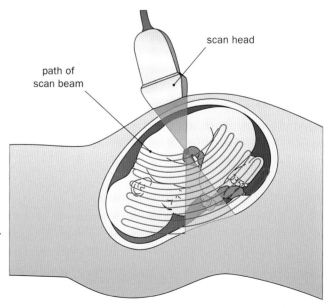

scan head

path of scan beam

▶ The ultrasound transducer

Have you heard of the **piezoelectric effect**?
When some crystals are deformed, a potential difference (p.d.) is produced across the crystal. This can be large enough to produce a spark.
You may have seen this working in a gas lighter.

These diagrams illustrate the effect:

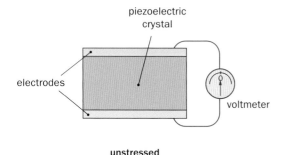

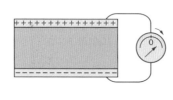

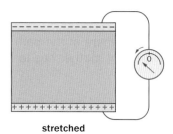

The direction of the p.d. changes as the crystal is squashed or stretched.

The process also works in reverse.
The crystal deforms when a p.d. is applied to it.
If this p.d. is alternating, the crystal will alternately contract and expand.

This allows the piezoelectric crystal to act as a transmitter and a receiver.

To transmit: an alternating p.d. is applied to the crystal. It vibrates rapidly, producing an ultrasound wave with the same frequency as the applied p.d.

To receive: the pressure variations in the incoming sound wave force the crystal to oscillate, creating a p.d. that alternates at the wave frequency.

The structure of the transducer is shown in the diagram:
The crystal faces are coated with silver and act as electrodes.

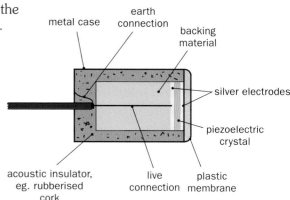

The thickness of the crystal is made so that the frequency of the applied alternating p.d. and the incoming ultrasound wave match one of the natural frequencies of oscillation of the crystal. The crystal then **resonates**: it vibrates with a large amplitude.

Can you think why the backing material is needed?
The transducer must produce short pulses of waves. The backing damps the vibrations after the applied p.d. is stopped.

Example 6
An ultrasonic transducer operates at a frequency f of 2.0 MHz.
a) Each pulse is made up of just 4.0 cycles of the wave. How long does each pulse last?
b) The wave travels in the body at an average speed of 1540 m s^{-1}. What is its wavelength?

a) Each cycle lasts for time T, the period of the wave:

$$T = \frac{1}{f} = \frac{1}{2.0 \times 10^6 \text{ Hz}} = 5.0 \times 10^{-7} \text{ s}$$

4 cycles last for: $4 \times 5.0 \times 10^{-7} \text{ s} = \underline{2.0 \times 10^{-6} \text{ s}}$

b) For the wavelength, use the wave equation:

$$\lambda = \frac{c}{f} = \frac{1540 \text{ m s}^{-1}}{2.0 \times 10^6 \text{ Hz}} = \underline{7.7 \times 10^{-4} \text{ m}}$$

This can also be written as $\underline{0.77 \text{ mm}}$

▶ Doppler ultrasound

Doppler ultrasound is used to see how well your blood flows.

It works by bouncing ultrasound off the moving red blood cells, and measuring the change in frequency: the Doppler shift.

In the diagram, a red blood cell is moving along the blood vessel towards the probe. Can you see how the wavelength of the reflected wave is reduced ahead of the moving blood cell?

The change in frequency Δf is used to compute the velocity v of the cell using the following equation:

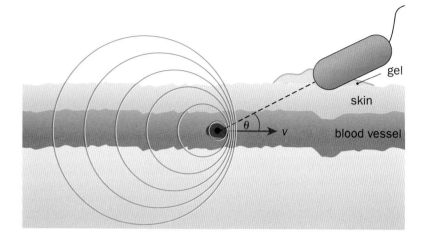

$$\boxed{\frac{\Delta f}{f} = \frac{2v \cos \theta}{c}}$$

where: f is the ultrasound frequency,
c is the velocity of the ultrasound,
θ is the angle between the ultrasound beam and the blood vessel.

The sonographer applies the probe at different angles and an average value for the blood velocity is calculated.

Look at the size of the Doppler shift in the example below.

Do you agree that 920 Hz is in the range of audible frequencies? The blood flow can be 'listened to' using speakers!

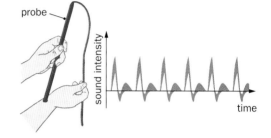

Sometimes both types of ultrasound imaging are combined to give pictures that represent the flow of blood through the blood vessels.

Example 7
An ultrasound Doppler test uses waves of frequency 5.0 MHz that move at 1540 m s^{-1} through the body. With the probe applied at 45°, the change in frequency is 920 Hz. What is the speed of the blood?

$$\frac{\Delta f}{f} = \frac{2v \cos \theta}{c}$$

$$\frac{920 \text{ Hz}}{5.0 \times 10^6 \text{ Hz}} = \frac{2 \times v \times \cos 45°}{1540 \text{ m s}^{-1}}$$

$$v = \frac{920 \text{ Hz} \times 1540 \text{ m s}^{-1}}{2 \times \cos 45° \times 5 \times 10^6 \text{ Hz}} = \underline{0.20 \text{ m s}^{-1}} \quad (2 \text{ s.f.})$$

How is this equation derived?

From page 415 we know that $\dfrac{\Delta f}{f} = \dfrac{v}{c}$

As it reflects the sound, the blood cell acts like a moving mirror.

The image in a moving mirror moves at twice the speed of the mirror: the formula becomes:

$$\frac{\Delta f}{f} = \frac{2v}{c}$$

The component of the cell's velocity parallel to the probe is $v \cos \theta$: the formula becomes:

$$\frac{\Delta f}{f} = \frac{2v \cos \theta}{c}$$

Advantages of ultrasound imaging	Disadvantages of ultrasound imaging
Ultrasound does not involve ionising radiation. Ultrasound energy is absorbed by the tissue through which it passes and can cause a heating effect, but the intensity is kept low to avoid this potential hazard.	Ultrasound does not penetrate bone because of its high acoustic impedance. Scanning the brain is very difficult. Ultrasound does not pass through air spaces, and this is a problem for imaging the lungs.
Ultrasound is very useful for imaging soft tissues, and real-time images can be obtained.	A typical wavelength for ultrasound waves in the body is 1 mm. This is far longer than the wavelength of X-rays, and so the resolution is poor and the scans cannot reveal fine detail. The higher the frequency, the better the resolution, but higher frequencies of ultrasound are more readily absorbed by the body tissues.
Ultrasound devices are relatively cheap and portable and they do not require a specialist room.	The sonographer must be skilled at operating the probe in order to get a good image.

▶ Magnetic resonance imaging (MRI)

The photograph shows an MRI scanner:
An MRI scan is able to create clear images of the spinal cord, the brain and most other organs. It can also detect tumours.

An MRI scan takes a long time. This is because the machine scans the body point by point, identifying the type of tissue at each point. It then uses the data to create 2D or 3D images.

So how does the scanner work?
The body contains billions of hydrogen atoms.
It is the behaviour of hydrogen nuclei in strong magnetic fields that is the key to MRI.

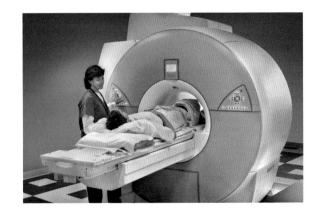

Precession

The hydrogen nucleus is made up of just one proton.
The proton possesses a quantum property called **spin**, which makes it behave like a tiny magnet.
We say the proton has a **magnetic moment**.
This is represented on the diagram by the **purple** arrow.

The proton has a magnetic moment.

Usually, the protons are orientated randomly, so that their magnetic moments cancel out. But in a *strong* magnetic field the protons *align* with the magnetic field lines.

Look at the diagram:
Here the magnetic field lines are blue.
Can you see that the nuclei do not align perfectly in the field?
They rotate or **precess** around the field lines, a bit like the way a child's top wobbles or precesses as it spins in the Earth's gravitational field.

Can you also see that there are two possible alignments?

Protons in the *parallel* alignment precess so that their moments are, on average, pointing in the field direction.
Protons in the *antiparallel* alignment precess so that their magnetic moments are in the *opposite* direction to the field.

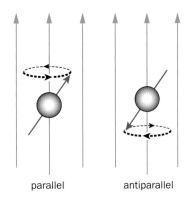

parallel   antiparallel

All the protons precess at the **Larmor frequency, f_L.**
Its value is proportional to the strength of the magnetic field.
For protons, the value of the Larmor frequency is given by:

$$\begin{array}{c|ccc} f_L & = & 4.258 \times 10^7 & \times & \boldsymbol{B} \\ \text{(Hz)} & & \text{(Hz T}^{-1}) & & \text{(T)} \end{array}$$

where B is the magnetic flux density at the proton.

Example 8
A typical magnetic flux density in an MRI scanner is 1.50 T. What value does this give for the Larmor frequency?

$f_L = 4.258 \times 10^7 \times \boldsymbol{B}$

$f_L = 4.258 \times 10^7 \times 1.5 \text{ T} = \underline{6.39 \times 10^7 \text{ Hz, or } 63.9 \text{ MHz.}}$

What region of the electromagnetic spectrum corresponds to a frequency of 60 MHz?

This is a typical frequency for radio waves.

438

Nuclear resonance

The patient lies inside the MRI scanner and the billions of hydrogen nuclei in his body line up in the magnetic field. The magnetic field runs along the centre of the machine, and so the protons line up so that their magnetic moments are pointing either to the patient's feet or to the head.

Roughly half go each way, but there are slightly more in the parallel than the antiparallel alignment, as shown here:

Why is there this mismatch?
The two alignments correspond to different energy levels, and the parallel alignment is the lower-energy state.

The MRI scanner forces some protons to **_flip_** from the parallel to the antiparallel alignment. How does it do this?

Remember, the protons are precessing at the Larmor frequency. A pulse of radio waves at this **_exact_** frequency will make the protons **_resonate_** (see page 110).
The protons **_absorb_** energy from the radio wave pulse, and flip from the lower to the higher energy level.

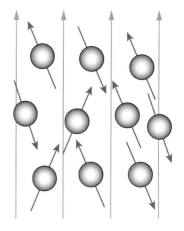

There are slightly more protons in the parallel alignment.

Relaxation times

The flipped or excited protons are now in a semi-stable state. They will **_relax_** (return) to the lower level by losing energy. This energy is emitted as a radio wave: the **_MRI signal_**.

The time it takes for an excited proton to return to the lower energy state is called the **relaxation time**.

The relaxation time varies for different types of tissue because it depends on the molecules surrounding the protons.
For example, relaxation times for hydrogen nuclei in water are around 2 s, but for brain tissue they are shorter, about 200 ms. The relaxation time for tumour tissue is between these two times.

If we know the relaxation time of the excited protons at any point in the body, we can identify the tissue type at that point.

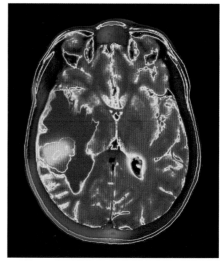

A false-colour MRI scan of the brain showing a tumour (yellow).

Advantages of MRI	Disadvantages of MRI
MRI does not involve ionising radiation, and there are no known side effects.	The machine is very noisy and a scan can take up to 45 minutes. Some people suffer from claustrophobia in the scanner.
The quality of the imaging is high, with good distinction between different tissue types.	The machines are very expensive (millions of pounds) and can scan relatively few people each day.
An image can be made for any slice in any orientation of the body.	MRI cannot be carried out on people with cochlear implants or pacemakers. The strong magnetic field can move, or affect the function of, devices that have ferrous materials in them.
Bone is not a barrier to radiofrequency waves, so high-quality images of the brain can be obtained.	The equipment must not have any radio waves from external sources within it, and so it must be in a screened room.

▶ The MRI scanner

Main magnet

The main magnet produces a stable, *uniform* magnetic field. This field is extremely strong, between 0.5 and 2.0 tesla. You can see how strong this is if you compare it to the strength of a typical fridge magnet, which is about 5×10^{-3} T!

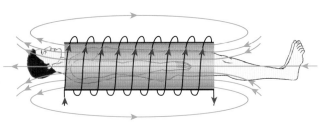

This field is produced by coils carrying huge electric currents. The coils are kept at close to absolute zero, at 4.2 K. At this temperature the coils are superconductors and so have no resistance. The coils must be kept cool with liquid helium, and this one of the reasons why the scanner is so expensive.

Gradient magnets

How is each point in the body pinpointed? Every point within a very thin slice of the body is allocated its own precise magnetic field value by use of the gradient magnets. Each magnet has its own magnetic field, which is superimposed on the main magnetic field. The resultant flux density varies from point to point, but is known accurately at every point. This is like each point on a map having its own grid reference.

The use of three gradient magnets allows the resultant magnetic field to be adjusted in any direction. This means that slices of the body can be scanned at any orientation in the 3D space. The loud hammering sound that the MRI scanner produces is caused by the gradient magnets switching off and on.

Radiofrequency coils

The radiofrequency (RF) coils *transmit* pulses of radio waves at a *precise* Larmor frequency. This is determined by the value of the flux density at a particular point in the body slice. After each radio wave pulse is switched off, the excited protons relax and emit radiofrequency waves. The relaxation time is measured from when the pulse stops.

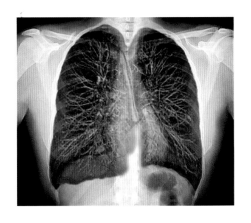

The same coils act as a *receiver* and pick up the RF waves from the patient. The signal is sent to a computer. The diagram shows that there are different coil designs for different anatomical regions.

Computer and monitor

The computer has to process a huge amount of data. It computes the point from which the RF signal has arisen, measures the relaxation time for that point, and from this identifies the tissue type. It integrates all the data to generate an image of the body.

The display can be either a slice through the patient or a 3D view of an organ. This can be rotated by the operator. False colours can be added to show different tissue types.

Summary

X-rays are produced when very fast-moving electrons collide with a target, losing their kinetic energy.

The minimum wavelength λ_{min} of the X-rays produced depends on the voltage V across the tube: $\lambda_{min} = \dfrac{h\,c}{e\,V}$

The intensity I of an X-ray beam attenuates with the thickness x of the material: $I = I_0\,e^{-\mu x}$

$x_{1/2}$ is the half-value thickness: the thickness of material needed to reduce the X-ray intensity by half: $x_{1/2} = \dfrac{\ln 2}{\mu}$

μ is the attenuation coefficient, which is a constant for a given material.

A medical tracer is a gamma-emitting radioisotope, chemically bound to a substance that will transport it around the body.

The effective half-life T_E, the time for the activity of a tracer in the body to be reduced to half, is: $\dfrac{1}{T_E} = \dfrac{1}{T_B} + \dfrac{1}{T_P}$
T_B is the biological half-life and T_P is the physical half-life.

The acoustic impedance Z of a material depends on the speed c of sound in the material, and its density ρ: $Z = \rho\,c$

At a tissue boundary Z_1 to Z_2, the fraction of the ultrasound wave intensity that is reflected is $\dfrac{I_r}{I_0} = \dfrac{(Z_2 - Z_1)^2}{(Z_2 + Z_1)^2}$

The speed v of blood is measured from the Doppler shift Δf in the ultrasound frequency f: $\dfrac{\Delta f}{f} = \dfrac{2v \cos \theta}{c}$
c is the velocity of the ultrasound in the blood.

Protons in a magnetic field B are forced to resonate by radio waves at the Larmor frequency: $f_L = 4.258 \times 10^7 \times B$
X-rays and medical tracers involve ionising radiation; ultrasound and MRI do not.

▶ Questions

You may need the following data:
Planck's constant is 6.6×10^{-34} J s
Charge on an electron is 1.6×10^{-19} C

1. Can you complete these sentences?
   a) An X-ray spectrum consists of a .... spectrum superimposed on a .... spectrum.
   b) In order to keep the .... low, a radioactive tracer must have either a short biological .... or a short .... half-life.
   c) Ultrasound waves have frequencies higher than ....
   d) In a strong magnetic field, protons .... around the magnetic .... at the .... frequency.

2. An X-ray tube operates at 120 kV. Calculate:
   a) the kinetic energy gained by the electrons in
      i) eV, ii) J,
   b) the minimum wavelength of the X-rays produced.

3. The attenuation coefficient of aluminium is 300 m^{-1}.
   Calculate: a) the half-value thickness of aluminium,
   b) the thickness of aluminium required to reduce the intensity of the X-ray beam to 10% of its initial value.

4. Name three measures that can be taken to give a high-quality X-ray image while keeping the radiation dose low.

5. Strontium-90 is a beta-emitter with a physical half-life of 1.1×10^4 days and a biological half-life of 1.8×10^4 days.
   a) Calculate the effective half-life of strontium-90.
   b) Give two reasons why this is not used as a tracer.

6. a) What type of radiation does a PET tracer emit?
   b) What happens to this radiation in body tissues?

7. The density of oil is 9.50×10^2 kg m^{-3} and the velocity of sound in oil is 1.50×10^3 m s^{-1}
   a) Calculate the acoustic impedance of oil.
   b) The acoustic impedance of skin is 1.99×10^6 kg m^{-2} s^{-1}.
      What fraction of ultrasound is reflected at an oil-to-skin boundary?

8. What is a coupling medium, in ultrasound scans?

9. The time-base on a CRO is set to 20 μs cm^{-1}.
   Reflected pulses from the front and back of the head of a foetus are 5.0 cm apart on the screen.
   Sound travels through the body at 1.54×10^3 m s^{-1}.
   What value does this give for the diameter of the head?

10. When an ultrasound Doppler probe is applied at 60° to the skin, the frequency shift is 780 Hz. The test uses 4.0 MHz waves, which move through the body at 1540 m s^{-1}. What is the speed of the blood?

11. a) What is the Larmor frequency for protons at a point where the magnetic flux density is 1.42 T?
    b) In MRI: i) What is meant by 'precession'?
       ii) What does the term 'relaxation time' mean?
    c) Give two advantages of MRI imaging compared to X-ray imaging.

Further questions on pages 444, 445.

Further Questions on Modern Physics

▶ Special Relativity

1. The diagram represents the Michelson–Morley interferometer.

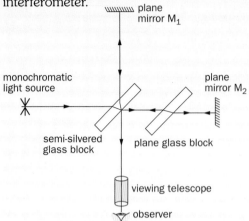

a) i) What was the principal purpose for which Michelson and Morley designed this apparatus?
   ii) Explain why interference fringes are observed through the telescope. [3]
b) Michelson and Morley expected the interference fringes would shift when the apparatus was rotated through 90°.
   i) Why was it thought that a fringe shift would be observed?
   ii) What conclusion did Michelson and Morley draw from the observation that the fringes did not shift? [3] (AQA)

2. π mesons, travelling in a straight line at a speed of $0.95c$, pass two detectors 34 m apart, as shown.

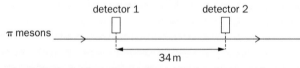

a) Calculate the time taken, in the frame of reference of the detectors, for a π meson to travel between the two detectors.
b) π mesons are unstable and decay with a half-life of 18 ns when at rest. Show that about 75% of the π mesons passing the first detector decay before they reach the second detector. [5] (AQA)

3. a) Light has a dual wave–particle nature. Outline a piece of evidence for the wave nature of light and a piece of evidence for its particle nature. For each piece of evidence, outline a characteristic feature that has been observed or measured. [6]
b) An electron is travelling at a speed of $0.890c$, where c is the speed of light in free space.
   i) Show that the electron has a de Broglie wavelength of 1.24×10^{-12} m. [2]
   ii) Calculate the energy of a photon of wavelength 1.24×10^{-12} m. [1]
   iii) Calculate the kinetic energy of an electron with a de Broglie wavelength 1.24×10^{-12} m. [2] (AQA)

4. a) Calculate the speed of a particle at which its mass is twice its rest mass. [2]
b) Use a copy of the axes below to show how the mass m of a particle changes from its rest mass m_0 as its speed v increases from zero. Mark and label on the graph the point **P** where the mass of the particle is twice its rest mass. [3]

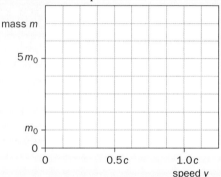

c) By considering the relationship between the energy of a particle and its mass, explain why the theory of special relativity does not allow a matter particle to travel as fast as light. [2] (AQA)

5. In an experiment, a beam of protons moving along a straight line at a constant speed of $1.8 \times 10^8 \, \text{ms}^{-1}$ took 95 ns to travel between two detectors at a fixed distance d_0 apart.

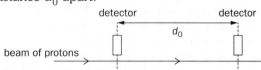

a) i) Calculate distance d_0 between the 2 detectors in the frame of reference of the detectors. [1]
   ii) Calculate the distance between the 2 detectors in the frame of reference of the protons. [2]
b) A proton is moving at a speed of $1.8 \times 10^8 \, \text{ms}^{-1}$. Calculate the ratio (kinetic energy of the proton) / (rest energy of the proton). [5] (AQA)

6. An experiment was performed in 1962 by W. Bertozzi to measure the speeds of electrons accelerated up to measured energies of 1.5 MeV. The idea was to make direct measurements of the time of flight of electrons travelling in a vacuum over a distance of 8.4 m. A direct measurement of the kinetic energy was also made.

Kinetic energy	T /MeV	0.25	0.50	1.00	1.50
Flight time	t /10^{-8} s	3.85	3.28	3.03	2.92

a) Calculate the speed v of the electrons for each value of the flight time given in the table.
b) Draw a graph of v^2 against T, the kinetic energy of the electrons in MeV.
c) Does your graph support the prediction of Newtonian mechanics that $v^2 \propto T$? Justify your answer.
d) What would you expect the best-fit graph line to do as T rises to 10 MeV and beyond? [8] (Edex)

▶ Astrophysics

7. Different types of telescope are used to detect the various parts of the electromagnetic spectrum. Discuss, with reference to three different parts of the electromagnetic spectrum, the factors which should be taken into account when deciding the siting and size of telescopes. [6] (AQA)

8. According to astronomers in Denmark and Australia a common type of active galactic nucleus (AGN) could be used as an accurate 'standard candle' for measuring cosmic distances.
a) i) State what is meant by a standard candle. [1]
ii) Explain how a standard candle is used to measure cosmic distances. [2]
b) i) State what is meant by redshift.
ii) Calculate the distance to a galaxy with a redshift $z = 0.12$. ($H_0 = 2.1 \times 10^{-18}\,s^{-1}$) [2]
c) Discuss how astronomers were led to propose the existence of dark matter and the consequences of its existence for the ultimate fate of the universe. [3]
d) Explain why the observable universe has a finite size. [2] (Edex)

9. In 1999 a planet was discovered orbiting a star in the constellation of Pegasus.
a) State one reason why it is difficult to make a direct observation of this planet. [1]
b) The initial discovery of the planet was made using the radial velocity method, which involved measuring a Doppler shift in the spectrum of the star. Explain how an orbiting planet causes a Doppler shift in the spectrum of a star. [2]
c) The discovery was confirmed by measuring the variation in the apparent magnitude of the star over a period of time. Explain how an orbiting planet causes a change in the apparent magnitude of a star. Sketch a graph of apparent magnitude against time (a light curve) as part of your answer. [3] (AQA)

10. a) Define the absolute magnitude of a star. [1]
b) Sketch the axes of a Hertzsprung–Russell (H–R) diagram, with absolute magnitude on the y-axis and temperature /K on the x-axis.
i) On each axis indicate a suitable range of values.
ii) Label with an S the current position of the Sun.
iii) Label the positions of the following stars:
(1) star W, which is significantly hotter and brighter than the Sun,
(2) star X, which is significantly cooler and larger than the Sun,
(3) star Y, which is the same size as the Sun, but significantly cooler,
(4) star Z, which is much smaller than the Sun and has molecular bands as an important feature in its spectrum. [7] (AQA)

11. SN 2008sr was a type 1a supernova detected in the Antennae Galaxies. This is the light curve of a type 1a supernova.

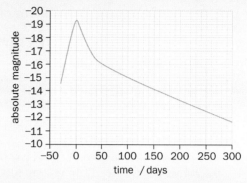

a) With reference to the diagram, explain why type 1a supernovae can be used as standard candles to determine distances. [2]
b) The peak value for the apparent magnitude of this supernova was 12.9. Using this measurement and information from the diagram, calculate the distance to the Antennae Galaxies in Mpc. [2]
c) Why is it important for astronomers to have several independent methods of determining the distance to galaxies? [1] (AQA)

12. a) Calculate the distance of 1 light-year (ly) in metres. [1]
b) The diagram shows an incomplete drawing by a student to show what is meant by a distance of 1 parsec (pc).

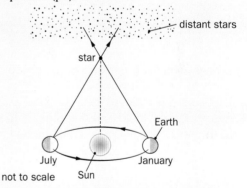

Copy and complete the diagram by showing distances of 1 pc and 1 AU, and the parallax angle of 1 second of arc (1"). [1]
c) A recent supernova, SN2011fe, in the Pinwheel Galaxy, M101, released 10^{44} J of energy. The supernova is 2.1×10^7 ly away.
i) Calculate the distance of this supernova in pc. $1\,pc = 3.1 \times 10^{16}$ m [2]
ii) Our Sun radiates energy at a rate of 4×10^{26} W. Estimate the time in years that it would take the Sun to release the same energy as the supernova SN2011fe. [2]
d) One of the possible remnants of a supernova event is a black hole. State two properties of a black hole. [2] (AQA)

Further Questions on Modern Physics

▶ Medical Physics

13. a) Describe briefly how X-rays are produced in an X-ray tube. [2]

b) Describe the Compton Effect in terms of an X-ray photon. [2]

c) A beam of X-rays of intensity $3.0 \times 10^9 \, \text{W m}^{-2}$ is used to target a tumour in a patient. The tumour is situated at a depth of 1.7 cm in soft tissue. The attenuation (absorption) coefficient μ of soft tissues is $6.5 \, \text{cm}^{-1}$.
   i) Show that the intensity of the X-rays at the tumour is about $5 \times 10^4 \, \text{W m}^{-2}$. [2]
   ii) The cross-sectional area of the X-ray beam at the tumour is 5 mm². The energy required to destroy the malignant cells of the tumour is 200 J. The tumour absorbs 10% of the energy from the X-rays. Calculate the total exposure time needed to destroy the tumour. [3]

d) Describe the operation of a computerised axial tomography (CAT) scanner. State one of the advantages of a CAT scan image over a conventional X-ray image. [5] (OCR)

14. a) The diagram shows an ultrasound transducer used in an A-scan.

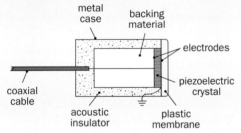

Outline, with reference to the diagram, the process by which the transducer produces a short pulse of ultrasound. [4]

b) Ultrasound is incident on the boundary between two materials. Some of the ultrasound is reflected at the boundary and the remainder is transmitted across the boundary. The ratio of the intensity of the reflected ultrasound, I_r, to the intensity of the incident ultrasound, I_i, is given by the equation

$$\frac{I_r}{I_i} = \left(\frac{Z_2 - Z_1}{Z_2 + Z_1} \right)^2$$

where Z_1 and Z_2 are the acoustic impedances of the two materials.
   i) Calculate the percentage of the incident ultrasound which would be transmitted into the skin when incident on an air–skin boundary. [2]
   acoustic impedance of air $= 4.29 \times 10^2 \, \text{kg m}^{-2} \text{s}^{-1}$
   acoustic impedance of skin $= 1.65 \times 10^6 \, \text{kg m}^{-2} \text{s}^{-1}$
   ii) When obtaining the ultrasound image of an unborn foetus, a coupling gel is used. Explain why a coupling gel is needed and state the property of the gel that ensures a good quality image. [2] (AQA)

15. Technetium-99m is a common medical tracer injected into patients before they have a scan with a gamma camera. Technetium-99m is a gamma emitter with a half-life of about 6 hours. Each gamma ray photon has energy $2.2 \times 10^{-14} \, \text{J}$. A patient is given a dose with an initial activity of 500 MBq.
a) Explain what is meant by *activity*. [1]
b) Calculate the initial rate of energy emissions from the dose of Technetium-99m. [2]
c) Name and describe the function of the main components of a gamma camera. In your answer you should make clear how a good quality image can be achieved with these components. [5] (OCR)

16. a) Explain why the effective half-life of a radionuclide in a biological system is always less than the physical half-life. [2]
b) The physical half-life of a radionuclide is 20 days. The nuclide was administered to a patient. Initially the corrected count rate at the patient's body was 2700 counts s⁻¹. Five days later, the corrected count rate at the same place on the patient was 1200 counts s⁻¹. Calculate the biological half-life of the nuclide. [4]
c) The table gives the properties of 2 radionuclides:

	Technetium-99m	Iodine-131
Emitted radiation	Gamma	Beta and gamma
Half-life / h	6.0	190
Energy of gamma ray / keV	140	610

By considering this information suggest which of these nuclides is more suitable for use as a tracer in medical diagnosis. [4] (AQA)

17. a) State one reason for using non-invasive techniques in medical diagnosis. [1]
b) Describe the use of medical tracers to diagnose the condition of organs. [2]
c) Describe the principles of positron emission tomography (PET). [5] (OCR)

18. a) A magnetic resonance imaging (MRI) scanner is a valuable item of diagnostic equipment found in most hospitals. It is capable of generating a three-dimensional image of the patient. The following terms are used in the description of MRI scanners:
   • Larmor frequency,
   • resonance of the protons,
   • relaxation times of the protons.
   Describe the operation of the MRI scanner with particular reference to these terms. [6]
b) An MRI scan can take a long time and it does produce an unpleasant loud noise. State one other disadvantage and one advantage of an MRI scan. [2] (OCR)

▶ **More questions on Modern Physics**
(including work from earlier chapters)

Special Relativity

19. a) One of the two postulates of Einstein's theory of special relativity is that the speed of light in free space, c, is invariant.
Explain what is meant by this statement. [1]

b) A beam of identical particles moving at a speed of $0.98c$ is directed along a straight line between two detectors 25 m apart.

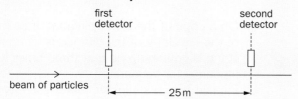

The particles are unstable and the intensity of the beam at the second detector is a quarter of the intensity at the first detector. Calculate the half-life of the particles in their rest frame. [4] (AQA)

20. a) In a science fiction film, a space rocket travels away from the Earth at a speed of $0.994c$, where c is the speed of light in free space. A radio message of duration 800 s is transmitted by the space rocket.
i) Calculate the duration of the message when it is received at the Earth.
ii) Calculate the distance moved by the rocket in the Earth's frame of reference in the time taken to send the message. [4]

b) A student claims that a twin who travels at a speed close to the speed of light from Earth to a distant star and back would, on return to Earth, be a different age to the twin who stayed on Earth. Discuss whether or not this claim is correct. [3] (AQA)

Astrophysics

21. a) Draw the ray diagram for a Cassegrain telescope. Your diagram should show the paths of two rays, initially parallel to the principal axis, as far as the eyepiece. [2]

b) A telescope design very similar to the Cassegrain was first proposed by James Gregory in 1663. His telescope design was also the first to include a parabolic primary reflector.
i) The use of a parabolic reflector overcomes the problem of spherical aberration. Draw a ray diagram to show how spherical aberration is caused by a concave spherical mirror. [1]
ii) The first telescope constructed to this design had a primary mirror of diameter 0.15 m. Calculate the minimum angular separation which could be resolved by this telescope when observing point sources of light of wavelength 630 nm. [2] (AQA)

22. Treated as a single source, the Andromeda Galaxy has an apparent magnitude of 3.54 and an absolute magnitude of -20.62.
a) Calculate the distance to Andromeda Galaxy. [2]
b) The Andromeda Galaxy is believed to be approaching the Milky Way at a speed of $105 \, \text{km s}^{-1}$. Calculate the wavelength of the radio waves produced by atomic hydrogen which would be detected from a source approaching the observer at a speed of $105 \, \text{km s}^{-1}$.
Wavelength of atomic hydrogen measured in a laboratory = 0.211 21 m [2]
c) Some astronomers believe Andromeda Galaxy may collide with the Milky Way in the distant future. Estimate a time, in s, which will elapse before a possible impact with the Milky Way. [2]
(AQA)

Medical Physics

23. An ECG trace is to be obtained for a patient.
a) State and explain the procedure and some design features of the equipment needed to ensure that a good trace is obtained. [6]
b) The diagram shows an ECG trace for a healthy person.

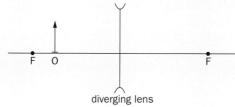

i) Suggest a suitable scale and unit for the potential axis. [2]
ii) Suggest a suitable scale for the time axis. [1]
iii) State the electrical events which give rise to the points **P**, **R** and **T**. [3] (AQA)

24. a) State what is meant by the principal focus and the power of a converging lens. [2]
b) Copy and complete the ray diagram below to show the formation of an image of a real object O by a diverging lens. Label the image clearly. [2]

F O   F
diverging lens

c) State the defect of vision that would be corrected using a diverging lens. [1]
d) A diverging lens of focal length -0.33 m is used to view a real object placed 0.25 m from the lens. Calculate the distance from the lens to the image. [2]
e) Two point sources of light are viewed by a normal eye and their images are formed at the fovea. State, in terms of the active receptors, the conditions necessary for two separate images to be seen. [2] (AQA)

▷ Synoptic Questions for A2

'Synoptic' questions may cover a wide range of topics within one question.

1. An electric wheelchair powered by a 12V battery has a maximum speed of $1.5\,\mathrm{m\,s^{-1}}$ on level ground. The maximum useful driving power of the motor and transmission system is 30W.
   The efficiency of the system is 70%.
   a) Explain briefly why all modes of transport have a maximum speed which depends on the power output of the propulsion system. [3]
   b) For the maximum speed, calculate the power supplied by the battery, and the current in the motor circuit. [4]
   c) Air resistance is negligible at $1.5\,\mathrm{m\,s^{-1}}$.
   Show that the resistance due to friction in the moving parts of the wheelchair is 20 N. [2]
   d) Ramps are often used to give wheelchair users access to buildings. One such ramp has a gradient that rises 0.12m for every metre moved along the ramp, as shown in the diagram. The total mass of a wheelchair and a student is 110 kg.

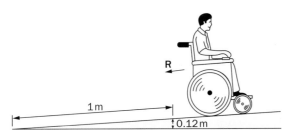

   i) Calculate the component force in the direction shown by the arrow R because of the weight of the wheelchair and student. [2]
   ii) Assuming that the resistance due to frictional forces is still 20 N, calculate the maximum speed of the wheelchair up the ramp. [2] (AQA)

2. This question is about forces on charged particles in electric and magnetic fields.
   Charged particles experience forces when moving in electric and magnetic fields. In the space below are four diagrams. Each diagram shows a charged particle entering a region in which a uniform electric or magnetic field exists.

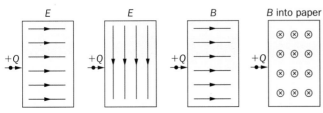

   On a copy of each diagram, draw a line to represent the continuing path of the particle in the field.
   Describe the effect on the velocity of the particle in each case. [8] (OCR)

3. The diagram shows the envelope of the vibrations of a stretched string that is emitting a note at its fundamental frequency.

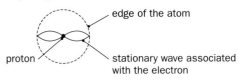

vibrating string

   a) On a copy of the lower diagram draw the shape of the envelope of the vibrations when the string is emitting a note at three times its fundamental frequency. [1]
   b) Explain how a stationary wave, such as that shown in the diagram, is produced. [2]
   c) A simple model of the hydrogen atom assumes that the wave associated with an electron in the atom is a stationary wave as shown below. The nodes correspond to opposite edges of the atom and the centre of the atom. The radius of the hydrogen atom is $1.0 \times 10^{-10}\,\mathrm{m}$.

   i) State the de Broglie wavelength of the electron in the diagram. [1]
   ii) Calculate the momentum of an electron in a hydrogen atom. [2]
   iii) State and explain briefly where, using this model, the electron in a hydrogen atom is most likely to be found. [2] (AQA)

4. A partially inflated balloon containing hydrogen carries meteorological equipment up through the atmosphere. Over a short distance its speed is almost constant.

   a) Copy and complete and label the diagram to show the forces acting on the balloon. [3]
   b) Explain what can be deduced about the resultant force on the balloon. [2]
   c) Explain why the balloon would not rise if the hydrogen were replaced by an identical volume of air at the same temperature and pressure. [2]
   d) Explain what happens to
   i) the pressure,
   ii) the volume of the hydrogen as the balloon rises a significant distance through the atmosphere. [4] (Edex)

5. The diagrams show a cross-section and plan view through a moving-coil loudspeaker. A coil is situated in the magnetic field provided by a permanent magnet. The coil is fixed to the diaphragm but is otherwise free to oscillate vertically. Oscillations of the coil cause the diaphragm to vibrate.

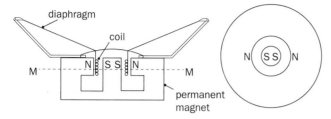

a) On a copy of the plan view, draw the magnetic field pattern between the poles of the magnet. [2]

b) The magnetic flux density at the position of the coil is 0.60 T. There are 50 turns on the coil, which has a diameter of 40 mm. Calculate the force acting on the coil when the current in it is 1.1 A. [4]

c) The effective mass of the diaphragm and coil is 12 g. Calculate the initial acceleration of the diaphragm and coil when the current in the coil is 1.1 A. [2] (AQA)

6. A hydrogen lamp is found to produce red light and blue light. The wavelengths of the light are 660 nm and 490 nm.

a) State which wavelength corresponds to red light. Explain why light of specific wavelengths is produced by the lamp. Calculate the energy change in an atom associated with the emission of a photon of red light. [5]

b) The refractive index of glass for the red light is 1.510 and for the blue light, 1.521. Light from the hydrogen lamp is incident at an angle of 30° on the glass–air boundary, as shown.

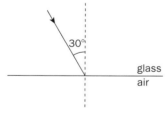

Calculate the angles of refraction for red and blue light and the angle between the red and blue refracted rays. [4]

c) The light from the hydrogen lamp is now directed normally on to a diffraction grating having 4.0×10^5 lines per metre. Calculate the angle between the red and the blue light in the first-order spectrum. [3]

d) A metal surface has a work function energy of 1.80 eV. By reference to your answer to (a) determine whether photo-emission of electrons from this surface is possible with red light. [3] (OCR)

7. The diagram shows a simple device to release excess steam in a model steam engine boiler. A light rod is hinged at one end, A. The safety valve is a plug at point B, 10 mm from A. C is a counterweight of 0.20 N which can be moved along the rod. When the pressure in the boiler increases above a critical value the upward force on plug B tilts the rod about A and releases steam.

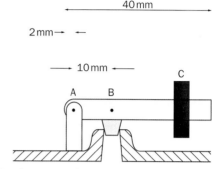

a) The force to lift the safety valve is 0.60 N. Calculate the position of the counterweight for this to happen. [2]

b) The area of the hole in the boiler is 4.0×10^{-6} m². Find the pressure difference between atmospheric pressure and boiler pressure to lift the safety valve. [2]

c) The situation is oversimplified. The uniform rod of length 40 mm has a weight of 0.10 N. Taking this into consideration, how far must the counterweight C be moved from the position calculated in (a) for the valve to open at the required pressure? [3] (OCR)

8. a) Suggest two reasons why an α particle causes more ionisation than a β particle of the same initial kinetic energy. [2]

b) A radioactive source has an activity of 3.2×10^9 Bq and emits α particles, each with kinetic energy of 5.2 MeV. The source is enclosed in a small aluminium container of mass 2.0×10^{-4} kg which absorbs the radiation completely.

i) Calculate the energy, in J, absorbed from the source each second by the container.

ii) Estimate the temperature rise of the aluminium container in 1 minute, assuming no energy is lost from the aluminium. Specific heat capacity of aluminium = 900 J kg⁻¹K⁻¹ [5] (AQA)

9. What is the correct unit for specific heat capacity?

A $m^2 s^{-2} K^{-1}$ B $m s^{-2} K^{-1}$
C $m^2 s^{-1} K^{-1}$ D $m^2 s^{-2} K$ [1] (OCR)

10. Which of the following is a standing wave?

A light emitted as a line spectrum
B ripples on water from a stone thrown into a pond
C sound from an opera singer in a theatre
D vibrations on a violin string as it is played.
[1] (Edex)

Synoptic Questions

11. The force between two masses and the force between two charges can be modelled in a similar way, using gravitational and electric fields.
A difference between these models is that
A an electric field is always a radial field,
B an electric field is always the stronger field,
C a gravitational field cannot be shielded,
D a gravitational field extends over an infinite range.
[1] (Edex)

12. A glass U-tube is constructed from hollow tubing having a square cross-section of side 2.0 cm, as shown in the diagram.
The U-tube has vertical arms and a horizontal section between the arms. Electrodes are set in the upper and lower faces of the horizontal section. Each electrode is length 5.0 cm and width 2.0 cm. The U-tube contains liquid sodium of density 960 kg m^{-3} and of resistivity $4.8 \times 10^{-8} \Omega \text{m}$.
In this question you may assume that the liquid sodium outside the electrodes has no effect on the resistance between the electrodes.

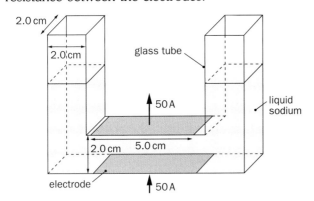

a) Calculate the resistance of the liquid sodium between the electrodes, and the p.d. V between the electrodes required to maintain a current of 50 A in the liquid sodium. [4]

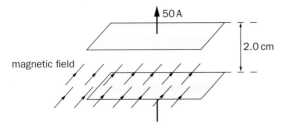

b) A uniform horizontal magnetic field of flux density 0.12 T is now applied at right angles to the horizontal section of the U-tube in the region between the electrodes (second diagram). A force is exerted on the liquid due to the magnetic field. State and explain the direction of the force and calculate its magnitude. [4]

c) By considering the pressure difference which the force in (b) causes, determine the difference in height of the surfaces of the liquid sodium in the vertical arms of the U-tube. [4] (OCR)

13. When gas is used as a fuel, the principal steps in providing electrical power to the consumer are
a) burning the gas to boil water at high pressure,
b) using the high pressure steam to drive a turbine,
c) using the turbine to drive a generator,
d) using current to carry power to the consumer.

For each of stages (a), (c) and (d) state briefly how and why the major 'waste' of energy occurs. [6] (Edex)

14. A potential difference is applied across the metal filament of a light bulb and charge flows.
a) By referring to the mean drift velocity of the electrons, explain what happens to the current in the metal filament if the potential difference is unchanged and the temperature of the metal increases. [3]
b) The graph shows the variation of current I with potential difference V for two components X and Y. X is a filament bulb and Y is a fixed resistor.

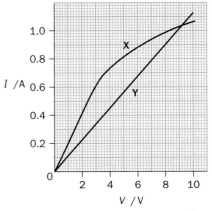

A potential divider circuit consisting of components X and Y is connected to a 9.0 V supply in series with a fixed resistor R as shown. The supply has a negligible internal resistance.

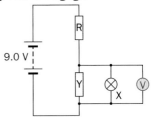

The reading on the voltmeter is 3.0 V.
i) Determine the current in the fixed resistor R. [2]
ii) Component X is removed from the circuit. Explain, without further calculation, how this would change the voltmeter reading. [3] (Edex)

15. Helium is a monatomic gas for which all the internal energy of the molecules may be considered to be translational kinetic energy.
At what temperature will the internal energy of 1.0 g of helium gas be equal to the kinetic energy of a tennis ball of mass 60 g travelling at 50 m s^{-1}? (molar mass of helium = 4×10^{-3} kg) [5] (AQA)

16. The diagram shows a positively charged oil drop held at rest between two parallel conducting plates A and B. The oil drop has a mass of 9.79×10^{-15} kg. The p.d. between the plates is 5000 V and plate B is at a potential of 0 V.

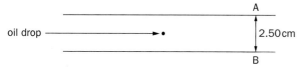

a) Is plate A positive or negative? [1]
b) Draw a labelled free-body force diagram which shows the forces acting on the oil drop. (You may ignore upthrust due to the air.) [2]
c) Calculate the electric field strength between the plates. [2]
d) Calculate the magnitude of the charge on the oil drop. How many electrons would have to be removed from a neutral oil drop for it to acquire this charge? [3] (Edex)

17. a) Explain what is meant by the concept of work. Hence derive the equation $E_p = mgh$ for the potential energy change of a mass m moved through a vertical distance h near the Earth's surface. [4]
b) Using the equation $pV = nRT$ and the kinetic theory of gases expression for the pressure exerted by a gas, $p = \frac{1}{3}(Nm/V)<c^2>$, show that the internal energy of one mole of an ideal gas is $\frac{3}{2}RT$. [7]
c) In a certain waterfall, water falls through a vertical distance of 24 m as shown in the diagram.

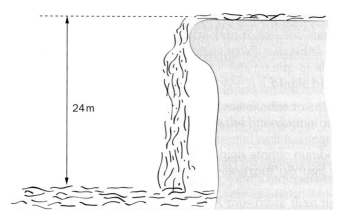

The water is brought to rest at the base of the waterfall. Calculate:
i) the change in gravitational potential energy of 18 kg of water when it descends the waterfall,
ii) the difference in temperature between the top and the bottom of the waterfall if all of the potential energy is converted into thermal energy. The specific heat capacity of water is $4.2 \text{ kJ kg}^{-1}\text{K}^{-1}$. [3] (OCR)

18. a) Explain what is meant by a field in physics. [2]
b) State two similarities between electric and gravitational fields. [2]
c) State two differences between electric and magnetic fields. [2]
d) A beam of electrons is accelerated from rest in an electric field of strength $8.5 \times 10^5 \text{ N C}^{-1}$. Calculate the force on each electron and the acceleration of each electron. [3]
e) Draw a labelled diagram showing what would happen if this beam then entered a uniform magnetic field at right angles to the direction of the beam. Show on your diagram the direction of the magnetic field and of the force acting on the electron. [3] (Edex)

19. a) Sketch stress–strain graphs from zero stress up to breaking stress for
i) a length of rubber cord,
ii) a cast iron rod. [3]
b) The stress–strain graphs can be explained in terms of the microstructure of the materials. Explain briefly the shape of the graph you have drawn for rubber. You may draw diagrams to help you explain if you wish. [4]
c) The diagram shows a rubber cord being used to secure luggage on the roof rack of a car. The tension in the cord is the same throughout. The natural length of the rubber cord is 0.80 m and measurements show that it obeys Hooke's Law for extensions as in the diagram.

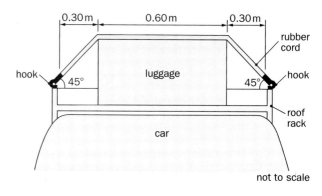

i) Calculate the extension of the rubber cord. [2]
ii) When the tension in the rubber cord is 20 N, it extends by 0.32 m. Calculate the tension in the rubber cord in the situation shown in the diagram. [2]
d) i) Calculate the energy stored in the rubber cord used to secure luggage on the roof rack. [2]
ii) A hook, of mass 0.15 kg, becomes detached from the roof rack. Neglecting the mass of the rubber cord, estimate the resulting speed of the hook. [2] (AQA)

20. Atmospheric Electricity

Lightning was probably the cause of the first fire observed by humans and today it still leads to danger and costly damage. It is now known that most lightning strokes bring negative charge to ground and that thunderstorm electric fields cause positive charges to be released from pointed objects near the ground.

Worldwide thunderstorm activity is responsible for maintaining a small negative charge on the surface of the Earth. An equal quantity of positive charge in the atmosphere leads to a typical p.d. of 300 kV between the Earth's surface and a conducting ionosphere layer at about 60 km. The resulting fair-weather electric field decreases with height because of the increasing conductivity of the air. Across the lowest metre there is a p.d. of about 100 V.

The 2000 thunderstorms estimated to be active at any one time each produce an average current of 1 A bringing negative charge to ground.
The resulting fair-weather field thus causes a leakage current of around 2000 A in the reverse direction, so that charge flows are in equilibrium.
The charge on the Earth and the fair-weather field are too small to cause us problems in everyday life. With an average current per storm of only 1 A, there is no scope for tapping into thunderstorms as an energy source.

The long-range sensing of lightning depends on detecting radio waves which lightning produces. Different frequency bands are chosen for different distances. The very high frequency (VHF) band at 30–300 MHz can only be used up to about 100 km because the Earth's curvature defines the radio horizon. Greater ranges, of several thousand kilometres, are achieved in the very low frequency (VLF) band of frequencies of 10–16 kHz. These signals bounce with little attenuation within the radio duct formed between the Earth and ionospheric layers at heights of 50–70 km.

A further system senses radio waves in the extremely low frequency (ELF) band around 1 kHz. ELF waves are diffracted in the region between the Earth's surface and the ionosphere and propagate up to several hundred kilometres. Horizontally polarised ELF waves do not propagate to any significant extent, hence this system avoids the polarisation error of conventional direction-finding systems.

a) Explain the meaning of the following terms as used in the passage:
   i) *to ground* (para. 1),
   ii) *leakage current* (para. 3),
   iii) *horizontally polarised* (para. 5). [5]

b) i) What is the electric field strength at the Earth's surface?
   ii) Calculate the average electric field strength between the Earth's surface and the conducting ionospheric layer.
   iii) Sketch a graph to show the variation of the Earth's fair-weather electric field with distance above the Earth's surface to a height of 60 km. [7]

c) The *power* associated with a lightning stroke is extremely large.
   Explain why *there is no scope for tapping into thunderstorms as an energy source* (para. 3). [3]

d) The diagram shows a lightning stroke close to the surface of the Earth.

   Copy the diagram and add:
   i) rays to it to illustrate the propagation of radio waves in the VLF band, and
   ii) wavefronts to illustrate the propagation of waves in the ELF band.
   Explain the meaning of the term *radio horizon* with reference to VHF radio waves. [7]

e) List the frequency ranges of VHF and ELF radio waves. Calculate the wavelength of
   i) a typical VHF signal,
   ii) an ELF signal. [4] (Edex)

21. An apparatus to demonstrate electromagnetic levitation is shown in the diagram. When there is an alternating current in the 400-turn coil, the aluminium ring rises a few centimetres above the coil.

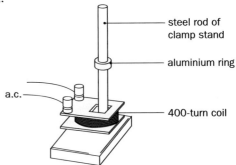

a) Explain why, when there is a varying current in the coil, there is an induced current in the aluminium ring. Suggest why the ring experiences an upward force. [6]

b) In one experiment the power transfer to the aluminium ring is 1.6 W. The induced current is then 140 A. Calculate the resistance of the aluminium ring. [2]

c) The dimensions of the ring are given on the diagram below. Use your value for the resistance to find the resistivity of aluminium.

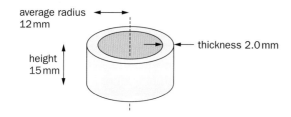

[3]

d) The aluminium ring becomes hot when the alternating current is left on for a few minutes.

In order to try to measure its temperature it is removed from the steel rod and then dropped into a small plastic cup containing cold water.

State what measurements you would take and what physical properties of water and aluminium you would need to look up in order to calculate the initial temperature of the hot aluminium ring. [3] (Edex)

22. A leaf electrometer consists of an aluminium leaf which hangs at an angle θ to the vertical when a p.d. V_{XY} is applied between the metal cap X and the conducting case Y of the electrometer.

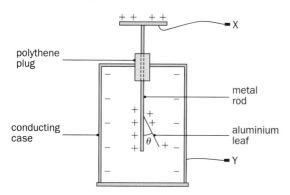

It is calibrated by connecting different values of V_{XY} across XY, with the following results:

V_{XY} /V	100	200	300	400
θ /°	11	14	18	23

V_{XY} /V	500	600	700	800
θ /°	30	37	45	53

V_{XY} /V	900	1000	1100	1200
θ /°	62	69	74	78

a) Plot a graph of θ against V_{XY}. [4]

b) The rate of change of θ with V_{XY} is known as the sensitivity of the electrometer. Over what range of p.d. is the sensitivity approximately constant? Calculate the sensitivity over this range. Show all your working. [5]

c) It is suggested that, between 300 V and 700 V, the p.d. is proportional to $\sin \theta$. Draw up a suitable table of values and plot a graph to test this suggestion. Does your graph support the suggestion? [7] (Edex)

23. The mass of a retort stand and clamp is 1.6 kg and their combined centre of mass lies on the line XY. A spring of negligible mass is attached to the clamp and supports a mass of 0.90 kg as shown. The spring requires a force of 6.0 N to stretch it 100 mm.

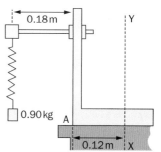

a) Calculate the extension of the spring. [2]

b) Show that this arrangement will not tip (ie. will not rotate about point A) when the 0.9 kg mass is at rest in its equilibrium position. [2]

c) If the mass is lifted and released, it will vibrate about the equilibrium position. Explain, without calculation, why the stand will tip over if the amplitude exceeds a certain value. [3] (AQA)

24. A hot-air balloon, tethered to the ground, is ready for release. The tension in the rope connecting it to the ground is 400 N. The total weight of the balloon (including the hot air within it) is 16 500 N.

a) i) Draw a free-body force diagram of the balloon and hence calculate the upthrust force U on the balloon.

ii) Explain how the upthrust force arises. What factors determine its magnitude during the flight of the balloon? [6]

b) Use the equation $pV = mRT/M$, where m is the total mass of air and M is the mass of 1 mole, to find a relationship between the density ρ of the air in the balloon and its kelvin temperature T. Under what circumstances will ρ be inversely proportional to T? [3]

c) Using the data below, show that the hot air in the balloon has a temperature of 81°C. [2]

Density of cold air outside balloon = 1.29 kg m^{-3}
Density of hot air inside balloon = 1.05 kg m^{-3}
Temperature of cold air outside balloon = 15 °C
Radius of balloon, assumed spherical = 5.5m (Edex)

25. A set of tuning forks is used to find a value for the speed of sound in air. A tuning fork is struck and then held near to the end of an air column formed by a moveable tube. The moveable tube is used to adjust the length, L, of the air column until a standing wave is set up in the tube and the loudest sound is heard. The experiment is repeated for different tuning forks.

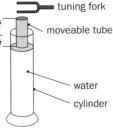
tuning fork
moveable tube
water
cylinder

The following results are obtained by a student.

Fork frequency /Hz	Length, L /cm	Speed of sound /m s^{-1}
256	31.9	327
320	25.6	328
512	16.1	330

Student A says "These results show that the speed of sound increases as the frequency of the sound increases". Student B says "The speed of sound should be the same for each frequency".
By estimating the uncertainties in these results from the precision of the data, conclude which of the statements is valid. [4] (Edex)

26. A length x is 50 mm $\pm$ 2 mm. A length y is 100 mm $\pm$ 6 mm. The length z is given by $z = y - x$.
What is the best estimate of the uncertainty in z?
A $\pm$1 mm B $\pm$4 mm C $\pm$5 mm D $\pm$8 mm   (OCR)

27. The photographs show a toy known as a popper. It is a hollow hemisphere made of rubber. When the top of the popper is pushed down, it changes shape as in the second photograph.

It remains in this shape for two to three seconds. It then returns to its original shape and is launched from the surface, rising nearly a metre.
a) A student concludes that the material of the popper should be classed as a plastic material rather than an elastic material because it remains inverted. Explain whether you think this conclusion is correct. [3]
b) The initial speed of the popper can be determined using only a metre rule to measure the maximum height reached by the popper.
   i) Describe how the maximum height measurement can be used to determine the launch speed. [3]
   ii) Comment on using the maximum height measurement as a means of determining an accurate value for the launch speed. [3] (Edex)

28. A motorcyclist riding on a level track is told to stop via a radio microphone in his helmet. The distance d travelled from this instant and the initial speed v are measured from a video recording.
A student is investigating how the stopping distance of a motorcycle with high-performance brakes varies with the initial speed.
a) Explain why the student predicts that v and d are related by the equation
$$d = \frac{v^2}{2a} + vt$$
where a is the magnitude of the deceleration of the motorcycle and t is the thinking time of the rider. [1]
b) The student decides to plot a graph of d/v on the y-axis against v on the x-axis. Explain why this is a sensible decision. [2]
c) The measured values of v and d are in the table.

v /m s^{-1}	d /m	d/v /s
10 $\pm$ 1	13.0 $\pm$ 0.5	
15 $\pm$ 1	24.5 $\pm$ 0.5	1.63 $\pm$ 0.14
20 $\pm$ 1	39.5 $\pm$ 0.5	1.98 $\pm$ 0.12
25 $\pm$ 1	57.5 $\pm$ 0.5	2.30 $\pm$ 0.11
30 $\pm$ 1	79.0 $\pm$ 0.5	2.63 $\pm$ 0.10
35 $\pm$ 1	103.0 $\pm$ 0.5	2.94 $\pm$ 0.09

i) Calculate the missing value of d/v in the table, including the absolute uncertainties. Use the data to complete a graph like the one shown. [2]

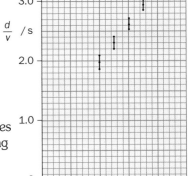

ii) Use your graph to determine the values of a and t, including their absolute uncertainties. [4]

d) It was suspected that the method used to determine the distance d included a zero error. The distance recorded by the student was larger than it should have been. Discuss how this would affect the actual value of t obtained in (c). [3] (OCR)

29. To find the density ρ of a metal wire, a student makes the following measurements:
length $l = 100 \pm 1$ mm
diameter $d = 2.50 \pm 0.05$ mm
mass $m = 4.00 \pm 0.02$ g
The equation $\rho = 4m/\pi d^2 l$ is used to calculate the density of the metal. What is the percentage uncertainty in the answer?
A $\pm$2.5% B $\pm$3.5% C $\pm$4.5% D $\pm$5.5% [1] (OCR)

30. A student undertakes an experiment to measure the acceleration of free fall, g. The diagram shows a steel sphere attached by a string to a steel bar. The bar is hinged at the top and acts as a pendulum. When the string is burnt through with a match, the sphere falls vertically from rest and the bar swings clockwise. As the bar reaches the vertical position, the sphere hits it and makes a mark on a sheet of pressure-sensitive paper that is attached to the bar.

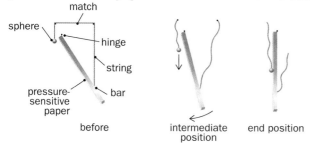

The student needs to measure the distance d fallen by the sphere in the time t taken for the bar to reach the vertical position. To measure t the student sets the bar swinging without the string attached and determines the time for the bar to swing through 10 small-angle oscillations.

a) The figure below shows the strip of paper after it has been removed from the bar. Mark, on a copy of the figure, the distance that the student should measure in order to determine d. [1]

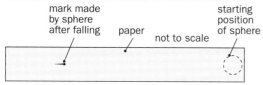

b) The student repeats the procedure several times. Data for the experiment are shown here:

d /m	0.752	0.758	0.746	0.701	0.772	0.769

Time for bar to swing through 10 oscillations as measured by a stop clock = 15.7 s. Calculate the time for one oscillation and hence the time t for the bar to reach the vertical position. [1]

c) Determine the percentage uncertainty in the time t suggested by the precision of the data. [2]

d) Use the data to calculate a value for d. [2]

e) Calculate the absolute uncertainty in your value of d. [1]

f) Determine a value for g and the absolute uncertainty in g. [3]

g) Discuss one change that could be made to reduce the uncertainty in the experiment. [2]

h) The student modifies the experiment by progressively shortening the bar so that the time for an oscillation becomes shorter. The student collects data of distance fallen s and corresponding times t over a range of times. Suggest, giving a clear explanation, how these data should be analysed to obtain a value for g. [3] (AQA)

31. Two point charges of $-6.0\,\mu C$ and $+6.0\,\mu C$ are arranged at points A and B, respectively, as in the diagram. Point X lies as shown, with ABX being an equilateral triangle.

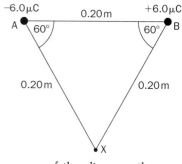

a) Indicate clearly on a copy of the diagram the directions of
   i) the electric field at X due to the charge at A (label it E_A), [1]
   ii) the electric field at X due to the charge at B (label it E_B), [1]
   iii) the resultant net electric field at X due to the charges at A and B (label it E_R). [1]

b) Calculate the magnitude of the resultant electric field at X. [3]

c) Point Y is at a distance 0.40 m to the right of B.

   i) Determine the electric potential at Y. [3]
   ii) Calculate the work required to bring a small charge of $+2.0\,\mu C$ from a distant point to Y. [2]
   iii) The small charge has a mass of 5.0×10^{-3} kg. If it is released from rest at point Y, determine its speed when it returns to a distant point. [2] (WJEC)

32. A container made of insulating material contains $1.7 \times 10^{-3}\,m^3$ of water. The water is heated by a 3 kW electric immersion heater. A student records the water temperature at 0.5 minute intervals.

Time /min	0.5	1.0	1.5	2.0	2.5	3.0
Temperature of water / °C	32.5	45.0	57.5	70.5	83.0	95.5

a) The density of water is 1.0×10^3 kg m^{-3}. Calculate the mass of the water. [1]

b) Plot a graph of the water temperature against time in minutes. [2]

c) Estimate the original temperature of the water. [1]

d) If the heating continues, how long after the start of heating will the water boil? [1]

e) The power of the heater is 3 kW. Determine a value for the specific heat capacity of the water in the insulating container. [3]

f) The student repeats the experiment but uses a container that is not such a good insulator. Readings are obtained at the same time intervals as before. State what happens to the
   i) values of temperature,
   ii) gradient of the graph,
   iii) value obtained for the specific heat capacity. Calculations are not required. [3] (WJEC)

Synoptic Questions

33. a) Describe the similarities and the differences between the gravitational field of a point mass and the electric field of a point charge. [3]

b) The diagram shows two identical negatively charged conducting spheres.

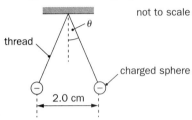

The spheres are tiny and each is suspended from a nylon thread. Each sphere has mass 6.0×10^{-5} kg and charge -4.0×10^{-9}C. The separation of the centres of the spheres is 2.0 cm.

i) Explain why the spheres are separated as shown. [2]

ii) Calculate the angle θ made by each thread with the vertical. [4]

c) This diagram shows two parallel vertical metal plates connected to a battery.

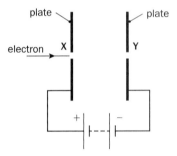

The plates are placed in a vacuum and have a separation of 1.2 cm. The uniform electric field strength between the plates is 1500 V m^{-1}. An electron travels through holes X and Y in the plates. The electron has a horizontal velocity of 5.0×10^{6} m s^{-1} when it enters hole X.

i) Draw 5 lines on a copy of the diagram to show the electric field between the plates. [2]

ii) Calculate the final speed of the electron as it leaves hole Y. [3] (OCR)

34. a) i) Draw a labelled diagram of the apparatus you could use to find the relationship between the resistance and length of a metal wire. [3]

ii) Sketch a graph to show the relationship you would expect to find. [1]

iii) Describe how you would use your graph to find the resistivity of the metal. You should describe the additional measurement you need to make and how you would use it. [3]

b) A metal wire has resistance R and is a cylinder of length l and uniform cross-sectional area A. The wire is now stretched to 3 times its original length while keeping the volume constant. Show that the resistance of the wire increases to $9R$. [3] (W)

35. a) Define the e.m.f. of a cell. [2]

b) A student carries out an experiment to determine the e.m.f. and internal resistance of a cell. The p.d. across the cell is measured when it is supplying various currents. The following readings are obtained. Plot these results on a grid (p.d. on the y-axis and current on the x-axis) and draw a line through your points. [3]

Current /A	0.20	0.42	0.66	0.96	1.20
p.d. /V	1.31	1.13	0.93	0.68	0.48

c) Use your graph to determine:
i) the e.m.f. of the cell, [1]
ii) the internal resistance of the cell. [2]

d) The cell is then connected to a torch bulb of resistance $6.0\,\Omega$ for 20 minutes. Calculate the charge that flows through the bulb in this time. Assume the e.m.f. remains constant. [4] (W)

36. A student obtains the following diffraction pattern on a wall by shining a red laser beam through a single narrow slit. The corresponding graph of intensity against position is shown below.

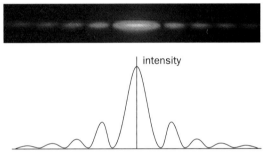

a) Explain how the diffraction pattern is created. [3]
b) Explain how the pattern would differ if green laser light were used instead of red laser light. [3]
c) A student replaces the single slit with a diffraction grating and obtains the pattern shown in the photograph.

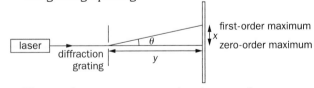

The photograph shows the zero-order maximum and the first and second orders on either side. The student takes measurements to determine the grating spacing.

The student measures x, the distance between the zero-order maximum and the first-order maximum, and y, the distance between the slit and the screen. $x = 23$ cm $y = 1.5$ m
Number of lines per millimetre = 300
Calculate the wavelength of light from the laser. [3] (Edex)

37. The graph shows how the refractive index n of a type of glass varies with the wavelength of light λ passing through the glass. The data for plotting the graph were determined by experiment.

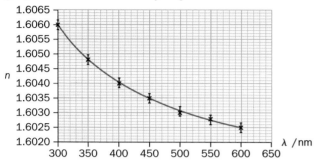

a) A student says that it resembles that of the decay of radioactive atomic nuclei with time and that it shows half-life behaviour. Comment on whether the student is correct. [1]
b) The dispersion D of glass is defined as the rate of change of its refractive index with wavelength. At a particular wavelength $D = \Delta n / \Delta \lambda$. Determine D at a wavelength 400 nm. State an appropriate unit for your answer. [3]
c) It is suggested that the relationship between n and λ is of the form $n = a + b/\lambda^2$ where a and b are constants.

λ /nm	300	350	400	450	500	550	600
n	1.6060	1.6048	1.6040	1.6035	1.6030	1.6028	1.6025

Plot a graph of n against $1/\lambda^2$. [3]
d) Use your graph to determine a. [1]
e) State the significance of a. [1]
f) Another suggestion for the relationship between n and λ is that $n = c\lambda^d$ where c and d are constants. Explain how d can be determined graphically. Do not attempt to carry out this analysis. [3] (AQA)

38. A wire is held under tension. A standing wave is set up by an oscillator at one end.

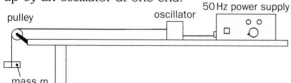

a) The wire is oscillated at a constant frequency. Measurements are taken to determine the wavelength for different values of the mass m. The following data are obtained. Draw a straight-line graph to test the relationship $\lambda^2 = km$. [4]

m /kg	0.100	0.150	0.200	0.250	0.300	0.350
λ /m	0.641	0.776	0.905	1.012	1.103	1.196

b) Use your graph to find a value for k. [2]
c) It is suggested that $k = g/f^2\mu$ where $g = 9.81$ N kg^{-1}, frequency $f = 50.0$ Hz and μ = the mass per unit length of the wire. Use your value for the gradient to calculate a value for μ. [3] (Edex)

39. The apparatus shown in the diagram is used to investigate the variation of the product PV with temperature in the range 20°C to 100°C. The pressure exerted by the air is P and the volume of air inside the flask is V.

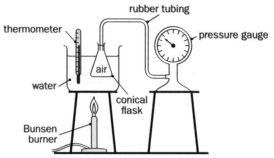

a) Describe how this apparatus can be set up and used to ensure accurate results. [4]
b) An investigation similar to that shown gives measurements of the pressure P, volume V and temperature θ in degrees Celsius of a fixed mass of gas. The results are used to plot the graph of PV against θ.

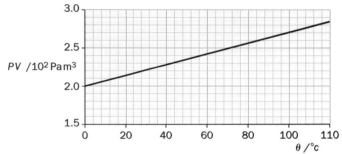

i) Explain, in terms of motion of particles, why the graph does not go through the origin. [2]
ii) The mass of a gas particle is 4.7×10^{-26} kg. Use the graph to calculate
1 the mass of the gas,
2 the internal energy of the gas at a temperature of 100°C. [4] (OCR)

40. A small loudspeaker emitting sound of constant frequency is positioned a short distance above a long glass tube containing water. When water is allowed to run slowly out of the tube, the intensity of the sound heard increases whenever the length l takes certain values.

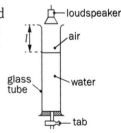

a) Explain these observations by reference to the physical principles involved. [4]
b) With the loudspeaker emitting sound of frequency 480 Hz, the effect described in (a) is noticed first when $l = 168$ mm. It next occurs when $l = 523$ mm. Use both values of l to calculate
i) the wavelength of the sound waves in the air column,
ii) the speed of these sound waves. [4] (AQA)

Study Skills

Making progress in Physics

Are you finding the Advanced Physics course quite hard?
Do you sometimes have difficulty with homework?
Or do you just feel that you want to do better?

This section has some ideas that should help you.
Of course they will require effort from you, but as they begin to help you, you will start to gain confidence and satisfaction.

Understanding work

It is most important that you try to understand the work which is being covered in class **at that time**, or later **the same day**. This is because:

- Much of the new work in Advanced Physics builds on the basic ideas that have been learned earlier in the course. If you don't understand the basic ideas really well then you will find it much harder to understand the new work.

- It is very difficult, when you are revising work, to remember facts which you don't really understand.

- You can prepare a list of questions to ask your teacher in the next lesson to help improve your understanding.

In class

- Try to concentrate as much as you can, and try to join in class discussions and group work. Research suggests that talking about your ideas helps you to understand Physics.

- Be brave enough to ask your teacher if you are not sure of anything! Either in class, or afterwards if you prefer. Your teacher will be able to explain an idea in several different ways in order to help you understand it.

Other ways to make progress

- It is really helpful if you can spend just 10 minutes at the end of the day reading notes from the lesson to remind yourself of work covered. It helps your brain to consolidate ideas. The more often that you read and think about new ideas, the more likely you are to remember and understand them.

- Talk to your friends about the key knowledge learned during the lesson. You can try formal questions or just discuss the ideas until you can explain them clearly to each other.

- Finding out about *everyday* applications of the Physics which you are learning will help you to understand it. It is a good idea to ask your teacher about applications as they explain new ideas to you. Reading books, watching TV or internet videos about Physics is also helpful.

Homework

Some people say that students do not learn Physics in lessons, but away from lessons, when they are doing homework.

Although you are introduced to new ideas in class, you are unlikely to learn and understand them fully without thinking about the ideas and *using* them. This is why work at home is so important.

Organising yourself for homework

Sometimes you may find it difficult to organise your time so that you complete your homework. There are a lot of distractions! Here are some ideas to help:

- *Make a routine!* Plan your time so that you always do your Physics at the same times each week – in the same free periods or on the same evenings when you have no other activities.

- *Do homework early!* This is so that you can get help with difficulties, from either friends, family or your teacher.

- *Take breaks!* People concentrate better in short bursts, so after about 30 minutes of work it is best to have a short break.

- *Deal with distractions!*
  There are lots of other people and other activities which can draw you away from Physics homework. You need to be strong-willed and avoid these distractions while you work. Ensure everybody knows you are busy and not to disturb you. Tell friends when you will and when you won't be available. Turn off your phone and keep it off while you work. Give yourself deadlines and reward yourself when you meet them.

Making progress from your homework

You will put many hours of effort into your homework during your Physics course. Here are some ideas on how to get the most out of the effort that you make.

1. Look up anything you are not sure about, especially definitions. Doing this will help your understanding of Physics, and may also help you to revise work you did earlier in the course.

2. Read your answer after you have written it. Is it clear? Does it answer the *exact* question asked? Will the person reading it understand what you have written?

3. Download a Mark Scheme and an Examiner's Report from the exam board to check examination questions. These let you judge your own answers and show you where you will score marks and where you might have missed some.

4. Aim to get a lot of detail into your explanations with a small number of words. This is quite a skill – it takes time to improve.

5. Use precise Physics ideas in your answers rather than everyday language. So, for example, instead of using the 'size' of an object, be clearer – is it mass, volume, diameter or length?

6. Read your teacher's feedback after your work has been marked. This should give you a clear idea about how to improve, and help to make sure that you don't make the same mistakes again.

▷ Becoming more confident with Maths

There are two parts to being good at Maths:

1. You must have knowledge and understanding of mathematical **skills**, such as solving equations or using trigonometry.

   The skills which you need are explained in more detail in the **Check Your Maths** section on page 472.

2. You must also be good at mathematical **problem-solving**. What does this mean?

To be good at mathematical problem-solving you need to be able to spot quickly what **type** of question you are answering.
For example, is it a 'constant acceleration' question or would it be better to use 'conservation of energy'?
This will allow you to think rapidly of a **plan** to find the answer.

The way to become good at Maths is by:

- **doing lots of questions**,
- getting them right,
- remembering how you answered them.

Improving your mathematical skills

It is very important that you get help when you don't understand some of the Maths used in a lesson.
You can ask your teacher to explain the Maths again, look for extra support from your mathematics department or ask a friend who is stronger at Maths than you to help you out.
The **Check Your Maths** section should help you as well.

Taking care with calculations is important. Check them carefully. It is best to put down all your working, especially if you are not confident with a calculation.

Improving your problem-solving ability

1. When you get stuck on a problem, look back through this book to see if there is a worked example which is similar.

2. If this does not help, then ask for help from friends, family or a teacher. Get them to explain how to do it, then go away and try to do it **on your own**. Don't just copy out their explanation and working!

3. It can help to do homework with friends, but don't just copy out their work – you have got to work it out **for yourself**!

4. When you make progress on a problem, think carefully whether there is anything new that you tried which might be useful for other questions.

5. Keep your answers well organised so that you can use them to help with revision later.

The opposite page shows the 10 steps to success in tackling mathematical questions in Physics, with an example.

▷ How to tackle mathematical homework questions

Here are 10 tips to help you improve your exam question skills, particularly if you use them over a long period of time.

1. *Read the question carefully*
A good idea, to help you to read a question carefully, is to underline any important words or numbers in the question.

2. *Imagine the situation in the question and draw a diagram*
Put as much on the diagram as you can, including numbers given in the question and anything that you can add (for example, the forces acting on an object).

3. *Note down the information clearly*
This means all the values of quantities that you are given including their units. You can also put a letter down with a '?' next to it, to remind you of what you need to calculate.

If you write down the information then there will be more 'space' left in your brain for thinking!

4. *Have a plan before you start*
This will help you to think carefully about the problem rather than just rushing in! Make sure that you are clear about what quantity you are trying to calculate.

Try not to just look for an equation to solve the problem, but think about what Physics principle applies to it. Eg. is it about Newton's 2nd Law or the Conservation of Energy?

5. *Think about which formulas might help solve the problem*
Write these down and check to see which formula has the quantity that you need to calculate and also has the quantities that are given to you in the question.

You may need to use more than one equation.

6. *Keep checking that you are on the right track*
As you go through the problem, think about whether your method seems to be working.

If you get totally stuck on part of a long question, then move on to the next part. You will get marks even if you have to use the wrong answer from the earlier part.

7. *Look at your answer and check that it is sensible*
Is the answer that you get about the right size?
Check that it isn't ridiculously large or small.

8. *Remember units!*
You will lose marks in exams if you forget units, so get into the habit of writing them in, always.

9. *Check significant figures*
Always give your answer to the correct number of significant figures. See page 9 for more details.

10. *Reflect on your success*
When you succeed, think about the method you used, and whether there was anything different that you tried, which may be useful for solving other problems.

An example

A ball is thrown <u>upwards</u> into the air and reaches a <u>height</u> of <u>30.0 m</u>. What was its <u>initial velocity</u>?

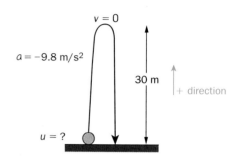

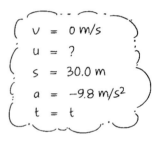

$v = 0$ m/s
$u = ?$
$s = 30.0$ m
$a = -9.8$ m/s²
$t = t$

$v = u + at$ ✗
$s = ut + \frac{1}{2}at^2$ ✗
$v^2 = u^2 + 2as$ ✓

$u = 5000$ m/s
seems too big —
where did I go wrong?

Doing the diagram seemed to help.

Revision skills

When you revise, you need to balance your time between:

- learning your notes,
- practising questions from past papers (for this you can use the eight Further Questions sections in this book).

This section concentrates mainly on how you can learn your notes *effectively*, so that you have a good knowledge and understanding of Physics when you go into the exam room.

Using a recorder may help.

Before you start:

- Get a copy of the syllabus 'specification' and any supporting materials from your teacher or the exam board website.
- Be clear about exactly which topics you need to revise for the exam you are about to take. The website at:
  www.oxfordsecondary.co.uk/advancedforyou
  tells you exactly which parts of this book you need for your particular exam syllabus. See also page 506.
- Work out which are your strong topics and which topics you will need to spend more time on.
  Use your test scores to help make a judgement.

Some helpful ideas

1. *Work out your best way of learning*
   Some people learn best from diagrams or videos, while some prefer listening (perhaps to taped notes) and making up rhymes and phrases. Others prefer to do something **active** with the information, like answering questions or making a poster on a topic. If you know which way you prefer, then this will help you to get the most out of your revision.

2. *Test yourself*
   Merely reading through notes will not make them stick.
   Get someone to test you on a section of work, or write down some questions testing your knowledge of the topic.
   Past paper questions are readily available from exam board websites, but ensure that you select the ones which are relevant.

3. *Teach others*
   If you can find someone to teach a topic to (friends or family?) then this is sometimes the best way to learn. You and some friends might take it in turns to present a topic to each other.

4. *Learn formulae and definitions*
   Check which formulae are **not** going to be given to you in the exam, and learn these by heart. Most people find it best to learn the formulae which are given, as well, so that they can be quickly remembered when needed for a question.

   Your specification will also have **key definitions** you will need to be able to recall. Make a list and memorise them.

5. *Make sure that you try to understand the work*
   This will help you to remember it better.

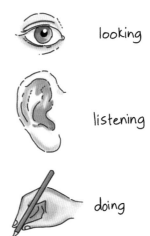

Three ways of learning:

looking

listening

doing

▷ Eight revision techniques

1. *Get an overall view of the topic first*

Before you start revising a topic, quickly read through the whole topic so that you have a general understanding of it and of how the different bits of it connect with each other.

A good place to start is the Summary shown at the end of each chapter in this book.

2. *Highlight* key *words* and phrases in your notes using highlighter pens in different colours.

3. *Make notes*

Rewriting and condensing notes is a good and ***active*** way of reading and then understanding what you need to learn.

You can also annotate exam questions you have done.

4. *Make yourself 'flash cards'*

These have questions on one side and answers with notes on the other. Test yourself by trying to answer the questions before checking with the notes. Add new cards to your pack as you reach new revision topics and remove cards you are consistently getting correct, to keep the pack size manageable.

5. *Make a Poster*

You can summarise a topic with a poster which you can put on the wall of your bedroom. Include important words and phrases in large, bold letters. If you are a visual learner, then include bright, colourful diagrams which illustrate the ideas.

6. *Create a Mind Map*

This is a poster that summarises a topic by emphasising the links between the different concepts which make it up: Making a Mind Map forces you to think about the links in a topic and can help you to understand it.

The more often you ***redraw*** and enhance the Mind Map, the better your recall and understanding will be.

7. *Revision Apps*

There are many revision applications available for tablets, smartphones and PCs. These can have high levels of animation, interactivity and testing. Used often, but for short periods of time, they can be very effective.
Eg. for ten minutes each day on the bus to school/college.

8. *Past Paper Questions*

The final examination is going to consist of examination questions. So no matter what form of revision you try, you also need to make sure you tackle as many of these as possible. The more you try, the better prepared you will be.

The eight ***Further Questions*** sections in this book contain past paper questions, and all exam boards have a selection along with Mark Schemes to help you check your answers.

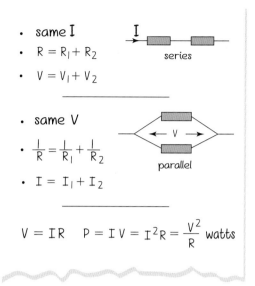

- same I
- $R = R_1 + R_2$
- $V = V_1 + V_2$

series

- same V
- $\frac{1}{R} = \frac{1}{R_1} + \frac{1}{R_2}$
- $I = I_1 + I_2$

parallel

$$V = IR \quad P = IV = I^2R = \frac{V^2}{R} \text{ watts}$$

Part of a Poster on circuits

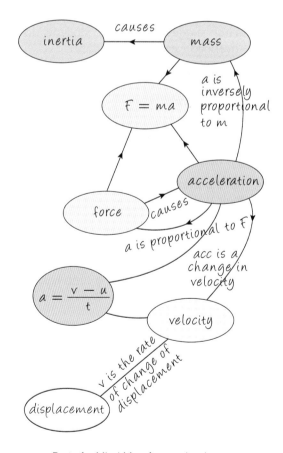

inertia — *causes* — mass

$F = ma$

a is inversely proportional to m

acceleration

force *causes*

a is proportional to F

acc is a change in velocity

$a = \dfrac{v - u}{t}$

velocity

v is the rate of change of displacement

displacement

Part of a Mind Map for mechanics

▷ Organising your revision

Do you have difficulty starting revision and feel that there is so much to do that you will never complete it?
Do you constantly put off revising and find other things to do?

Here are some ways in which you can help yourself:

- **_Think about the positive effects of starting_**
  When you have finished a session you will feel good that you have made progress, and you'll feel less anxious about not getting enough work done.

- **_Think about the negative effects of delaying_**
  What will happen and who will be affected if you put off starting?

- **_Give yourself rewards_**
  Think of things with which you can reward yourself at the end of a session. Eg. a cup of tea or listening to some favourite music.

- **_Get help!_**
  Think of ways to involve friends, family and fellow students which will make revision easier and more enjoyable.

Making a Revision Timetable

Research suggests that some students do better than others at Advanced Level because they:

- revise topics **_throughout_** the whole course,

- start their examination revision earlier,

- use better techniques for learning work (such as testing themselves, rather than just reading their notes),

- get help from others rather than working alone,

- have a **_planned_** revision timetable which includes working on their weaknesses.

You can use these ideas to help you plan an efficient Revision Timetable.

1. Start your revision a long time before your exams (at least eight weeks).

2. Note down when you will cover each topic and stick to this!

3. Spend more time on your weaker topics, especially as you get close to the exam. Ask your teacher to give revision lessons and activities focussed on these to help you.

4. Do lots of past paper questions and get feedback from your teacher about how you can improve.

5. Arrange some revision periods to work with someone who can help you.

6. Do some social activities in between revision sessions, so that you don't go completely crazy!

What will you revise?

You will need to divide your time effectively so that you:

1. Learn the work which you have covered so that you have a good knowledge and understanding of it, including knowing formulae and definitions of quantities.

2. Practise past questions so that you know what to do when you get a similar question in the exam.
   This can be useful in learning work, because you will often need to read your notes to help you answer a question. You can also judge where your weaknesses are, to focus on them.

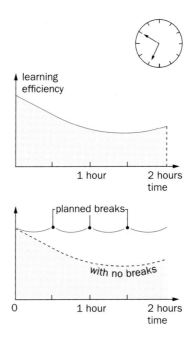

How long should you revise for?

Research suggests that if you start revising with no thought of when you will stop, then your learning efficiency just gets lower and lower.

However, if you decide on a **fixed** time, say 30 minutes, then you learn the most at the beginning of a session and *also* just before you have decided to stop, when your brain realises that it is coming to the end of a session (see the first graph).

Which would be better: one 2-hour session or four 25-minute sessions with a break of 10 minutes in between them?

The four sessions would be much better, as shown by the yellow area on the second graph:

However, it is important to stick to a definite *fixed* time, using a clock, and have a break, rather than just carrying on.

This way you get through more work *and* you feel less tired!

How often should you revise?

This graph shows how much information your memory can recall at different times after you have finished a revision session:

As it shows, you very quickly forget much of what you have revised.

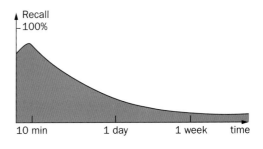

The next graph shows that if you briefly *review* the work again after 10 minutes, then the amount that you remember increases:

The graph stays higher!
You can do this review at the end of your 10-minute break.

So a good system of working is to:

- work against the clock for a definite time (eg. 25 minutes),
- have a 10-minute break and do something entirely different,
- review, briefly, the work you were revising in the last session,
- then begin the cycle again, by working against the clock...

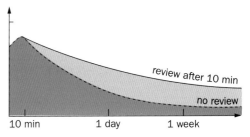

If you revise this material again after a **day**, and then again after a **week**, then you remember even more of the work, as shown by the last graph:

The trick is to briefly revise a topic again at regular intervals.

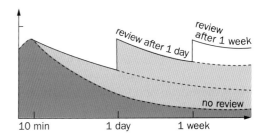

Examination technique

Before the exam

1. Make sure that you check carefully the dates and the times of all your exams, so that you are not late!

2. Make sure that you know which type of paper (eg. multiple-choice, short answer) is on which day, and which topics are being examined on which paper.

3. Make sure that you know how many questions you have to do on each paper and how long it is. Plan how long you should spend on each question on the paper.

4. Make sure that you are familiar with the Physics formula sheet which you will be given for the exam, and that you know what all the symbols mean. Make sure that you have memorised equations which are not on the sheet.

5. On the night before the exam, it may help you to steady your nerves if you look briefly through your notes. But don't do too much!
Make sure that you get a good night's rest before your exam!

On the day of the exam

- Aim to arrive early at the exam and try to get into the room as early as possible. This will help you to settle your nerves and give you time to prepare.

- Don't eat too much or too little food, and avoid caffeine.

- Make sure that you are properly equipped with pens and pencils (and spares), a rubber, ruler, protractor, calculator (check the batteries) and a watch. Don't take a phone or other banned items into the exam room.

During the exam

1. Don't waste time when you get the paper!
Fill in your details on the front of any sheets of paper or answer booklets that you are going to use.
Read the instructions on the front page of the exam paper.

2. Read each question very carefully and <u>underline</u> important parts. Make sure you know exactly what you have to do to score as many marks as possible.

3. If you have a choice of which questions to attempt, then read all of the questions on the paper first.
Never dive into a question without reading all of it first.

4. Write neatly and in short sentences which will be easier for the marker to understand.

5. Do not spend too long on any question! Pace yourself carefully with a watch so that you don't run out of time.

6. If you are stuck, just leave some space so that you can go back to it later. It is easier to get 50% on all the questions than 100% on half of them.

7. Sometimes you may be completely stuck on one part of the question. Look to see if there are any later parts which you can do easily and try them – you may need to 'make up' answers for the early parts to allow you to continue, but you will still score marks.

8. Check that you have done all the required questions. Turn over to check the last page of the booklet, just in case!

Ten hints on answering questions

1. If the exam paper shows that a question is worth 4 marks then put down (at least) 4 points in your answer.

2. Questions have **command words** such as 'State', 'List' or 'Explain' which tell you how much detail is required in your answer. Make sure that you know exactly what each one means – your exam board has a list.

3. Use correct scientific vocabulary as much as possible, and keep your spelling to a high standard.

4. Never explain something just because you know how to! You only earn marks for **exactly** what the question asks.

5. When you do calculation questions always show your working. There is often a mark for substituting data and manipulating the right equation. You can get your mark for this part, even if you don't finally get the right answer.

6. **Check** your answers! First check whether your answer is much too big or too small. Ensure that you check **units** and significant figures, or you'll lose marks.

7. Make your diagrams clear and neat, but do not spend a long time trying to make them perfect.

8. If you are asked to sketch a graph, label the axes (including units) and put as many values on the graph as you can, including numbers given in the question.

9. In a multiple-choice question, narrow down your options by crossing out those answers which can't possibly be right. If you are not sure – make an **educated guess**!

10. Even though you get no marks for doing working in multi-choice questions, you are more likely to get them right if you do careful calculations.

1 mark   1 mark

The loudspeaker works by having an alternating current in a coil. The current is in a magnetic field, so it has a force on it. The coil vibrates because the direction of the force constantly changes.

1 mark   1 mark

$$P = \frac{F}{A} \qquad F = 5.0\,N$$

1 mark for formula

$$A = \pi r^2$$
$$= \pi (0.010)^2$$
$$= 3.1415927 \times 10^{-4}\,m^2$$

1 mark for formula

$$\therefore P = \frac{5.0\,N}{3.1415927 \times 10^{-4}\,m^2}$$

$$P = 15915.494\,N\,m^{-2}$$

$$= 1.6 \times 10^4\,N\,m^{-2}\ (2\ s.f.)$$

1 mark
but only with units and to 2 s.f. (because F and r are to 2 s.f.)

Doing Your Practical Work

Practical work is an important aspect of AS and A-level Physics. During your course you will complete a set of practical activities designed to develop your skills and increase your understanding of measurement and analysis.
Your abilities will be assessed in two different ways:

- Using *examination questions* which will test your understanding of the experimental process.
  These will account for 15% of your final A-level grade!

- Through *Common Practical Assessment Criteria* which will assess how well you carry out experiments. Your teacher will decide whether you have *passed* based on the evidence you collect and keep in your practical portfolio.
  This result will not count towards your final A-level grade.

Developing your practical skills

There are 4 different aspects you need to develop during your course: **Planning**, **Implementing**, **Analysis** and **Evaluation**. The following sections will help you to understand the skills needed to succeed in practical work and exams.

▷ Planning

Your Exam Board will have a set of investigations you will need to carry out during the course, and your teacher may also include some extra ones to help you develop your skills. Before carrying out each practical, you need to develop a plan. Research from textbooks or the internet can help you with the experimental design.

Your plan should describe how you intend to investigate the relationships between variables. Key parts of the plan include:

- Selecting the **independent variable** (the one you will change) and the **dependent variable** (the one you will measure).
  For example, in a resistivity experiment you may change the diameter of the wire while measuring the effect on the resistance.

- Identifying **control variables**. These are any other variables which could affect the outcome of the experiment.
  In the resistivity experiment, you would need to control the length of the wire and its temperature.

- The apparatus you will use, including the *detailed specification* of any measuring instruments.

A ruler with a range of 0–300 mm and a resolution of 1 mm.

- A thorough *practical method*, explaining all of the steps you will take during your experiment. Numbered steps are best.

- A description explaining how you will process any data.

- You may also need a *risk assessment* to ensure hazards are identified and managed.

466

▷ Implementing

While carrying out your experiment you will need to:

- make sure you work safely,
- set up and use the apparatus correctly and skilfully,
- check for sources of error in your technique and take action to reduce them (see below),
- use the measuring instruments carefully, aiming for accurate and precise results,
- record all measurements in a column to the same number of decimal places, which should match the resolution of the measuring instrument used,
- repeat your readings when necessary and check any readings which don't fit with the others (**anomalous** results),
- be flexible and adapt your plan if necessary.

Showing the uncertainty in your measurements

Every measuring instrument has limitations, and there will always be an uncertainty in any readings taken using it.
For example, a reading of 1.0 V means that the voltage is between 0.95 V and 1.05 V but we cannot be certain of the exact value. You need to take this uncertainty into account when recording measurements by using an appropriate number of significant figures.

If you measured some wire as being 110 cm long to the nearest centimetre, how should this be recorded in metres?
You should write it as 1.10 m.
If you write it as 1.1 m, then this suggests you could only measure to the nearest 10 cm. If you record it as 1.100 m, it implies that you measured to the nearest millimetre. Both are misleading.

Systematic errors

These are errors in the experimental method or equipment, when readings are consistently too big or consistently too small.

For example, if your newton-meter reads 0.2 N with no weight on it, then your measurements of force will always be 0.2 N too large. Remember to check for these **zero errors** before using any equipment. In addition, some instruments, such as balances, can be **calibrated** by checking them against known standards.

By reducing systematic errors the data becomes **more accurate**, so it reflects the true value more closely.

Random errors

These are errors which mean that the readings are sometimes too big and sometimes too small. For example, when timing a pendulum, there is an error in your timing because of your reactions. Some of the measurements will be below the true value and some will be above it.

The effect of random errors can be reduced by taking repeat readings; the mean value will be **more precise** than individual measurements.

Measurement error

Systematic errors and random errors combine to cause a measured value to be different from the true value. Always check your equipment and refine your technique to keep measurement uncertainty to a minimum.

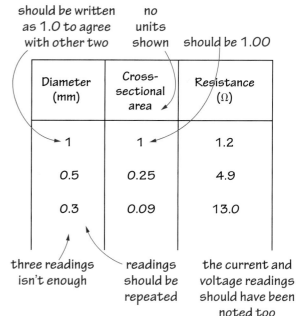

Diameter (mm)	Cross-sectional area	Resistance (Ω)
1	1	1.2
0.5	0.25	4.9
0.3	0.09	13.0

should be written as 1.0 to agree with other two

no units shown

should be 1.00

three readings isn't enough

readings should be repeated

the current and voltage readings should have been noted too

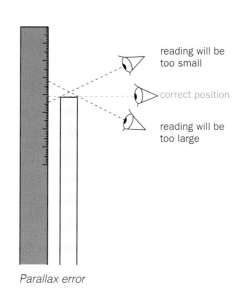

reading will be too small

correct position

reading will be too large

Parallax error

▷ Analysing evidence and drawing conclusions

The most common way of analysing your data is to produce a graph comparing variables and look for a relationship between them. Imagine that you had to investigate how the period of a **pendulum** is related to its length. Theory shows that the relationship between the period T and length l is $T = 2\pi\sqrt{l/g}$ (see page 106).

This means that T should be directly proportional to $\sqrt{l}$. A suitable graph should confirm this relationship.

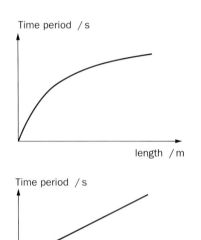

1. Think carefully about which graph to plot

For the pendulum, a graph of T against l would produce a curve. It would show there is a clear relationship between the variables, but it would not show whether T is directly proportional to $\sqrt{l}$.

A more suitable graph would be T against $\sqrt{l}$.
This should produce a straight line (which would pass through the origin of the graph) if your data matches the theory.

2. Make sure that you plot your graphs correctly

See page 476 for advice on plotting graphs. Remember to:

- Use a pencil to mark your points neatly, draw the axes and your line of best fit, so you can correct mistakes easily.
- Decide the form of best-fit line to draw. Does theory suggest a straight line or a curve (and would it go through the origin)?
- A best-fit line should have roughly the same number of offline points above it and below it.
- If some points are a long way away from the best-fit line, then these can be classed as *anomalous* results.
  Draw a circle round such results, as shown, and label them. Make sure you ignore them when drawing the line of best fit.

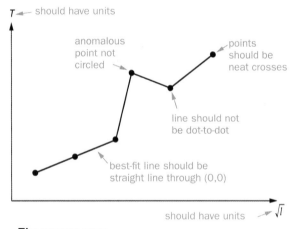

The wrong way . . .

. . . *and the right way:*

3. Find the gradient and intercepts

Often we can use the gradient of the graph or one of the intercepts to find a key value or relationship.
For the pendulum, the relationship between the period and length should be given by $T = 2\pi\sqrt{l/g}$.

This can be expressed as $T = \sqrt{l}\,(2\pi/\sqrt{g})$. Comparing this to the equation for a straight line ($y = mx + c$) tells us that the gradient of our graph should be $(2\pi/\sqrt{g})$.

So, if we find the gradient, we can determine a value for g, the acceleration due to gravity, as $g = 4\pi^2/\text{gradient}^2$.
Measurement of the gradient is explained on page 476.

4. State your result

Once you have analysed your data, you should clearly state your result or conclusion.
You can also estimate the amount of uncertainty in any final answer.

Pages 470 and 471 will help you to determine this.

You should compare the value you have found to relevant theory and any 'standard' values you have been given. For example, internet research will tell you the accepted value of g, and you can compare your measured value to this, and discuss any difference.

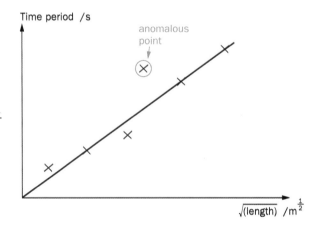

▷ Evaluation of procedures and results

Evaluating the quality of your evidence

Your experimental results will not always match your expectations and there will often be a difference between the value you find experimentally and an 'accepted value'.

You can assess and comment on the quality of your evidence based on these three factors:

- The uncertainty of your final answer. A smaller uncertainty in the result is clearly better than a large one.
- The number of anomalous results in your data. The fewer of these there are, the better your experimental technique was.
- The strength of any correlation in your graph. The closer the data points lie to the line of best fit, the better.

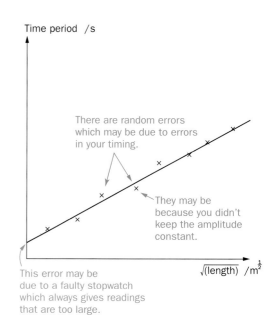

Identifying sources of error and uncertainty

In your evaluation you need to show that you understand what may have caused any difference between your conclusions and the expected result. Differences may be due to factors such as:

- Limitations in your experimental design or techniques which led to errors. (Eg. poor insulation in an experiment measuring specific heat capacity will give larger values than expected.)
- You may not have calibrated or zeroed your instrument (eg. a newton-meter or balance) and this results in systematic errors.
- The instrument may be faulty. (Eg. an old ammeter may not have a linear response to current, producing systematic error.)
- The resolution of the instrument may not be high enough, and this leads to a large uncertainty in the measurements and therefore in the final answer.

By finding the percentage difference between your answer and the accepted value, and comparing this to the combined percentage uncertainty caused by the instruments, you can evaluate how well you have carried out the experiment.

Suggesting improvements

There is always room to improve an experiment, and you need to be able to refine your original plan to give better results. These improvements should include suggestions about changes to the measuring instruments, and the use of techniques which are less prone to systematic or random error. You need to be able to explain why making these changes will give you a smaller level of uncertainty in your final answer. Suggestions could include:

- Repeating readings – this should help to reduce random errors.
- Using instruments with higher resolution – this will decrease uncertainty in measurements used to find the final answer.
- Adjusting the range of values used for the independent variable so that interesting areas of the dependent variable are explored in more detail (eg. near a sharp curve in your graph).
- New ideas for ensuring that the control variables cannot change during your experiment.

▷ Dealing with errors and uncertainty

Precision and accuracy

No reading is ever going to match exactly the true value of a variable, due to uncertainty of instruments and error in the experimental technique. Taking this into account, we try to produce results which are both **accurate** and **precise**:

- *Accurate measurements* lie close to the true value of a variable – sometimes a bit above and sometimes a bit below. An *accurate answer* will therefore be close to the true value.

- *Precise measurements* will all be grouped together, with all of them being nearly the same value. But note that this does not mean that they are anywhere near the true value, as they could all be affected by a systematic error.

How are uncertainties written down?

In an electrical experiment, you may get a reading of 1.1 V on your voltmeter. The meter has a **resolution** of 0.1 V; this is the smallest change it can display.

So you know that the voltage is about 1.1 V, not 1.0 V or 1.2 V.

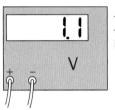

The uncertainty of the measurement is 0.05 V.

This means that it lies somewhere between 1.05 V and 1.15 V. This is because if it was 1.04 V the meter would display 1.0 V, and if it was 1.16 V then the meter would show 1.2 V.

The way to show the reading is between 1.05 V and 1.15 V is by writing it as: 1.10 ± 0.05 V.
In this example, the **uncertainty** is ± 0.05 V.

Often it's more important to calculate the **percentage uncertainty**. In this example:

$$\text{percentage uncertainty} = \frac{\text{uncertainty}}{\text{measurement}} = \frac{0.05}{1.10} \times 100 = \pm 4.5\%$$

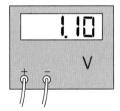

A meter with higher resolution. The uncertainty of the measurement is now 0.005 V.

The voltage must be between 1.095 V and 1.105 V.

Estimating uncertainty

Sometimes the uncertainty in a reading is greater than the resolution of the instrument because there are errors involved.

For example, the resolution for a stopwatch may be 0.01 s but the reaction time of the user means it is impossible to measure time to anywhere near this resolution. In this case, you may want to assume the resolution is 0.1 s and base the uncertainty on this.

If you have a set of repeat readings, such as the timing of a pendulum, you can calculate the mean value.
You can also estimate the uncertainty by looking at the **range** of the repeat values in the set.

For example, if the mean time was 10.3 s, and the lowest reading in the set was 9.9 s (0.4 s below the mean) and the highest was 10.9 s (0.6 s above the mean), then the average difference from the mean is 0.5 s. In this case we could state the time as (10.3 ± 0.5) s.

▷ Calculations involving uncertainties

When we calculate an answer using values which have uncertainties involved, we need to take those uncertainties into account in the final answer.

1. When adding or subtracting

When adding *or* subtracting quantities, just **add** their uncertainties.

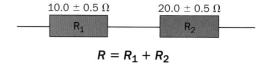

The diagram shows 2 resistors in series.
What is the uncertainty in the total resistance of this combination?

Nominal value for total resistance = 10.0 Ω + 20.0 Ω = 30.0 Ω
Total uncertainty = 0.5 Ω + 0.5 Ω = 1.0 Ω
So we state the resistance as (30.0 ± 1.0) Ω

You can confirm this by calculating the maximum and minimum possible resistances (31 Ω and 29 Ω respectively).

2. When multiplying or dividing

In these cases you need to **add** the **percentage** uncertainties.

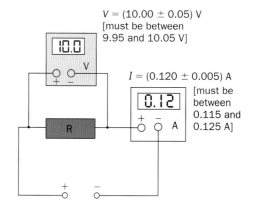

For the measurements shown in the diagram:
Resistance = V/I = 10.00 V/0.120 A = 83.3 Ω
Percentage uncertainty in V = (0.05/10.00) × 100 = 0.5%
Percentage uncertainty in I = (0.005/0.120) × 100 = 4.2%
Total percentage uncertainty = 0.5% + 4.2% = 4.7%
4.7% of 83.3 Ω = 0.047 × 83.3 = 3.9 Ω
So: Resistance = (83.3 ± 3.9) Ω = (83 ± 4) Ω

3. When dealing with powers

When squaring a quantity, the percentage uncertainty in the value doubles; when cubing, the percentage uncertainty is multiplied by 3. Finding a square root multiplies the percentage uncertainty by 0.5.

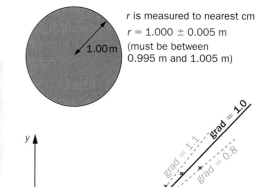

What is the area of the circle in the diagram?
Area = πr^2 = $\pi \times 1.00^2$ = 3.14 m^2

Percentage uncertainty in r = (0.005/1.000) × 100 = 0.5%
∴ Percentage uncertainty in r^2 is 1.0%
Uncertainty in the area is 1% of 3.14 = 0.03 m^2
∴ We can state the area as (3.14 ± 0.03) m^2

Uncertainty in gradients

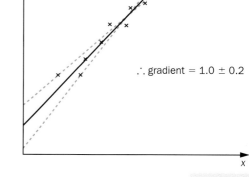

When you draw a line of best fit through some points, it is difficult to judge exactly where the line should go. This means the gradient is uncertain to some extent, and so is any value derived from it.

To estimate this uncertainty you can draw two other possible lines of best fit, one with the maximum possible gradient and the other with the minimum. Then find the gradients of these 2 lines as well.

For a more sophisticated treatment, you should draw **error bars**, as explained at: www.oxfordsecondary.co.uk/advancedforyou

Check Your Maths

Do you find Maths difficult?
Do you find it hard to remember?
If so, these pages can help you with some of the Maths that
you need in your Physics lessons.

Symbols used in Maths

Here are some of the symbols you may meet in Physics:

$\propto$	is proportional to	$=$	is equal to
$\sim$	about the same size as	$\approx$	is approximately equal to
$>$	is greater than	$<$	is less than
$>>$	is much greater than	$<<$	is much less than
$\geqslant$	greater or equal to	$\pm$	plus or minus
Δx	a change in x	δx	a small change in x
$x^{\frac{1}{2}}$	the square root of x	$\sqrt{x}$	the square root of x
$\bar{x}$	mean of the values of x	Σx	sum of all the values of x
$\therefore$	therefore	$\Rightarrow$	implies that

Significant figures

There is more detailed explanation of significant figures on page 9,
but here are some reminders:

- To find the number of significant figures (**s.f.**), count the number
  of digits starting from the ***first non-zero*** number on the left.
  Include zeros, once you have started counting. For example:

  2.7 (2 s.f.)   271 (3 s.f.)   271.0 (4 s.f.)

  1.200 (4 s.f.)   0.0120 (3 s.f.)   0.000 12 (2 s.f.)

- In a calculation, give the answer to the ***lowest*** number of s.f.
  of ***any*** of the numbers used to calculate the answer.
  Eg. $56.21 \times 3.1 = 174.251$ but the answer should be written
  as just: 170 (2 s.f.) because 3.1 is only 2 s.f.
  A better way of writing this answer would be 1.7×10^2 (2 s.f.).

Vectors

For work on vectors please see pages 10–16.

Prefixes

For prefixes (like milli, kilo, mega, etc.) see page 8.

▷ Powers

The power of a number is the small figure perched on the shoulder of the number. This is the 'index' (plural: 'indices'). It tells us how many times the number is multiplied by *itself*.

For example: $2^3 = 2 \times 2 \times 2$ and $5^4 = 5 \times 5 \times 5 \times 5$

Any number to the power 1 is itself, so $5^1 = 5$.
Any number to the power 0 is equal to 1, so $2^0 = 1$ and $7^0 = 1$.

Rules for powers

$y^a \times y^b = y^{a+b}$ eg. $10^2 \times 10^3 = 10^{2+3} = 10^5$

$y^a / y^b = y^{a-b}$ eg. $10^3 / 10^2 = 10^{3-2} = 10^1$

$(y^a)^b = y^{a \times b}$ eg. $(2^4)^3 = 2^4 \times 2^4 \times 2^4 = 2^{12}$

Negative Powers

If a number has a negative power, like 10^{-1}, what does this mean?
This is equal to $^1/$(10 to the positive power).

For example: $10^{-1} = {}^1/_{10^1} = {}^1/_{10} = 0.1$
$10^{-2} = {}^1/_{10^2} = {}^1/_{100} = 0.01$

Standard Form

Many quantities used in Physics are very small or very large.
Eg. mass of the Earth is 6 000 000 000 000 000 000 000 000 kg.
The diameter of an atom is about 0.000 000 000 1 metres.

Can you see the problem with using such large numbers?
It is very easy to make mistakes by missing one of the noughts off!

If we use *Standard Form* it helps us to avoid this.
This is when a number is written as a number between 1 and 10 and then *multiplied by a power of 10*.

For example: $1280 = 1.28 \times 1000 = 1.28 \times 10^3$
$0.0128 = 1.28 \times 0.01 = 1.28 \times 10^{-2}$

So the mass of the Earth $= 6 \times 10^{24}$ kg
The diameter of an atom $= 1 \times 10^{-10}$ m

The index tells you how many decimal places to move. Some examples:

$2.1 \times 10^6 = 2\,100\,000$

$0.21 \times 10^7 = 2\,100\,000$

$4 \times 10^{-3} = 0.004$

$4.1 \times 10^{-4} = 0.00041$

Using Standard Form on a calculator

It is important to put Standard Form into your calculator correctly!
The **EXP** button on your calculator means "$\times 10$ to the power of".

So what would you press to put 1.28×10^3 into your calculator?

You would press   then then

Notice that there is no need to put in the $\times 10$ bit, because this is taken care of by the EXP button.

If you saw $\boxed{2 \quad 03}$ on your calculator, would this mean 2^3?

No, it would mean $2 \times 10^3 = 2000$.
Remember that the EXP button means: "$\times 10$ to the power of".

Find the EXP button on your calculator.

▷ Geometry and Trigonometry

Triangles

The angles of a triangle add up to 180°.   Area of a triangle $= \frac{1}{2} \times$ base $\times$ height

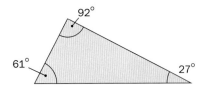

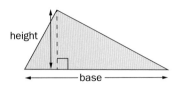

Right-angled triangles

θ ('theta') is an angle within a right-angled triangle.
The sides of the triangle are given the names
hypotenuse, *opposite* and *adjacent*, as shown:

Pythagoras' Theorem

This is useful when you know the lengths of 2 sides of
a right-angled triangle and you wish to find the third side.

$$a^2 = b^2 + c^2$$

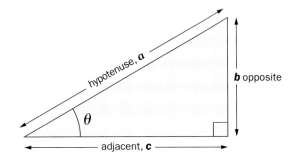

Sine, Cosine, Tangent

For a certain angle θ, the ratios b/a c/a and b/c will always
be the same whatever the size of the right-angled triangle.
These ratios are called the sine (sin), cosine (cos) and tangent (tan)
of the angle θ.

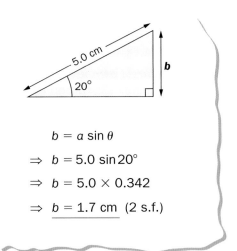

$\sin \theta = \dfrac{\text{opposite}}{\text{hypotenuse}} = \dfrac{b}{a}$	$\therefore$	$b = a \sin \theta$
$\cos \theta = \dfrac{\text{adjacent}}{\text{hypotenuse}} = \dfrac{c}{a}$	$\therefore$	$c = a \cos \theta$
$\tan \theta = \dfrac{\text{opposite}}{\text{adjacent}} = \dfrac{b}{c}$	$\therefore$	$b = c \tan \theta$

$b = a \sin \theta$

$\Rightarrow \quad b = 5.0 \sin 20°$

$\Rightarrow \quad b = 5.0 \times 0.342$

$\Rightarrow \quad b = 1.7$ cm (2 s.f.)

Using a calculator with sin, cos and tan

- Check that you are working in the right units for your question
  (degrees or radians, see opposite page).
- On some calculators you need to press the sin, cos or tan
  button *after* you have entered the angle.
- If you know the sine of an angle and want to know the angle
  itself, use the **sin⁻¹** button. See this example:
  You can use the $\cos^{-1}$ and $\tan^{-1}$ buttons in the same way.

$\sin \theta = 0.75$

$\Rightarrow \quad \theta = \sin^{-1} 0.75$

$\Rightarrow \quad \theta = 49°$ (2 s.f.)

More Geometry

Circumference of a circle $= 2\pi r$

Area of circle $= \pi r^2$

Surface area of sphere $= 4\pi r^2$

Volume of sphere $= {}^4/_3\pi r^3$

Volume of a cylinder $= \pi r^2 h$

Area of trapezium $= {}^1/_2 (a + b) h$

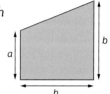

Radians

In higher level maths it is often necessary to measure angles in units called radians (rads). See also page 78.
Looking at the diagram opposite:

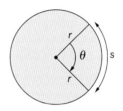

$$\text{the angle } \theta \text{ in radians } = \frac{\text{arc length}}{\text{radius of circle}} = \frac{s}{r}$$

Converting from radians to degrees

One complete revolution of a circle is 360°.
The arc length for one revolution is the whole circumference of the circle, $2\pi r$

So, the number of radians in one complete revolution $= \dfrac{\text{arc length}}{\text{radius}} = \dfrac{2\pi r}{r} = 2\pi$ rad

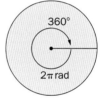

360° is equivalent to 2π rad   1 radian $= 57.3°$

180° is equivalent to π rad

90° is equivalent to $\pi/_2$ rad

Using a calculator with degrees and radians

Your calculator will work with either degrees *or* radians.

If it is working with degrees you will see **DEG** on the top of the screen.
RAD will show on the screen if you are working in radians.
You can change from one to another on most calculators by using the MODE button. Make sure you know how to do this.

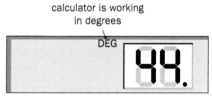

calculator is working in degrees

Calculator screen.

Small angles

It is sometimes useful to make the following approximations when an angle is *small*.

$\sin\theta \approx \tan\theta \approx \theta$ measured in radians.   Also, $\cos\theta \approx 1$

These work quite well for angles less than 10°.
The smaller the angle, the better the approximation!

▷ Plotting graphs

When you are constructing a graph after an experiment:

- Have at least 5 points to plot. If your graph curves, you may need more points to draw the curve more accurately.

- Scales must increase in equal steps. Choose scales that will make the graph as *large* as possible, without making the scale awkward for plotting points.

- Label your scales with the units being used. For example, time /s   means "time measured in seconds".

- As a general rule, the *independent* variable (the one you are deliberately changing) goes on the *x*-axis. The other variable which changes as a result (the *dependent* variable) goes on the *y*-axis. Time usually goes on the *x*-axis.

- Lines of best fit should be drawn. Sometimes they should be straight lines, sometimes curves. If you are not sure which to draw, look at the theory about your practical. Would you expect a curve or a straight line from your results? See also page 468.

- Think carefully about whether theory suggests that your graph should go through (0,0). Not all graphs should go through the origin!

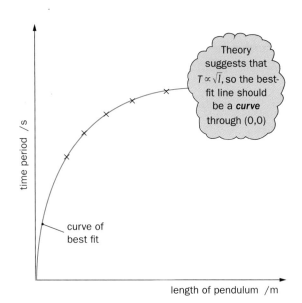

A graph showing how the period of a pendulum depends on its length.

▷ The gradient of a straight-line graph

The diagram shows you how to find the gradient of a line:

Δy means the change in the value of y.
Δx means the change in the value of x.

$$\textbf{Gradient} = \frac{\Delta y}{\Delta x}$$

In the example opposite:

$$\text{Gradient} = \frac{\Delta y}{\Delta x} = \frac{30\ \text{N} - 10\ \text{N}}{0.6\ \text{m} - 0.2\ \text{m}} = \frac{20\ \text{N}}{0.4\ \text{m}} = 50\ \text{N m}^{-1}$$

A gradient will normally have units. You can find them by dividing the units of the *y*-axis by the units of the *x*-axis. Notice that the units of the gradient above are N m^{-1}.

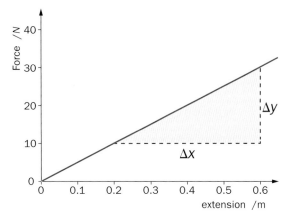

The gradient of a straight line is constant.
The gradient of a **curve** is always changing. We can measure the gradient of a curve at selected points.

To do this, a **tangent** to the curve is drawn and the gradient of this tangent is measured in the same way as for a straight line:
(A tangent is a straight line which just touches the curve at one point, as shown in the diagram.)

To find gradients accurately, make Δy and Δx as *large* as possible.

For the treatment of uncertainties, see page 471.

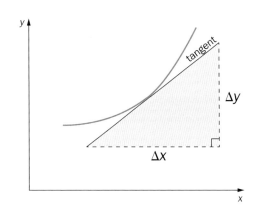

476

▷ Graphs and the area beneath a line

In many situations in Physics we need to use the product of two numbers to work out a quantity.
For example, to find the work done by a force, we can multiply the force F by the displacement s in the direction of the force:

$$W = Fs \qquad \text{(see page 68)}$$

However, this assumes that the force is constant throughout the process, and this is often not true.

One way to calculate the work done by a varying force is to find the **area beneath** a graph of force against displacement.

You can picture the area under the graph as lots of slices like the one shown in yellow:
The area of each thin slice is (height × width), which is force × displacement ($Fs = W$).
So the area of each slice represents a small amount of work done.
Adding up the area of all the slices gives you the total work done.

For an estimate of this area in a practical situation, we can use an approach which counts grid squares, as shown below.

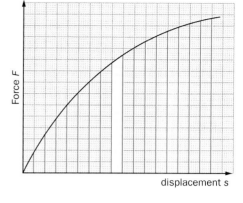

Adding up the area of these small strips will tell us the work done by the force

Worked example

The **red** line on the graph on the right shows the results of an experiment to measure the extension caused by a force stretching a thick elastic rope:

How much energy was stored by the elastic as it stretched? The graph is not a straight line. The rope is not obeying Hooke's Law and so we can't find the elastic energy stored by using $E_p = \frac{1}{2} F \Delta x$.

Instead we look at the area under the **red** line on the graph. We can use simple geometric shapes (rectangles and triangles) to calculate most of the squares, or we can just count them. Combining fragments of squares also helps our count; adding a small bit of a square to an almost full one gives us one complete square. Or see the simple rule in the yellow box:

The diagram shows how the squares have been counted. There is a total of 107 large squares beneath the red line.

Now we know the number of squares, we can use the scales on the graph to calculate the energy that these represent. In this case, each square represents 1 kN by 1 m, which is 1 kJ. This gives us a total energy stored of 107 kJ.

Page 193 explained that rubber stretches differently during loading and unloading (a process called hysteresis). Energy is lost due to heating of the rubber.
The **green** line shows the unloading curve for the elastic rope. Use the same technique as we used above to work out how much energy was lost in the loading–unloading cycle. You should find that this is approximately 20 kJ.

You will find that this technique is particularly useful for working out distance travelled during a journey when the speed is not constant. See page 41 for a further example.

See other examples on pages 60, 68, 187, 191, 274, 302–3.

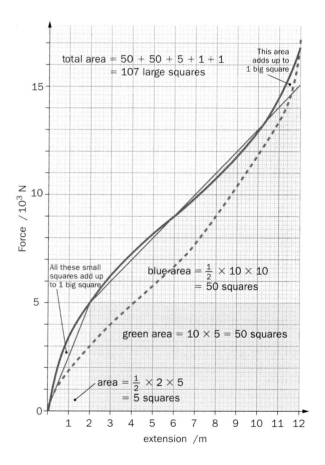

Count the squares under the graph.

A simple rule is: if more than half the square is under the graph, count it as a whole square; if less than half a square, don't count the square.

The errors in this method will tend to balance out.

▷ The equation of a straight-line graph

Look at the 4 graphs shown here and the equations
which they represent:
Notice that the number in front of the x in the equation
is the same as the **gradient** of the graph.
Notice that the number added on after the x is the same
as the place where the line cuts the y-axis (the **intercept**).

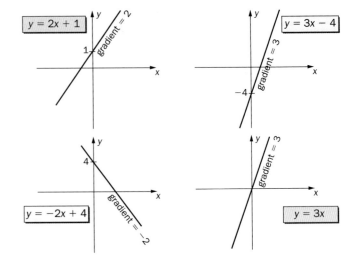

All **straight-line graphs** have an equation of the form

$$ y = m\,x + c $$

y and x are the quantities on the y- and x-axes.
m is the **gradient** of the graph.
c is the **intercept** of the graph on the y-axis.

▷ Some common graph shapes

In each case, k is a constant – a number which does not change.

Inverse proportion

$$ y = \frac{k}{x} \qquad \text{as in} \quad a = \frac{F}{m} \qquad \text{(see page 55)} $$

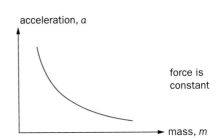

Inverse-square relationship

$$ y = \frac{k}{x^2} \qquad \text{as in} \quad E = \frac{k\,Q}{r^2} \qquad \text{(see page 293)} $$

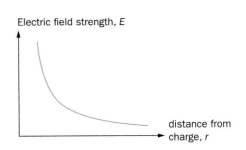

Notice that this has a very similar shape to the one above,
and so you need to investigate further when you get a graph
of this shape, to see which one it is.

A square relationship

$$ y = k\,x^2 \qquad \text{as in} \quad s = \tfrac{1}{2}\,a\,t^2 \quad \text{(see page 42)} $$

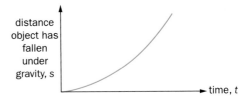

A sine graph

$$ y = k\,\sin x \qquad \text{as in} \quad s = A \sin \omega t \quad \text{(see page 103)} $$

A cosine graph

$$ y = k\,\cos x \qquad \text{as in} \quad s = A \cos \omega t \quad \text{(see page 103)} $$

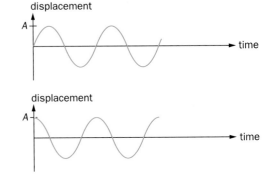

For exponential graphs, $y = A\,e^{-kx}$, see page 484.

478

▷ Direct proportion

A variable y is **directly** proportional to a variable x if

$$y = k\,x$$ (k is a constant number)

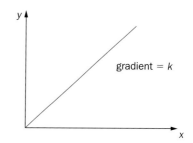

gradient $= k$

If this equation is true, then:

- a graph of y against x will be a straight line through $(0,0)$ with gradient k, as shown here:
- y/x will always be the same number (the constant k),
- if you double y, then x will double.

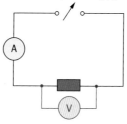

variable power supply

Look at the circuit shown here. If you changed the voltage and measured the current each time, then you might get results like this which show direct proportion:

Can you see that, as the voltage doubles from 1 V to 2 V, the current also doubles?
Also, when the voltage doubles from 2 V to 4 V then the current doubles again.

Voltage (V)	Current (A)
0.00	0.00
1.00	0.25
2.00	0.50
3.00	0.75
4.00	1.00

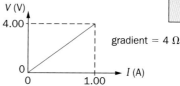

gradient $= 4\ \Omega$

Notice also that voltage divided by current is always the same number, 4. This is actually the resistance of the wire. Remember that $V/I = R$ (page 232).

▷ Inverse proportion

The two variables y and x are **inversely** proportional if

$$y = \frac{k}{x}$$ This means: $y\,x = k$ (a constant)

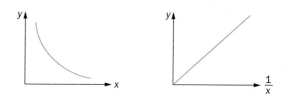

- to get a straight line through $(0,0)$, plot **y** against **$1/x$**:
- if you double y, then x will **halve**.

If you change the pressure on a gas and measure its volume (see page 210) then you could get results like the ones here, which show **inverse** proportion:

When the pressure is doubled from 1×10^5 N m^{-2} to 2×10^5 N m^{-2}, the gas volume halves from 16 m^3 to 8 m^3. Can you see the volume halves again when the pressure is doubled again from 2×10^5 N m^{-2} to 4×10^5 N m^{-2}?

What is the constant $P \times V$ equal to?
In this experiment it is always 16×10^5 N m.

Pressure (N m^{-2})	Volume (m^3)
1×10^5	16
2×10^5	8
4×10^5	4
8×10^5	2
16×10^5	1

Inverse-square relationship

$$y = \frac{k}{x^2}$$ and so $y\,x^2 = k$ (a constant)

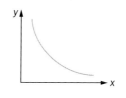
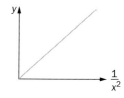

- to get a straight line through $(0,0)$, plot y against $1/x^2$:
- if you double x, then y falls to a quarter.

▷ Equations

Changing the subject of an equation

From page 42, there is an equation for acceleration, $a = \dfrac{v - u}{t}$

But what if you know a, u and t, and you want to find $\boldsymbol{v}$?
You will need to change the **subject** of the equation to v.

As long as you do the **same** thing to **both** sides of an equation,
it will still balance. For example:

First multiply by t: $\qquad a \times t = \dfrac{(v - u) \times \cancel{t}}{\cancel{t}}$

then cancel: $\qquad \therefore \ a\,t = v - u$

Now add u: $\qquad u + a\,t = v - \cancel{u} + \cancel{u}$

then cancel: $\qquad \therefore \ u + a\,t = v$

So v is now the subject. Often it is easier to put in your numbers
before you rearrange the equation, as shown here:

A more complex example

Change the subject of: $\qquad F = \dfrac{G\,m_1\,m_2}{r^2}$ (page 86) to be **G**.

First multiply by r^2: $\qquad F \times r^2 = \dfrac{G\,m_1\,m_2}{\cancel{r^2}} \times \cancel{r^2}$

$\qquad\qquad\qquad \therefore \ F\,r^2 = G\,m_1\,m_2$

Divide by m_2: $\qquad \dfrac{F\,r^2}{m_2} = \dfrac{G\,m_1\,\cancel{m_2}}{\cancel{m_2}}$

$\qquad\qquad \therefore \ \dfrac{F\,r^2}{m_2} = G\,m_1$

Divide by m_1: $\qquad \dfrac{F\,r^2}{m_1\,m_2} = \dfrac{G\,\cancel{m_1}}{\cancel{m_1}}$

$\qquad\qquad \therefore \ \dfrac{F\,r^2}{m_1\,m_2} = G$

Combining equations

Sometimes you need to combine two equations.
For example, in circuit work (Chapters 16, 17) if you know that
voltage is given by $V = IR$ and power by $P = IV$,
what is the power of a resistor R with a voltage V across it?

$P = I \times V$ $\qquad$ But: $V = I \times R$, $\quad$ so: $I = \dfrac{V}{R}$

Therefore: $P = \dfrac{V}{R} \times V$

So: $P = \dfrac{V^2}{R}$

$t = 10\,\text{s}$

$u = 0$

$a = 2\ \text{m/s}^2 \qquad V = ?$

$a = \dfrac{v - u}{t}$

$\therefore \ 2 = \dfrac{v - 0}{10}$

$\therefore \ 2 = \dfrac{v}{10}$

$\therefore \ V = 20\ \text{m/s}$

Usually equations are written
with the subject on the
left-hand side, so

$V = U + at$

and

$G = \dfrac{F\,r^2}{m_1\,m_2}$

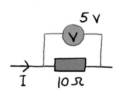

$5\,\text{V}$

$I \qquad 10\,\Omega$

$P = \dfrac{V^2}{R}$

$= \dfrac{5^2}{10} = \dfrac{25}{10}$

Power $= 2.5\ \text{W}$

▷ Simultaneous equations

In some equations there may be more than one unknown quantity.
In this case you need to solve 2 or more equations simultaneously.

For example, the diagram shows two different masses connected by
an inextensible string over a freely rotating pulley:
When released, the masses will accelerate.
Let's calculate the acceleration a, and the tension T in the string.

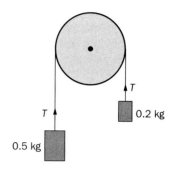

There are 2 unknowns (a, T), so we need to find 2 equations.
Ignoring friction in the pulley, the tension T in the string must be the
same throughout it (so T on the left is the same size as T on the right).

Looking at the mass on the left, we have an unbalanced resultant force
downwards, which produces an acceleration ($F = ma$).
The unbalanced force is the resultant of the weight of the mass ($m_1 g$)
and the tension in the spring (T) so we can write the equation:

$$m_1 g - T = m_1 a$$

$$0.5 \times 9.8 - T = 0.5a \qquad \text{So:} \quad a = 9.8 - 2T \quad [1]$$

This equation still has 2 unknown values and so cannot yet be solved.

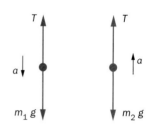

$$m_1 g - T = m_1 a \qquad T - m_2 g = m_2 a$$

Now do the same for the mass on the right of the diagram:

$$T - m_2 g = m_2 a$$

$$T - 0.2 \times 9.8 = 0.2a \qquad \text{So:} \quad 5T - 9.8 = a \quad [2]$$

Now we have two (simultaneous) equations which describe the same
acceleration. We can substitute the value of a from equation [1] into
equation [2] to get:

$$5T - 9.8 = 9.8 - 2T \quad \text{from which we get:} \quad T = 2.8 \text{ N}$$

Substituting this value for T into equation [1] gives $a = 4.2 \text{ m s}^{-2}$

▷ Quadratic equations

Quadratic equations are equations of the form $ax^2 + bx + c = 0$

They have solutions given by: $x = \dfrac{-b \pm \sqrt{b^2 - 4ac}}{2a}$

This type of equation is useful in solving some Physics problems,
for example in this worked example:

A ball is thrown from a cliff 200 m high. It is thrown downwards at a
vertical velocity of 10 m s^{-1}. How long will it take to reach the ground?

If we start with this equation: $s = ut + \frac{1}{2}at^2$ (from page 42)
we can write this like the quadratic above: $\frac{1}{2}at^2 + ut - s = 0$

Substituting values: $4.9t^2 + 10t - 200 = 0$

Now substitute the coefficients $\quad t = \dfrac{-10 \pm \sqrt{10^2 - 4 \times 4.9 \times -200}}{2 \times 4.9}$
in the equation to give:

This gives two solutions: $t = 5.4$ s and $t = -7.5$ s.
The solution we are looking for is $t = 5.4$ s.

(The negative solution is not physically possible.)

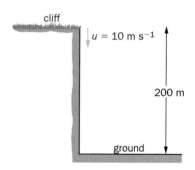

Taking downwards
direction as positive:

$s = +200$ m
$u = +10$ m s^{-1}
$v = ?$
$a = +9.8$ m s^{-2}
$t = ?$

481

▷ Using Spreadsheets

Spreadsheets are really useful tools for understanding some areas of Physics. The on-line resources for this book have several example spreadsheets, showing just how useful they can be in:
- modelling behaviour,
- analysis of data,
- simulations.

The on-line resources are at: www.oxfordsecondary.co.uk/advancedforyou

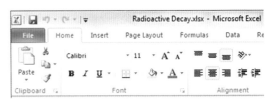

▷ Modelling behaviour

In some situations you need to model the behaviour of a system involving **rates of change**, where the rate of change is proportional to the 'amount' of something remaining.
This is expressed in the general equation:

$$\frac{\Delta x}{\Delta t} = -k\,x$$

where x is the amount remaining, and
k is the fraction of x which will decay per second.

The two most important examples of this behaviour are radioactive decay (see page 368) and the decay of charge on a capacitor discharging through a resistor (see page 307).

For **radioactive decay** we have the relationship:

$$A = \frac{\Delta N}{\Delta t} = -\lambda\,N$$

where A is the activity (the rate of decay),
N is the number of active nuclei remaining, and
λ is the decay constant (the fraction which decays each second).

For a **capacitor discharge** we have the relationship:

$$\frac{\Delta Q}{\Delta t} = -\frac{1}{C\,R}Q$$

where Q is the charge remaining on the capacitor,
C is the capacitance and R is the resistance.

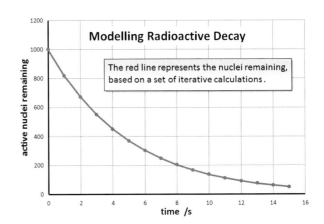

Using the initial charge, the capacitance and resistance, we can calculate the amount of charge ΔQ that will leave the capacitor in the first second. From this we know the charge remaining after one second. The example in the box shows the principle:

We can then use this new value for charge to calculate the charge leaving in the **next** second and thus the charge remaining after two seconds, and so on. Repeating this calculation 15 times to find the charge remaining after 15 seconds would be very time-consuming.

The relationship can be modelled more effectively by performing a series of repetitive (or iterative) calculations using a **spreadsheet**.

The spreadsheet approach also has the advantage of allowing you to change the **initial conditions** of a model quickly, and letting the spreadsheet instantly perform the new series of calculations and produce a new graph.

On-line at www.oxfordsecondary.co.uk/advancedforyou
you can find spreadsheets illustrating how this is achieved
for **radioactive decay** and for **capacitor discharge**.

For example, consider a capacitor discharging in a circuit with a time constant of 5.0 s.
Suppose at time $t = 0$ the charge on the capacitor is 1000 μC.

What charge flows off the capacitor in the first second? Using the formula in the main text:

$$\Delta Q = \frac{1}{C\,R} \times Q \times \Delta t$$

$$\Delta Q = \frac{1}{5.0\text{ s}} \times 1000\text{ μC} \times 1\text{ s} = 200\text{ μC}$$

So after 1 second the charge remaining on the capacitor is: 1000 μC − 200 μC =   800 μC

We can then use this value to calculate the charge leaving in the next second, and so on.

This repetition or **iteration** is what the spreadsheet does for you.

▷ Analysis of data

Spreadsheets are also very useful for the processing and analysis of large sets of data.

For example, think about an experiment to find out how the diameter of a wire affects the resistance.
You might collect 30 different measurements of current and potential difference and need to calculate resistances from them. You would then need to find mean resistance and plot a graph comparing resistance to cross-sectional area.

A suitably designed spreadsheet can be used to collect this data, perform the repetitive calculations and plot the required graphs.

An example spreadsheet called **Resistivity Analysis** is provided on-line at: www.oxfordsecondary.co.uk/advancedforyou

In it the data is processed using a simple formula ($R = V/I$). The spreadsheet is then used to calculate the mean resistance, cross-sectional area (A) and $1/A$ for wires of different diameters.

The spreadsheet plots a graph of R against A and shows that there is a clear relationship between them.
Then a second graph of R plotted against $1/A$ shows that R is inversely proportional to A.

The spreadsheet has also been used to determine the gradient of the line, and finally the resistivity.

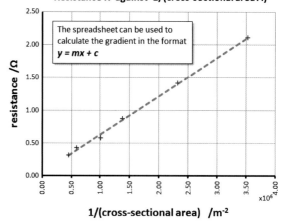

Resistance *R* against 1/(cross-sectional area *A*)

The spreadsheet can be used to calculate the gradient in the format
y = mx + c

1/(cross-sectional area) /m⁻²

▷ Simulations

Sometimes it can be difficult to understand how changes to a system can alter its behaviour. Spreadsheets can help in this area too, by performing simple simulations based on the relevant equations.

Three examples are provided on-line, for you to look at:

- The behaviour of three pendulums is simulated in the file called **SHM Pendulums**.
  Altering the properties of these pendulums within the spreadsheet allows you to compare pendulums of different lengths or in different gravitational fields:

- Another example of a simulation spreadsheet is called **SHM Masses**.
  The behaviour of 3 mass–spring systems is modelled for easy comparison. You can alter the mass and the spring constant for each system, to investigate what happens.

- The superposition of two waves is shown in the file called **Superposition Model**.
  Here the frequency, amplitude and phase of the two source waves can be altered and the results are shown on the graph.

We suggest that you download each of these spreadsheets from: www.oxfordsecondary.co.uk/advancedforyou
and 'play' with them by altering the parameters.

Look at how each spreadsheet is constructed, by reading the explanations that we have provided on each one.

Comparing Pendulums

You can alter the values for *g*, length, amplitude, phase to compare the motions of three different pendulums.

time /s

Learning how to use spreadsheets effectively can take some time.

Download and use the on-line worksheet **Projectile Motion** to construct your own spreadsheet to investigate launching projectiles.

▷ Exponentials

You will come across an **exponential decay curve** in the radioactivity topic (see page 369) and also when learning about the discharge of capacitors (page 307).

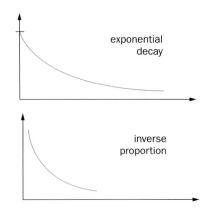

exponential decay

inverse proportion

Look at the curve and compare it with the graph showing inverse proportion (page 479). How is it different?

One thing which is different is that the curve touches the y-axis, but the inverse proportion curve doesn't.
The exponential curve has some very definite features.

How can you tell if a curve is exponential?

Look at the graph:
It's a graph of the amount of radiation given off by a radioactive substance each second, called the activity of the substance.

After each 70 seconds, the activity **halves**. After 70 secs it has **halved** from 400 to 200. Then, in the next 70 seconds, it has **halved** from 200 to 100. In the next 70 seconds it halves again.

One rule for exponential curves is that they always take the same time to halve (called the **half-life**), no matter how much you have to start with. The substance above has a half-life of 70 seconds.

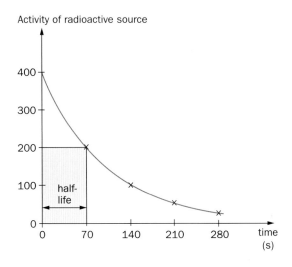

Activity of radioactive source

The exponential equation

The mathematical equation for any exponential decay is

$$y = A\,e^{-kt}$$

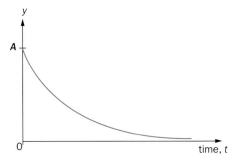

y is the quantity which is decaying (eg. the activity of the radioactive source).

As you can see from the graph, A is the starting value of y.
e is a special 'never-ending' number in maths (similar to π) with a value of 2.7128 to 4 decimal places.
k is a constant. It depends on how quickly the quantity decays.
t is the time that the quantity has been decaying for.

Logarithms

In order to understand exponentials better, we need to know about logarithms. What is a logarithm or 'log'?

Since **$100 = 10^2$** we say that:
the log of 100 (to the base 10) is equal to 2.

This can be written as: **$\log_{10} 100 = 2$**.

In the same way, since $1000 = 10^3$ then $\log_{10} 1000 = 3$.

Also, for example, $8241 = 10^{3.916}$.
So what is $\log_{10} 8241$? It is 3.916.

We can have logs to other bases. For example, $2^4 = 16$, so the log of 16 to the base 2 is 4. That is: $\log_2 16 = 4$.

You're lucky, son, I had to use log tables when I was at school. Yes, dad.

Two rules for using logs (to any base)

1. $\log(a \times b) = \log a + \log b$ so: $\log(4 \times 7) = \log 4 + \log 7$

2. $\log a^b = b \log a$ so: $\log 10^3 = 3 \times \log 10$

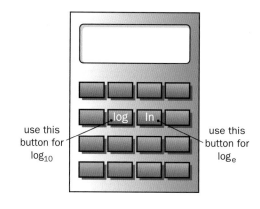

use this button for $\log_{10}$

use this button for $\log_e$

Taking logs of both sides of a power equation

In general, if you have an equation $y = A x^n$

then $\log y = \log(A x^n)$

But from rule 1 on logs above: $\log(ab) = \log a + \log b$

$\therefore$ $\log y = \log A + \log x^n$

But by rule 2: $\log a^b = b \log a$ so: $\log x^n = n \log x$

So if $y = A x^n$ then $\log y = \log A + n \log x$ . . . . . [1]

On many calculators you must put in the number first and then press the log/ln button.

Taking logs of both sides of an exponential equation

$y = A e^{-kt}$ and so, from [1] $\log_e y = \log_e A + \log_e e^{-kt}$

Since $\log_{10} 10^2 = 2$ and $\log_{10} 10^3 = 3$
can you see that $\log_e e^{-kt} = -kt$?

So we get: $\log_e y = -kt + \log_e A$ . . . . . [2]

This can be used to check whether a curve is exponential and to work out the constants A and k in the decay equation $y = A e^{-kt}$, as follows:

Plotting a log graph to check whether decay is exponential

Compare equation [2] above with the equation for a straight line, $y = m x + c$ (see page 478):

$$\log_e y = -k \, t + \log_e A$$
$$y = m \, x + c$$

What would a graph of **$\log_e y$** against **t** look like?

It will be a straight line as shown, with gradient $= -k$ and an intercept on the y-axis $= \log_e A$.

You will only get a straight-line log-graph if the decay is exponential! This is another way of checking whether a curve is exponential.

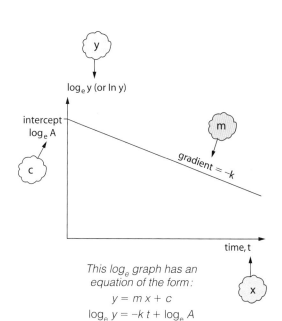

y

$\log_e y$ (or ln y)

intercept $\log_e A$

c

m

gradient $= -k$

time, t

x

This $\log_e$ graph has an equation of the form:
$y = m x + c$
$\log_e y = -k t + \log_e A$

The half-life equation

The constant k is useful because you can use it to work out the half-life of the decay.

It turns out that: $\text{half-life} = \dfrac{0.693}{k}$ (page 369)

▷ Worked example: Using logarithmic graphs and a spreadsheet

During an investigation of the properties of protactinium-234^m a student recorded the **corrected count rate** for a sample over a period of time, as shown in the table:
How can we use this data to determine the decay constant for protactinium-234^m?

t /s	C /s^{-1}
0	800
30	601
60	445
90	338
120	252
150	193
180	137
210	106

A plot of this data would show the expected exponential decay curve (as on page 484) but this does not help us find the decay constant from the data.

To find the decay constant, the exponential equation for radioactive decay has to be manipulated into a form which will produce a straight-line graph.
We start with the equation for the count rate C from page 369:

$$C = C_0 \, e^{-\lambda t}$$

Taking natural logarithms (base e) of both sides, we get:

$$\ln(C) = \ln(C_0 \, e^{-\lambda t})$$

Using log rule 1 from the previous page, this becomes:

$$\ln(C) = \ln(C_0) + \ln(e^{-\lambda t})$$

As ln (or log$_e$) and the exponential function are the reverse of each other, this becomes:

$$\ln(C) = \ln(C_0) - \lambda t$$

And, with a slight rearrangement, we get the same form as a straight-line graph:

$$\ln(C) = -\lambda \quad t \quad + \quad \ln(C_0)$$
$$y \quad = \quad m \quad x \quad + \quad c$$

With exponentials use natural logs.

So plotting $\ln(C)$ on the y-axis against t on the x-axis will give a straight line with a gradient of $-\lambda$.
The intercept on the y-axis will be $\ln(C_0)$.

Now, to find λ we process the original data, getting the spreadsheet to calculate a set of values for $\ln(C)$ as shown in the table:
We can then get the spreadsheet to plot the required graph of $\ln(C)$ against t, which will give a straight-line graph as shown:

t /s	C /s^{-1}	$\ln(C)$
0	800	6.685
30	601	6.399
60	445	6.098
90	338	5.824
120	252	5.529
150	193	5.262
180	137	4.923
210	106	4.663

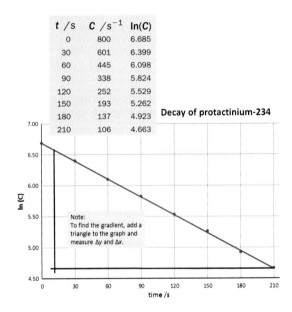

Decay of protactinium-234

Note:
To find the gradient, add a triangle to the graph and measure Δy and Δx.

This 'protactinium' spreadsheet can be downloaded from:
www.oxfordsecondary.co.uk/advancedforyou

The gradient λ is found using a large Δx Δy triangle as on p. 476.

▷ Worked example: Power functions

It was suggested (on page 357) that the relationship between the nuclear radius r and number of nucleons A is given by:

$$r = r_0 \, A^{\frac{1}{3}} \qquad \text{where } r_0 \text{ is a constant.}$$

Taking logs of both sides we get: $\log_{10}(r) = \log_{10}(r_0 \, A^{\frac{1}{3}})$

Applying both rules for logs (see page 485), can you get:

$$\log_{10}(r) = \frac{1}{3} \log_{10}(A) + \log_{10}(r_0)$$

nucleon number, A	nuclear radius, r /$\times 10^{-15}$ m
30	3.88
34	4.05
45	4.45
73	5.22
104	5.88

Can you see which straight-line graph to draw to verify the $\frac{1}{3}$ power formula and find r_0?
The data and the graph are on the same spreadsheet as above.

Hints and Answers

Give me a hint...

▷ **Mechanics**

Chapter 1 (Basic Ideas) page 17
1. a) unit b) base, derived c) size, direction, size (magnitude)
   d) resultant e) component f) component, vector, cosine
2. a) 2 GJ b) 5.9 kg c) 5 ms d) 345 kN e) 20 μm
3. a) F: $kg\,m\,s^{-2}$ A: m^2 v: $m\,s^{-1}$ ρ: $kg\,m^{-3}$
   b) See Example 5.
      i) $A^2\rho v = m^4\ kg\,m^{-3}\ ms^{-1} = kg\,m^2s^{-1}$ ✗
      ii) $A\rho^2 v = m^2\ kg^2m^{-6}\ ms^{-1} = kg^2m^{-3}s^{-1}$ ✗
      iii) $A\rho v^2 = m^2\ kg\,m^{-3}\ m^2s^{-2} = kg\,m\,s^{-2}$ ✔
4. a) 600 N ⟵————•————⟶ 475 N
   b) 125 N to the left
5. a)
   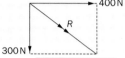
   b) $R = \sqrt{400N^2 + 300N^2} = 500\,N$
      $\tan\theta = 300/400$ ∴ $\theta = 36.9°$
      Resultant is 500 N at E 36.9° S
6.
   a) $v_V = 20\cos45° = 14\,ms^{-1}$
   b) $v_H = 20\cos45° = 14\,ms^{-1}$
7.
   a) $F_H = 120\cos25° = 109\,N$
   b) $F_V = 120\cos65° = 50.7\,N$
8. a) Redraw vectors head to tail using suitable scale.
      (A larger scale will give you a more precise result.)
   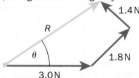

   Measuring R and θ directly should give the same
   answers as by calculation in (b).
   b) See Example 14. Two vectors are already at 90° so
      resolve the 3.0 N force in these directions:

   $R = \sqrt{3.92^2 + 0.72^2} = 4.0\,N$ (2 s.f.)
   $\tan\theta = \dfrac{3.92}{0.72}$ ∴ $\theta = 79.6°$
   Resultant is 4.0 N at 79.6° − 45° = 35° (2 s.f.)
   above the original 3.0 N force.

Chapter 2 (Looking at Forces) page 31
1. a) mass, kilograms b) gravity, newtons c) friction, drag
   d) depth, density, temperature e) upthrust, weight
2. 120 kg mass on Venus weighs more (1060 N).
3. See Example 3. Force = weight = 29.4 N
   a) Area = 70 cm² (0.007 m²) 0.42 N cm⁻²(4200 Pa)
   b) Area = 300 cm² (0.03 m²) 0.098 N cm⁻²(980 Pa)

4. a) High pressure. Pierces surfaces easily.
   b) Spreads weight over large area. Reduces pressure on ice.
   c) Low pressure. Don't sink into mud.
5. a)   b)
   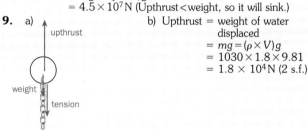
6. At limit, force up slope = force down slope
   $F = W\cos55° = 120 \times 9.81 \times \cos55° = 675\,N$
7. $\Delta p = (h_2 - h_1)\rho g$
   $= (3.5 - 1.0) \times 1000 \times 9.81 = 25\,kPa$ (2 s.f.)
8. Upthrust = weight of water displaced
   $= mg = (\rho \times V)g = 1030 \times 4500 \times 9.81$
   $= 4.5 \times 10^7\,N$ (Upthrust < weight, so it will sink.)
9. a)   b) Upthrust = weight of water
                                    displaced
   upthrust $= mg = (\rho \times V)g$
                                    $= 1030 \times 1.8 \times 9.81$
   weight $= 1.8 \times 10^4\,N$ (2 s.f.)
   tension

   c) Buoy not moving ∴ upthrust = weight + tension
      ∴ tension = upthrust − weight
         $= 1.8 \times 10^4 - (200 \times 9.81)$
         $= 1.6 \times 10^4\,N$ (2 s.f.)

Chapter 3 (Turning Effects of Forces) page 37
1. a) perpendicular, force b) newton-metres (Nm)
   c) equilibrium, clockwise, equal, anticlockwise
   d) weight e) size (magnitude), opposite, turning
2. a) Moment = 200 N × 0.20 m = 40 Nm.
   b) A longer spanner gives more turning moment for the
      same force. This makes it easier to use.
3. See Example 3.
   a) $X = 9\,N$ b) $Y = 4.5\,N$ c) $Z = 5\,cm$
4. a) Low centre of gravity keeps it stable.
   b) Low centre of gravity and wide base gives high stability.
   c) Large effective base area keeps him stable.
   d) Weight of wine raises overall centre of gravity.
   e) Load makes centre of gravity higher.
5. See Example 2.
   a) couple = 15 N × 0.75 m = 11 Nm (2 s.f.)
   b) couple = 24 N × 0.80 m = 19 Nm (2 s.f.)
6. a) Weight of ruler (1 N) acts 40 cm from pivot.
      $(X \times 10\,cm) = (1\,N \times 40\,cm)$ ∴ X = 4 N
   b) Weight of ruler (1 N) acts 35 cm from pivot.
      $(1\,N \times 35\,cm) + (5\,N \times Y) = (10\,N \times 15\,cm)$
      ∴ Y = 23 cm
7. a)
   b) See Example 6. Take moments about B to find
      $R_A = 110\,N$ (2 s.f.)
      $R_A + R_B = 250\,N$ ∴ $R_B = 140\,N$ (2 s.f.)

8. a)

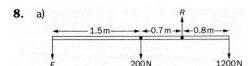

b) See Example 6. Take moments about pivot to find $F = 370\,\text{N}$ (2 s.f.)

c) Total force up = total force down ∴ $R = 1770\,\text{N}$

Chapter 4 (Describing Motion) pages 50–51

1. a) displacement, metres per second $(\text{m}\,\text{s}^{-1})$, velocity, metres per second squared $(\text{m}\,\text{s}^{-2})$

b) speed, direction   c) gradient, greater (faster, higher)

d) velocity, time, displacement   e) direction

f) gravity, air resistance, increases, terminal

g) same, horizontal, vertical

2. a) Distance $= \pi r = 3.1\,\text{m}$

b) Time = distance ÷ speed $= 5.2\,\text{s}$

c) Displacement $= 2.0\,\text{m}$ (diameter of circular track)

d) Average velocity = displacement ÷ time $= 0.38\,\text{m}\,\text{s}^{-1}$ (from A to B)

3. a) Acceleration $= \dfrac{\Delta v}{t} = \dfrac{12\,\text{m}\,\text{s}^{-1}}{4.0\,\text{s}} = 3.0\,\text{m}\,\text{s}^{-2}$

b) Time $= \dfrac{\Delta v}{a} = \dfrac{-20\,\text{m}\,\text{s}^{-1}}{-4.0\,\text{m}\,\text{s}^{-2}} = 5.0\,\text{s}$

4. a) 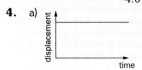   b)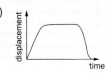

c)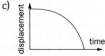

5. a) = (ii)   b) = (iv)   c) = (i)   d) = (iii)

6. a) Acceleration $= \dfrac{\Delta v}{t} = \dfrac{14\,\text{m}\,\text{s}^{-1}}{5.0\,\text{s}} = 2.8\,\text{m}\,\text{s}^{-2}$

b) Deceleration $= \dfrac{\Delta v}{t} = \dfrac{14\,\text{m}\,\text{s}^{-1}}{10\,\text{s}} = 1.4\,\text{m}\,\text{s}^{-2}$

c) Displacement = area under graph $= 35\,\text{m}$

d) Total displacement = total area under graph $= 245\,\text{m}$

7. a) $s = ut + \frac{1}{2}at^2 = 0 + (\frac{1}{2} \times 3.4 \times 3.0^2) = 15\,\text{m}$ (2 s.f.)

b) $v = u + at = 0 + (3.4 \times 3.0) = 10\,\text{m}\,\text{s}^{-1}$ (2 s.f.)

c) Time for last $85\,\text{m}$ = distance ÷ speed $= 8.3\,\text{s}$, so total time for race $= 8.3\,\text{s} + 3.0\,\text{s} = 11\,\text{s}$ (2 s.f.)

8. a) Acceleration $= \dfrac{\Delta v}{t} = \dfrac{2.0\,\text{m}\,\text{s}^{-1}}{3.0\,\text{s}} = 0.67\,\text{m}\,\text{s}^{-2}$

b) Slows to a stop

c) Displacement = area under graph $= 18\,\text{m}$

d) Lift accelerates downwards. It then decelerates to a stop.

e) Area DEF under graph $= 6\,\text{m}$. So overall displacement $= 18\,\text{m}$ up $+ 6\,\text{m}$ down $= 12\,\text{m}$ up

9. a) $u = 5.0\,\text{m}\,\text{s}^{-1}$ upwards

b) See Example 7. With down as positive: $s = 150\,\text{m}$, $a = 9.8\,\text{m}\,\text{s}^{-2}, u = -5.0\,\text{m}\,\text{s}^{-1}$. Use $v^2 = u^2 + 2as$ to find $v = 54\,\text{m}\,\text{s}^{-1}$.
Then use $v = u + at$ to find $t = 6.0\,\text{s}$.

10. See Example 8.

a) Vertically, with down as positive: $s = 0.65\,\text{m}$, $u = 0$, $a = 9.81\,\text{m}\,\text{s}^{-2}$. Use $s = ut + \frac{1}{2}at^2$ to find $t = 0.36\,\text{s}$.

b) Horizontally: $u = 2\,\text{m}\,\text{s}^{-1}$, $t = 0.36\,\text{s}$, $a = 0$
Horiz. distance = horiz. velocity × time $= 0.72\,\text{m}$

11. See Example 8.

a) $u_{\text{HORIZ}} = 60.0\,\text{m}\,\text{s}^{-1}$, $u_{\text{VERT}} = 0\,\text{m}\,\text{s}^{-1}$

b) Vertically, with down as positive: $s = 1000\,\text{m}$, $u = 0$, $a = 9.81\,\text{m}\,\text{s}^{-2}$. Use $s = ut + \frac{1}{2}at^2$ to find $t = 14.3\,\text{s}$

c) Horizontally: $u = 60\,\text{m}\,\text{s}^{-1}$, $t = 14.3\,\text{s}$, $a = 0$
Horiz. distance = horiz. velocity × time $= 858\,\text{m}$

d) Parcel takes longer to hit the ground as air resistance reduces the downward velocity. Plane should be closer to target on release as drag reduces the horizontal velocity, reducing the horizontal distance.

12. See Example 9.

a) Vertically from A to B with up as positive: $v = 0$, $u = 26\cos 45°$, $a = -9.81\,\text{m}\,\text{s}^{-2}$. Use $v = u + at$ to find t. Double this for total time = 3.7 s

b) Horizontally: $u = 26\cos 45°$, $t = 3.7\,\text{s}$, $a = 0$
Horiz. distance = horiz. velocity × time $= 68\,\text{m}$

c) Vertically from A to B with up as positive: $v = 0$, $t = 1.87\,\text{s}$
$u = 26\cos 45°$. Use $s = 1/2(u + v)t$ to give $s = 17\,\text{m}$.

d) The ball has a lower maximum height and a shorter range. This results in a shorter time in the air.

Chapter 5 (Newton's Laws, Momentum) pages 66–67

1. a) force, accelerate, velocity

b) mass, velocity, kilogram metres per second $(\text{kg}\,\text{m}\,\text{s}^{-1})$

c) mass

d) 1 kilogram (1 kg), 1 metre per second squared $(1\,\text{m}\,\text{s}^{-2})$

e) magnitude (size), directions

f) force, time, newton-seconds (N s), momentum, force–time

g) momentum, force

2. See Example 1. a) $6000\,\text{kg}\,\text{m}\,\text{s}^{-1}$ b) $6.0\,\text{kg}\,\text{m}\,\text{s}^{-1}$

3. $F = ma = 25\,000\,\text{kg} \times 4\,\text{m}\,\text{s}^{-2} = 100\,000\,\text{N}$

4. a) $a = F \div m = (9000 - 1500)\text{N} \div 1500\,\text{kg} = 5\,\text{m}\,\text{s}^{-2}$

b) Drag increases with velocity, reducing the resultant accelerating force.

5. a) $p = mv = 2.4 \times 10^{-3} \times 0.65 = 1.6 \times 10^{-3}\,\text{kg}\,\text{m}\,\text{s}^{-1}$

b) Δp = impulse $= F \times t$ ∴ $F = \Delta p \div t = 1.3\,\text{N}$

6. See page 57 and Example 5.

7. i)

(We have ignored the very small gravitational attraction between the man and the chair.)

ii)

iii)

8. a) $a = \dfrac{\Delta v}{t} = \dfrac{6.0\,\text{m}\,\text{s}^{-1}}{3.0\,\text{s}} = 2.0\,\text{m}\,\text{s}^{-2}$

b) $W = mg = 770\,\text{N}$ (2 s.f.)

c) Force down slope $= (W\cos 60° - D)$
Using $F = ma$, $770\cos 60° - D = 79\,\text{kg} \times 2.0\,\text{m}\,\text{s}^{-2}$
∴ $D = 227\,\text{N}$ (2 s.f.)

d) Force up slope = force down slope
∴ $D = 770\cos 60° = 390\,\text{N}$ (2 s.f.)

9. See Example 7.

a) $T = W = mg = 11\,800\,\text{N}$ (3 s.f.)

b) $T - W = ma \therefore T - 11\,800\,\text{N} = 1200\,\text{kg} \times 1.5\,\text{ms}^{-2}$
   $\therefore T = 13\,600\,\text{N}$ (3 s.f.)
c) $W - T = ma \therefore 11\,800\,\text{N} - T = 1200\,\text{kg} \times 2.0\,\text{ms}^{-2}$
   $\therefore T = 9400\,\text{N}$

10. a) $p = mv = 0.060\,\text{kg} \times 31\,\text{ms}^{-1}$
   $= 1.9\,\text{kg\,ms}^{-1}$ (2 s.f.)
b) $\Delta p = $ impulse $= 1.9\,\text{Ns}$ (2 s.f.)
c) Force $=$ impulse $\div$ time $= 74\,\text{N}$

11. a) $\Delta p = m\Delta v = 65 \times 25 = 1625$
   $= 1600\,\text{kg\,ms}^{-1}$ (2 s.f.)

b) Force $= \dfrac{\text{impulse}}{\text{time}} = \dfrac{\Delta p}{t} = 8100\,\text{N}$ (2 s.f.)

c) $W = mg = 640\,\text{N}$ $\therefore$ force $\sim 13W$

12. See Example 11.
a) Impulse $=$ area under graph $= 10\,\text{Ns}$
b) $\Delta p = $ impulse $= 10\,\text{kg\,ms}^{-1}$
c) $v = p \div m = 10\,\text{kg\,ms}^{-1} \div 3.0\,\text{kg} = 3.3\,\text{ms}^{-1}$

13. See Example 13.
a) Mass per second $= 0.09\,\text{kg}$
   $\therefore \Delta p$ per second $= m\Delta v = 1.1\,\text{kg\,ms}^{-1}$ (2 s.f.)
b) $F = 1.1\,\text{N}$ (2 s.f.)

14. Momentum before $=$ momentum after
$(0.20 \times 0.60) + 0 = (0.20 + 0.25) \times v$
$\therefore v = 0.27\,\text{ms}^{-1}$

15. See Example 14.
Momentum before $=$ momentum after
$(9000 \times 20) - (1500 \times 12) = (9000 \times 15) + (1500 \times v)$
$\therefore v = 18\,\text{ms}^{-1}$ in direction of lorry

16. See Example 15.
Momentum before $=$ momentum after
$(3500 \times 250) = (2300 \times v) - (1200 \times 20)$
$\therefore v = 390\,\text{ms}^{-1}$

17. See Example 15.
a) Momentum before $=$ momentum after
   $0 = (55 \times 35) - (m \times 2.5)$ $\therefore m = 770\,\text{kg}$
b) Momentum before $=$ momentum after
   $(55 \times 35) = (655\,\text{kg} \times v)$ $\therefore v = 2.9\,\text{ms}^{-1}$

18. Let $v_\text{E} = $ lorry's component of velocity to the east
$v_\text{N} = $ lorry's component of velocity to the north
In easterly direction:
momentum before collision $=$ momentum after collision
$(1200 \times 16.2) + 0 = (1200 \times 12\cos75°) + 3500\,v_\text{E}$
   $\therefore v_\text{E} = 4.49\,\text{ms}^{-1}$
In northerly direction:
momentum before collision $=$ momentum after collision
$(3500 \times 10.4) + 0 = (1200 \times 12\sin75°) + 3500\,v_\text{N}$
   $\therefore v_\text{N} = 6.43\,\text{ms}^{-1}$
Resultant velocity $v = \sqrt{4.49^2 + 6.43^2} = 7.84\,\text{ms}^{-1}$
$\tan\theta = 6.43/4.49$ $\therefore \theta = 55.1°$
So the resultant velocity is $7.84\,\text{ms}^{-1}$ in a direction E$55.1°$N.

Chapter 6 (Work, Energy and Power) page 75

1. a) energy, joules b) one newton (1 N), one metre (1 m)
c) rate, watts, joule, second d) kinetic, potential
e) internal f) kinetic, inelastic
2. $W = F \times d = 150\,\text{N} \times 80\,\text{m} = 12\,000\,\text{J}$
3. a) $F = P \div v = 54\,000\,\text{W} \div 30\,\text{ms}^{-1} = 1800\,\text{N}$
b) $1800\,\text{N}$ (no resultant force if $a = 0$)
4. $P = Fv = 15\,000\,\text{N} \times 3.0\,\text{ms}^{-1} = 45\,\text{kW}$
5.

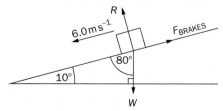

a) Weight acting down hill $= W\cos80° = 170\,\text{N}$
b) Force up hill $=$ force down hill, for steady speed
   $\therefore$ braking force $= 170\,\text{N}$ (2 s.f.)
c) $W = F \times d = 170\,\text{N} \times 18\,\text{m} = 3.1\,\text{kJ}$

6. $P = \dfrac{\Delta W}{\Delta t} = \dfrac{2000\,\text{N} \times 15\,\text{m}}{60\,\text{s}} = 500\,\text{W}$
a) Input power $= 500\,\text{W}$ b) Input power $= 770\,\text{W}$
7. a) PE lost $=$ KE gained
   $(0.060 \times 9.81 \times 1.5) = (1/2 \times 0.060 \times v^2)$
   $\therefore v = 5.4\,\text{ms}^{-1}$
b) KE at impact $= 0.875\,\text{J}$. KE on rebound $= 0.656\,\text{J}$
   $\therefore v = 4.7\,\text{ms}^{-1}$
c) KE lost $=$ PE gained
   $0.656\,\text{J} = (0.060 \times 9.81 \times h)$ $\therefore h = 1.1\,\text{m}$ (2 s.f.)
8. a) PE lost $= mgh = 10 \times 9.81 \times 3.0\sin30° = 150\,\text{J}$
b) $W = F \times d = 35\,\text{N} \times 3.0\,\text{m} = 105\,\text{J}$
c) PE lost $=$ work done against friction $+$ KE gained
   $150\,\text{J} = 105\,\text{J} +$ KE gained $\therefore$ KE $= 45\,\text{J}$
d) $45\,\text{J} = 1/2 \times 10 \times v^2$ $\therefore v = 3.0\,\text{ms}^{-1}$

Chapter 7 (Circular Motion) page 85

1. a) towards, centripetal, straight, first
b) acceleration, speed, radius
c) frequency, hertz (Hz), time period d) 2π
2. $360° = 2\pi\,\text{rad}$ $180° = \pi\,\text{rad}$
$90° = \pi/2\,\text{rad}$ $45° = \pi/4\,\text{rad}$
$30° = \pi/6\,\text{rad}$ $1° = \pi/180\,\text{rad} = 0.017\,\text{rad}$
3. $2\pi\,\text{rad} = 360°$ $\pi/2\,\text{rad} = 90°$
$\pi/3\,\text{rad} = 60°$ $1\,\text{rad} = 180°/\pi = 57.3°$
4. $w = \theta \div t = 6\,\text{rad} \div 0.2\,\text{s} = 30\,\text{rad\,s}^{-1}$
5. $v = rw = 0.20\,\text{m} \times 85\,\text{rad\,s}^{-1} = 17\,\text{ms}^{-1}$
6.

Period (s)	Frequency (Hz)	Angular speed (rad s^{-1})
0.04	25	160
4.2	0.24	1.5
0.52	1.9	12

(all to 2 s.f.)
7. $T = 2\pi / \omega = 2\pi/10.5\,\text{rad\,s}^{-1} = 0.60\,\text{s}$
8. a) $a = v^2/r = 8.0^2/30 = 2.1\,\text{ms}^{-2}$
b) $F = ma = 800\,\text{kg} \times 2.1\,\text{ms}^{-2} = 1700\,\text{N}$
9. Centripetal force $=$ resultant towards centre
$= 3300\cos20° + 11\,000\cos70° = 6900\,\text{N}$ (2 s.f.)
10. a) $f = 1/T = 1/(365 \times 24 \times 60 \times 60\,\text{s})$
   $= 3.2 \times 10^{-8}\,\text{Hz}$
b) $\omega = 2\pi f = 2\pi \times 3.2 \times 10^{-8}\,\text{Hz} = 2.0 \times 10^{-7}\,\text{rad\,s}^{-1}$
c) $v = r\omega = 1.5 \times 10^{11} \times 2.0 \times 10^{-7} = 3.0 \times 10^4\,\text{ms}^{-1}$

11. a) $F = \dfrac{mv^2}{r} = \dfrac{80 \times 60^2}{100} = 2880\,\text{N} = 2900\,\text{N}$ (2 s.f.)
b)

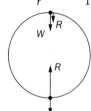

c) i) At top: $W + R = 2880\,\text{N}$
      $\therefore R = 2080\,\text{N}$
   ii) At bottom:
      $R = 2880 + W = 3680\,\text{N}$
d) i) $2080\,\text{N} = 2.6 \times W$
   ii) $3680\,\text{N} = 4.6 \times W$

Chapter 8 (Gravitational Forces, Fields) pages 98–99

1. a) masses, separation b) square, quarters
c) field, strength, direction, radial d) acceleration, 9.81
e) geostationary (geosynchronous)
f) equipotential   g) potential, mass
2. a) $W = mg = 19.62 = 20\,\text{N}$ (2 s.f.)
b) $F = Gm_1m_2/r^2 = 19.54 = 20\,\text{N}$ (2 s.f.)
c) Same answer to 2 s.f.
3. See Example 2. Use $F = Gm_1m_2/r^2$
a) $9.49 \times 10^{-37}\,\text{N}$ b) $8.5 \times 10^{-22}\,\text{N}$ c) $1.20 \times 10^{13}\,\text{N}$
4. a) $F = Gm_1m_2/r^2 = 3.56 \times 10^{22}\,\text{N}$
b) Mass increase is greater than increase in (radius)^2 so a larger gravitational force is exerted.
5. Both stars exert equal gravitational field strength
$(g = GM/r^2)$ at P:
$$\frac{G \times 6.2 \times 10^{32}}{(1.4 \times 10^{16} - 1.5 \times 10^{15})^2} = \frac{GM}{(1.5 \times 10^{15})^2}$$
$\therefore M = 8.9 \times 10^{30}\,\text{kg}$

6. a) $g = GM/r^2 = 9.2\,\mathrm{Nkg^{-1}}$
b) $W = mg$ i) $690\,\mathrm{N}$ (2 s.f.) ii) $640\,\mathrm{N}$ (2 s.f.)
7. a) Use $g = GM/r^2$

Distance	R	$2R$	$3R$	$4R$	$5R$
g $(\mathrm{Nkg^{-1}})$	9.77	2.44	1.09	0.61	0.39

b)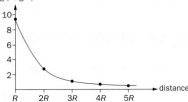

8. Use $g = GM/r^2$:
a) $g \times 1/4$ when radius doubles $\therefore r_E$ above surface.
$6.4 \times 10^6\,\mathrm{m}$
b) $g \times 1/9$ when radius triples $\therefore 2 \times r_E$ above surface.
$12.8 \times 10^6\,\mathrm{m}$
9. Rearrange equation for g: $r^2 = GM/g$
$r = 3.39 \times 10^6\,\mathrm{m}$
10. a) $g = F/m = 16.6\,\mathrm{N}/5\,\mathrm{kg} = 3.32\,\mathrm{Nkg^{-1}}$
b) Rearrange equation for g: $M = gr^2/G$
$r = (3\,390\,000 + 200\,000)\,\mathrm{m}$
$\therefore M = 6.42 \times 10^{23}\,\mathrm{kg}$
11. a) $T = 24$ hours $= 86\,400\,\mathrm{s}$
b) $\omega = 2\pi/T = 7.27 \times 10^{-5}\,\mathrm{rad\,s^{-1}}$
$r = v/\omega = 4.24 \times 10^7\,\mathrm{m}$
$\therefore$ height above Earth $= 3.6 \times 10^7\,\mathrm{m}$
12. See Example 5.
Grav. force of attraction = mass × centripetal acceleration
$\dfrac{GMm}{r^2} = \dfrac{mv^2}{r}$ $\therefore$ $v^2 = \dfrac{GM}{r}$
$\therefore$ $v = 3.52 \times 10^4\,\mathrm{ms^{-1}}$
13. $v_E = \sqrt{\dfrac{2GM}{R}} = \sqrt{\dfrac{2 \times 6.67 \times 10^{-11} \times 5.6 \times 10^{26}}{6.0 \times 10^7}} = 3.5 \times 10^4\,\mathrm{ms^{-1}}$
14. See Example 10. $T^2 \propto r^3$
$\therefore \dfrac{T_1^2}{r_1^3} = \dfrac{T_2^2}{r_2^3}$ $\therefore \dfrac{(3.4 \times 10^9)^2}{(2.2 \times 10^{17})^3} = \dfrac{T_2^2}{(3.6 \times 10^{17})^3}$
$\therefore T_2 = 7.1 \times 10^9$ years
15. a) Using $V = -\dfrac{GM_E}{R_E} = -6.25 \times 10^7\,\mathrm{Jkg^{-1}}$
b) $2 \times$ radius, so R_E above surface
16. See Example 8.
a) Using $V = -GM/r$, $V_P = -6.16 \times 10^7\,\mathrm{Jkg^{-1}}$,
$V_Q = -6.06 \times 10^7\,\mathrm{Jkg^{-1}}$
b) $\Delta E = \Delta Vm = 3.3 \times 10^{10}\,\mathrm{J}$
c) No change in potential (no motion in direction of force).
17. a) Rearranging equation for V gives $M = -Vr/G$
$= 8.8 \times 10^{25}\,\mathrm{kg}$
b) See page 92. $V = 0$ at infinity. Work must be done to pull an object away from the planet to infinity.
18. a) $V = -GM/r = -2.818 \times 10^6\,\mathrm{Jkg^{-1}}$
b) At $1\,\mathrm{km}$, $r = 1741\,\mathrm{km}$
$V = -GM/r = -2.816 \times 10^6\,\mathrm{Jkg^{-1}}$
$\Delta E = \Delta Vm = -200\,\mathrm{Jkg^{-1}} \times 100\,\mathrm{kg}$
$= -2 \times 10^5\,\mathrm{J}$
c) PE lost $(2 \times 10^5\,\mathrm{J})$ = KE gained $(1/2 \times 100\,\mathrm{kg} \times v^2)$
$\therefore v = 60\,\mathrm{ms^{-1}}$ (1 s.f.)
19. See Example 5.
a) Grav. force of attraction = mass × centripetal accel.
$\dfrac{GMm}{r^2} = \dfrac{mv^2}{r}$ $\therefore$ $v^2 = \dfrac{GM}{r}$
$\therefore$ $v = 7.6 \times 10^3\,\mathrm{ms^{-1}}$
b) KE $= 1/2\,mv^2 = 5.7 \times 10^{11}\,\mathrm{J}$
c) $V = -GM/r = -5.8 \times 10^7\,\mathrm{Jkg^{-1}}$
d) V at surface $= -6.253 \times 10^7\,\mathrm{Jkg^{-1}}$
$\Delta E = \Delta V\,m = 4.5 \times 10^6\,\mathrm{Jkg^{-1}} \times 19\,500\,\mathrm{kg}$
$= 8.8 \times 10^{10}\,\mathrm{J}$
e) Total energy needed = KE + PE $= 6.6 \times 10^{11}\,\mathrm{J}$

Chapter 9 (Simple Harmonic Motion) pages 112–113
1. a) period, amplitude   b) equilibrium, displacement
c) acceleration   d) displacement
e) centre (equilibrium position), zero
2. a) $f = 1/T = 1/3.0\,\mathrm{s} = 0.33\,\mathrm{Hz}$
b) $a = -(2\pi f)^2 x$ $\therefore a_{MAX} = -(2\pi f)^2 A$
$= -(2\pi \times 0.33\,\mathrm{Hz})^2 \times 0.10\,\mathrm{m} = -0.44\,\mathrm{ms^{-2}}$
3. a), b)
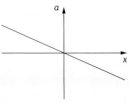
c) Phase difference = 1/4 cycle = $\pi/2$ rad
4. a) Using $a = -(2\pi f)^2 x$ b)

x (m)	a $(\mathrm{ms^{-2}})$
± 0.01	± 0.77
± 0.02	± 1.55
± 0.03	± 2.32
± 0.04	± 3.10
± 0.05	± 3.87

5. At A and C: a = max, $v = 0$, KE = 0, PE = max
At B: $a = 0$, v = max, KE = max, PE = 0
6. a) $T = 2\pi\sqrt{m/k} = 2\pi\sqrt{10/250} = 1.3\,\mathrm{s}$
b) $f = 1/T = 0.80\,\mathrm{Hz}$
7. $T = 2\pi\sqrt{l/g} = 2\pi\sqrt{2.0/9.81} = 2.8\,\mathrm{s}$
8. Rearranging equation for period of a pendulum gives
$l = (T^2 g)/(2\pi) = (2^2 \times 9.81)/4\pi^2 = 0.994\,\mathrm{m}$
9. Rearranging equation for a mass on a spring gives
$k = (m \times 4\pi^2)/T^2 = (9 \times 4\pi^2)/1.2^2 = 250\,\mathrm{Nm^{-1}}$
10. Period of pendulum changes (increases) since g is smaller on the Moon.
11. a) $A = 6\,\mathrm{m}$ b) $T = 8\,\mathrm{s}$ c) $f = 1/T = 1/8 = 0.125\,\mathrm{Hz}$
d) $a_{MAX} = (2\pi f)^2 A = (2\pi \times 0.125)^2 \times 6 = 3.7\,\mathrm{ms^{-2}}$
12. a) $v_{MAX} = 2\pi f A = 2\pi \times 0.55\,\mathrm{Hz} \times 0.10\,\mathrm{m} = 0.35\,\mathrm{ms^{-1}}$
b) $v = 2\pi f\sqrt{(A^2 - x^2)} = 2\pi \times 0.55\sqrt{(0.1^2 - 0.08^2)}$
$= 0.21\,\mathrm{ms^{-1}}$
13. a) $A = 1.0\,\mathrm{cm} = 0.01\,\mathrm{m}$
b) $a_{MAX} = (2\pi f)^2 A = (2\pi \times 2.4)^2 \times 0.01 = 2.3\,\mathrm{ms^{-2}}$
14. a)

T (s)	0.90	1.25	1.80	2.53	3.59	See
l (m)	0.20	0.40	0.80	1.60	3.20	page 106.
$\sqrt{l}$	0.45	0.63	0.89	1.26	1.79	

b) Straight line through the origin.
c) gradient $= 2.0$ $(\mathrm{s\,m^{-1/2}})$
d) $g = (2\pi / \text{gradient})^2 = 9.87\,\mathrm{ms^{-2}}$
15. a) $a = 5.0\,\mathrm{m}$ b) $f = 1/T = 1/2.0 = 0.50\,\mathrm{Hz}$
c) $v_{MAX} = 2\pi f A = 2\pi \times 0.50\,\mathrm{Hz} \times 5.0\,\mathrm{m} = 16\,\mathrm{ms^{-1}}$
d) $a_{MAX} = (2\pi f)^2 A = (2\pi \times 0.50)^2 \times 5.0 = 49\,\mathrm{ms^{-2}}$
e)

energy $(\times 10^{-3}\mathrm{J})$

2.4

-5.0 0 $+5.0$ displacement (m)

16. a) $v_{MAX} = 2\pi f A = 2\pi \times 2.6\,\mathrm{Hz} \times 0.45\,\mathrm{m} = 7.4\,\mathrm{ms^{-1}}$
b) $\mathrm{KE}_{MAX} = \frac{1}{2}mv^2 = 1/2 \times 0.60 \times 7.4^2 = 16\,\mathrm{J}$
c) Lost in overcoming air resistance. Transferred as internal energy of surrounding air.
17. Mirror is resonating. Engine vibrations at this speed must equal the natural frequency of vibration of the mirror.

Basic Ideas (page 114)

1. length, ampere, energy, weight
2. The value to the nearest mm is 0.392 m.
3. In base units, density is $kg\,m^{-3}$ and force is $kg\,m\,s^{-2}$.
4. Answer A
5. k is in $N/(kg\,m^{-3})(m^2\,s^{-2})$ or m^2.
6. a) 120 N/equal to weight; vertically upwards/opposite direction to weight.
   b) Correct diagram:
      Use of $120^2 = 70^2 + T^2$ to give $T = 97\,N$

7. a) 5 N using Pythagoras' theorem.
   b) Perpendicular components can be treated independently.
8. $(5.8\,m\,s^{-1}) \times \cos 45° = 4.1\,m\,s^{-1}$

Looking at Forces (page 114)

9. A = his/her weight,
   B = reaction from the ground, he/she is decelerating to come to rest.
10. a) Use $m = V\rho = 0.010 \times 5.0 \times 10^{-3} \times 1000 = 0.050\,kg$
    and weight $= mg = 0.050 \times 9.81 = 0.49\,N$
    b) Weight of cork = upthrust or weight of displaced water $= 0.49\,N$
11. a) The forces perpendicular to OX must balance.
    $30\,kN\cos(90° - 15°)$ balances $18\,kN\cos(90° - \theta)$
    $= 18\,kN\sin\theta$ $\theta = 25°$
    b) Components parallel to OX add up.
    $F = (30\,kN\cos 15°) + (18\,kN\cos 25°) = 45\,kN$
12. a) Tension $T = 8g$;
    $T + T\cos 40° = 138\,N$
    b) $T + T\cos 50° = 129\,N$
    c) The force in (a) is balanced by the frictional force between the patient and the bed.

Turning Effects of Forces (page 115)

13. a) air resistance, friction (backwards)
    b) Motive force from friction between tyres and road, (forwards). Wheels turn due to torque from engine.
    c) weight (down), reaction from road (up)
14. Using $p = F/A$,
    $A = F/p = $ weight$/p = 120\,m^2$
15. a) moment = force × perpendicular distance from point/elbow
    b) i) The point where the weight appears to act.
       ii) Moment $= (0.150 \times 18) + (0.460 \times 30) = 16.5\,Nm$
       iii) 1) Same/equal to 16.5 Nm/equal to clockwise moment.
            2) Perpendicular distance between elbow and line of action of F decreases or the vertical force $F\cos\theta$ is the same or $F\cos\theta = 412.5\,N$ or $F \propto 1/\cos\theta$.
16. a) Centre of gravity is 1/3 of way from A to B.
    b) No, because c of g is not at the centre of the beam.
17. Answer C
18. a) Take moments about the edge of the step.
    $F \times 0.25\,m = W \times 0.25\sin 45°$
    $\therefore F = 590\,N$
    b) $F \times 0.25\sin 45° = W \times 0.25 \times \sin 45°$
    $\therefore F = W = 830\,N$
    c)

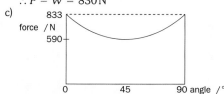

19. Answer B
20. a) Take moments about S. $WS = 0.03\,m$
    b) S is just nearer A than B because the moment of A about the centre > moment of B about the centre.

Describing Motion (page 116)

21. a) initial acceleration = initial gradient $= 1.7\,m\,s^{-2}$
    b) total distance = area under graph $= 400\,m$
    c) average speed = total distance/total time $= 13\,m\,s^{-1}$
22. a) Constant acceleration from $t = 0$ to $5\,s$ then constant deceleration to $t = 25\,s$.   b) 2500 m
    c) area = displacement
       At $t = 5\,s$ displacement $= 1/2 \times 5 \times 200 = 500\,m$
       Graph starts at origin and curves up in a parabola to pass $x = 500\,m$ at $t = 5\,s$; it then continues upwards with a decreasing gradient to finish at $x = 2500\,m$ at $t = 25\,s$.
23. a) Terminal speed is horiz. asymptote at $57\,m\,s^{-1}$.
    b) Area under first 13s of graph $\approx 500\,m$.
    c) Constant speed implies zero resultant force, weight is balanced by resistive forces.
24. a) $0.3 + 1.6 + 0.3 = 2.2\,s$
    b) time at max. height $= 1.1\,s$ and $v = 0$ $v = u + at$,
       $0 = u - g \times 1.1$ gives $u = 10.8\,m\,s^{-1}$
    c) $x = ut + \frac{1}{2}at^2 = 10.8 \times 0.3 - \frac{1}{2} \times 9.81 \times 0.3^2 = 2.8\,m$
25. a) Using $s = \frac{1}{2}gt^2$ for the vertical motion ($u = 0$), $t = 3.2\,s$
    b) $s = vt$ for the horiz. motion ($a = 0$) $\therefore s = 63\,m$
    c) vertical component: $v_V = gt = 31\,m\,s^{-1}$
       horizontal component: $v_H = 20\,m\,s^{-1}$
       velocity of stone $= 37\,m\,s^{-1}$
    d) Both would decrease because resistive forces would produce deceleration / reduce acceleration.
26. a) i) Using $v = at$, $t = 30\,s$ (ii) $s = \frac{1}{2}at^2 = 1100\,m$ (2 s.f.)
    b) In 2.5 s before braking starts, it travels 190 m, decelerating at $4\,m\,s^{-2}$ it stops in 700 m. Total distance to stop (890 m) $< 1100\,m$ remaining so stops safely.
27. a) Using $v^2 = 2gs$, $v = 4.2\,m\,s^{-1}$
    b) new speed $= 2.8\,m\,s^{-1}$, change in speed $= 1.4\,m\,s^{-1}$
       change in velocity $= 2.8 - (-4.2) = 7.0\,m\,s^{-1}$ upwards
    c) Use $v = u + at$ to find time to max. height after bounce
       ($v = 0$, $u = 2.8$, $a = g$) $\therefore t = 0.29\,s$

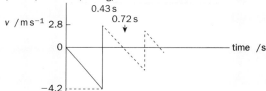

Newton's Laws and Momentum (page 117)

28. a) Use of $v = u + at$ gives $a = (55 - 0)/2.2 = 25\,m\,s^{-2}$
    b) Use of $s = ut + \frac{1}{2}at^2$
       $s = 0 + 1/2 \times 25 \times 2.2^2 = 60.5\,m$
    c) $F = ma = 3.2 \times 10^4 \times 25 = 8.0 \times 10^5\,N$
29. a) before: $(0.3\,kg \times 0.2\,m\,s^{-1}) - (0.2\,kg \times 0.5\,m\,s^{-1})$
       after: $-(0.3\,kg \times 0.2\,m\,s^{-1}) + (0.2\,kg \times v_B)$
       giving $v_B = +0.1\,m\,s^{-1}$ (left to right)
    b) No external forces acting on system.
    c) Collision is elastic if kinetic energy E_K conserved.
       Using $E_K = \frac{1}{2}mv^2$, before: $E_K = 0.031\,J$,
       after: $E_K = 0.007\,J$, therefore collision is inelastic.
30. a) $V = \pi r^2 h = 1500\pi$ or $4710\,m^3$ and $m = V\rho = 6130\,kg$
    b) i) Momentum $= 7.4 \times 10^4\,kg\,m\,s^{-1}$
       ii) Force $= 14\,700\,N$
       iii) Mass $= 1500\,kg$
31. a) Using $F = \Delta(mv)/t = 33\,MN$
    b) $F = 33\,MN - $ (weight of rocket) $= 3.6\,MN$
32. a) Using $F = \Delta(mv)/t = 2400\,N$ (2 s.f.)
    b) Momentum is transferred to parachutist plus Earth; the Earth is very massive compared to person.
33. a)
       ↑ reaction from floor

       ↓ weight
    b) force exerted on floor of lift by person, gravitational pull of person on Earth

c) The force exerted on the person by the lift increases as upward acceleration increases.
(Resultant force up = ma)
34. a) i) $27\,\text{ms}^{-2}$ ii) using $F = ma = 2.3 \times 10^4\,\text{N}$
  b) using $E = \frac{1}{2}mv^2$ gives maximum speed $v = 33\,\text{ms}^{-1}$
  c) using $m_1u = (m_1 + m_2)\,v$ gives $v = 3.1\,\text{ms}^{-1}$
35. The answer is B. The two forces act on different objects.
36. a) Collision is elastic when both momentum and kinetic energy are conserved:
  mom. before = mom. after = $1.2\,\text{kgms}^{-1}$
  kinetic energy before = E_K after = $3.6\,\text{J}$
  b) Impulse = change in momentum of $L = 2.0\,\text{Ns}$

Work, Energy and Power (page 118)
37. a) Use $\Delta E = mg\Delta h$ to give $490\,\text{J}$
  b) $P = E/t = 8.2\,\text{W}$. Rate of energy conversion $= 41\,\text{W}$ allowing for efficiency.
38. a) See page 69.
  b) Using $P = Fv$, $F = 40\,\text{N}$
  c) i) $F = kv$, $P = Fv = kv^2$
    ii) Using $P \propto v^2$, $290\,\text{W}$
  d) extra $= mg\Delta h/t = 163\,\text{W}$
    ($\Delta h = 5\,\text{m} \times \sin\theta = 0.167\,\text{m in 1s}$).
    Total power = $363\,\text{W}$
39. a) $F = mg\cos 80° = 2600\,\text{N}$ (2 s.f.)
  b) Work done $= Fvt = 61\,\text{kJ}$
40. a) Work done = area under graph $\approx 25\,\text{J}$
  b) $E_K = \frac{1}{2}mv^2$, $v = 25\,\text{ms}^{-1}$
41. a) loss of KE $= 600 - 37.5 = 560\,\text{J}$
  b) Using work done = loss in KE and force = work/distance $= 3.7 \times 10^4\,\text{N}$
42. a) mass $= V\rho = 7.0 \times 10^6 \times 1000 = 7.0 \times 10^9\,\text{kg}$
    Use PE $= mgh$ to get $= 4.1 \times 10^{13}\,\text{J}$
  b) Time = energy/power $= 0.88 \times 4.1 \times 10^{13}/$
    $6 \times 300 \times 10^6 = 2.0 \times 10^4\,\text{s}$
  c) i) mass rate of flow = mass/time $= 7.0 \times 10^9/$
    $2.0 \times 10^4 = 3.5 \times 10^5\,\text{kgs}^{-1}$
    ii) $0.10 \times 4.1 \times 10^{13}/2.0 \times 10^4 = 2.0 \times 10^8\,\text{W}$
    (note 10% of total power is wasted not 10% of rated power)
43. a) Use $a = \Delta v/t$.
  b) Using $F = ma$, $F = 3.0 \times 10^{-4}\,\text{N}$
  c) $P = F \times$ (average v) $= F \times 0.40\,\text{ms}^{-1}$
    $= 1.2 \times 10^{-4}\,\text{W}$
44. a) i) work done $= Fx$ ii) $F = ma$
  b) KE = work done $= Fx = max$
    $v^2 = 2ax$ substituting gives KE $= \frac{1}{2}mv^2$
  c) The braking distance is more than $50\,\text{m}$.
    As the mass is greater, the KE is greater; the force is the same and KE $= Fx$, so x must be greater.

Circular Motion (page 119)
45. a) $W = mg = 0.50 \times 9.81 = 4.9\,\text{N}$
  b) i) $\omega = v/r = 7.1\,\text{rads}^{-1}$
    ii) $a = v^2/r = 50\,\text{ms}^{-2}$
  c) i) top of the circle
    ii) $F = ma = 25\,\text{N}$
    $\therefore$ least $T = 25\,\text{N} -$ weight $= 20\,\text{N}$
46. a) $\omega = 2\pi/T = 0.60\,\text{rads}^{-1}$
  b) i) Lift = weight $= 1.5\,\text{N}$, ii) $F = mr\omega^2 = 0.54\,\text{N}$
    Horiz. component provides centripetal force.
47. a) i)   tension
    weight
    ii) Tension provides component forces to balance weight of bung and provide centripetal force.
    iii) Resultant acts towards centre of circle. Keeps bung moving at constant speed in a circle.
  b) i) Using $T \cos\theta = mg$ $T = 2.3\,\text{N}$
    ii) $T \cos 15° = 2.2\,\text{N}$
    iii) Using $2.2\,\text{N} = mv^2/r$ $3.8\,\text{ms}^{-1}$

Gravitational Forces and Fields (page 119)
48. a) Work = force $\times$ distance $= 6500 \times 600 = 3.9\,\text{MJ}$
  b)   Work = power $\times$ time
      $= 320\,000 \times 55 = 17.6\,\text{MJ}$
    Efficiency = useful work out / total work in
      $= 3.9/17.6 = 22\%$
49. a) Force per unit mass
  b) i) $g = Gm/r^2 = 6.67 \times 10^{-11} \times 7.35 \times 10^{22} /$
    $(1.74 \times 10^6)^2 = 1.62\,\text{Nkg}^{-1}$
    ii) $F = GMm/d^2 = 6.67 \times 10^{-11} \times 5.98 \times 10^{24} \times$
    $7.35 \times 10^{22}/(3.84 \times 10^8)^2 = 1.99 \times 10^{20}\,\text{N}$
    iii) $m\omega^2 r = $ force $= 1.99 \times 10^{20}$ and $\omega = 2\pi/T$ giving
    $T = 27.4\,\text{days}$
  c) Same period/angular velocity and direction as Earth's rotation and above the equator.
50. a) i) zero along an equipot. (ii) $\Delta V \times m = +1500\,\text{MJ}$
  b) i) force is attraction, $V = 0$ at infinity
    ii) Use $V = -Gm/r$ so all points at same radius have same potential.

Simple Harmonic Motion (page 120)
51. a) i) $f = 1/\text{period} = 0.83\,\text{Hz}$,   (ii) same period $\propto (m/k)$
    iii) show that variation is sinusoidal
  b) i) parabola, inverted, mirrored in PE, peak at $0.16\,\text{J}$
    ii) parabola, inverted as (i) but peak at $0.04\,\text{J}$
    (1/2 ext $\rightarrow$ 1/4 PE stored since energy $\propto A^2$)
  c) Using PE $= \frac{1}{2}kx^2$, $k = 8.0\,\text{Nm}^{-1}$
52. a) check restoring force $\propto$ displacement etc.
  b) smaller mass, smaller period, mean position higher up; smaller wings, less damping
53. a) i) Using $\omega^2 = k/m$ and $\omega = 2\pi f$ or $T = 1/f$
    $m = T^2 k/4\pi^2 = 1.6^2 \times 2640/4\pi^2 = 171\,\text{kg}$
    ii) Using $\frac{1}{2}mv^2 = 2150$ gives $v = 5.01\,\text{ms}^{-1}$
    iii) $2.15\,\text{KJ}$ quoting conservation of energy or all KE transferred to PE
    iv) $v = \omega A$ gives $A = 1.28\,\text{m}$
    v) Using $x = \pm A \sin(2\pi ft)$ and $a = -\omega^2 x = 13.9\,\text{ms}^{-2}$
  b) Resonance gives max. amplitude as system is at the natural frequency:
    $f = 1/T = 1/0.625 = 1.6\,\text{Hz}$
54. Answer B
55. Answer D
56. a) $a = -(2\pi f)^2 x$
    $\therefore a_{max} = (-)(2\pi \times 2.4 \times 10^3)^2 \times 1.8 \times 10^{-3}$
    $= (-)4.1 \times 10^5\,\text{ms}^{-2}$
  b) Work done = mean force $\times$ distance moved
    For 1/4 oscillation distance moved $= 1.8\,\text{mm}$
    Work done $= 0.25 \times 1.8 \times 10^{-3} = 4.5 \times 10^{-4}\,\text{J}$
    Time taken $\Delta t = 1/4\,T = 1/4 \times (1/2.4 \times 10^{-3})$
      $= 1.04 \times 10^{-4}\,\text{s}$
    Power = work done/Δt
      $= 0.25 \times 1.8 \times 10^{-3}/1.04 \times 10^{-4} = 4.3\,\text{W}$
57. a) $E_K = 1/2\,mv^2 = 0.16\,\text{J}$; $F = E_K/d = 1.5\,\text{N}(d = 11\,\text{cm})$
  b) E_K of ball $= 0.90\,\text{J}$ before; after, $v = 8.6\,\text{ms}^{-1}$
  c) Conservation of momentum shows can could not fall off.
58. a) Component of weight perpendicular to board balances normal contact force.
    Resultant = component of weight down slope
  b) $mg\cos 60° = 0.49\,\text{N}$
  c) KE gained = PE lost $= 0.098\,\text{J}$
  d) KE gained = PE lost $-$ work done against friction
    $= 0.098 - (0.19 \times 0.20) = 0.060\,\text{J}$
  e) no work is done against friction
59. a) area under graph $= 1550\,\text{m}$
  b) i) $a = \Delta v/t = 1.5\,\text{ms}^{-2}$ (ii) $F = ma = 1800\,\text{N}$
    iii) work $= Fd = 540\,\text{kJ}$
  c) i) $P = Fv = 15\,\text{kW}$
    ii) extra power = PE gained/sec $= mg \times 1.5\,\text{m} = 18\,\text{kW}$
  d) E_K to internal energy
  e) i) $F = mv^2/r = 8.0\,\text{kN}$ (ii) No, F perpendicular to r
60. a) Using PE gained = KE lost, $v = 1.7\,\text{ms}^{-1}$
  b) conserve momentum to get $u = 340\,\text{ms}^{-1}$
61. a) momentum $= 5000\,\text{kgms}^{-1}$; $E_K = 1250\,\text{J}$
  b) $F = \Delta mv$ per second $= V\Delta m$ per second $= 50\,\text{N}$

▷ Waves

Chapter 10 (Wave Motion) page 137

1. a) energy
   b) period (time period), wavelength
   c) amplitude, centre (midpoint, middle)
   d) wavefronts
   e) difference, degrees
   f) longitudinal, transverse
   g) transverse

2. See Example 3. $0.5\,Hz$, $50\,Hz$, $2.3\,ms$, $1.09 \times 10^{-8}\,s$

3. a) Definition of frequency. $4\,Hz$
   b) See Example 1. $0.06\,m$ ($6\,cm$)
   c) $0.24\,ms^{-1}$ ($24\,cm\,s^{-1}$)

4. Use $v = x/t$ to find t for each wave separately. $71\,s$ (2 s.f.)

5. a) Definition of dispersion on page 125.
      Speed v depends on wavelength λ.
   b) i) $8.8\,ms^{-1}$
      ii) $12\,ms^{-1}$ (2 s.f.)
   c) See Question 4 above.
      $660\,s$ (11 minutes) (2 s.f.)

6. See page 126. $2.25\,MJ$ ($2250\,kJ$)

7. a) $180°$ is $\frac{1}{2}\lambda$ (ie. anti-phase)
   b) $\pi/2$ radians is $\frac{1}{4}\lambda$

8. See page 132.

9. See Example 5. a) $20\,W$ b) $0.1\,Wm^{-2}$ (1 s.f.)

Chapter 11 (Reflection and Refraction) pages 156–157

1. a) equal   b) Snell
   c) refractive, vacuum (free space)
   d) away, critical, reflected
   e) convex (converging), virtual
   f) focal, principal   g) power, dioptres (D), negative
   h) lens, mirror, convex (eyepiece)
   i) cone, rod, intensities
   j) distant, diverging (concave), close, converging (convex)
   k) toric, curvatures, directions

2. See Example 1.
   By rows: $59°$, $22°$, $45°$ (2 s.f.)

3. a) See page 140. 1.61, 1.52 (3 s.f.)
   b) See Example 5. $70.8°$ (3 s.f.)

4. See Example 5. $76.9°$ (3 s.f.)

5. See Example 6. By rows: $+2.0\,D$, $-4.0\,D$,
   converging $5.0\,m$, diverging $2.0\,m$

6. See Example 6.
   $+5.5\,D$, $0.18\,m$ ($18\,cm$) (2 s.f.)

7. See page 147 and Example 7.
   By rows: $30\,cm$, $23\,cm$, $-33\,cm$, $-20\,cm$, $15\,cm$ (2 s.f.)

8. a) See page 147.
      $13\,cm$ from the lens on the same side as the object, $1.5\,cm$ in diameter (2 s.f.)
   b) See page 147. $40\,cm$ from the lens on the same side as the object, $4.4\,cm$ in diameter (2 s.f.)

9. a) Light from a very distant object is parallel as it reaches the lens.
   b) $5.0\,cm$ (2 s.f.)
   c) $5.5\,cm$ (2 s.f.)
   d) See Example 7. $55\,cm$ (2 s.f.)

10. a) See Example 2. $1.24 \times 10^8\,ms^{-1}$ (3 s.f.)
   b) See Example 4. $24.4°$ (3 s.f.)
   c) Light entering the diamond undergoes many internal reflections, eventually escaping in random directions.
   d) As the ring is turned, the 'sparkles' will change in colour due to the different refractions of light entering and leaving the diamond, making the ring more attractive.

11. $40\times$, $30\times$

12. a) myopic
   b) the left eye
   c) Both eyes are astigmatic, and the values inform you of the curvature and orientation of the cylindrical component.

Chapter 12 (Interference, Diffraction) pages 178–179

1. a) phase, phase, constructive
   b) anti, destructive
   c) standing, nodes, anti-nodes d) small, large
   e) diffraction, minimum f) Young, grating

2. a) See pages 160 and 161.
      nodes (destructive interference)
   b) See Example 2. $56\,cm$
   c) $610\,Hz$ (2 s.f.)

3. a) Distance between nodes $= \frac{1}{2}\lambda$. $1.3\,m$ (2 s.f.)
   b) $85\,Hz$ (2 s.f.)
   c) Wavelength is the same. $440\,ms^{-1}$ (2 s.f.)

4. $f_1 = \dfrac{1}{2l} \times \sqrt{\dfrac{T}{\mu}} = \dfrac{1}{2 \times 0.42} \times \sqrt{\dfrac{110\,N}{8.2 \times 10^{-4}\,kg\,m^{-1}}}$
   $= 440\,Hz$ (2 s.f.)

5. Distance between nodes $= \frac{1}{2}\lambda$ $10\,ms^{-1}$

6. See pages 164 and 165.
   a) $236\,Hz$ and $472\,Hz$ (3 s.f.)
   b) $118\,Hz$ and $354\,Hz$

7. See page 165. $340\,ms^{-1}$ (3 s.f.)

8. Compare wavelengths with a typical doorway width, (about $1\,m$).
   $0.34\,m$, $0.34\,cm$ (2 s.f.)

9. a) $0.024\,m$ ($2.4\,cm$) (2 s.f.)
   b) See Example 6.
      $0.015\,rad$ ($0.86°$) (2 s.f.)
   c) Use angle from (b) and trigonometry or small-angle approximation. $5.4 \times 10^5\,m$, $9.2 \times 10^{11}\,m^2$

10. a) See Example 8. $0.0095\,m$ ($9.5\,mm$) (2 s.f.)
   b) The fringe separation would decrease.

11. a) See Example 8. $0.56\,m$
   b) s is not very much smaller than D (page 173)

12. See Example 7, use small-angle approximation.
   $3.3 \times 10^{-7}\,rad$ ($1.9 \times 10^{-5}°$)

13. See Example 9. (4 s.f. in data)
   $5.878 \times 10^{-7}\,m$ ($587.8\,nm$) (4 s.f.)

14. a) See Example 9. $2.00 \times 10^{-6}\,m$ (3 s.f.)
   b) See Example 9. $17.1°$ ($0.299\,rad$) (3 s.f.)
   c) As in (b). $36.1°$ ($0.630\,rad$) (3 s.f.)
   d) As in (b). $62.1°$ ($1.08\,rad$) (3 s.f.)
   e) Maximum possible path difference $=$ grating spacing, which must be 4λ for the 4th-order maximum and $2.00 \times 10^{-6}\,m < 4\lambda$ ($4\lambda = 2.36 \times 10^{-6}\,m$)

▷ Further Questions on Waves

Wave Motion (page 180)

1. c) $c = f\lambda = 0.8\,ms^{-1}$

2. a) i) BD ii) ACE iii) AE iv) any adjacent pair
   b) $10\,mm$, $2.0\,m$

3. c) You might observe the trace on an oscilloscope and measure the period.

4. a) $f = 1/\text{period} = 170\,Hz$ (2 s.f.)
   b) $90°$ or $\pi/2$
   c) longitudinal waves
   d) A (larger amplitude)
   e) Using $c = f\lambda$, $\lambda = 2.0\,m$

5. Answer B

6. b) Dims to almost zero at $90°$ and brightens again towards $180°$.
   c) radio, microwaves, ultraviolet

Reflection and Refraction (page 181)

7. a) Using Snell's Law, i) $r = 28°$ ii) $c = 42°$
   b) t.i.r. at A, escapes symmetrically at right side
   c) critical angle between glass and liquid $>$ critical angle between glass and air (velocity ratio is less)

8. b) magnification $= v/u = \times 4$

9. b) $c = 41°$, $c < 45°$ so t.i.r. at hypotenuse
   c) No light loss on t.i.r., no silvering needed, no 'ghost images' from reflection off mirror glass.

10. a) Core is transmission medium for em waves to progress (by total internal reflection). Cladding provides lower refractive index so that total internal reflection takes place and offers protection of boundary from scratching which could lead to light leaving the core.

b) Blue travels slower than red due to the greater refractive index. Red reaches end before blue, leading to material pulse broadening.
Alternative calculations for first mark
Time for blue = $d/v = d/(c/n)$
$= 1200/(3 \times 10^8/1.467) = 5.87 \times 10^{-6}$ s
Time for red = $d/v = d/(c/n)$
$= 1200/(3 \times 10^8/1.459) = 5.84 \times 10^{-6}$ s
Time difference = $5.87 \times 10^{-6} - 5.84 \times 10^{-6}$
$= 3(.2) \times 10^{-8}$ s

c) Discussions to include:
use of monochromatic source so speed of pulse constant,
use of shorter repeaters so that the pulse is reformed before significant pulse broadening has taken place,
use of monomode fibre to reduce multipath dispersion.

11. a) $a \approx 0.05$ mm, $b \approx 0.1$ mm to 1 mm

b) Using Snell's Law, $r = 9.4°$, $i = 80.6°$

c) $_{clad}n_{core} = 1.014$, $n_{clad} = n_{core}/_{clad}n_{core} = 1.46$

d) Light would escape into the cladding.

12. a) $c = 81°$ b) Similar shape graph with cladding at 1.45, fibre at 1.47 but with fibre only 1 μm diameter.

c) Using $s = vt$ where v is speed of light in core, $t = 25$ μs

Interference and Diffraction (page 182)

13. a) $s \approx 0.5$ mm, $D \approx 1$ m

b) Light from the 2 sources has arrived in phase (bottom slit is λ further away than top slit).

c) i) reduces x (ii) reduces x d) Same frequency, always in phase or always have same phase difference.

14. Answer D

15. b) What has happened to λ? Spacing of nodes and anti-nodes halved.

d) Near the metal sheet the amplitudes of the incident and reflected waves are more comparable.

16. a) Length of string must be an exact multiple of $\frac{1}{2}\lambda$, so only certain values of λ are possible.

b) $c = f\lambda = 96$ m s^{-1}

c) length = $n \times 10 = (n + 2) \times 8$, so $n = 8$ loops

17. a) use $d \sin\theta = n\lambda$, $1/d = 5.8 \times 10^5$ m^{-1}

b) 650 nm c) $\sin\theta < 1$ when $n \le 2$

18. a) $\lambda = 9$ mm (diagram not full size), $f = 9$ Hz (1 s.f.)

b) i) Join intersections of crests which are equidistant from A and B, λ further from A than B etc.

ii) Join lines half way between lines in (i).

Synoptic Questions on Waves (page 183)

19. a) 500 mm (image is 225 mm behind mirror)

b) Rays leaving the spectacle lens still diverge, so the image is virtual with $v = -250$ mm.
Converging lens of power $+2$D

20. b)

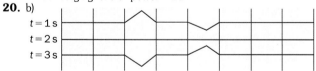

21. a) at P, path difference = zero; at Q, $\lambda/2$

b) $\lambda = 0.027$ m, $f = 11$ GHz

22. a) movement is parallel to wave direction

b) C where atoms close together; λ apart; R midway between

c) same pattern as sketch Y but shifted $\frac{1}{4}\lambda$ to the right

23. a) Use of $v = d/t$ to give $v = 340 \pm 10$ m s^{-1}

b) i) $\lambda = (0.30 \times 1.6)/1.2 = 0.040$ (m)

ii) $v = 332$ (m s^{-1})

iii) $\lambda = 1.1$ (m) or $\lambda > a$ or $\lambda > 0.3$ (m) or $\lambda > S_1 S_2$
Maximum path difference possible for waves from S_1 and S_2 is the slit separation, which is only 0.30 m **or** path difference can never be large enough. Or Young's fringes equation gives 'first' maximum at 4.4 m from central dot.

▶ Matter and Molecules

Chapter 13 (Materials) page 197

1. a) mass, volume b) proportional, limit, elastic

c) force, area, extension, length (original length), Young

d) ductile (plastic, malleable), brittle, chains

2. a) See Example 3. 50 N m^{-1} (2 s.f.)

b) See Example 4. 0.25 J (2 s.f.)

3. a) Gradient of linear part. 20 N m^{-1} (2 s.f.)

b) Area under curve up to extension of 0.6 m, between 3.5 J and 3.6 J (2 s.f.)

c) Where curve bends over. 10 N to 10.5 N

4. See Example 5. 5.0×10^7 Pa (N m^{-2}) (2 s.f.)

5. See Example 6. 25 N (2 s.f.)

6. See Example 6. 7.9 N (2 s.f.)

7. See Example 7. 1.2 m, 6.0 (600%)

8. See Example 8. 1.3×10^{11} Pa (N m^{-2}) (2 s.f.)

9. See Example 1. 15 kg (2 s.f.), not very rapidly

10. See Example 1 and diagram below it.

11. See *Physics at Work* on tennis rackets on page 196. Glass alone would shatter, resin alone is not stiff enough.

Chapter 14 (Thermodynamics) page 208

1. a) energy, working (work)

b) energy, molecules, cold

c) melting, melt, state d) bonds e) boiling, energy, gas

2. a) See Example 1. -20 J

b) See Example 1. $+50$ J

c) See Example 1. $+30$ J

3. What fraction of the distance between 0 °C and 100 °C is X at? 75 °C (2 s.f.)

4. See Example 2.
By rows: -253 °C, 195 K, 6273 K (about 6000 K)

5. See Example 4. 3.3×10^5 J (2 s.f.)

6. As in Example 4, rearranging the equation.
1.9 K (1.9 °C) (2 s.f.)

7. a) See Example 4.
5.2×10^5 J (2 s.f.)

b) Use $P =$ energy/t. 260 s (2 s.f.) (about 4 minutes) The kettle itself has no significant increase in internal energy.

8. a) See Example 7. 1.2×10^6 J (2 s.f.)

b) Use $P =$ energy/t. 580 s (about 10 minutes) (2 s.f.)

9. a) See Example 7. 1.7×10^4 J (2 s.f.)

b) Use answer to (a) as approximate value of energy removed from lemonade and see Example 4.
$\Delta T = 9.8$ °C, so final temperature = 10.2 °C (≈ 10 °C)

Chapter 15 (Gases and Kinetic Theory) page 221

1. a) solids, molecules, molecules, pressure

b) inversely, Boyle

c) ideal, gas, moles

d) r.m.s. (root-mean-square), kelvin (K)

2. See Example 1. 8000 Pa (2 s.f.)

3. See Example 2. 5000 J (2 s.f.)

4. See Example 3. 150 kPa (2 s.f.)

5. Find how many moles (0.004 kg) there are in 1 kg.
1.5×10^{26} molecules (2 s.f.)

6. Calculate exactly as for pressure in Example 4.
0.0036 m^3 (2 s.f.)

7. a) See Example 1 in Chapter 13.
50 m^2, 60 kg (2 s.f.)

b) See question 6 above. 2000 mol (2 s.f.)

8. a) See Example 5.
3.0×10^7 Pa (2 s.f.)

b) As part (a). 3.7 m^3 (2 s.f.)

9. See Example 6. 8400 Pa (2 s.f.)

10. a) Find the volume of 1 mol of helium (4.0×10^{-3} kg) and see Example 6.

b) The mass of air molecules is 1.2/0.17 $\times$ greater than that of helium molecules; what does this tell you about their r.m.s. speed, c? (See Example 8.)

Materials (page 222)

1. eg. metal (crystalline), polyethene (polymer), glass (amorphous)
2. a) $E = 1.0 \times 10^{10}$ Pa, UTS $= 2.6 \times 10^8$ Pa, ductile
   b) straight line through origin ending at 3.2×10^8 Pa, 0.094 strain
3. b) area of cross-section, original length
4. a) $E = 2.1 \times 10^{11}$ Pa, UTS $= 3.0 \times 10^9$ Pa
   b) 53 kN, 0.17 m, energy $= \frac{1}{2}F\Delta x = 45\,000$ J
5. b) See graph in question 6, page 222.
   c) cross-sectional area decreases, ductile fracture
6. a) $F = k\Delta x$, point of departure from straight line
   b) $E = Fl/A\Delta x = 2.0 \times 10^{11}$ Pa
   c) energy $= \frac{1}{2}F\Delta x = 0.65$ J
7. Answer A   8. Answer C
9. a) i) It has maximum <u>stress</u> at this point.
   ii) The tape has (permanent) extension or deformation when the force or stress is removed.
   b) **Measurement:** Diameter, and any <u>two</u> from: original or initial length (not: final length)
   extension or initial <u>and</u> final lengths;   weight or mass
   Equipment: micrometer or vernier (calliper) (for the diameter of the wire), and any <u>two</u> from: Ruler or (metre) rule or tape measure (for measuring the original length or extension). Travelling microscope (for measuring extension) Scales or balance or newtonmeter (for the weight of hanging masses) or 'known' weights used
   Young modulus = stress/strain or Young modulus is equal to the gradient from stress–strain graph (in the linear region)

Thermodynamics and Gases (pages 223–224)

10. a) Heat capacity is the energy needed to raise the temperature of the whole system (0.525 kg of water) by 1 K,   2.2 kJK^{-1} b) initial gradient = 0.073 Ks^{-1}
    c) 160 W   d) increasing energy lost to surroundings
11. a) $P = mc\Delta\theta/t = 0.82$ W   b) using $E = mL$, $m = 36$ g
    c) water vapour above ice condensed to water
12. a) eg. use a smoke cell b) i) spacing slightly greater in ice
    ii) ice more ordered iii) vibrational energy only in ice, translational energy in water
13. b) At 100 °C, translational KE of steam molecules = vibrational KE + PE of water molecules.
14. a) i) $\Delta W = p\Delta V = 23$ kJ   ii) 177 kJ
    b) ideal gas when V/T is constant: yes, nearly
15. a) i) using $E = Pt$, 10.3 MJ   ii) 1.1 MJ
    b) i) conduction   ii) evaporation and convection
16. a) $E_K = \frac{1}{2}mv^2 = 3.7$ J   b) Gas molecules hitting moving disc rebound with speed reduced, hence cool down.

Gases and Kinetic Theory (page 224)

17. b) faster particles, more collisions, bigger momentum change per collision   c) 478 ms^{-1} d) no change
18. a) assume linear scale, 41 °C
    b) i) both extrapolate to -270 °C   ii) 0.006 mol
19. a) i) $pV = nRT$ $n = pV/RT = \dfrac{1.2 \times 10^7 \times 0.05}{8.31 \times (273 + 21)} = 250$
    ii) mass $= n \times 0.029 = 250 \times 0.029 = 7.1$ kg
    b) $n_{\text{air added}} = pV/RT = \dfrac{1.0 \times 10^5 \times 1.5}{8.31 \times (273 + 21)} = 61.4$
    $n_{\text{total}} = n_{\text{initial}} + n_{\text{air added}} = 246 + 61.4$
    $p_{\text{final}} = n_{\text{total}} \times (RT/V)$
    $= 307 \times [(8.31 \times (273 + 21))]/0.050$
    $p_{\text{final}} = 1.5 \times 10^7$ Pa
20. a) use $pV = nRT$ b) 900 K   c) 600 kPa
    d) A → B, ΔU positive, ΔQ positive, ΔW negative
    B → C, ΔU zero, ΔQ negative, ΔW positive
    e) Approx. the area of the 'triangle' ABC, 4 kJ, decrease
21. $\Delta U = 0$; $\Delta Q = 48$ J; $\Delta W = -48$ J
22. b) $\Delta W = p\Delta V = 14$ kJ   c) use $\Delta U = \Delta Q + p\Delta V$
23. a) $2p_0$, $3p_0$, $1.5p_0$, p_0
    c) ΔW done on gas, ΔQ extracted, $\Delta W + \Delta Q = 0$
24. a) ΔW positive, ΔQ negative, ΔU zero
    b) ΔW zero, ΔQ positive, ΔU positive

Chapter 16 (Current and Charge) pages 242–243

1. a) electrons (charge), electrons, energy
   b) coulombs, amperes, time, seconds
   c) resistance, p.d. (voltage)
   d) current, proportional, p.d. (voltage), temperature
   e) increases, increases, decreases
   f) watts, amperes, p.d. (voltage), volts
2. a) very large numbers of free electrons
   b) very few free electrons
   c) semiconductors have fewer ($\sim 10^{10}$ m^{-3} fewer) free electrons
   d) vibration of metal ions increases, more opposition to electron flow
   e) number of free electrons increases as temperature rises
3. Using $Q = It$, a) $Q = 35$ C b) $Q = 36$ C
4. Using $Q = It$, a) $I = 0.15$ A b) $I = 20$ A c) $I = 2.0$ mA
5. a) 4.0×10^{-3} C
   b) Using $Q = It$, $I = 80\,\mu$A
   c) I is the number of coulombs per second; number of electrons s^{-1} = $I/e = 5.0 \times 10^{14}$
6. Using $Q = It$, $Q = 360\,\mu$C; number of electrons $= Q/e = 2.3 \times 10^{15}$
7. Charge = area under graph = 70 C
8. Using $I = nAve$, $v = 1.3 \times 10^{-4}$ ms^{-1}
9. a) Using $W = QV$, $W = 4800$ J
   b) Using $W = VIt$, $W = 900$ J
10. Using $W = VIt$, $W = 1.4 \times 10^5$ J
11. a) Using $W = QV$, $V = 12$ V
    b) Using $W = VIt$, $V = 45$ V
12. Using $W = VIt$, $I = 10$ A
13. Using $V = IR$ a) $V = 12$ V, b) $R = 24\,\Omega$, c) $I = 0.20$ A
14. a) Using $R = \rho l/A$, $R = 0.34\,\Omega$
    b) Using $V = IR$, $V = 4.4$ V
15. Using $R = \rho l/A$ and $A = \pi d^2/4$, $R = 1.2\,\Omega$
16. Using $R = \rho l/A$ and $A = \pi d^2/4$, a) 0.23 m b) 0.57 m
17. a) Using $R = V/I$, $R = 10\,\Omega$ at θ_1, and $R = 13\,\Omega$ at θ_2
    b) θ_2 is the higher temperature
18. Using $R = V/I$,   a) $120\,\Omega$ b) $16\,\Omega$
19. Using $W = Pt$, a) W (in J) $= 3000$ W $\times 1800$ s $= 5.4$ MJ
    b) W (in kWh) $= 3.0$ kW $\times 0.5$ h $= 1.5$ kWh
20. 14p (0.06 kW $\times 30$ h $= 1.8$ kWh)
21. Using $P = VI$,   a) $P = 1.7$ kW   b) $I = 4.0$ A
22. Using $P = I^2R$,   a) $P = 40$ W   b) $P = 160$ W
    $2 \times$ current gives $4 \times$ power
23. Using $P = I^2R$, $P = 3.1 \times 10^{-2}$ W
24. a) Using $P = IV$, $I = 625$ A
    b) Using $P = I^2R$, $P = 2.0$ MW
25. Using $P = V^2/R$, $P = 1.1$ kW
26. a) Using $R = \rho l/A$ and $A = \pi r^2$, $R = 760\,\Omega$
    b) Using $P = V^2/R$, $P = 70$ W
    c) i) resistance increases, ii) power falls

Chapter 17 (Electric Circuits) pages 254–255

1. a) current, same b) p.d. (voltage), sum
   c) current, sum
   d) p.d. (voltage), resistor (component)
   e) closed, sum, p.d.s.
2. a) Less opposition to current because more than one path for electrons to take.
   b) 3.0 V when connected in the same direction; 0 V when in opposite directions.
   c) Energy is wasted within the cell itself due to its internal resistance.
   d) See page 253.
3. a) Lamp not lit, high resistance voltmeter reduces p.d. across lamp and current through it.
   b) Lamp not lit, low resistance ammeter short circuits lamp.
4. a) 2.5 kΩ b) 3 Ω c) 7.6 Ω d) 6 Ω
5. a) 3 Ω, half of 6 Ω
   b) 4 Ω
   c) 3 Ω, one–third of 9 Ω

6. a) Using $V = IR$, $I = 0.5$A
   b) 0.5 A, current for resistors in series is the same
   c) Using $V = IR$, $V = 4.0$V
7. See Example 2. a) $2\,\Omega$ b) 6V c) 2A d) 1A
8. a) $6.0\,\Omega$ (parallel $R = 4.5\,\Omega$, $+ 1.5\,\Omega$ in series)
   b) Using $V = IR$, $I = 2.0$A
   c) Using $V = IR$, $V_1 = 3.0$V, $V_2 = 9.0$V
   d) Using $V = IR$, $I_1 = 0.5$A (9V/18Ω),
      $I_2 = 1.5$A (9V/6.0Ω)
9. From definition of e.m.f., a) 2.0J b) 5000J
10. a) Using $W = Q\epsilon$, $W = 15$J
   b) Using $W = Q\epsilon$ and $Q = It$, $W = 18$J
11. Using $\epsilon = I(R + r)$, $I = 0.4$A
12. i) See Example 4. a) $18\,\Omega$ b) 0.3A c) 5.4V
   d) $0.6\,\mathrm{JC^{-1}}$
   ii) See Example 5. a) $4\,\Omega$ b) 1A c) 4V d) $2\,\mathrm{JC^{-1}}$
13. Using $\epsilon = I(R + r)$, $I = 1.5$A
14. a) 1.5V (e.m.f. = terminal p.d. when current through cell is zero)
   b) Using $v = Ir$ and $v = 1.5$V $- 1.2$V, $r = 1.0\,\Omega$
   c) Using $V = IR$, $R = 4.0\,\Omega$
15. See Example 6. 3.0A
16. See page 248. i) 2V ii) 100V
17. See pages 248 and 249. 0 V to 10V
18. See Example 3. a) 5V b) 1V

Chapter 18 (Magnetic Fields) pages 268–269
1. a) field, stronger b) magnetic, current, circles (cylinders), wire
   c) parallel, spaced, uniform (constant), bar
   d) force, force, 90°, field e) turns, current
2. a) Direction of vertical force on wire changes with every half-cycle of a.c.
   b) See page 265. Force on charged particle is always at 90° to its velocity.
   c) See page 266. Each wire experiences a repulsive force due to the field of the other wire.
3. a) See page 257. b) See page 258.
   c) See page 266. d) radial field, inwards from N to S
4. Currents around b and a are in same direction.
5. Using $B = \mu_0 I/2\pi r$, $B = 1.6 \times 10^{-5}$T
6. Using $B = \mu_0 I/2\pi r$, $r = 2.0$cm
7. Using $B = \mu_0 n I$ ($n = 1200/0.4$), $B = 9.4 \times 10^{-3}$T
8. a) i), ii), and iv)
   b) i) perpendicular to I, towards bottom right
      ii) vertically downwards iv) out of page
9. Using $B = \mu_0 n I$, $n = 1.6 \times 10^4$ turns per metre
10. ii) and iv)
11. a) See Example 7. Using $F = \mu_0 I_1 I_2 l/2\pi r$,
      $F = 8.0 \times 10^{-5}$N
   b) 8 mm from P; current in P is $4 \times$ current in Q, so fields are equal at a point $4 \times$ further from P than Q.
12. Using $F = BIl\sin\theta$, a) $F = 0.12$N b) $F = 0.10$N
   c) Into the page in both cases.
13. Using $F = BIl$ (from balance $F = 2.2 \times 10^{-3} \times 10$N),
   $B = 0.11$T
14. See page 264. a) X positively charged,
   b) See page 265. i) and ii), radius of path decreases,
   iii) X circles clockwise
15. a) Using $F = BQv$, $F = 8.0 \times 10^{-14}$N
   b) Using $F = mv^2/r$, $r = 4.6$mm
   c) $F = BQv\sin\theta$, $\sin 30° = 0.5$, so $F = 4.0 \times 10^{-14}$N
   The radius of the path becomes $2 \times 4.6 = 9.2$mm.
   The particle spirals round the field lines – see page 265.

Chapter 19 (Electromagnetic Induction) page 279
1. a) flux, induced b) size, e.m.f., rate, flux
   c) direction, e.m.f., oppose d) speed, strength, area, turns
2. a) The wire is not 'cutting' through lines of flux.
   b) See page 276, magnetic braking
   c) See page 276, magnetic braking
3. Using $\Phi = BA$, $\Phi = 110$Wb
4. Using $N\Phi = NBA$, $N\Phi = 1.5$Wb
5. Using $\epsilon = N\Delta\Phi/\Delta t$, and $\Delta\Phi = B\Delta A$, $\epsilon = 0.68$V
6. Using $\epsilon = N\Delta\Phi/\Delta t$, and $\Delta\Phi = B\Delta A$, $B = 0.50$T

7. See Example 5. $\epsilon = 8.0$mV
8. See Example 5. $B = 2 \times 10^{-5}$T
9. a) Need anticlockwise couple to oppose clockwise motion; applying Fleming's left-hand rule, induced current must flow ABCD.
   b) Using $\epsilon_0 = 2\pi f BAN$, $\epsilon_0 = 1.4$V
10. a) Total change in flux linkage $N\Delta\Phi$ = area under graph,
      $N\Delta\Phi = 0.12$Wb
   b) Using $N\Delta\Phi = NB\Delta A$, $B = 0.32$T

Chapter 20 (Alternating Current) page 287
1. a) $\sqrt{2}$, r.m.s
   b) stepped down, ratio, 100%
   c) p.d. (voltage), time base
2. a) Laminations reduce induced eddy currents; reduces heating of core.
   b) d.c. produces steady magnetic flux in iron core; no change in flux linkage of secondary coil, so no induced secondary voltage.
   c) i) High voltage, low current, minimises power losses in cables.
      ii) a.c. so that transformers can be used to step voltages up and down.
3. Using $V_0 = 2V_{\mathrm{rms}}$, $V_0 = 17$V
4. See page 280. a) 50ms b) 20Hz c) 6.0V d) 4.2V
5. See Example 5. a) Using $P = IV$, $I = 30$A
   b) Using $P = I^2 R$, $P = 4500$W
6. Secondary p.d. = $23\,V_{\mathrm{rms}}$, step down
   Primary turns = 400, step up
   Secondary turns = 12 000, step up
7. See Example 2. a) 3.0V b) 1.1V c) 40ms d) 25Hz
8. 75 beats per minute (1 beat in 4×200ms)
9. a) Using $N_S/N_P = V_S/V_P$, $N_S = 250$ turns
   b) i) $I_S = 0.60$A ii) $V_S \times I_S = V_P \times I_P$, $I_P = 0.030$A

Chapter 21 (Electric Fields) pages 298–299
1. a) charge, separation (distance apart)
   b) coulombs, coulombs, electrons
   c) charge, force, strength, direction
   d) parallel, equally spaced
   e) equipotential f) potential, charge
2. a) Spray gives positive paint droplets and induces a negative charge on bike frame; droplets attracted to bike.
   b) Paint droplets follow field lines to the frame.
3. See page 291.
4. a) See page 289.
   b) See page 295. The potential energy of the charge does not change.
   c) Positive ball induces a negative charge on the earthed plate; ball and plate attract.
5. Using $F = kQ_1Q_2/r^2$, a) $2F$ b) $4F$ c) $F/4$
6. Using $F = kQ_1Q_2/r^2$, $F = 2.3 \times 10^{-8}$N
7. a) Using $F = kQ_1Q_2/r^2$, $Q = 2.0$nC
   b) Number of electrons = charge Q /charge on electron
      $= 1.3 \times 10^{10}$
8. Using $E = kQ/r^2$, a) $E = 36 \times 10^4\,\mathrm{NC^{-1}}$,
   b) $E = 9 \times 10^4\,\mathrm{NC^{-1}}$
   The radial field is an inverse square field;
   when r doubles, E drops to 1/4.
9. a) Using $E = kQ/r^2$ for each charge and subtracting,
      $E = 2.0 \times 10^4\,\mathrm{NC^{-1}}$
   b) Using $E = kQ/r^2$ for each charge and adding,
      $E = 6.0 \times 10^4\,\mathrm{NC^{-1}}$
10. a) Using $E = V/d$, $E = 4.0 \times 10^4\,\mathrm{Vm^{-1}}$
   b) From plate Z to plate Y.
   c) Twice as many field lines. $E = 8.0 \times 10^4\,\mathrm{Vm^{-1}}$
   d) Using $E = V/d$, $V = 4.0$kV
11. Using $E = V/d$, a) $V = 2400$V b) $d = 6.2$cm
12. See section on lightning, page 295.
13. Using $F = QE$, $F = 9.6 \times 10^{-6}$N
14. a) Uniform field, so $V_X = +300$V, $V_Y = +600$V,
      $V_Z = +900$ V
   b) See Example 4. $W = -9.6 \times 10^{-14}$J
   c) Lost potential energy

15. a) Weight of oil drop (mg) downwards; force due to electric field (QE) upwards.
b) Using $mg = QE$ and $E = V/d$, $Q = 3.2 \times 10^{-19}$C
16. a) Using $E = kQ/r^2$, $Q = 7.5 \times 10^{-6}$C
b) Using $V = kQ/r$, $V = 4.5 \times 10$V
17. Using $V = kQ/r$, $\Delta V = kQ(1/r_1 - 1/r_2) = 24$ V
Using $\Delta W = Q\Delta V$, $\Delta W = 8.6 \times 10^{-11}$J (2 s.f.)
18. a) Using $E = V/d$, $E = 1.0 \times 10^5$Vm^{-1}
b) Using $F = QE$, $F = 1.6 \times 10^{-14}$N

Chapter 22 (Capacitors) page 311
1. a) charge, negative b) p.d. (voltage), supply (battery)
c) energy   d) parallel   e) series
2. a) See page 301. There is current only while charging.
b) See page 306. As capacitor loses its charge, its p.d. falls and so the current falls.
3. Using $C = Q/V$, $C = 0.50\,\mu$F
4. a) Using $Q = CV$, $Q = 900\,\mu$C
b) Using $Q = It$, $t = 18$s
5. Using $W = \frac{1}{2}QV$, $W = 3.6 \times 10^{-4}$J
6. a) Using $C = C_1 + C_2$, $C = 7.5\,\mu$F
b) Using $1/C = 1/C_1 + 1/C_2$, $C = 2.0\,\mu$F
c) Add capacitors in parallel, then use $1/C = 1/C_1 + 1/C_2$, $C = 100\,\mu$F
7. a) Two $10\,\mu$F capacitors in parallel.
b) Two $10\,\mu$F capacitors in series.
c) Two pairs of $10\,\mu$F in series, connected in parallel.
8. See Example 3. a) $Q = 60\,\mu$C b) $C = 25\,\mu$F
c) $V = 2.4$V.
9. See Example 4. a) $C = 20\,\mu$F b) $Q = 120\,\mu$C
c) $V_1 = 1.2$V $V_2 = 4.8$V.
10. See Example 6. a) Using $Q_0 = CV_0$, $Q_0 = 60\,\mu$C
b) Using $\tau = CR$, $\tau = 3.2$s
c) Using $Q = Q_0 e^{-t/CR}$, $Q = 13\,\mu$C
11. See Example 7.
a) Using $I = V/R$, $I_0 = 6/20000 = 3 \times 10^{-4}$A
b) Using $V = V_0(1 - e^{-t/CR})$, $V = 2.4$V (2 s.f.)
12. Using $V = V_0 e^{-t/CR}$ ($V = 2.5$V, $V_0 = 9.0$V), $t = 420$s

▷ Further Questions on Electricity

Current and Charge (pages 312–314)
1. a) 960W, 2880W (3×960W), 1920W (2×960W) (all elements in parallel)
b) Using $P = IV$ and $V = IR$, $I = 4.0$A, $R = 60\,\Omega$
c) 3 equal resistors in parallel, $R_T = 20\,\Omega$
2. a) Lamps and resistor in series, using $V = IR$ and $P = I^2R$, $I = 0.8$A, $P = 3.2$W
b) Lamps in parallel and together are in series with resistor, $I = 0.8$A (total current = 1.6A)
c) Same current and resistance, so same brightness. Battery lasts longer because total current from battery is less in 1 (0.8A) than in 2 (1.6A).
d) Remaining lamp becomes brighter as current through it increases (to 1.2A).
3. a) Resistance of the filament rises as its temperature rises. At $V = 4.0$V, $I = 0.8$A and $R = 5.0\,\Omega$
b) Current in $3\,\Omega$ resistor is 1.33A, so total current is 2.13A. Using $P = IV$ gives $P = 8.5$W
c) Using $R = \rho l/A$ and $A = \pi d^2/4$ gives $d = 3.9 \times 10^{-5}$m
4. a) 3 in series gives $30\,\Omega$, two $30\,\Omega$ in parallel gives $15\,\Omega$
Using $P = V^2/R$, $P = 9.6$W.
b) Each set of 3 in series is a potential divider.
$V_{CD} = 4$V $V_{FD} = 8$V $V_{CF} = 4$V
5. a) i) electron   ii) ion   Both required for 1 mark.
b) i) $I = Q/t = 650/5 = 130$A
ii) $n = I/e = 130/1.6 \times 10^{-19} = 8.13 \times 10^{20}$
iii) $I = 10^{29}Aev$ giving $8.13 \times 10^{20} = 10^{29}Av$
$v = 8.13 \times 10^{20} / 10^{29} \times 3.0 \times 10^{-4}$
$= 2.7 \times 10^{-5}$ (ms^{-1}).
c) i) Because of Kirchhoff's first law or statement of this law.
ii) Using $I = nAev$ so v is proportional to $1/A$ giving 5.4×10^{-5}ms^{-1}.

6. a) Place the ammeter in series and the voltmeter in parallel with the lamp.
b) Using $P = V^2/R$, $R = 6.0\,\Omega$
c) For 12V across $24\,\Omega$ potentiometer and lamp, slider must be at top of potentiometer.
Current in lamp is 2.0A, 0.5A in potentiometer, so current of 2.5A in R.
P.d. across R is 8V with current of 2.5A, so $R = 3.2\,\Omega$.
7. a) i) Using $V = IR$, $I = 6.0$A
ii) Using $P = IV$ gives $P = 1260$W
iii) When $I = 6.0$A, p.d. across cables = 20V, so $3.3\,\Omega$
b) The 20V drop in voltage is the voltage across the supply cables to maintain a current of 6A.
c) The greater the current drawn by an appliance in the cottage, the smaller the voltage across it; energy is lost as heat in the cables.
8. a) Current in each lamp = 2A giving a total current from the battery of 4A.
b) Using $I = Q/t$ gives $t = 3 \times 10^4$s = 8.3h
c) i) No, they are in a separate parallel circuit.
ii) No, from $P = V^2/R$, power decreases because resistance increases and voltage is the same.
d) All four lamps are in parallel, so failure of any one lamp does not extinguish any other lamp.
9. b) For potential divider, $V_{AB} = (5/9) \times 12$V $= 6.7$V
c) Extra resistor 'loads' the potentiometer. In the lower section $5\,\Omega$ is in parallel with $20\,\Omega$, giving combined resistance of $4\,\Omega$. Thus p.d. across lower section is 6V, giving 3V across AB.
d) Using $I = V/R$ gives $I = 0.3$A
10. b) Total $R = 3.0\,\Omega$, ammeter reads 2.0A
$8.0\,\Omega$ and $4.0\,\Omega$ make 2:1 potential divider, voltmeter reads 2V.
c) i) terminal p.d. = p.d. across the $8.0\,\Omega$ and $4.0\,\Omega$ potential divider, $= 3 \times$ voltmeter reading.
ii) Using $V = E - Ir$, $r = 0.75\,\Omega$
11. a) $160\,\Omega$
b) i) R_{TH} drops in relation to $300\,\Omega$, so V_{TH} drops.
ii) Using $V_{OUT}/V_{IN} = R_1/(R_1 + R_2)$, $V_{OUT} = 3.9$V
12. a) 5.56V and data point plotted correctly to $\pm 1/2$ small square.
b) Best-fit straight line drawn through the last 4 data points. Allow a maximum of 2 marks if the line of best fit is drawn through all 5 data points.
Gradient $= 14 \pm 1$m^{-1}
Resistivity = gradient $\times A = (1.1 \pm 0.1) \times 10^{-6}\,\Omega$ m
c) The actual resistance values will be smaller. The gradient of the graph will be lower. Hence resistivity of the metal will be smaller than the value in (b).
13. The answer is D (remember, the resistance is NOT the gradient of the line or 1/gradient).

Magnetic Fields (page 315)
14. a) Using $F = BIl$, $F = 6 \times 10^{-5}$N
b) Using Fleming's LH rule, force is to the right.
15. a) Right to left along axis of solenoid.
b) B is uniform within solenoid remote from ends.
c) weakest W Z Y X strongest
16. a) $F = BIl$
b) they form a Newton III pair of forces
c) Graph of m against I is a straight line of gradient $0.14g$A^{-1}, or (since $g = 9.8$Nkg^{-1}) 1.4×10^{-3}NA^{-1}
d) Force is reversed and acts upwards on the magnet and current reading decreases.
17. a) The magnetic field (must be) at right angles to the current.
b) All three units for force, length and current clearly identified. (The unit of force is kgms^{-2}, the unit of current is A, the unit of length is m) T $= $ kgA^{-1}s^{-2}
18. Answer A
19. a) $\sin \theta = 0$ or $\theta = 0°$ or $\theta = 180$, π etc.
Travels along field lines
b) $F = BQv\sin\theta$ understood: $\theta = 30°/0.52$ radian

Electromagnetic Induction (page 316)

20. a) Aircraft is metal moving in the Earth's field.
b) See Example 5, page 275.
   i) $\Delta A = 3.3 \times 10^3 \, \text{m}^2$ ii) $\epsilon = 0.13 \, \text{V}$
21. a) See page 272. Note that there is an induced e.m.f. only when flux is changing.
b) i) along line XX from right to left,   ii) induced e.m.f. drives current round Q to oppose flux change.
22. Answer D
23. Answer B
24. a) Top side is a N-pole, arrow at X points left.
b) See page 274. Magnet moving faster, larger e.m.f. for a shorter time.
25. a) (magnetic) flux linkage
b) Lenz's law / conservation of energy induced current/e.m.f. (direction)
   Opposes the change (that produced it)
26. a) Using $N\Phi = NBA$, $N\Phi = 0.015 \, \text{Wb}$
b) Using $\epsilon = N\Delta\Phi/\Delta t$, $\epsilon = 5.9 \, \text{mV}$

Electric Fields (page 317)

27. a) Using $E = Q/4\pi\varepsilon_0 r^2$, $E = 4.5 \times 10^{10} \, \text{NC}^{-1}$ towards B
b) zero   c) towards P on either side of P
28. a)

b) tension, $T = 23 \, \text{mN}$ c) force, $F = T\sin\theta$
d) Using $F = Q^2/4\pi\varepsilon_0 r^2$, $Q = 220 \, \text{nC}$
29. a) parallel vertical lines with arrows down page
b) i) $E = V/d = 50 \, \text{kVm}^{-1}$ ii) $F = QE = 8.0 \times 10^{-15} \, \text{N}$
30. a) $V = Q/4\pi\varepsilon_0 r$
   gives $Q \, (= 4\pi\varepsilon_0 rV)$
   $Q = 4\pi \times 8.85 \times 10^{-12} \times 0.30 \times 3.0$
   $= 1.0 \times 10^{-10} \, (\text{C})$
b) use of $V \propto 1/r$ gives $V_N = V_L/3 = +1.0 \, \text{V}$
c) $E = Q/4\pi\varepsilon_0 r^2$
   $= 1.0 \times 10^{-10}/(4\pi \times 8.85 \times 10^{-12} \times 0.60^2)$
   $(= 2.50 \, \text{Vm}^{-1})$
31. Answer C
32. a) See page 292. Parallel vertical lines with arrows down page, at edges of plates lines curve outwards.
b) See page 295.
c) See page 294. Electric potential is a scalar.
33. a) $E = V/d = 2.0 \times 10^6 \, \text{Vm}^{-1}$
b) Ball is charged by contact, moves in the E-field and is earthed as it makes contact with P_2.

Capacitors (page 318)

34. a) straight line through the origin, $V \propto q$
b) energy stored = area under graph
35. a) energy = power × time = $8 \, \text{J}$
b) Using $W = \frac{1}{2}CV^2$, $V = 580 \, \text{V}$
c) $Q = CV = 27 \, \text{mC}$ d) $I = Q/t = 14 \, \text{A}$
36. a) Using $V = IR$, $I = 4.0 \, \text{mA}$ b) zero
c) $Q = CV = 5.6 \, \mu\text{C}$ d) $W = \frac{1}{2}CV^2 = 34 \, \mu\text{J}$
37. a) $Q = CV = 5.6 \, \text{mC}$
b) See pages 306 and 307. At $t = 0$, $I = 0.55 \, \text{mA}$; at $t = CR = 10 \, \text{s}$, $I = 0.20 \, \text{mA}$
38. a) 600 V each in series   b) $200 \, \mu\text{F}$ c) $360 \, \text{J}$
39. See pages 304 and 305.
a) 550 pF to 1000 pF   b) 45 pF to 250 pF
40. a) charge = $0.040 \times 60 = 2.4 \, \text{C}$
b) energy = $\frac{1}{2}Q^2/C = \frac{1}{2} \times 2.4^2/0.1 = 29 \, \text{J}$
41. a) $d = \varepsilon_r\varepsilon_0 A/C$
   $= 8.9 \times 10^{-12} \times 2.3 \times 250 \times 10^{-4}/370 \times 10^{-12}$
   $= 1.4 \times 10^{-3} \, \text{m} \, (1.4 \, (1.38) \, \text{mm})$
b) i) New capacitance = 161 pF
   New $V = 13 \, \text{nC}/161 \, \text{pF} = 81 \, \text{V}$
   ii) Energy stored = $\frac{1}{2} \times 161 \times 10^{-12} \times 81^2$
   $= 0.53 \, \mu\text{J}$

b) iii) Energy increases because:
   In the polar dielectric, molecules align in the field with positive charged end towards the negative plate (or words to that effect).
   Work is done on the capacitor separating the positively charged surface of the dielectric from the negatively charged plate (or vice versa).

Alternating Current (page 319)

42. a) 10.0 (V)
b) $V_{\text{rms}} = 10.0/\sqrt{2} = 7.1 \, \text{V}$
c) time period = $3 \times 2 = 6 \, \text{ms}$
d) frequency = $1/6 \, \text{ms} = 1/0.006 = 167 \, \text{Hz}$
43.

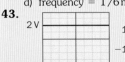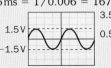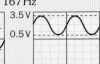

44. a) $\approx 500 \, \text{kV}$ b) adding ordinates at any time gives zero

Synoptic Questions on Electricity

45. a) period = $1/f = 20 \, \text{ms}$
b) i) half wave rectified waveform (see page 286), amplitude 2.5 cm, repeat every 4 cm
   ii) same direction as diode symbol
c) horizontal trace 4.5 cm above zero line
46. a) Using $V = Ir$, $r = 1.6 \, \text{M}\Omega$
b) i) Using $V = V_0 \, e^{-t/CR}$, where $CR = 70 \, \text{s}$, $t = 290 \, \text{s}$
   ii) $CR = 70 \, \text{s}$, $R = 17.5 \, \text{M}\Omega$ iii) no, less safe
c) $Q_0 = CV = 13 \, \text{mC}$; exponential decrease; at $t = 0$, $Q = Q_0$; at $t = 70 \, \text{s}$, $Q = 0.37 Q_0 = 4.8 \, \text{mC}$

▶ Quantum Physics

Chapter 23 (Electrons and Photons) page 336

1. a) electrons, thermionic, light (electromagnetic radiation), photoelectric
b) electron-volt (eV), 1.6×10^{-19}
c) emitted, threshold, higher (larger)
d) frequency (wavelength), photon, work   e) particle, waves
2. a) See Example 1. $1.6 \times 10^{-18} \, \text{J}$, 10 eV (2 s.f.)
b) $1.6 \times 10^{-15} \, \text{J}$, 10 keV ($10^4 \, \text{eV}$) (2 s.f.)
3. a) See Example 6. $6.6 \times 10^{-26} \, \text{J}$, $4.1 \times 10^{-7} \, \text{eV}$ (2 s.f.)
b) Use $v = f\lambda$, then as (a). $2.0 \times 10^{-24} \, \text{J}$, $1.2 \times 10^{-5} \, \text{eV}$
c) $2.0 \times 10^{-22} \, \text{J}$, $1.2 \times 10^{-3} \, \text{eV}$ (2 s.f.)
d) $2.0 \times 10^{-18} \, \text{J}$, 12 eV (2 s.f.)
e) $2.0 \times 10^{-17} \, \text{J}$, 120 eV (2 s.f.)
4. a) See page 329.
b) See page 329. $3.0 \times 10^{-19} \, \text{J}$ to $3.1 \times 10^{-19} \, \text{J}$ (1.9 eV)
c) $6.6 \times 10^{-34} \, \text{Js}$ (2 s.f.) (It is the Planck constant.)
5. a) Compare photon energy with work function.
   Yes, as 3.1 eV > 2.8 eV
b) See Example 7. $2.9 \times 10^{-7} \, \text{m}$ (290 nm) (2 s.f.)
6. a) Kinetic energy = $\frac{1}{2}mv^2$. $1.3 \times 10^7 \, \text{ms}^{-1}$ (2 s.f.)
b) Momentum = mv. $1.2 \times 10^{-23} \, \text{kg m s}^{-1}$ (2 s.f.)
c) See Example 10. $5.5 \times 10^{-11} \, \text{m}$ (2 s.f.)

Chapter 24 (Spectra and Energy Levels) pages 350–351

1. a) gamma, wavelength, radioactive (unstable), X-rays, visible (visible light)
b) interference, radio, gamma
c) all, line, absorption   d) excited, levels
e) higher, lower, absorption, ground
f) ionised, ionisation
2. See pages 321, 328 and 337. By rows: visible light 2.5 eV ($4.0 \times 10^{-19} \, \text{J}$), ultraviolet $2.5 \times 10^{-8} \, \text{m}$ (25 nm), infrared $4.0 \times 10^{-5} \, \text{m}$ (40 μm) (2 s.f.)
3. See Example 3.
a) 6400 K   b) $1.3 \times 10^{10} \, \text{m}$
4. See page 328. $8.3 \times 10^{-7} \, \text{m}$ (830 nm) (2 s.f.)

5. a) See page 344. $-3.6\,eV$ = ground state, $0\,eV$ = ionisation level.
  b) See 'A potential well' on page 344. The zero level is defined for a free electron, so an electron trapped at a lower potential energy in the potential well of an atom has negative potential energy.
  c) Ionisation takes the electron from the ground state to the zero energy level. $3.6\,eV$
  d) See pages 344 and 346. $3.6\,eV$
  e) $0.4\,eV$
6. a) See page 337 for photon energies in eV. $2.7\,eV$ is visible (wavelength of $460\,nm$)
  b) $3.2\,eV$ is ultraviolet (wavelength of $390\,nm$, $<400\,nm$)
  c) $0.5\,eV$ is infrared (wavelength of $2.5\,\mu m$, $>700\,nm$)
7. a) See pages 321 and 328.
  b) Transition must $= 1.96\,eV$. From the $-4.03\,eV$ level to the $-5.99\,eV$ level.
8. a) See photon energies on pages 321, 328 and 337. $91\,keV$ photons are emitted or absorbed in transitions
  b) See pages 321 and 328. $1.4 \times 10^{-11}\,m$ (2 s.f.)
  c) See page 337. Gamma
9. a) See page 344. $10.4\,eV$
  b) See Example 5. $1.8\,eV$, $4.9\,eV$ and $6.7\,eV$
  c) See pages 321, 328 and 337. $6.9 \times 10^{-7}\,m$ ($690\,nm$) is visible, $2.5 \times 10^{-7}\,m$ ($250\,nm$) is ultraviolet, $1.85 \times 10^{-7}\,m$ ($180\,nm$) is ultraviolet (2 s.f.)
  d) Which of the transitions in part (b) involved the ground state? All transitions from the ground state require photons of ultraviolet energy.

▶ Further Questions on Quantum Physics

Electrons and Photons (page 352)
1. d) Using $E = hc/\lambda$ and $E = \phi + 0.76\,eV$ gives $\phi = 3.2 \times 10^{-19}\,J$, $f = 4.8 \times 10^{14}\,Hz$
2. c) $\phi = hf$ when $E_{max} = 0$, $\phi = 2.6 \times 10^{-19}\,J$, $\lambda = 750\,nm$
  d) more electrons released but no change in graph
3. a) One photon interacts with one electron, intensity $\propto$ photons/s increases. so number of electrons/s increases.
  b) Photon energy $\propto$ frequency, which is constant and $hf = \phi + $ max KE, where ϕ is constant therefore max KE is constant.
4. a) Drag force = weight = $6\pi\eta rv = 4\pi r^3 \rho g/3$ giving $r = (9\eta v/2\rho g)^{1/2}$
  b) Calculate r using above equation and then mass = density × volume
  c) Vertical forces balance, $QV/d = mg$ gives $Q = 3.2 \times 10^{-9}\,C$
5. a) $280\,s^{-1}$ b) absorption, reflection off surface of eye
6. a) use $qV = \frac{1}{2}mv^2$ b) $\lambda = h/mv = 3.2 \times 10^{-11}\,m$
  c) λ is smaller than the atomic spacing of graphite
  d) nuclear radii $\approx 10^{-14}\,m$, so $\lambda \approx 10^{-14}\,m$ required
7. Force due to electric field (eE) balances force due to magnetic field (Bev) giving $v = E/B$

Spectra and Energy Levels (page 353)
8. c) ultraviolet, using $E = hf$, $E = 16.5 \times 10^{-19}\,J = 10.3\,eV$
9. $16 \pm 1\,\mu m$; use $\lambda_{max}T = 2.898 \times 10^{-3}$ gives range $T = 170 - 207\,K$
10. Total power $= \sigma AT^4$ and surface area $= 4\pi r^2$ so ratio $= (r_R)^2 (T_R)^4/(r_S)^2 (T_S)^4$ ratio $= 70^2 \times 2^4 = 80000$
11. $A = P/\sigma T^4 = 1.3 \times 10^{22}\,m^2$ so diameter $= 6.4 \times 10^{10}\,m$
12. c) Using $E = hc/\lambda$, $\lambda = 91\,nm$
13. a) $1.8\,eV$
  b) Using $E = hc/\lambda$, $\lambda = 120\,nm$
14. a) $1.7 \times 10^{-18}\,J$,
  b) 4 routes
  c) $600\,nm$ equates to $2.1\,eV$, $-1.6\,eV$ to $-3.7\,eV$
15. b) i) Using $d\sin\theta = n\lambda$ for red and blue, $\Delta\theta = 3.78° - 2.81° = 0.97°$
  ii) Using $\sin i/\sin r = n$ for red and blue, $\Delta r = 36.6° - 36.1° = 0.50°$

▶ Nuclear Physics

Chapter 25 (Radioactivity) page 373
1. a) proton (atomic), nucleon (mass)
  b) particle, positively, helium
  c) particles, negatively, high d) alpha, beta, magnetic, gamma
2. a) See page 363. β less strongly ionising than α.
  b) See page 364.
  c) α- and β-particles are absorbed by the medium.
  d) See pages 366 and 368 on nuclear stability.
  e) i) α-scattering is due to interactions with single nuclei, and to avoid absorption of α-particles.
    ii) path of incident α-particles is known and so scattering angle can be measured accurately.
3. 55 protons and 82 neutrons in nucleus
4. a) Using $r = r_0 A^{1/3}$, $r = 7.0 \times 10^{-15}\,m$
  b) $1.4 \times 10^{-42}\,m^3$ c) density $= m/V = 2.3 \times 10^{17}\,kg\,m^{-3}$
5. Using inverse-square law pages 364 and 365, $I/9$
6. Using inverse-square law pages 364 and 365, $2.5\,m$
7. See page 361, but with lines of two different lengths.
8. a) $^{19}_{8}O \rightarrow ^{19}_{9}F + ^{0}_{-1}e + ^{0}_{0}\bar{v}$
  b) $^{212}_{84}Po \rightarrow ^{208}_{82}Pb + ^{4}_{2}He$
  c) $^{56}_{27}Co \rightarrow ^{56}_{26}Fe + ^{0}_{+1}e + ^{0}_{0}v$
9. Using $\lambda = 0.693/t_{1/2}$ ($t_{1/2} = 10.8 \times 60\,s$), $\lambda = 1.1 \times 10^{-3}\,s^{-1}$
10. Using $A = \lambda N$ and $\lambda = 0.693/t_{1/2}$ ($t_{1/2} = 5 \times 24 \times 3600\,s$), $N = 2.5 \times 10^{11}$
11. a) $A_0 = 3340\,Bq$
  b) $A_0 = \lambda N_0$, $\lambda = 2.2 \times 10^{-6}\,s^{-1}$
  c) Using $t_{1/2} = 0.693/\lambda$, $t_{1/2} = 3.1 \times 10^5\,s = 3.6\,days$
12. Graph passes through: (0, 800), (5, 400), (10, 200), (15, 100), (20, 50).
13. a) 4800 years is $3 \times$ half-life. $1/8$ $(1/2^3)$ remains and so 7/8 decayed.
  b) 6400 years is $4 \times$ half-life. $1/16$ $(1/2^4)$ remains.
14. See page 371. b) Half-life ~ $50\,s$
15. a) Using $N = N_0 e^{-\lambda t}$, $N = 3300$
  b) Using $A = \lambda N$, $A = 50\,Bq$
16. a) Using $\lambda = 0.693/t_{1/2}$, $\lambda = 1.1 \times 10^{-2}\,min^{-1}$
  b) Using $A = A_0 e^{-\lambda t}$, $A = 44\,Bq$
17. a) Strontium in bones emits ionising β-particles for a long period of time.
  b) Using $A = A_0 e^{-\lambda t}$, and $\lambda = \ln 2/t_{1/2}$ ($A = 5\%$, $A_0 = 100\%$), $t = 120$ years

Chapter 26 (Nuclear Energy) page 383
1. a) less, defect (difference)   b) binding
  c) nucleon, nucleus   d) less, energy
  e) fission, fusion, increases
2. See Example 1.
  a) $0.042\,167\,u$, $7.00 \times 10^{-29}\,kg$
  b) $0.254\,06\,u$, $4.22 \times 10^{-28}\,kg$
  c) $0.592\,23\,u$, $9.83 \times 10^{-28}\,kg$
  d) $1.675\,00\,u$, $2.78 \times 10^{-27}\,kg$
3. a) See Example 3.   b) See Example 4.
  Li: $39.28\,MeV$ Li: $5.61\,MeV/$ nucleon
  Si: $236.7\,MeV$ Si: $8.45\,MeV/$ nucleon
  Cu: $551.7\,MeV$ Cu: $8.76\,MeV/$ nucleon
  Au: $1560\,MeV$ Au: $7.92\,MeV/$ nucleon
4. a) Yes. More mass on l.h.s. than r.h.s. so reaction occurs spontaneously, releasing energy.
  b) See α-decay example on p. 351. Mass difference of $0.005\,29\,u$, $4.93\,MeV$
5. Mass difference between l.h.s. and r.h.s. is $4.25 \times 10^{-29}\,kg$. Use $E = mc^2$ then convert from joules to eV. $23.9\,MeV$.
6. Mass difference between l.h.s. and r.h.s. is $0.204\,86\,u$. Convert to MeV using $931.5\,MeV$ per u. $190.8\,MeV$.
7. Mass on l.h.s. $= 63.916\,81\,u$. Mass on r.h.s. $= 63.929\,17\,u$. Extra mass/energy needed on l.h.s. is $0.012\,36\,u$ or $11.51\,MeV$ (assuming products have zero KE).

Chapter 27 (Particle Physics) page 393

1. a) mass, charge, spin b) annihilate, energy c) pair
   d) up, down, charm, strange, top, bottom
   e) electron (e^-), electron-neutrino (ν_e), muon (μ^-), muon-neutrino (ν_μ), tau (τ^-), tau-neutrino (ν_τ) f) strong
   g) baryons, mesons h) protons, neutrons, pions, kaons
   i) baryon, lepton j) gluon, weak, photons

2. a) $p = uud$ b) $n = udd$
   c) $\pi^0 = u\bar{u}$ or $d\bar{d}$, $\pi^+ = u\bar{d}$, $\pi^- = \bar{u}d$
   d) $K^0 = d\bar{s}$ or $\bar{d}s$, $K^+ = u\bar{s}$, $K^- = \bar{u}s$

3.

	p	n	π^0	π^+	π^-	K^0	K^+	K^-
Q	+1	0	0	+1	−1	0	+1	−1
B	+1	+1	0	0	0	0	0	0
S	0	0	0	0	0	+1	+1	−1

4. a) $\bar{p} = \bar{u}\bar{u}\bar{d}$ b) $\bar{n} = \bar{u}\bar{d}\bar{d}$
   c) $\bar{\pi}^0 = \bar{u}u$ or $\bar{d}d$, $\bar{\pi}^+ = \bar{u}d$, $\bar{\pi}^- = u\bar{d}$
   d) $\bar{K}^0 = \bar{d}s$ or $d\bar{s}$, $\bar{K}^+ = \bar{u}s$, $\bar{K}^- = u\bar{s}$
   π^0 is its own antiparticle (particle and antiparticle identical)

5. Mass of $p + \bar{p} = 2 \times 1.67 \times 10^{-27}\,\text{kg} = 3.34 \times 10^{-27}\,\text{kg}$
   $E = mc^2 = 3.34 \times 10^{-27} \times (3.00 \times 10^8)^2$
   $= 3.01 \times 10^{-10}\,\text{J}$ $(= 1.88\,\text{GeV})$
   (assuming created with zero kinetic energy)

6. a)

	μ^-	$\rightarrow$	e^-	+	$\bar{\nu}_e$	+	ν_μ	
Charge:	−1		−1		0		0	✔
B:	0		0		0		0	✔
L:	+1		+1		−1		+1	✔
S:	0		0		0		0	✔

   This can happen.

   b)

	p	+	e^-	$\rightarrow$	n	+	ν_e	
Charge:	+1		−1		0		0	✔
B:	+1		0		+1		0	✔
L:	0		+1		0		+1	✔
S:	0		0		0		0	✔

   This can happen.

   c)

	K^+	$\rightarrow$	π^+	+	π^0	
Charge:	+1		+1		0	✔
B:	0		0		0	✔
L:	0		0		0	✔
S:	+1		0		0	✔

   This can happen. Strangeness can change by ±1 in weak interactions.

7. a) Electron capture b) Repulsion

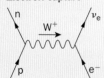
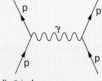

   c) Strong interaction d) β^+-decay

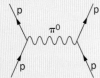

8. Only S quarks have strangeness.
   Possible quark combinations in mesons & antimesons ($q\bar{q}$):
   strange + antistrange: $S = -1 + 1 = 0$
   strange + other antiquark: $S = -1 + 0 = -1$
   antistrange + other quark: $S = +1 + 0 = +1$

9. a) See Example 3.
   p.d. $= 6.0 \times 10^3\,\text{V} \times 12 = 7.2 \times 10^4\,\text{V}$
   ∴ energy gained $= 7.2 \times 10^4\,\text{eV}$
   b) Need to gain $50 \times 10^9\,\text{eV} - 50 \times 10^6\,\text{eV}$
   $= 49.95 \times 10^9\,\text{eV}$
   No. of laps $= \dfrac{49.95 \times 10^9\,\text{eV}}{7.2 \times 10^4\,\text{eV/lap}} = 6.9 \times 10^5\,\text{laps}$

▶ Further Questions on Nuclear Physics

Radioactivity (page 394)

1. a) Using $A = \lambda N$, $\lambda = 1.05 \times 10^{-11}\,\text{s}^{-1}$
   $t_{1/2} = 0.693/\lambda = 6.6 \times 10^{10}\,\text{s}$ (about 2000 y)
   b) ratio of activities $= 32:1$, 5 half-lives, $= 10^4\,\text{y}$.

2. a) Boron-11 has 6 neutrons, Boron-12 has 7.
   b) N_A atoms in 12 g; $(600/12 \times 10^{-3}) \times N_A$ atoms
   $= 3.0 \times 10^{26}$ atoms in 6.0 kg.

3. a) Graph levels out to constant activity.
   Activity of C = 100 cpm (counts per minute).
   b) 3000 falls to 1500 cpm or 2000 falls to 1000 cpm in approx 1.3 min

4. a) insert very thin absorber (eg. sheet of paper)
   b) i) γ are uncharged,
   ii) β have a range of energies.

5. a) $^{235}_{92}\text{U}$ has 92 p, 143 n and 92 e
   $^{238}_{92}\text{U}$ has 92 p, 146 n and 92 e
   b) $^{238}_{92}\text{U}$ has halved in given time; while $^{235}_{92}\text{U}$ has gone through 6.43 half-lives, a factor of 86 ($1/2^{6.43}$).
   Ratio is $2/86 = 0.023$.
   c) Earth is older than $4.5 \times 10^9\,\text{y}$.

6. a) Causes ionisation which kills living cells.
   b) i) ^{52}Mn, short half-life, γ detected through earth
   ii) ^{60}Co, long half-life (constant dose rate), γ can penetrate to tumour
   iii) ^{241}Am, long life, α gives strong ionisation with low penetration (no shielding problems)

7. a) 8 protons, 10 neutrons
   b) $(9 \times 1.6 \times 10^{-19})/(18 \times 1.67 \times 10^{-27}) = 4.79 \times 10^7\,\text{Ckg}^{-1}$
   c) show similar calculation to give $4.26 \times 10^7\,\text{Ckg}^{-1}$ or compare 9/18 with 8/18

8. Using $A = A_0 e^{-\lambda t}$, where $\lambda = 0.693/t_{1/2}$,
   $A = 11.7\,\text{min}^{-1}$

9. a) X is a proton
   b) N_A atoms in 14 g; $(7.5/14 \times 10^{-3}) \times N_A$ atoms
   $= 3.2 \times 10^{26}$ atoms in 7.5 kg of ^{14}C.
   c) Using $\lambda = 0.693/t_{1/2}$, $\lambda = 1.2 \times 10^{-4}\,\text{y}^{-1}$
   d) Using $A = \lambda N$, $N = 2.5 \times 10^{30}$ atoms
   e) Activity falls exponentially from 0.8 to 0.1 Bq after 3×5700 years; 0.3 Bq occurs after approx 8000 y
   f) activity $< 1\,\text{s}^{-1}$, similar to background count rate

10. a) N_A atoms in 238 g; $(2.0/238 \times 10^{-3}) \times N_A$ atoms
    $= 5.0 \times 10^{24}$ atoms in 2.0 kg plutonium-238.
    b) Using $\lambda = 0.693/t_{1/2}$, $\lambda = 2.5 \times 10^{-10}\,\text{s}^{-1}$
    c) Using $A = \lambda N$, $A = 1.24 \times 10^{15}\,\text{Bq}$
    d) Using $A = A_0 e^{-\lambda t}$, $t = 9.2 \times 10^9\,\text{s}$ (about 290 y)

11. a) Using $A = A_0 e^{-\lambda t}$, $\lambda = 1.0 \times 10^{-6}\,\text{s}^{-1}$
    b) Using $t_{1/2} = 0.693/\lambda$, $t_{1/2} = 6.9 \times 10^5\,\text{s}$

12. See page 355 on Rutherford scattering experiment.

13. a) α-particle does work against repulsive force – kinetic energy transferred to potential energy.
    b) 1.7 MeV ($2.7 \times 10^{-13}\,\text{J}$)
    c) α-particle moves directly towards centre of nucleus.

14. See page 355 on Rutherford scattering experiment.

Nuclear Energy (page 396)

15. The mass of the constituents is greater to allow for the binding energy.

16. See pages 376–379.

17. The mass of the reactants is 0.001 29 u (1.2 MeV) less than the mass of the products. E_K for the incident α-particle must be much greater than 1.2 MeV.

18. a) Fission, a large nucleus splits into two smaller nuclei with release of energy triggered by absorption of a neutron. E represents E_K for the products (the energy liberated).
    b) mass difference $= 0.21\,\text{u} = 3.5 \times 10^{-28}\,\text{kg}$, use $\Delta E = c^2\Delta m$ to get $3.1 \times 10^{-11}\,\text{J}$ ($= 196\,\text{MeV}$)

19. See pages 378, 380–381.

20. See pages 380–381.

21. a) A in range 54 to 64
   Stability increases as binding energy per nucleon increases (or, large binding energy per nucleon shows nucleus is difficult to break apart).
   b) Binding energy per nucleon increases from about 7.6 to 8.5; increase of about 0.9 MeV for 235 nucleons; hence 210 MeV (approx 200 MeV) in total.
22. mass change $= 5.0 \times 10^{-30}$ kg $\equiv 4.5 \times 10^{-13}$ J $(E = mc^2)$
23. a) See page 376. b) See pages 378–379.
24. a) mass change $= 0.02561$ u, energy release $= 3.8 \times 10^{-12}$ J
   b) energy required s^{-1} = 9.6 MW which requires 2.5×10^{18} fusions s^{-1}, a burn rate of 8.4×10^{-6} mol s^{-1}. 1000 mol would last 1.2×10^8 s ($\sim$3.8y)

Particle Physics (page 397)
25. a) See page 359.   b) using $s = vt$, track length $= 70$ mm
26. a) Time or sequence of events.
   b) A = proton or u quark; B = neutron or d quark; C = W^+; D = neutrino. c) Weak interaction.
27. a) Must be neutral and a lepton, v_e.
   b) Can't be a lepton or baryon and must be positive, π^+.
   c) Yes in (a) because of the neutrino or the change in quark flavour.
   No in (b) because of absence of neutrino or no change in overall quarks.
28. Answer D
29. a) Composed of quarks, feel strong nuclear force, have a baryon number which is always conserved.
   b) meson = $q\bar{q}$, baryon = qqq or $\bar{q}\bar{q}\bar{q}$
   c) Conserving values for Q, B and S for each gives
   i) X has $Q = 0$, $B = 0$, $S = 0$
   ii) Y has $Q = 0$, $B = +1$, $S = 0$
   iii) Z has $Q = -1$, $B = +1$, $S = -1$
30. a) $^3_1\text{H} \rightarrow \,^3_2\text{X} + \,^{0}_{-1}\beta$, X is helium-3
   b) $\lambda = 0.693 / t_{1/2} = 0.056$ y^{-1}
   c) 12.5% is 3 half-lives, so age $= 36.9$ y
   d) $\Delta m = 0.0086$ u $= 1.42 \times 10^{-29}$ kg
   Using $E = mc^2 = 1.28 \times 10^{-12}$ J
31. a) B and D (curve opposite way to electrons)
   b) Uncharged so they create little ionisation.
   c) Electron takes some photon energy, so less for pair production in X. Lower-energy particles are more deflected by magnetic field.

▷ **Modern Physics**

Chapter 28 (Special Relativity) page 407
1. a) source, observer
   b) inertial, inertial
   c) slowly, greater/higher/faster, dilation
   d) parallel   e) increases
   f) rest, kinetic
2. a) 24 m s^{-1}
   b) Both record the speed as 3.00×10^8 m s^{-1} due to the invariance of the speed of light.
3. a) This is due to time dilation. According to special relativity, time on the moving rocket runs slower, so 1 s on the rocket is a much longer time on Earth.
   b) The Earth is assumed to be an inertial frame of reference.
4. The proper time t_0 is the time recorded by the person holding the torch on the train. Here $t = 5.0$ s and t_0 is our unknown.
   $5.0 = t_0(1 - 0.85^2 c^2/c^2)^{-1/2}$ $\therefore 5.0 = t_0 (0.2775)^{-1/2}$
   $\therefore t_0 = 5.0/(0.2775)^{-1/2} = 2.6$ s
5. a) increased   b) increased   c) unchanged
6. The proper length of a metre rule is 1.00 m.
   $l = l_0(1 - v^2/c^2)^{1/2}$ $\therefore 0.80 = 1.00 (1 - v^2/c^2)^{1/2}$
   $\therefore 0.80^2 = (1 - v^2/c^2)$ $\therefore v^2/c^2 = (1 - 0.80^2) = 0.36$
   $\therefore v = \sqrt{0.36 c^2} = 1.8 \times 10^8$ m s^{-1}

7. a) $m = m_0 (1 - v^2/c^2)^{-1/2}$
      $= 1.67 \times 10^{-27} (1 - 0.950^2 c^2/c^2)^{-1/2} = 5.35 \times 10^{-27}$ kg
   b) $E = m_0 c^2 (1 - v^2/c^2)^{-1/2}$
      $= 1.67 \times 10^{-27} c^2 (1 - 0.950^2 c^2/c^2)^{-1/2} = 4.81 \times 10^{-10}$ J
   c) Rest energy $E_0 = m_0 c^2$
      $= 1.67 \times 10^{-27} \times (3.00 \times 10^8)^2$
      $= 1.50 \times 10^{-10}$ J
   d) KE = total energy − rest energy
      $= 4.81 \times 10^{-10} - 1.50 \times 10^{-10} = 3.31 \times 10^{-10}$ J
   e) KE gained = work done = eV
      $\therefore V = \dfrac{\text{KE}}{e} = \dfrac{3.31 \times 10^{-10}}{1.60 \times 10^{-19}} = 2.07 \times 10^9$ V
8. As the speed increases, the increase in KE comes from an increase in mass. As $v \rightarrow c$, $m \rightarrow \infty$ and so KE $\rightarrow \infty$. An object's speed can never reach the speed of light as this would require infinite energy.

Chapter 29 (Astrophysics) page 422
1. a) visible light, radio or microwaves, infrared, ultraviolet
   b) CCD   c) light, resolve   d) parsec
   e) magnitude or Hipparcos, dimmer
   f) Hertzsprung–Russell or HR
   g) blue, Doppler   h) expanding
2. See Example 1 and page 170; 1.3×10^{-7} rad (or $7.6 \times 10^{-6\,\circ}$)
3. megametre, astronomical unit, light-year, kiloparsec
4. See Example 4; $M = -0.1$
5. a) A   b) B   c) E
6. HR diagram as on page 413; label point near left of Main Sequence, then in supergiant region; it will become a neutron star or black hole.
7. See page 413; red giants are formed by Sun-like stars contracting and throwing off material before expanding and cooling.
8. White dwarfs are small, dense and very hot, formed by small stars at the end of their lives. Neutron stars are much more massive, and are extremely dense. They are left after a red supergiant undergoes a supernova. Black holes are formed like neutron stars but are even more massive so their mass contracts them down to have no size (singularities).
9. a) 2.3×10^7 pc $= 23$ Mpc See Example 9. b) See page 419.
10. Sun is rotating anticlockwise, with left-hand edge coming towards us at 3900 m s^{-1}.
11. v (y-axis) against d (x-axis) graph has best-fit straight line through origin of gradient $H_0 = 68$ km s^{-1} Mpc^{-1}
12. See page 171. Arecibo has $3.9\times$ the resolving power and $16\times$ the area, so $16\times$ the collecting power when compared with the Lovell telescope.
13. See Example 7. K line gives $z = 0.071$, H line gives $z = 0.069$, so mean $z = 0.070$.
14. Dark matter and dark energy.
15. See page 419; $R_S = 300\,000$ m
16. See page 420.
17. See page 421.

Chapter 30 Medical Imaging (page 441)
1. a) line, continuous   b) dose, half-life, physical
   c) 20 kHz   d) precess, field lines, Larmor
2. a) See page 425. i) 120 keV   ii) 1.92×10^{-14} J
   b) Using $E_{max} = hc/\lambda_{min}$, $\lambda_{min} = 1.0 \times 10^{-11}$ m (2 s.f.)
3. a) See Example 2. $x_{1/2} = 2.3 \times 10^{-3}$ m or 2.3 mm (2 s.f.)
   b) $I_r/I_0 = e^{-\mu x}$, $0.1 = e^{-300 \times x}$, $x = 7.7 \times 10^{-3}$ m or 7.7 mm (2 s.f.)
4. See page 428.
5. a) See Example 3. $T_E = 6.8 \times 10^3$ days (2 s.f.)
   b) Effective half-life is too long; it is a beta-emitter.
6. a) positron emission
   b) positron annihilates with an electron
7. a) Using $Z = \rho c$, $Z = 1.43 \times 10^6$ kg m^2 s^{-1}
   b) See Example 4. $I_r/I_0 = 0.16$ or 16% (2 s.f.)
8. See page 435.
9. See Example 5. Diameter $= 7.7 \times 10^{-2}$ m or 7.7 cm (2 s.f.)
10. See Example 7. $v = 0.30$ m s^{-1} (2 s.f.)
11. a) See Example 8. $f_L = 6.05 \times 10^7$ Hz (3 s.f.)
   b) i) See page 438. ii) See page 439.
   c) See page 439.

▷ Further Questions on Modern Physics

Special Relativity (page 442)

1. a) i) To detect Earth's absolute motion.
   ii) Light reaches observer via each mirror; there is a phase difference between the two light beams. Bright fringes are seen where the two light beams are in phase (or, dark fringes where in antiphase).
   b) i) Earth's motion was thought to affect speed of light. Distance travelled by each beam did not change. The difference in the time taken by each beam would change so phase difference would change.
   ii) Earth's motion does not affect the speed of light.

2. a) time taken = distance/speed
   $= 34/(0.95 \times 3.0 \times 10^8) = 1.19 \times 10^{-7}$s
   b) use of $t = t_0/(1 - v^2/c^2)^{\frac{1}{2}}$ where $t_0 = 18$ns and t is half-life in the detectors' frame of reference
   so $t = (18 \times 10^{-9})/(1 - 0.95^2)^{\frac{1}{2}} = 57.6 \times 10^{-9}$s
   Time take for π meson to pass from one detector to the other = 2 half-lives (approx) (in the detectors' frame of reference). After 2 half-lives only 25% of the π mesons remain, so 75% decay before reaching the second detector.

3. a) Wave-like property: interference or diffraction; observed in bright and dark fringes of double slit interference or single slit diffraction or spectra of diffraction grating; explained with ref to path or phase difference; particle theory predicts two bright fringes for double slits or a single bright fringe for single slit or no diffraction for diffraction grating. Particle-like property: photoelectricity; observed in threshold frequency or instant emission of electrons; explained with ref to photon energy hf, the work function and '1 photon absorbed by 1 electron'; wave theory predicts emission at all light frequencies or delayed emission for (very) low intensity.
   b) i) $m = m_0 (1 - v^2/c^2)^{-0.5}$
   $= 9.11 \times 10^{-31} (1 - 0.890^2)^{-0.5} = 2.0 \times 10^{-30}$kg
   $\lambda = h/mv = (6.36 \times 10^{-34})/(2.0 \times 10^{-30} \times 0.89 \times 3.0 \times 10^8) = 1.24 \times 10^{-12}$ m
   ii) $E_{ph} = hf = hc/\lambda = (6.63 \times 10^{-34} \times 3.00 \times 10^8)/(1.24 \times 10^{-12}) = 1.6 \times 10^{-13}$J
   iii) $E_K = (m - m_0)c^2$
   $= (1.998 \times 10^{-30} - 9.11 \times 10^{-31}) \times (3.0 \times 10^8)^2$
   $= 9.78 \times 10^{-14}$J (3 s.f.)

4. a) Using $m = m_0/\sqrt{(1 - v^2/c^2)}$ gives
   $2 = 1/\sqrt{(1 - v^2/c^2)}$ or $\sqrt{(1 - v^2/c^2)} = 0.5$
   $v = \sqrt{(1 - 0.5)^2}c = 0.866c = 2.6 \times 10^8$ms^{-1}
   b) curve starts at $v = 0$, $m = m_0$ and rises smoothly
   curve passes through $2m_0$, at $v = 0.87c$
   curve gets closer to $v = c$ (but does not touch $v = c$ or curve back)
   c) Energy $= mc^2$ so (as $v \to c$) energy of particle increases as mass increases. Mass $\to$ infinity as $v \to c$ so energy $\to$ infinity which is (physically) impossible.

5. a) i) $d_0 = $ speed $\times$ time $= 1.8 \times 10^8 \times 95 \times 10^{-9} = 17(.1)$m
   ii) $d = d_0 (1 - v^2/c^2)^{\frac{1}{2}}$
   $= 17.1 \times (1 - (1.8 \times 10^8/3.0 \times 10^8)^2)^{\frac{1}{2}} = 14$m
   b) $m = m_0 (1 - v^2/c^2)^{-\frac{1}{2}}$
   $= 1.67 \times 10^{-27} \times (1 - (1.8 \times 10^8/3.0 \times 10^8)^2)^{-\frac{1}{2}}$
   $= 2.09 \times 10^{-27}$kg
   Kinetic energy $= (m - m_0)c^2$
   (kinetic energy)/(rest energy) $= (m - m_0)c^2/m_0c^2$
   $= (2.09 - 1.67) \times 10^{-27}/1.67 \times 10^{-27} = 0.25$

6. a) 2.18×10^8ms^{-1}; 2.56×10^8ms^{-1}; 2.77×10^8ms^{-1}; 2.88×10^8ms^{-1}
   b) Graph: attempt to calculate v^2; scales and units; points plotted; best-fit line

   c) No, it is not a straight line.
   d) The line would level out; KE increase due to increase in mass not speed as it approaches speed of light.

Astrophysics (page 443)

7. Siting: Apart from visible and some radio, em spectrum is significantly absorbed by atmosphere. IR telescopes in dry areas and/or very high up; UV and X-ray telescopes are generally put into orbit. To avoid atmospheric distortion and light pollution, visible telescopes are often high up and away from centres of population.

Due to interference from terrestrial sources, radio telescopes are situated away from centres of population.

Size: Telescopes have large diameters to increase the collecting power, which is proportional to the diameter^2, and to improve the resolving power, as minimum angle resolved is proportional to 1/diameter.

8. a) i) (A standard candle is) an object of known luminosity.
   ii) Flux or brightness or intensity of standard candle is measured. Inverse square law used to calculate distance to standard candle.
   b) i) An increase in the wavelength (of radiation) received from a receding source.
   ii) Use of $z = v/c$ and $v = H_0d$ [$z = H_0d/c$]
   $v = zc = 0.12 \times 3 \times 10^8ms^{-1} = 3.6 \times 10^7$ms^{-1}
   $d = v/H = 3.6 \times 10^7ms^{-1}/2.1 \times 10^{-18}$s^{-1}
   $= 1.71 \times 10^{25}$m
   c) Dark matter has mass but does not emit em-radiation. Observations of galaxies indicated that they must contain more matter than could be seen. The existence of dark matter will increase the density of the universe. This may make it more likely that the universe is closed. (Accept will contract or end with a 'Big Crunch'.)
   d) The universe started from a small initial point. The universe has a finite age. We can only see as far as (speed of light) $\times$ (age of universe).

9. a) Star much brighter than reflected light from planet. OR Planet very small and distant – subtends very small angle compared to resolution of telescopes.
   b) Planet and star orbit around common centre of mass that means the star moves towards or away from Earth as planet orbits. Causes shift in wavelength of light received from star.
   c) Light curve showing constant value with dip. When planet passes in front of star (as seen from Earth), some of the light from star is absorbed and therefore the amount of light reaching Earth is reduced. Apparent magnitude is a measure of the amount of light reaching Earth from the star.

10. a) brightness (or apparent magnitude) of star from a distance of 10 pc
   b) i) temperature from 50 000 K to 2500 K absolute magnitude from +15 to −10
   ii) S at 6000 K, and abs mag 5
   iii) W above and to left of S
   X above and to right of S
   Y below and to right of S
   Z below and to right of S

11. a) All type 1a supernovae have same peak absolute magnitude. Apparent magnitude can be measured. Reference to $m - M \log(d/10)$ or inverse-square law.
   b) Use of $m - M = 5 \log(d/10)$
   $12.9 - (-19.3) = 5 \log(d/10)$
   gives $\log(d/10) = 6.44$ $d = 27.5$ (Mpc)
   c) To make the accepted value for the distance more reliable.

12. a) distance $= 3.0 \times 10^8 \times 3.16 \times 10^7 = 9.48 \times 10^{15}$ (m)
   $\approx 9.5 \times 10^{15}$ (m)
   b) Correct labelling of 1 pc, 1 AU and 1"
   c) i) distance $= 9.5 \times 10^{15} \times 2.1 \times 10^7$m or 2.0×10^{23}m
   distance in pc $= 2.0 \times 10^{23}/3.1 \times 10^{16}$
   $= 6.4 \times 10^6$ (pc)
   ii) time $= 10^{44}/4 \times 10^{26}$s $= 2.5 \times 10^{17}$s
   $= 2.5 \times 10^{17}/3.16 \times 10^7 = 7.9 \times 10^9$y
   d) Any one from: Very dense / infinite density / very small / singularity
   Any one from: Very strong gravitational field therefore light cannot escape from it / curves space / slows down time / emits Hawking radiation

Medical Imaging Techniques (page 444)

13. a) Fast-moving electrons hit an anode. The kinetic energy of the electrons is transferred into X-rays.

b) An X-ray photon interacts with an electron within the atom. The electron is ejected and the frequency of the scattered photon is reduced.

c) i) $I = I_0 e^{-ux} = 3.0 \times 10^9 \times e^{-(6.5 \times 1.7)}$
   Intensity $= 4.8 \times 10^4 \, \mathrm{W\,m^{-2}}$

   ii) power of beam $= 4.8 \times 10^4 \times 5.0 \times 10^{-6} = 0.24 \, \mathrm{W}$
   2000 J needs to be delivered to tumour
   so $t = 2000/0.24 = 8.3 \times 10^3 \, \mathrm{s}$

d) X-ray beam passes through the patient at different angles or X-ray tube rotates around the patient. A thin fan-shaped beam is used. Images of 'slices' through the patient in one plane are produced with the help of computer software. X-ray tube or detectors are moved along the patient for the next slice through the patient. Advantage: 3D image; better contrast between different soft tissues.

14. a) Alternating p.d. applied across the crystal causes it to expand and contract creating pressure waves in the crystal / plastic membrane. Frequency of alternating p.d. equal to resonant frequency of crystal which is above 20 kHz. Short application of a.c. to produce short pulse. Use of backing material to damp and stop vibration of crystal.

b) i) correct calculation of ratio $I_r/I_i = 0.99896$
   Subtract from 1 and multiply by 100 to give 0.10%

   ii) Gel is between the probe and the skin to exclude air. Gel should have acoustic impedance equal to that of the skin to ensure maximum transmission into the body.

15. a) number of γ photons emitted per unit time

b) rate of energy $= 500 \times 10^6 \times 2.2 \times 10^{-14}$
   rate of energy emission $= 1.1 \times 10^{-5} \, \mathrm{J\,s^{-1}}$

c) See page 431.

16. a) The physical half-life depends only on the properties of the radioactive nuclide. Biological removal of the nuclide also occurs thus removing the nuclide more quickly overall.

b) $A_t = A_0 e^{-\lambda t}$ $1200 = 2700 \, e^{-5\lambda}$
   ($\lambda_E = \ell n \, (2700/1200)/5 = 0.1622$)
   $T_E = \ell n \, (2) / 0.1622 = 4.273$
   $1/4.273 = 1/20 + 1/T_B$ $T_B = 5.4$ days

c) Beta more strongly ionising than gamma so ^{131}I more likely to damage cells.
   Gamma rays for ^{131}I are over $4 \times$ more energetic which can cause problems when imaging with a gamma camera.
   190 h $\gg$ 6.0 h so with ^{131}I body will remain radioactive for longer, posing a greater danger to patient and others.
   Half-life of ^{99}T^{m} may be too short for certain types of diagnosis to be undertaken.
   Sensible conclusion based on above points.

17. a) Less chance of infection

18. b) Disadvantages: No metallic objects; patient has to remain still for a long time; confined space so difficult for patient suffering from claustrophobia.
   Advantages: Non-ionising; non-invasive; better contrast between soft tissues.

More Questions (page 445)

19. a) c is the same, regardless of the speed of the light source or the observer.

b) distance between detectors in rest frame of particles $= 25 \times (1 - 0.98^2)^{1/2} = 5.0 \, \mathrm{m}$
   time taken in rest frame of particles
   ($= $ distance/time $= 5.0/0.98c) = 1.7 \times 10^{-8} \, \mathrm{s}$
   time taken to decrease by 1/4 = 2 half-lives
   half-life ($= 1.7 \times 10^{-8}/2) = 8.5 \times 10^{-9} \, \mathrm{s}$

20. a) i) $t_0 = 800$ (s); use of $t = t_0 \, (1 - v^2/c^2)^{-1/2}$
   gives $t = 800(1 - 0.994^2)^{-1/2} = 7300 \, \mathrm{s}$

   ii) distance $= 0.994ct = 0.994 \times 3 \times 10^8 \times 7300$
   $= 2.2 \times 10^{12} \, \mathrm{m}$

b) space twin's travel time = proper time (or t_0)
   time on Earth, $t = t_0 \, (1 - v^2/c^2)^{-1/2}$
   $t > t_0$ so the time for traveller slows down compared with Earth twin; space twin ages less than Earth twin; travelling in non-inertial frame of reference

21. a) Concave primary and convex secondary; both rays correct to eyepiece.

b) i) Diagram to show two pairs of parallel rays being brought to a focus, those further from the axis being focused closer to the mirror.

   ii) use of $\theta = \lambda/D$ to give $\theta = 630 \times 10^{-9}/0.15$
   $= 4.2 \times 10^{-6}$ rad.

22. a) Use of $m - M = 5 \log(d/10)$ gives
   $3.54 - (-20.62) = 5 \log(d/10)$
   $d = 6.7(9) \times 10^5 \, \mathrm{pc}$

b) Use of $\Delta \lambda / \lambda = -v/c$
   $\Delta \lambda = -(0.21121 \times 10^5 \times 10^3)/(3.0 \times 10^8) = -7.4 \times 10^{-5}$
   $\lambda = 0.21121 - 7.4 \times 10^{-5} = 0.21114 \, \mathrm{m}$

c) $t = d/v = (6.79 \times 10^5 \times 3.08 \times 10^{16})/(105 \times 10^3)$
   $= 2.0 \times 10^{17} \, \mathrm{s}$

23. a) To reduce contact resistance: Sandpaper skin to remove hairs and some dead skin; Apply conducting gel; Securely attach more than one electrode.
   To remove unwanted signals: Electrodes should be non-reactive; Patient to remain relaxed and still; Shielded leads reducing interference from a.c. sources. Properties of amplifier: Amplifier has large input impedance, high gain and low noise.

b) i) starts at 0 to maximum of 1; unit mV
   ii) starts at 0 to 0.7 at end of T wave
   iii) **P** – depolarisation of atria; **R** – depolarisation of ventricles and repolarisation of atria;
   T – repolarisation of ventricles

24. a) See page 145. b) See page 147.

c) myopia or short sight

d) use of $1/f = 1/u + 1/v$; $1 - 0.33 = 1/0.25 + 1/v$;
   $v = (-)0.14 \, \mathrm{m}$

e) Cones active or stimulated; Cones stimulated by images must be separated by at least 1 unstimulated cone.

$\triangleright$ Synoptic Questions for A2

1. a) no force to produce acceleration when max speed $\times$ resistive forces = power available

b) power from battery $= 30 \mathrm{W}/0.70 = 43 \mathrm{W}$,
   $I = 43\mathrm{W}/12\mathrm{V} = 3.6 \, \mathrm{A}$

c) Using $P = Fv$, $F = 20 \, \mathrm{N}$

d) i) $R = 110g \, (0.12/1.0) = 130 \, \mathrm{N}$
   ii) total force down ramp $= 150 \, \mathrm{N}$,
   $v = P/F = 30\mathrm{W}/150\mathrm{N} = 0.20 \, \mathrm{m\,s^{-1}}$

2. Referring to diagrams from left to right:
1. moves parallel to E, accelerates
2. curves downwards, horiz: const. vel. down: const. accel.
3. moves parallel to B, no change in velocity
4. curves upwards in circle, constant speed

3. a) 3 loops

c) i) $2.0 \times 10^{-10} \, \mathrm{m}$,
   ii) Using $mv = h/\lambda$ gives $mv = 3.3 \times 10^{-24} \, \mathrm{Ns}$
   iii) where standing wave has largest amplitude

4. a)

b) zero because velocity is constant

c) no density difference, so no upthrust

d) Pressure drops because (i) atmospheric pressure drops and there is no pressure difference across the balloon fabric; (ii) volume increases.

5. a) a radial field in gap, arrows from N to S

b) Using $F = BIl$, where $l = 50 \times \pi \times d$, $F = 4.1 \, \mathrm{N}$

c) Using $F = ma$, $a = 350 \, \mathrm{m\,s^{-2}}$

6. a) the red line is at 660 nm, $E = hc/\lambda = 3.0 \times 10^{-19} \, \mathrm{J}$

b) $r_{\mathrm{red}} = 49.03°$, $r_{\mathrm{blue}} = 49.51°$, difference $= 0.48°$

c) $\theta_{\mathrm{red}} - \theta_{\mathrm{blue}} = 15.3 - 11.3 = 4.0°$

d) $\phi = 2.9 \times 10^{-19} \, \mathrm{J}$, so just possible

7. a) $0.60\,N \times 10\,mm = 0.20\,N \times X$
$X = 30\,mm$ from pivot A
b) $p = F/A = 0.60\,N/4.0 \times 10^{-6}\,m^2 = 150\,kPa$
c) $0.60 \times 10\,mm = (0.2\,N \times X) + (0.10\,N \times 18\,mm)$
$X = 21\,mm$ ∴ move counterweight 9 mm left
8. a) i) α-particle: has much more mass or momentum than β-particle; has twice as much charge as a β-particle; travels much slower than a β-particle.
b) i) Energy absorbed per sec $= 3.2 \times 10^9 \times 5.2 \times 10^6$
$\times 1.6 \times 10^{-19} = 2.7 \times 10^{-3}\,J$
ii) Temperature rise in 1 min $=$ energy absorbed in
1 min/mass × sp. ht. capacity
$= (2.7 \times 10^{-3} \times 60)/(0.20 \times 10^{-3} \times 900) = 0.90\,K$
9. Answer A
10. Answer D
11. Answer C
12. a) Using $R = \rho l/A$, $R = 0.96\,\mu\Omega$, $V = 48\,\mu V$
b) force to left – Fleming's left–hand rule,
Using $F = BIl$, $F = 0.12\,N$
c) Using $P = F/A$, mass $=$ density, $\rho \times$ volume, Ah
pressure due to BIl force $=$ pressure due to liquid
$BIl/A = \rho\,Ah\,g/A$, $h = 3.2\,cm$
13. See pages 305–306, 18–19, 247.
14. a) Amplitude of lattice vibration increases, resulting in more frequent collisions of electrons with lattice ions, resulting in a smaller mean drift velocity $I = nAve$ and the current decreases.
b) i) Read current values at 3 V for X and Y.
Values are 0.33 A and 0.61 A
Current through fixed resistor R $= 0.33 + 0.61$
$= 0.94\,A$
ii) Resistance of Y will be greater than resistance of parallel combination. Y will have a greater share of the p.d. so the reading on the voltmeter will increase.
15. $E_k = 75\,J = \frac{1}{2}N_Amc^2$ $\frac{1}{3}N_Amc^2 = pV = RT$ $T = 24\,K$
16. a) negative b) electrical force up balances weight downwards
c) Using $E = V/d$, $E = 2.0 \times 10^5\,Vm^{-1}$
d) charge $= 4.8 \times 10^{-19}\,C$ or 3 electronic charges
17. c) i) Using $E_p = mg\Delta h$, $E_p = 4.2\,kJ$
ii) Using $E = mc\Delta\theta$, $\Delta\theta = 0.056\,°C$
18. a), b) and c) See pages 88, 256, 291 and 292.
d) Using $F = QE$, $F = 1.4 \times 10^{-13}\,N$
Using $F = ma$, $a = 1.5 \times 10^{17}\,ms^{-2}$
e) See page 265.
19. c) i) stretched length $= 1.45\,m$, extension $= 0.65\,m$
ii) tension $= 41\,N$
d) i) Using $E = \frac{1}{2}F\Delta x$, $E = 13\,J$
ii) Using $E_k = \frac{1}{2}mv^2$, $v = 13\,ms^{-1}$
20. a) i) to the Earth's surface
ii) Current produced by the fairweather field in opposite direction to lightning.
iii) Waves (E-field) oscillating in the horizontal plane.
b) i) $100\,Vm^{-1}$ ii) Using $E = V/d = 5.0\,Vm^{-1}$
iii) scales marked on axes; E in Vm^{-1}, h in km, negative gradient curve, starting from (0, 100) getting less steep as h increases.
c) ideas of storms spread out in space, and time; low average current (only 1 A); strikes last for a very short time
d) i) straight lines reflecting off Earth or ionosphere with more than one bounce
ii) wavefronts equally spaced and curved; idea of waves from above the Earth's surface meeting the surface and creating a shadow
e) VHF at 30–300 MHz, ELF about 1 kHz.
i) 1–10 m ii) about $3 \times 10^5\,m$.
21. a) See page 276. Field produced by induced ring current opposes field produced by coil current.
b) Using $P = I^2R$, $R = 8.2 \times 10^{-5}\,\Omega$
c) Using $R = \rho l/A$, $l = 2\pi \times 12\,mm$,
$A = 15\,mm \times 2.0\,mm$, gives $\rho = 3.3 \times 10^{-8}\,\Omega\,m$.
d) Measure mass of ring and mass of water, rise in temperature of water and look up specific heat capacities of water and aluminium.

22. a) x-axis labelled V_{XY} in V, y-axis θ in degrees; points plotted correctly; smooth curve through points
b) sensitivity constant when gradient of graph is constant: about 450 V to 850 V, about $0.077\,deg\,V^{-1}$
c) Produce table as below. Plot $\sin\theta$ against V_{XY}, close to a straight line through origin.

V_{XY}	300	400	500	600	700
θ	18	23	30	37	45
$\sin\theta$	0.309	0.391	0.500	0.602	0.707

23. a) $k = 6\,N/0.1\,m = 60\,Nm^{-1}$
$\Delta l = F/k = 0.90 \times g/60\,Nm^{-1} = 0.15\,m$
b) clockwise moment $= 1.6 \times g \times 0.12\,m = 1.88\,Nm$
anti-cw moment $= 0.90 \times g \times 0.18\,m = 1.59\,Nm$
therefore no tipping.
c) As mass vibrates, tension in spring changes. Maximum tension increases as amplitude increases. Tipping occurs when anticlockwise moment of tension exceeds clockwise moment of weight of clamp.
24. a) i) 16 900 N
ii) displaced cold air weighs more than balloon + hot air
b) $p(V/m) = p/\rho$ so $\rho \propto 1/T$ if p constant
c) $\rho \propto 1/T$ so: $T_{hot}/T_{cold} = \rho_{cold}/\rho_{hot}$ (T in K).
25. %uncertainty: in L is $(0.1/25.6) \times 100\% = 0.4\%$;
in F is $(1/320) \times 100\% = 0.3\%$; so uncertainty in speed is
$0.4 + 0.3 = 0.7\%$
Uncertainty in speed $= 328 \times 0.007 = 2\,ms^{-1}$
so speed $= 328 \pm 2\,ms^{-1}$
All three results are within the calculated uncertainty so student B is correct.
26. Answer D
27. a) Student's conclusion is incorrect because the popper returns to its original shape, even though there is a time delay; elastic material returns to its original shape when the deforming force is removed but a plastic material would suffer a permanent deformation.
b) i) Refer to $v^2 = u^2 + 2as$ where s is height reached, v is zero, $a = -g$; so $u = \sqrt{(2gs)}$
ii) Air resistance will act on the popper as a decelerating force so the initial speed will be lower than in the absence of air resistance, so the suggestion is not correct.
28. a) for thinking time t rider moves $s = vt$
for (constant) deceleration from v to 0, $v^2 = 2as$
so total $s = d = v^2/2a + vt$
b) Using $y = mx + c$: $d/v = v/2a + t$ gives an equation resulting in a straight-line graph as a and t are constants.
c) i) 1.30 ± 0.18 entered in table; two points correctly plotted on graph with error bars
Line of best-fit: lower end of line should pass between (9.5, 1.3) and (10.5, 1.3) and upper end of line should pass between (34.0, 2.9) and (35.5, 2.9).
ii) Gradient should be about 0.065
$a = 1/(2 \times$ gradient) giving $a = 7.7\,ms^{-2}$;
y-intercept of best-fit line; should be about 0.65;
$t = $ y-intercept so should be about 0.65 s;
uncertainty in gradient; should be about 0.010 to 0.012;
giving uncertainty in a to be about ± 1.1 to ± 1.2:
uncertainty in y-intercept and t should be about ± 0.3
d) actual d/v values will be lower; so the y-intercept will be lower; hence the actual t ($=$ y-intercept) will be smaller.
29. Answer D:
%error $l + 2 \times$ %error $d + $ %error m
30. a) Clear identification of distance from centre of sphere to r.h. end of mark, or near r.h. end of mark.
b) 0.393 (s)
c) for 10 oscillations % uncertainty $= 0.1 / 15.7 = 0.006\,37 = 0.64\% = $ same for the 1/4 period
d) identifies anomaly [0.701] and calculates mean distance $= 0.759$ (m)
e) largest to smallest variation $= 0.026$ (m)
absolute uncertainty $= 0.013$ (m)

f) Use of $g = 2d/t^2$ leading to $9.83\,\text{ms}^{-2}$
percentage uncertainty in distance = 1.7%
Total percentage uncertainty = $1.7 + 2 \times 0.64 = 3.0\%$
Absolute uncertainty = $0.30\,\text{ms}^{-2}$. $g = 10 \pm 03\,\text{ms}^{-2}$

g) Suggests one change and sensible comment about change or its impact on uncertainty. E.g. use pointed mass not sphere because this will give better defined mark OR because the distance determination has most impact on uncertainty.
OR Time more swings/oscillations as this reduces the percentage uncertainty in timing.
OR Longer/heavier bar would take a greater time to the vertical, increasing t and s and reducing the percentage uncertainty in each.

h) [$s = \frac{1}{2}gt^2$] plot graph of s against t^2 or $s^{1/2}$ against t. Calculate the gradient; the gradient is $g/2$ or $(g/2)^{1/2}$.

31. a) i) along line XA   ii) along line BX
iii) 'horizontal' and to the left

b) $E = 2\dfrac{1}{4\pi\varepsilon_0}\dfrac{6 \times 10^{-6}}{(0.2)^2}\cos 60° = 1.35 \times 10^6\,\text{NC}^{-1}$

c) i) $V = -\dfrac{1}{4\pi\varepsilon_0}\dfrac{6 \times 10^{-6}}{(0.6)} + \dfrac{1}{4\pi\varepsilon_0}\dfrac{6 \times 10^{-6}}{(0.4)}$
$= 4.5 \times 10^4\,\text{V}$

ii) $W = q\,\Delta V = (2 \times 10^{-6})(4.5 \times 10^4) = 0.09\,\text{J}$

iii) $\frac{1}{2}mv^2 = 0.09$
$v = \sqrt{\dfrac{2(0.09)}{5 \times 10^{-3}}} = 6\,\text{ms}^{-1}$

32. a) $m = \rho V = 10^3(1.7 \times 10^{-3}) = 1.7\,\text{kg}$
b) all points correct, straight line, sensible scales   c) $20\,°\text{C}$
d) 3.2 min or 192 s
e) Heat supplied to water in eg. 2.5 min (Q)
$= (3 \times 10^3)(2.5 \times 60) = 4.5 \times 10^5\,\text{J}$
$\Delta\theta = 95.5 - 32.5 = 63\,°\text{C}$
$c = Q/m\Delta\theta = (4.5 \times 10^5)/((1.7) \times (63))$
$= 4.2 \times 10^3\,\text{Jkg}^{-1}\text{K}^{-1}$

f) i) All temperature measurements lower because heat taken by container.
ii) gradient of graph shallower   iii) c would be larger

33. a) Similarity: field strength or force $\propto 1/\text{separation}^2$ or both produce a radial field.
Differences: gravitational field is linked to mass and electric field is linked to charge; gravitational field is always attractive whereas electric field can be either attractive or repulsive.

b) i) The charges repel each other (because they have like charges). Each charge is in equilibrium under the action of the three forces: downward weight, a horizontal electrical force and an upwardly inclined force due to the tension in the string.

ii) $F = (4.0 \times 10^{-9})^2 / (4\pi\varepsilon_0 \times 0.02^2)$
$= 3.596.... \times 10^{-4}\,\text{N}$
weight $W = 6.0 \times 10^{-5} \times 9.81 = 5.886 \times 10^{-4}\,\text{N}$
$\tan\theta = (3.596 \times 10^{-4})/(5.886 \times 10^{-4})$
angle $\theta = 31°$

c) i) parallel and equidistant field lines
field direction is correct (from left to right)
ii) work done $= 1500 \times 1.6 \times 10^{-19} \times 1.2 \times 10^{-2}$
$= 2.88 \times 10^{-18}$
$\frac{1}{2}mv^2 = 1/2 \times 9.11 \times 10^{-31} \times (5.0 \times 10^6)^2$
$- 2.88 \times 10^{-18}$
speed $= 4.3 \times 10^6\,\text{ms}^{-1}$

34. a) Ruler and wire; moving crocodile clip; ohmmeter connected with no power supply or voltmeter and ammeter positioned correctly with power supply
ii) straight line through origin
iii) gradient $= R/l$ or pair of R and l values from graph; measure diameter to calculate area
resistivity, $\rho = \text{gradient} \times \text{area}$

b) Volume $= Al = \frac{1}{3}A \times 3l$ so cross-sectional area is $1/3$ of original.
new $R = (\rho \times 3 \times l)/(A/3) = 9\rho l/A$
so new $R = 9 \times$ original R

35. a) Electrical energy transferred to whole of circuit per unit charge.
b) Sensible scale and axes labelled with unit. All points correct. Line of best fit.
c) i) $E = \text{intercept} = 1.48\,\text{V}$
ii) $r = (-)\text{gradient} = (E - V)/I = 0.83\,\Omega$
d) $I = E/(R + r) = 1.48/(6 + 0.83) = 0.22\,\text{A}$
$t = 20 \times 60 = 1200\,\text{s}$
$Q = It = 0.22 \times 1200 = 264\,\text{C}$

36. a) Waves passing through a narrow gap spread out; light reaches the wall from each part of the slit with differing phase (OR path lengths); when the waves meet superposition takes place and if the waves are in antiphase it results in destructive interference so a dark spot is seen (OR if in phase it results in constructive interference so a bright(er) region is seen).

b) Green light has a shorter wavelength than red light so green light diffracts less than red light so the dark points would be closer to the centre, more dark points would be seen in the same space on the wall and the central fringe would be narrower.

c) $\theta = \tan^{-1}(0.23\,\text{m}/1.5\,\text{m}) = 8.7°$
$d = 10^{-3}/300 = 3.3 \times 10^{-6}\,\text{m}$
$\lambda = 3.3 \times 10^{-6}\,\text{m} \times \sin 8.7° = 5.1 \times 10^{-7}\,\text{m} = 510\,\text{nm}$

37. a) n changes by 4 units, 2 units, 1 unit for each change in 100 nm OR the magnitude of n does not halve every interval

b) Sensible long (>8 cm) tangent drawn, correct read-off for points from triangle at least half length of line and readings taken.
Substitution correct: $(-)(1.5 \pm 0.2) \times 10^4\,\text{m}^{-1}$

c) Column heading correct, all calculations correct, appropriate (3) s.f.

$1/\lambda^2$ /$10^{-12}\,\text{m}^{-2}$	11.1	8.16	6.25	4.94	4.00	3.31	2.78

Graph axes labelled correctly and sensible axes, plots correct to within half a square, best-fit line.

d) Intercept correct to within half a square (1.6014).
e) The value of refractive index at infinite or very long wavelength.
f) $\log n = \log c + d \log \lambda$; plot $\log n$ versus $\log \lambda$, d is the gradient of the graph

38. a) Graph of λ^2 against m to give straight line through origin.

m /kg	0.100	0.150	0.200	0.250	0.300	0.350
λ^2 /m^2	0.411	0.602	0.819	1.024	1.217	1.430

Axes labelled with suitable scales, quantities and units.
b) large triangle to include half the plotted length
$k = 4.08\,\text{m}^2\text{kg}^{-1}$
c) $\mu = g/f^2k$; so $\mu = 9.81\,\text{Nkg}^{-1}/((50.0\,\text{Hz})^2 \times 4.08\,\text{m}^2\text{kg}^{-1})$
$= 9.62 \times 10^{-4}\,\text{kgm}^{-1}$

39. a) Ensure largest possible proportion of flask is immersed. Make volume of tubing small compared to volume of flask. Remove heat source and stir water to ensure water at uniform temperature throughout. Allow time for heat energy to conduct through glass to air before reading temperature.

b) i) Pressure is caused by collisions of particles with sides. Velocity of particles (and volume of gas) are not zero at $0\,°\text{C}$.

ii) 1. Gradient of graph $0.75 \times 10^2/100 = 0.75$
Number of moles of gas = gradient/R = $0.75/8.31 = 0.09$
Mass of gas $= 0.09 \times 6.02 \times 10^{23} \times 4.7 \times 10^{-27}$
$= 2.5 \times 10^{-4}$ (kg)

2. Internal energy $= 3/2 \times NkT$
$= 1.5 \times 0.09 \times 6.02 \times 10^{23} \times 1.38 \times 10^{-23} \times (100 + 273) = 410\,\text{J}$

40. a) Air set into vibration at frequency of loudspeaker; resonance when driving frequency = natural frequency of air column; wave reflected from surface; interference/superposition between transmitted and reflected waves; maximum intensity when path difference is $n\lambda$; maxima observed when l changes by $\lambda/2$

b) i) $\lambda/2 = 523 - 168 = 355\,\text{mm}$; $\lambda = 710\,\text{mm}$
ii) $c = f\lambda = 480 \times 0.71 = 341\,\text{ms}^{-1}$

	Examination Boards' Specifications (syllabuses) for Physics						
	Page numbers needed for each one: **Blue** = AS-Level **Red** = A-Level						
	✔ = whole chapter needed, except ( ) [ ] = optional topic						
Chapter	**AQA**	**OCR**	**Edexcel**	**CCEA**	**WJEC**	**CIE**	**Edexcel Int'nat'l**
1 Basic Ideas	✔	✔	✔	✔ (16)	✔ (14)	✔	✔ (16)
2 Looking at Forces	✔ (26-9)	✔ 19	✔ (19, 26)	18-9, 20-1	18-21, 24	✔ (23, 25)	18, 21-4
3 Turning Effects of Forces	✔	✔	✔	33-34	✔ [32]	✔	35
4 Describing Motion	✔	✔	✔	✔ (45-49)	✔	✔ (48)	✔ (48)
5 Newton's Laws, Momentum	✔ (64)	✔	✔	✔(60-5), 62	✔ [60-62]	✔ (60-61)	54-9, 60-2
6 Work, Energy and Power	✔	✔	✔	✔ (74)	✔	✔ (74)	✔
7 Circular Motion	✔	✔	✔	✔	✔	✔	✔
8 Gravitational Forces, Fields	✔ (94)	✔	✔ (93-6)	✔ (92-4)	✔	✔ (92-6)	86-9
9 Simple Harmonic Motion	✔	✔	✔	✔ (105-8)	✔	✔	✔
10 Wave Motion	✔ [130-5]	✔ (134-5)	✔ (134-5)	✔ (126-8)	✔ (134-5)	✔ (127-35)	✔ (134-5)
11 Reflection and Refraction	✔[145-155]	138-142	✔ (149-55)	✔ (149-51)	139-142		138-142
12 Interference and Diffraction	✔[167, 171]	✔ (167)	✔ (164-5)	✔ (175-7)	✔ (164-5)	✔(162-7)	158-9, 168-9
13 Materials	✔ (185)	✔ (185)	✔ (185)	186-191	✔ (185)	✔ (193-6)	✔ (193-6)
14 Thermodynamics	✔	✔	198, 201-4	198, 204-5	✔ (203)	201 ✔	198, 201-4
15 Gases and Kinetic Theory	✔	✔	212, 217-8	✔ (209)	✔	209, 218	214, 218
16 Current and Charge	✔ (229)	✔	✔	✔ (236-9)	✔	✔ (237-41)	✔ (237-41)
17 Electric Circuits	✔	✔	✔	✔	✔	✔	✔
18 Magnetic Fields	✔ (261, 266)	✔ (266)	✔ (266)	✔ (266)	✔	✔	✔ (266-7)
19 Electromagnetic Induction	✔	✔	✔	✔	✔ (272)	✔	✔ (272)
20 Alternating Current	✔	282-3	280-1	282-3 ✔	[280-3]	282-3 ✔	
21 Electric Fields	✔ (288-9)	✔ (289)	✔	290-3	✔ (288-9)	292-4 ✔	✔
22 Capacitors	✔ (304-5)	✔ (309)	✔ (304-5)	✔ (309)	✔	✔ (307-8)	✔ (304-5)
23 Electrons and Photons	✔ 320-4	✔ 322-3	326-36	326-33	✔ (320-5)	321, 326-9	320-9 ✔
24 Spectra and Energy Levels	✔ [337-41]	✔ (337-8)	344-6, 337	344-7	✔ (341)	337, 345-6	344-6, 339
25 Radioactivity	355-66 ✔	✔	354-71	354-369	✔ (358-9)	354-66 ✔	✔
26 Nuclear Energy	✔	✔	374-8	✔	✔ (380-2)	✔ (380-2)	✔ (379-82)
27 Particle Physics	✔	✔	384-9	✔	✔ (391), 392	386-8, 391	384-5, 392
28 Special Relativity	[✔]		400				✔
29 Astrophysics	[✔]	✔ (411)	✔ (408-9)		409, 411-7	414	414-6, 411-7
30 Medical Imaging	[✔]	✔ (438-40)	433-5	✔ (431-2)	440 [✔]	437 ✔ (430)	433-7

See the much more detailed analysis for your specification at: **www.oxfordsecondary.co.uk/advancedforyou**

Acknowledgements

Cover: sportstock / iStock. p6(T): Stockbyte/Getty Images; p6(C): Wavebreak Media Ltd/Corbis; p6(B): Al Bello/Allsport/Getty Images; p8(T): Jacek Chabraszewski/Shutterstock; p8(B): Larry Mulvehill/Getty Images; p9: Stockbyte/Getty Images; p10: David J Phillip/AP Imgaes; p13(T): StockbrokerXtra/Alamy; p13(B): Spencer Platt/Getty Images; p15(T): Gary M Prior/Allsport/Getty Images; p15(B): Susan Leggett/Alamy; p18: Stockbyte/Getty Images; p19(T): Ted Kinsman/Science Photo Library; p19(B): Jean Collombet/Science Photo Library; p21(T): Custom Life Science Images/Alamy; p22(T): Matt9122/Shutterstock; p22(B): culture-images GmbH/Alamy; p23: Charles D Winters/Science Photo Library; p25: Craig Prentis /Allsport/Getty Images; p26: Anze Bizjan/Shutterstock; p27: Loic Lagarde/Getty Images; p28: Gary S Chapman / Getty Images; p29: ZUMA/Rex Features; p21(B): Szefei/Shutterstock; p30(T): Action Press/Rex Features; p30(B): Alexander Klein/AFP/Getty Images; p34(T): Peter Beavis/Getty Images; p34(B): Ken Reid/Getty Images; p36: Paolo Bona/Shutterstock; p38(T): David E Myers/Getty Images; p38(a): Anatoliy Lukich/Shutterstock; p38(b): Natursports/Shutterstock; p38(c): Kim Walker/Robert Harding World Imagery/Corbis; p38(d): Frederic Haslin/TempSport/Corbis; p38(e): olcha/Shutterstock; p39: Pete Saloutos/Getty Images; p41: George Impey/Alamy; p43(T): Jonathan & Angela Scott/JAI/Corbis; p43(C): Pete Saloutos/Getty Images; p43(B): Image Source/Rex Features; p45: Hector Mandel/Getty Images; p46: EPA; p47: Ddp USA/Rex Features; p48: Alex Coppel/Newspix/Rex Features; p49(T): Hennessey/Splash News/Corbis; p49(C): Kieran Doherty/Reuters; p49(B): Peter Luckhurst/Rex Features; p53(T): NASA; p53(B): TRL Ltd/Science Photo Library; p55(T): Susan Cazenove/Getty Images; p55(B): Omikron/Getty Images; p57: I T A L O/Shutterstock; p59: Patrice Latron/Look at Sciences/Science Photo Library; p60(T): Tony Bowler/Shutterstock; p60(B): Chris Brandis/AP Photo; p61: ID1974/Shutterstock; p63(L): Almonfoto/Shutterstock; p63(R): NASA; p64: Motorradcbr/Shutterstock; p65(T): Minerva Studio/Shutterstock; p65(C): Peter Dazeley/Photographer's Choice/Getty Images; p65(B): FStop Images/Caspar Benson/Getty Images; p69: Peter Bernik/Shutterstock; p71: ChameleonsEye/Shutterstock; p72: Mitch Gunn/Shutterstock; p74(T): John Smith/Corbis; p74(B): Robert Crum/Shutterstock; p76(C): Gray Mortimore/ALLSPORT/Getty Images; p76(T): Eky Studio/Shutterstock; p76(B): Martin Rogers/Getty Images; p79(T): Digital Vision/Getty Images; p79(B): Peter Kramer/NBC NewsWire/Getty Images; p80(T): NASA/Science Photo Library; p80(B): Ria Novosti/Science Photo Library; p81: Mattia Terrando/Shutterstock; p82: Chad Slattery/Getty Images; p83: NASA; p84: Al Bello/Allsport/Getty Images; p88: NASA; p90: Kevin A Horgan/Science Photo Library; p95: Tristan3D/Shutterstock; p97(T): NASA; p97(B): SSPL/Getty Images; p94: Universal History Archive/Getty Images; p100: Eye Ubiquitous/Alamy; p101: Zak Kendal/Cultura/Science Photo Library; p103: Martyn F Chillmaid; p104: Dyoma/Shutterstock; p105: Wally Stemberger/Shutterstock; p107: StockBrazil/Alamy; p109(L): Carloscastilla/iStockphoto; p109(C): Andrew Lambert Photography/Science Photo Library; p109(R): Ghislain & Marie David De Lossy/Science Photo Library; p111(T): AP Images; p111(C): Fred de Noyelle/Godong/Corbis; p111(B): Michael Neelon(misc)/Alamy; p115: Soundsnaps/Shutterstock; p122: Mark Downey/Masterfile; p130(T): Monty Rakusen/Cultura/Science Photo Library; p130(B): Ivan Smuk/Shutterstock; p132: Artsmela/Shutterstock; p133: Arletta Cwalina/Alamy; p135: Patrick Lane/Somos Images/Corbis; p136: Rod Haestier/Alamy; p138: Howard Kingsnorth/Getty Images; p139: Jan van der Hoeven/Shutterstock; p150: Dorling Kindersley/UIG/Science Photo Library; p155(T): NASA; p155(B): NASA; p144: BSIP, VEM/Science Photo Library; p158: North Wind Picture Archives/Alamy; p159: Martyn F Chillmaid/Science Photo Library; p162: Harvard Natural Science Lecture Demonstrations, FAS Science Division; p163: Vladimir Koletic/Shutterstock; p164: Venus Angel/Shutterstock; p166(T): Geoffrey Swaine/Rex Features; p166(B): Fabrice Coffrini/Keystone/AP Photo; p167: SPARK: Museum of Electrical Invention; p168: Peter Lewis/Getty Images; p169: onlinetuition.com.my; p171(T): Bill Meadows/Mary Evans Picture Library; p171(C): David Parker/Science Photo Library; p171(B): David Ducros/Science Photo Library; p172: Russell Kightley/Science Photo Library; p174(T): Source: http://www.itp.uni-hannover.de/~zawischa/ITP/multibeam.html; p174(B): Sue Holt; p175(T): Dr Keith Wheeler/Science Photo Library; p175(B): Science Museum/Science & Society Picture Library; p176: Source: http://www.itp.uni-hannover.de/~zawischa/ITP/multibeam.html; p177: Giphotostock/Science Photo Library; p184(L): NREL/US Department of Energy/Science Photo Library; p184(R): Peter Thorne, Johnson Matthey/Science Photo Library; p185(B): Drs A Yazdani & DJ Hornbaker/Science Photo Library; p187: Leslie Garland Picture Library/Alamy; p192(T): Martyn F Chillmaid/Science Photo Library; p192(B): Martyn F Chillmaid; p193(L): Stockphoto mania/Shuttertstock; p193(R): Jodie Johnson/Shuttertstock; p194: G Muller, Struers Gmbh/Science Photo Library; p196(T): Sorin Colac/Alamy; p196(C): Marwan Naamani/AFP/Getty Images; p196(B): Buzz Pictures/Alamy; p198: Jean-Loup Charmet/Science Photo Library; p206(T): Rob Wilson/Shutterstock; p206(B): RL Christiansen/Corbis; p209: DenisNata/Shutterstock; p219(T): Charles Falco/Science Photo Library; p219(B): David Parker/Science Photo Library; p220(TL): Time Life Pictures/NASA/The LIFE Picture Collection/Getty Images; p220(TR): Digital Vision/Getty Images; p220(C): Andrey Nekrasov/Alamy; p220(B): Top-Pics TBK/Alamy; p235: Image courtesy of IBM Research - Zurich; p236: John Foxx/Getty Images; p237(T): Tracy Hebden/Alamy; p237(B): Sheila Terry/Science Photo Library; p238(T): National Motor Museum/Rex Features; p238(B): Yasuyoshi Chiba/AFP/Getty Images; p239: "Long Island Power Authority (LIPA) is utilizing a cable system manufactured by Nexans that utilizes AMSC's HTS wire and an Air Liquide cooling system. Energized in April of 2008, this is the world's first superconductor transmission-voltage cable system and is capable of transmitting up to 574 megawatts (MW) of electricity and powering 300,000 homes. Photo courtesy of AMSC."; p244: Percom/Shutterstock; p249: Andrew Lambert Photography/Science Photo Library; p250: Saturno dona'/Alamy; p253(T): Martyn F Chillmaid/Science Photo Library; p253(B): Rick Rusing/Transtock/Corbis; p259(T): David R Frazier/Science Photo Library; p259(B): Alfred Eisenstaedt/The LIFE Picture Collection/Getty Images; p264: Pekka Parviainen/Science Photo Library; p265: Sciencephotos/Alamy; p267: Alex Bartel/Science Photo Library; p270: Pedrosala/Shutterstock; p272: Heritage Images/Corbis; p278: Nielskliim/Shutterstock; p280: Masterfile; p283: James King-Holmes/Science Photo Library; p285: Stockbyte/Getty Images; p288: Weatherstock/Superstock; p289: Martyn F Chillmaid/Science Photo Library; p290: Sciencephotos/Alamy; p300: Martyn F Chillmaid/Science Photo Library; p309(T): Ted Kinsman/Science Photo Library; p309(B): NOAA/GSFC/Suomi NPP/VIIRS/Norman Kuring/NASA; p310(T): Stu49/iStockphoto; p310(B): Stu49/Shutterstock; p320(T): Umberto Shtanzman/Shutterstock; p320(B): Science Photo Library; p327(R): Science Photo Library; p327(L): Science Source/Science Photo Library; p331(T): David Ducros/Science Photo Library; p331(C): Hank Morgan/Science Photo Library; p331(B): Martin Bond/Science Photo Library; p337: Gordon Garradd/Science Photo Library; p338: Dr Ray Clark & Mervyn Goff/Science Photo Library; p339: Djgis/Shutterstock; p341(T): Pascal Goetgheluck/Science Photo Library; p341(B): Oleg Golovnev/Shutterstock; p342(T): Savas Keskiner/iStockphoto; p342(C): Colomboriccardo/iStockphoto; p342(B): Jerry Mason/Science Photo Library; p346(T): Physics Dept, Imperial College/Science Photo Library; p346(B): NOAO/Science Photo Library; p347: Philippe Plailly/Science Photo Library; p348: Pedro Custodio/EyeEm/Getty Images; p349(T): Pavel Vakhrushev/Shutterstock; p349(B): Mikael Karlsson/Alamy; p358: Time Life Pictures/Mansell/The LIFE Picture Collection/Getty Images; p359: Martyn F Chillmaid; p362(L): Steve Allen/Science Photo Library; p362(R): Courtesy Argonne National Laboratory; p364: Patrick Blackett/Science Photo Library; p367: Martyn F Chillmaid; p371(T): NASA/Science Photo Library; p371(C): www.BibleLandPictures.com/Alamy; p371(B): Gianni Tortoli/Getty Images; p372: Age Fotostock/Alamy; p374: Steve Allen/Science Photo Library; p377(T): Bettmann/CORBIS; p377(B): Alfred Pasieka/Science Photo Library; p378: Interfoto/Alamy; p379(T): Science Photo Library/Getty Images; p379(B): EUROfusion; p381: David Paul Morris/Bloomberg/Getty Images; p382(T): Photofusion/Getty Images; p382(B): Phil Noble/Reuters; p382(C): Ho New/Reuters; p384: Roger-Viollet/Rex Features; p385(T): Lawrence Berkeley Laboratory/Science Photo Library; p385(B): Lawrence Berkeley Laboratory/Science Photo Library; p386: Photo Inc/Getty Images; p391: Shelley Gazin/CORBIS; p392(T): Elise Amendola/AP Images/Press Association; p392(C): CERN, P Loiez/Science Photo Library; p392(B): Maximilien Brice, CERN/Science Photo Library; p398: Courtesy of the Leo Baeck Institute, New York; p399: Huntington Library/Superstock; p400: Science Photo Library; p405: Massachusetts Institute of Technology; p408: New York Public Library/Science Photo Library; p409: NASA/Science Photo Library; p413: European Southern Observatory/Science Photo Library; p414: Dundee Photographics/Alamy; p415(T): Victor De Schwanberg/Science Photo Library; p415(B): NASA; p416(T): Emilio Segre Visual Archives/American Institute of Physics/Science Photo Library; p416(B): Detlev Van Ravenswaay/Science Photo Library; p417: Emilio Segre Visual Archives/American Institute of Physics/Science Photo Library; p418(T): NASA/Science Photo Library; p418(C): Mark Garlick/Science Photo Library; p418(B): Celestial Image Picture Co/Science Photo Library; p419: Mark Garlick/Science Photo Library; p420(T): Emilio Segre Visual Archives/American Institute of Physics/Science Photo Library; p420(B): NASA/Science Photo Library; p421: NASA; p423(T): P Saada/Eurelios/Science Photo Library; p423(C): C J Tavin/Everett/Rex Features; p423(BR): Stammers/Thompson/Science Photo Library; p423(BL): Science Photo Library; p427: Praisaeng/Shutterstock; p430: Burger/Phanie/Alamy; p432(T): Oak Ridge Associated Universities; p432(B): Dr Robert Friedland/Science Photo Library; p433: Samuel Ashfield/Science Photo Library; p435(T): Medic Image/Universal Images Group/Getty Images; p435(B): www.accutome.com; p436: Bork/Shutterstock; p438: Lester Lefkowitz/Photographer's Choice/Getty Images; p439: Simon Fraser/Royal Victoria Infirmary, Newcastle Upon Tyne/Science Photo Library; p440: Zephyr/Science Photo Library; p454(T): Edward Kinsman/Science Photo Library; p454(B): OUP.

Examination Boards

We would like to thank the following Examination Boards for permission to use questions from past examination papers. They are labelled in the Further Questions sections as follows:

AQA	Assessment and Qualifications Alliance	www.aqa.org.uk
Edex	Pearson Education Ltd	http://qualifications.pearson.com
OCR	Oxford, Cambridge and RSA	www.ocr.org.uk
W	Welsh Joint Education Committee	www.wjec.co.uk
CCEA	Northern Ireland Council for the Curriculum, Examinations and Assessment	www.rewardinglearning.org.uk

Answers are entirely the responsibility of the authors and publisher and have not been endorsed by the examination boards.

You can find more details about syllabus coverage at: www.oxfordsecondary.co.uk/advancedforyou

Further reading

Looking for further reading? Don't know what presents you want for Christmas or your birthday?

- ***The Cartoon Guide to Physics*** by Larry Gonick & Art Huffman (HarperCollins, ISBN 006-2731009)
- ***Mr Tomkins in Paperback*** by George Gamow & Roger Penrose (Cambridge University Press, ISBN 0521-447712)
- ***The Flying Circus of Physics*** (with answers) by Jearl Walker (John Wiley, ISBN 0471-762733)

Index

D

Dalton, John 354
damping 109–12
dark energy/dark matter 417
dating, radioactive 371
d.c. (direct-current) 283, 286
de Broglie equations 332–4, 336
decay, radioactive 364–72, 377, 482-4
decay constant 368–9, 372
decibels 134–5
deep inelastic scattering 386
Democritus 354
density 7, 27–9, 31, 184–5, 197
derived quantities 7, 42–3, 84
destructive interference 158–9, 172
dielectric materials 300, 310
diffraction 158, 168–78, 185, 330–2, 343
diffraction gratings 176–8, 330–2, 343
digital signals 143–4
diodes 234, 286–7
dioptres 145, 153
dip, angle of 257
Dirac, Paul 384
direct-current (d.c.) 283, 286
discharging capacitors 300–1, 306–7, 482
dispersion 125, 143
displacement 38–43, 68–9, 75, 102–12, 126, 129
displacement–time graphs 40, 50, 102–3, 129
dissipation, energy 72
diverging lens 145, 147, 152–4, 156
Doppler effect 414–15, 421, 437, 441
Doppler ultrasound 437, 441
double-slit experiments 169, 173–5, 178
drag, forces 22–3, 31, 49, 50
drift velocity 227, 229, 231
drum brakes 30
ductile materials 187, 192

E

ears 134–5
earthing 294, 309
earthquakes 136
Earth's magnetic fields 257, 264
ECG (electrocardiograms) 241
eddy currents 276, 278, 285
efficiency, energy 73, 75
Einstein, Albert 323, 327, 375–6, 383, 398–407
elastic collisions 74–5, 209
elastic limit 186–7, 189–90, 192
elastic materials 107–12, 186–91, 197
elastic (strain) potential energy 187
elasticity modulus 189–91, 197
electric charges 226–43, 264–8, 289–93, 302–11, 321–5
electric circuits 244–55
electric conduction 229, 235, 239
electric currents 6, 226–55, 258–68, 277, 280–7, 290–3
electric field strength 292–3, 296–8
electric fields 288–99, 303, 321, 322–5

electric motors 19, 73, 123
electric plugs 237
electric potential 294–6, 298
electric power 286
electric resistance 7–8, 232–43, 246–55, 285–7, 301–8
electric shocks 238
electric vehicles 238
electricity 226–43, 285–7
electrocardiograms (ECG) 241
electrolytes 229
electromagnetic forces 19–20, 31, 390
electromagnetic induction 270–9
electromagnetic spectrum 130–1, 328, 337–50, 409
electromagnetic waves 130–3, 137–44, 156–79, 326–43, 398–9, 407
electromagnets 259, 267
electromotive force (e.m.f.) 250–4, 270–9
electron capture 366
electron degeneracy pressure 419
electron guns 282, 321, 333
electron microscopes 335
electron-volt 321, 336–7, 350, 375, 383
electrons 77, 226–55, 265, 282, 320–36, 354, 370
electroscopes 293, 326, 358
electrostatics 288–99
e.m.f. (electromotive force) 250–4, 270–9
emission spectra 338, 345, 347–8, 350
energy
    binding 376–9, 383
    conservation of 72, 74–5
    efficiency 73, 75
    Einstein's equation 375, 383, 405–7
    electrical 226–43, 245, 285–7
    gravitational potential 93, 98
    ionisation 344–5, 350
    levels 343–7, 350
    mass 375–6, 383, 405–7
    photons 328–31, 336–7, 343–50
    simple harmonic motion 108–12
    thermodynamics 198–208
    transfer diagrams 72
    transmutation 377
    waves 126–7, 137
    work 7, 68–75, 198–200, 208, 211
equations
    help with 478–81, 484–6
    of motion 42–3, 50
equilibrium 14, 32–6, 101, 200, 208, 211
equipotentials 93, 295
errors, dealing with 467, 470–1
escape velocity 96–7
event horizons 419
exams, advice about 464–5
    syllabus coverage 460, 506
exchange particles 390, 393
excited states 344–8, 350, 365
exoplanets 421
exponential decay 367–9, 372, 426, 484–5
exponential discharge 307, 311
exponential equation 307–8, 311, 484–5
exponential growth 308, 311
eye, lenses 151–3

F

Faraday, Michael, law 270–9
farad 302
ferromagnetic materials 259
Feynman diagrams 391, 393
Feynman, Richard 391
fibre optics 142–4, 156
fields, electric 288–99, 303, 322–5
    gravitational 83, 88–9
    lines 88–9, 98, 256–68
    magnetic 256–79
first harmonic 162–5
first laws, Kepler's 94
    Kirchhoff's 244
    Newton's 53–4, 65–6, 76
    thermodynamics 200, 208, 211
fission, nuclear 378, 380–3
Fizeau's calculation 131
flavours, quarks 387
Fleming's left hand rule 262, 264–8
flexibility, materials 197
floating 29
fluid pressure 26–7
fluid resistance (drag) 22–3, 31
fluorescence 348, 429
flux lines 256–68, 271, 273, 275–6
flux linkage 271–9
FM (frequency modulation) 159, 168
focal length 145–7, 149, 152–3, 156
focus, principal 145–50, 152, 154
force–extension graphs 186–7, 193, 197
forces 7, 18–37, 52–85, 186
    centripetal 76–85, 265–6
    contact forces 19–20
    drag 22–3, 31, 49, 50
    electric currents 266, 290–3
    electromagnetic 19–20, 31, 390
    fluids 22–3, 26–7, 31
    free-body diagrams 24–5
    friction 20–1, 31
    gravitational 18, 31, 86–99, 390
    lift 23
    magnetic 262–8, 322, 335
    moments of 32–4, 36
    motive 21, 52–67
    nuclear 19, 31, 364, 389–91, 393
    pressure 7, 26–31, 209–19, 221
    tension 20, 195–6
    turning effects 32–7
    work 68–75
force–time graphs 60–1
Foucault, Leon 111
fracture, crystal 192
frames of reference 399–407
Franklin, Benjamin 288
Fraunhofer lines 346, 412
free fall 44–5, 50
free-body force diagrams 24–5
frequency
    alternating currents 280, 283
    circular motion 78, 85
    Doppler effect 414–15
    simple harmonic motion 100–3
    waves 124–5, 137, 159, 162–8